APPLIED MATHEMATICS SERIES

Edited by

I. S. Sokolnikoff

Ordinary Differential Equations

APPLIED MATHEMATICS SERIES

The Applied Mathematics Series is devoted to books dealing with mathematical theories underlying physical and biological sciences, and with advanced mathematical techniques needed for solving problems of these sciences.

Ordinary Differential Equations

WILLIAM T. REID
Phillips Professor of Mathematics
University of Oklahoma, Norman, Oklahoma

JOHN WILEY & SONS, INC.
New York · London · Sydney · Toronto

Library of Congress Catalog Card Number: 74-123745

ISBN 0-471-71499-2

Printed in the United States of America

10 9 8 7 6 5 4 3 2 1

To my wife, Idalia

Preface

This book is based upon lecture notes of courses presented in past years at the University of Chicago, Northwestern University, the State University of Iowa, and the University of Oklahoma. More material is included than was ever given in a single academic year. In a two-semester, or three-quarter, course the major portion of the material in Chapters I, III, IV was systematically covered, together with additional material from one of the following three alternatives, depending upon the maturity and interests of the particular group of students: (a) Chapters V, VI; (b) Chapter VIII; (c) Chapters II, VII.

It is felt that particularly strong points of the present manuscript are as follows.

(1) Existence theorems, and the dependence of solutions of ordinary differential equations upon initial data, are treated in sufficient generality and depth to provide a foundation for the rigorous discussion of fields of extremals for simple integral variational problems, and for the study of first-order partial differential equations by the method of characteristic strips.

(2) As presented in Chapters III and IV, the general theory of vector differential equations and associated differential systems with two-point boundary conditions deals with a vector-valued differential operator of such generality that whenever the involved matrix coefficients are merely continuous the corresponding adjoint operator is of the same form as the given differential operator. Moreover, when these matrix coefficients satisfy still weaker conditions, and "solution" is taken in the Carathéodory sense, this general formulation permits the automatic incorporation of certain types of interface conditions.

(3) There is presented a comprehensive discussion of the Sturmian theory, both for the classical case of a second-order scalar linear differential equation {Chapters V, VI}, and for self-adjoint differential systems {Chapter VII}. In particular, the student is introduced to the theory of M. Morse for such systems. In general, the treatment is based upon underlying variational principles, although no previous acquaintance with variational theory is required for a complete understanding of the discussion.

(4) For generality, and also for applications in the "control formulation" of certain variational problems, the general system treated in Chapter VII is written in terms of "canonical variables." Moreover, in this general setting

the spectral theory for self-adjoint boundary problems in canonical form, and involving two-point boundary conditions, is derived by elementary means.

(5) The inverse of a vector differential operator, and the associated Green's matrix, is presented in the setting of a general linear problem involving two-point boundary conditions, utilizing the E. H. Moore generalized inverse of an algebraic matrix.

(6) Both in text and problem sets there is extensive discussion of the Riccati matrix differential equation, properties of its solutions, and the relation of the theory of such an equation to disconjugacy criteria for self-adjoint differential systems.

Appendices contain prefatory results on vector spaces, algebra of matrices, quadratic and hermitian forms, the Jordan canonical form for matrices, and elementary convergence theorems; in particular, Section 3 of Appendix B is devoted to a concise discussion of the E. H. Moore generalized inverse of an $m \times n$ matrix. In general, these appendices provide the students with a common reference for basic concepts, definitions and notations, and should be a source for initiating review and possible further study in these areas.

To facilitate usage and reference, a superscript "*s*" on a problem number is employed to indicate that the result of the problem will be used at some later place in another problem; superscript "*ss*" on a problem number signifies usage of the result in the body of the text, perhaps in addition to usage also in a later problem.

For the material of Chapters II and VII, acquaintance with the Lebesgue integral in one dimension is presupposed. For the remainder of the text, however, the requisite elements of real analysis are to be found in most books of "Advanced Calculus" level. The material of a "usual" one-semester course in complex analysis provides the student with sufficient background for the work of Chapter I on differential equations in the complex plane, and for various results of Chapters III, IV.

The author acknowledges his indebtedness to many authors and books, and at best the listed Bibliography and terminal chapter sections on Notes and Comments provide limited records of comparative results. In particular, however, the manuscript has been profoundly influenced by the early association of the author with G. A. Bliss. This influence is especially strong in the discussion of the differentiability of solutions in Section 10 of Chapter I. Also, the presentation of the material of Chapters III, IV, VII owes much to the basic work of Bliss on "definitely self-adjoint differential systems," and the relation of such systems to accessory boundary problems of the calculus of variations.

During many of the several years of writing this manuscript the author has been the recipient of research grants from the Air Force Office of Scientific Research. In particular, considerable portions of the text and problem material of Chapter VII are to be found in papers written by him in this period. To the AFOSR grateful acknowledgment is made, both for personal support and also for the support of graduate students as Research Assistants, who in the course of their work provided helpful comments and aid to the author on various portions and editions of the lecture notes. In this category belong, in particular, Dr. John S. Bradley, (University of Iowa, 1963–64), Dr. Calvin D. Ahlbrandt, (University of Oklahoma, 1967–68), and Dr. Churl S. Kim (University of Oklahoma, 1968–70). Profound thanks are extended to Dean John S. Ezell, College of Arts and Sciences, and Dean Carl D. Riggs, Graduate College, of the University of Oklahoma for their support in providing secretarial help. The author is also deeply grateful to Mrs. Carolyn Johnson for her typing of the final version of this manuscript.

Norman, Oklahoma
September 1970

WILLIAM T. REID

Contents

PROLOGUE 1

1. Historical remarks 1
2. Prospectus 5
3. Comments 6

I
EXISTENCE THEOREMS FOR DIFFERENTIAL EQUATIONS 9

1. Introduction 9
2. Auxiliary results 13
3. Existence theorems of Cauchy-Peano type 16
4. Extreme solutions for real scalar differential equations 27
5. Uniqueness theorems 32
6. Existence theorems by the method of iterations 47
7. Application of the contraction mapping theorem to existence theorems 54
8. Differential equations of higher order 58
9. Linear vector differential equations 61
10. Differentiability of solutions of real differential systems 70
11. Differential equations in the complex plane 80
12. Notes and remarks 86

II
EXTENSION OF THE CONCEPT OF A SOLUTION OF A DIFFERENTIAL EQUATION 90

1. Introduction 90
2. Preliminary definitions and results 91
3. Existence theorems 92
4. Differentiability of solutions 104
5. Notes and remarks 108

III
LINEAR DIFFERENTIAL SYSTEMS 109

1. Introduction 109
2. Adjoint vector differential equations 110
3. Adjoint n-th order differential equations 117
4. Homogeneous differential systems involving two-point boundary conditions 126
5. Nonhomogeneous differential systems involving two-point boundary conditions 129
6. Adjoint differential systems 132
7. The Green's matrix 138
8. Differential systems involving a single n-th order linear differential equation 143
9. Differential systems involving an n-th order linear vector differential equation 152
10. The generalized Green's matrix 156
11. Equivalent differential systems 165
12. Notes and remarks 172

IV
DIFFERENTIAL SYSTEMS INVOLVING LINEARLY A PARAMETER 174

1. Formulation of the problem 174
2. Elementary properties of boundary value problems 175
3. Properties of the Green's matrix 183
4. Boundary problems involving an n-th order linear vector differential equation 190
5. Self-adjoint boundary problems 196
6. Definite boundary problems 200
7. Notes and remarks 217

V
SECOND ORDER LINEAR DIFFERENTIAL EQUATIONS 218

1. Introduction 218
2. Preliminary properties of solutions of (1.1) 221
3. An associated functional 224
4. The associated Riccati differential equation 231
5. Oscillation criteria 233
6. Comparison theorems 238

7. Differential systems involving a real parameter 244
8. Fundamental quadratic forms for conjugate and focal points 253
9. Notes and remarks 260

VI
SELF-ADJOINT BOUNDARY PROBLEMS ASSOCIATED WITH SECOND ORDER LINEAR DIFFERENTIAL EQUATIONS 262

1. Canonical forms for self-adjoint boundary problems 262
2. Extremum properties for self-adjoint systems (B) 267
3. Existence of proper values 272
4. Comparison theorems 279
5. Expansion theorems 288
6. Notes and remarks 300

VII
SELF-ADJOINT DIFFERENTIAL SYSTEMS 302

1. Introduction 302
2. Preliminary results 303
3. Normality and abnormality 313
4. An associated functional 322
5. Disconjugacy criteria 337
6. Comparison theorems 354
7. Morse fundamental hermitian forms for differential systems 356
8. Self-adjoint differential systems 368
9. Normality and abnormality of boundary problems ($\mathscr{B}$) 372
10. Preliminary existence theorems 374
11. Existence of proper values 381
12. Comparison theorems 390
13. Notes and remarks 398

VIII
STABILITY AND ASYMPTOTIC BEHAVIOR OF DIFFERENTIAL EQUATIONS 401

1. Introduction 401
2. Linear differential equations 403
3. Preliminary results for nonhomogeneous linear systems 421
4. Stability of related homogeneous differential systems 423
5. Further results on the asymptotic behavior of solutions of homogeneous differential equations 430

6. Differential equations with periodic coefficients 440
7. Stability of nonlinear differential equations 447
8. The direct method of Liapunov 454
9. Notes and remarks 459

APPENDIX A
VECTOR SPACES 463

1. Introduction 463
2. Linear independence 464
3. The conjugate space of linear functionals 466
4. Linear transformations 469
5. Hermitian forms 471

APPENDIX B
MATRICES, SYSTEMS OF LINEAR EQUATIONS 475

1. Algebra of matrices 475
2. Solvability theorems for a system of linear algebraic equations 478
3. The E. H. Moore generalized inverse of a matrix 480
4. Notes and remarks 484

APPENDIX C
DETERMINANTS 485

1. Definition 485
2. Properties 485

APPENDIX D
GEOMETRIC AND ANALYTIC ASPECTS OF EUCLIDEAN SPACE 488

1. A metric for $\mathbf{C}_n$ 488
2. Sequences of vectors and matrices 492

APPENDIX E
SPECTRAL PROPERTIES OF MATRICES 496

1. Elementary properties of linear transformations of $\mathbf{C}_n$ into $\mathbf{C}_n$ 496
2. Hermitian matrices and forms 500
3. The Jordan canonical form 509
4. Notes and remarks 517

APPENDIX F
MATRIX AND VECTOR FUNCTIONS 518

1. Preliminary concepts 518
2. Convergence theorems 525
3. Notes and remarks 528

BIBLIOGRAPHY 529

AUTHOR INDEX 543

SUBJECT INDEX 546

Prologue

1. HISTORICAL REMARKS

The study of ordinary differential equations dates from the works of G. W. Leibniz (1646–1716) and Isaac Newton (1642–1727) in the development of the calculus during the eighth decade of the seventeenth century. Indeed, some authors have assigned the genesis of this subject to November 11, 1675, when Leibniz wrote the equation $\int y\, dy = \frac{1}{2}y^2$, which may properly be considered the solution of a simple differential equation (e.g., see, Ince [1, p. 529]; here, as elsewhere throughout this manuscript, numbers in square brackets refer to the bibliography at the end of this volume). Before the invention of the calculus, however, specific problems in the determination of plane curves whose tangents satisfied a prescribed condition were solved by René Descartes (1596–1650), Isaac Barrow (1630–1677), and James Gregory (1638–1675). At an early stage there emerged the idea of obtaining the equation of a curve that possessed a given geometrical, kinematical, or dynamical property by expressing its characterization in the form of a differential equation and then finding its solution under given initial conditions. One outstanding example of this procedure was the problem formulated by Newton of the determination of the shape of the solid of revolution which will encounter a minimum resistance when moved in the direction of its axis through a resisting medium. Another was the famous brachistochrone problem, solved in 1696 by John Bernoulli (1667–1748), which involved the determination of the curve in a vertical plane down which a particle would fall from one given point to another in the shortest time. In passing, it is to be remarked that these two cited problems are among the earliest examples of the calculus of variations. It is to be noted that this latter problem was also solved by Leibniz, Newton, G. F. A. l'Hospital (1661–1704), and James Bernoulli (1654–1705).

By the end of the seventeenth century most elementary methods of solution of ordinary differential equations of the first order had been discovered. In the eighteenth century mathematics provided the basis for a rational outlook on the universe through dynamical astronomy, analytical mechanics, and potential theory. In each of these disciplines a central role was played by differential equations, both ordinary and partial. In the earlier part of this century there was continued work by James and John Bernoulli, and other members of their family so illustrious in science, notably Daniel Bernoulli (1700–1782), who has been called the founder of mathematical physics. Other outstanding mathematicians predominantly associated with this century, who contributed greatly to mathematical analysis in general and differential equations in particular, were Brook Taylor (1685–1731), Count Jacopo Riccati (1676–1754), Leonhard Euler (1707–1783), Jean-le-Rond d'Alembert (1717–1783), A. C. Clairaut (1713–1765), J. L. Lagrange (1736–1813), P. S. Laplace (1749–1827), and A.-M. Legendre (1752–1833), although the scientific activities of the last three extended into the nineteenth century.

The work of the mathematical physicist J. B. J. Fourier (1758–1830) and the mathematical astronomer F. W. Bessel (1784–1846) in the first quarter of the nineteenth century was basic to the theory of boundary problems, which involves the determination of solutions of differential equations that satisfy prescribed boundary conditions. By the application of the method of separation of variables to partial differential equations of mathematical physics, one was lead to the expansion problem of an "arbitrary" function in terms of a system of functions appearing as the "proper functions" or "eigenfunctions" of a differential equation for corresponding "proper values" or "eigenvalues" of an involved parameter.

The early decades of the nineteenth century were marked by the introduction of greater rigor in mathematics, and in the area of mathematical analysis the greatest contributor in this regard was A.-L. Cauchy (1789–1857). In particular, in his lectures at the École Polytechnique in the 1820's he developed an existence theorem for differential equations, which today is still referred to as the "Cauchy polygon method"; a refinement of this method, known as the "Cauchy-Lipschitz method," is due to R. Lipschitz (1832–1903). Perhaps for differential equations one of the most important developments of the nineteenth century was a growing awareness of the fact that in general the solutions of a differential equation were not representable in closed form in terms of a limited number of "elementary functions", and that the really pertinent problem was the determination of the qualitative nature of solutions; that is, the specification of basic properties of functions that derive from the fact that they are solutions of differential equations of some prescribed form. A simple illustration of this procedure appears in Problem I.9:10 (i.e., problem 10 of Section 9 in Chapter I), which deals with the trigonometric functions sine and cosine and shows how the basic aspects of analytical trigonometry are direct consequences of properties of solutions of the differential equation

$x'' + x = 0$. Another example is presented by the differential system

$$x''(t) = g(t)[1 + h(t)\,x(t)], \qquad x(1) = 0,$$

where $g(t)$ and $h(t)$ are real-valued continuous functions on $[0, 1]$ with $g(t) > 0$ on this interval. To establish the qualitative result that $x(t) < 0$ for $t \in [0, 1)$, we might actually determine the solution of this system as

$$x(t) = -\int_t^1 g(\tau)\left[\exp\left\{-2\int_t^\tau g(s)\,h(s)\,ds\right\}\right] d\tau, \qquad t \in [0, 1],$$

from which the conclusion is immediate. The conclusion, however, that $x(t) < 0$ for $t \in [0, 1)$ is obtainable by an elementary argument involving no solution formulas whatsoever and using only the facts that $x(1) = 0$ and that if s is any value on $[0, 1]$ at which $x(s) = 0$ then $x'(s) = g(s) > 0$. An outstanding example of a study of the qualitative nature of solutions was the work of J. C. F. Sturm (1803–1855) in his famous 1836 paper dealing with oscillation and comparison theorems for linear homogeneous second-order ordinary differential equations. The closely allied work of J. Liouville (1809–1882) in the period between 1835 and 1841 on ordinary linear differential equations of the second order led to a type of boundary problem known as a "Sturm-Liouville problem," and was an introduction to the study of the asymptotic behavior of solutions of differential equations by the use of integral equations.

The work on ordinary differential equations during the nineteenth and twentieth centuries has been so extensive that it would be utterly presumptious to attempt a comprehensive survey here. For contributions made in the nineteenth century and the first decade of the twentieth century the reader is referred to the encyclopedic articles of M. Bôcher [1], P. Painlevé [1, 2], and E. Vessiot [1, 2]. For more general historical surveys the reader is referred to F. Cajori [1], E. T. Bell [1, 2], and also the admirable concise historical note that forms Appendix A in E. L. Ince [1].

In particular, however, mention is made of the following aspects of the theory of ordinary differential equations whose initial development and flowering may be assigned to the nineteenth century.

1. The theory of differential equations in the complex plane, dating from an existence theorem given by Cauchy in the 1820's and including major contributions by N. H. Abel (1802–1829), C. G. J. Jacobi (1804–1851), K. Weierstrass (1815–1897), A. A. Briot (1817–1882), J. C. Bouquet (1819–1895), G. F. B. Riemann (1826–1866), L. Fuchs (1833–1902), G. Frobenius (1849–1917) and P. Painlevé (1863–1933).

2. The application of group theory to the theory of differential equations, first in the theory of continuous groups, as developed by S. Lie (1842–1899), and second by É. Picard (1856–1941) in initiating the development for linear

differential equations of an analog of the Galois theory for algebraic equations.

3. The development of the theory of integral equations, first in the work of V. Volterra (1860–1940) and then in the work of I. Fredholm (1866–1927).

4. The study of the concept of stability of solutions of a differential equation and the attendant treatment of the qualitative nature of solutions, notably in the extensive and fundamental investigations of H. Poincaré (1854–1912) and the classic 1893 memoir of A. M. Liapunov (1857–1918).

5. The development of more general existence theorems dating from the basic work of G. Peano (1858–1932) in the 1880's.

The twentieth century has seen extensive considerations of the geometric theory of differential equations, emanating from the basic contributions of Poincaré and Liapunov and owing much to the fundamental work of G. D. Birkhoff (1884–1944) on dynamical systems. Also, for the important generalization of Fourier series to expansions in terms of the proper functions of a boundary problem involving a linear homogeneous differential equation, and which in general is not self-adjoint, the contributions of Birkhoff and J. D. Tamarkin (1888–1945) have been basic for the solution of the problem of asymptotic developments of solutions of differential equations. The modern theory of singular differential operators dates from the 1909 memoir of H. Weyl (1885–1955) on singular self-adjoint linear differential operators of the second order, with later development of the general spectral theory of linear differential operators progressing through the work of M. H. Stone (1903–), J. von Neumann (1903–1957), K. Friedrichs (1901–), K. Kodaira (1915–), and many others. The early 1900's also saw the elaboration and extension of the oscillation, separation, and comparison theorems of Sturm by M. Bôcher (1867–1918), both in his many papers on the subject and his lectures at the Sorbonne in 1913–1914, which were subsequently published under the title *Leçons sur les méthodes de Sturm dans la théorie des équations différentielles linéaires et leurs développements modernes*.

For the study of boundary problems associated with self-adjoint differential systems the work of David Hilbert (1862–1941) in the first decade of the twentieth century has been fundamental, both in regard to the development of the theory of integral equations and in connection with the calculus of variations and the associated variational characterization of proper values and proper solutions of these systems. The significance of the calculus of variations for such boundary problems has also been emphasized by Gilbert A. Bliss (1876–1951) and Marston Morse (1892–). In particular, the latter showed in his basic 1930 paper in *Mathematische Annalen* that variational principles provided the appropriate environment for the extension to self-adjoint differential systems of the classical Sturmian theory for linear homogeneous ordinary differential equations of the second order.

A particular aspect of the calculus of variations that bears the name of optimal control theory has within the last two decades stimulated the introduction of generalizations of the concept of a differential equation. This subject, as well as the general area of functional analysis has lead to an ever-increasing study of differential equations in abstract spaces.

2. PROSPECTUS

From even a cursory reading of the comments of the preceding section it is apparent that no one volume on the subject of ordinary differential equations could begin to cover the whole field. In particular, the scope of the present work is most modest and is now briefly described.

The first two chapters are concerned with existence theorems for vector ordinary differential equations and the properties of solutions of these equations as functions of initial values. In Chapter I the major portion of the discussion is concerned with ordinary differential equations which involve continuous functions of a real variable, although Section 10 of this chapter presents an introduction to differential equations in the complex plane. Chapter II is devoted to the generalization of a differential equation due to C. Carathéodory, in which the solutions are merely absolutely continuous functions and the derivative condition of the differential equation is required to hold almost everywhere in the sense of Lebesgue.

Chapter III treats the algebraic properties of solutions of a linear vector differential system that involves two-point boundary conditions, whereas Chapter IV is concerned with the theory of boundary problems for related differential systems which contain linearly a characteristic parameter. In particular, Section 6 of Chapter IV is concerned with self-adjoint boundary problems which satisfy a condition of definiteness and which historically arose from the theory of accessory boundary problems of the calculus of variations.

Chapters V, VI, and VII may be considered as a unit in which the central feature is the systematic use of variational principles applied to certain quadratic and hermitian functionals, although it is to be emphasized that the treatment is entirely self-contained and no previous acquaintance with the calculus of variations is required for a thorough understanding of the material. Stated briefly, Chapter V is devoted to the Sturmian theory for a real scalar linear homogeneous differential equation of the second order, which deals primarily with the oscillation, separation, and comparison theorems for such an equation. The last section of Chapter V introduces the reader to the fundamental quadratic form due to Marston Morse. Chapter VI is concerned with the treatment of boundary problems which involve two-point boundary conditions and are associated with a scalar differential equation of the sort discussed in Chapter V; the proofs of the existence of proper values and associated expansion theorems are based on the extremal properties of these proper values and associated proper solutions. In particular, the boundary problems of this chapter include the classical Sturm-Liouville problems.

Chapter VII may be described briefly as the extension of the Sturmian theory of Chapters V and VI to the case of self-adjoint differential systems and associated boundary problems. Sections 1 to 7 provide the extension of the results of Chapter V, and the corresponding generalizations of the results of Chapter VI are given in Sections 8 to 12. Section 7 is devoted to the Morse fundamental quadratic form for such systems and its use in the derivation of results on the distribution of conjugate and focal points. It is to be remarked that the desire to present the material of this chapter in a unified form was my chief motivating factor in undertaking the preparation of this manuscript.

It is to be mentioned that throughout this work there are numerous instances of a Riccati differential equation and its generalizations. Some of the basic properties of the solution of a scalar Riccati differential equation are presented in Problem I.9:16 and Section V.4 (i.e., Section 4 of Chapter V). Basic properties of solutions of a matrix Riccati differential equation appear in Problems II.3:8, II.3:9, and II.3:10, whereas hermitian matrix Riccati differential equations appear frequently in the text and problem sets of Chapter VII.

Finally, Chapter VIII provides a relatively brief introduction to the concept of stability for an ordinary differential equation and some of the basic theorems on the qualitative nature of solutions.

3. COMMENTS

As indicated in the limited historical resumé of Section 1, the subject of differential equations is of long duration and a discipline that has continually expanded in scope and contact with various areas of knowledge. With the advent of integral equations, integro-differential equations, and other types of functional equation the mathematician is ever reminded that a differential equation is but one kind of functional equation that may be of use in interpreting a given phenomenon mathematically, and that in the larger sense differential equation theory is but one special example of the comprehensive area of functional analysis. These comments are not meant to disparage the study of differential equations per se, for indeed it is by such study that we gain insight into the dominant characteristics of these special problems and the motivation for the introduction of various concepts in the larger realm of functional analysis.

From the standpoint of applications of differential equations, and as in other disciplines, in the construction of a model we ignore some aspects of the situation to be interpreted and select other aspects as basic. Consequently, the validity of the results is actually established only for that model, and the question of its usefulness depends on how the results of the mathematical theory compare with those observed. Viewed in this fashion, the ability to interpret a given phenomenon by a differential equation is a most special situation. If the differential equation fails to possess a unique solution satisfying given initial data, we usually decide that the mathematical formulation is one that does not really describe the considered situation. On the other hand, if the considered differential equation does possess a unique solution that satisfies given initial data, the knowledge of the solution within any nondegenerate interval completely determines the solution throughout its entire interval of existence and this determinism in many instances points to the ineptness of differential equations as the vehicle for mathematical interpretation. Thus we might be led to the use of an integral equation, an integro-differential equation, a differential equation with retarded argument, or some other type of functional equation. It is to be emphasized, however, that the theory of differential equations still maintains a significant position for the study of each of these types of related functional equation. Consequently, the user of mathematical analysis should be familiar with the essential aspects of the theory of differential equations and realize that certain situations can be described by such equations, whereas other situations are not amenable to such description.

In the present volume there is no discussion of the numerical solution of differential equations. Moreover, there is no discussion of specific applications, since it is felt that this treatment comes most properly from the specialist in the involved discipline. However, the material presented is on aspects of the subject that appear repeatedly in applications, notably in the physical and biological sciences, and in engineering. Therefore, it has been my goal to present the material in a comprehensive and mathematically rigorous fashion and in the spirit of certain aspects of functional analysis, but without requiring extensive mathematical prerequisites. It is felt that such treatment would be useful to both the individual pursuing graduate study in mathematics and the individual desiring a basic knowledge of the field for applications.

I

Existence Theorems for Differential Equations

1. INTRODUCTION

If $f(t, y)$ is a real-valued continuous function in an open region $\mathfrak{R}$ of the (t, y)-plane, then for the first-order ordinary differential equation

$$\frac{dy}{dt} = f(t, y) \tag{1.1}$$

a function $y = y(t)$ with $(t, y(t))$ in $\mathfrak{R}$ for t on a nondegenerate (open or closed) interval I is termed a *solution* if $y(t)$ is continuous and has a continuous derivative $dy/dt = y'(t)$ such that $y'(t) = f(t, y(t))$ for $t \in I$. If $t = a$ is an end point of I which belongs to this interval, then in the above definition of solution of (1.1) it is to be understood that $y'(a)$ signifies the one-sided derivative of $y(t)$ at $t = a$ determined by the functional values of $y(t)$ on I.

Geometrically, (1.1) determines for each point (t_0, y_0) of $\mathfrak{R}$ a direction with angle of inclination θ_0, $(-\pi/2 < \theta_0 < \pi/2)$, specified by $\tan\theta_0 = f(t_0, y_0)$. The triple of numbers $t_0, y_0, m_0 = f(t_0, y_0)$ is called a *lineal element* issuing from (t_0, y_0), and the totality of such lineal elements with (t_0, y_0) in $\mathfrak{R}$ is called a *slope field*, (or *direction field*), for this differential equation. Clearly a curve $\Gamma: y = y(t)$ with $(t, y(t))$ in $\mathfrak{R}$ for t on I is a *solution-curve*, or *integral curve*, if and only if $y(t)$ is continuously differentiable and at each point on Γ the tangent line has the slope prescribed by the slope field. As a notion of the slope field is afforded by plotting a selected number of points (t_0, y_0) with attendant short line segments of corresponding slopes $m_0 = f(t_0, y_0)$, a qualitative picture of the solution curves of (1.1) may be obtained by sketching curves on such a plot of the slope field, exercising care to insure that at each point the sketched curve has approximately the prescribed slope of the field. Frequently the construction of a plot of the slope field is facilitated by finding first the curves of constant slope, or *isoclines*, which are the level curves $f(t, y) = c$ for the function of the right-hand member of (1.1).

It is to be emphasized that any such graphical study of solutions has its limitations. For example, it is doubtful that application of the above-described procedure to the equations

$$(1.2) \qquad \text{(a) } y' = \sqrt{1 + y^2}, \qquad \text{(b) } y' = 1 + y^2$$

would cause one to suspect that solutions of these two equations are quite different in character. Actually, the general solution of (a) is $y = \sinh(t - c)$, which for each value of c is a solution on the infinite interval $-\infty < t < \infty$, whereas for (b) the general solution is $y = \tan(t - c)$ so that the solution of this equation passing through a given point (t_0, y_0) may be continued along the t-axis only over a corresponding interval $a < t < b$ of length $b - a = \pi$. In particular, an example of the type (1.2b) illustrates the fact that in general the existence of solutions of (1.1) is a *local* problem, and that the interval on which a solution of such an equation is defined depends not only upon the particular equation considered but also upon the initial point (t_0, y_0) through which the solution passes.

It is to be emphasized, moreover, that although the slope field of an equation (1.1) prescribes a unique slope at each individual point it is not true that the continuity of $f(t, y)$ implies the existence of a unique solution of (1.1) passing through a given initial point. For example, although $f(t, y) = 3y^{2/3}$ is continuous throughout the (t, y)-plane, the equation $y' = 3y^{2/3}$ has the distinct solutions $y \equiv 0$ and $y = t^3$ passing through the origin $(0, 0)$; indeed, if $c_1 \leq 0 \leq c_2$ and $y(t; c_1, c_2) = (t - c_1)^3$ on $t \leq c_1$, $y(t; c_1, c_2) \equiv 0$ on $c_1 \leq t \leq c_2$, and $y(t; c_1, c_2) = (t - c_2)^3$ on $t \geq c_2$, then $y = y(t; c_1, c_2)$ is also a solution of this differential equation.

It is to be noted that if $y = y(t)$ is a solution of (1.1) for t on the interval I, and $y(\tau) = \eta$ for a point $t = \tau$ of I, then

$$(1.3) \qquad y(t) = \eta + \int_\tau^t f(s, y(s))\, ds, \quad \text{for} \quad t \in I.$$

On the other hand, suppose that $y(t)$ is a function defined on I and such that $y(t)$ is a solution of (1.3) in the sense that $f(s, y(s))$ is integrable on each closed subinterval of I, and this equation holds on I. As the integral is a continuous function of its limits of integration it then follows that $y(t)$ is continuous on I, and whenever $f(t, y)$ is a continuous function of (t, y) in a neighborhood of the curve $\{(t, y(t)) \mid t \in I\}$, then $f(s, y(s))$ is continuous on I and at each point of this interval $y(t)$ has a continuous derivative $y'(t) = f(t, y(t))$. That is, for $f(t, y)$ continuous throughout a region $\mathfrak{R}$ of the (t, y)-plane the condition that a curve $\{(t, y(t)) \mid t \in I\}$ in $\mathfrak{R}$ be a solution curve of (1.1) passing through a given point (τ, η) of this region is equivalent to the condition that $y(t)$ be a solution of (1.3). In view of these remarks (1.3) presents a means for extending in a natural fashion the concept of a solution of (1.1) whenever $f(t, y)$ is discontinuous. As noted above, if $y(t)$ is such that $f(s, y(s))$ is integrable on each closed subinterval of I containing τ, and (1.3) holds on I, then $y(t)$ is necessarily continuous. In particular, if $f(t, y)$ is such that $f(s, y(s))$ is integrable on each closed subinterval of I whenever $y(t)$ is continuous on I, then the condition that $y(t)$ is a solution of (1.3) in the sense stated above is equivalent to the condition that $y(t)$ is a continuous function such that (1.3) holds. For such functions $f(t, y)$ the statement that "$y = y(t)$ is a solution of (1.1) on I satisfying $y(\tau) = \eta$," or "$\{(t, y(t)) \mid t \in I\}$ is a solution curve passing through the point (τ, η)," will be understood to mean that "$\tau \in I$, and $y(t)$ is a continuous function on this interval satisfying (1.3)." For example, if $f(t, y) = -y$ on $t \leq 1$, and $f(t, y) = 2y$ on $t > 1$, then the solution on $-\infty < t < \infty$ of $y' = f(t, y)$, $y(0) = 1$ is the solution of $y(t) = 1 + \int_0^t f(s, y(s))\, ds$, $-\infty < t < \infty$, which is $y(t) = e^{-t}$ on $t \leq 1$, $y(t) = e^{2t-3}$ on $t \geq 1$.

In the above comments it has been assumed implicitly that the involved integration procedure is that of the Riemann integral. Such need not be the case, however, and the following chapter is devoted to comments on the case when the integral in (1.3) is taken in the sense of Lebesgue.

The majority of the results of this chapter are concerned with a system of n ordinary differential equations

$$(1.4) \qquad y'_\alpha = f_\alpha(t, y_1, \ldots, y_n), \qquad (\alpha = 1, \ldots, n),$$

involving one real independent variable t and n dependent variables $y_1, \ldots, y_n$, and where derivatives with respect to t are denoted by accents. In the spirit of generality the y_α are taken as complex variables and the f_α are allowed to be complex-valued. If the functions $f_\alpha(t, y_1, \ldots, y_n)$, $(\alpha = 1, \ldots, n)$, are continuous in a region $\mathfrak{R}$ of $(t, y_1, \ldots, y_n)$-space then $y_\alpha = y_\alpha(t)$, $(\alpha = 1, \ldots, n)$, with $(t, y_1(t), \ldots, y_n(t))$ in $\mathfrak{R}$ for t on a nondegenerate interval I, is a solution of (1.4) if the functions $y_\alpha(t)$ are continuous and have continuous derivatives $y'_\alpha(t)$ such that $y'_\alpha(t) = f_\alpha(t, y_1(t), \ldots, y_n(t))$ on this interval. As in the case of one real dependent variable discussed above, for discontinuous $f_\alpha(t, y_1, \ldots, y_n)$ the concept of a solution of (1.4) may be generalized. In particular, if the functions f_α are such that $f_\alpha(s, y_1(s), \ldots, y_n(s))$ is integrable on each closed subinterval of I, then the statement "$y_\alpha = y_\alpha(t)$, $(\alpha = 1, \ldots, n)$, is a solution of (1.4) on I satisfying $y_\alpha(\tau) = \eta_\alpha$, $(\alpha = 1, \ldots, n)$," or "$\{(t, y_1(t), \ldots, y_n(t)) \mid t \in I\}$ is a solution curve passing through the point $(\tau, \eta_1, \ldots, \eta_n)$," is understood to mean that "$\tau \in I$, and $y_\alpha(t)$, $(\alpha = 1, \ldots, n)$, are continuous functions on this interval satisfying

$$(1.5) \quad y_\alpha(t) = \eta_\alpha + \int_\tau^t f_\alpha(s, y_1(s), \ldots, y_n(s))\, ds, \qquad (\alpha = 1, \ldots, n;\, t \in I)."$$

It is to be commented that the consideration of systems (1.4) involving complex-valued y_α and f_α is actually no more general than the consideration of such systems involving real-valued quantities. Indeed, if $y_\alpha = u_\alpha + iu_{n+\alpha}$, and $f_\alpha(t, y_1, \ldots, y_n) = g_\alpha(t, u_1, \ldots, u_{2n}) + ig_{n+\alpha}(t, u_1, \ldots, u_{2n})$, $(\alpha = 1, \ldots, n)$, where u_α, $u_{n+\alpha}$ and g_α, $g_{n+\alpha}$ are real-valued, then (1.4) is equivalent to the system

$$(1.4') \qquad u'_\sigma = g_\sigma(t, u_1, \ldots, u_{2n}), \qquad (\sigma = 1, \ldots, 2n),$$

of $2n$ real differential equations in the real dependent variables $u_1, \ldots, u_{2n}$.

On the other hand, if $g_\alpha(t, u_1, \ldots, u_n)$, $(\alpha = 1, \ldots, n)$, are given real-valued functions defined for the real variables $t, u_1, \ldots, u_n$ in a region $\mathfrak{R}_1$ of $(t, u_1, \ldots, u_n)$-space, then the system of real differential equations

$$(1.6) \qquad u'_\alpha = g_\alpha(t, u_1, \ldots, u_n), \qquad (\alpha = 1, \ldots, n),$$

may be considered as a special case of (1.4) involving the complex-valued $y_\alpha = u_\alpha + iu_{n+\alpha}$, with $f_\alpha(t, y_1, \ldots, y_n) = g_\alpha(t, u_1, \ldots, u_n)$ for $(t, y_1, \ldots, y_n)$ in the region $\mathfrak{R}$ of (t, y)-space specified by $(t, u_1, \ldots, u_n)$ in $\mathfrak{R}_1$ and $u_{n+1}, \ldots, u_{2n}$ arbitrary. For the corresponding system (1.4′) the functions $g_{n+1}, \ldots, g_{2n}$ are identically zero, and this system has a solution $y_\alpha(t)$, $(\alpha = 1, \ldots, n)$, satisfying $y_\alpha(\tau) = \eta_\alpha = \zeta_\alpha + i\zeta_{n+\alpha}$, $(\alpha = 1, \ldots, 2n)$, with $\zeta_1, \ldots, \zeta_{2n}$ real-valued, if and only if $y_\alpha(t) \equiv u_\alpha(t) + i\zeta_{n+\alpha}$ and $u_\alpha(t)$, $(\alpha = 1, \ldots, n)$, is a real-valued solution of (1.6) satisfying $u_\alpha(\tau) = \zeta_\alpha$.

At certain points in the following treatment of general differential systems attention will be restricted to real solutions of a system of differential equations involving given real-valued functions; in such instances we shall use for the considered system the notation (1.6), or a corresponding vector equation in $u = (u_\alpha)$.

Throughout most of the subsequent discussion vector and matrix notation will be employed, and basic relations on the differentiation and integration of vector and matrix functions will be used in the text without comment. For a brief resumé of the involved concepts the reader is referred to the Appendices, especially Appendix A and Appendix F.

Problems I.1

1. Find the totality of real solutions of the differential equation $u' = g(t, u)$ passing through the point $(0, 0)$, where the real-valued function $g(t, u)$ is defined in the (t, u)-plane as follows:

(i) $g(t, u) = -(\frac{3}{2})u^{1/3}$;
(ii) $g(t, u) = u^{2/3}$ for $|u| \leq 1$, $g(t, u) = |u|$ for $|u| \geq 1$;
(iii) $g(t, u) = 1 + u^2$ for $|u| \leq 1$, $g(t, u) = 2$ for $|u| \geq 1$.
(iv) $g(t, u) = -1$ for $|u| < 1$, $g(t, u) = 0$ for $|u| = 1$, $g(t, u) = 1$ for $|u| > 1$.

2. AUXILIARY RESULTS

Consider the first order linear scalar differential equation

$$v' = r(t) + k(t)v, \tag{2.1}$$

where $r(t)$ and $k(t)$ are piecewise continuous functions on a finite closed interval $[a, b]$ of the real line. As (2.1) implies

$$\left[\exp\left\{-\int_\tau^t k(s)\,ds\right\} v(t)\right]' = \exp\left\{-\int_\tau^t k(s)\,ds\right\} r(t),$$

where τ is an arbitrary point of $[a, b]$, it follows that the general solution of this differential equation is given by

$$v(t) = \exp\left\{\int_\tau^t k(s)\,ds\right\} v(\tau) + \int_\tau^t \exp\left\{\int_s^t k(s_1)\,ds_1\right\} r(s)\,ds. \tag{2.2}$$

In particular, this explicit relation clearly implies the following monotoneity principle for solutions of (2.1).

Monotoneity Principle A. *If $k(t)$, $r_1(t)$ and $r_2(t)$ are real-valued piecewise continuous functions on $[a, b]$, and $v_j(t)$, $(j = 1, 2)$, are the solutions of*

$$v_j' = r_j(t) + k(t)\,v_j, \qquad v_j(\tau) = \gamma, \qquad (j = 1, 2),$$

where γ is a given real quantity and τ is a value on $[a, b]$, then the condition $r_1(t) \geq r_2(t)$ on $[a, b]$ implies

$$v_1(t) \geq v_2(t), \quad \textit{for} \quad t \in [\tau, b];\ v_1(t) \leq v_2(t), \quad \textit{for} \quad t \in [a, \tau].$$

This monotoneity principle leads to the result of the following theorem on linear inequalities; this result, together with the conclusions of the corollaries of this theorem, will be used frequently in the subsequent treatment of various aspects of the theory of differential equations.

Theorem 2.1. *Suppose that $\phi(t)$, $\kappa(t)$ and $\rho(t)$ are real-valued piecewise continuous functions on $[a, b]$, and $\kappa(t) \geq 0$ on this interval. If*

$$\rho(t) \leq \phi(t) + \int_\tau^t \kappa(s)\,\rho(s)\,ds, \quad \text{for} \quad t \in [\tau, b], \tag{2.3}$$

for some τ on $[a, b]$, then

$$\rho(t) \leq \phi(t) + \int_\tau^t \exp\left\{\int_s^t \kappa(s_1)\,ds_1\right\} \kappa(s)\,\phi(s)\,ds, \quad \text{for} \quad t \in [\tau, b]; \tag{2.4}$$

moreover, if $\phi(s) \leq \Phi[\tau, t]$ for $\tau \leq s \leq t \leq b$, then

$$\rho(t) \leq \Phi[\tau, t]\exp\left\{\int_\tau^t \kappa(s)\,ds\right\}, \quad \text{for} \quad t \in [\tau, b]. \tag{2.5}$$

Correspondingly, if

$$\rho(t) \leq \phi(t) + \int_t^\tau \kappa(s)\,\rho(s)\,ds, \quad \text{for} \quad t \in [a, \tau], \tag{2.3'}$$

then

$$\rho(t) \leq \phi(t) + \int_t^\tau \exp\left\{\int_t^s \kappa(s_1)\,ds_1\right\} \kappa(s)\,\phi(s)\,ds, \quad \text{for} \quad t \in [a, \tau]; \tag{2.4'}$$

moreover, if $\phi(s) \leq \Phi[t, \tau]$ *for* $a \leq t \leq s \leq \tau$, *then*

$$\rho(t) \leq \Phi[t, \tau]\exp\left\{\int_t^\tau \kappa(s)\,ds\right\}, \quad \text{for} \quad t \in [a, \tau]. \tag{2.5'}$$

If $\rho(t)$ satisfies (2.3) then $\theta(t) = \rho(t) - \int_\tau^t \kappa(s)\,\rho(s)\,ds$ is such that $\theta(t) \leq \phi(t)$ on $[\tau, b]$ and $v(t) = \int_\tau^t \kappa(s)\,\rho(s)\,ds$ is the solution of

$$v' = \kappa(t)\,\theta(t) + \kappa(t)v, \qquad v(\tau) = 0.$$

As the condition $\kappa(t) \geq 0$ implies that $\kappa(t)\,\theta(t) \leq \kappa(t)\,\phi(t)$, relation (2.4) is an immediate consequence of the above Monotoneity Principle A. Relation (2.5) then follows from the fact that the right-hand member of (2.4) is not greater than

$$\begin{aligned}\Phi[\tau, t]\left(1 + \int_\tau^t \exp\left\{\int_s^t \kappa(s_1)\,ds_1\right\} \kappa(s)\,ds\right) &= \Phi[\tau, t]\left(1 - \left[\exp\left\{\int_s^t \kappa(s_1)\,ds_1\right\}\right]_{s=\tau}^{s=t}\right)\\ &= \Phi[\tau, t]\exp\left\{\int_\tau^t \kappa(s)\,ds\right\}.\end{aligned}$$

Correspondingly, if $\rho(t)$ satisfies (2.3′), then $\theta(t) = \rho(t) - \int_t^\tau \kappa(s)\,\rho(s)\,ds$ is such that $\theta(t) \leq \phi(t)$ on $[a, \tau]$, and $v(t) = \int_t^\tau \kappa(s)\,\rho(s)\,ds$ satisfies

$$v' = -\kappa(t)\,\theta(t) - \kappa(t)v, \qquad v(\tau) = 0.$$

Since $\kappa(t) \geq 0$ implies that $-\kappa(t)\,\theta(t) \geq -\kappa(t)\,\phi(t)$, it follows from the Monotoneity Principle A that

$$v(t) \leq \int_\tau^t \exp\left\{-\int_s^t \kappa(s_1)\,ds_1\right\}[-\kappa(s)\,\phi(s)]\,ds = \int_t^\tau \exp\left\{\int_t^s \kappa(s_1)\,ds_1\right\} \kappa(s)\,\phi(s)\,ds;$$

hence inequality (2.4′) holds. Finally (2.5′) follows from (2.4′) in the same manner that (2.5) results from (2.4).

***Corollary* 1.** *If* $\rho(t)$ *is a non-negative real-valued piecewise continuous function satisfying with non-negative real constants* κ_0, κ_1, κ *the inequality*

$$\rho(t) \leq \kappa_0 + \kappa_1\,|t - \tau| + \left|\int_\tau^t \kappa\rho(s)\,ds\right|, \quad \text{for} \quad t \in [a, b], \tag{2.6}$$

then on this interval,

$$(2.7)\quad \rho(t) \leq \kappa_0 \exp\{\kappa\,|t-\tau|\} + \frac{\kappa_1}{\kappa}\,[\exp\{\kappa\,|t-\tau|\} - 1] \quad \text{if} \quad \kappa > 0.$$
$$\leq \kappa_0 + \kappa_1\,|t-\tau|, \quad \text{if} \quad \kappa = 0.$$

If $\kappa > 0$, then (2.6) implies that (2.3) and (2.3′) hold for $\phi(t) = \kappa_0 + \kappa_1\,|t-\tau|$, $\kappa(t) \equiv \kappa$ and the corresponding relations (2.4) and (2.4′) reduce on integration by parts to the corresponding inequality of (2.7) on $[\tau, b]$ and $[a, \tau]$, respectively. In case $\kappa = 0$ inequality (2.6) reduces to the second relation of (2.7).

***Corollary* 2.** *If the components of the vector* $y(t) \equiv (y_\alpha(t))$, $(\alpha = 1, \ldots, n)$, *are continuous and have piecewise continuous derivatives on* $[a, b]$ *and there are non-negative real constants* κ, κ_1 *such that on* $[a, b]$

$$(2.8)\qquad |y'(t)| \leq \kappa\,|y(t)| + \kappa_1,$$

then for arbitrary τ, t *on this interval,*

$$(2.9)\quad |y(t)| \leq |y(\tau)| \exp\{\kappa\,|t-\tau|\} + \frac{\kappa_1}{\kappa}\,[\exp\{\kappa\,|t-\tau|\} - 1] \quad \text{if} \quad \kappa > 0,$$
$$\leq |y(\tau)| + \kappa_1\,|t-\tau|, \quad \text{if} \quad \kappa = 0.$$

As $y(t) = y(\tau) + \int_\tau^t y'(s)\,ds$, the triangle inequality implies $|y(t)| \leq |y(\tau)| + |\int_\tau^t| y'(s)\,|ds|$ on $[a, b]$, and in view of (2.8) it follows that (2.6) holds with $\kappa_0 = |y(\tau)|$, κ and κ_1 the constants of (2.8), and $\rho(t) = |y(t)|$, so that the conclusion (2.7) of Corollary 1 becomes (2.9).

***Corollary* 3.** *If* $\eta^{(j)} \equiv (\eta_\alpha^{(j)})$, $(\alpha = 1, \ldots, n; j = 0, 1, \ldots, p)$, *are vectors of* $\mathbf{C}_n$ *which satisfy with constants* $\kappa \geq 0$, $\kappa_1 \geq 0$, $\delta_h > 0$, $(h = 0, 1, \ldots, p-1)$, *the inequalities*

$$(2.10)\quad |\eta^{(j+1)} - \eta^{(j)}| \leq [\kappa\,|\eta^{(j)}| + \kappa_1]\,\delta_j, \qquad (j = 0, 1, \ldots, p-1),$$

then for $j = 0, 1, \ldots, p-1$, *we have*

$$(2.11)\quad |\eta^{(j+1)}| \leq |\eta^{(0)}| \exp\left\{\kappa \sum_{h=0}^{j} \delta_h\right\} + \frac{\kappa_1}{\kappa}\left[\exp\left\{\kappa \sum_{h=0}^{j} \delta_h\right\} - 1\right] \quad \text{if} \quad \kappa > 0,$$
$$\leq |\eta^{(0)}| + \kappa_1 \sum_{h=0}^{j} \delta_h \quad \text{if} \quad \kappa = 0.$$

Since the triangle inequality implies

$$|\eta^{(j+1)}| \leq |\eta^{(0)}| + \sum_{h=0}^{j} |\eta^{(h+1)} - \eta^{(h)}|, \qquad (j = 0, 1, \ldots, p-1),$$

from the inequality (2.10) it follows that

$$(2.12)\quad |\eta^{(j+1)}| \le \left(|\eta^{(0)}| + \kappa_1 \sum_{h=0}^{j} \delta_h\right) + \kappa \sum_{h=0}^{j} |\eta^{(h)}|\, \delta_h, \quad (j = 0, 1, \ldots, p-1).$$

For $b = \sum_{h=0}^{p-1} \delta_h$, let $\rho(t)$ be defined on $[0, b]$ as follows: $\rho(t) = |\eta^{(0)}|$ on $0 \le t < \delta_0$; $\rho(t) = |\eta^{(j)}|$ on $\sum_{h=0}^{j-1} \delta_h \le t < \sum_{h=0}^{j} \delta_h$, $(j = 1, \ldots, p-1)$; $\rho(b) = |\eta^{(p)}|$. From inequality (2.12) it then follows that (2.6) holds on $[0, b]$ with $\tau = 0$, $\kappa_0 = |\eta^{(0)}|$, and κ, κ_1 the constants in (2.12), so that (2.11) follows from the conclusion (2.7) of Corollary 1 for $t = \sum_{h=0}^{j} \delta_h$, $(j = 0, 1, \ldots, p-1)$.

3. EXISTENCE THEOREMS OF CAUCHY-PEANO TYPE

Attention will now be given to a system of ordinary differential equations (1.4), which for $y(t) \equiv (y_\alpha(t))$ and $f(t, y) \equiv (f_\alpha(t, y)) \equiv (f_\alpha(t, y_1, \ldots, y_n))$, $(\alpha = 1, \ldots, n)$, may be written in vector form as

$$(3.1)\qquad y' = f(t, y).$$

Although the same type letters are used for the vector functions $y(t)$ and $f(t, y)$ in (3.1) as for the corresponding scalar functions in the scalar equation (1.1), it is felt that no ambiguity will arise for the reader. Throughout the present work the majority of attention is devoted to systems of equations, and all equations of the form (3.1) are to be considered as vector equations unless restrictive hypotheses on the nature of the equation are specifically stated in a given chapter or section.

If $y(t)$ is a continuously differentiable vector function on $[a, b]$ such that $f(t, y)$ is continuous in a region of (t, y)-space containing the curve

$$\{(t, y(t)) \mid t \in [a, b]\},$$

then as in the introductory discussion of Section 1 the vector function $y(t)$ is a solution of (3.1) on $[a, b]$ satisfying $y(\tau) = \eta$ if and only if $y(t)$ satisfies the vector integral equation

$$(3.2)\qquad y(t) = \eta + \int_\tau^t f(s, y(s))\, ds, \quad \text{for} \quad t \in [a, b].$$

For a general differential equation (3.1) all existence theorems involve some sort of a limiting process to pass from an "approximate solution" to an actual solution of this equation. For the present discussion the measure of approximation to an actual solution is phrased in terms of the integral operator of (3.2). A vector function $y(t)$ defined on $[a, b]$ is said to be an *ε-approximate solution* of (3.1) on $[a, b]$ if on this interval $y(t)$ is continuous, $f(t, y(t))$ is integrable, and for a suitable constant vector $\eta \equiv (\eta_\alpha)$ and value

with mesh $|\Pi|$ equal to the maximum length of subinterval $t_{r+1} - t_r$, $(r = -q, \ldots, p-1)$, and define $y(t) = y(t; \Pi)$ as follows:

$$\begin{aligned}
&\text{(a) } y(\tau) = \eta, \\
(3.10)\quad &\text{(b) } y(t) = y(t_j) + f(t_j, y(t_j))(t - t_j), \quad \text{for} \quad t \in (t_j, t_{j+1}], \\
&\qquad\qquad (j = 0, 1, \ldots, p-1), \\
&\text{(c) } y(t) = y(t_k) + f(t_k, y(t_k))(t - t_k), \quad \text{for} \quad t \in [t_{k-1}, t_k), \\
&\qquad\qquad (k = 0, -1, \ldots, -q+1).
\end{aligned}$$

It will be established first that $|y(t_r)| \leq \nu$, $(r = -q, \ldots, p)$, where

$$(3.11)\quad \begin{aligned}
&\nu = |\eta| \exp\{\kappa(b-a)\} + \frac{\kappa_1}{\kappa}[\exp\{\kappa(b-a)\} - 1], \quad \text{if} \quad \kappa > 0, \\
&\nu = |\eta| + \kappa_1(b-a), \quad \text{if} \quad \kappa = 0.
\end{aligned}$$

Indeed, if $\delta_h = t_{h+1} - t_h$, $(h = 0, 1, \ldots, p-1)$, then, in view of (3.8), it follows from (a), (b) of (3.10) that $\eta^{(j)} = y(t_j)$, $(j = 0, 1, \ldots, p)$, satisfies (2.10) with the constants κ, κ_1 of (3.8) and from Corollary 3 of Theorem 2.1 it follows that $|y(t_j)| \leq \nu$, $(j = 0, 1, \ldots, p)$. In a similar fashion it follows from (a), (c) of (3.10) that $|y(t_k)| \leq \nu$, $(k = 0, -1, \ldots, -q)$. As $y(t)$ is linear on each subinterval of Π it then follows that $|y(t)| \leq \nu$ on $[a, b]$. It is to be remarked that the bound ν of (3.11) is independent of the value τ on $[a, b]$, and depends only upon $|\eta|$, $b - a$, and the constants κ, κ_1 of (3.8).

Now in view of (3.8) the constant $\mu = \kappa\nu + \kappa_1$ is such that $|f(t, y)| \leq \mu$ on the set $\Delta_\nu = \{(t, y) \mid t \in [a, b], |y| \leq \nu\}$, and consequently for arbitrary partitions (3.9) of $[a, b]$ the vector function $y(t) = y(t; \Pi)$ defined by (3.10) satisfies on $[a, b]$ a Lipschitz condition with constant μ. From the continuity of $f(t, y)$ it follows that for a given $\delta > 0$ the function $|f(t^1, y^1) - f(t^2, y^2)|$ has a maximum $\omega_\nu(\delta)$ on the set of values t^1, y^1, t^2, y^2 with $(t^1, y^1) \in \Delta_\nu$, $(t^2, y^2) \in \Delta_\nu$, $|t^1 - t^2| \leq \delta$, $|y^1 - y^2| \leq \mu\delta$, and $\omega_\nu(\delta) \to 0$ as $\delta \to 0$.

Let the step function $g(t; \Pi)$ be defined in terms of $y(t) = y(t; \Pi)$ as follows:

$$\begin{aligned}
g(t; \Pi) &= f(t_j, y(t_j)), \quad \text{for} \quad t \in [t_j, t_{j+1}), \\
&\qquad (j = 0, 1, \ldots, p-1); \; g(b; \Pi) = f(b, y(b)); \\
g(t; \Pi) &= f(t_k, y(t_k)), \quad \text{for} \quad t \in (t_{k-1}, t_k], \\
&\qquad (k = 0, -1, \ldots, -q+1); \; g(a; \Pi) = f(a, y(a)).
\end{aligned}$$

Then $y(t) = y(t; \Pi)$ has the integral representation

$$y(t) = \eta + \int_\tau^t g(s; \Pi)\, ds \quad \text{for} \quad t \in [a, b]$$

Moreover, $|g(t; \Pi) - f(t, y(t))| \leq \omega_\nu(|\Pi|)$ on $[a, b]$, and

$$\left|\int_\tau^t [g(s; \Pi) - f(s, y(s))]\, ds\right| \leq |t - \tau|\, \omega_\nu(|\Pi|),$$

so that uniformly on $[a, b]$,

$$(3.12) \qquad |y(t) - \eta - \int_\tau^t f(s, y(s))\, ds| \leq (b - a)\, \omega_\nu(|\Pi|).$$

Consequently, if Π is a partition (3.9) of $[a, b]$ with $|\Pi|$ so small that $\omega_\nu(|\Pi|)$ is less than $\varepsilon/(b - a)$, then the polygonal vector function $y(t) = y(t; \Pi)$ defined by (3.10) is an ε-approximate solution of (3.1) on $[a, b]$.

Second proof of conclusion (i). The device used in the following determination of approximate solutions was introduced by Tonelli in the proof of existence theorems for functional equations of Volterra type.

For δ a positive number so small that $a < \tau - \delta$ and $\tau + \delta < b$, define $y(t) = y(t; \delta)$ as follows:

$$(3.13) \qquad \begin{aligned} &\text{(a)} \quad y(t) = \eta, && \text{for} \quad t \in [\tau - \delta, \tau + \delta]; \\ &\text{(b)} \quad y(t) = \eta + \int_\tau^{t-\delta} f(s, y(s))\, ds, && \text{for} \quad t \in [\tau + \delta, b]; \\ &\text{(c)} \quad y(t) = \eta + \int_\tau^{t+\delta} f(s, y(s))\, ds, && \text{for} \quad t \in [a, \tau - \delta]. \end{aligned}$$

It is to be noted that the definition of $y(t)$ is iterative in character on the individual subintervals $[\tau, b]$ and $[a, \tau]$; for example, from (a), (b) of (3.13) the knowledge of $y(t)$ on a subinterval $[\tau, \tau + j\delta]$ of $[\tau, b)$ provides the value of $y(t)$ on a larger interval $[\tau, \tau + (j + 1)\delta]$ or $[\tau, b]$, according as $\tau + (j + 1)\delta < b$ or $\tau + (j + 1)\delta \geq b$.

The function $y(t)$ defined by (3.13) satisfies $|y(t)| \leq \nu$ on $[a, b]$, where ν is the value (3.11). Indeed, since

$$\left|\int_\tau^{t-\delta} f(s, y(s))\, ds\right| \leq \int_\tau^t |f(s, y(s))|\, ds, \quad \text{for} \quad t \in [\tau + \delta, b],$$
$$\left|\int_\tau^{t+\delta} f(s, y(s))\, ds\right| \leq \int_t^\tau |f(s, y(s))|\, ds, \quad \text{for} \quad t \in [a, \tau - \delta],$$

it follows from (3.8) that $\rho(t) = |y(t)|$ satisfies (2.6) with $\kappa_0 = |\eta|$ and κ, κ_1 the constants of (3.8), and the inequality $|y(t)| \leq \nu$ results from Corollary 1 of Theorem 2.1. In view of (3.8) the constant $\mu = \kappa\nu + \kappa_1$ is such that $|f(t, y(t))| \leq \mu$ for $t \in [a, b]$. Now if $\zeta(t) = \zeta(t; \delta)$ is defined as $\zeta(t) = \tau$ on $[\tau - \delta, \tau + \delta]$, $\zeta(t) = t - \delta$ on $[\tau + \delta, b]$, $\zeta(t) = t + \delta$ on $[a, \tau - \delta]$, then for $t \in [a, b]$ we have $|\zeta(t) - t| \leq \delta$ and

$$\left|\int_t^{\zeta(t)} |f(s, y(s))|\, ds\right| \leq \mu\, |\zeta(t) - t| \leq \mu\delta.$$

Consequently, the vector function $y(t) = y(t; \delta)$ defined by (3.13) satisfies for $t \in [a, b]$ the inequality

$$(3.14) \qquad \left| y(t) - \eta - \int_\tau^t f(s, y(s))\, ds \right| = \left| \int_t^{\zeta(t)} f(s, y(s))\, ds \right| \leq \mu\delta,$$

and $y(t)$ is an ε-approximate solution of (3.1) satisfying (3.3) whenever $0 < \delta < \min[b - \tau, \tau - a, \varepsilon/\mu]$.

The above theorem establishes the existence of a solution of (3.1) that is defined on the interval $[a, b]$ and passes through the initial point (τ, η). Now there may be more than one solution of (3.1) passing through the given point (τ, η), and for any one such solution there is the possibility that it may be impossible to extend its interval of definition to the whole interval $[a, b]$. In Theorem 3.3 below it is proved that under the hypotheses of Theorem 3.2 such extension of the interval of definition is always possible.

Theorem 3.3. *If the vector function $f(t, y)$ is continuous and satisfies* (3.8) *on the infinite strip $\Delta = [a, b] \times \mathbf{C}_n$, and $y = \phi(t)$ is a solution of* (3.1) *on a nondegenerate subinterval I of $[a, b]$, then there exists on $[a, b]$ a solution $y = y(t)$ of* (3.1) *such that $y(t) \equiv \phi(t)$ for $t \in I$.*

Let a_1 and b_1, $(a_1 < b_1)$, be the end-points of I, and for τ an arbitrarily chosen point of (a_1, b_1) let δ be such that $0 < \delta < \delta_0 = \min[\tau - a_1, b_1 - \tau]$. The desired result will be obtained by applying the method of the second proof of conclusion (i) of Theorem 3.2 to the initial point $(b_1 - \delta, \phi(b_1 - \delta))$ on the interval $[b_1 - \delta, b]$, and to the initial point $(a_1 + \delta, \phi(a_1 + \delta))$ on the interval $[a, a_1 + \delta]$. Now if we set $\eta = \phi(\tau)$ then

$$\phi(t) = \eta + \int_\tau^t f(s, \phi(s))\, ds, \quad \text{for} \quad t \in I,$$

and the involved procedure may be described concisely as the determination of the unique continuous vector function $y(t, \delta)$ on $[a, b]$ such that

$$y(t, \delta) \equiv \phi(t) \quad \text{for} \quad t \in [a_1 + \delta, b_1 - \delta],$$

$$y(t, \delta) = \eta + \int_\tau^{\zeta(t)} f(s, y(s, \delta))\, ds, \quad \text{for} \quad t \in [a, b],$$

where the function $\zeta(t) = \zeta(t, \delta)$ is defined as follows:

$$\zeta(t) = t \quad \text{on} \quad [a_1 + \delta, b_1 - \delta]; \qquad \zeta(t) = b_1 - \delta \quad \text{on} \quad [b_1 - \delta, b_1];$$
$$\zeta(t) = t - \delta \quad \text{on} \quad [b_1, b]; \qquad \zeta(t) = a_1 + \delta \quad \text{on} \quad [a_1, a_1 + \delta];$$
$$\zeta(t) = t + \delta \quad \text{on} \quad [a, a_1].$$

Since $\tau \leq \zeta(t) \leq t$ on $[\tau, b]$, $t \leq \zeta(t) \leq \tau$ on $[a, \tau]$, and $|\zeta(t) - t| \leq \delta$ on $[a, b]$, as in the proofs of the corresponding results for the vector function

(3.13) it follows that for ν the value (3.11) and $\mu = \kappa\nu + \kappa_1$ we have

$$(3.15)\quad |y(t, \delta)| \leq \nu, \qquad |y(t, \delta) - \eta - \int_\tau^t f(s, y(s, \delta))\, ds| \leq \mu\delta,$$

$$\text{for} \quad t \in [a, b].$$

Consequently, if $0 < \delta_m < \delta_0$, $(m = 1, 2, \ldots)$, and $\{\delta_m\} \to 0$ as $m \to \infty$, Theorem 3.1 implies that there is a subsequence $\{y(t, \delta_{m_k})\}$, $(m_1 < m_2 < \cdots)$, which converges uniformly on $[a, b]$ to a solution $y(t)$ of (3.1) satisfying $y(\tau) = \eta$. As $y(t, \delta) = \phi(t)$ for $t \in [a_1 + \delta, b_1 - \delta]$, it follows that the limit vector function $y(t)$ is identical with $\phi(t)$ on the open interval (a_1, b_1); moreover, if t_0 is an end-point of I which is contained in I, then by continuity it follows that $y(t_0) = \phi(t_0)$, so that $y(t) = \phi(t)$ for $t \in I$.

As a consequence of Theorem 3.2 we have the following more classical local existence theorem.

Theorem 3.4. *If the vector function $f(t, y)$ is continuous on an $(n + 1)$-dimensional interval $\square = \{(t, y) \mid |t - \tau| \leq \varepsilon, |y - \eta| \leq \delta\}$ and μ is such that $|f(t, y)| \leq \mu$ on $\square$, then for $\varepsilon_1 = \min[\varepsilon, \delta/\mu]$ there is a solution $y = y(t)$ of* (3.1) *on $[\tau - \varepsilon_1, \tau + \varepsilon_1]$ satisfying $y(\tau) = \eta$ and $|y(t) - \eta| < \delta$ on $(\tau - \varepsilon_1, \tau + \varepsilon_1)$. If $y = \phi(t)$ is a solution of* (3.1) *on a nondegenerate subinterval I of $[\tau - \varepsilon_1, \tau + \varepsilon_1]$ containing τ and $\phi(\tau) = \eta$, then there exists on $[\tau - \varepsilon_1, \tau + \varepsilon_1]$ a solution $y(t)$ of* (3.1) *such that $y(t) = \phi(t)$ on I, and for any such extension the inequality $|y(t) - \eta| < \delta$ holds for $t \in (\tau - \varepsilon_1, \tau + \varepsilon_1)$.*

Define the vector function $h(t, y)$ on $\Delta = [\tau - \varepsilon, \tau + \varepsilon] \times \mathbf{C}_n$ as follows: $h(t, y) = f(t, y)$ on $\square$; $h(t, y) = f(t, \eta + \delta(y - \eta)/|y - \eta|)$ if $|t - \tau| \leq \varepsilon$, $|y - \eta| > \delta$. Then $h(t, y)$ is continuous and satisfies $|h(t, y)| \leq \mu$ on Δ, so that $h(t, y)$ satisfies on Δ an inequality of the form (3.8) with $\kappa_1 = \mu$ and κ an arbitrary positive number. From Theorem 3.2 it then follows that the vector differential equation $y' = h(t, y)$ has on $[\tau - \varepsilon, \tau + \varepsilon]$ a solution $y = y(t)$ satisfying $y(\tau) = \eta$. Moreover, since

$$|y(t) - \eta| \leq \left| \int_\tau^t |h(s, y(s))|\, ds \right| \leq \mu\, |t - \tau|,$$

on $[\tau - \varepsilon, \tau + \varepsilon]$, and $f(t, y) \equiv h(t, y)$ on $\square$, it follows that $y(t)$ is a solution of (3.1) on $[\tau - \varepsilon_1, \tau + \varepsilon_1]$ satisfying $y(\tau) = \eta$ and $|y(t) - \eta| < \delta$ on $(\tau - \varepsilon_1, \tau + \varepsilon_1)$. If $y = \phi(t)$ is a solution of (3.1) on a nondegenerate subinterval I of $[\tau - \varepsilon_1, \tau + \varepsilon_1]$ which contains τ, then on this subinterval $y = \phi(t)$ is also a solution of the auxiliary differential equation $y' = h(t, y)$. Application of the result of Theorem 3.3 to this latter equation, and the fact that the resulting vector function $y(t)$ has graph $\{(t, y(t)) \mid t \in [\tau - \varepsilon_1, \tau + \varepsilon_1]\}$ in $\square$, then implies the final conclusion of the theorem.

If $\square = \{(t, y) \mid |t - \tau| \leq \varepsilon, |y - \eta| \leq \delta\}$, and $0 < \varepsilon^0 < \varepsilon$, $0 < \delta^0 < \delta$, then $\square^0 = \{(t, y) \mid |t - \tau^0| \leq \varepsilon - \varepsilon^0, |y - \eta^0| \leq \delta - \delta^0\}$ is contained in $\square$

whenever $|\tau^0 - \tau| \leq \varepsilon^0$, $|\eta^0 - \eta| \leq \delta^0$. Moreover, if $\varepsilon_1 = \min[\varepsilon, \delta/\mu]$ as in Theorem 3.4, and $0 < \varepsilon_2 < \varepsilon_1$, $\varepsilon_3 = \min[\varepsilon - \varepsilon^0, (\delta - \delta^0)/\mu]$, then by elementary inequalities it follows that if ε^0, δ^0 are so small that $\varepsilon^0 < (\varepsilon - \varepsilon_2)/2$ and $\varepsilon^0 + \delta^0/\mu < \delta/\mu - \varepsilon_2$, then $\tau^0 - \varepsilon_3 < \tau - \varepsilon_2$ and $\tau + \varepsilon_2 < \tau^0 + \varepsilon_3$ whenever $\tau^0 \in [\tau - \varepsilon^0, \tau + \varepsilon^0]$. Consequently, as a corollary to Theorem 3.4 one has the following result.

Corollary. *Suppose that on* $\Box = \{(t, y) \mid |t - \tau| \leq \varepsilon, |y - \eta| \leq \delta\}$ *the vector function* $f(t, y)$ *is continuous,* μ *is such that* $|f(t, y)| \leq \mu$ *on* $\Box$, *and* $\varepsilon_1 = \min[\varepsilon, \delta/\mu]$. *If* $0 < \varepsilon_2 < \varepsilon_1$, *then there exist positive constants* ε^0, δ^0 *such that* $\varepsilon^0 < \varepsilon_2$ *and if* $|\tau^0 - \tau| \leq \varepsilon^0$, $|\eta^0 - \eta| \leq \delta^0$ *then there is a solution* $y = y(t)$ *of* (3.1) *on* $[\tau - \varepsilon_2, \tau + \varepsilon_2]$ *satisfying* $y(\tau^0) = \eta^0$, *and* $|y(t) - \eta| < \delta$ *for* $t \in (\tau - \varepsilon_2, \tau + \varepsilon_2)$. *If* $y = \phi(t)$ *is a solution of* (3.1) *on a nondegenerate subinterval* I *of* $[\tau - \varepsilon_2, \tau + \varepsilon_2]$ *containing* τ^0 *and* $\phi(\tau^0) = \eta^0$, *then there exists on* $[\tau - \varepsilon_2, \tau + \varepsilon_2]$ *a solution* $y(t)$ *of* (3.1) *such that* $y(t) = \phi(t)$ *for* $t \in I$; *moreover, for any such extension the inequality* $|y(t) - \eta| < \delta$ *holds for* $t \in (\tau - \varepsilon_2, \tau + \varepsilon_2)$.

Now suppose that the vector function $f(t, y)$ of (3.1) is continuous in an open region $\mathfrak{R}$ of (t, y)-space, and that $y = \phi(t)$ is a solution of this differential equation on a nondegenerate interval I, with graph $\{(t, \phi(t)) \mid t \in I\}$ in $\mathfrak{R}$. The interval I is said to be a *right-hand maximal interval of existence* of ϕ if there does not exist a solution $y(t)$, $t \in I_0$, of (3.1) with graph $\{(t, y(t)) \mid t \in I_0\}$ in $\mathfrak{R}$, where $I \subset I_0$, $y(t) = \phi(t)$ for $t \in I$, and there is a $t_0 \in I_0$ such that $t < t_0$ for every $t \in I$. A *left-hand maximal interval of existence* for a solution of (3.1) is defined similarly.

If I is a right-hand maximal interval of existence for a solution $\phi(t)$ of (3.1), and b is the right-hand endpoint of I, then b cannot be finite and belong to I, since in this case $(b, \phi(b))$ would belong to $\mathfrak{R}$ and Theorem 3.4 would imply the existence of an extension of this solution to an interval with right-hand endpoint $b + \varepsilon_1 > b$. Similarly, if I is a left-hand maximal interval of existence for a solution of (3.1), then the left-hand endpoint of I cannot be finite and belong to I.

If an open interval (a, b) is an interval of existence of a vector function $y(t)$ with graph $\{(t, y(t)) \mid t \in (a, b)\}$ in $\mathfrak{R}$, then $y(t)$ is said to *tend to the boundary of* $\mathfrak{R}$ *as* $t \to b$ if either $b = +\infty$, or $b < +\infty$ and for S an arbitrary compact subset of $\mathfrak{R}$ there is a corresponding $b_0 < b$ such that if $t \in (b_0, b)$ then $(t, y(t)) \notin S$. An alternate specification of this latter condition is that there is no point (b, η) in $\mathfrak{R}$ which is a limit point of the graph $\{(t, y(t)) \mid t \in (a, b)\}$. The concept of $y(t)$ tending to the boundary of $\mathfrak{R}$ as $t \to a$ is defined similarly.

Theorem 3.5. *If* $f(t, y)$ *is a continuous vector function in an open region* $\mathfrak{R}$ *of* (t, y)*-space, and* $y^{(0)}(t)$, $t \in (a_0, b_0)$, *is a solution of* (3.1) *with graph in* $\mathfrak{R}$, *then there exists a solution* $y(t)$, $t \in (a, b)$, *of* (3.1) *which is an extension of* $y^{(0)}(t)$ *and such that* (a, b) *is a maximal interval of existence for* $y(t)$; *moreover,* $y(t)$ *tends to the boundary of* $\mathfrak{R}$ *as* $t \to a$ *and as* $t \to b$.

Let the rational real numbers greater than b_0 be ordered as a sequence $\{b_m\}$, $(m = 1, 2, \ldots)$. If (a_0, b_0) is not a right-hand maximal interval of existence of the solution $y^{(0)}(t)$, then there exists a smallest integer m_1 such that there is a solution $y^{(1)}(t)$, $t \in (a_0, b_{m_1})$, of (3.1) which is an extension of $y^{(0)}(t)$. Now suppose that integers m_j, $(j = 1, \ldots, k)$, have been determined, where $m_1 < m_2 < \cdots < m_k$, and there are solutions $y^{(j)}(t)$, $t \in (a_0, b_{m_j})$, of (3.1) such that $y^{(j)}(t)$ is an extension of $y^{(j-1)}(t)$ for $j = 1, \ldots, k$. If (a_0, b_{m_k}) is not a right-hand maximal interval of existence of $y^{(k)}(t)$, let m_{k+1} be the smallest integer such that $b_{m_{k+1}} > b_{m_k}$, and there is a solution $y^{(k+1)}(t)$, $t \in (a_0, b_{m_{k+1}})$, which is an extension of $y^{(k)}(t)$. Proceeding in this fashion, there is obtained a finite or denumerably infinite sequence of solutions $y^{(j)}(t)$, $t \in (a_0, b_{m_j})$. If $I = (a_0, b)$ is the union of the intervals (a_0, b_{m_j}), then for $t_0 \in I$ and k such that $t_0 \in (a_0, b_{m_k})$, the condition $y(t_0) = y^{(k)}(t_0)$ unambiguously defines a solution $y(t)$ of (3.1) on (a_0, b) with graph in $\mathfrak{R}$. Moreover, the manner of choice of the b_{m_j} implies that (a_0, b) is a right-hand maximal interval of existence of $y(t)$.

Now if $b < +\infty$ it will be shown that the assumption that $y(t)$ does not tend to the boundary of $\mathfrak{R}$ as $t \to b$ leads to a contradiction. Indeed, if $(b, \eta) \in \mathfrak{R}$ and (b, η) is a limit point of the graph $\{(t, y(t)) \mid t \in (a, b)\}$, let ε, δ be positive constants such that the set $\square = \{(t, y) \mid |t - b| \leq \varepsilon, |y - \eta| \leq \delta\}$ is in $\mathfrak{R}$, and let μ be a constant such that $|f(t, y)| \leq \mu$ on $\square$. If $0 < \varepsilon_2 < \min[\varepsilon, \delta/\mu]$, and ε^0, δ^0 are positive constants such that $\varepsilon^0 < (\varepsilon - \varepsilon_2)/2$ and $\varepsilon^0 + \delta^0/\mu < \delta/\mu - \varepsilon_2$ then, as there exists a $\tau^1 \in [b - \varepsilon^0, b)$ for which $|y(\tau^1) - \eta| \leq \delta^0$, it follows from the Corollary to Theorem 3.4 that there is a solution $y^{(1)}(t)$, $t \in [\tau^1, b + \varepsilon_2]$ of (3.1) such that $y^{(1)}(t) = y(t)$ for $t \in [\tau^1, b)$. Consequently, there is a solution of (3.1) which is an extension of $y(t)$ to the interval $(a_0, b + \varepsilon_2)$, contrary to the fact that (a_0, b) is a right-hand maximal interval of existence of $y(t)$.

A similar argument establishes the existence of a solution $y(t)$, $t \in (a, b_0)$, of (3.1) which is an extension of $y^{(0)}(t)$, such that (a, b_0) is a left-hand maximal interval of existence of $y(t)$, and $y(t)$ tends to the boundary of $\mathfrak{R}$ as $t \to a$.

For a vector function $f(t, y)$ that is continuous in an open region $\mathfrak{R}$ of (t, y)-space Theorems 3.4 and 3.5 imply that if $(\tau, \eta) \in \mathfrak{R}$ there exists a solution $y(t)$, $t \in (a, b)$ such that $y(\tau) = \eta$, (a, b) is the maximal interval of existence of $y(t)$, and the graph $\{(t, y(t)) \mid t \in (a, b)\}$ of $y(t)$ "extends from boundary to boundary of $\mathfrak{R}$." It is to be emphasized that in general such a solution is not unique and that in each case the maximal interval of existence is associated with an individual solution. From Theorem 3.4 it follows, however, that for every such solution of (3.1) passing through (τ, η) the corresponding maximal interval of existence contains the interval $[\tau - \varepsilon_1, \tau + \varepsilon_1]$, with ε_1 defined as in the statement of that theorem.

Problems I.3

1. If the vector function $f(t, y)$ is continuous on the infinite strip $\Delta = [a, b] \times \mathbf{C}_n$, show that the hypothesis that there are constants $\kappa > 0$, $\kappa_1 \geq 0$, $\beta > 1$ such that

$$|f(t, y)| \leq \kappa\, |y|^\beta + \kappa_1, \quad \text{for} \quad (t, y) \in \Delta,$$

is not strong enough to imply that the interval of definition of a solution of (3.1) passing through a given point (τ, η) of Δ may be extended to the whole interval $[a, b]$.

Hint. Consider the specific differential equation $y' = |y|^\beta$.

2.[ss] If $f(t, y)$ is continuous and satisfies condition (3.8) on the infinite strip $\Delta = [a, b] \times \mathbf{C}_n$, and $\{y^{(m)}(t)\}$ is a sequence of continuous vector functions on $[a, b]$ satisfying the condition

$$\lim_{m\to\infty} \left| y^{(m)}(t) - \eta^{(m)} - \int_{\tau^{(m)}}^{t} f(s, y^{(m)}(s))\, ds \right| = 0, \quad \text{uniformly for} \quad t \in [a, b],$$

where $\{(\tau^{(m)}, \eta^{(m)})\}$ is a sequence of points in Δ converging to a limit point (τ, η), then there exists a subsequence $\{y^{(m_k)}(t)\}$, $(m_1 < m_2 < \cdots)$, which converges uniformly on $[a, b]$ to a continuous vector function $y(t)$ that is a solution of $y' = f(t, y)$ on $[a, b]$ passing through the initial point (τ, η).

Hint. Modify suitably the proof of Theorem 3.1.

3.[ss] Suppose that $f(t, y), f^{(m)}(t, y)$, $(m = 1, 2, \ldots)$, are continuous vector functions on a compact set S in (t, y)-space, that $\{f^{(m)}(t, y)\} \to f(t, y)$ uniformly on S, and that there is an interval $[a, b]$ such that for $m = 1, 2, \ldots$ there is a solution $y = y^{(m)}(t)$ of

$$(3.16_m) \qquad y' = f^{(m)}(t, y), \qquad t \in [a, b],$$

with graph $\{(t, y^{(m)}(t)) \mid t \in [a, b]\}$ in S. Prove that there exists a subsequence $\{y^{(m_k)}(t)\}$ which is uniformly convergent on $[a, b]$ to a vector function $y(t)$, and that $y(t)$ is a solution of $y' = f(t, y)$ with graph $\{(t, y(t)) \mid t \in [a, b]\}$ in S. In particular, if there exists a sequence of values $t_m \in [a, b]$ such that $\{(t_m, y^{(m)}(t_m))\}$ converges to a point (τ, η), and there is a unique solution $y(t)$ of $y' = f(t, y)$ on $[a, b]$ which passes through (τ, η) and has graph in S, then the sequence $\{y^{(m)}(t)\}$ converges uniformly on $[a, b]$ to $y(t)$.

Hint. Show that there exists a constant κ_1 such that $|f^{(m)}(t, y)| \leq \kappa_1$ for $t, y) \in S$ and $m = 1, 2, \ldots$. Deduce that the vector functions $y^{(m)}(t)$ are uniformly bounded and equi-continuous on $[a, b]$, and apply the Ascoli selection theorem.

4.[s] Suppose that on the rectangular region $\Delta_\nu = \{(t, u) \mid t \in [a, b], u \in [-\nu, \nu]\}$ in the real (t, u)-plane the real-valued functions $g_1(t, u)$, $g_2(t, u)$ are continuous and satisfy the inequality $g_2(t, u) > g_1(t, u)$. If $u_1(t)$ and $u_2(t)$ are real-valued solutions of the differential equations

$$u'_\alpha = g_\alpha(t, u_\alpha), \qquad (\alpha = 1, 2), \tag{3.17}$$

with graphs $\{(t, u_\alpha(t)) \mid t \in [a, b]\}$ in Δ_ν, and there is a value τ on $[a, b]$ such that $u_1(\tau) = u_2(\tau)$, show that $u_2(t) > u_1(t)$ for $t \in (\tau, b]$, and $u_2(t) < u_1(t)$ for $t \in [a, \tau)$. Conclude that if there is a value $\tau_0 \in [a, b]$ such that $u_2(\tau_0) \geq u_1(\tau_0)$ then $u_2(t) > u_1(t)$ for $t \in (\tau_0, b]$; correspondingly, if $u_1(\tau_0) \geq u_2(\tau_0)$, then $u_1(t) > u_2(t)$ for $t \in [a, \tau_0)$.

Hint. Note that if t_0 is a value such that $u_1(t_0) = u_2(t_0)$ then $u_1'(t_0) < u_2'(t_0)$, and consequently there exists an $\varepsilon > 0$ such that if $t \in [a, b] \cap [t_0 - \varepsilon, t_0 + \varepsilon]$ then $u_2(t) > u_1(t)$ for $t > t_0$ and $u_2(t) < u_1(t)$ for $t < t_0$.

5. If $u(t)$ is a real-valued solution of the differential equation $u' = 1 + t^2 + u^2$, show that $u(t)$ cannot be a solution on the whole infinite interval $(-\infty, \infty)$. Determine a more precise bound on the length of an interval on which there exists a solution of this differential equation.

Hint. Use the result of the preceding problem to compare a solution of the given equation with a solution of an equation $u' = a^2 + u^2$, where a is a positive constant.

6.[s] Suppose that the vector function $f(t, y)$ is continuous in an open region $\mathfrak{R}$ of (t, y)-space, and for $(\tau, \eta) \in \mathfrak{R}$ let $\mathcal{F}(\tau, \eta)$ denote the set of all solutions $y(t)$, $t \in I$, of (3.1) that pass through the point (τ, η) and have graph $\{(t, y(t)) \mid t \in I\}$ in $\mathfrak{R}$. Moreover, suppose that there is a compact interval $[a, b]$ which contains $t = \tau$ and is such that each solution of $\mathcal{F}(\tau, \eta)$ has an extension whose interval of existence contains $[a, b]$. Prove that: (i) there exists a compact set S in $\mathfrak{R}$ such that if $y(t)$, $t \in I$, is a member of $\mathcal{F}(\tau, \eta)$ and $I \supset [a, b]$, then $\{(t, y(t)) \mid t \in [a, b]\} \subset S$; (ii) if $s \in [a, b]$, and $\mathcal{F}(\tau, \eta; s)$ is the set of points (s, ξ) for which there exists a solution $y(t)$, $t \in I$, of $\mathcal{F}(\tau, \eta)$ with $s \in I$ and $y(s) = \xi$, then $\mathcal{F}(\tau, \eta; s)$ is a closed set.

Hints. (i) Let $\{T_n\}$, $(n = 1, 2, \ldots)$, be a sequence of nondegenerate $(n + 1)$-dimensional intervals $\{(t, y) \mid |t - t_n^0| \leq \varepsilon_n^0, |y - y_n^0| \leq \delta_n^0\}$ which lie in $\mathfrak{R}$, and such that $\mathfrak{R}$ is the union of the sets T_n; for example, $\{T_n\}$ may be chosen as the collection of $(n + 1)$-dimensional intervals lying in $\mathfrak{R}$, with centers (t_n^0, y_n^0) having rational values, and with the corresponding dimensions ε_n^0, δ_n^0 positive rational numbers, ordered as a sequence. If $S_m = \bigcup_{n=1}^m T_n$ then each S_m is a compact subset of $\mathfrak{R}$, and if S is any compact subset of $\mathfrak{R}$ then there exists an m such that $S \subset S_m$. For $\tau < s \leq b$, let $P(s)$ denote the property that there exists a positive integer m such that if $y(t)$,

$t \in I$, is a solution of $\mathcal{F}(\tau, \eta)$ with $[\tau, s] \subset I$ then $\{(t, y(t)) \mid t \in [\tau, s]\} \subset S_m$. Note that if $\square = \{(t, y) \mid |t - \tau| \leq \varepsilon, |y - \eta| \leq \delta\} \subset \mathfrak{R}$, and ε_1 is determined as in Theorem 3.4, then property $P(s)$ holds if $s \in (\tau, \tau + \varepsilon_1]$. If $c \in (\tau, b]$ and for each $s \in (\tau, c)$ property $P(s)$ holds, use the result of Problem 3 above and the Cantor diagonal process to show that if $\{y^{(j)}(t), t \in I_j\}$ is any sequence of solutions of $\mathcal{F}(\tau, \eta)$ for which $[\tau, c) \subset I_j$, $(j = 1, 2, \ldots)$, then there exists a solution $y(t)$, $t \in [\tau, c)$, of (3.1) and a subsequence $\{y^{(j_k)}(t)\}$ which converges to $y(t)$ uniformly on each closed subinterval of $[\tau, c)$. Note that the hypotheses of the problem imply that the solution $y(t)$ may be extended to an interval $[\tau, c + \varepsilon_1]$, and with the aid of the Corollary to Theorem 3.4 conclude that there is a $\square_1 = \{(t, y) \mid |t - c| \leq \varepsilon_2, |y - y(c)| \leq \delta_2\}$ such that $\square_1 \subset \mathfrak{R}$ and for k sufficiently large all solutions $y^{(j_k)}(t)$ may be extended to the interval $[\tau, c + \varepsilon_2]$, and $\{(t, y^{(j_k)}(t)) \mid |t - c| \leq \varepsilon_2\} \subset \square_1$. With the aid of this result show that if c_1 is the supremum of values $c \in (\tau, b]$ such that property $P(s)$ holds for $s \in (\tau, c)$, then $c_1 = b$, and there exists a positive integer m such that $\{(t, y(t)) \mid t \in [\tau, b]\} \subset S_m$ for all $y(t)$, $t \in [\tau, b]$ that belong to $\mathcal{F}(\tau, \eta)$.

(ii) Use the results of Part (i) above, and of Problem 3.

4. EXTREME SOLUTIONS FOR REAL SCALAR DIFFERENTIAL EQUATIONS

Suppose that $g(t, u)$ is a real-valued continuous function of (t, u) in an open region $\mathfrak{R}$ of the real plane, and that $(\tau, \xi) \in \mathfrak{R}$. From Theorem 3.5 it follows that there is at least one solution of the scalar differential equation

$$u' = g(t, u) \tag{4.1}$$

passing through the initial point (τ, ξ), and that any such solution may be continued to a maximal interval of existence on which the solution curve extends from boundary to boundary of $\mathfrak{R}$. In general, there is more than one solution of (4.1) passing through the point (τ, ξ), and, as in Problem 6 above, the totality of such solutions will be denoted by $\mathcal{F}(\tau, \xi)$. In this section the character of $\mathcal{F}(\tau, \xi)$ for scalar equations (4.1) will be considered in more detail. As a preliminary result, we have the following lemma.

***Lemma* 4.1.** *Suppose that $g(t, u)$ is a real-valued continuous function in an open region $\mathfrak{R}$ of real (t, u)-space, and that $u_1(t)$ and $u_2(t)$ are two solutions of $\mathcal{F}(\tau, \xi)$ for* (4.1) *with a common interval of existence I containing $t = \tau$. If $u^+(t) = \max[u_1(t), u_2(t)]$ and $u^-(t) = \min[u_1(t), u_2(t)]$ for $t \in I$, then $u = u^+(t)$ and $u^-(t)$ are solutions of $\mathcal{F}(\tau, \xi)$ for* (4.1).

If $u_2(t) - u_1(t)$ does not change sign on I, then the result of the lemma is obvious. In general, if $t_0 \in I$ and $u_2(t_0) > u_1(t_0)$, then there is a maximal open subinterval of I containing t_0, and throughout which $u_2(t) > u_1(t)$. If there is only a finite number of such subintervals, then again the result of the lemma is clearly true. Now if there are infinitely many such subintervals, then they are denumerably infinite in number, and may be ordered as a sequence $\{I_j\}$, $(j = 1, 2, \ldots)$. For each positive integer m, let $w_m(t) = u_2(t)$ for $t \in I_j$, $(j = 1, 2, \ldots, m)$, $w_m(t) = u_1(t)$ for other values of t on I. Each $w_m(t)$, $t \in I$, is then a solution of $\mathcal{F}(\tau, \xi)$ for (4.1), and

$$u_1(t) \leq w_m(t) \leq w_{m+1}(t) \leq u^+(t), \quad \text{for} \quad t \in I, \quad \text{and} \quad m = 1, 2, \ldots;$$
$$\lim_{m\to\infty} w_m(t) = u^+(t), \quad \text{for} \quad t \in I.$$

Now by elementary argument it follows that $u^+(t) = \max[u_1(t), u_2(t)]$ is a continuous function on I, and thus $\{w_m(t)\}$ is a sequence of continuous functions which converges monotonically on I to the continuous limit function $u^+(t)$. From the Dini criterion, (see Appendix F), it then follows that the convergence of $\{w_m(t)\}$ to $u^+(t)$ is uniform on each compact subinterval of I. Now if I_0 is a compact subinterval of I containing $t = \tau$, then $\{(t, u_\alpha(t)) \mid t \in I_0, \alpha = 1, 2\}$ is a compact subset of $\mathfrak{R}$, and from Problem I.3:3 it follows that $u^+(t)$, $t \in I_0$, is a solution of (4.1) passing through the point (τ, ξ). Consequently, as I_0 may be an arbitrary compact subinterval of I containing $t = \tau$, it follows that $u^+(t)$, $t \in I$, is a solution of $\mathcal{F}(\tau, \xi)$ for (4.1). The corresponding result for $u^-(t)$, $t \in I$, may be established by a similar argument.

If $(\tau, \xi) \in \mathfrak{R}$, a solution $u = u_0(t)$, $t \in I_0$, of (4.1) that passes through the point (τ, ξ) is said to be a *maximal* {*minimal*} *solution* of (4.1) relative to (τ, ξ) if for all solutions $u(t)$, $t \in I$, of $\mathcal{F}(\tau, \xi)$ it is true that $u(t) \leq u_0(t)$, $\{u(t) \geq u_0(t)\}$, for all $t \in I \cap I_0$.

Theorem 4.1. *If $g(t, u)$ is a real-valued continuous function in an open region $\mathfrak{R}$ of real (t, u)-space, and $(\tau, \xi) \in \mathfrak{R}$, then for (4.1) relative to (τ, ξ) there is a unique maximal solution $u_M(t)$, $t \in (a_M, b_M)$, and a unique minimal solution $u_m(t)$, $t \in (a_m, b_m)$, such that (a_M, b_M) and (a_m, b_m) are maximal intervals of existence for $u_M(t)$ and $u_m(t)$, respectively. Moreover, if $a \geq$* $\max[a_M, a_m]$, $b \leq \min[b_M, b_m]$, *and such that the set*

$$\text{(4.2)} \qquad \{(t, u) \mid t \in [a, b], \quad u_m(t) \leq u \leq u_M(t)\}$$

belongs to $\mathfrak{R}$, then for $\tau^0 \in (a, b)$ and $u_m(\tau^0) \leq \xi^0 \leq u_M(\tau^0)$ there is a solution $u(t)$ of $\mathcal{F}(\tau, \xi)$ such that $u(\tau^0) = \xi^0$.

Let $\square = \{(t, u) \mid |t - \tau| \leq \xi, \ |u - \xi| \leq \delta\}$ be such that $\square \subset \mathfrak{R}$, μ a constant such that $|g(t, u)| \leq \mu$ on $\square$, and $\varepsilon_1 = \min[\varepsilon, \delta/\mu]$. From Theorem 3.4 it then follows that all solutions $u(t)$ of $\mathcal{F}(\tau, \xi)$ may be continued to the interval $[\tau - \varepsilon_1, \tau + \varepsilon_1]$, and that $|u(s_1) - u(s_2)| \leq \mu\,|s_2 - s_1|$ for arbitrary s_1, s_2 on this interval. For $t \in [\tau - \varepsilon_1, \tau + \varepsilon_1]$, let $u_M(t)$ and $u_m(t)$ denote, respectively, the supremum and infimum of the values $u(t)$ for solutions of $\mathscr{F}(\tau, \xi)$ for (4.1).

Now let $\{t_j\}$, $(j = 1, 2, \ldots)$, be a sequence of values on $[\tau - \varepsilon_1, \tau + \varepsilon_1]$ that is everywhere dense on this interval. In view of the definition of $u_M(t)$, for each pair of positive integers j and k there exists a solution $u_{j,k}(t)$ of $\mathscr{F}(\tau, \xi)$ whose interval of definition contains $[\tau - \varepsilon_1, \tau + \varepsilon_1]$, and is such that $u_{j,k}(t_j) > u_M(t_j) - (1/k)$. Consequently, if $u_k(t)$ denotes the maximum of $u_{j,k}(t)$ for $j = 1, \ldots, k$, it follows from Lemma 4.1 that $u_k(t)$ is a solution of $\mathscr{F}(\tau, \xi)$ whose interval of definition contains $[\tau - \varepsilon_1, \tau + \varepsilon_1]$, and is such that $u_k(t_j) > u_M(t_j) - (1/k)$ for $j = 1, \ldots, k$. Since $|u_k(s_1) - u_k(s_2)| \leq \mu\, |s_1 - s_2|$ for arbitrary s_1, s_2 on $[\tau - \varepsilon_1, \tau + \varepsilon_1]$, and $\{(t, u_k(t)) \mid t \in [\tau - \varepsilon_1, \tau + \varepsilon_1]\} \subset \square$, the functions $u_k(t)$ are uniformly bounded and equicontinuous on $[\tau - \varepsilon_1, \tau + \varepsilon_1]$. With the aid of the Ascoli selection theorem it then follows that there is a function $v(t)$ which is continuous on $[\tau - \varepsilon_1, \tau + \varepsilon_1]$, and a subsequence $\{u_{k_j}(t)\}$ such that $\{u_{k_j}(t)\}$ converges to $v(t)$ uniformly on this interval. Since $u_{k_j}(\tau) = \xi$, for $k = 1, 2, \ldots$, we have $v(\tau) = \xi$, and from Problem I.3:3 it follows that $v(t)$ is a solution of $\mathscr{F}(\tau, \xi)$ for (4.1). As $v(t_j) = u_M(t_j)$ for $j = 1, 2, \ldots$, it follows that if $u(t)$ is any solution of $\mathscr{F}(\tau, \xi)$ then $u(t) \leq v(t)$ throughout $[\tau - \varepsilon_1, \tau + \varepsilon_1]$, and consequently $u_M(t) = v(t)$ for $t \in [\tau - \varepsilon_1, \tau + \varepsilon_1]$, and $u_M(t)$, $t \in [\tau - \varepsilon_1, \tau + \varepsilon_1]$, is a maximal solution of (4.1) relative to (τ, ξ). The existence of a minimal solution $u_m(t)$, $t \in [\tau - \varepsilon_1, \tau + \varepsilon_1]$, of (4.1) relative to (τ, ξ) may be established by a similar argument.

If $u = u_\alpha(t)$, $t \in I_\alpha$, $(\alpha = 1, 2)$, are two maximal solutions of (4.1) relative to (τ, ξ), from the definition of maximal solution it follows that $u_1(t) = u_2(t)$ for $t \in I_1 \cap I_2$. Consequently, if $U(\tau, \xi)$ is the totality of maximal solutions $u(t)$, $t \in I$, of (4.1) relative to (τ, ξ), and with interval of definition I an open interval, the union I_0 of all intervals I belonging to solutions of $U(\tau, \xi)$ is an open interval (a_0, b_0) on which there is unambiguously defined a maximal solution $u_0(t)$ of (4.1) relative to (τ, ξ). Now $u_0(t)$ tends to the boundary of $\mathfrak{R}$ as $t \to b_0$, since otherwise, as in the proof of Theorem 3.4, this solution would have an extension to an interval $(a_0, b_0 + \varepsilon_1)$. In particular, $u_0(b_0) = \lim_{t \to b_0} u_0(t)$ would be such that $(b_0, u_0(b_0)) \in \mathfrak{R}$, and in view of the proof given above there would exist a $b_1 > b_0$ such that on $[b_0, b_1]$ there exists a maximal solution $u_1(t)$ of (4.1) relative to $(b_0, u_0(b_0))$. If $u(t)$, $t \in I$, is a solution of $\mathscr{F}(\tau, \xi)$, then $u(t) \leq u_0(t)$ for $t \in I \cap [\tau, b_0)$. Moreover, if $I \cap [b_0, b_1)$ is nonempty, and there is a point t_0 on this interval such that $u(t_0) > u_1(t_0)$, then by continuity there would exist a value $t_1' \in [b_0, t_0)$ such that $u(t_1') = u_1(t_1')$. The function $u_2(t)$ defined by $u_2(t) = u_1(t)$ for $t \in [b_0, t_1']$, and $u_2(t) = u(t)$ for $t \in (t_1', b_1) \cap I$, would then be a solution of $\mathscr{F}(b_0, u_0(b_0))$ that is not dominated by $u_1(t)$ on $[b_0, b_1) \cap I$, contrary to the definition of $u_1(t)$. Hence $u(t) \leq u_1(t)$ for $t \in [b_0, b_1) \cap I$, and the function $u_3(t) = u_0(t)$

for $t \in [\tau, b_0)$, $u_3(t) = u_1(t)$ for $t \in [b_0, b_1)$, would be a maximal solution of (4.1) relative to (τ, ξ), contrary to the definition of $u_0(t)$ and the interval (a_0, b_0). Consequently, (a_0, b_0) is a right-hand maximal interval of existence for $u_0(t)$ as a solution of (4.1). In a similar fashion it may be established that (a_0, b_0) is also a left-hand maximal interval of existence for $u_0(t)$. Finally, since two maximal solutions of (4.1) relative to (τ, ξ) must agree on the intersection of their intervals of existence, if $u_M(t)$, $t \in (a_M, b_M)$, is a maximal solution of (4.1) relative to (τ, ξ), and such that (a_M, b_M) is a maximal interval of existence of $u_M(t)$ as a solution of (4.1), it follows that $(a_M, b_M) = (a_0, b_0)$ and $u_M(t) \equiv u_0(t)$.

The conclusions of the theorem concerning the existence of a unique minimal solution $u_m(t)$, $t \in (a_m, b_m)$, of (4.1) relative to (τ, ξ), and such that (a_m, b_m) is the maximal interval of existence for $u_m(t)$, may be established in a similar fashion.

Now suppose that $a \geq \max[a_M, a_m]$, $b \leq \min[b_M, b_m]$, and such that the set (4.2) belongs to $\mathfrak{R}$, while $\tau^0 \in (a, b)$, and $u_m(\tau^0) \leq \xi^0 \leq u_M(\tau^0)$; for preciseness of statement we shall assume $\tau^0 > \tau$, since the case of $\tau^0 < \tau$ may be treated in an analogous fashion. If $\xi^0 = u_M(\tau^0)$ or $\xi^0 = u_m(\tau^0)$, then the respective solution $u_M(t)$ or $u_m(t)$ satisfies the final conclusion of the theorem. If $u_m(\tau^0) < \xi^0 < u_M(\tau^0)$, then there exist positive constants ε^0, δ^0 such that $\square^0 = \{(t, u) \mid |t - \tau^0| \leq \varepsilon^0, |u - \xi^0| \leq \delta^0\}$ is in $\mathfrak{R}$, and neither of the sets $\{(t, u_M(t)) \mid t \in [\tau^0 - \varepsilon^0, \tau^0]\}$, $\{(t, u_m(t)) \mid t \in [\tau^0 - \varepsilon^0, \tau^0]\}$ has a point in common with $\square^0$. Theorem 3.5 implies the existence of a solution $u_1(t)$, $t \in (a_1, b_1)$, of (4.1) such that $u_1(\tau^0) = \xi^0$, and (a_1, b_1) is a maximal interval of existence of $u_1(t)$ as a solution of (4.1). Now if $|g(t, u)| \leq \mu^0$ for $(t, u) \in \square^0$, and $\varepsilon_1{}^0 = \min\{\varepsilon^0, \delta^0/\mu^0\}$, then $u_m(t) < u_1(t) < u_M(t)$ for $t \in [\tau^0 - \varepsilon_1{}^0, \tau^0]$. As $S = \{(t, y) \mid t \in [\tau, \tau^0], u_m(t) \leq y \leq u_M(t)\}$ is a compact subset of $\mathfrak{R}$, there exists a smallest value s on $[\tau, \tau^0)$ such that $(t, u_1(t)) \in S$ for $t \in [s, \tau^0]$, and either $u_1(s) = u_M(s)$ or $u_1(s) = u_m(s)$. If $u(t)$, $t \in [\tau, \tau^0]$, is then defined as equal to $u_1(t)$ for $t \in [s, \tau^0]$, and $u(t)$ equal to the respective $u_M(t)$ or $u_m(t)$ for $t \in [\tau, s]$, then $u(t)$ is a solution of $\mathcal{F}(\tau, \xi)$ for (4.1) such that $u(\tau^0) = \xi^0$, thus completing the proof of Theorem 4.1.

In the examples presented explicitly in the problems of this and preceding sections, the set of points (τ, ξ) for which the aggregate of solutions $\mathcal{F}(\tau, \xi)$ consists of more than a single curve is most special in character, consisting either of a single point or the set of points lying on some curve. In turn, such examples may be used to construct others in which the phenomenon occurs at a finite number of isolated points, or the set of points lying on a finite number of curves. There do exist examples of extreme nonuniqueness, however, in which the continuous function $g(t, u)$ is such that for every point (τ, ξ) the corresponding aggregate of solutions $\mathcal{F}(\tau, \xi)$ consists of more than a single curve.

Problems I.4

1. If $g(t, u)$ is defined in the real (t, u)-plane as $g(0, 0) = 0$, $g(t, u) = 4t^3u/(t^4 + u^2)$ for $(t, u) \neq (0, 0)$, show that:

(i) $g(t, u)$ is a continuous function of (t, u);
(ii) for (4.1) relative to $(0, 0)$ the maximal solution is $u = t^2$, $t \in (-\infty, \infty)$, and the minimal solution is $u = -t^2$, $t \in (-\infty, \infty)$.

Hint. (ii) By elementary means show that $u' = g(t, u)$ has as solutions $u(t) = c \pm \sqrt{t^4 + c^2}$.

2.[s] If $u(t)$, $t \in (a, b)$, is a maximal solution of (4.1) relative to $(a, u(a))$, and $c \in (a, b)$, prove that $u(t)$, $t \in (c, b)$ is a maximal solution of (4.1) relative to $(c, u(c))$.

3.[s] Suppose that $g(t, u)$ is a real-valued continuous function in an open region $\mathfrak{R}$ of the real (t, u)-plane which contains the origin $(0, 0)$, and let $\mathfrak{R}_0$ denote the reflection of $\mathfrak{R}$ in the origin; that is $\mathfrak{R}_0 = \{(t, u) \mid (-t, -u) \in \mathfrak{R}\}$. If $g_0(t, u) = g(-t, -u)$ for $(t, u) \in \mathfrak{R}_0$, show that $u(t)$, $t \in (a, b)$, is a solution of (4.1) with graph in $\mathfrak{R}$ if and only if $u_0(t) = -u(-t)$, $t \in (-b, -a)$, is a solution of $u_0' = g_0(t, u_0)$ with graph in $\mathfrak{R}_0$. In particular, if $u(t)$, $t \in (a, b)$, is a maximal solution of (4.1) relative to (a, η), then $u_0(t) = -u(-t)$, $t \in (-b, -a)$, is a minimal solution of $u_0' = g_0(t, u_0)$ relative to $(-a, -\eta)$.

4.[s] Suppose that $g(t, u)$ is a real-valued continuous function in an open region $\mathfrak{R}$ of the real (t, u)-plane, ε and δ are positive constants such that $\square = \{(t, u) \mid |t - \tau| \leq \varepsilon,\ |u - \xi| \leq \delta\} \subset \mathfrak{R}$, and $0 < \varepsilon_2 < \min[\varepsilon, \delta/\mu]$, where μ is a constant such that $|g(t, u)| \leq \mu$ on $\square$. Show that there exists a positive value h_0 such that if $0 < h \leq h_0$ then any solution $v = v(t, h)$ of

$$v' = g(t, v) + h, \qquad v(\tau) = \xi,$$

is extensible to the interval $[\tau - \varepsilon_2, \tau + \varepsilon_2]$, and that there is a solution $u_1(t)$, $t \in [\tau - \varepsilon_2, \tau + \varepsilon_2]$, of (4.1) such that $v(t, h) \to u_1(t)$ uniformly on $[\tau - \varepsilon_2, \tau + \varepsilon_2]$ as $h \to 0$; moreover, $u_1(t)$, $t \in [\tau, \tau + \varepsilon_2]$ is a maximal solution of (4.1) relative to (τ, ξ), and $u_1(t)$, $t \in [\tau - \varepsilon_2, \tau]$ is a minimal solution of (4.1) relative to (τ, ξ).

Hint. Use the result of the preceding problem to limit attention to the interval $[\tau, \tau + \varepsilon_2]$; for this modified problem use results of the Corollary to Theorems 3.4, and of Problems I.3:3,4.

5.[s] Suppose that $g(t, u)$ and $g^0(t, u)$ are real-valued continuous functions of (t, u) in an open region $\mathfrak{R}$ of the real (t, u)-plane, and that $g(t, u) \geq g^0(t, u)$ for $(t, u) \in \mathfrak{R}$. For $(\tau, \xi) \in \mathfrak{R}$, let $u_M(t)$, $t \in (a_M, b_M)$ and $u_m(t)$, $t \in (a_m, b_m)$, denote the maximal and minimal solutions of (4.1) relative to

(τ, ξ); similarly, let $u_M{}^0(t)$, $t \in (a_M{}^0, b_M{}^0)$ and $u_m{}^0(t)$, $t \in (a_m{}^0, b_m{}^0)$, denote the maximal and minimal solutions of $u' = g^0(t, u)$ relative to (τ, ξ). Prove that: (i) $b_M{}^0 \geq b_M$, and $u_M(t) \geq u_M{}^0(t)$ for $t \in [\tau, b_M)$; (ii) $a_m{}^0 \leq a_m$, and $u_m(t) \geq u_m{}^0(t)$ for $t \in (a_m, \tau]$.

Hint. Use results of Problem I.3:4 and of above Problems 2, 3, 4.

6. Suppose that $g(t, u)$ is a real-valued continuous function on $(-\infty, \infty) \times (-\infty, \infty)$ which is non-negative, and such that for $j = 1, 2, \ldots$ there exists a finite positive constant M_j such that

$$g(t, u) \leq M_j, \quad \text{for} \quad (t, u) \in (-\infty, \infty) \times [j-1, j). \tag{4.3}$$

If $u_M(t)$, $t \in (a_M, b_M)$, denotes the maximal solution of $u' = g(t, u)$ relative to the initial point $(0, 0)$, prove that $b_M = \infty$ if $\sum_{j=1}^{\infty} (1/M_j) = \infty$. If in addition to (4.3) there exist positive constants m_j such that

$$m_j \leq g(t, u), \quad \text{for} \quad (t, u) \in (-\infty, \infty) \times [j-1, j), \qquad (j = 1, 2, \ldots), \tag{4.4}$$

and the infinite series $\sum_{j=1}^{\infty} (1/m_j)$ converges, prove that $\sum_{j=1}^{\infty} (1/M_j) \leq b_M \leq \sum_{j=1}^{\infty} (1/m_j)$. Formulate, and establish, corresponding results concerning a_M. Also formulate and establish corresponding results concerning the minimum solution $u_m(t)$, $t \in (a_m, b_m)$, of $u' = g(t, u)$ relative to $(0, 0)$.

5. UNIQUENESS THEOREMS

The results of the Section I.3 present sufficient conditions for the existence of a solution of (3.1) passing through a given initial point, but provide no information on the number of such solutions. We shall now consider the important problem of conditions which insure the uniqueness of solution for the initial value problem, that is, conditions which insure that there is *at most one* solution of (3.1) passing through a given initial point (τ, η).

A vector function $f(t, y)$ defined for (t, y) in a set $\mathfrak{D}$ of (t, y)-space is said to satisfy a *Lipschitz condition* with respect to y on $\mathfrak{D}$, or to be *Lipschitzian* in y on $\mathfrak{D}$, if there is a (Lipschitz) constant κ such that

$$|f(t, z) - f(t, y)| \leq \kappa |z - y|, \quad \text{for} \quad (t, y) \in \mathfrak{D}, \qquad (t, z) \in \mathfrak{D}. \tag{5.1}$$

If $f(t, y)$ is Lipschitzian in y on $\mathfrak{D}$, then for each fixed $t = \tau$ the vector function $f(t, y)$ is uniformly continuous on the set $\mathfrak{D}_\tau$ of values y belonging to points (τ, y) of $\mathfrak{D}$. Of particular interest is the case in which each $\mathfrak{D}_\tau$ is an open set in y-space. In this case, if $y_\beta = u_\beta + iu_{n+\beta}$, $(\beta = 1, \ldots, n)$, with the u_γ, $(\gamma = 1, \ldots, 2n)$, real, then whenever the component functions $f_\alpha(t, y) = g_\alpha(t, u_1, \ldots, u_{2n}) + ig_{n+\alpha}(t, u_1, \ldots, u_{2n})$ are differentiable with respect to the real variables u_γ the Lipschitz condition (5.1) implies that on $\mathfrak{D}$ each of the partial derivatives $f_{\alpha u_\gamma} = \partial f_\alpha / \partial u_\gamma = \partial g_\alpha / \partial u_\gamma + i \partial g_{n+\alpha} / \partial u_\gamma$ has absolute value not exceeding κ.

A converse of this latter result holds if $\mathfrak{D}$ is convex in the y-variables, that is, if for arbitrary $(t, y) = (t, u_1 + iu_{n+1}, \ldots, u_n + iu_{2n})$ and $(t, z) = (t, v_1 + iv_{n+1}, \ldots, v_n + iv_{2n})$ of $\mathfrak{D}$ all the points $(t, [1 - \theta]y + \theta z)$, $0 \leq \theta \leq 1$, are also in $\mathfrak{D}$. If $\mathfrak{D}$ is convex in the y-variables and each $\mathfrak{D}_\tau$ is an open set, then the condition that each partial derivative $f_{\alpha u_\gamma}$ is a continuous function on $\mathfrak{D}_\tau$ permits the evaluation

$$\begin{aligned} f_\alpha(\tau, z) - f_\alpha(\tau, y) &= f_\alpha(\tau, [1 - \theta]y + \theta z)\Big|_{\theta=0}^{\theta=1}, \\ &= \int_0^1 \frac{df_\alpha}{d\theta}\, d\theta, \\ &= \sum_{\gamma=1}^{2n} \left[\int_0^1 f_{\alpha u_\gamma}\, d\theta \right] (v_\gamma - u_\gamma), \end{aligned}$$

where in the above integrals the arguments of f_α and $f_{\alpha u_\gamma}$ are $(\tau, [1 - \theta]y + \theta z)$. Consequently, if on such a set $\mathfrak{D}$ the functions f_α possess partial derivatives $f_{\alpha u_\gamma}$ which are continuous on every non-null cross section $\mathfrak{D}_\tau$, and there is a constant κ_0 such that $|f_{\alpha u_\gamma}| \leq \kappa_0$, $(\alpha = 1, \ldots, n;\ \gamma = 1, \ldots, 2n)$ on $\mathfrak{D}$, then by the Lagrange-Cauchy inequality we have that

$$|f_\alpha(t, z) - f_\alpha(t, y)| \leq \sqrt{2n}\, \kappa_0 \left[\sum_{\gamma=1}^{2n} (v_\gamma - u_\gamma)^2 \right]^{1/2} = \sqrt{2n}\, \kappa_0\, |z - y|,$$

$$(\alpha = 1, \ldots, n)$$

and (5.1) holds with $\kappa = \sqrt{2}\, n\kappa_0$.

In particular, if on an open region $\mathfrak{R}$ of (t, y)-space the functions f_α have partial derivatives with respect to the real and pure imaginary parts of each y_β that are continuous functions of (t, y) on $\mathfrak{R}$, then $f(t, y)$ is *locally Lipschitzian in y on $\mathfrak{R}$*; that is, if $(\tau, \eta) \in \mathfrak{R}$ there is a neighborhood of this point on which $f(t, y)$ is Lipschitzian in y.

It is to be remarked that if $\mathfrak{R}$ is an open region in (t, y)-space and the component functions $f_\alpha(t, y)$ are differentiable, (holomorphic), functions of the individual complex variables $y_\beta = u_\beta + iu_{n+\beta}$, $(\beta = 1, \ldots, n)$, then the existence of the partial derivative $f_{\alpha y_\beta}$ implies the existence of the partial derivatives $f_{\alpha u_\beta}$, $f_{\alpha u_{n+\beta}}$ and $f_{\alpha u_\beta} = -if_{\alpha u_{n+\beta}}$. In particular if $f(t, y)$ is continuous in (t, y) on an open region $\mathfrak{R}$ of (t, y)-space, and differentiable in each of the individual complex variables y_α, $(\alpha = 1, \ldots, n)$, then an application of the Cauchy integral formula yields the result that each partial derivative f_{y_β} is continuous in (t, y) on $\mathfrak{R}$, and differentiable in each of the complex variables y_α.

Theorem 5.1. *If on the infinite strip* $\Delta = [a, b] \times \mathbf{C}_n$ *the vector function* $f(t, y)$ *is continuous and satisfies a Lipschitz condition* (5.1), *then for each point* (τ, η) *of* Δ *there is a unique solution* $y(t)$ *of* (3.1) *on* $[a, b]$ *satisfying* $y(\tau) = \eta$.

It is to be remarked that Theorem 5.1 is both an existence theorem and a uniqueness theorem. The condition that $f(t, y)$ is Lipschitzian in y on Δ implies that

$$|f(t, y)| \leq |f(t, 0)| + |f(t, y) - f(t, 0)| \leq |f(t, 0)| + \kappa\, |y|,$$

and hence on Δ the vector function $f(t, y)$ satisfies (3.8) with κ the Lipschitz constant and κ_1 an upper bound on $[a, b]$ of the continuous function $|f(t, 0)|$. Consequently, by Theorem 3.2 the hypotheses of Theorem 5.1 imply that (3.1) has a solution $y(t)$ on $[a, b]$ satisfying $y(\tau) = \eta$.

Now suppose that $y = y(t)$ and $y = z(t)$ are both solutions of (3.1) on $[a, b]$ passing through the initial point (τ, η). Then

$$z(t) - y(t) = \int_\tau^t [f(s, z(s)) - f(s, y(s))]\, ds, \tag{5.2}$$

and as $|f(s, z(s)) - f(s, y(s))| \leq \kappa\, |z(s) - y(s)|$ the function $\rho(t) = |z(t) - y(t)|$ satisfies (2.6) on $[a, b]$ with κ the Lipschitz constant and $\kappa_0 = 0 = \kappa_1$. From Corollary 1 of Theorem 2.1 it follows that $\rho(t) \equiv 0$, and hence $z(t) \equiv y(t)$ on $[a, b]$, thus establishing the portion of Theorem 5.1 dealing with the uniqueness of solution.

Corresponding to Theorem 3.4 we have also the following result.

Theorem 5.2. *If on* $\square = \{(t, y) \mid |t - \tau| \leq \varepsilon,\ |y - \eta| \leq \delta\}$ *the vector function* $f(t, y)$ *is continuous in* (t, y), *Lipschitzian in* y, *and* μ *is such that* $|f(t, y)| \leq \mu$ *on* $\square$, *then for* $\varepsilon_1 = \min[\varepsilon, \delta/\mu]$ *there is a unique solution* $y = y(t)$ *of* (3.1) *on* $[\tau - \varepsilon_1, \tau + \varepsilon_1]$ *satisfying* $y(\tau) = \eta$, *and* $|y(t) - \eta| < \delta$ *on* $(\tau - \varepsilon_1, \tau + \varepsilon_1)$.

If $f(t, y)$ is continuous on $\Delta = [a, b] \times \mathbf{C}_n$ and $\tau \in (a, b)$, then the condition that there is at most one solution of (3.1) on $[a, b]$ passing through the point (τ, η) is equivalent to the following two statements: (i) there is at most one solution of (3.1) on $[\tau, b]$ passing through the point (τ, η); (ii) there is at most one solution of (3.1) on $[a, \tau]$ passing through the point (τ, η). We shall proceed to show that each of the individual "unilateral" conditions (i) and (ii) is valid under conditions weaker than the Lipschitz condition requisite for the applicability of Theorem 5.1.

Theorem 5.3. *If* $f(t, y)$ *is continuous on* $\Delta_1 = [\tau, b] \times \mathbf{C}_n$, *and there is a non-negative constant* λ *such that*

$$\Re\{[z^* - y^*][f(t, z) - f(t, y)]\} \leq \lambda\, |z - y|^2 \tag{5.3}$$

for arbitrary points (t, y), (t, z) *of* Δ_1, *then there is at most one solution of* (3.1) *on* $[\tau, b]$ *passing through a given initial point* (τ, η).

If $y = y(t)$ and $y = z(t)$ are solutions of (3.1) on $[\tau, b]$ passing through the initial point (τ, η), then $\rho(t) = |z(t) - y(t)|^2$ is such that $\rho(\tau) = 0$, and

$$\rho'(t) = 2\Re e\{[z^*(t) - y^*(t)][f(t, z(t)) - f(t, y(t))]\} \leq 2\lambda\rho(t)$$

on $[\tau, b]$, so that $\rho(t)$ satisfies (2.3) with $\phi(t) \equiv 0$, $\kappa(t) \equiv 2\lambda$, and hence from Theorem 2.1 it follows that $\rho(t) \equiv 0$ and $z(t) \equiv y(t)$ on $[\tau, b]$.

In a similar fashion one establishes the corresponding result of the following theorem.

Theorem 5.4. *If* $f(t, y)$ *is continuous on* $\Delta_2 = [a, \tau] \times \mathbf{C}_n$, *and there is a non-negative constant* λ *such that*

$$\Re e\{[z^* - y^*][f(t, y) - f(t, z)]\} \leq \lambda\, |z - y|^2 \tag{5.4}$$

for arbitrary points (t, y), (t, z) *of* Δ_2, *then there is at most one solution of* (3.1) *on* $[a, \tau]$ *passing through a given initial point* (τ, η).

It is to be emphasized that condition (5.3) or (5.4) is weaker than the condition that $f(t, y)$ is Lipschitzian in y on the respective infinite strip Δ_1 or Δ_2. Indeed, if $f(t, y)$ is Lipschitzian in y on a set $\mathfrak{D}$ with Lipschitz constant κ then in view of the Lagrange-Cauchy inequality it follows from (5.1) that $|[z^* - y^*][f(t, z) - f(t, y)]| \leq \kappa\, |z - y|^2$ and conditions (5.3) and (5.4) both hold for $\lambda = \kappa$ and (t, y), (t, z) arbitrary points of $\mathfrak{D}$. On the other hand, in the case of the real scalar equation $u' = g(t, u)$ the condition (5.3) holds on Δ_1 if for each t on $[\tau, b]$ the real-valued function $g(t, u)$ is monotone nonincreasing on $(-\infty, \infty)$, and clearly such a condition of monotoneity imposes no restriction on the order of growth of $g(t, u)$ as u tends to $\pm\infty$, as would be implied by a Lipschitz condition. Correspondingly, for the real scalar equation $u' = g(t, u)$ condition (5.4) holds on Δ_2 if for each t on $[a, \tau]$ the function $g(t, u)$ is monotone nondecreasing on $(-\infty, \infty)$.

Theorem 5.5. *Suppose that* $f(t, y)$ *is continuous and satisfies* (3.8) *on the infinite strip* $\Delta = [a, b] \times \mathbf{C}_n$, *and that for a point* (τ^0, η^0) *of* Δ *there is a unique solution* $y = \phi(t)$ *of* (3.1) *which is defined on* $[a, b]$ *and such that* $\phi(\tau^0) = \eta^0$. *Then for an arbitrary* $\zeta > 0$ *there are positive constants* ε'_ζ, ε''_ζ *such that if*

$$\tau \in [a, b], \qquad |\tau - \tau^0| < \varepsilon'_\zeta, \qquad |\eta - \eta^0| < \varepsilon'_\zeta, \tag{5.5}$$

and $y(t)$ *is a continuous vector function satisfying*

$$\left| y(t) - \eta - \int_\tau^t f(s, y(s))\, ds \right| < \varepsilon''_\zeta \quad \text{for} \quad t \in [a, b],$$

then $|y(t) - \phi(t)| < \zeta$ *on* $[a, b]$.

The theorem will be established by indirect argument. On the assumption that the conclusion of the theorem is not valid there is some positive value $\zeta = \zeta^0$ for which there are no corresponding values ε'_ζ, ε''_ζ satisfying the conclusion of the theorem. In particular, this conclusion does not hold for $\varepsilon'_\zeta = \varepsilon''_\zeta = 1/m$, so that for each $m = 1, 2, \ldots$ there is a point (τ^m, η^m) of Δ with $|\tau^m - \tau^0| < 1/m$, $|\eta^m - \eta^0| < 1/m$, and a corresponding continuous vector function $y^{(m)}(t)$ satisfying

$$\left|y^{(m)}(t) - \eta^m - \int_{\tau^m}^{t} f(s, y^{(m)}(s))\, ds\right| < \frac{1}{m} \quad \text{for} \quad t \in [a, b],$$

while there is a point $t = t_m$ of this interval such that $|y^{(m)}(t_m) - \phi(t_m)| \geq \zeta^0$. In view of Problem I.3:2 there is a subsequence $\{y^{(m_k)}(t)\}$ which converges uniformly on $[a, b]$ to a solution $y(t)$ of (3.1) satisfying $y(\tau^0) = \eta^0$. Now, by hypothesis, $y = \phi(t)$ is the unique solution of (3.1) on $[a, b]$ passing through the initial point (τ^0, η^0), so that $y(t) \equiv \phi(t)$, and the uniform convergence on $[a, b]$ of the subsequence $\{y^{(m_k)}(t)\}$ to $\phi(t)$ contradicts the above condition that for each m there is a value t_m on $[a, b]$ such that $|y^{(m)}(t_m) - \phi(t_m)| \geq \zeta^0$.

The following result is clearly contained in the above theorem.

Corollary. *Under the hypotheses of Theorem* 5.5 *a solution* $y(t)$, $t \in [a, b]$, *of* (3.1) *that passes through an initial point* (τ, η) *satisfying* (5.5) *is such that* $|y(t) - \phi(t)| < \zeta$ *for* $t \in [a, b]$.

Theorem 5.6. *Suppose that* $f(t, y)$ *is continuous in* (t, y) *on an open region* $\mathfrak{R}$ *of* (t, y)*-space, and that for a given point* (τ^0, η^0) *of* $\mathfrak{R}$ *there is a nondegenerate interval* $[a, b]$ *with* $\tau^0 \in (a, b)$ *and a solution* $y = \phi(t)$ *of* (3.1) *that is the unique solution of this differential equation on* $[a, b]$ *with graph in* $\mathfrak{R}$, *and passing through the initial point* (τ^0, η^0). *If* ζ *is a positive constant such that the set* $S_\zeta = \{(t, y) \mid t \in [a, b], |y - \phi(t)| \leq \zeta\}$ *is in* $\mathfrak{R}$, *then:*

(i) *there exist positive constants* ε'_ζ, δ'_ζ *such that if* Π *is a partition* (3.9) *of* $[a, b]$ *with norm* $|\Pi| < \delta'_\zeta$, *and* (τ, η) *satisfies* (5.5), *then the polygonal function* $y = y(t; \Pi)$ *of* (3.10) *satisfies the inequality* $|y(t; \Pi) - \phi(t)| < \zeta$ *on* $[a, b]$;

(ii) *there exist positive constants* ε'_ζ, δ''_ζ *such that if* $0 < \delta < \delta''_\zeta$ *and* (τ, η) *satisfies* (5.5), *then the function* $y = y(t; \delta)$ *of* (3.13) *satisfies* $|y(t; \delta) - \phi(t)| < \zeta$ *on* $[a, b]$;

(iii) *there is a* $\delta_\zeta > 0$ *such that if*

$$(5.6) \qquad |\tau - \tau^0| < \delta_\zeta, \qquad |\eta - \eta^0| < \delta_\zeta,$$

and $y = y(t)$ *is a solution of* (3.1) *passing through* (τ, η), *then the interval of definition of* $y(t)$ *may be extended to* $[a, b]$, *and for any such extension* $|y(t) - \phi(t)| < \zeta$ *for* $t \in [a, b]$.

Let the vector function $h(t, y)$ be defined on $\Delta = [a, b] \times \mathbf{C}_n$ as follows: $h(t, y) = f(t, y)$ on S_ζ, $h(t, y) = f(t, \phi(t) + \zeta[y - \phi(t)]/|y - \phi(t)|)$ if $t \in [a, b]$ and $|y - \phi(t)| > \zeta$. On Δ the function $h(t, y)$ is continuous and there is a constant κ_1 such that $|h(t, y)| \leq \kappa_1$, so that $h(t, y)$ satisfies on Δ condition (3.8) with the above determined κ_1 and $\kappa = 0$. For a given $\zeta > 0$ we shall denote by ε', ε'' the quantities of Theorem 5.5 determined for the differential equation

$$y' = h(t, y). \tag{5.7}$$

Denote by $y = y^0(t; \Pi)$ the polygonal vector function determined by (3.10) with $f(t, y)$ replaced by the $h(t, y)$ of (5.7). In view of (3.12) it follows from Theorem 5.5 that if $(b - a)\omega_\nu(|\Pi|) < \varepsilon''_\zeta$ then $|y^0(t; \Pi) - \phi(t)| < \zeta$ for $t \in [a, b]$; consequently, the graph $\{(t, y^0(t; \Pi)) \mid t \in [a, b]\}$ lies in S_ζ and $y^0(t; \Pi) \equiv y(t; \Pi)$. Therefore, conclusion (i) of Theorem 5.6 holds for ε'_ζ thus determined, and δ'_ζ a positive constant such that $(b - a)\omega_\nu(\delta'_\zeta) < \varepsilon''_\zeta$. A similar argument involving the vector function $y^0(t; \delta)$ determined by (3.13) with $f(t, y)$ replaced by the $h(t, y)$ of (5.7) implies in view of (3.14) that conclusion (ii) holds for ε'_ζ as determined above, and δ''_ζ such that $0 < \delta''_\zeta < \min[b - \tau, \tau - a, \varepsilon''_\zeta/\kappa_1]$.

In order to establish conclusion (iii), it is noted first that if $y(t)$, $t \in [a, b]$, is a solution of (5.7) satisfying $y(\tau) = \eta$, then from the Corollary to Theorem 5.5 it follows that $|y(t) - \phi(t)| < \zeta$ for $t \in [a, b]$ whenever (τ, η) satisfies (5.5), where again ε'_ζ, ε''_ζ are to denote the constants of Theorem 5.5 for the differential equation (5.7). We shall proceed to show that conclusion (iii) holds for δ_ζ a positive constant for which $\delta_\zeta \leq \varepsilon'_\zeta$, and such that if $|\tau - \tau^0| < \delta_\zeta$, $|\eta - \eta^0| < \delta_\zeta$, then $\tau \in [a, b]$ and $|\eta - \phi(\tau)| < \zeta$. Suppose that (τ, η) satisfies (5.6) with such a δ_ζ, and $y(t)$, $t \in I$, is a solution of (3.1) passing through (τ, η), where I is a subinterval of $[a, b]$. If it is not true that $|y(t) - \phi(t)| < \zeta$ on I, then there exists a subinterval I_0 of I which contains $t = \tau$, and such that $|y(t) - \phi(t)| < \zeta$ for values t interior to I_0, while $|y(t_1) - \phi(t_1)| = \zeta$ for t_1 one of the endpoints of I_0. By Theorem 3.3 there is a solution $y^0(t)$, $t \in [a, b]$, of (5.7) such that $y^0(t) = y(t)$ for $t \in I_0$. As (5.5) and the condition $\delta_\zeta \leq \varepsilon'_\zeta$ imply that $|y^0(t) - \phi(t)| < \zeta$ for $t \in [a, b]$, one has the contradictory result $|y(t_1) - \phi(t_1)| < \zeta$. Hence $|y(t) - \phi(t)| < \zeta$ for $t \in I$, and if $y^0(t)$, $t \in [a, b]$ is a solution of (5.7) such that $y^0(t) = y(t)$ for $t \in I$ then $|y^0(t) - \phi(t)| < \zeta$ for $t \in [a, b]$, and $y^0(t)$ is a solution of (3.1) on the interval $[a, b]$.

If the vector function $f(t, y)$ is defined on an open region $\mathfrak{R}$ of (t, y)-space, and $(t^0, \eta^0) \in \mathfrak{R}$, then (t^0, η^0) is said to be a *point of local uniqueness* for (3.1) if there exists a positive $\varepsilon^0 = \varepsilon^0(t^0, \eta^0)$ such that if $0 < \varepsilon < \varepsilon^0$ then there is at most one solution $y(t)$, $t \in [t^0 - \varepsilon, t^0 + \varepsilon]$, of this differential equation with graph in $\mathfrak{R}$, and passing through the initial point (t^0, η^0). If each point of $\mathfrak{R}$

is a point of local uniqueness for (3.1), then $\mathfrak{R}$ is called a *region of local uniqueness* for this differential equation. If $\mathfrak{R}$ is a region of local uniqueness for (3.1), and also for each $(t^0, \eta^0) \in \mathfrak{R}$ there exists a corresponding positive $\varepsilon^1 = \varepsilon^1(t^0, \eta^0)$ such that (3.1) has a solution $y(t)$, $t \in [t^0 - \varepsilon^1, t^0 + \varepsilon^1]$ with graph in $\mathfrak{R}$, and passing through (t^0, η^0), then for brevity we say that the *solutions of* (3.1) *are locally unique* in $\mathfrak{R}$. For example, if $f(t, y)$ is continuous in an open region $\mathfrak{R}$, and locally Lipschitzian in y on $\mathfrak{R}$, then in view of Theorems 3.4 and 5.2 the solutions of (3.1) are locally unique in $\mathfrak{R}$.

Theorem 5.7. *If $f(t, y)$ is continuous on an open region $\mathfrak{R}$ of (t, y)-space, and the solutions of* (3.1) *are locally unique in $\mathfrak{R}$, then the solution $y = \phi(t; \tau, \eta)$ of this differential equation passing through the initial point (τ, η), and with maximal interval of existence*

$$a(\tau, \eta) < t < b(\tau, \eta) \tag{5.8}$$

has the following properties:

(i) *if $a(\tau, \eta) \leq a < \tau < b \leq b(\tau, \eta)$, then $y = \phi(t; \tau, \eta)$ is the unique solution of* (3.1) *defined on (a, b), with graph in $\mathfrak{R}$, and passing through the point (τ, η);*

(ii) *$\phi(t; \tau, \eta)$ tends to the boundary of $\mathfrak{R}$ as $t \to a(\tau, \eta)$, and as $t \to b(\tau, \eta)$;*

(iii) *$\phi(t; \tau, \eta)$ is continuous in (t, τ, η) on the set*

$$(\tau, \eta) \in \mathfrak{R}, \qquad a(\tau, \eta) < t < b(\tau, \eta). \tag{5.9}$$

Conclusions (i) and (ii) of the theorem are direct consequences of Theorem 3.5 and the assumption that the solutions of (3.1) are locally unique in $\mathfrak{R}$. In order to establish conclusion (iii), for a given point (τ^0, η^0) of $\mathfrak{R}$ let t^0 be a value on $a(\tau^0, \eta^0) < t < b(\tau^0, \eta^0)$, and consider values a, b satisfying $a(\tau^0, \eta^0) < a < b < b(\tau^0, \eta^0)$, and such that $t = t^0$ and $t = \tau^0$ are values on the open interval (a, b). In view of the continuity of $\phi(t; \tau^0, \eta^0)$ as a function of t, for a given $\varepsilon > 0$ there exists a $\delta_\varepsilon{}^0 > 0$ such that $\delta_\varepsilon{}^0 < \min[b - t^0, t^0 - a]$ and $|\phi(t; \tau^0, \eta^0) - \phi(t^0; \tau^0, \eta^0)| < \varepsilon/2$ whenever $|t - t^0| < \delta_\varepsilon{}^0$. Now consider the set $S_\zeta = \{(t, y) \mid t \in [a, b], |y - \phi(t; \tau^0, \eta^0)| \leq \zeta\}$ for $\zeta = \zeta(\varepsilon)$ so small that $0 < \zeta < \varepsilon/2$ and S_ζ is in the region $\mathfrak{R}$. From conclusion (iii) of Theorem 5.6 it then follows that there exists a positive $\delta_\varepsilon{}^1 = \delta_{\zeta(\varepsilon)}$ such that if $|\tau - \tau^0| < \delta_\varepsilon{}^1$, $|\eta - \eta^0| < \delta_\varepsilon{}^1$, then the solution $y = \phi(t; \tau, \eta)$ exists for $t \in [a, b]$, and $|\phi(t; \tau, \eta) - \phi(t; \tau^0, \eta^0)| < \zeta(\varepsilon) < \varepsilon/2$ for $t \in [a, b]$. Consequently, if (t, τ, η) belongs to the set (5.9) and $|t - t^0| < \delta_\varepsilon{}^0$, $|\tau - \tau^0| < \delta_\varepsilon{}^1$, $|\eta - \eta^0| < \delta_\varepsilon{}^1$, then

$$\begin{aligned}|\phi(t; \tau, \eta) - \phi(t^0; \tau^0, \eta^0)| &\leq |\phi(t; \tau, \eta) - \phi(t; \tau^0, \eta^0)| \\ &\quad + |\phi(t; \tau^0, \eta^0) - \phi(t^0; \tau^0, \eta^0)| \\ &< \frac{\varepsilon}{2} + \frac{\varepsilon}{2} = \varepsilon,\end{aligned}$$

thus establishing the continuity of $\phi(t; \tau, \eta)$ at the element (t^0, τ^0, η^0) of the set (5.9).

The preceding theorem admits various extensions to the case of vector differential equations involving parameters; for brevity, attention will be restricted to a particular result that is an easy consequence of Theorem 5.7. Let $f(t, y; c) \equiv (f_\alpha(t, y_1, \ldots, y_n; c_1, \ldots, c_k))$, $(\alpha = 1, \ldots, n)$, be a vector function involving the complex parameter vector $c \equiv (c_\nu)$, $(\nu = 1, \ldots, k)$, which satisfies the following condition:

($\mathfrak{H}$) *$f(t, y; c)$ is continuous in (t, y, c) in an open region $\mathfrak{D}$ of (t, y, c)-space; moreover, if $\mathfrak{D}_c$ denotes the section of $\mathfrak{D}$ consisting of all (t, y) such that (t, y, c) is in $\mathfrak{D}$, then for each c such that $\mathfrak{D}_c$ is nonvacuous the solutions of the vector differential equation*

$$y' = f(t, y; c) \tag{5.10}$$

are locally unique in $\mathfrak{D}_c$.

Theorem 5.8. *If the vector function $f(t, y; c)$ satisfies the above hypothesis ($\mathfrak{H}$), then the solution $y = \phi(t; \tau, \eta, c)$ of (5.10) passing through the initial point (τ, η) in $\mathfrak{D}_c$, and with maximal interval of existence*

$$a(\tau, \eta, c) < t < b(\tau, \eta, c) \tag{5.11}$$

has the following properties:

(i) *if $a(\tau, \eta, c) \leq a < \tau < b \leq b(\tau, \eta, c)$, then $y = \phi(t; \tau, \eta, c)$ is the unique solution of (5.10) defined on (a, b), with graph in $\mathfrak{D}_c$, and passing through the initial point (τ, η);*

(ii) *$\phi(t; \tau, \eta, c)$ tends to the boundary of $\mathfrak{D}_c$ as $t \to a(\tau, \eta, c)$, and as $t \to b(\tau, \eta, c)$;*

(iii) *$\phi(t; \tau, \eta, c)$ is continuous in (t, τ, η, c) on the set*

$$(\tau, \eta, c) \in \mathfrak{D}, \qquad a(\tau, \eta, c) < t < b(\tau, \eta, c). \tag{5.12}$$

If $(\tau, \eta, c) \in \mathfrak{D}$ then $\mathfrak{D}_c$ is a nonvacuous open set of (t, y)-space containing the point (τ, η), and for this fixed value of c the vector function $f(t, y; c)$ of (5.10) satisfies the hypotheses of Theorem 5.7 on the maximal connected subset of $\mathfrak{D}_c$ containing (τ, η). Application of Theorem 5.7 to (5.10) for fixed values of c yields consequences (i), (ii) of Theorem 5.8; on the other hand, the result (iii) of Theorem 5.7 when applied to (5.10) with c fixed does not yield the full result of conclusion (iii) of Theorem 5.8, but merely that for fixed c the vector function $\phi(t; \tau, \eta, c)$ is a continuous function of (t, τ, η) on the set $\{(t, \tau, \eta) \mid (\tau, \eta) \in \mathfrak{D}_c, a(t, \eta, c) < t < b(\tau, \eta, c)\}$.

In order to establish the full result of conclusion (iii) of Theorem 5.8, consider the $(n + k)$-dimensional vector $w \equiv (w_\gamma)$, $(\gamma = 1, \ldots, n + k)$, and set $g_\alpha(t, w) = f_\alpha(t, w_1, \ldots, w_n; w_{n+1}, \ldots, w_{n+k})$, $(\alpha = 1, \ldots, n)$, $g_{n+\nu}(t, w) \equiv 0$, $(\nu = 1, \ldots, k)$. Then the vector function $g(t, w) \equiv (g_\gamma(t, w))$, $(\gamma = 1, \ldots, n + k)$, is continuous in (t, w) in the open region $\mathfrak{D}$ of (t, w)-space, and $w(t) \equiv (w_\gamma(t))$ is a solution of the vector differential equation

$$w' = g(t, w), \tag{5.13}$$

with graph in $\mathfrak{D}$, and passing through an initial point (τ, w^0) of this region, if and only if $w_{n+\nu}(t) \equiv w^0_{n+\nu}$, $(\nu = 1, \ldots, k)$, and for $c_\nu = w^0_{n+\nu}$, $(\nu = 1, \ldots, k)$, the n-dimensional vector function $y \equiv (w_\alpha(t))$ is a solution of (5.10) passing through the initial point $(\tau, \eta) = (\tau, w_1{}^0, \ldots, w_n{}^0)$ of the corresponding section $\mathfrak{D}_c$ of $\mathfrak{D}$. In particular, the condition that the solutions of (5.10) are locally unique in each nonvacuous $\mathfrak{D}_c$ implies that the solutions of (5.13) are locally unique in $\mathfrak{D}$. The desired result then follows from applying Theorem 5.7 to the vector equation (5.13); indeed, in view of the above remarks each conclusion (i), (ii), (iii) of Theorem 5.7 for the differential equation (5.13) is equivalent to the correspondingly numbered conclusion of Theorem 5.8 for the differential equation (5.10).

Problems I.5

1. Show that if on $\Delta = [a, b] \times \mathbf{C}_n$ the vector function $f(t, y)$ is Lipschitzian in y, and continuous in t for each fixed y, then $f(t, y)$ is continuous in (t, y) on Δ.

2.[ss] If the vector function $f(t, y)$ is Lipschitzian in y on

$$\square = \{(t, y) \mid |t - \tau| \leq \varepsilon, |y - \eta| \leq \delta\},$$

with Lipschitz constant κ, and $h(t, y) \equiv f(t, y)$ on $\square$, $h(t, y) = f(t, \eta + \delta(y - \eta)/|y - \eta|)$ for $|t - \tau| \leq \varepsilon$, $|y - \eta| > \delta$, show that $h(t, y)$ is Lipschitzian in y on $\Delta = [\tau - \varepsilon, \tau + \varepsilon] \times \mathbf{C}_n$ with the same Lipschitz constant κ.

Hint. Let $x\{y\} = y$ for $|y - \eta| \leq \delta$, and $x\{y\} = \eta + \delta(y - \eta)/|y - \eta|$ for $|y - \eta| > \delta$, and by elementary consideration of $|y - z|^2 = |(y - \eta) - (z - \eta)|^2$ proceed to show that $|y - z|^2 \geq |x\{y\} - x\{z\}|^2$ for arbitrary vectors y, z. Note that for arbitrary $(t, y) \in \Delta$ we have $(t, x\{y\}) \in \square$ and $h(t, y) = f(t, x\{y\})$.

3. Show that $u(t) \equiv 0$ is the unique solution of the real-valued scalar differential equation $u' = -tu^{1/3}$ passing through the point $(0, 0)$, but that for $\tau \neq 0$ there are infinitely many solutions of the differential equation passing through the point $(\tau, 0)$.

Hint. Note that if $h > 0$, and $u(t, h) \equiv 0$ for $|t| > h$,

$$u(t, h) = \left[\frac{(h^2 - t^2)}{3}\right]^{3/2} \quad \text{for} \quad t \in [-h, h],$$

then $u = u(t, h)$ is a solution of the given differential equation.

4.[s] Suppose that $x(t) \equiv (x_\alpha(t))$, $z(t) \equiv (z_\alpha(t))$, $(\alpha = 1, \ldots, n)$, are continuous and have piecewise continuous derivatives on $[a, b]$, and $f(t, y) \equiv (f_\alpha(t, y))$, $(\alpha = 1, \ldots, n)$, is continuous in (t, y) and satisfies a Lipschitz condition (5.1) in a region of (t, y)-space containing the graphs of $x(t)$ and $z(t)$. Show that if

$$(5.14) \quad |[x'(t) - f(t, x(t))] - [z'(t) - f(t, z(t))]| \leq \kappa_1 \quad \text{for} \quad t \in [a, b],$$

then for $(t, \tau) \in [a, b] \times [a, b]$,

$$|x(t) - z(t)| \leq |x(\tau) - z(\tau)| \exp\{\kappa\,|t - \tau|\} + \frac{\kappa_1}{\kappa}[\exp\{\kappa\,|t - \tau|\} - 1].$$

Hint. Show that $y(t) = x(t) - z(t)$ satisfies the conditions of Corollary 2 of Theorem 2.1, with κ the Lipschitz constant of (4.1) and κ_1 the constant of (5.14).

5.[s] Suppose that $x(t) \equiv (x_\alpha(t))$, $z(t) \equiv (z_\alpha(t))$, $(\alpha = 1, \ldots, n)$, are continuously differentiable on $[a, b]$, and that in a region of (t, y)-space containing the graphs of $x(t)$ and $z(t)$ the vector functions $f(t, y) \equiv (f_\alpha(t, y))$, $g(t, y) \equiv (g_\alpha(t, y))$, $(\alpha = 1, \ldots, n)$, are continuous in (t, y) and $f(t, y)$ satisfies a Lipschitz condition (5.1). Show that if $x'(t) = f(t, x(t))$, $z'(t) = g(t, z(t))$ on $[a, b]$, and there are constants μ_1, κ_1 such that

$$|f(t, x(t))| \leq \mu_1, \; |f(t, z(t)) - g(t, z(t))| \leq \kappa_1, \quad \text{for} \quad t \in [a, b],$$

then for $(t, \tau, \tau^0) \in [a, b] \times [a, b] \times [a, b]$,

$$|x(t) - z(t)| \leq [|x(\tau^0) - z(\tau)| + \mu_1\,|\tau^0 - \tau|]\exp\{\kappa\,|t - \tau|\} + \frac{\kappa_1}{\kappa}[\exp\{\kappa\,|t - \tau|\} - 1].$$

Hint. Use the result of Problem 4, noting that $|x(\tau^0) - x(\tau)| \leq \mu_1\,|\tau^0 - \tau|$.

6. Show that the result of Theorem 5.3 remains valid whenever (5.3) is replaced by the condition

$$\Re e\{[z^* - y^*][f(t, z) - f(t, y)]\} \leq \lambda(t)\Omega(|z - y|^2)$$

for arbitrary (t, y), (t, z) of Δ_1, with $\lambda(t)$ integrable on $[\tau, b]$, and $\Omega(s)$ a real-valued continuous function on $[0, \infty)$ such that $\Omega(0) = 0$, $\Omega(s) > 0$ for

$s \in (0, \infty)$, and for $\sigma > 0$, $\int_\varepsilon^\sigma [1/\Omega(s)]\, ds \to \infty$ as $\varepsilon \to 0^+$. State and prove a corresponding generalization of the result of Theorem 5.4.

7. If on $\square = \{(t, y) \mid |t - \tau| \leq \varepsilon,\ |y - \eta| \leq \delta\}$ the vector function $f(t, y)$ is continuous in (t, y), $|f(t, y)| \leq \mu$, and

$$|t - \tau|\, |f(t, z) - f(t, y)| \leq |z - y|, \quad \text{for} \quad (t, y) \in \square, \qquad (t, z) \in \square,$$

then on $|t - \tau| \leq \varepsilon_1 = \min[\varepsilon, \delta/\mu]$ the differential equation (3.1) has a unique solution passing through (τ, η).

Hint. Use Theorem 3.4. Note also that if $y = y(t)$ and $y = z(t)$ are solutions of (3.1) on $[\tau - \varepsilon_1, \tau + \varepsilon_1]$ with graphs in $\square$, and passing through the initial point (τ, η), then the real-valued function $\theta(t) = |[z(t) - y(t)]/(t - \tau)|$ for $0 < |t - \tau| \leq \varepsilon_1$, $\theta(\tau) = 0$, is continuous on $[\tau - \varepsilon_1, \tau + \varepsilon_1]$, and

$$0 \leq \theta(t) \leq \left| \frac{1}{t - \tau} \int_\tau^t \theta(s)\, ds \right| \quad \text{for} \quad 0 < |t - \tau| \leq \varepsilon_1. \tag{5.15}$$

Proceed to show that under the assumption that $\theta(t)$ has a nonzero maximum θ_M on $[\tau - \varepsilon_1, \tau + \varepsilon_1]$ the inequality (5.15) implies the contradictory relation $\theta_M < \theta_M$.

8. Suppose that on $\square = \{(t, y) \mid |t - \tau| \leq \varepsilon,\ |y - \eta| \leq \delta\}$ the vector function $f(t, y)$ is continuous in (t, y), Lipschitzian in y, and $|f(t, y)| \leq \mu$. Show that for $0 < \varepsilon_2 < \min[\varepsilon, \delta/\mu]$ there are positive constants ε^0, δ^0 such that if $|\tau^0 - \tau| \leq \varepsilon^0$, $|\eta^0 - \eta| \leq \delta^0$ then there is a unique solution $y = y(t)$ of (3.1) on $[\tau - \varepsilon_2, \tau + \varepsilon_2]$ satisfying $y(\tau^0) = \eta^0$, and $|y(t) - \eta| < \delta$ for $t \in (\tau - \varepsilon_2, \tau + \varepsilon_2)$.

Hint. Use results of Theorem 5.2 and the Corollary to Theorem 3.4.

9. Show that for k a positive integer the pair of differential equations

$$x' = \frac{1}{2k} x(x^2 + y^2)^k - \frac{1}{k} y(x^2 + y^2)^{1+k},$$

$$y' = \frac{1}{2k} y(x^2 + y^2)^k + \frac{1}{k} x(x^2 + y^2)^{1+k},$$

has for solutions the curve $x \equiv 0, y \equiv 0, -\infty < t < +\infty$, and the family of curves

$$x = (c_2 - t)^{-1/(2k)} \cos[(c_2 - t)^{-1/k} + c_1],$$
$$y = (c_2 - t)^{-1/(2k)} \sin[(c_2 - t)^{-1/k} + c_1], \qquad -\infty < t < c_2.$$

10.[s] Suppose that on $\Delta = [a, b] \times \mathbf{C}_n$ the vector function $h(t, y)$ is continuous in (t, y), bounded, and uniformly continuous in y; let μ be a constant such that $|h(t, y)| \leq \mu$ for $(t, y) \in \Delta$, and for $\varepsilon > 0$ let $\delta(\varepsilon)$ be such that $|h(t, y^1) - h(t, y^2)| < \varepsilon$ if $(t, y^1) \in \Delta$, $(t, y^2) \in \Delta$, and $|y^1 - y^2| < \delta(\varepsilon)$. Prove that for a given $\sigma > 0$ there exists a function $h(t, y; \sigma)$ which on Δ is continuous in (t, y), Lipschitzian in y, is such that $|h(t, y; \sigma)| \leq \mu$, and $|h(t, y; \sigma) - h(t, y)| < \sigma$.

Hint. Consider first the case of a real-valued vector function $g(t, u) \equiv (g_\alpha(t, u_1, \ldots, u_n))$, $(\alpha = 1, \ldots, n)$, which on $\Delta = [a, b] \times \mathbf{R}_n$ is continuous in (t, u), bounded and uniformly continuous in u; let μ be such that $|g(t, u)| \leq \mu$ for $(t, u) \in \Delta$, and for $\varepsilon > 0$ let $\delta(\varepsilon)$ be such that $|g(t, u^1) - g(t, u^2)| < \varepsilon$ if $(t, u^1) \in \Delta$, $(t, u^2) \in \Delta$, and $|u^1 - u^2| < \delta(\varepsilon)$. For $r > 0$, and $(t, u) \in \Delta$, set

$$g(t, u; r) = (2r)^{-n} \int_{-r}^{r} \cdots \int_{-r}^{r} g(t, u_1 + s_1, \ldots, u_n + s_n)\, ds_n \cdots ds_1,$$
$$= (2r)^{-n} \int_{u_1 - r}^{u_1 + r} \cdots \int_{u_n - r}^{u_n + r} g(t, v_1, \ldots, v_n)\, dv_n \cdots dv_1.$$

Show that $|g(t, u; r)| \leq \mu$ on Δ, and that $|g(t, u; r) - g(t, u)| < \sigma$ if $0 < r < \delta(\sigma)/\sqrt{n}$. Moreover, $g(t, u; r)$ has partial derivatives with respect to the u_β which are continuous functions of (t, u) on Δ, and $|g_{u_\beta}(t, u; r)| \leq \mu/r$, $(\beta = 1, \ldots, n)$. Indeed, if $L(t, w_1, \ldots, w_n, w) = g(t, w_1 + w, w_2, \ldots, w_n) - g(t, w_1 - w, w_2, \ldots, w_n)$, then

$$g_{u_1}(t, u; r) = (2r)^{-n} \int_{u_2 - r}^{u_2 + r} \cdots \int_{u_n - r}^{u_n + r} L(t, u_1, v_2, \ldots, v_n, r)\, dv_n \cdots dv_2,$$

with similar expressions for $g_{u_\beta}(t, u; r)$, $(\beta = 2, \ldots, n)$. Reduce the case of complex-valued vector functions to that of real-valued vector functions by the method given in Section 1.

11.[s] Suppose that the vector function $f(t, y)$ is continuous in (t,y) on $\Delta = [a, b] \times \mathbf{C}_n$. As in Problem I.3:6, let $\mathcal{F}(a, \eta)$ denote the set of all solutions of (3.1) passing through the point (a, η), and for $s \in [a, b]$ let $\mathcal{F}(a, \eta; s)$ be the set of all points (s, ξ) such that there is a solution $y(t)$, $t \in [a, s]$, of $\mathcal{F}(a, \eta)$ satisfying $y(s) = \xi$. If all solutions of $\mathcal{F}(a, \eta)$ are extensible to the interval $[a, b]$, prove that for $s \in [a, b]$ the set $\mathcal{F}(a, \eta; s)$ is a bounded continuum; that is, it is bounded, closed, and connected.

Hint. Note that in view of the result of Problem I.3:6 the extensibility of all solutions of $\mathcal{F}(a, \eta)$ to $[a, b]$ implies that there exists a $\nu > 0$ such that all solutions $y(t)$, $t \in [a, b]$, of $\mathcal{F}(a, \eta)$ lie in the set $\Delta_{\nu-1} = [a, b] \times \{y \mid |y| \leq \nu - 1\}$. Moreover, for $s \in [a, b]$ the set $\mathcal{F}(a, \eta; s)$ is closed, so that it remains only to prove that $\mathcal{F}(a, \eta; s)$ is connected. It is to be remarked that although the hypotheses of the cited Problem 6 require $f(t, y)$ to be continuous on an open region the results are applicable to an equation (3.1) with $f(t, y)$

continuous on a strip Δ, since such an $f(t, y)$ may be extended continuously on $(-\infty, \infty) \times \mathbf{C}_n$ by setting $f(t, y) = f(a, y)$ for $t < a$, $f(t, y) = f(b, y)$ for $t > b$. Now if $s \in (a, b]$, and $\mathcal{F}(a, \eta; s)$ is not connected, there exist two nonempty closed sets S^0, S^1 of $\{(s, \xi) \mid |\xi| \leq \nu - 1\}$ such that $\mathcal{F}(a, \eta; s) = S^0 \cup S^1$ and $d(S^0, S^1) > 0$, where $d(S^0, S^1)$ is the distance between S^0 and S^1 defined as the infimum of $|\xi^0 - \xi^1|$ for $\xi^0 \in S^0$ and $\xi^1 \in S^1$. For a given fixed pair ξ^0, ξ^1 with $\xi^\alpha \in S^\alpha$, $(\alpha = 0, 1)$, let $y^\alpha(t)$, $t \in [a, b]$, be solutions of $\mathcal{F}(a, \eta)$ such that $y^\alpha(s) = \xi^\alpha$. Now let $h(t, y)$ be defined on Δ as $h(t, y) = f(t, y)$ for $(t, y) \in \Delta_\nu$, $h(t, y) = f(t, \nu y/|y|)$ for $(t, y) \in \Delta$, $|y| > \nu$. For $j = 1, 2, \ldots$, let $h^j(t, y)$ be the function $h(t, y; 1/j)$ corresponding to $h(t, y)$ as in Problem 10 above. In particular, on Δ the function $h^j(t, y)$ is continuous in (t, y), Lipschitzian in y, and $|h^j(t, y) - h(t, y)| < 1/j$, while $|h^j(t, y)| \leq \mu$, where μ is a constant that dominates $|h(t, y)| = |f(t, y)|$ on Δ_ν. Define $h^{(j,\alpha)}(t, y)$ and $h^j(t, y \mid \theta)$ as follows:

$$h^{(j,\alpha)}(t, y) = h^j(t, y) + h(t, y^\alpha(t)) - h^j(t, y^\alpha(t)), \qquad (\alpha = 0, 1; j = 1, 2, \ldots);$$
$$h^j(t, y \mid \theta) = (1 - \theta)h^{(j,0)}(t, y) + \theta h^{(j,1)}(t, y), \qquad \theta \in [0, 1]; j = 1, 2, \ldots.$$

Note that for $\theta \in [0, 1]$, and $j = 1, 2, \ldots$, on Δ the vector function $h^j(t, y \mid \theta)$ is continuous in (t, y), Lipschitzian in y, $|h^j(t, y \mid \theta) - h(t, y)| < 1/j$, and $|h^j(t, y \mid \theta)| \leq \mu + 1/j \leq \mu + 1$. Moreover, $h^{(j,\alpha)}(t, y^\alpha(t)) = h(t, y^\alpha(t)) = f(t, y^\alpha(t))$, and therefore $y = y^\alpha(t)$, $(\alpha = 0,1)$, is a solution of

$$y' = h^j(t, y \mid \theta), \tag{5.16}$$

for the respective values $\theta = 0$ and $\theta = 1$. Show that the solutions of (5.16) are locally unique in Δ, and if $y = y^j(t, \theta)$ is the solution of this equation passing through the initial point (τ, η) then $y^j(s, \theta)$ is continuous in θ on $[0, 1]$. By a continuity argument conclude that there exists a value θ_j on $(0, 1)$ such that $\xi^j = y^j(s; \theta_j)$ satisfies the equality $d(S^0, \xi^j) = d(S^1, \xi^j)$. On a suitable set $\Delta_\rho = [a, b] \times \{y \mid |y| \leq \rho\}$ apply the result of Problem I.3:3 to the differential equations $y' = h(t, y)$, $y' = h^j(t, y \mid \theta_j)$, and show that there exists a solution $y(t)$, $t \in [a, b]$ of $y' = h(t, y)$ passing through the initial point (τ, η) and such that $d(S^0, y(s)) = d(S^1, y(s))$. Finally, from the definitive property of $\Delta_{\nu-1}$ conclude that $y(t)$ is a solution of $y' = f(t, y)$ on $[a, b]$, thus obtaining the contradictory result that the point $(s, y(s))$ of $\mathcal{F}(a, \eta; s)$ does not belong to $S^0 \cup S^1$.

12.[s] Suppose that $\omega(t, r)$ is a continuous non-negative real-valued function on a domain $\{(t, r) \mid t \in (0, q], r \geq 0\}$, with the following properties:

(i) $\omega(t, 0) = 0$ for $t \in (0, q]$;
(ii) for each $c \in (0, q]$, the only solution of the scalar differential equation

$$r' = \omega(t, r) \tag{5.17}$$

on the interval $(0, c]$ which is such that

$$(5.18) \qquad r(t) \to 0, \qquad \frac{r(t)}{t} \to 0, \quad \text{as} \quad t \to 0^+$$

is $r(t) \equiv 0$.

If on $\square_1 = \{(t, y) \mid t \in [\tau, \tau + q], |y - \eta| \le d\}$ the vector valued function $f(t, y)$ is continuous and satisfies the inequality

$$(5.19) \qquad |f(t, z) - f(t, y)| \le \omega(t - \tau, |z - y|)$$

for every pair of points (t, y), (t, z) on $\square_1$ with $t \in (\tau, \tau + q]$, prove that if $b \in (\tau, \tau + q]$ then there is at most one solution $y(t)$, $t \in [\tau, b]$, of (3.1) passing through the initial point (τ, η), and having graph in $\square_1$.

Hint. If for $\alpha = 1, 2$ the vector functions $y = y^\alpha(t)$, $t \in [\tau, b]$, are solutions of $\mathcal{F}(\tau, \eta)$ for (3.1) with graphs in $\square_1$, and $\rho(t) = |y^1(t) - y^2(t)|$ is not identically zero on $(\tau, b]$, note that there exists a $c \in (\tau, b]$ and a value $a_0 \in [\tau, c)$ such that $0 = \rho(a_0) < \rho(t)$ for $t \in (a_0, c]$, and from (5.19) conclude that

$$\rho'(t) \le \omega(t - \tau, \rho(t)), \quad \text{for} \quad t \in (a_0, c].$$

If $r_m(t)$ is the minimal solution of $r' = \omega(t - \tau, r)$ passing through the point $(c, \rho(c))$, with the aid of the result of Problem I.4:5 conclude that $\rho(t) \ge r_m(t)$ on any subinterval $(a_1, c]$ of $(a_0, c]$ on which $r_m(t)$ exists. Proceed to show that the existence of a value c_1 on $(a_1, c]$ such that $r_m(c_1) = 0$ implies that $r(t) = r_m(t + \tau)$, $t \in [c_1 - \tau, c - \tau]$, $r(t) = 0$ for $t \in (0, c_1 - \tau]$, is a solution of (5.17) on $(0, c - \tau]$ which satisfies (5.18), but is not identically zero. In view of conditions (i) and (ii), show in turn that either of the possibilities $a_0 > \tau$, $a_0 = \tau$ leads to a similar contradiction, and hence $y^1(t) = y^2(t)$ for $t \in (\tau, b]$.

13. For (3.1) formulate, and establish, a criterion for uniqueness of solution on an interval $[\tau - q, \tau]$ that is analogous to the result of Problem 12. In particular, show that the result of Theorem 5.2, and also that of Problem 7 above, are consequences of the criterion of Problem 12 and the analogous criterion of this problem.

14. Suppose that $\omega(t, r)$ is a continuous non-negative real-valued function satisfying on $\{(t, r) \mid t \in (0, q], r \ge 0\}$ the conditions (i), (ii) of Problem 12, and that $V(t, u)$ is a non-negative real-valued function that is continuous and continuously differentiable in (t, u) on $[0, q] \times \mathbf{R}_n$, and such that if $t \in [0, q]$ then $V(t, u) = 0$ if and only if $u = 0$. If on $\square_1 = \{(t, u) \mid t \in [\tau, \tau + q],$ $|u - \xi| \le d\}$ the real-valued vector function $g(t, u) = (g_\alpha(t, u_1, \ldots, u_n))$ is continuous and satisfies the inequality

$$(5.20)\quad V_t(t, v-u) + \sum_{\alpha=1}^{n} V_{u_\alpha}(t, v-u)[g_\alpha(t, v) - g_\alpha(t, u)] \le \omega(t-\tau, V(t, v-u))$$

for points (t, u), (t, v) in $\square_1$ with $t \in (\tau, \tau + q]$, prove that if $b \in (\tau, \tau + q]$ then there is at most one solution $u(t)$, $t \in [\tau, b]$, of

$$(5.21)\qquad u' = g(t, u)$$

passing through the initial point (τ, ξ) and having graph in $\square_1$.

Hint. If $u = u^\alpha(t)$, $t \in [\tau, b]$, $\alpha = 1, 2$, are solutions of (5.21) with graphs in $\square_1$, and $u^1(t) - u^2(t)$ is not identically zero on $[\tau, b]$, proceed in a manner similar to that described in the Hint for Problem 12. In particular, if I is a subinterval of $[\tau, b]$ throughout which $u^1(t) - u^2(t) \neq 0$, then $\psi(t) = V(t, u^1(t) - u^2(t))$ is such that

$$\psi'(t) \le \omega(t-\tau, \psi(t)), \quad \text{for} \quad t \in I.$$

15.[s] Suppose that $\chi(t, r)$ is a continuous, non-negative real-valued function on a domain $\{(t, r) \mid t \in (0, q], r \ge 0\}$ with the following properties:

(i) $\chi(t, 0) = 0$ for $t \in (0, q]$;
(ii) for each $c \in (0, q]$, the only solution of the scalar differential equation

$$(5.22)\qquad r' = \chi(t, r)$$

on the interval $(0, c]$ which is such that

$$(5.23)\qquad r(t) \to 0, \ \frac{r(t)}{t^2} \to 0 \quad \text{as} \quad t \to 0^+$$

is $r(t) \equiv 0$.

If on $\square_1 = \{(t, y) \mid t \in [\tau, \tau + q], |y - \eta| \le d\}$ the vector valued function $f(t, y)$ is continuous and satisfies the inequality

$$(5.24)\qquad 2\Re\{[z^* - y^*][f(t, z) - f(t, y)]\} \le \chi(t - \tau, |z - y|^2),$$

prove that if $b \in (\tau, \tau + q]$ then there is at most one solution $y(t)$, $t \in [\tau, b]$, of (3.1) passing through the initial point (τ, η) and having graph in $\square_1$. Show that the criterion of Theorem 5.3, and also that of Problem 6 above, are special cases of the criterion of this problem.

16. For (3.1) formulate, and establish, a criterion for uniqueness of solution on an interval $[\tau - q, \tau]$ that is analogous to that of Problem 15, and which includes as a special instance the criterion of Theorem 5.4.

17. Suppose that $\sigma(t, r)$ is a continuous non-negative real-valued function on the domain $Q = \{(t, r) \mid t \in (0, q], r \geq 0\}$, and that each solution of the differential equation

$$r' = \sigma(t, r) \tag{5.25}$$

passing through a point (t^0, r^0) of Q is extensible to the whole interval $[t^0, q]$. If the vector function $f(t, y)$ is continuous and satisfies the inequality

$$|f(t, y)| \leq \sigma(t - \tau, |y|), \quad \text{for} \quad (t, y) \in (\tau, \tau + q] \times \mathbf{C}_n, \tag{5.26}$$

prove that each solution $y(t)$ of (3.1) is extensible to the interval $[\tau, \tau + q]$.

Hint. If $y^0(t)$ is a solution of (3.1) passing through (τ, η), and $y^0(t)$ is not extensible to the entire interval $[\tau, \tau + q]$, note that in view of Theorem 3.5 there exists a $c \in (\tau, \tau + q]$ and an extension $y(t)$ of $y^0(t)$ to $[\tau, c)$ with $|y(t)| \to \infty$ as $t \to c$. In particular, there is a $c_0 \in (\tau, c)$ such that $|y(t)| \geq 1$ on $[c_0, c)$; moreover, from (5.26) it follows that $|y(t)|' \leq \sigma(t - \tau, |y(t)|)$ on $[c_0, c)$. Consequently, $\rho(t) = |y(\tau + t)|$ satisfies $\rho'(t) \leq \sigma(t, \rho(t))$ for $t \in [c_0 - \tau, c - \tau)$, $\rho(t) \geq 1$ on this interval, and $\rho(t) \to \infty$ as $t \to c - \tau$. If $r_M(t)$ is the maximal solution of (5.25) passing through the point $(c_0 - \tau, \rho(c_0 - \tau))$, conclude with the aid of the result of Problem I.4:5 that $\rho(t) \leq r_M(t)$ on any interval $[c_0 - \tau, d)$ common to the intervals of existence of these two functions, and thus obtain a contradiction to the extensibility of $r_M(t)$ to $[c_0 - \tau, q]$.

18. If $\lambda(t)$ is continuous on $(-\infty, \infty)$, and $\Phi(r)$ is a non-negative real-valued continuous function for $r \geq 0$ which is positive for $r > 0$ and such that

$$\int_1^\infty \frac{dr}{\Phi(r)} = \infty,$$

show that for arbitrary real τ, ξ each solution of the differential equation

$$r' = \lambda(t)\Phi(r)$$

is extensible to the interval $(-\infty, \infty)$. Consequently, with the aid of the result of the preceding problem, and an analogous extensibility result for an interval $[\tau - q, \tau]$, establish the following criterion of extensibility: If on $(-\infty, \infty) \times \mathbf{C}_n$ the vector function $f(t, y)$ is continuous in (t, y), and $|f(t, y)| \leq \lambda(t)\Phi(|y|)$, where $\lambda(t)$ and $\Phi(r)$ satisfy the above conditions, then each solution of (3.1) is extensible to the interval $(-\infty, \infty)$.

6. EXISTENCE THEOREMS BY THE METHOD OF ITERATIONS

The result of Theorem 3.2 on the existence of a solution of (3.1) passing through an initial point is of limited computational value, as it does not in general provide a constructive method for determining a solution of the initial value problem. Under the hypothesis of Theorem 5.6 to the effect that there is on an interval $[a, b]$ a unique solution $y = \phi(t)$ of (3.1) passing through a given initial point (τ, η), it does follow from Theorem 5.6 that this solution is on $[a, b]$ the uniform limit of the approximate solutions (3.10) as $|\Pi| \to 0$, and also the uniform limit of the approximate solutions (3.13) as $\delta \to 0$. Even in this case, however, for an ε-approximate solution $y(t)$ of (3.1) satisfying (3.3) there is not available in general a computable upper bound for $|y(t) - \phi(t)|$ on $[a, b]$. We shall proceed now to discuss an iterative procedure which under the hypothesis of a Lipschitz condition may be used to establish for the initial value problem the existence of the unique solution, and which also provides estimates for the order of convergence of the iterates to the solution.

The methods of iterations is also an old device in the history of differential equations. In particular, it was used by Liouville in 1838 for the case of a homogeneous second order linear differential equation. The proof of existence theorems by this method for nonlinear differential equations was given in 1890 by Picard, whose method was improved in detail by Lindelöf in 1894. Frequently the procedure is referred to as the Picard-Lindelöf method of "successive approximations" or "iterations."

For our preliminary consideration of the iteration procedure, suppose that the vector function $f(t, y) \equiv (f_\alpha(t, y))$, $(\alpha = 1, \ldots, n)$, is continuous in (t, y) on the infinite strip $\Delta = [a, b] \times \mathbf{C}_n$, and for $y^{(0)}(t)$ an arbitrary continuous vector function on $[a. b]$ define the sequence of vector functions $\{y^{(j)}(t)\}$, $(j = 0, 1, \ldots)$, by the iterative process

$$(6.1) \qquad y^{(m+1)}(t) = \eta + \int_\tau^t f(s, y^{(m)}(s))\, ds, \qquad (m = 0, 1, \ldots).$$

By induction it follows readily that each $y^{(j)}(t)$ is continuous on $[a, b]$.

Theorem 6.1. *If on the infinite strip* $\Delta = [a, b] \times \mathbf{C}_n$ *the vector function* $f(t, y)$ *is continuous and satisfies a Lipschitz condition* (5.1), *then for* $y^{(0)}(t)$ *an arbitrary continuous vector function on* $[a, b]$ *the sequence* $\{y^{(j)}(t)\}$ *defined by* (6.1) *converges uniformly on* $[a, b]$ *to the unique solution of* (3.1) *passing through the initial point* (τ, η) *of* Δ.

The uniform convergence on $[a, b]$ of the sequence $\{y^{(j)}(t)\}$ defined by (6.1) is equivalent to the uniform convergence on $[a, b]$ of the vector series

$$(6.2) \quad y^{(0)}(t) + [y^{(1)}(t) - y^{(0)}(t)] + \cdots + [y^{(m+1)}(t) - y^{(m)}(t)] + \cdots.$$

As $|y^{(1)}(t) - y^{(0)}(t)|$ is a continuous function on $[a, b]$, there is a $\lambda \geq 0$ such that $|y^{(1)}(t) - y^{(0)}(t)| \leq \lambda$ on this interval. It will be shown by induction that in terms of such a constant λ and the Lipschitz constant of (5.1) we have

$$(6.3)\qquad |y^{(m+1)}(t) - y^{(m)}(t)| \leq \frac{\lambda(\kappa\,|t-\tau|)^m}{m!}, \qquad (m = 1, 2, \ldots).$$

Indeed, as $|f(s, y^{(1)}(s)) - f(s, y^{(0)}(s))| \leq \kappa\,|y^{(1)}(s) - y^{(0)}(s)| \leq \lambda\kappa$ on $[a, b]$, on this interval we have

$$|y^{(2)}(t) - y^{(1)}(t)| = \left|\int_\tau^t [f(s, y^{(1)}(s)) - f(s, y^{(0)}(s))]\, ds\right| \leq \lambda\kappa\,|t-\tau|.$$

Correspondingly, if (6.3) holds for $m = j$, then

$$\begin{aligned} |f(s, y^{(j+1)}(s)) - f(s, y^{(j)}(s))| &\leq \kappa\,|y^{(j+1)}(s) - y^{(j)}(s)|, \\ &\leq \frac{\lambda\kappa(\kappa\,|s-\tau|)^j}{j!}, \quad \text{for} \quad s \in [a, b]. \end{aligned}$$

Hence

$$\begin{aligned} |y^{(j+2)}(t) - y^{(j+1)}(t)| &= \left|\int_\tau^t [f(s, y^{(j+1)}(s)) - f(s, y^{(j)}(s))]\, ds\right|, \\ &\leq \lambda\left|\int_\tau^t \frac{\kappa(\kappa\,|s-\tau|)^j}{j!}\, ds\right| = \frac{\lambda(\kappa\,|t-\tau|)^{j+1}}{(j+1)!}, \end{aligned}$$

so that (6.3) holds for $m = j + 1$, thus completing the inductive proof of this relation. In particular, (6.3) implies that uniformly on $[a, b]$,

$$(6.4)\qquad |y^{(m+1)}(t) - y^{(m)}(t)| \leq \frac{\lambda(\kappa[b-a])^m}{m!}, \qquad (m = 1, 2, \ldots).$$

If λ_0 is a constant such that the continuous function $|y^{(0)}(t)|$ satisfies $|y^{(0)}(t)| \leq \lambda_0$ on $[a, b]$, then it follows that this vector series converges uniformly on $[a, b]$. Moreover, if $y(t)$ denotes the sum of this series then for $t \in [a, b]$ we have

$$(6.5)\qquad \begin{aligned} |y(t)| &\leq \lambda_0 + \lambda \exp\{\kappa(b-a)\}, \\ |y(t) - y^{(m)}(t)| &\leq \lambda \sum_{k=0}^{\infty} \frac{(\kappa\,|t-\tau|)^{m+k}}{(m+k)!}. \end{aligned}$$

As $(\kappa\,|t-\tau|)^{m+k}/(m+k)!$ is dominated by

$$\left\{\frac{(\kappa\,|t-\tau|)^m}{m!}\right\}\left[\frac{(\kappa(b-a))^k}{k!}\right],$$

from the second inequality of (6.5) it follows that for $m = 1, 2, \ldots,$ and uniformly in t on $[a, b]$, we have

$$
(6.6)\qquad
\begin{aligned}
|y(t) - y^{(m)}(t)| &\leq \frac{\lambda(\kappa\,|t - \tau|)^m}{m!}\exp\{\kappa(b-a)\},\\
&\leq \frac{\lambda(\kappa(b-a))^m}{m!}\exp\{\kappa(b-a)\}.
\end{aligned}
$$

In particular, (6.6) implies that the sequence $\{y^{(j)}(t)\}$ converges uniformly on $[a, b]$ to $y(t)$, and hence this vector function is continuous on $[a, b]$. As

$$
\begin{aligned}
\left|\int_\tau^t f(s, y(s))\,ds - \int_\tau^t f(s, y^{(m)}(s))\,ds\right| &\leq \left|\int_\tau^t |f(s, y(s)) - f(s, y^{(m)}(s))|\,ds\right|,\\
&\leq \left|\int_\tau^t \kappa|y(s) - y^{(m)}(s)|\,ds\right|,
\end{aligned}
$$

it also follows from (6.6) that $\int_\tau^t f(s, y^{(m)}(s))\,ds \to \int_\tau^t f(s, y(s))\,ds$ as $m \to \infty$. Upon letting $m \to \infty$ in (6.1), it then follows that

$$
(6.7)\qquad y(t) = \eta + \int_\tau^t f(s, y(s))\,ds \quad \textit{for} \quad t \in [a, b],
$$

and consequently $y(t)$ is a solution of (3.1) on $[a, b]$ satisfying $y(\tau) = \eta$. In view of Theorem 5.1 there is a unique solution of (3.1) on $[a, b]$ passing through the initial point (τ, η), so the proof of the theorem is complete.

For the local existence theorem one has the following result using the method of iterations.

Theorem 6.2. *Suppose that on* $\square = \{(t, y) \mid |t - \tau| \leq \varepsilon, |y - \eta| \leq \delta\}$ *the vector function* $f(t, y)$ *is continuous in* (t, y), *Lipschitzian in* y, μ *is such that* $|f(t, y)| \leq \mu$ *on* $\square$, *and* $\varepsilon_1 = \min[\varepsilon, \delta/\mu]$. *If* $y^{(0)}(t)$ *is an arbitrary continuous vector function on* $[\tau - \varepsilon_1, \tau + \varepsilon_1]$ *satisfying* $|y^{(0)}(t) - \eta| \leq \delta$, *then the sequence* $\{y^{(j)}(t)\}$ *defined by* (6.1) *for* $t \in [\tau - \varepsilon_1, \tau + \varepsilon_1]$ *converges uniformly on this interval to the unique solution* $y(t)$ *of* (3.1) *satisfying* $y(\tau) = \eta$.

One may show by an inductive argument that each of the vector functions $y^{(j)}(t)$ of (6.1) satisfies $|y^{(j)}(t) - \eta| \leq \delta$ on $[\tau - \varepsilon_1, \tau + \varepsilon_1]$, and then proceed as in Theorem 6.1 with the strip Δ of that theorem replaced by $\square$. An alternate procedure is to use the result of Problem 2 of Section I.5, and from Theorem 6.1 deduce the result of Theorem 6.2 in the same fashion that the result of Theorem 5.2 was derived from Theorem 5.1.

Problems I.6

1. For the following real-valued scalar differential equations find the solution passing through the indicated initial point, using the method of iterations and starting with the given function $u^{(0)}(t)$ in each case:

(i) $u' = t + u$, $(t = 0, u = 1)$, $u^{(0)}(t) \equiv 1$;
(ii) $u' = u^2 - t^2 - 1$, $(t = 0, u = 0)$, (a) $u^{(0)}(t) \equiv 0$, (b) $u^{(0)}(t) = t$.

2. If on $\Delta = [a, b] \times \mathbf{C}_n$ the vector function $f(t, y)$ is continuous and satisfies a Lipschitz condition (5.1), show that the solution $y = y(t; \tau, \eta)$ of (3.1) passing through the point (τ, η) is such that:

(i) $|y(t; \tau, \eta) - \eta| \leq |\int_\tau^t| f(s, \eta)\,|ds|\, \exp\{\kappa\,|t - \tau|\}$;
$\leq (\int_a^b |f(s, \eta)|\, ds)\exp\{\kappa(b - a)\}$;

(ii) if $\mu(\tau, \eta)$ is such that $|f(t, y(t; \tau, \eta))| \leq \mu(\tau, \eta)$ for $t \in [a, b]$, then

$$|y(t^1; \tau^1, \eta^1) - y(t; \tau, \eta)| \leq \mu(\tau, \eta)\,|t^1 - t| + [|\eta^1 - \eta| + \mu(\tau, \eta)\,|\tau^1 - \tau|]\exp\{\kappa(b - a)\}.$$

Hints. (i) Use estimate (6.5), for $y^{(0)}(t) \equiv \eta$.
(ii) Use the result of Problem 1.5:5.

3.[s] Suppose that the vector function $f(t, y; c) \equiv (f_\alpha(t, y_1, \ldots, y_n; c_1, \ldots, c_k))$ $(\alpha = 1, \ldots, n)$, is continuous in (t, y, c) for (t, y) on the infinite strip $\Delta = [a, b] \times \mathbf{C}_n$, and $c \equiv (c_\nu)$, $(\nu = 1, \ldots, k)$, on a set Γ of $\mathbf{C}_k$; moreover, suppose that there exists a constant $\kappa > 0$ such that the Lipschitz condition

$$|f(t, y; c) - f(t, z; c)| \leq \kappa\,|y - z|$$

holds for arbitrary t, y, z, c with (t, y), (t, z) on Δ and c on Γ. If $y = y(t; \tau, \eta, c)$ is the solution of $y' = f(t, y; c)$ passing through the initial point (τ, η), and $\mu(\tau, \eta, c)$, $\kappa_1(\tau, \eta, c, c^1)$ are such that for $t \in [a, b]$ we have

$$|f(t, y(t; \tau, \eta, c); c)| \leq \mu(\tau, \eta, c),$$
$$|f(t, y(t; \tau, \eta, c); c) - f(t, y(t; \tau, \eta, c); c^1)| \leq \kappa_1(\tau, \eta, c, c^1),$$

show that

$$|y(t^1; \tau^1, \eta^1, c^1) - y(t; \tau, \eta, c)| \leq \mu(\tau, \eta, c)\,|t^1 - t| + \mu_1(\tau, \eta, c, \tau^1, \eta^1, c^1)\exp\{\kappa(b - a)\}$$

for arbitrary t^1, t on $[a, b]$, (τ^1, η^1) and (τ, η) on Δ, and c^1, c on Γ, where

$$\mu_1(\tau, \eta, c, \tau^1, \eta^1, c^1) = |\eta^1 - \eta| + \mu(\tau, \eta, c)\,|\tau^1 - \tau| + \kappa_1(\tau, \eta, c, c^1)(b - a).$$

Hint. Use the result of Problem I.5:5.

4. For $\theta(t)$ a real-valued continuously differentiable function on $(-\infty, \infty)$ satisfying $\theta(t) \equiv 0$ on $(-\infty, 0]$, $\theta'(t) > 0$ on $(0, \infty)$, let $g(t, u)$ be defined as follows:

$$\begin{aligned} g(t, u) &= 0, \quad \text{for} \quad t \le 0, u \in (-\infty, \infty), \\ &= \theta'(t), \quad \text{for} \quad t > 0, u \in (-\infty, 0], \\ &= \theta'(t) - \frac{\theta'(t)}{\theta(t)} u, \quad \text{for} \quad t > 0, 0 < u \le \theta(t), \\ &= 0, \quad \text{for} \quad t > 0, u > \theta(t). \end{aligned}$$

Show that $g(t, u)$ is continuous throughout the entire (t, u) plane, and that the scalar differential equation $u' = g(t, u)$ has the unique solution $u = \theta(t)/2$ passing through the point $(0, 0)$. However, when the method of iterations is employed with initial function $u^{(0)}(t) \equiv 0$ the sequence of iterates does not converge.

5. Suppose that $\omega(t, r)$ satisfies the conditions of Problem 1.5:12, and in addition is a nondecreasing function of r on $[0, \infty)$ for each $t \in (0, q]$. Moreover, suppose that on $\square_1 = \{(t, y) \mid t \in [\tau, \tau + q], |y - \eta| \le d\}$ the vector valued function $f(t, y)$ is continuous and satisfies the inequality (5.19) for each pair of points (t, y), (t, z) on $\square_1$ with $t \in (\tau, \tau + q]$. If μ is such that $|f(t, y)| \le \mu$ on $\square_1$, and $\varepsilon_1 = \min\{q, d/\mu\}$, then the successive approximations

$$y^{(0)}(t) = \eta, \; y^{(m+1)}(t) = \eta + \int_\tau^t f(s, y^{(m)}(s))\, ds, \qquad (m = 0, 1, \ldots),$$

converge uniformly on $[\tau, \tau + \varepsilon_1]$ to the unique solution $y(t)$ of (3.1) on this interval satisfying $y(\tau) = \eta$.

Hint. Note that Problem I.5:12 implies that there is at most one solution $y(t)$, $t \in [\tau, \tau + \varepsilon_1]$, of (3.1) satisfying $y(\tau) = \eta$. Moreover, on $[\tau, \tau + \varepsilon_1]$ each of the approximations $y^{(m)}(t)$ satisfies $|y^{(m)}(t) - \eta| \le \mu[t - \tau] \le d$, and $|y^{(m)}(t_1) - y^{(m)}(t_2)| \le \mu\, |t_2 - t_1|$ for $t_\alpha \in [\tau, \tau + \varepsilon_1]$, $(\alpha = 1, 2)$; in particular, the vector functions $y^{(m)}(t)$, $t \in [\tau, \tau + \varepsilon_1]$ are equi-continuous. Let $\{m_k\}$ be a subsequence of positive integers such that $\{y^{(m_k)}(t)\}$ converges uniformly on $[\tau, \tau + \varepsilon_1]$ to a limit vector function $y(t)$, which is necessarily continuous. Moreover, if $r_k = m_k + 1$ then the subsequence $\{y^{(r_k)}(t)\}$ converges uniformly on $[\tau, \tau + \varepsilon_1]$ to $\eta + \int_\tau^t f(s, y(s))\, ds$, and hence the conclusion that $y(t)$ is a solution of (3.1) satisfying $y(\tau) = \eta$ follows if it can be established that the sequence $\{y^{(m+1)}(t) - y^{(m)}(t)\}$, $(m = 0, 1, \ldots)$, converges uniformly on $[\tau, \tau + \varepsilon_1]$ to the zero vector function.

In order to prove this latter result, for $t \in [\tau, \tau + \varepsilon_1]$ and $m = 1, 2, \ldots$, let $w^{(m)}(t) = \sup\{|y^{(j+1)}(t) - y^{(j)}(t)| \mid j = m, m + 1, \ldots\}$. Show that $|w^{(m)}(t_2) - w^{(m)}(t_1)| \le 2\mu\, |t_2 - t_1|$ for $t_\alpha \in [\tau, \tau + \varepsilon_1]$, $(\alpha = 1, 2)$; in particular, $|w^{(m)}(t)| \le 2\mu[t - \tau]$. From the fact that $0 \le w^{(m+1)}(t) \le w^{(m)}(t)$, and the equi-continuity of the functions $w^{(m)}(t)$ on $[\tau, \tau + \varepsilon_1]$, conclude that there is a continuous non-negative function $w(t)$ such that the sequence $\{w^{(m)}(t)\}$ converges uniformly to $w(t)$ on $[\tau, \tau + \varepsilon_1]$; the function $w(t)$ is the *limit superior* of the sequence $\{w^{(m)}(t)\}$. Use the following argument to show that existence of an ε_2, $0 < \varepsilon_2 \leq \varepsilon_1$ such that $w(\tau + \varepsilon_2) > 0$ leads to a contradiction. Note that condition (5.19), together with the monotoneity of ω in r, implies that if $\tau < \nu < \tau + \varepsilon_2$ then

$$(6.8) \quad w^{(m)}(t) \leq w^{(m)}(\nu) + \int_\nu^t \omega(s - \tau, w^{(m-1)}(s))\, ds, \quad for \quad t \in [\nu, \tau + \varepsilon_2],$$

and conclude that

$$(6.9) \qquad w(t) \leq w(\nu) + \int_\nu^t \omega(s - \tau, w(s))\, ds, \quad for \quad t \in [\nu, \tau + \varepsilon_2].$$

If $v(t, \nu) = w(\nu) + \int_\nu^t \omega(s - \tau, w(s))\, ds$, $t \in [\nu, \tau + \varepsilon_1]$, show that if $\tau < \nu_2 < \nu_1 < \tau + \varepsilon_2$ then (6.9) implies $v(t, \nu_1) \leq v(t, \nu_2)$ for $t \in [\nu_1, \tau + \varepsilon_2]$; in particular, $v(\tau + \varepsilon_2, \nu_1) \leq v(\tau + \varepsilon_2, \nu_2)$. Moreover $v'(t, \nu) = \omega(t - \tau, w(t)) \leq \omega(t - \tau, v(t, \nu))$, in view of the monotoneity of ω in r. Now let $r(t, \nu)$ be the minimal solution of $r' = \omega(t - \tau, r)$ passing through the point $(\tau + \varepsilon_2, v(\tau + \varepsilon_2, \nu))$; if $r(t, \nu)$ exists on a subinterval $(t_1, \tau + \varepsilon_2]$ of $(\tau, \tau + \varepsilon_2]$, then Theorem 5.4 implies that $r = r(t, \nu)$ is the unique solution of this differential equation on $(t_1, \tau + \varepsilon_2]$ satisfying the initial condition $r(\tau + \varepsilon_2) = v(\tau + \varepsilon_2, \nu)$. As in the proof of the result of Problem I.5:12, show that $v(t, \nu) \geq r(t, \nu) > 0$ throughout a subinterval $(t_1, \tau + \varepsilon_2]$ of $[\nu, \tau + \varepsilon_2]$ on which $r(t, \nu)$ exists, and proceed to conclude that $r(t, \nu)$ exists throughout the entire interval $[\nu, \tau + \varepsilon_2]$, and $0 < r(\nu, \nu) \leq v(\nu, \nu) = w(\nu)$. If $\tau < \nu_2 < \nu_1 < \tau + \varepsilon_2$, note that the condition $v(\tau + \varepsilon_2, \nu_2) \geq v(\tau + \varepsilon_2, \nu_1)$ implies that $r(t, \nu_2) \geq r(t, \nu_1)$ throughout a subinterval $(t_1, \tau + \varepsilon_2]$ of $[\nu_2, \tau + \varepsilon_2]$ on which $r(t, \nu_1)$ exists. Proceed to show that if $\tau < \nu < \tau + \varepsilon_2$ then $r(t, \nu)$ exists on $(\tau, \tau + \varepsilon_2]$, and $0 < r(t, \nu) \leq w(t)$ for $t \in (\tau, \nu]$. Now for $t \in [\tau, \tau + \varepsilon_2]$ let

$$\lambda(t) = \max\{|f(t, y^1) - f(t, y^2)| \mid |y^\alpha - \eta| \leq \mu[t - \tau], \alpha = 1, 2\}.$$

Then $\lambda(t)$ is continuous on $[\tau, \tau + \varepsilon_2]$, $\lambda(\tau) = 0$, and

$$|y^{(m+1)}(t) - y^{(m)}(t)| \leq \int_\tau^t \lambda(s)\, ds, \qquad m = 0. 1, \ldots,$$

from which is follows that $w^{(m)}(t) \leq \int_\tau^t \lambda(s)\, ds$ and $w(t) \leq \int_\tau^t \lambda(s)\, ds$; in particular, $w(t)/(t - \tau) \to 0$ as $t \to \tau^+$. Conclude that if $\tau < \nu < \tau + \varepsilon_2$ then $r(t) = r(\tau + t, \nu)$ is a solution of $r' = \omega(t, r)$ which is not identically zero on $(0, \varepsilon_2]$, and for which $r(t) \to 0$ and $r(t)/t \to 0$ as $t \to 0^+$, in contradiction to the hypothesis of the problem.

7. APPLICATION OF THE CONTRACTION MAPPING THEOREM TO EXISTENCE THEOREMS

One of the basic steps in the proof of Theorem 6.1 was the fact that if the vector function $f(t, y)$ is continuous on the strip $\Delta = [a, b] \times \mathbf{C}_n$, $(\tau, \eta) \in \Delta$, and y belongs to the class $\mathfrak{C}_n[a, b]$ of n-dimensional vector functions which are continuous on $[a, b]$, then the function z defined by

$$(7.1) \qquad z(t) = \eta + \int_\tau^t f(s, y(s))\, ds, \qquad t \in [a, b],$$

is also an element of $\mathfrak{C}_n[a, b]$. That is, the relation (7.1) defines a transformation $z = Ty$ on $\mathfrak{C}_n[a, b]$ into $\mathfrak{C}_n[a, b]$. Moreover, the condition that a particular $y \in \mathfrak{C}_n[a, b]$ is a solution of (3.1) passing through the initial point (τ, η) means that

$$(7.2) \qquad y = Ty;$$

that is, y is a "fixed element" or "fixed point" of the transformation T.

With this interpretation in mind, we proceed to establish a fixed point theorem for a certain type of abstract space.

Suppose that $\mathfrak{M}$ is a "space" of "elements" or "points," together with a distance function $d(x, y)$ defined on $\mathfrak{M} \times \mathfrak{M}$, and possessing the following properties:

(7.3)
1°. $d(x, y) \geq 0$, *and* $d(x, y) = 0$ *if and only if* $x = y$;
2°. $d(x, y) = d(y, x)$;
3°. $d(x, z) \leq d(x, y) + d(y, z)$;
4°. *If* $\{y^{(k)}\}$, $(k = 1, 2, \ldots)$, *is a sequence of points of* $\mathfrak{M}$ *satisfying* $\lim_{h,k\to\infty} d(y^{(h)}, y^{(k)}) = 0$, *then there exists a* $y \in \mathfrak{M}$ *such that* $\lim_{k\to\infty} d(y^{(k)}, y) = 0$.

A space $\mathfrak{M}$ with distance function $d(x, y)$ satisfying conditions 1°, 2°, 3° is called a *metric space*, and whenever condition 4° holds also $\mathfrak{M}$ is a *complete* metric space. Condition 4° may be phrased as: "in $\mathfrak{M}$ every Cauchy sequence is a convergent sequence." Examples of complete metric spaces are:

(a) Real Euclidean n-dimensional space $\mathbf{R}_n$ of real n-tuples $x = (x_\alpha)$, with $d(x, y) = [\sum_{\alpha=1}^n (x_\alpha - y_\alpha)^2]^{1/2}$, or complex Euclidean n-dimensional space $\mathbf{C}_n$ of complex n-tuples $x = (x_\alpha)$, with $d(x, y) = [\sum_{\alpha=1}^n |x_\alpha - y_\alpha|^2]^{1/2}$.

(b) The set $\mathfrak{C}_n[a, b]$ of complex-valued n-dimensional vector functions $y = (y_\alpha(t))$, $(\alpha = 1, \ldots, n)$, which are continuous on $[a, b]$, with

$$d(x, y) = \max(|x(t) - y(t)|\, \theta(t) \mid t \in [a, b]),$$

where $\theta(t)$ is a given continuous scalar function satisfying $\theta(t) > 0$ for $t \in [a, b]$.

For example (b), the condition that y and $y^{(k)}$, $(k = 1, 2, \ldots)$, are such that $\lim_{k\to\infty} d(y^{(k)}, y) = 0$ is equivalent to the condition that the sequence $y^{(k)}(t)$ converges to $y(t)$ uniformly on I.

Theorem 7.1. (Contraction Mapping Theorem) *Suppose that $\mathfrak{M}$ is a complete metric space with distance function $d(x, y)$, and that T is a transformation of $\mathfrak{M}$ into $\mathfrak{M}$. If there exists a λ such that $0 < \lambda < 1$, and*

$$(7.4) \qquad d(Tx, Ty) \leq \lambda\, d(x, y), \quad \textit{for} \quad x \in \mathfrak{M},\ y \in \mathfrak{M},$$

then there exists a unique $y \in \mathfrak{M}$ such that

$$(7.5) \qquad y = Ty.$$

Let $y^{(0)}$ be an arbitrary element of $\mathfrak{M}$, and consider the sequence defined recursively by

$$(7.6) \qquad y^{(m+1)} = Ty^{(m)}, \qquad (m = 0, 1, \ldots).$$

Now

$$d(y^{(2)}, y^{(1)}) = d(Ty^{(1)}, Ty^{(0)}) \leq \lambda\, d(y^{(1)}, y^{(0)}),$$
$$d(y^{(3)}, y^{(2)}) = d(Ty^{(2)}, Ty^{(1)}) \leq \lambda\, d(y^{(2)}, y^{(1)}) \leq \lambda^2\, d(y^{(1)}, y^{(0)}),$$

and by mathematical induction it follows that

$$(7.7) \qquad d(y^{(m+1)}, y^{(m)}) \leq \lambda^m\, d(y^{(1)}, y^{(0)}), \qquad (m = 0, 1, \ldots).$$

In view of property (7.3-3°) of the distance function it follows that if $k > h$ then $d(y^{(k)}, y^{(h)}) \leq \sum_{j=h}^{k-1} d(y^{(j+1)}, y^{(j)})$, and hence with the aid of (7.7) we obtain

$$(7.8) \qquad d(y^{(k)}, y^{(h)}) \leq d(y^{(1)}, y^{(0)}) \sum_{j=h}^{k-1} \lambda^j.$$

Since $0 < \lambda < 1$, by hypothesis, the series $\sum_{j=0}^{\infty} \lambda^j$ is convergent, so that (7.8) implies $\lim_{h,k\to\infty} d(y^{(k)}, y^{(h)}) = 0$, and hence from the completeness property (7.3-4°) it follows that there is a $y \in \mathfrak{M}$ such that $\lim_{k\to\infty} d(y^{(k)}, y) = 0$. In view of (7.6) we have

$$(7.9) \quad 0 \leq d(y, Ty) \leq d(y, y^{(m+1)}) + d(Ty^{(m)}, Ty),$$
$$\leq d(y, y^{(m+1)}) + \lambda d(y^{(m)}, y), \qquad (m = 0, 1, \ldots),$$

and as the terms $d(y, y^{(m+1)})$ and $d(y^{(m)}, y)$ tend to zero as $m \to \infty$ it follows that y is a solution of (7.5).

Finally, the solution of (7.5) is unique since if y and z are elements of $\mathfrak{M}$ satisfying $y = Ty$ and $z = Tz$, then

$$d(y, z) = d(Ty, Tz) \leq \lambda\, d(y, z),$$

and as $0 < \lambda < 1$ it follows that $d(y, z) = 0$ and $y = z$.

Now suppose that $f(t, y)$ is an n-dimensional vector function which on $\Delta = [a, b] \times \mathbf{C}_n$ is continuous in (t, y), and satisfies the Lipschitz condition (5.1), while (τ, η) is a given point of Δ. Let $\mathfrak{M}$ denote the space of Example (b), above, with $\theta(t)$ the function $\exp\{-(\kappa + 1)\,|t - \tau|\}$. If T is the transformation on $\mathfrak{M}$ into $\mathfrak{M}$ defined by (7.1), and $x \in \mathfrak{M}$, $y \in \mathfrak{M}$, then

$$d(Tx, Ty) \\ = \max\left(\left|\int_\tau^t [f(s, x(s)) - f(s, y(s))]\, ds\right| \exp\{-(\kappa + 1)\,|t - \tau|\} \,\middle|\, t \in [a, b]\right).$$

Moreover, in view of the inequalities

$$\begin{aligned}\left|\int_\tau^t [f(s, x(s)) - f(s, y(s))]\, ds\right| &\le \left|\int_\tau^t |f(s, x(s)) - f(s, y(s))|\, ds\right|, \\ &\le \kappa \left|\int_\tau^t |x(s) - y(s)|\, ds\right|, \\ &\le \kappa\, d(x, y) \left|\int_\tau^t \exp\{(\kappa + 1)\,|s - \tau|\}\, ds\right|, \\ &\le \frac{\kappa}{\kappa + 1}\, d(x, y)[-1 + \exp\{(\kappa + 1)\,|t - \tau|\}],\end{aligned}$$

it follows that

$$d(Tx, Ty) \le \frac{\kappa}{\kappa + 1}\, d(x, y).$$

That is, condition (7.4) holds with $\lambda = \kappa/(\kappa + 1)$, and hence the result of Theorem 6.1 is a direct consequence of Theorem 7.1.

In a similar fashion the result of Theorem 6.2 may be established as a special instance of the Contraction Mapping Theorem, where now $\mathfrak{M}$ denotes the set of n-dimensional vector functions $y(t)$ which are continuous and satisfy $|y(t) - \eta| \le \delta$ on $[\tau - \varepsilon_1, \tau + \varepsilon_1]$, and with

$$d(x, y) = \max(|x(t) - y(t)| \exp\{-(\kappa + 1)\,|t - \tau|\} \mid t \in [\tau - \varepsilon_1, \tau + \varepsilon_1]).$$

Problems I.7

1. Find a constant $\rho > 0$ such that if $|A_{\alpha\beta}| \le \rho$, $(\alpha, \beta = 1, 2, \ldots, n)$, then for arbitrary $v = (v_\alpha)$, $(\alpha = 1, \ldots, n)$, the system of linear algebraic equations

$$u_\alpha = \sum_{\beta=1}^{n} A_{\alpha\beta} u_\beta + v_\alpha, \qquad (\alpha = 1, \ldots, n),$$

has a unique solution $u \equiv (u_\alpha)$.

2. If $k(t, s)$ is a continuous function of (t, s) on $S = \{(t, s) \mid a \leq s \leq t \leq b\}$, and $f(t)$ is a continuous function of t on $[a, b]$, prove that there is a unique solution of the (*Volterra*) *integral equation*

$$y(t) = \int_a^t k(t, s)\, y(s)\, ds + f(t), \quad \text{for} \quad t \in [a, b].$$

Hint. If κ is a constant such that $|k(t, s)| \leq \kappa$ on S, let $\mathfrak{M}$ be the space of Example (b), with $\theta(t) = \exp\{-(\kappa + 1)(t - a)\}$.

3. If $k(t, s)$ is a continuous function of (t, s) on $[a, b] \times [a, b]$, and $f(t)$ is a continuous function on $[a, b]$, prove that there is a unique solution of the (*Fredholm*) *integral equation*

$$y(t) = \int_a^b k(t, s)\, y(s)\, ds + f(t), \qquad t \in [a, b],$$

in case there is a positive continuous scalar function $\theta(t)$ on $[a, b]$ such that

$$\theta(t) \int_a^b \frac{|k(t, s)|}{\theta(s)}\, ds < 1, \qquad t \in [a, b].$$

As an example of this general result, prove that if $f(t)$ is continuous on $[0, 1]$ then for $|\lambda| < 1$ the equation

$$y(t) = f(t) + \lambda \int_0^1 e^{t-s}\, y(s)\, ds, \qquad t \in [0, 1],$$

has a unique solution.

4. From Theorem 7.1 conclude that if $\alpha > 0$, $\beta > 0$, and $\alpha + \beta < 1$, then there is a unique continuously differentiable function $y(t)$ on $0 \leq t \leq 1$ such that

$$y' = \alpha y + \beta y(1) + 1, \qquad y(0) = 0, \qquad 0 \leq t \leq 1.$$

Proceed by elementary methods to show that this solution $y(t)$ is

$$y(t) = \frac{e^{\alpha t} - 1}{\alpha + \beta - \beta e^{\alpha}}.$$

5. Prove that a real-valued function $d(x, y)$ defined on $\mathfrak{M} \times \mathfrak{M}$ possesses properties 1°, 2°, 3° of (7.3) if and only if $d(x, y)$ has the following two properties:

(a) $d(x, y) = 0$ if and only if $x = y$;

(b) $d(x, z) \leq d(x, y) + d(z, y)$.

6. Suppose that $\mathfrak{M}$ is a complete metric space with distance function $d(x, y)$, $x_0 \in \mathfrak{M}$, $N_\delta = \{y \mid y \in \mathfrak{M}, d(y, x_0) < \delta\}$ is a neighborhood of x_0, and

T is a transformation on N_δ into $\mathfrak{M}$ for which there is a constant λ such that $0 < \lambda < 1$ and

$$d(Tx, Ty) \leq \lambda d(x, y) \quad \text{for} \quad x \in N_\delta, \qquad y \in N_\delta.$$

If $d(Tx_0, x_0) < (1 - \lambda)\delta$, and the sequence $\{y^{(m)}\}$, $m = 0, 1, 2, \ldots$, is defined as $y^{(0)} = x_0$, $y^{(m+1)} = Ty^{(m)}$, $(m = 0, 1, \ldots)$, prove by induction that each $y^{(m)}$ is in N_δ and $\{y^{(m)}\}$ converges to a y which is the unique element of $\mathfrak{M}$ in N_δ such that $y = Ty$.

8. DIFFERENTIAL EQUATIONS OF HIGHER ORDER

Suppose that $p(t, y) = p(t, y_1, \ldots, y_n)$ is a complex-valued function of a real variable t and n complex variables $y_1, \ldots, y_n$ which is continuous in a region $\mathfrak{R}$ of (t, y)-space. A function $x(t)$, $t \in I$, is said to be a solution of the n-th *order (scalar) differential equation*

$$x^{[n]} = p(t, x, x', \ldots, x^{[n-1]}) \tag{8.1}$$

on a nondegenerate interval I if $x(t)$ is continuous and has continuous derivatives of the first n orders on this interval, the elements $(t, x(t), \ldots, x^{[n-1]}(t))$, $t \in I$, are all in $\mathfrak{R}$, and $x^{[n]}(t) = p(t, x(t), \ldots, x^{[n-1]}(t))$ on this interval. In view of Taylor's formula with integral form of remainder, such a solution $x(t)$ of (8.1) satisfies the initial condition $x^{[\alpha-1]}(\tau) = \eta_\alpha$, $(\alpha = 1, \ldots, n)$, for τ a point on I if and only if

$$x(t) = \eta_1 + \eta_2(t - \tau) + \cdots + \eta_n \frac{(t - \tau)^{n-1}}{(n - 1)!} + \int_\tau^t \frac{(t - s)^{n-1}}{(n - 1)!} p(s, x(s), \ldots, x^{[n-1]}(s))\, ds, \tag{8.2}$$

for t on I, in which case for $\alpha = 2, \ldots, n$ we have,

$$x^{[\alpha-1]}(t) = \eta_\alpha + \eta_{\alpha+1}(t - \tau) + \cdots + \eta_n \frac{(t - \tau)^{n-\alpha}}{(n - \alpha)!} + \int_\tau^t \frac{(t - s)^{n-\alpha}}{(n - \alpha)!} p(s, x(s), \ldots, x^{[n-1]}(s))\, ds. \tag{8.3}$$

From the above discussion it is clear that under the substitution

$$x = y_1, \qquad x' = y_2, \ldots, x^{[n-1]} = y_n \tag{8.4}$$

the n-th order equation (8.1) is equivalent to the first order system

$$\begin{aligned} y'_\alpha &= y_{\alpha+1}, \qquad (\alpha = 1, \ldots, n - 1), \\ y'_n &= p(t, y_1, \ldots, y_n), \end{aligned} \tag{8.5}$$

whenever $p(t, y)$ is a continuous function of its arguments. For discontinuous functions $p(t, y)$ the generalization of the concept of a solution of (8.5) provided by the corresponding system of integral equations (1.5) leads to the generalization of the concept of a solution of (8.1) that is equivalent to the existence of a function $x(t)$ which is continuous and has continuous derivatives of the first $n - 1$ orders on I, and such that equations (8.2), (8.3) are valid on this interval.

Theorems on the existence, uniqueness, and properties of solutions of (8.1) are immediate consequences of the results of the preceding sections for the system (8.5).

In particular, (8.5) is of the form (3.1) with $f_\alpha(t, y) = y_{\alpha+1}$, $(\alpha = 1, 2, \ldots, n - 1)$, $f_n(t, y) = p(t, y)$, and it follows readily that this vector function $f(t, y)$ satisfies a Lipschitz condition (5.1) on a set $\mathfrak{D}$ of (t, y)-space if and only if the scalar function $p(t, y)$ satisfies a corresponding Lipschitz condition

$$(8.6) \qquad |p(t, z) - p(t, y)| \leq \kappa^0 |z - y|$$

on $\mathfrak{D}$; moreover if κ^0 is such that (8.6) holds then (5.1) is valid with $\kappa = \kappa^0 + 1$.

It is to be remarked that although the n-th order equation (8.1) is equivalent to a first order system (8.5) there are special procedures that are applicable to (8.1), and which are not mere applications to (8.5) of procedures that have been introduced for general first order differential systems. For example, if $p(t, y) \equiv p(t, y_1, \ldots, y_n)$ is continuous in (t, y) on the infinite strip $\Delta = [a, b] \times \mathbf{C}_n$, and $x_0(t)$ is a given function which is continuous and has continuous derivatives of the first $n - 1$ orders on $[a, b]$, one might consider the sequence of functions $\{x_j(t)\}$, $(j = 0, 1, 2, \ldots)$, defined by

$$(8.7) \quad x_{m+1}^{[n]}(t) = p(t, x_m(t), \ldots, x_m^{[n-1]}(t)), \quad x_{m+1}^{[\alpha-1]}(\tau) = \eta_\alpha, \quad (\alpha = 1, \ldots, n).$$

The above iterative procedure is not identical with that obtained by applying the method of Section 6 to the system (8.5). It is highly similar, however, and it is to be expected that in case a Lipschitz condition (8.6) holds on Δ the sequence $\{x_j(t)\}$ defined by (8.7) would be such that $\{x_j^{[\alpha-1]}(t)\} \to x^{[\alpha-1]}(t)$, $(\alpha = 1, \ldots, n)$, uniformly on $[a, b]$, where $x(t)$ is the unique solution of (8.1) on $[a, b]$ satisfying $x^{[\alpha-1]}(\tau) = \eta_\alpha$, $(\alpha = 1, \ldots, n)$. Indeed, for $\alpha = 1, \ldots, n$,

$$x_{m+1}^{[\alpha-1]}(t) = \eta_\alpha + \eta_{\alpha+1}(t - \tau) + \cdots + \eta_n \frac{(t - \tau)^{n-\alpha}}{(n - \alpha)!}$$
$$+ \int_\tau^t \frac{(t - s)^{n-\alpha}}{(n - \alpha)!} p(s, x_m(s), \ldots, x_m^{[n-1]}(s))\, ds,$$

and the vectors $y^{(m)}(t) \equiv (y_\alpha^{(m)}(t)) \equiv (x_m^{[\alpha-1]}(t))$, $(\alpha = 1, \ldots, n)$, satisfy with the vector $w(u) \equiv (w_\alpha(u)) \equiv (u^{n-\alpha}/(n-\alpha)!)$, $(\alpha = 1, \ldots, n)$, the iterative relations

$$y^{(m+1)}(t) - y^{(m)}(t) = \int_\tau^t w(t-s)[p(s, y^{(m)}(s)) - p(s, y^{(m-1)}(s))]\, ds.$$

Consequently, if (8.6) holds for (t, y), (t, z) in Δ, and κ' is such that $|w(b-a)| \leq \kappa'$, then for $m = 1, 2, \ldots$ we have

$$|y^{(m+1)}(t) - y^{(m)}(t)| \leq \left| \int_\tau^t \kappa' \kappa^0 | y^{(m)}(s) - y^{(m-1)}(s) | ds \right|,$$

and the above stated convergence properties of the sequence $\{x_j(t)\}$ follow by an argument similar to that of Section 6.

Finally, it is to be commented that the theory of a system of differential equations of the form

$$z_\gamma^{[n_\gamma]} = p_\gamma(t, z_1, \ldots, z_1^{[n_1-1]}, z_2, \ldots, z_2^{[n_2-1]}, \ldots, z_k, \ldots, z_k^{[n_k-1]}),$$
$$(\gamma = 1, \ldots, k),$$

where $n_\gamma \geq 1$, $(\gamma = 1, \ldots, k)$, is equivalent under the transformation

$$y_1 = z_1, \ldots, y_{n_1} = z_1^{[n_1-1]}, y_{n_1+1} = z_2, \ldots, y_{n_1+n_2} = z_2^{[n_2-1]}, \ldots$$
$$y_{n_1+\cdots+n_{k-1}+1} = z_k, \ldots, y_{n_1+\cdots+n_k} = z_k^{[n_k-1]}$$

to a system of $n = n_1 + \cdots + n_k$ first order equations $y'_\alpha = f_\alpha(t, y_1, \ldots, y_n)$, $(\alpha = 1, \ldots, n)$, where $f_\alpha = y_{\alpha+1}$, $(\alpha \neq n_1, n_1 + n_2, \ldots, n_1 + \cdots + n_k)$, $f_{n_1+\ldots+n_j} = p_j(t, y_1, \ldots, y_n)$, $(j = 1, \ldots, k)$.

Problems I.8

1. For the differential equation $x'' + x = 0$ determine the solution satisfying $x(0) = 1$, $x'(0) = 0$, using (a) the method of iterations of Section I.6 for the equivalent first-order system $u_1' = u_2$, $u_2' = -u_1$, $u_1(0) = 1$, $u_2(0) = 0$, with $u_1^{(0)}(t) \equiv 1$, $u_2^{(0)}(t) \equiv 0$; (b) the method of iterations given by (8.7), with $x_0(t) \equiv 1$.

2. If $p(t, y) \equiv p(t, y_1, \ldots, y_n)$ is continuous on $\Delta_1 = [\tau, b] \times \mathbf{C}_n$ and there is a constant $\lambda \geq 0$ such that

$$\Re\{(\bar{z}_n - \bar{y}_n)[p(t, z_1, \ldots, z_n) - p(t, y_1, \ldots, y_n)]\} \leq \lambda \sum_{\alpha=1}^n |z_\alpha - y_\alpha|^2$$

for arbitrary (t, y), (t, z) of Δ_1, show that there is at most one solution of (8.1) on $[\tau, b]$ that satisfies $x^{[\alpha-1]}(\tau) = \eta_\alpha$, $(\alpha = 1, \ldots, n)$. Formulate and prove a corresponding criterion for the existence on an interval $[a, \tau]$ of at most one solution of (8.1) that satisfies $x^{[\alpha-1]}(\tau) = \eta_\alpha$, $(\alpha = 1, \ldots, n)$.

Hint. Use the results of Theorems 5.3 and 5.4.

3. Two sterling students, Mr. Z. Yx and Mr. W. Vut, were considering the following proposition:

Suppose that the real-valued function $g(t, u, v)$ of the real variables (t, u, v) is continuous and has continuous partial derivatives $g_u(t, u, v)$ and $g_v(t, u, v)$ for (t, u) in an open region $\mathfrak{R}$ of the (t, u)-plane and $-\infty < v < \infty$. Then for an arbitrary point (t_0, u_0) of $\mathfrak{R}$ and arbitrary v_0 there is a unique solution $u = u(t)$ of the second-order differential equation

$$u'' = g(t, u, u')$$

such that $u(t_0) = u_0$, $u'(t_0) = v_0$; moreover, the graph of this solution in the (t, u)-plane may be extended to the boundary of $\mathfrak{R}$.

Mr. Yx insisted that the proposition was a consequence of Theorems 5.2 and 5.7 of this Chapter. Mr. Vut disagreed with Mr. Yx on the possibility of extending the graph of the solution to the boundary of $\mathfrak{R}$, and countered with the remark: "Look at the equation $u'' = \frac{1}{2}(u')^3$."

Indicate whether you agree with Mr. Yx or with Mr. Vut and give the reasons for your answer.

9. LINEAR VECTOR DIFFERENTIAL EQUATIONS

Throughout this section it will be supposed that $A_{\alpha\beta}(t)$, $g_\alpha(t)$, $(\alpha, \beta = 1, \ldots, n)$, are continuous functions of t on the interval $[a, b]$. Then $f_\alpha(t, y) = \sum_{\beta=1}^{n} A_{\alpha\beta}(t)y_\beta + g_\alpha(t)$, $(\alpha = 1, \ldots, n)$, are continuous in (t, y) on the strip $\Delta = [a, b] \times \mathbf{C}_n$ and the vector function $f(t, y) \equiv (f_\alpha(t, y))$ satisfies on Δ a Lipschitz condition (5.1) with Lipschitz constant κ not exceeding the maximum on $[a, b]$ of the continuous function $(\sum_{\alpha,\beta=1}^{n} |A_{\alpha\beta}(t)|^2)^{1/2}$. From Theorem 5.1 it follows that there is a unique solution of the linear vector differential equation

$$y' = A(t)y + g(t), \qquad t \in [a, b], \tag{9.1}$$

which passes through a given point (τ, η) of Δ.

For $g_\alpha(t) \equiv 0$, $(\alpha = 1, \ldots, n)$, the equation (9.1) reduces to the homogeneous linear vector differential equation

$$y' = A(t)y, \qquad t \in [a, b]. \tag{9.2}$$

The totality of solutions $y(t) \equiv (y_\alpha(t))$, $(\alpha = 1, \ldots, n)$, of (9.2) is clearly a linear vector space of n-dimensional vector functions defined on $[a, b]$; that is, if $y = y^{(j)}(t)$, $(j = 1, \ldots, k)$, are solutions of (9.2) then for arbitrary scalars $\xi_1, \ldots, \xi_k$ the vector function $y = y^{(1)}(t)\xi_1 + \cdots + y^{(k)}(t)\xi_k$ is also a solution of (9.2). The particular vector function $y(t) \equiv 0$ is referred to as the *trivial* solution of (9.2). From the uniqueness of solutions of this differential equation passing through a given initial point it follows that a solution $y(t)$ of (9.2) is the trivial solution of this equation on $[a, b]$ whenever there is a single point $t = \tau$ of this interval at which $y(\tau)$ is the null vector. Moreover, solutions $y^{(1)}(t), \ldots, y^{(k)}(t)$ are linearly independent vector functions on $[a, b]$ if and only if for some $t = \tau$ of this interval the constant vectors $y^{(1)}(\tau), \ldots, y^{(k)}(\tau)$ are linearly independent vectors of $\mathbf{C}_n$.

Theorem 9.1. *If* $y = y^{(j)}(t) \equiv (y_\alpha^{(j)}(t))$, $(j = 1, \ldots, k)$, *are solutions of* (9.2) *then the* $n \times k$ *matrix* $[y_\alpha^{(j)}(t)]$, $(\alpha = 1, \ldots, n; j = 1, \ldots, k)$, *has constant rank on* $[a, b]$. *In particular, if* $Y(t) \equiv [Y_{\alpha\beta}(t)]$, $(\alpha, \beta = 1, \ldots, n)$, *is a square matrix whose column vectors are solutions of* (9.2), *and* $Y(t)$ *is nonsingular for a value* $t = \tau$ *on* $[a, b]$, *then* $Y(t)$ *is nonsingular throughout this interval and an arbitrary solution* $y(t)$ *of* (9.2) *is of the form* $y(t) = Y(t)\xi$, *where* $\xi \equiv (\xi_\alpha)$, $(\alpha = 1, \ldots, n)$, *is a constant vector of* $\mathbf{C}_n$.

If the matrix $[y_\alpha^{(j)}(\tau)]$ has rank $k - q$ for a value $t = \tau$ on $[a, b]$ then there exist q linearly independent vectors $\xi^{(\rho)} \equiv (\xi_j^{(\rho)})$, $(\rho = 1, \ldots, q)$, of $\mathbf{C}_k$ for which the vector functions $w^{(\rho)}(t) = y^{(1)}(t)\xi_1^{(\rho)} + \cdots + y^{(k)}(t)\xi_k^{(\rho)}$ are such that $y = w^{(\rho)}(t)$, $(\rho = 1, \ldots, q)$, is a solution of (9.2) that reduces to the null vector for $t = \tau$. Consequently, each $w^{(\rho)}(t)$ is the trivial solution of (9.2) on $[a, b]$, and for arbitrary $t \in [a, b]$ the rank of $[y_\alpha^{(j)}(t)]$ does not exceed the rank of $[y_\alpha^{(j)}(\tau)]$. As τ may be chosen arbitrarily on $[a, b]$, it follows that the matrix $[y_\alpha^{(j)}(t)]$ has constant rank on this interval. In particular, if $k = n$ and there is a value τ on $[a, b]$ such that $[y_\alpha^{(j)}(\tau)]$ is nonsingular, then this matrix is nonsingular throughout this interval. If $Y(t)$ is an $n \times n$ nonsingular matrix whose n column vectors are solutions of (9.2), and $y^{(0)}(t) \equiv (y_\alpha^{(0)}(t))$ is any solution of this equation, then for a fixed τ on $[a, b]$ there is a unique constant vector ξ of $\mathbf{C}_n$ such that $y^{(0)}(\tau) = Y(\tau)\xi$. Then $y(t) = y^{(0)}(t) - Y(t)\xi$ is a solution of (9.2) satisfying $y(\tau) = 0$, and hence $y(t) \equiv 0$ and $y^{(0)}(t) \equiv Y(t)\xi$ on $[a, b]$. In geometric terminology the last result of this theorem states that the linear vector space of solutions of (9.2) has dimension n.

A nonsingular $n \times n$ matrix $Y(t)$ whose column vectors are solutions of (9.2) on $[a, b]$ is termed a *fundamental matrix* for this differential equation; clearly a fundamental matrix $Y(t)$ for (9.2) on $[a, b]$ may be characterized as a nonsingular solution of the *matrix differential equation*

$$Y' = A(t)Y, \qquad t \in [a, b]. \tag{9.3}$$

If $Y_1(t)$ is a fundamental matrix for (9.2) on $[a, b]$, then from the existence and uniqueness theorem it follows immediately that an $n \times n$ matrix $Y(t)$ is a solution of (9.3) if and only if $Y(t) = Y_1(t)C$, where C is a constant matrix given by $C = Y_1^{-1}(\tau)Y(\tau)$ for arbitrary τ of $[a, b]$; moreover, $Y(t)$ is a fundamental matrix for (9.2) if and only if the constant matrix C is nonsingular. In particular, if $Y(t) = Y(t; \tau)$ is the fundamental matrix for (9.2) satisfying $Y(\tau) = E$ then $Y(t, \tau) = Y_1(t)\,Y_1^{-1}(\tau)$, where $Y_1(t)$ is an arbitrary fundamental matrix for (9.2). From the iterative procedure of Picard-Lindelöf it follows that

$$(9.4)\quad Y(t; \tau) = E + \int_\tau^t A(s_1)\,ds_1 + \int_\tau^t A(s_2)\left[\int_\tau^{s_2} A(s_1)\,ds_1\right] ds_2 + \cdots.$$

The expression (9.4) has been called the *matrizant* of $A(t)$. In particular, if $A(t)$ is a constant matrix A on $[a, b]$, then (9.4) becomes

$$(9.5)\quad E + (t-\tau)A + \frac{(t-\tau)^2}{2!}A^2 + \cdots + \frac{(t-\tau)^m}{m!}A^m + \cdots = \exp\{(t-\tau)A\}.$$

From Theorem 9.1 it follows that if $Y(t)$ is an $n \times n$ matrix satisfying (9.3) then $\det Y(t) \neq 0$ throughout $[a, b]$ whenever there is a single point τ of this interval at which $\det Y(\tau) \neq 0$. The specific relation between $\det Y(t)$ and $\det Y(\tau)$ is afforded by the following theorem.

Theorem 9.2. *If $Y(t)$ is an $n \times n$ matrix satisfying* (9.3) *then for arbitrary $t \in [a, b]$, $\tau \in [a, b]$,*

$$(9.6)\qquad \det Y(t) = [\det Y(\tau)] \exp\left\{\int_\tau^t \sum_{\alpha=1}^{n} A_{\alpha\alpha}(s)\,ds\right\}.$$

To establish (9.6) it suffices to note that $[\det Y(t)]' = [\det U^{(1)}(t)] + \cdots + [\det U^{(n)}(t)]$, where for $k = 1, 2, \ldots, n$,

$$\begin{aligned} U^{(k)}_{k\beta}(t) &= Y'_{k\beta}(t), \qquad (\beta = 1, 2, \ldots, n),\\ U^{(k)}_{\alpha\beta}(t) &= Y_{\alpha\beta}(t), \qquad (\alpha \neq k;\ \alpha, \beta = 1, 2, \ldots n). \end{aligned}$$

Then $U^{(k)}_{k\beta}(t) = Y'_{k\beta}(t) = \sum_{\gamma=1}^{n} A_{k\gamma}(t)\,Y_{\gamma\beta}(t)$, and $\det U^{(k)}(t) = A_{kk}(t)[\det Y(t)]$, $(k = 1, 2, \ldots, n)$. Consequently, the scalar function $\theta(t) = \det Y(t)$ satisfies the first order linear differential equation $\theta'(t) = \theta(t)\sum_{\alpha=1}^{n} A_{\alpha\alpha}(t)$, and (9.6) is immediate.

Theorem 9.3. *If $Y(t)$ is a fundamental matrix for* (9.2) *on $[a, b]$, then the solution of* (9.1) *satisfying $y(\tau) = \eta$ is given by*

$$(9.7)\qquad y(t) = Y(t)Y^{-1}(\tau)\eta + \int_\tau^t Y(t)Y^{-1}(s)g(s)\,ds, \quad \textit{for} \quad t \in [a, b].$$

Indeed, if we set $y(t) = Y(t)u(t)$, then $y(t)$ is the solution of (9.1) satisfying $y(\tau) = \eta$ if and only if $u'(t) = Y^{-1}(t)g(t)$, $u(\tau) = Y^{-1}(\tau)\eta$, so that $u(t) = Y^{-1}(\tau)\eta + \int_\tau^t Y^{-1}(s)\,g(s)\,ds$. In view of the above definition of the matrizant

$Y(t;\tau)$, it follows that (9.7) may be written as

$$(9.7') \qquad y(t) = Y(t;\tau)\eta + \int_\tau^t Y(t;s)g(s)\,ds, \quad \textit{for} \quad t \in [a, b].$$

For brevity, let $l_n[x]$ denote the n-th order linear scalar differential expression

$$(9.8) \qquad l_n[x] \equiv p_n(t)x^{[n]} + \cdots + p_1(t)x' + p_0(t)x,$$

where $p_0(t), \ldots, p_n(t)$ are continuous on $[a, b]$ and $p_n(t) \neq 0$ on this interval. If $r(t)$ is a given continuous scalar function on $[a, b]$, then the nonhomogeneous n-th order linear scalar differential equation

$$(9.9) \qquad l_n[x] = r(t), \qquad t \in [a, b],$$

is equivalent under the substitution (8.4) to a vector equation (9.1) with

$$(9.10) \qquad A(t) = \begin{bmatrix} 0 & 1 & 0 & \cdots & 0 \\ 0 & 0 & 1 & \cdots & 0 \\ \cdot & \cdot & \cdot & \cdots & \cdot \\ 0 & 0 & 0 & \cdots & 1 \\ \multicolumn{3}{c}{\dfrac{-p_0(t)}{p_n(t)}} & \cdots & \dfrac{-p_{n-1}(t)}{p_n(t)} \end{bmatrix}; \qquad g(t) = \begin{bmatrix} 0 \\ \cdot \\ \cdot \\ \cdot \\ \dfrac{r(t)}{p_n(t)} \end{bmatrix}.$$

For such a matrix $A(t)$ the corresponding homogeneous vector equation (9.2) is equivalent under the substitution (8.4) to the homogeneous n-th order linear scalar differential equation

$$(9.11) \qquad l_n[x] = 0;$$

moreover, an $n \times n$ matrix $Y(t)$ is a solution of the corresponding matrix equation (9.3) if and only if $Y(t)$ is of the form

$$(9.12) \qquad \begin{bmatrix} x_1(t) & \cdots & x_n(t) \\ x_1'(t) & \cdots & x_n'(t) \\ \cdots & \cdots & \cdots \\ x_1^{[n-1]}(t) & \cdots & x_n^{[n-1]}(t) \end{bmatrix},$$

where $x_1(t), \ldots, x_n(t)$ are solutions of (9.11). The matrix (9.12) is called the *Wronskian matrix* of $x_1(t), \ldots, x_n(t)$ and is denoted by $W(t; x_1, \ldots, x_n)$.

Equation (9.7) is called the *variation of constants* or *variation of parameters* formula for the solution of the first-order vector differential equation (9.1). In particular, $y_0(t) = \int_\tau^t Y(t)Y^{-1}(s)g(s)\,ds$ is the solution of this equation that satisfies the initial condition $y_0(\tau) = 0$. For (9.1), with $A(t)$, $g(t)$ given by (9.10), and $Y(t)$ the Wronskian matrix (9.12), the vector equation

$Y(t)u'(t) = g(t)$ may be written as the system of scalar equations

$$\sum_{\beta=1}^{n} x_\beta^{[\alpha-1]}(t)\, u'_\beta(t) = 0, \qquad (\alpha = 1, \ldots, n-1),$$

$$\sum_{\beta=1}^{n} x_\beta^{[n-1]}(t)\, u'_\beta(t) = \frac{r(t)}{p_n(t)}.$$

Therefore, if $C_\beta(t)$ denotes the cofactor of the element in the n-th row and β-th column of (9.12) and $D(t) = \det W(t; x_1, \ldots, x_n)$, then $u'_\beta(t) = [C_\beta(t)\, r(t)]/[D(t)\, p_n(t)]$, $(\beta = 1, \ldots, n)$. Consequently the particular solution $x = x_0(t)$ of (9.9) which satisfies the initial condition $x^{[\alpha-1]}(\tau) = 0$, $(\alpha = 1, \ldots, n)$, is given by

$$x_0(t) = \int_\tau^t \frac{K(t, s)}{W(s)} \frac{r(s)}{p_n(s)}\, ds, \qquad t \in [a, b],$$

where

$$K(t, s) = \det \begin{bmatrix} x_1(s) & \cdots & x_n(s) \\ \cdots & \cdots & \cdots \\ x_1^{[n-2]}(s) & \cdots & x_n^{[n-2]}(s) \\ x_1(t) & \cdots & x_n(t) \end{bmatrix}.$$

Problems I.9

1. If $A \equiv [A_{\alpha\beta}]$, $(\alpha, \beta = 1, \ldots, n)$, is a constant matrix, show that the solution of (9.1) satisfying $y(\tau) = \eta$ is given by

$$y(t) = Y(t - \tau)\eta + \int_\tau^t Y(t - s) g(s)\, ds,$$

where $Y(t)$ is the solution of $Y' = AY$ satisfying $Y(0) = E$.

2. Find the solution of the differential system

$$y_1' = y_2 + \sin t,$$
$$y_2' = -y_1 + 1,$$

satisfying the initial conditions $y_1(0) = 0$, $y_2(0) = 1$.

3. For $A = [A_{\alpha\beta}]$, $B = [B_{\alpha\beta}]$, $(\alpha, \beta = 1, \ldots, n)$, constant matrices, show that $\exp\{t(A + B)\} = [\exp\{tA\}][\exp\{tB\}]$ for all t if and only if $AB = BA$.

4. For $A = [A_{\alpha\beta}]$, $(\alpha, \beta = 1, \ldots, n)$, a constant matrix, show that (a) $\det[\exp\{tA\}] = \exp\{t \sum_{\alpha=1}^{n} A_{\alpha\alpha}\}$; (b) $[\exp\{tA\}]^{-1} = \exp\{-tA\}$.

5.[s] If $A \equiv [A_{\alpha\beta}]$, $(\alpha, \beta = 1, \ldots, n)$, is a constant matrix and $c_0, \ldots, c_{r-1}$ are constants such that $A^r + c_{r-1}A^{r-1} + \cdots + c_1 A + c_0 E = 0$, then $\exp\{tA\} = x_1(t)E + x_2(t)A + \cdots + x_r(t)A^{r-1}$, where $x_\alpha(t)$, $(\alpha = 1, \ldots, r)$, is the solution of the r-th order scalar differential equation $x^{[r]} + c_{r-1}x^{[r-1]} + \cdots + c_1x' + c_0x = 0$ satisfying $x_\alpha^{[\beta-1]}(0) = \delta_{\alpha\beta}$, $(\alpha, \beta = 1, \ldots, r)$.

Hint. Show that $Y(t)$ is uniquely determined by the condition that $Y^{(r)} + c_{r-1}Y^{(r-1)} + \cdots + c_1Y' + c_0Y = 0$ on $[a, b]$ and $Y(0) = E$, $Y^{(j)}(0) = A^j$, $(j = 1, \ldots, r-1)$.

6. Use the result of Problem 5 to compute $\exp\{tA\}$ for:

$$\text{(a) } A = \begin{bmatrix} 0 & 1 \\ -1 & 0 \end{bmatrix}, \qquad \text{(b) } A = \begin{bmatrix} -2 & 1 \\ -3 & 2 \end{bmatrix}, \qquad \text{(c) } A = \begin{bmatrix} 1 & 1 & 0 \\ 0 & 1 & 0 \\ 0 & 0 & -1 \end{bmatrix}.$$

7.[s] Show that if $x_1(t), \ldots, x_n(t)$ are solutions of (9.11) then the Wronskian matrix $W(t; x_1, \ldots, x_n)$ given by (9.12) satisfies the relation

$$\det W(t; x_1, \ldots, x_n) = \det W(\tau; x_1, \ldots, x_n) \exp\left[-\int_\tau^t \frac{p_{n-1}(s)}{p_n(s)}\,ds\right].$$

8. If $x_1(t), \ldots, x_n(t)$ are functions which are continuous and have continuous derivatives of the first n orders on a nondegenerate interval I, and $W(t; x_1, \ldots, x_n)$ is nonsingular on this interval, show that there exists a differential equation of the form (9.11) with $p_n(t) \neq 0$ on I and such that $x_1(t), \ldots, x_n(t)$ are solutions of this equation.

9.[s] If $c_0, c_1, \ldots, c_n$ are constants with $c_n \neq 0$, and $x(t)$ is a solution of the n-th order homogeneous linear scalar differential equation

$$l_n[x] \equiv c_nx^{[n]} + \cdots + c_0x = 0, \quad \textit{for} \quad t \in (-\infty, \infty),$$

prove that each of the following functions is also a solution of this equation: (a) $x_1(t) = x(t + t_0)$, where t_0 is an arbitrary real constant; (b) $x_2(t) = x'(t)$; (c) $x_3(t) = (2r)^{-1}\int_{-r}^{r} x(t + s)\,ds = (2r)^{-1}\int_{t-r}^{t+r} x(\sigma)\,d\sigma$.

10. If $c(t)$ and $s(t)$ denote the solutions of the differential equation $x'' + x = 0$ determined by the initial conditions

$$c(0) = 1, \qquad c'(0) = 0, \qquad s(0) = 0, \qquad s'(0) = 1,$$

prove that

(a) $s'(t) = c(t)$, $c'(t) = -s(t)$;
(b) $c^2(t) + s^2(t) = 1$;
(c) $c(t) = 1 - t^2/2! + \cdots + (-1)^m t^{2m}/(2m)! + \cdots$;
$s(t) = t - t^3/3! + \cdots + (-1)^m t^{2m+1}/(2m+1)! + \cdots$;
(d) $c(t)$ has positive zeros, and if h denotes the smallest positive zero of $c(t)$, then $h < 3$;
(e) if h is defined as in (d), then $s(h + t) = c(t)$, $c(h + t) = -s(t)$;
(f) if h is defined as in (d), and $\pi = 2h$, then

$$s(\pi + t) = -s(t), \qquad c(\pi + t) = -c(t),$$
$$s(2\pi + t) = s(t), \qquad c(2\pi + t) = c(t).$$

Consequently, the functions $c(t)$ and $s(t)$ thus defined are the familiar trigonometric functions $\cos t$ and $\sin t$, and the above relations embody the basic results of analytic trigonometry.

Hints. (a) Note that in view of (b) of Problem 9 the functions $x(t) = s'(t)$ and $x(t) = c'(t)$ are solutions of $x'' + x = 0$, and identify these solutions by the initial values of x and x' at $t = 0$.

(b) Apply the result of Problem 7.

(c) Solve for $c(t)$ and $s(t)$ by the method of iterations.

(d) Note that for $t = 3$ the series for $c(t)$ is an alternating series with the sequence of terms having nonincreasing absolute value, and that $c(3) < 1 - (\frac{9}{2}) + (\frac{27}{8}) = -\frac{1}{8}$.

(e) Note that $1 \geq c(t) > 0$ for $t \in [0, h)$, and conclusion (a) leads to the result that on this interval $c(t)$ is decreasing, and $s(t)$ is increasing. With the aid of conclusion (b), obtain that $s(h) = 1$, and proceed to consider the initial values at $t = 0$ of the solutions $x(t) = s(h + t)$ and $x(t) = c(h + t)$ of $x'' + x = 0$.

11. Suppose that $A = [A_{\alpha\beta}]$, $(\alpha, \beta = 1, \ldots, n)$, is a constant matrix such that the equation $\det[A - \lambda E] = 0$ has distinct roots $\lambda_1, \ldots, \lambda_n$, and denote by $\eta^{(\beta)}$ a non-null vector satisfying $[A - \lambda_\beta E]\eta^{(\beta)} = 0$. Show that $Y(t) = [\eta_\alpha^{(\beta)} \exp\{\lambda_\beta(t - \tau)\}]$, $(\alpha, \beta = 1, \ldots, n)$, is a fundamental matrix for the differential equation $y' = Ay$.

Hint. Show that if $\sum_{\beta=1}^{n} \eta_\alpha^{(\beta)} c_\beta = 0$, $(\alpha = 1, \ldots, n)$, then $\sum_{\beta=1}^{n} \eta_\alpha^{(\beta)} \lambda_\beta^{\,k} c_\beta = 0$, $(\alpha = 1, \ldots, n;\ k = 1, 2, \ldots)$. Use the fact that the distinct character of $\lambda_1, \ldots, \lambda_n$ implies that the Vandermonde determinant

$$\det \begin{bmatrix} 1 & \cdots & 1 \\ \lambda_1 & \cdots & \lambda_n \\ \cdot & \cdots & \cdot \\ \cdot & \cdots & \cdot \\ \lambda_1^{n-1} & \cdots & \lambda_n^{n-1} \end{bmatrix}$$

is different from zero, and proceed to show that the constant matrix $[\eta_\alpha^{(\beta)}]$ is nonsingular.

12. If $A = [A_{\alpha\beta}]$, $(\alpha, \beta = 1, \ldots, n)$, is a constant matrix, and for a given λ there exist non-null vectors $\zeta^1, \ldots, \zeta^r$ such that

$$[A - \lambda E]\zeta^1 = 0, \qquad [A - \lambda E]\zeta^{j+1} = \zeta^j, \qquad (j = 1, 2, \ldots, r - 1),$$

show that

$$y(t) = \left[\zeta^r + \frac{t}{1!}\zeta^{r-1} + \cdots + \frac{t^{r-1}}{(r-1)!}\zeta^1\right] \exp\{\lambda t\}$$

is a solution of $y' = Ay$. Using this result, and the Jordan canonical form of the constant matrix A, (see Appendix E), obtain an expression for a fundamental matrix that generalizes the result of Problem 11.

13.[ss] If $A_{\alpha\beta}(t, c) = A_{\alpha\beta}(t, c_1, \ldots, c_k)$, $g_\alpha(t, c) = g_\alpha(t, c_1, \ldots, c_k)$ are continuous in (t, c) on $[a, b] \times \Gamma$, where Γ is a subset of $\mathbf{C}_k$, and $y = y(t; \tau, \eta, c)$ is the solution of the linear vector differential equation

$$y' = A(t, c)y + g(t, c)$$

passing through the initial point (τ, η), show that $y(t; \tau, \eta, c)$ is continuous in (t, τ, η, c) on $[a, b] \times [a, b] \times \mathbf{C}_n \times \Gamma$.

Hint. Use the result of Problem I.6:3.

14. If the elements of the constant matrix $A = [A_{\alpha\beta}]$, $(\alpha, \beta = 1, \ldots, n)$, are real-valued, prove that a necessary and sufficient condition for all elements of $\exp\{tA\}$ to be non-negative on $[0, \infty)$ is that $A_{\alpha\beta} \geq 0$ for $\alpha \neq \beta$; $\alpha, \beta = 1, \ldots, n$.

Hint. In establishing the sufficiency of the stated condition, note that if $Y(t)$ is a solution of the matrix differential equation $Y' = AY$, and $Y_1(t) = [\exp\{\lambda t\}]\, Y(t)$, where λ is a scalar, then $Y_1' = [A + \lambda E]\, Y_1$.

15. If $A(t)$ and $B(t)$ are $n \times n$ matrix functions which are continuous on $[a, b]$, and $W(t)$ is an $n \times n$ matrix function that is a solution of the matrix differential equation

$$W' + A(t)W + WB(t)W = 0, \quad \textit{for} \quad t \in [a, b],$$

prove that $W(t)$ is of constant rank on $[a, b]$.

16.[s] Suppose that on $[a, b]$ the functions $p_0(t), p_1(t)$ are continuous, the function $q(t)$ is continuously differentiable and different from zero on this interval, and $\tau \in [a, b]$. Prove that:

(i) $w(t)$ is a solution of the *Riccati differential equation*

$$w' + p_0(t) + p_1(t)w + q(t)w^2 = 0 \tag{9.13}$$

on $[a, b]$ if and only if $u(t) = \exp\{\int_\tau^t q(s)\, w(s)\, ds\}$ is a solution on $[a, b]$ of the linear homogeneous second order differential equation

$$q(t)u'' + [q(t)\, p_1(t) - q'(t)]u' + q^2(t)\, p_0(t)u = 0; \tag{9.14}$$

(ii) if $w = w_\alpha(t)$, $(\alpha = 0, 1)$ are solutions of (9.13) on $[a, b]$, and $w_1(t) - w_0(t) \neq 0$ for $t \in [a, b]$, then $s(t) = 1/[w_1(t) - w_0(t)]$ is a solution of the first order linear differential equation

$$s' = [p_1(t) + 2q(t)\, w_0(t)]s + q(t) \tag{9.15}$$

on $[a, b]$. Also, if $s_1(t)$, $s_2(t)$ are distinct solutions of (9.15), then for any solution $s(t)$ of this equation there is a constant k such that $s(t) - s_1(t) = k[s_2(t) - s_1(t)]$

(iii) if $w = w_\alpha(t)$, $\alpha = 1, 2, 3, 4$, are solutions of (9.13) on an interval $[a, b]$, and $w_4(t) - w_1(t)$ and $w_3(t) - w_2(t)$ are nonzero on this interval, then the "cross ratio," or "anharmonic ratio"

$$\{w_1, w_2, w_3, w_4\} = (w_3 - w_1)(w_3 - w_2)^{-1}(w_4 - w_2)(w_4 - w_1)^{-1}$$

is constant on $[a, b]$.

Hint. (i) Show that (9.13) and (9.14) are equivalent under the substitution $u' = qwu$.

17. If $y^{(1)}(t), \ldots, y^{(k)}(t)$, $(1 \leq k < n)$, are n-dimensional vector functions which are continuously differentiable on $[a, b]$, and such that the $n \times k$ matrix function $[y_\alpha^{(j)}(t)]$, $(\alpha = 1, \ldots, n; j = 1, \ldots, k)$ has rank k on this interval, show that there exist continuous vector functions $y^{(k+1)}(t), \ldots, y^{(n)}(t)$ such that the $n \times n$ matrix $[y_\alpha^{(\beta)}(t)]$ is nonsingular on $[a, b]$. Moreover, the $y^{(k+1)}(t), \ldots, y^{(n)}(t)$ may be chosen to be continuously differentiable, and in this case there exists a continuous $n \times n$ matrix function $A(t)$ such that $Y(t) = [y_\alpha^{(\beta)}(t)]$ is a fundamental matrix for $y' = A(t)y$.

Hint 1. In order to prove the first conclusion, note that a basic inductive step is established by showing that there is a continuous vector function $y^{(k+1)}(t)$ such that the $n \times (k + 1)$ matrix $[y_\alpha^{(m)}(t)]$, $(\alpha = 1, \ldots, n;\ m = 1, \ldots, k + 1)$, is of rank $k + 1$ on $[a, b]$. Suppose that on a subinterval $[a, t']$, $t' \in [a, b)$, the vector function $y^{(k+1)}(t)$ is continuous and such that the $n \times (k + 1)$ matrix $[y_\alpha^{(m)}(t)]$ is of rank $k + 1$ on $[a, t']$. Let α_i, $(i = 1, \ldots, k)$, be indices such that $1 \leq \alpha_1 < \alpha_2 < \cdots < \alpha_k \leq n$, and the $k \times k$ matrix $[y_{\alpha_i}^{(j)}(t)]$, $(i, j = 1, \ldots, k)$, is nonsingular on a subinterval $[t', t'']$ of $[t', b]$. If $\alpha_0 \neq \alpha_i$, $(i = 1, \ldots, k)$, and for $\alpha \neq \alpha_0$ the component functions $y_\alpha^{(k+1)}(t)$ are defined on $[t', t'']$ in any manner such that they are continuous on $[a, t'']$, show that the component function $y_{\alpha_0}^{(k+1)}(t)$ may then be defined on $[t', t'']$ in a manner so that it is also continuous on $[a, t'']$ and the $(k + 1) \times (k + 1)$ matrix

$$\begin{bmatrix} y_{\alpha_1}^{(1)}(t) & \cdots & y_{\alpha_1}^{(k+1)}(t) \\ \cdots & \cdots & \cdots \\ y_{\alpha_0}^{(1)}(t) & \cdots & y_{\alpha_0}^{(k+1)}(t) \end{bmatrix}$$

is nonsingular on $[t', t'']$.

To establish the final result of the problem, approximate the continuous vector functions $y^{(k+1)}(t), \ldots, y^{(n)}(t)$ by vector functions that are continuously differentiable. One such method of approximation is based upon the fact that if $y(t)$ is a continuous vector function on an interval $[a - \delta_0, b + \delta_0]$,

($\delta_0 > 0$), then for $0 < \delta < \delta_1 < \delta_0$ the vector function

$$y^{(\delta)}(t) = (2\delta)^{-1}\int_{-\delta}^{\delta} y(t+s)\,ds$$

is continuously differentiable on $(a - [\delta_0 - \delta_1],\ b + [\delta_0 - \delta_1])$, and $y^{(\delta)}(t) \to y(t)$ uniformly on $[a, b]$ as $\delta \to 0$.

Hint 2. Note that $Y(t) = [y_\alpha^{(j)}(t)]$, $(\alpha = 1, \ldots, n; j = 1, \ldots, k)$ is such that the $k \times k$ matrix $U(t) = Y^*(t)Y(t)$ is nonsingular on $[a, b]$, and hence $Y'(t) = A_1(t)Y(t)$, where $A_1(t) = Y'(t)U^{-1}(t)Y^*(t)$. For $\tau \in [a, b]$, and $\eta^1, \ldots, \eta^{n-k}$ vectors such that $Y_1 = [Y(\tau) \quad \eta^1 \cdots \eta^{n-k}]$ is nonsingular, consider the solution $Y_0(t)$ of $Y_0' = A_1(t)Y_0$, $Y_0(\tau) = Y_1$.

18. If K is an $n \times n$ matrix, prove that the solution of the second-order linear vector differential equation $y'' + K^2 y = 0$ satisfying the initial conditions $y(0) = \eta$, $y'(0) = K\zeta$, is $y(t) = [\cos tK]\eta + [\sin tK]\zeta$. If I is an interval containing $t = 0$ throughout which $\cos tK$ is nonsingular, show that $T(t) = [\sin tK][\cos tK]^{-1}$ is the solution of the matrix differential equation $T' = K + TKT$ on I for which $T(0) = 0$; moreover, if K is an hermitian matrix then $T(t)$ is hermitian for $t \in I$.

10. DIFFERENTIABILITY OF SOLUTIONS OF REAL DIFFERENTIAL SYSTEMS

In the present section the considered differential system is of the form

$$u' = g(t, u), \tag{10.1}$$

where the component functions $g_\alpha(t, u) = g_\alpha(t, u_1, \ldots, u_n)$, $(\alpha = 1, \ldots, n)$, are real-valued functions of the real variables $(t, u) = (t, u_1, \ldots, u_n)$ on the region of consideration. For brevity, the vector function $g(t, u)$ is said to be *of class* $\mathfrak{C}^{(q)}$ *in* u on a region $\mathfrak{R}$ of (t, u)-space if g and all of its partial derivatives with respect to the components u_β of u up to and including those of order q are defined and continuous in (t, u) throughout $\mathfrak{R}$.

Theorem 10.1. *Suppose that the real-valued vector function $g(t, u)$ is of class $\mathfrak{C}^{(1)}$ in u on an open region $\mathfrak{R}$ of real (t, u)-space and that $u = u(t, h)$ is a one-parameter family of solutions of* (10.1) *with graph in $\mathfrak{R}$ for $t \in [a, b]$, $h \in (-\delta, \delta)$. If for a fixed τ on $[a, b]$ the vector function $\eta(h) = u(\tau, h)$, $h \in (-\delta, \delta)$ is continuous and has a finite derivative at $h = 0$, then on $[a, b]$ the derivative vector function $z(t) = u_h(t, 0) \equiv (u_{\alpha h}(t, 0))$ exists and is a solution of the linear vector differential equation*

$$z' = g_u(t, u(t, 0))z, \tag{10.2}$$

where $g_u(t, u(t, 0))$ is the $n \times n$ matrix function $[g_{\alpha u_\beta}(t, u(t, 0))]$.

The linear vector differential equation (10.2) is called the *equation of variation*, or *perturbation equation* of (10.1) along the solution $u = u(t, 0)$ of (10.1).

The hypotheses of the theorem imply that $g(t, u)$ is locally Lipschitzian in u on $\mathfrak{R}$, and hence the solutions of (10.1) are locally unique in $\mathfrak{R}$. If $u = \phi(t; \tau, \eta)$, $a(\tau, \eta) < t < b(\tau, \eta)$, is the solution of (10.1) passing through a point (τ, η) of $\mathfrak{R}$, as determined in Theorem 5.7, then $u(t, h) = \phi(t; \tau, \eta(h))$ and from that theorem it follows that $u(t, h)$ is continuous in (t, h) on $[a, b] \times (-\delta, \delta)$. If we set $\Delta u = u(t, h) - u(t, 0)$, then the elements

$$A_{\alpha\beta}(t, h) = \int_0^1 g_{\alpha u_\beta}(t, u(t, 0) + \theta\, \Delta u)\, d\theta, \qquad (\alpha, \beta = 1, \ldots, n),$$

of the matrix $A(t, h) \equiv [A_{\alpha\beta}(t, h)]$ are continuous in (t, h), and from the mean value theorem it follows that for $h \in [-\delta, \delta]$ the vector function $z = z(t, h) = (1/h)\,\Delta u$ is a solution of the homogeneous linear vector differential equation

$$\text{(10.3)} \qquad z' = A(t, h)z, \qquad t \in [a, b].$$

Since $A_{\alpha\beta}(t, 0) = g_{\alpha u_\beta}(t, u(t, 0))$, $(\alpha, \beta = 1, \ldots, n)$, for $h = 0$ the vector equation (10.3) reduces to (10.2). Now by Problem I.9:13 the solution $z = z(t; \tau, \eta, h)$ of (10.3) passing through the initial point (τ, η) is continuous in (t, τ, η, h) on $[a, b] \times [a, b] \times \mathbf{R}_n \times (-\delta, \delta)$. In particular, for $h \in (-\delta, \delta)$ we have $(1/h)\,\Delta u = z(t; \tau, [\eta(h) - \eta(0)]/h, h)$, and since

$$[\eta(h) - \eta(0)]/h \to \eta'(0)$$

as $h \to 0$ it follows that on $[a, b]$ we have $(1/h)\Delta u \to u_h(t, 0) = z(t; \tau, \eta'(0), 0)$ as $h \to 0$. Consequently, on this interval $z = (u_{\alpha h}(t, 0))$ is the solution of (10.2) passing through the initial point $(\tau, \eta'(0))$.

Now consider an n-dimensional real vector differential equation

$$\text{(10.4)} \qquad u' = g(t, u, c),$$

where $g(t, u, c) \equiv (g_\alpha(t, u_1, \ldots, u_n, c_1, \ldots, c_k))$, $(\alpha = 1, \ldots, n)$, is a real-valued vector function of the real variables $(t, u, c) = (t, u_1, \ldots, u_n, c_1, \ldots, c_k)$ on an open region $\mathfrak{D}$ of (t, u, c)-space. The section of $\mathfrak{D}$ consisting of all (t, u) such that (t, u, c) is in $\mathfrak{D}$ will be denoted by $\mathfrak{D}_c$. If $g(t, u, c)$ is of class $\mathfrak{C}^{(1)}$ in u on $\mathfrak{D}$ then for each c such that $\mathfrak{D}_c$ is nonvacuous the vector function $g(t, u, c)$ is locally Lipschitzian in u on $\mathfrak{D}_c$, and hence the solutions of (10.4) are locally unique in $\mathfrak{D}_c$. If (τ, η, c) is in $\mathfrak{D}$, then from Theorem 5.8 it follows that the solution

$$\text{(10.5)} \qquad u = \phi(t; \tau, \eta, c) = (\phi_\alpha(t; \tau, \eta, c))$$

of (10.4) passing through the point (τ, η), and with graph in $\mathfrak{D}_c$, has a maximal interval of definition

$$(10.6) \qquad I(\tau, \eta, c) = \{t \mid a(\tau, \eta, c) < t < b(\tau, \eta, c)\}$$

and that the vector function $\phi(t; \tau, \eta, c)$ is continuous in (t, τ, η, c) on

$$(10.7) \qquad \{(t, \tau, \eta, c) \mid (\tau, \eta, c) \in \mathfrak{D}, t \in I(\tau, \eta, c)\}.$$

The following Theorems 8.2, 8.3, and 8.4 provide further results on the differentiability properties of the solutions (10.5) of the differential equation (10.4).

Theorem 10.2. *If* $g(t, u, c)$ *is of class* $\mathfrak{C}^{(q)}$, $(q \geq 1)$, *in* u *on* $\mathfrak{D}$ *then for the solutions* (10.5) *of* (10.4) *the partial derivatives*

$$(10.8) \qquad \phi_{\alpha\eta_\gamma}(t; \tau, \eta, c),\ \phi'_{\alpha\eta_\gamma}(t; \tau, \eta, c),\ (\alpha, \gamma = 1, \ldots, n),$$

exist and are of class $\mathfrak{C}^{(q-1)}$ *in* η *on the open region of* (t, τ, η, c)*-space defined by* (10.7). *Moreover,* $z(t) = (\phi_{\alpha\eta_\gamma}(t; \tau, \eta, c))$, $(\gamma = 1, \ldots, n)$, *are solutions of the linear vector homogeneous equation*

$$(10.9) \qquad z' = g_u(t, \phi(t; \tau, \eta, c), c)z$$

on $I(\tau, \eta, c)$ *satisfying the initial conditions*

$$(10.10) \qquad \phi_{\alpha\eta_\gamma}(\tau; \tau, \eta, c) = \delta_{\alpha\gamma}, \qquad (\alpha, \gamma = 1, \ldots, n).$$

Suppose that the hypotheses of the theorem hold for $q = 1$, and for a given (τ, η, c) of $\mathfrak{D}$ consider an interval $[a, b]$ with $a(\tau, \eta, c) < a < \tau < b < b(\tau, \eta, c)$. As the graph of $u = \phi(t; \tau, \eta, c)$, $t \in [a, b]$, is in the open subset $\mathfrak{D}_c$ of (t, u)-space, from the continuity of the vector function $\phi(t; \tau, \eta, c)$ on (10.7) it follows that there is a $\delta > 0$ such that if $|\eta^1 - \eta| < \delta$ then $[a, b]$ is a subinterval of $I(\tau, \eta^1, c)$; that is, on $[a, b]$ the solution $u = \phi(t; \tau, \eta^1, c)$ is defined and has graph in $\mathfrak{D}_c$. In particular, for $h \in (-\delta, \delta)$ and $\gamma = 1, \ldots, n$ the vector $u = u(t, h) = \phi(t; \tau, \eta + he^{(\gamma)}, c)$ is a solution of (10.4) on $[a, b]$ with graph in $\mathfrak{D}_c$; moreover, on $(-\delta, \delta)$ the vector function $\eta(h) = u(\tau, h) = \eta + he^{(\gamma)}$ is continuous and $\eta'(0) = e^{(\gamma)}$. Application of Theorem 10.1 to the equation (10.4) then establishes the result that $z(t) = u_h(t, 0) = (\phi_{\alpha\eta_\gamma}(t; \tau, \eta, c))$ is the solution of (10.9) on $[a, b]$ satisfying the initial conditions $z(\tau) = e^{(\gamma)}$, $(\gamma = 1, \ldots, n)$, that is, $\phi_{\alpha\eta_\gamma}(\tau; \tau, \eta, c) = \delta_{\alpha\gamma}$. Since $[a, b]$ may be chosen as an arbitrary closed subinterval of $I(\tau, \eta, c)$ containing $t = \tau$ as an interior point, it follows that $z(t) = (\phi_{\alpha\eta_\gamma}(t; \tau, \eta, c))$, $(\gamma = 1, \ldots, n)$, is a solution of (10.9) on $I(\tau, \eta, c)$. Now the functions appearing as the right-hand members of (10.9) are continuous in the open region

$$\mathfrak{D}^1 = \{(t, z, \tau, \eta, c) \mid (\tau, \eta, c) \in \mathfrak{D},\ t \in I(\tau, \eta, c), |z| < \infty\}$$

of (t, z, τ, η, c)-space, and the solutions of this linear differential equation are locally unique on the sections $\mathfrak{D}^1(\tau, \eta, c) = \{(t, z) \mid t \in I(\tau, \eta, c), |z| < \infty\}$ of $\mathfrak{D}^1$. Therefore one may apply Theorem 5.8 to the linear equation (10.9) involving the parameters τ, η, c, and conclude that the functions $\phi_{\alpha\eta_\gamma}(t; \tau, \eta, c)$ are continuous in (t, τ, η, c) on (10.7). From (10.9) it then follows that the functions

$$\phi'_{\alpha\eta_\gamma}(t; \tau, \eta, c) = \sum_{\beta=1}^{n} g_{\alpha u_\beta}(t, \phi(t; \tau, \eta, c), c)\phi_{\beta\eta\gamma}(t; \tau, \eta, c)$$

are also continuous in (t, τ, η, c) on (10.7).

To establish by mathematical induction the full result of the theorem, suppose that the result of the theorem holds for $q = p$, $(p \geq 1)$, and for a vector differential equation (10.4) with $g(t, u, c)$ of class $\mathfrak{C}^{(p+1)}$ in u on an open region $\mathfrak{D}$ of real (t, u, c)-space consider the differential system

$$\text{(10.11)} \qquad \begin{aligned} z'_\alpha &= \sum_{\beta=1}^{n} g_{\alpha u_\beta}(t, \phi(t; s, w, c), c) z_\beta, \\ w'_\alpha &= 0, \qquad\qquad (\alpha = 1, \ldots, n). \end{aligned}$$

Since $g(t, u, c)$ is of class $\mathfrak{C}^{(p+1)}$ in u on $\mathfrak{D}$, it follows from the inductive hypothesis that $\phi(t; s, w, c)$ is of class $\mathfrak{C}^{(p)}$ in w on the region $\{(t, s, w, c) \mid (s, w, c) \in \mathfrak{D},\ t \in I(s, w, c)\}$ of (t, s, w, c)-space, and consequently the functions appearing as right-hand members of the equations (10.11) are of class $\mathfrak{C}^{(p)}$ in $(z, w) = (z_1, \ldots, z_n, w_1, \ldots, w_n)$ on the open region

$$\text{(10.12)} \quad \{(t, z, w, s, c) \mid (s, w, c) \in \mathfrak{D}, a(s, w, c) < t < b(s, w, c), |z| < \infty\}$$

of real (t, z, w, s, c)-space. For a point $(t, z, w, s, c) = (\tau, \zeta, \eta, s, c)$ of (10.12) let $z_\alpha = \theta_\alpha(t; \tau, \zeta, \eta, s, c)$, $w_\alpha = \theta_{n+\alpha}(t; \tau, \zeta, \eta, s, c) \equiv \eta_\alpha$, $(\alpha = 1, \ldots, n)$, be the solution of (10.11) such that $z_\alpha = \zeta_\alpha$, $w_\alpha = \eta_\alpha$, $(\alpha = 1, \ldots, n)$, for $t = \tau$. From the inductive hypothesis it follows that the functions $\theta_{\alpha\zeta_\gamma}$ and $\theta_{\alpha\eta_\gamma}$ are of class $\mathfrak{C}^{(p-1)}$ in (ζ, η) on (10.12); that is, the functions $\theta_\alpha(t; \tau, \zeta, \eta, s, c)$ are of class $\mathfrak{C}^{(p)}$ in (ζ, η) on (10.12). Now $\phi_{\alpha\eta_\gamma}(t; \tau, \eta, c) = \theta_\alpha(t; \tau, e^{(\gamma)}, \eta, \tau, c)$, and consequently $\phi_{\alpha\eta_\gamma}(t; \tau, \eta, c)$, $(\gamma = 1, \ldots, n)$, are of class $\mathfrak{C}^{(p)}$ in η on (10.7). From the fact that $z = (\phi_{\alpha\eta_\gamma}(t; \tau, \eta, c))$, $(\gamma = 1, \ldots, n)$, are solutions of (10.9) it follows that the functions $\phi'_{\alpha\eta_\gamma}$ are also of class $\mathfrak{C}^{(p)}$ in η on (10.7), thus completing the proof of the inductive step that the validity of Theorem 10.2 for $q = p$ implies the validity of the theorem for $q = p + 1$.

Theorem 10.3. *If $g(t, u, c)$ is of class $\mathfrak{C}^{(q)}$, $(q \geq 1)$, in u on $\mathfrak{D}$ then for the solutions* (10.5) *of* (10.4) *the partial derivatives*

$$\text{(10.13)} \qquad \phi_{\alpha\tau}(t; \tau, \eta, c), \qquad \phi'_{\alpha\tau}(t; \tau, \eta, c), \qquad (\alpha = 1, \ldots, n),$$

exist and are of class $\mathfrak{C}^{(q-1)}$ *in* η *on the open region of* (t, τ, η, c)*-space defined by* (10.7). *Moreover,* $z(t) = (\phi_{\alpha\tau}(t; \tau, \eta, c))$ *is the solution of the linear vector homogeneous equation* (10.9) *on* $I(\tau, \eta, c)$ *satisfying the initial conditions*

$$\phi_{\alpha\tau}(\tau; \tau, \eta, c) = -g_\alpha(\tau, \eta, c), \qquad (\alpha = 1, \ldots, n). \tag{10.14}$$

For a given (τ, η, c) of $\mathfrak{D}$ consider a subinterval $[a, b]$ of $I(\tau, \eta, c)$ such that $\tau \in (a, b)$, and choose $\delta > 0$ such that for $h \in (-\delta, \delta)$ we have $\tau + h \in [a, b]$, and the solution $u(t, h) = \phi(t; \tau + h, \eta, c)$ is defined and has graph in $\mathfrak{D}_c$ for $t \in [a, b]$. In view of the continuity of $\phi(t; \tau, \eta, c)$ on (10.6), it follows that $u(t, h)$ is continuous in (t, h) on $[a, b] \times (-\delta, \delta)$. The relation

$$\frac{1}{h}[u(\tau, h) - u(\tau, 0)] = \frac{1}{h}\int_{\tau+h}^{\tau} g(t, u(t, h), c)\, dt$$

then implies that $\eta(h) = u(\tau, h)$ has at $h = 0$ the derivative $\eta'(0) = -g(\tau, u(\tau, 0), c) = -g(\tau, \eta, c)$ From Theorem 10.1 it then follows that on $[a, b]$ the vector function $z(t) = u_h(t, 0) = (\phi_{\alpha\tau}(t; \tau, \eta, c))$ exists and is a solution of (10.9). Again, since $[a, b]$ may be chosen as an arbitrary closed subinterval of $I(\tau, \eta, c)$ containing $t = \tau$ as an interior point, it follows that $z(t) = (\phi_{\alpha\tau}(t; \tau, \eta, c))$ is the solution of (10.9) on $I(\tau, \eta, c)$ satisfying $z(\tau) = -g(\tau, \eta, c)$. As the column vectors of $[\phi_{\alpha\eta_\gamma}(t; \tau, \eta, c)]$, $(\alpha, \gamma = 1, \ldots, n)$, are solutions of (10.9) on $I(\tau, \eta, c)$, and this matrix is the identity matrix for $t = \tau$, it follows from Theorem 9.1 that each solution $z(t)$ of (10.9) is of the form $z_\alpha(t) = \sum_{\gamma=1}^{n} \phi_{\alpha\eta_\gamma}(t; \tau, \eta, c)\xi_\gamma$ for suitable constants ξ_γ. In particular, $\phi_{\alpha\tau}(\tau; \tau, \eta, c) = -\sum_{\gamma=1}^{n} \phi_{\alpha\eta_\gamma}(\tau; \tau, \eta, c) g_\gamma(\tau, \eta, c)$, and hence for $(\tau, \eta, c) \in \mathfrak{D}$, $t \in I(\tau, \eta, c)$, we have

$$\phi_{\alpha\tau}(t; \tau, \eta, c) = -\sum_{\gamma=1}^{n} \phi_{\alpha\eta_\gamma}(t; \tau, \eta, c) g_\gamma(\tau, \eta, c), \qquad (\alpha = 1, \ldots, n), \tag{10.15}$$

so that $z(t) = (\phi_{\alpha\tau}(t; \tau, \eta, c))$ is the solution of (10.9) on $I(\tau, \eta, c)$ satisfying the initial conditions (10.14). This fact, together with the results of Theorem 10.2 on the continuity and differentiability of the functions (10.8), imply the stated results on the continuity and differentiability of the functions (10.13).

Theorem 10.4. *If* $g(t, u, c)$ *is of class* $\mathfrak{C}^{(q)}$, $(q \geq 1)$, *in* (u, c) *on* $\mathfrak{D}$, *then for the solutions* (10.5) *of* (10.4) *the partial derivatives* (10.8), (10.13) *and also the partial derivatives*

$$\phi_{\alpha c_\nu}(t; \tau, \eta, c),\ \phi'_{\alpha c_\nu}(t; \tau, \eta, c), \qquad (\alpha = 1, \ldots, n;\ \nu = 1, \ldots, k), \tag{10.16}$$

exist and are of class $\mathfrak{C}^{(q-1)}$ *in* (η, c) *on* (10.7).

Under the hypotheses of Theorem 10.4 the real differential system

$$(10.17)\qquad \begin{aligned} u'_\alpha &= g_\alpha(t, u, c), \qquad (\alpha = 1, \ldots, n),\\ c'_\nu &= 0, \qquad (\nu = 1, \ldots, k), \end{aligned}$$

in the $n + k$ real variables $u_1, \ldots, u_n, c_1, \ldots, c_k$ satisfies the hypotheses of Theorems 10.2 and 10.3, and consequently Theorem 10.4 is an immediate corollary to these preceding theorems.

Theorem 10.5. *If $g(t, u, c)$ is of class $\mathfrak{C}^{(q)}$, $(q \geq 1)$, in (t, u, c) on $\mathfrak{D}$, then the functions $\phi_\alpha(t; \tau, \eta, c)$ and $\phi'_\alpha(t; \tau, \eta, c)$, $(\alpha = 1, \ldots, n)$, are of class $\mathfrak{C}^{(q)}$ in (t, τ, η, c) on (10.7).*

The result of Theorem 10.5 for $q = 1$ is a ready consequence of the preceding theorems. Indeed, if $g(t, u, c)$ is of class $\mathfrak{C}^{(1)}$ in (t, u, c) on $\mathfrak{D}$ then Theorems 10.2, 10.3 and 10.4 imply that the functions $\phi_\alpha(t; \tau, \eta, c)$ are of class $\mathfrak{C}^{(1)}$ in (t, τ, η, c) on (10.7), and from the equation

$$(10.18)\qquad \phi'(t; \tau, \eta, c) = g(t, \phi(t; \tau, \eta, c), c),$$

it follows that the functions $\phi'_\alpha(t; \tau, \eta, c)$ are also of class $\mathfrak{C}^{(1)}$ in (t, τ, η, c) on (10.7).

Now suppose that the result of Theorem 10.5 holds for $q = p$, $(p \geq 1)$, and consider a vector equation (10.4) with $g(t, u, c)$ of class $\mathfrak{C}^{(p+1)}$ in (t, u, c) on an open region $\mathfrak{D}$ of real (t, u, c)-space. In particular, the inductive hypothesis implies that the functions $\phi_\alpha(t; \tau, \eta, c)$ are of class $\mathfrak{C}^{(p)}$ in (t, τ, η, c) on (10.7). Then the functions occurring as right-hand members of the linear vector differential equations

$$(10.19)\qquad z' = g_u(t, \phi(t; s, \eta, c), c)z,$$

$$(10.20)\qquad z' = g_u(t, \phi(t; s, \eta, c), c)z + g_{c_\nu}(t, \phi(t; s, \eta, c), c),$$

where $g_{c_\nu}(t, u, c) = (g_{\alpha c_\nu}(t, u, c))$, are of class $\mathfrak{C}^{(p)}$ in (t, z, s, η, c) on the open region

$$(10.21)\qquad \{(t, z, s, \eta, c) \mid (s, \eta, c) \in \mathfrak{D}, |z| < \infty, t \in I(s, \eta, c)\}$$

of (t, z, s, η, c)-space. Let $z = (\theta_\alpha(t; \tau, \zeta, s, \eta, c))$ and $z = (\psi_{\alpha;\nu}(t; \tau, \zeta, s, \eta, c))$ be the solutions of (10.19) and (10.20), respectively, such that $z = \zeta$ when $t = \tau$. In view of the inductive hypothesis that the result of Theorem 10.5 holds for $q = p$, it follows that the functions $\theta_\alpha, \psi_{\alpha;\nu}$ are of class $\mathfrak{C}^{(p)}$ in $(t, \tau, \zeta, s, \eta, c)$ on the region

(10.22)

$$\{(t, \tau, \zeta, s, \eta, c) \mid (s, \eta, c) \in \mathfrak{D}, |\zeta| < \infty, \tau \in I(s, \eta, c), t \in I(s, \eta, c)\}.$$

Since $\phi_{\alpha\eta_\gamma}(t; \tau, \eta, c) = \theta_\alpha(t; \tau, e^{(\gamma)}, \tau, \eta, c)$, $\phi_{\alpha\tau}(t; \tau, \eta, c) = \theta_\alpha(t; \tau, -g(\tau, \eta, c), \tau, \eta, c)$, and $\phi_{\alpha c_\nu}(t; \tau, \eta, c) = \psi_{\alpha;\nu}(t; \tau, 0, \tau, \eta, c)$, it follows that the functions $\phi_{\alpha\eta_\gamma}$, $\phi_{\alpha\tau}$, $\phi_{\alpha c_\nu}$ are of class $\mathfrak{C}^{(p)}$ in (t, τ, η, c) on (10.7). Equation (10.18) then implies that the functions ϕ'_α are of class $\mathfrak{C}^{(p)}$ in (t, τ, η, c) on (10.7), and consequently the functions ϕ_α are of class $\mathfrak{C}^{(p+1)}$ in (t, τ, η, c) on (10.7). In turn, equation (10.18) implies that the functions ϕ'_α are of class $\mathfrak{C}^{(p+1)}$ in (t, τ, η, c) on (10.7), thus completing the proof of the inductive step that the validity of Theorem 10.5 for $q = p$ implies the validity of this theorem for $q = p + 1$.

For the case of vector differential equations

$$y' = f(t, y, c),$$

in which t is a real variable, the $y = (y_\alpha)$ and $c = (c_\nu)$ are complex-valued, and the components of the vector function $f(t, y, c) = (f_\alpha(t, y, c))$ are differentiable functions of the individual y_α, c_ν, the results of the above theorems remain valid with the understanding that the occurring partial derivatives are now with respect to the involved complex variables. This may be established, either by noting that the individual steps leading to the above results remain valid under the new hypothesis, or by reducing the equation to an associated real-valued vector differential equation in the manner given in Section I.1, using the above results for this real-valued equation, and obtain locally the differentiability with respect to the involved complex variable by verifying the validity of the associated Cauchy-Riemann partial differential equations.

The special case of a linear vector differential equation

$$y' = [A(t) + \lambda B(t)]y, \tag{10.23}$$

in which the coefficient matrix is linear in a complex parameter λ, is of special importance for the subsequent discussion of boundary value problems. Problem 4 below is concerned with an independent direct proof that for (10.23) one may choose a fundamental matrix to be an entire function of λ.

For a vector differential equation (10.1) with $g(t, u)$ of class $\mathfrak{C}^{(1)}$ in u on an open region $\mathfrak{R}$ of (t, u)-space, a scalar function $W(t, u)$ is called a *first integral* of (10.1) if this function is of class $\mathfrak{C}^{(1)}$ in (t, u) on an open subset $\mathfrak{R}_0$ of $\mathfrak{R}$, is not constant on $\mathfrak{R}_0$, and is such that for every solution $u(t)$, $t \in I$, of (10.1) with graph in $\mathfrak{R}_0$ the function $W(t, u(t))$, $t \in I$, is constant. The following result is immediate.

Theorem 10.6. *If $g(t, u)$ is of class $\mathfrak{C}^{(1)}$ in u on an open region $\mathfrak{R}$ of (t, u)-space, then a function $W(t, u)$ which is nonconstant and of class $\mathfrak{C}^{(1)}$ in (t, u) on an open subset $\mathfrak{R}_0$ of $\mathfrak{R}$ is a first integral of* (10.1) *if and only if on $\mathfrak{R}_0$ it is a solution of the partial differential equation*

$$W_t + \sum_{\alpha=1}^{n} g_\alpha(t, u) W_{u_\alpha} = 0. \tag{10.24}$$

If $u = \phi(t; \tau, \eta) = (\phi_\alpha(t; \tau, \eta))$, $a(\tau, \eta) < t < b(\tau, \eta)$, denotes the solution of (10.1) passing through (τ, η), and with graph in $\mathfrak{R}$, then it follows from Theorems 10.2 and 10.3 that in a suitable neighborhood of a point (t^0, u^0) of $\mathfrak{R}$ each of the component functions $\phi_\alpha(t^0; t, u)$, $(\alpha = 1, \ldots, n)$, is a first integral of (10.1). Indeed, this result is an immediate consequence of the relation $\phi_\alpha(t^0; t, \phi(t; \tau, \eta)) = \phi_\alpha(t^0; \tau, \eta)$, which holds whenever t is such that the interval with endpoints t^0 and t is a subinterval of $(a(\tau, \eta), b(\tau, \eta))$.

Similarly, for an n-th order scalar or vector differential equation

$$u^{[n]} = g(t, u, u', \ldots, u^{[n-1]}), \tag{10.25}$$

a function $W(t, u_1, \ldots, u_n)$ is called a first integral whenever this function is a first integral for the first order differential system to which (10.25) is reducible under the transformations presented in Section I.8. For example, $W(t, u, u') = u'^2 + u^2$ is a first integral for the differential equation $u'' + u = 0$.

Problems I.10

1. Suppose that $g(t, u) = (g_\alpha(t, u_1, \ldots, u_n))$, $(\alpha = 1, \ldots, n)$, is a real valued continuous vector function of (t, u) which is of class $\mathfrak{C}^{(1)}$ in u on an open region $\mathfrak{R}$ of real (t, u)-space, and that $u = \phi(t; \tau, \eta)$, $a(\tau, \eta) < t < b(\tau, \eta)$, denotes the solution of (10.1) passing through the point (τ, η) and with graph in $\mathfrak{R}$, as determined above. Show that

(i) $\det[\phi_\eta(t; \tau, \eta)] = \exp\left\{\int_\tau^t \sum_{\alpha=1}^n g_{\alpha u_\alpha}(s, \phi(s; \tau, \eta))\, ds\right\}$;

(ii) $\nu[\phi_\eta(t; \tau, \eta)] \leq \exp\{\operatorname{sgn}(t - \tau) \int_\tau^t \lambda(s)\, ds\}$, where $\lambda(t)$ is the greatest proper value of the hermitian matrix $\frac{1}{2}(\operatorname{sgn}(t - \tau))(g_u + g_u^*)$, and the arguments of g_u and g_u^* are $(t, \phi(t; \tau, \eta))$.

(iii) if $(\tau, \eta(\theta)) = (\tau, (1 - \theta)\eta^0 + \theta\eta^1) \in \mathfrak{R}$ and $[a, b] \subset (a(\tau, \eta(\theta)), b(\tau, \eta(\theta)))$ for $\theta \in [0, 1]$, while $\lambda_0(t)$ majorizes the greatest proper value of the hermitian matrix

$$\tfrac{1}{2}[\operatorname{sgn}(t - \tau)][g_u(t, \phi(t; \tau, \eta(\theta))) + g_u^*(t, \phi(t; \tau, \eta(\theta)))]$$

for $0 \leq \theta \leq 1$, then for $t \in [a, b]$,

$$|\phi(t; \tau, \eta^1) - \phi(t; \tau, \eta^0)| \leq |\eta^1 - \eta^0| \exp\left\{\operatorname{sgn}(t - \tau)\int_\tau^t \lambda_0(s)\, ds\right\}.$$

Hints. (i), (ii) Use the fact that $U(t) = \phi_\eta(t; \tau, \eta)$ is the fundamental matrix of $U' = g_u(t, \phi(t; \tau, \eta))U$, such that $U(\tau) = E$.

(iii) Note that

$$\phi(t;\tau,\eta^1)-\phi(t;\tau,\eta^0)=\int_0^1 (d/d\theta)\phi(t;\tau,\eta(\theta))\,d\theta$$
$$=\left[\int_0^1 \phi_\eta(t;\tau,\eta(\theta))\,d\theta\right](\eta^1-\eta^0),$$

and use the result of conclusion (ii).

2. Suppose that $g(t, u) = (g_\alpha(t, u_1, \ldots, u_n))$ and $h(t, u) = (h_\alpha(t, u_1, \ldots, u_n))$ are real-valued functions of (t, u) which are of class $\mathfrak{C}^{(1)}$ in u on an open region $\mathfrak{R}$ of real (t, u)-space, and that $u = \phi(t; \tau, \eta)$, $a_g(\tau, \eta) < t < b_g(\tau, \eta)$, and $u = \psi(t; \tau, \eta)$, $a_h(\tau, \eta) < t < b_h(\tau, \eta)$, are the solutions of the respective vector differential equations $u' = g(t, u)$, $u' = h(t, u)$, passing through the point (τ, η) and with graphs in $\mathfrak{R}$. Show that if t is such that the compact interval I with endpoints τ and t is a subinterval of each of the intervals $(a_h(\tau, \eta), b_h(\tau, \eta))$ and $(a_g(s, \psi(s; \tau, \eta)), b_g(s, \psi(s; \tau, \eta)))$ for $s \in I$, then

$$\psi(t;\tau,\eta)-\phi(t;\tau,\eta)$$
$$=\int_\tau^t \phi_\eta(t;s,\psi(s;\tau,\eta))[h(s,\psi(s;\tau,\eta))-g(s,\psi(s;\tau,\eta))]\,ds.$$

Hint. Compute $(d/ds)\,\phi(t; s, \psi(s; \tau, \eta))$.

3. Suppose that $g(t, u_1, u_2)$ is real-valued and continuous in (t, u_1, u_2) on $\Delta = [a, b] \times \mathbf{R}_1 \times \mathbf{R}_1$, and that there exists a constant μ such that

$$|g(t,u_1,u_2)|\le\mu$$

on Δ. If ξ^a and ξ^b are given real constants, prove that there exists at least one solution $u(t)$ of

$$u''=g(t,u,u'),\qquad t\in[a,b], \tag{10.26}$$

which satisfies the boundary conditions $u(a) = \xi^a$, $u(b) = \xi^b$.

Hint. If $u = u(t; m)$ is a solution of (10.26) satisfying the initial conditions $u(a) = \xi^a$, $u'(a) = m$, show that $u(t; m)$ is extensible to $[a, b]$ and for $t \in [a, b]$,

$$u(t;m)=\int_a^t (t-s)\,g(s,u(s;m),u'(s;m))\,ds+m(t-a)+\xi^a.$$

Conclude that

$$\xi^a+m(b-a)-\frac{\mu(b-a)^2}{2}\le u(b;m)\le \xi^a+m(b-a)+\frac{\mu(b-a)^2}{2},$$

and hence that there exist values m_1, m_2 such that $u(b; m) < \xi^b$ if $m < m_1$, and $u(b; m) > \xi^b$ if $m > m_2$. If $g(t, u_1, u_2)$ satisfies the additional property of having partial derivatives with respect to the u_α, $(\alpha = 1, 2)$, which are continuous in (t, u_1, u_2) on Δ, use a result of Theorem 10.2 to conclude that there is a value $m = m_b$ such that $u(b; m_b) = \xi^b$. For the general case in which this additional condition is not satisfied, employ an approximation procedure similar to that suggested for the proof of Problem I.5:11.

4.[ss] If $A(t)$ and $B(t)$ are continuous $n \times n$ matrix functions on $[a, b]$, and λ is a complex parameter, show that the fundamental matrix $Y = Y(t, \tau; \lambda)$ of (10.23) satisfying the initial condition $Y(\tau, \tau; \lambda) = E$ at a given $\tau \in [a, b]$, is an entire function of λ.

Hint. If the matrix functions $Y_j(t, \tau)$, $(j = 0, 1, \ldots)$, are defined recursively as solutions of the matrix differential systems

$$Y_0'(t, \tau) = A(t)Y_0(t, \tau), \qquad Y_0(\tau, \tau) = E,$$

$$Y_m'(t, \tau) = A(t)Y_m(t, \tau) + B(t)Y_{m-1}(t, \tau), \qquad Y_m(\tau, \tau) = 0, \quad (m = 1, 2, \ldots),$$

show inductively that $\nu[Y_m(t, \tau)] \leq \kappa_0(\kappa_1 |t - \tau|)^m/m!$ for $(t, \tau) \in [a, b] \times [a, b]$, where κ_0, κ_1 are constants such that $\nu[Y_0(t, \tau)] \leq \kappa_0$ for $(t, \tau) \in [a, b] \times [a, b]$, and $\nu[Y_0(t, \tau) Y_0^{-1}(s, \tau) B(s)] \leq \kappa_1$ for $(t, \tau, s) \in [a, b] \times [a, b] \times [a, b]$. Conclude that on each compact set $\{(t, \tau, \lambda) \mid (t, \tau) \in [a, b] \times [a, b],\ |\lambda| \leq q\}$ the series $Y_0(t, \tau) + \sum_{m=1}^{\infty} \lambda^m Y_m(t, \tau)$ converges uniformly, and the sum $Y(t, \tau; \lambda)$ of this series is such that

$$Y(t, \tau; \lambda) = E + \int_\tau^t [A(s) + \lambda B(s)]\, Y(s, \tau; \lambda)\, ds$$

for $(t, \tau) \in [a, b] \times [a, b]$ and arbitrary complex λ.

5.[ss] Suppose that $A(t)$, $B(t)$, and λ are as in Problem 4, $g(t)$ is a continuous n-dimensional vector function on $[a, b]$, and $y = y^{(0)}(t)$ is a solution of

$$y'(t) = [A(t) + \lambda B(t)]y(t) + g(t), \qquad t \in [a, b], \tag{10.27}$$

for $\lambda = \lambda_0$. Prove that the solution $y(t; \lambda)$ of (10.27) which satisfies the initial condition $y(a; \lambda) = y^{(0)}(a)$ is for fixed $t \in [a, b]$ an entire function of λ, and $y(t; \lambda_0) = y^{(0)}(t)$.

Hint. Use results of Problem 4, observing that, in the notation of that problem,

$$y(t; \lambda) = Y(t, a; \lambda)y^{(0)}(a) + \int_a^t Y(t, s; \lambda)\, g(s)\, ds.$$

6. Show that the following functions W are first integrals of the accompanying differential equations:

(i) $u'' + g \sin u = 0, \qquad W = u'^2 - 2g \cos u;$

(ii) $r'' - r\theta'^2 + \dfrac{k^2}{r^2} = 0, \qquad W_1 = r'^2 + r^2\theta' - \dfrac{2k^2}{r};$

$r\theta'' + 2r'\theta' = 0, \qquad W_2 = r^2\theta'.$

11. DIFFERENTIAL EQUATIONS IN THE COMPLEX PLANE

This section will be devoted to a brief discussion of a vector differential equation of the form

$$w' = f(z, w) \tag{11.1}$$

where $f(z, w) = (f_\alpha(z, w_1, \ldots, w_n))$, $(\alpha = 1, \ldots, n)$, is a complex-valued vector function defined on an open region $\mathfrak{D}$ in complex (z, w)-space. It will be supposed that f is single-valued and analytic in (z, w) on $\mathfrak{D}$, in the sense that f is continuous on $\mathfrak{D}$ and for fixed values of n of the $n + 1$ variables $z, w_1, \ldots, w_n$ the components of f are single-valued and analytic, (holomorphic), in the remaining variable on the nonempty sections of $\mathfrak{D}$ determined by the fixed variables. In view of the Cauchy integral formula for holomorphic functions of a single complex variable, it follows that, in the terminology of the preceding section, f is of class $\mathfrak{C}^{(q)}$, $(q = 1, 2, \ldots)$, in (z, w) on $\mathfrak{D}$. An equivalent definition is that for each point $(z^0, w^0) \in \mathfrak{D}$ the vector function $f(z, w)$ is representable in some neighborhood of this point as the sum of a power series in $(z - z^0, w_1 - w_1^0, \ldots, w_n - w_n^0)$.

Theorem 11.1. *Suppose that $f(z, w) = (f_\alpha(z, w_1, \ldots, w_n))$, $(\alpha = 1, \ldots, n)$, is single-valued and analytic on an open region $\mathfrak{D}$ in complex (z, w)-space, and $(z^0, w^0) \in \mathfrak{D}$. If ε, δ are positive constants such that $\square = \{(z, w) \mid |z - z^0| \leq \varepsilon, |w - w^0| \leq \delta\}$ lies in $\mathfrak{D}$, μ is a constant such that $|f(z, w)| \leq \mu$ on $\square$, and $\varepsilon_1 = \min[\varepsilon, \delta/\mu]$, then for $0 < \varepsilon_2 \leq \varepsilon_1$ there exists on $\mathfrak{N}_2 = \{z \mid |z - z^0| < \varepsilon_2\}$ a unique holomorphic solution $w(z) = (w_\alpha(z))$ of (11.1) which satisfies the initial condition*

$$w(z^0) = w^0, \tag{11.2}$$

and

$$|w(z) - w^0| \leq \mu|z - z^0| < \delta \quad \textit{for} \quad z \in \mathfrak{N}_2.$$

As the compact convex set $\square$ lies in $\mathfrak{D}$, it follows that the partial derivatives $f_{\alpha w_\beta}$, $(\alpha, \beta = 1, \ldots, n)$, are bounded on $\square$, and f is Lipschitzian in w on $\square$. The method of iterations of Section I.6 is applicable here also, if one considers the successive approximations

$$w^{(0)}(z) = w^0,$$

$$w^{(m+1)}(z) = w^0 + \int_{z^0}^{z} f(\zeta, w^{(m)}(\zeta))\, d\zeta, \qquad (m = 0, 1, \ldots), \tag{11.3}$$

where for definiteness, the integrals in (11.3) are taken along the linear segment from z^0 to z. It may be established by induction that on $\mathfrak{N}_1 = \{z \mid |z - z^0| < \varepsilon_1\}$ each $w^{(j)}(z)$ is holomorphic, $|w^{(j)}(z) - w^0| < \delta$, and

$$|w^{(m+1)}(z) - w^{(m)}(z)| \leq \frac{\mu\kappa^{m}|z - z_0|^{m+1}}{(m+1)!}, \qquad (m = 0, 1, \ldots),$$

where κ is a Lipschitz constant such that $|f(z, w^1) - f(z, w^2)| \leq \kappa\, |w^1 - w^2|$ if $(z, w^1) \in \square$, $(z, w^2) \in \square$. Consequently, the sequence $\{w^{(j)}(z)\}$ converges uniformly on $\mathfrak{N}_1$ to a limit function $w(z)$, and in view of the basic convergence theorem on sequences of holomorphic functions it follows that $w(z)$ is holomorphic on $\mathfrak{N}_1$. Moreover,

$$w(z) = w^0 + \int_{z^0}^{z} f(\zeta, w(\zeta))\, d\zeta, \quad \textit{for} \quad z \in \mathfrak{N}_1, \tag{11.4}$$

so that on $\mathfrak{N}_1$ the vector function $w(z)$ is a solution of (11.1) satisfying (11.2).

In general, it is to be noted that if $w(z)$ is a holomorphic solution of (11.1) on $\mathfrak{N}_r = \{z \mid |z - z^0| < r\}$ which satisfies (11.2), and $|w(z) - w^0| \leq \delta$ for $z \in \mathfrak{N}_r$, then also $|w(z) - w^0| \leq \mu\, |z - z^0| < \mu r$ for $z \in \mathfrak{N}_r$. Consequently, if $0 < \varepsilon_2 \leq \varepsilon_1$ and on $\mathfrak{N}_2 = \{z \mid |z - z^0| < \varepsilon_2\}$ a vector function $w = w^1(z)$ is any holomorphic solution of (11.1) satisfying (11.2), then $|w^1(z) - w^0| \leq \mu\, |z - z^0| < \mu\varepsilon_2 \leq \delta$ for $z \in \mathfrak{N}_2$. Therefore, if $w(z)$ is the solution obtained as the limit of the iterates defined by (11.3), it follows that $v(z) = w(z) - w^1(z)$ satisfies for $z \in \mathfrak{N}_2$ the inequality

$$\begin{aligned} |v(z)| &\leq \int_{z^0}^{z} |f(\zeta, w(\zeta)) - f(\zeta, w^1(\zeta))|\, |d\zeta|, \\ &\leq \int_{z^0}^{z} \kappa\, |v(\zeta)|\, |d\zeta|, \end{aligned}$$

where, as in the preceding argument, κ is a Lipschitz constant for f on $\square$, and the integrals are taken along the linear segment from z^0 to z. When these integrals are written in terms of the arc length parameter along the paths of integration, it follows from the basic inequality result of Theorem 2.1 in Section I.2 that $v(z) = 0$ and $w(z) = w^1(z)$ for $z \in \mathfrak{N}_2$.

Of particular interest is the case of a linear vector differential equation

$$w' = A(z)w + g(z), \tag{11.5}$$

for which one has the following important result.

Theorem 11.2. *If the $n \times n$ matrix function $A(z)$ and the n-dimensional vector function $g(z)$ are holomorphic in a simply connected open region $\mathfrak{R}$ of the complex z-plane, and $z^0 \in \mathfrak{R}$, then for arbitrary w^0 there is a unique vector function $w(z)$ which is a holomorphic solution of* (11.5) *in $\mathfrak{R}$, and satisfies the initial condition* (11.2).

The result of this theorem is a consequence of the local Existence Theorem 11.1, and the Monodromy Theorem of complex function theory. It may be established directly by the method of successive approximations, however, as will be presented now. Let $\mathfrak{S}_n$, $(n = 1, 2, \ldots)$, be a sequence of simply connected open subsets of $\mathfrak{R}$ such that $z^0 \in \mathfrak{S}_n$, $\mathfrak{S}_n \subset \mathfrak{S}_{n+1}$, the closure $\overline{\mathfrak{S}}_n$ of $\mathfrak{S}_n$ is a subset of $\mathfrak{R}$, and $\bigcup_{n=1}^{\infty} \mathfrak{S}_n = \mathfrak{R}$. For example, with the aid of the Heine-Borel Theorem it can be established readily that $\mathfrak{S}_n$ may be chosen as the interior of a simply closed polygonal curve Γ_n, where each component arc of Γ_n is parallel to one of the coordinate axes. In particular, since $A(z)$ and $g(z)$ are holomorphic on $\mathfrak{R}$, there exist positive constants κ_n, $\kappa_n{}^0$, λ_n such that $\nu[A(z)] \leq \kappa_n$, $|g(z)| \leq \kappa_n{}^0$ on $\mathfrak{S}_n \cup \Gamma_n$, and for any $z \in \mathfrak{S}_n$ there is a polygonal path $\gamma_n(z)$ in $\mathfrak{S}_n$ which joins z^0 to z, and has length not exceeding λ_n.

Now for $z \in \mathfrak{S}_n$, let

$$\begin{aligned} w^0(z) &= w^0, \\ w^{(m+1)}(z) &= w^0 + \int_{z^0}^{z} [A(\zeta)w^{(m)}(\zeta) + g(\zeta)]\, d\zeta, \qquad (m = 1, 2, \ldots), \end{aligned} \tag{11.6}$$

where the path of integration is a rectifiable path in $\mathfrak{S}_n$ which joins z^0 to z. By induction, it may be established that each $w^{(j)}(z)$ is holomorphic in z on $\mathfrak{S}_n$, so that the value of the integral in (11.6) is independent of the path of integration. For $z^1 \in \mathfrak{S}_n$, one may use a polygonal path $\gamma_n(z^1)$ of length not exceeding λ_n, and on this path by induction one obtains the estimates

$$|w^{(1)}(z) - w^{(0)}(z)| \leq (\kappa_n |w^0| + \kappa_n{}^0)s(z),$$

$$|w^{(j+1)}(z) - w^{(j)}(z)| \leq \frac{(\kappa_n |w^0| + \kappa_n{}^0)s^j(z)}{j!}, \qquad (j = 1, 2, \ldots),$$

where $s(z)$ denotes the arc length of the initial segment of $\gamma_n(z^1)$ joining z^0 to z. In particular, for $z \in \mathfrak{S}_n$ we have

$$|w^{(j+1)}(z) - w^{(j)}(z)| \leq \frac{(\kappa_n |w^0| + \kappa_n{}^0)\lambda_n{}^j}{j!}, \qquad (j = 0, 1, \ldots),$$

and therefore the sequence $\{w^{(j)}(z)\}$ converges uniformly on $\mathfrak{S}_n$. Consequently, if $w(z)$ denotes the limit of $\{w^{(j)}(z)\}$ for $z \in \mathfrak{S}_n$, the vector function thus defined is holomorphic on $\mathfrak{S}_n$, and

$$w(z) = w^0 + \int_{z^0}^{z} [A(\zeta)\, w(\zeta) + g(\zeta)]\, d\zeta, \quad \text{for} \quad z \in \mathfrak{S}_n,$$

so that $w(z)$ is a solution of (11.5) on $\mathfrak{S}_n$ satisfying (11.2). In a manner similar to that employed in the proof of Theorem 11.1, it may be established that $w(z)$ is the unique holomorphic solution of (11.5) on $\mathfrak{S}_n$ satisfying the initial condition (11.2). Moreover, since $\mathfrak{S}_n \subset \mathfrak{S}_{n+1}$ it follows that the solution vector determined for $\mathfrak{S}_{n+1}$ by the above described method of successive approximations is an extension of the solution vector obtained for $\mathfrak{S}_n$ by this method. As $\bigcup_{n=1}^{\infty} \mathfrak{S}_n = \mathfrak{R}$ we have that the sequence $\{w^{(j)}(z)\}$ defined on $\mathfrak{R}$ by the iterates (11.6) converges on $\mathfrak{R}$ to the holomorphic solution $w(z)$ of (11.5) on $\mathfrak{R}$ satisfying the initial condition (11.2), and the convergence of this sequence is uniform on any compact subset of $\mathfrak{R}$.

***Corollary* 1.** *If the $n \times n$ matrix function $A(z)$ is holomorphic on a simply connected region $\mathfrak{R}$ of the complex z-plane, and $z^0 \in \mathfrak{R}$, then the matrix differential equation*

$$\Omega' = A(z)\Omega \tag{11.7}$$

has a unique holomorphic solution $\Omega = \Omega(z; z^0)$ satisfying the initial condition $\Omega(z^0) = E$. Moreover, $\Omega(z; z^0)$ is nonsingular for $(z, z^0) \in \mathfrak{R} \times \mathfrak{R}$, and if $w(z)$ is any holomorphic solution of

$$w' = A(z)w \tag{11.8}$$

on $\mathfrak{R}$, then $w(z) = \Omega(z; z^0)\, w(z^0)$.

It is to be remarked that for the general nonlinear differential equation (11.1) the vector function $f(z, w)$ may be single-valued and analytic on an open region $\mathfrak{D}$ consisting of all (z, w) with z in a given simply connected region $\mathfrak{R}$ of the z-plane, and the solutions of (11.1) not be holomorphic functions of z on $\mathfrak{R}$. For example, for $n = 1$ the scalar functions $f(z, w) = 1 + w^2$ and $f(z, w) = e^w$ are single-valued analytic functions on the entire (z, w)-space, while the general solutions of the differential equations $w' = 1 + w^2$ and $w' = e^w$ are $w = \tan(z + c)$ and $w = \log[1/(c - z)]$, respectively.

For the linear differential equation (11.5), if the $A(z)$ and $g(z)$ are holomorphic in an open region $\mathfrak{R}$ of the z-plane which is not simply connected then a local holomorphic solution of this equation will determine by analytic continuation an analytic vector function which in general is multiple-valued, and each of whose elements is a local holomorphic solution of this differential equation. For example, for $n = 1$ the scalar matrix $A(z) = 1/(2z)$ is holomorphic on the punctured plane $\{z \mid 0 < |z| < \infty\}$, and the differential equation $w' = [1/(2z)]w$ has for its general solution $w = cz^{1/2}$, which possesses a branch point at $z = 0$ except in the special case $c = 0$.

The following theorem presents a basic result for linear homogeneous differential equations (11.7) involving matrix functions $A(z)$ which are holomorphic in a punctured neighborhood of a given point.

Theorem 11.3. *If the $n \times n$ matrix function $A(z)$ is holomorphic in a punctured neighborhood $\mathfrak{N} = \{z \mid 0 < |z - z^0| < \rho\}$ of z^0, then any fundamental matrix $\Omega(z)$ of (11.7) on $\mathfrak{N}$ is of the form*

$$\Omega(z) = \Omega^0(z)(z - z^0)^K, \tag{11.8}$$

where $\Omega^0(z)$ is holomorphic on $\mathfrak{N}$, K is a constant matrix, and

$$(z - z^0)^K = \exp\{K \log(z - z_0)\} = \sum_{m=0}^{\infty} \frac{[K \log(z - z_0)]^m}{m!}. \tag{11.9}$$

For $z^1 \in \mathfrak{N}$, let θ be such that there is a value $\arg(z^1 - z^0)$ of the argument of $z^1 - z^0$ satisfying the inequality $\theta < \arg(z^1 - z^0) < \theta + \pi$, and let $\mathfrak{N}_1$, $\mathfrak{N}_2$ denote the simply connected regions specified by the conditions

$$\begin{aligned} \mathfrak{N}_1 &= \{z \mid z \in \mathfrak{N},\ \theta < \arg(z - z^0) < \theta + 2\pi\}, \\ \mathfrak{N}_2 &= \{z \mid z \in \mathfrak{N},\ \theta - \pi < \arg(z - z^0) < \theta + \pi\}. \end{aligned}$$

Then $z^1 \in \mathfrak{N}_1 \cap \mathfrak{N}_2$, and, by the above Corollary to Theorem 11.2, for $k = 1, 2$ there exists on $\mathfrak{N}_k$ a unique holomorphic fundamental matrix solution $\Omega = \Omega_k(z; z^1)$ of (11.7) which at $z = z^1$ is equal to the identity matrix. If $\mathfrak{N}^+$ and $\mathfrak{N}^-$ denote the simply connected regions specified by the conditions

$$\begin{aligned} \mathfrak{N}^+ &= \{z \mid z \in \mathfrak{N},\ \theta < \arg(z - z^0) < \theta + \pi\}, \\ \mathfrak{N}^- &= \{z \mid z \in \mathfrak{N},\ \theta + \pi < \arg(z - z^0) < \theta + 2\pi\}, \\ &= \{z \mid z \in \mathfrak{N},\ \theta - \pi < \arg(z - z^0) < \theta\}, \end{aligned}$$

then $z^1 \in \mathfrak{N}^+$ and $\Omega_1(z; z^1) = \Omega_2(z; z^1)$ on $\mathfrak{N}^+$. Moreover, since each of the $\Omega_k(z; z^1)$ is a fundamental solution of (11.7) on $\mathfrak{N}^-$, there exists a nonsingular constant $n \times n$ matrix C such that $\Omega_2(z; z^1)C = \Omega_1(z; z^1)$ for $z \in \mathfrak{N}^-$. Consequently, $\Omega_2(z; z^1)C$ provides an analytic continuation of $\Omega_1(z; z^1)$ across the segment $z = z^0 + r \exp\{i\theta\}$, $0 < r < \rho$, and on $\mathfrak{N}^+$ this analytic continuation is equal to $\Omega_1(z; z^1)C$. In other words, if $z^2 \in \mathfrak{N}^+$, and $\gamma(z^2)$ denotes the circular arc $z = z^2 + (z^2 - z^0)\exp(is)$, $0 \leq s \leq 2\pi$, then analytic continuation of $\Omega_1(z; z^1)$ along $\gamma(z^2)$ results in the terminal functional value $\Omega_1(z^2; z^1)C$ at $z = z^2$. Now let K_1 be an $n \times n$ constant matrix such that $C = \exp\{2\pi i K_1\}$; the existence of such a matrix is established in Appendix E. If $\Omega^0(z; z^1) = \Omega_1(z; z^1)\exp\{-K_1 \log(z - z^0)\}$, it then follows that if $z^2 \in \mathfrak{N}^+$ and $\Omega^0(z; z^1)$ is analytically continued along the above defined circular arc $\gamma(z^2)$, then the terminal value of this matrix function at z^2 is the same as its initial value, so that $\Omega^0(z; z^1)$ defines a holomorphic function of z on $\mathfrak{N}$, and

$$\Omega_1(z; z^1) = \Omega^0(z; z^1)\exp[K_1 \log(z - z^0)] = \Omega^0(z; z^1)(z - z^0)^{K_1}.$$

In general, if $\Omega(z)$ is a fundamental matrix solution of (11.7) and one of its determinations at z^1 is equal to M, then $\Omega(z)$ is of the form (11.8) with $\Omega^0(z) = \Omega^0(z; z^1)M$ and $K = M^{-1}K_1M$.

Problems I.11

1. If $A(z)$ and $g(z)$ are holomorphic functions on a simply connected region $\mathfrak{R}$ of the z-plane, and $\Omega(z; z^0)$ is the fundamental matrix for (11.7) as determined in Corollary 1 to Theorem 11.2, show that the solution of (11.5) satisfying the initial condition (11.2) at a point z^0 of $\mathfrak{R}$ is given by the integral formula

$$w(z) = \Omega(z; z^0)w^0 + \int_{z^0}^{z} \Omega(z; \zeta)\, g(\zeta)\, d\zeta, \tag{11.10}$$

where the path of integration in (11.10) is any rectifiable arc in $\mathfrak{R}$ joining z^0 to z.

2. For each of the following 2×2 matrices $A(z)$ show that the accompanying matrix $\Omega(z)$ is a fundamental matrix for (11.7) on the indicated region $\mathfrak{R}$:

(i) $A(z) = \begin{bmatrix} 0 & 1 \\ -(z-1)^{-1} & z(z-1)^{-1} \end{bmatrix}$; $\mathfrak{R} = \{z \mid 0 < |z| < 1\}$;

$$\Omega(z) = \begin{bmatrix} z & e^z \\ 1 & e^z \end{bmatrix}.$$

(ii) $A(z) = \begin{bmatrix} 0 & 1 \\ -2z^{-2} & 2z^{-1} \end{bmatrix}$; $\mathfrak{R} = \{z \mid 0 < |z| < \infty\}$; $\Omega(z) = \begin{bmatrix} z^2 & z \\ 2z & 1 \end{bmatrix}$.

(iii) $A(z) = \begin{bmatrix} 0 & 1 \\ -2[z(2-z)]^{-1} & 0 \end{bmatrix}$; $\mathfrak{R} = \{z \mid 0 < |z| < 2\}$;

$$\Omega(z) = \begin{bmatrix} z(2-z) & 2(z-1) + z(2-z)\log\dfrac{z}{2-z} \\ 2(1-z) & 4 + 2(1-z)\log\dfrac{z}{2-z} \end{bmatrix}.$$

3. If $w(z)$ is a nonconstant meromorphic function of z in a region $\mathfrak{R}$ of the z-plane, then the associated meromorphic function

$$\left(\frac{w''}{w'}\right)' - \frac{1}{2}\left(\frac{w''}{w'}\right)^2 = \frac{w'''}{w'} - \frac{3}{2}\left(\frac{w''}{w'}\right)^2 \tag{11.11}$$

is denoted by $\{w, z\}$, and called the *Schwarzian derivative*. Show that the Schwarzian derivative is a differential invariant for the group of all fractional linear transformations; that is, if $W = (Aw + B)/(Cw + D)$, with $AD - BC \neq 0$, then $\{W, z\} = \{w, z\}$. Also if $Z = (az + b)/(cz + d)$, with $ad - bc \neq 0$, then $\{w, z\} = [Z'(z)]^2\{w, z\}$.

4. Suppose that $\mathfrak{R}$ is a simply connected region of the z-plane, and that $p(z)$ is holomorphic in $\mathfrak{R}$. If $\phi_1(z)$ and $\phi_2(z)$ are linearly independent solutions of the second order linear differential equation

$$\phi'' + p(z)\phi = 0 \tag{11.12}$$

on $\mathfrak{R}$, prove that: (a) $\phi_1\phi_2' - \phi_1'\phi_2$ is equal to a nonzero constant on $\mathfrak{R}$; (b) $w(z) = \phi_1(z)/\phi_2(z)$ is a solution of the third order nonlinear differential equation

$$\{w, z\} = 2p(z) \tag{11.13}$$

on $\mathfrak{R}$. Conversely, if $z^0 \in \mathfrak{R}$ and $w(z)$ is holomorphic in a neighborhood of z^0, with $w'(z^0) \neq 0$, and which satisfies the differential equation (11.13) in this neighborhood, then there exist solutions $\phi_1(z)$, $\phi_2(z)$ of (11.12) such that $w(z) = \phi_1(z)/\phi_2(z)$ in this neighborhood, and these solutions are uniquely determined if $\phi_2(z)$ is such that $\phi_2(z^0) = 1$.

Hint. Use appropriate existence theorems, noting that if $\phi_1(z)$, $\phi_2(z)$ are solutions of (11.12) with $\phi_2(z^0) = 1$, and $w(z) = \phi_1(z)/\phi_2(z)$, then $\phi_1(z^0) = w(z^0)$, $\phi_1'(z^0) = w'(z^0) - w(z^0)\, w''(z^0)/[2w'(z^0)]$, where the derivative symbol means differentiation with respect to z, $\phi_2'(z^0) = -w''(z^0)/[2w'(z^0)]$.

12. NOTES AND REMARKS

Suggested collateral references for this chapter are Bliss [5, Appendix], Kamke [1, Chs. II, IV], Sansone [1-I, Ch. I; 1-II, Ch. VIII], Graves [1, Ch. X], Coddington and Levinson [2, Ch. I], Birkhoff and Rota [1, Ch. V], Hartman [7, Chs. III, IV, V], and Hille [4, Chs. 2, 3, 6, Appendix D].

The basic concept of the integral inequality of Theorem 2.1 is quite old; in particular, it is essentially present in Peano [1]. The specific case of Theorem 2.1 in which $\phi(t)$ and $\kappa(t)$ are constants was established by Gronwall [1], and frequently an inequality of the general character of this theorem is called "Gronwall's inequality," or a "Gronwall-type inequality." The more general case in which $\phi(t)$ and $\psi(t)$ are merely assumed to be Lebesgue integrable was evidently explicitly presented first in Reid [1, p. 296]. This general result has been re-discovered from time to time, and frequently its introduction has been erroneously credited to Bellman [1]. Various extensions of this result have been obtained; in particular, the reader is referred to Giuliano [3].

The definition of an ε-approximate solution in Section 3 is at variance with that frequently used, (e.g., see Coddington and Levinson [2, Ch. 1]), in that this concept is presented here in integral form. Also, it is felt that the presentation of results under hypotheses of the form appearing in Theorems 3.1, 3.2 has an advantage over the presentation of similar results under the more restrictive condition of boundedness, since the more general condition permits automatic inclusion of nonlocal results for linear differential equations.

The approximating functions appearing in the first proof of conclusion (i) of Theorem 3.2 are known as "Cauchy polygonal vector functions," since for the treatment of a scalar equation such functions were used by Cauchy in the third decade of the nineteenth century. Refinements of the method were made by Lipschitz [1], and existence theorems based on this process and the Ascoli convergence theorem date from Peano [1]. The second proof of Conclusion (i) of Theorem 3.2 employs a method introduced by Tonelli [1] in the proof of the existence of solutions for functional equations of Volterra type. For references to existence theorems of the type given in Theorem 3.5, see Kamke [1, Th. 2, p. 75 and Th. 2, p. 135], Graves [1, Th. 5, p. 159], and Hartman [7, Th. 3.1, p. 12].

For a real, scalar, continuous function $g(t, u)$ the local existence of a solution of the differential equation (4.1), as well as the concepts of maximal and minimal solutions for such an equation, dates from Peano [1; 3]; in this connection, the reader is also referred to Perron [1]. For associated results, extending to differential systems the concept of an extremal solution through the notion of a solution curve that lies wholly on the boundary of the set of points belonging to the loci of the curves comprising the aggregate of solutions $\mathcal{F}(\tau, \eta)$ passing through an initial point (τ, η), see Hukuhara [2] and Kamke [2].

The local result of Theorem 5.2 is frequently referred to as the "Cauchy-Lipschitz theorem." The special case of Theorem 5.3 presented by a real scalar differential equation in which the function $f(t, y)$ is monotone non-decreasing in y is due to Peano [3]; for the general case of real vector differential equations, and under hypotheses of the "Carathéodory type" as considered in Chapter II, see McShane [1; 2, Chapter IX].

The fact that uniqueness of solution through an initial point implies a continuity property of the sort established in Theorem 5.5 was first proved by Carathéodory [1, Ch. XI, Th. 5], again under conditions on the involved functions of the generality to be considered in Chapter II.

The presented examples of differential equations which exhibit non-uniqueness are most elementary in character. It is to be remarked that examples have been given of continuous $f(t, y)$ which are such that the condition of local uniqueness holds at no point (τ, η) of the domain of existence of $f(t, y)$. The first such example was given by Lavrientieff in 1925; a greatly simplified example of this phenomenon has been given by Hartman [6; 7, Ch. II, Sec. 5].

The contraction mapping theorem, as given in Theorem 7.1, is the most elementary "fixed point theorem" and historically dates from the dissertation (1922) of S. Banach.

The concept of a fundamental set of solutions of a linear differential equation is due to Lagrange in about 1765, while the specific term "fundamental set of solutions" was introduced by Fuchs in 1866. The formula (9.6) is associated with the names of Abel, Liouville, and Jacobi. The analogous formula for a scalar second order equation was given by Abel in about 1826, and by Liouville in about 1838; Theorem 9.2 of the text was established by Jacobi in 1845. The method of variation of constants, or variation of parameters, is essentially due to Lagrange.

Paper [2] of Peano is noteworthy in that the treatment of first-order linear differential systems presented therein is entirely by "multipartite number" or vector methods. In particular, (9.4), (9.5), and (9.7) are all present in this paper in notation entirely equivalent to the vector and matrix notation of the present text. In this connection attention is directed also to papers [1] and [2] of Bliss, both for the historical fact that they were relatively early papers on differential equations to be written in vector notation and for the modern character of their mathematical contents. In particular, in [2] it is established that under suitable continuity and Lipschitz conditions for $f(t, y)$ the solutions $y = \phi(t; \tau, \eta \,|\, f)$ of a real vector differential equation (3.1) are continuous and have differentials as functions of (t, τ, η, f) and that the differentials which involve the variations of the functions f are in a sense analogous to those introduced by Fréchet and Volterra in the theory of functions of curves.

The discussion in Section 10 of the differentiability properties of solutions follows closely the treatments of Bliss [5, Appendix] and Graves [1, Ch. IX]. This section presents a rigorous and complete derivation of those properties that are needed for the construction of a field of extremals for a problem of the calculus of variations and those properties of solutions that are requisite for a rigorous treatment of partial differential equations of the first order by the method of characteristic strips.

For more elaborate treatment of differential equations in the complex plane than the brief introduction in Section 11 the reader is referred to Ince

[1, Part II], Sansone [1-I, Ch. III], Hartman [7, Ch. IV-Appendix], and Hille [4; Ch. 2-Sec. 2.5, Appendix B, Ch. VI].

RELATED REFERENCES AND COMMENTS FOR SPECIFIC PROBLEMS

I.3:4 Kamke [1, Ch. II-Sec. 10].

I.4:4,5 Kamke [1, Ch. II-Sec. 10].

I.5:6 Giuliano [1; 2].

I.5:7 Nagumo [1].

I.5:11 The result of this problem is an extension due to Kamke [2] of a local result of H. Kneser [1], in which the suggested pattern of proof is that employed by Müller [2] to establish the Kneser result; see also Hukuhara [1].

I.5:12 Kamke [1, Ch. IV-Sec. 16]; the result is an extension of earlier results of Perron [2]. The result of this problem includes the criterion of Osgood [1] for $\omega(t, r) = \lambda(t)\,\Phi(r)$, where $\lambda(t)$ is real-valued, non-negative and continuous on $(0, q]$ with $\lim_{\varepsilon\to 0}\int_\varepsilon^q \lambda(t)\,dt < +\infty$, while on $0 \le r < \infty$ the real-valued function $\Phi(r)$ is continuous, positive for $r > 0$, with $\Phi(0) = 0$ and

$$\lim_{\varepsilon\to 0}\int_\varepsilon^{r_0} [\Phi(r)]^{-1}\,dr = +\infty.$$

I.5:14,15 Brauer and S. Sternberg [1].

I.5:17 Wintner [7]; Brauer and S. Sternberg [1].

I.5:18 Wintner [1].

I.6:4 The example of this problem is a slight modification of an example due to Müller [1].

I.6:5 Coddington and Levinson [1; 2, Ch. 2-Sec. 2]. It has been established by Olech that the result of this problem remains true when the assumption of monotoneity is dropped; see Hartman [7, Ch. III-Th. 9.1].

I.9:16 Ince [1, Ch. I], Hille [4, Appendix C].

I.9:17 Bliss [5-Sec. 79]. The method of the second hint was suggested to the author by C. D. Ahlbrandt.

I.10:2 Brauer [1, Lemma 3]; this result is attributed by Brauer to V. M. Alekseev.

I.10:3 Sansone [1-II, Ch. VIII-Sec. 7].

I.11:3,4 For a more extensive discussion of the Schwarzian derivative, and some of its applications, the reader is referred to Hille [4, Appendix D]; see also, Nehari [1] and Hille [2].

II

Extension of the Concept of a Solution of a Differential Equation

1. INTRODUCTION

In the existence theorems of the preceding chapter it has been emphasized that the stated hypotheses on f are such that if τ is a value on an interval I then the initial value problem

$$y' = f(t, y), \tag{1.1}$$

$$y(\tau) = \eta, \tag{1.2}$$

has a solution $y(t)$ on I if and only if $y(t)$ is a solution of the integral equation

$$y(t) = \eta + \int_{\tau}^{t} f(s, y(s))\, ds, \qquad t \in I. \tag{1.3}$$

Indeed, the concept of an approximate solution, and the details of proofs of the existence theorems, dealt directly with the integral equation (1.3). Whereas in the preceding discussion the hypotheses were such that the integral in this equation existed as a Riemann integral, we shall now consider the situation when the integral is taken in the sense of Lebesgue.

If $y(t)$ is such that $f(t, y(t))$ is Lebesgue integrable on arbitrary compact subintervals of I containing the value τ, and (1.3) holds, then $y(\tau) = \eta$ and $y(t)$ is a.c., (absolutely continuous), on each compact subinterval of I containing τ, so that a.e., (almost everywhere, i.e., except for a set of Lebesgue measure zero), on I the derivative $y'(t)$ exists and satisfies the differential equation (1.1). Conversely, if $y(t)$ is a vector function which satisfies $y(\tau) = \eta$ and is a.c., on compact subintervals of I containing τ, while the differential equation (1.1) holds a.e. on I, then $y(t)$ is a solution of the integral equation (1.3). Thus when the integral of (1.3) is considered in the sense of Lebesgue the concept of a solution of (1.1) is extended to mean a *vector function $y(t)$ which is a.c. on arbitrary compact subintervals of its interval I of existence, and such that the derivative vector $y'(t)$ satisfies* (1.1) *a.e. on I.* The first person to consider differential equations with this general concept was C. Carathéodory, and frequently such solutions are referred to as "solutions in the sense of Carathéodory."

2. PRELIMINARY DEFINITIONS AND RESULTS

For a given set J on the real line, the symbols $\mathfrak{L}_{nr}(J)$, $\mathfrak{L}_{nr}{}^{p}(J)$, $\mathfrak{L}_{nr}{}^{\infty}(J)$ will be used to denote the class of $n \times r$ matrix functions $M(t) = [M_{\alpha\beta}(t)]$, $(\alpha = 1, \ldots, n;\ \beta = 1, \ldots, r)$, which on J are respectively (Lebesgue) integrable, (Lebesgue) measurable and $|M_{\alpha\beta}(t)|^p$ integrable, $(\alpha = 1, \ldots, n;$ $\beta = 1, \ldots, r)$, measurable and essentially bounded, i.e., there exists a finite constant k such that each of the sets $\{t \mid |M_{\alpha\beta}(t)| > k\}$ has measure zero. For brevity, $\mathfrak{L}_n(J)$, $\mathfrak{L}_n{}^{p}(J)$, $\mathfrak{L}_n{}^{\infty}(J)$ will be used for $\mathfrak{L}_{n1}(J)$, $\mathfrak{L}_{n1}{}^{p}(J)$, $\mathfrak{L}_{n1}{}^{\infty}(J)$, respectively, and $\mathfrak{L}(J)$, $\mathfrak{L}^{p}(J)$, $\mathfrak{L}^{\infty}(J)$ for the sets designated by $n = 1$, $r = 1$; also, when there is no ambiguity as to the set J under consideration, the symbol J will be omitted. Moreover, if J is an interval $[a, b]$ or $[a, b)$ the symbols $\mathfrak{L}_{nr}([a, b])$, $\mathfrak{L}_{nr}([a, b))$ will be abbreviated to $\mathfrak{L}_{nr}[a, b]$, $\mathfrak{L}_{nr}[a, b)$, with similar abbreviations for other types of intervals and other integrable classes. If T is a subset of the real line then $\phi(t; T)$ will be used to denote the characteristic function of T; that is, $\phi(t; T) = 1$ for $t \in T$, $\phi(t; T) = 0$ for $t \notin T$.

If the matrix function $F(t, y) = (F_{\alpha\beta}(t, y_1, \ldots, y_n))$, $(\alpha = 1, \ldots, n;$ $\beta = 1, \ldots, r)$ is defined for (t, y) on a set $Q = I \times S$, where I is an interval on the real line and S is a subset of $\mathbf{C}_n$, then F is said to be *of class* $\Gamma(Q)$ if:

(i) *for $y \in S$, F is measurable in t on I;*

(ii) *for $t \in I$, F is continuous in y on S.*

Also, F is said to be *of class* $\Gamma L(Q)$ if F is of class $\Gamma(Q)$, and there exist non-negative real-valued functions $\kappa(t)$, $\kappa_1(t)$ which are integrable on arbitrary compact subsets of I and such that

$$\nu[F(t, y)] \leq |y|\, \kappa(t) + \kappa_1(t), \quad \text{for} \quad (t, y) \in Q = I \times S. \tag{2.1}$$

In particular, for a vector function $f(t, y) = (f_\alpha(t, y))$, this latter condition is

$$|f(t, y)| \leq |y|\, \kappa(t) + \kappa_1(t), \quad \text{for} \quad (t, y) \in Q = I \times S. \tag{2.1'}$$

The following result is basic for the results of this section.

***Lemma* 2.1.** *Suppose that* $\Delta = [a, b] \times \mathbf{C}_n$ *and* $f(t, y) = (f_\alpha(t, y_1, \ldots, y_n))$ *belongs to* $\Gamma(\Delta)$. *If* $y(t) \in \mathfrak{L}_n^\infty[a, b]$, *then* $f(t, y(t))$ *is measurable on* $[a, b]$; *moreover, if* $f \in \Gamma L(\Delta)$ *and* $y(t) \in \mathfrak{L}_n^\infty[a, b]$ *then* $f(t, y(t)) \in \mathfrak{L}_n[a, b]$.

Consider first the case where $y(t)$ is a *step-function*, or *simple function* on $[a, b]$; that is, there exist a finite number of mutually disjoint subintervals I_j, $(j = 1, \ldots, m)$, of $[a, b]$ such that $\bigcup_{j=1}^m I_j = [a, b]$, and constants v_j such that $y(t) = v_j$ for $t \in I_j$. Since the condition that $f \in \Gamma(\Delta)$ implies that each $f(t, v_j)$ is measurable on $[a, b]$, and the characteristic functions $\phi(t; I_j)$ are measurable, the measurability of $f(t, y(t))$ follows from the relation $f(t, y(t)) = \sum_{j=1}^m f(t, v_j)\, \phi(t; I_j)$. Secondly, consider the case of $y(t)$ continuous on $[a, b]$, and let $I_{j,m}$, $(j = 1, \ldots, m)$, be mutually disjoint subintervals of $[a, b]$, with each of length $(b - a)/m$ and $\bigcup_{j=1}^m I_{j,m} = [a, b]$, and denote by $t_{j,m}$ the midpoint of $I_{j,m}$. Then $y_m(t) = \sum_{j=1}^m y(t_{j,m})\, \phi(t; I_{j,m})$, $m = 1, 2, \ldots$, is a sequence of step functions such that $\{y_m(t)\} \to y(t)$ for each $t \in [a, b]$. Since each $f(t, y_m(t))$ is measurable, and condition (ii) of the definition of $\Gamma(\Delta)$ implies that $\{f(t, y_m(t))\} \to f(t, y(t))$ for each $t \in [a, b]$, it then follows that $f(t, y(t))$ is measurable on $[a, b]$. Now suppose that $y(t) \in \mathfrak{L}_n^\infty[a, b]$, and let k be a constant such that $|y(t)| \leq k$ a.e. on $[a, b]$. If $y_0(t) = y(t)$ for $t \in \{s \mid s \in [a, b], |y(s)| \leq k\}$, and $y_0(t) = 0$ elsewhere on the real line, then for $h > 0$ and $t \in (-\infty, \infty)$ the function

$$y_h(t) = \frac{1}{2h} \int_{t-h}^{t+h} y_0(s)\, ds = \frac{1}{2h} \int_{-h}^{h} y_0(t + \sigma)\, d\sigma,$$

is continuous, $|y_h(t)| \leq k$, and $\lim_{h\to 0} y_h(t) = y_0(t)$ a.e. on $(-\infty, \infty)$. Then $\lim_{h\to 0} y_h(t) = y(t)$ a.e. on $[a, b]$, and since for $h > 0$ the vector function $f(t, y_h(t))$ is measurable on $[a, b]$ it follows that $f(t, y(t))$ is measurable on this interval.

Finally, if $f \in \Gamma L(\Delta)$ and $y(t) \in \mathfrak{L}_n^\infty[a, b]$, then for k as above we have

$$|f(t, y_h(t))| \leq k\kappa(t) + \kappa_1(t), \quad \text{for} \quad t \in [a, b], \qquad h > 0,$$

and as $k\kappa(t) + \kappa_1(t)$ is integrable on $[a, b]$ it follows by the dominated convergence theorem of Lebesgue that $f(t, y(t))$ is also integrable on this interval, moreover, $\int_T f(t, y(t))\, dt = \lim_{h\to 0} \int_T f(t, y_h(t))\, dt$ for T an arbitrary measurable subset of $[a, b]$.

3. EXISTENCE THEOREMS

We shall proceed to present results that correspond to results established in Chapter I under stronger hypotheses. Since in each case the proofs may be given in steps which closely parallel those of earlier results, only the pertinent differences in details will be mentioned. In particular, *we shall refer to relations that appeared in earlier discussion, understanding that any integrals which appear are now to be taken in the sense of Lebesgue, and that "solution" is to be understood in the extended sense defined above.*

In this chapter, as subsequently in this work, references to numbered theorems and formulas in a chapter other than the one in which the statement occurs include an adjoined Roman numeral designating the chapter of reference, while references to similar items in the current chapter do not contain the designating Roman numeral. For example, in this chapter a reference to Theorem 3.1 or formula (1.1) of Chapter I would be made by referring to Theorem I.3.1 or formula (I.1.1).

***Lemma* 3.1.** *Suppose that $\phi(t)$, $\kappa(t)$, $\rho(t)$ are real-valued functions with $\phi(t) \in \mathfrak{L}^\infty[a, b]$, $\rho(t) \in \mathfrak{L}^\infty[a, b]$, $\kappa(t) \in \mathfrak{L}[a, b]$ and $\kappa(t) \geq 0$ on this interval. If* (I.2.3) *holds for some $\tau \in [a, b]$, then conclusions* (I.2.4), (I.2.5) *are valid; if* (I.2.3′) *holds for some $\tau \in [a, b]$, then conclusions* (I.2.4′), (I.2.5′) *are valid.*

Theorem 3.1. *For $\Delta = [a, b] \times \mathbf{C}_n$, suppose that the vector function $f(t, y)$ belongs to $\Gamma(\Delta)$ and $(\tau, \eta) \in \Delta$. If ε_m, $(m = 1, 2, \ldots)$, is a sequence of positive constants converging to zero, and $y^{(m)}(t)$ is a corresponding sequence of approximate solutions satisfying* (I.3.3) *with $\varepsilon = \varepsilon_m$, and for which there are non-negative real-valued functions $\kappa(t)$, $\kappa_1(t)$ in $\mathfrak{L}[a, b]$ such that*

(3.1)
$$|f(t, y^{(m)}(t))| \leq |y^{(m)}(t)|\, \kappa(t) + \kappa_1(t), \qquad t \in [a, b], \qquad m = 1, 2, \ldots,$$

then there is subsequence $y^{(m_k)}(t)$, $(m_1 < m_2 < \cdots)$, which converges uniformly on $[a, b]$ to a solution $y(t)$ of (1.1), (1.2).

If $r^{(m)}(t)$ is defined by (I.3.5), then as in the proof of Theorem I.3.1 there exists a non-negative constant γ such that $|\eta + r^{(m)}(t)| \leq |\eta| + \gamma$ on $[a, b]$. In view of (3.1) it follows that $\rho(t) = |y^{(m)}(t)|$ satisfies (I.2.3) and (I.2.3′) with $\phi(t) = |\eta| + \gamma + |\int_\tau^t \kappa_1(s)\, ds|$ and $\kappa(t)$. As

$$|\phi(t)| \leq |\eta| + \gamma + \int_a^b \kappa_1(s)\, ds,$$

it follows from Lemma 3.1 that

$$(3.2) \quad |y^{(m)}(t)| \leq \nu, \quad \textit{for} \quad \nu = \left[|\eta| + \gamma + \int_a^b \kappa_1(s)\, ds\right] \exp\left\{\int_a^b \kappa(s)\, ds\right\},$$

and hence

$$(3.3) \quad |f(t, y^{(m)}(t))| \leq \mu(t), \quad \textit{with} \quad \mu(t) = \nu\kappa(t) + \kappa_1(t), \quad \text{for} \quad t \in [a, b].$$

Consequently the vector functions $z^{(m)}(t) = \eta + \int_\tau^t f(s, y^{(m)}(s))\,ds$ are such that

$$z^{(m)}(\tau) = \eta, \; |z^{(m)}(t_2) - z^{(m)}(t_1)| \leq \left| \int_{t_1}^{t_2} \mu(s)\,ds \right| = |\lambda(t_2) - \lambda(t_1)|,$$

for arbitrary t_1, t_2 on $[a, b]$, where $\lambda(t)$ is the continuous function $\lambda(t) = \int_a^t \mu(s)\,ds$, $t \in [a, b]$. By the uniform continuity of $\lambda(t)$ on $[a, b]$ it then follows that the sequence $\{z^{(m)}(t)\}$ is equicontinuous on $[a, b]$ and $\{|z^{(m)}(t)|\}$ is bounded for $t \in [a, b]$. Therefore as in the proof of Theorem I.3.1, there is a subsequence $\{y^{(m_k)}(t)\}$ that converges uniformly on $[a, b]$ to a limit function $y(t)$, and in view of (3.3) the limit relation (I.3.7) now follows from the dominated convergence theorem of Lebesgue.

Corresponding to the proof of Theorem I.3.2 by the method of Tonelli, and now utilizing the result of the above theorem, we may establish the following result.

Theorem 3.2. *For* $\Delta = [a, b] \times \mathbf{C}_n$, *suppose that* $f(t, y) \in \Gamma L(\Delta)$. *If* $(\tau, \eta) \in \Delta$, *then:* (i) *for each* $\varepsilon > 0$ *there is a corresponding approximate solution of* (1.1) *satisfying* (I.3.3)*;* (ii) *there is a solution* $y(t)$ *of* (1.1) *on* $[a, b]$ *satisfying* (1.2).

As in the proof of Theorem I.3.2, conclusion (ii) is a ready consequence of conclusion (i), now in view of the result of Theorem 3.1, and with simple modifications the second proof of (i) of Theorem I.3.2 yields the result of (i) of the present theorem. Indeed, if $y(t) = y(t, \delta)$ is defined as in (I.3.13), and $\kappa(t)$, $\kappa_1(t)$ are the functions of (2.1), it then follows that $\rho(t) = |y(t, \delta)|$ satisfies (I.2.3) and (I.2.3′) with $\phi(t) = |\eta| + \left| \int_\tau^t \kappa_1(s)\,ds \right|$ and $\kappa(t)$. Lemma 3.1 then implies the inequality

$$|y(t, \delta)| \leq \nu_0, \quad \textit{with} \quad \nu_0 = \left[|\eta| + \int_a^b \kappa_1(s)\,ds \right] \exp\left\{ \int_a^b \kappa(s)\,ds \right\},$$

and, therefore,

$$|f(t, y(t, \delta))| \leq \mu_0(t), \quad \textit{with} \quad \mu_0(t) = \nu_0 \kappa(t) + \kappa_1(t), \quad \text{for} \quad t \in [a, b].$$

Consequently,

$$\begin{aligned} |y(t, \delta) - \eta - \int_\tau^t f(s, y(s, \delta))\,ds| &= \left| \int_t^{\zeta(t)} f(s, y(s, \delta))\,ds \right| \\ &\leq \left| \int_t^{\zeta(t)} \mu_0(s)\,ds \right| = |\lambda_0(\zeta(t)) - \lambda_0(t)|, \end{aligned}$$

where $\lambda_0(t)$ is the continuous function $\lambda_0(t) = \int_a^t \mu_0(s)\,ds$. As $|\zeta(t) - t| \leq \delta$ for $t \in [a, b]$, it follows from the uniform continuity of $\lambda_0(t)$ on $[a, b]$ that for $\varepsilon > 0$ there is a $\delta_\varepsilon > 0$ such that if $0 < \delta < \delta_\varepsilon$ then $y(t, \delta)$ is an ε-approximate solution satisfying (I.3.3).

Corresponding to Theorem I.3.3., one now has the following result.

Theorem 3.3. *If $\Delta = [a, b] \times \mathbf{C}_n, f(t, y) \in \Gamma L(\Delta)$, and $y = \phi(t)$ is a solution of* (1.1) *on a nondegenerate subinterval I of $[a, b]$, then there exists on $[a, b]$ a solution $y(t)$ of this equation such that $y(t) = \phi(t)$ for $t \in I$.*

Also, in view of the result of Problem 1 below, by a method of proof analogous to that of Theorem I.3.4 the following theorem may be established. This theorem also has as corollary a result corresponding to that of the Corollary to Theorem I.3.4, which for brevity will not be stated explicitly here.

Theorem 3.4. *If the vector function $f(t, y)$ is of class $\Gamma(\square)$ on $\square = \{(t, y) \mid |t - \tau| \leq \varepsilon, |y - \eta| \leq \delta\}$, and $\kappa_1(t)$ is integrable on $[\tau - \varepsilon, \tau + \varepsilon]$ and such that $|f(t, y)| \leq \kappa_1(t)$ for $(t, y) \in \square$, then for*

$$\varepsilon_1 = \sup\left\{\xi \mid 0 < \xi < \varepsilon, \int_\tau^{\tau+\xi} \kappa_1(s)\, ds < \delta, \int_{\tau-\xi}^{\tau} \kappa_1(s)\, ds < \delta\right\},$$

there is a solution $y(t)$ of (1.1) *on $[\tau - \varepsilon_1, \tau + \varepsilon_1]$ satisfying $y(\tau) = \eta$ and $|y(t) - \eta| < \delta$ on $(\tau - \varepsilon_1, \tau + \varepsilon_1)$. If $y = \phi(t)$ is a solution of* (1.1) *on a nondegenerate subinterval I of $[\tau - \varepsilon_1, \tau + \varepsilon_1]$ containing τ with $\phi(\tau) = \eta$, then there exists on $[\tau - \varepsilon_1, \tau + \varepsilon_1]$ a solution $y(t)$ of* (1.1) *such that $y(t) = \phi(t)$ on I, and for any such extension the inequality $|y(t) - \eta| < \delta$ holds for $t \in (\tau - \varepsilon_1, \tau + \varepsilon_1)$.*

A matrix function $F(t, y)$ will be said to be *locally of class ΓL on an open region* $\mathfrak{R}$ of (t, y)-space if for each (τ, η) in $\mathfrak{R}$ there exists a

$$\square = \{(t, y) \mid |t - \tau| \leq \varepsilon, \quad |y - \eta| \leq \delta\}$$

with $\square \subset \mathfrak{R}$, and a non-negative real-valued function $\kappa_\square(t)$ integrable on $[\tau - \varepsilon, \tau + \varepsilon]$, such that $F(t, y) \in \Gamma L(\square)$ with $\nu[F(t, y)] \leq \kappa_\square(t)$ for $(t, y) \in \square$. In particular, for a vector function $f(t, y)$ this latter condition is $|f(t, y)| \leq \kappa_\square(t)$ for $(t, y) \in \square$.

If the vector function $f(t, y)$ is locally of class ΓL on an open region $\mathfrak{R}$ of (t, y)-space, then for a solution of (1.1) one has the extension of this solution to a maximal interval of existence, precisely as in Theorem I.3.5. Moreover, if the real-valued scalar function $g(t, u)$ is locally of class ΓL on an open region $\mathfrak{R}$ of real (t, u)-space, then for the differential equation $u' = g(t, u)$ the existence and properties of maximal and minimal solutions may be established by the same methods as those employed in Section I.4 for the case of continuous $g(t, u)$. For these and other extensions of the results of Chapter I to differential equations with vector functions $f(t, y)$ which are locally of class ΓL in an open region of (t, y)-space, the basic modifications in details or argument are of the following types:

(a) Whereas continuity of $f(t, y)$ on a set $\square = \{(t, y) \mid |t - \tau| \leq \varepsilon, |y - \eta| \leq \delta\}$ in $\mathfrak{R}$ implies the boundedness of $|f(t, y)|$ on this set, and the uniform Lipschitzian character of integrals of the form $I(t) = \int_\tau^t f(s, y(s))\, ds$, $t \in [\tau - \varepsilon, \tau + \varepsilon]$, for $y(t)$ a continuous vector function with graph in $\square$, domination of $|f(t, y)|$ on $\square$ by an integrable function $\kappa_\square(t)$ implies for such integrals $I(t)$ the equi-continuity condition $|I(t_2) - I(t_1)| \leq |\lambda(t_2) - \lambda(t_1)|$, where $\lambda(t)$ is the absolutely continuous function $\lambda(t) = \int_\tau^t \kappa_\square(s)\, ds$.

(b) If $y(t)$, $t \in [a, b]$, is a continuous vector function with graph in $\mathfrak{R}$, and $\zeta > 0$ is such that the set $S_\zeta = \{(t, y) \mid t \in [a, b], |y - y(t)| \leq \zeta\}$ is in $\mathfrak{R}$, then continuity of $f(t, y)$ on $\mathfrak{R}$ implies boundedness of $|f(t, y)|$ on S_ζ, while in the case of $f(t, y)$ locally of class ΓL a simple application of the Heine-Borel Theorem assures the existence of a $\kappa_0(t)$ which is Lebesgue integrable on $[a, b]$, and such that $|f(t, y)| \leq \kappa_0(t)$ for $(t, y) \in S_\zeta$.

A vector function $f(t, y)$ defined on a set $\mathfrak{D}$ of (t, y)-space is said to be *of class* Lip$[\mathfrak{D}; \kappa(t)]$ *in* y if

$$|f(t, z) - f(t, y)| \leq \kappa(t)\, |z - y|, \quad \text{for} \quad (t, z) \in \mathfrak{D},\ (t, y) \in \mathfrak{D}.$$

Such a vector function is said to be *locally of class* Lip *in* y *on an open region* $\mathfrak{R}$ of (t, y)-space if for each (τ, η) in $\mathfrak{R}$ there is a corresponding $\square = \{(t, y) \mid |t - \tau| \leq \varepsilon, |y - \eta| \leq \delta\}$ with $\square \subset \mathfrak{R}$ and a non-negative real-valued function $\kappa_\square(t)$ integrable on $[\tau - \varepsilon, \tau + \varepsilon]$ such that $f(t, y) \in$ Lip$\{\square; \kappa_\square(t)\}$.

The following theorem combines results corresponding to those of Theorems I.5.1 and I.6.1. Also, with the aid of the result of Problem 2 below one establishes the result corresponding to that of Theorem I.5.2, where now ε_1 has the value appearing in the above Theorem 3.4.

Theorem 3.5. *If* $\Delta = [a, b] \times \mathbf{C}_n$, *and* $f(t, y)$ *is of class* Lip$[\Delta; \kappa(t)]$ *in* y *with* $\kappa(t) \in \mathfrak{L}[a, b]$, *and for fixed* $y \in \mathbf{C}_n$ *the vector function* $f(t, y) \in \mathfrak{L}_n[a, b]$, *then for* $(\tau, \eta) \in \Delta$ *there is a unique solution* $y(t)$ *of* (1.1) *on* $[a, b]$ *satisfying* (1.2). *Moreover, if* $y^{(0)}(t)$ *is an arbitrary continuous vector function on* $[a, b]$, *and the sequence* $\{y^{(j)}(t)\}$, $(j = 0, 1, \ldots)$, *is defined by the iterative process* (I.5.1), *then* $\{y^{(j)}(t)\}$ *converges uniformly on* $[a, b]$ *to the solution* $y(t)$ *of* (1.1), (1.2).

The hypotheses of this theorem clearly imply that $f(t, y) \in \Gamma L(\Delta)$, with $\kappa(t)$ the function in the condition Lip$[\Delta; \kappa(t)]$ and $\kappa_1(t) = |f(t, 0)|$, and the existence of a solution of (1.1), (1.2) on $[a, b]$ follows from Theorem 3.2; the uniqueness of this solution may be established in a fashion similar to that of Theorem I.5.1. The proof of the result of the theorem by successive approximations parallels that of Theorem I.6.1, where it is to be noted now that if λ_0, λ are constants such that $|y^{(0)}(t)| \leq \lambda_0$, $|y^{(1)}(t) - y^{(0)}(t)| \leq \lambda$ on $[a, b]$, then by induction one may establish the following inequalities for $t \in [a, b]$,

$$|y^{(m+1)}(t) - y^{(m)}(t)| \leq \lambda \left| \int_\tau^t \kappa(s)\, ds \right|^m \Big/ m!, \qquad (m = 0, 1, \ldots);$$

$$|y(t)| \leq \lambda_0 + \lambda \exp\left\{ \left| \int_\tau^t \kappa(s)\, ds \right| \right\};$$

$$\begin{aligned} |y(t) - y^{(m)}(t)| &\leq \lambda \left[\left| \int_\tau^t \kappa(s)\, ds \right|^m \Big/ m! \right] \exp\left\{ \left| \int_\tau^t \kappa(s)\, ds \right| \right\}, \\ &\leq \lambda \left[\left[\int_a^b \kappa(s)\, ds \right]^m \Big/ m! \right] \exp\left\{ \int_a^b \kappa(s)\, ds \right\}. \end{aligned}$$

Corresponding to Theorem I.5.5, one now has the following result.

Theorem 3.6. *Suppose that* $\Delta = [a, b] \times \mathbf{C}_n$, $f(t, y) \in \Gamma L(\Delta)$, *and that for a point* $(\tau^0, \eta^0) \in \Delta$ *there is a unique solution* $y = \phi(t)$ *of* (1.1) *which is defined on* $[a, b]$ *and such that* $\phi(\tau^0) = \eta^0$. *Then for arbitrary* $\zeta > 0$ *there are positive constants* ε'_ζ, ε''_ζ *such that if*

(3.4) $$\tau \in [a, b], \qquad |\tau - \tau^0| < \varepsilon'_\zeta, \qquad |\eta - \eta^0| < \varepsilon'_\zeta,$$

and $y(t)$ *is a continuous vector function satisfying*

$$|y(t) - \eta - \int_\tau^t f(s, y(s))\, ds| < \varepsilon''_\zeta \quad \textit{for} \quad t \in [a, b],$$

then $|y(t) - \phi(t)| < \zeta$ *on* $[a, b]$.

With the aid of Theorem 3.6, and the result of Problem 3 below, an argument which parallels that used to establish conclusions (ii) and (iii) of Theorem I.5.6 now yields the following result.

Theorem 3.7. *Suppose that* $f(t, y)$ *is locally of class* ΓL *in an open region* $\mathfrak{R}$ *of* (t, y)*-space, and that for a given point* (τ^0, η^0) *of* $\mathfrak{R}$ *there is a nondegenerate interval* $[a, b]$ *with* $\tau^0 \in (a, b)$ *and a solution* $y = \phi(t)$ *of* (1.1) *that is the unique solution of this differential equation on* $[a, b]$ *with graph in* $\mathfrak{R}$, *and passing through the initial point* (τ^0, η^0). *If* ζ *is a positive constant such that the set* $S_\zeta = \{(t, y) \mid t \in [a, b], |y - \phi(t)| \leq \zeta\}$ *is in* $\mathfrak{R}$, *then:*

(a) *there exist positive constants* ε'_ζ, δ''_ζ *such that if* $0 < \delta < \delta''_\zeta$ *and* (τ, η) *satisfies* (3.4), *then the function* $y = y(t, \delta)$ *of* (I.3.13) *satisfies*

$$|y(t, \delta) - \phi(t)| < \zeta \quad \textit{on} \quad [a, b];$$

(b) *there is a* $\delta_\zeta > 0$ *such that if*

$$|\tau - \tau^0| < \delta_\zeta, \qquad |\eta - \eta^0| < \delta_\zeta,$$

and $y = y(t)$ *is a solution of* (1.1) *passing through* (τ, η), *then the interval of definition of* $y(t)$ *may be extended to* $[a, b]$ *and for any such extension* $|y(t) - \phi(t)| < \zeta$ *for* $t \in [a, b]$.

With the aid of the result of the above theorem one may then proceed as in Section I.4 to establish the following analogue of Theorem I.5.7.

Theorem 3.8. *If* $f(t, y)$ *is locally of class* ΓL *in an open region* $\mathfrak{R}$ *of* (t, y)-*space, and the solutions of* (1.1) *are locally unique in* $\mathfrak{R}$, *then the solution* $y = \phi(t; \tau, \eta)$ *of this differential equation passing through the initial point* (τ, η), *and with maximal interval of existence*

$$a(\tau, \eta) < t < b(\tau, \eta) \tag{3.5}$$

possesses the properties (i), (ii), (iii) *of Theorem* I.5.7.

If $f(t, y; c) \equiv (f_\alpha(t, y_1, \ldots, y_n; c_1, \ldots, c_k))$, $(\alpha = 1, \ldots, n)$, is a vector function involving the complex parameter vector $c = (c_\nu)$, $(\nu = 1, \ldots, k)$, then the phrase "f is locally of class ΓL on an open region $\mathfrak{D}$ of (t, y, c)-space" is to be interpreted as meaning that the $(n + k)$-dimensional vector function $g(t, w)$ in $(t, w) = (t, w_1, \ldots, w_{n+k})$, with $g_\alpha(t, w) = f_\alpha(t, w_1, \ldots, w_n; w_{n+1}, \ldots, w_{n+k})$, $(\alpha = 1, \ldots, n)$, and $g_{n+\nu}(t, w) \equiv 0$, $(\nu = 1, \ldots, k)$, is locally of class ΓL on $\mathfrak{D}$ according to the definition following the statement of Theorem 3.4. That is, in $\mathfrak{D}$ the vector function $f(t, y; c)$ is measurable in t for fixed (y, c), continuous in (y, c) for fixed t, and locally $|f(t, y; c)|$ is dominated by an integrable function of t. Corresponding to Theorem I.5.8, one now considers a vector differential equation

$$y' = f(t, y; c), \tag{3.6}$$

involving a vector function f satisfying the following condition.

($\mathfrak{H}'$) $f(t, y; c)$ *is locally of class* ΓL *on an open region* $\mathfrak{D}$ *of* (t, y, c)-*space; moreover, if* $\mathfrak{D}_c$ *denotes the section of* $\mathfrak{D}$ *consisting of all* (t, y) *such that* (t, y, c) *is in* $\mathfrak{D}$, *then for each* c *such that* $\mathfrak{D}_c$ *is nonvacuous the solutions of* (3.6) *are locally unique in* $\mathfrak{D}_c$.

The following theorem is deducible from the above Theorem 3.8 in the same fashion that Theorem I.5.8 was deduced from Theorem I.5.7.

Theorem 3.9. *If the vector function* $f(t, y; c)$ *satisfies the above hypothesis* ($\mathfrak{H}'$), *then the solution* $y = \phi(t; \tau, \eta, c)$ *passing through the initial point* (τ, η) in $\mathfrak{D}_c$, *and with maximal interval of existence*

$$a(\tau, \eta, c) < t < b(\tau, \eta, c), \tag{3.7}$$

possesses the properties (i), (ii), (iii) *of Theorem* I.5.8.

Clearly the above results on first order vector differential equations are directly applicable to higher order scalar and vector differential equations by means of the transformations introduced in Section I.8, and no detailed account of these results will be given here.

If the $n \times n$ matrix function $A(t) = [A_{\alpha\beta}(t)]$ and the n-dimensional vector function $g(t) = (g_\alpha(t))$ are integrable on $[a, b]$, then on $\Delta = [a, b] \times \mathbf{C}_n$ the vector function $f(t, y) = A(t)y + g(t)$ is of class $\Gamma L(\Delta)$, satisfying inequality (2.1) with $\kappa_1(t) = |g(t)|$, $\kappa(t) = [\sum_{\alpha,\beta} |A_{\alpha\beta}(t)|^2]^{1/2}$, and of class $\text{Lip}[\Delta, \kappa(t)]$ in y with this same $\kappa(t)$. Indeed, a better dominating function $\kappa(t)$ is given by $\kappa(t) = \nu[A(t)]$. The reader is referred to Appendix F for a discussion of pertinent relations, and, in particular, to Problem 2 of F.1 for the integrability of $\nu[A(t)]$.

Consequently under the hypothesis that $A(t) \in \mathfrak{L}_{nn}[a, b]$ and $g(t) \in \mathfrak{L}_n[a, b]$, the linear vector differential equation

$$y' = A(t)y + g(t) \tag{3.8}$$

has a unique solution passing through a point (τ, η) of Δ, and this solution exists throughout the interval $[a, b]$. Moreover, the results of Theorems I.9.1, I.9.2, I.9.3 hold for such linear vector differential equations without any change in wording. Also, these results for linear systems of the first order imply results for higher order scalar equations of the form (I.9.9) whenever the coefficient functions $p_0(t), \ldots, p_n(t)$ are such that $p_n(t) \neq 0$ a.e. on $[a, b]$ and $r(t)/p_n(t)$, $p_0(t)/p_n(t), \ldots, p_{n-1}(t)/p_n(t)$ are integrable on this interval. Corresponding results for higher order linear vector differential equations are immediate, by the same methods as presented at the end of Section I.8.

Problems II.3

1.[ss] Suppose that the vector function $f(t, y)$ is of class $\Gamma(\square)$, and $|f(t, y)| \leq \kappa_1(t)$ for $(t, y) \in \square$, where $\square = \{(t, y) \mid |t - \tau| \leq \varepsilon, |y - \eta| \leq \delta\}$, and $\kappa_1(t)$ is integrable on $[\tau - \varepsilon, \tau + \varepsilon]$. If $h(t, y)$ is defined on $\Delta = [\tau - \varepsilon, \tau + \varepsilon] \times \mathbf{C}_n$ as $h(t, y) = f(t, y)$ on $\square$,

$$h(t, y) = f\left(t, \eta + \frac{\delta(y - \eta)}{|y - \eta|}\right)$$

if $|t - \tau| \leq \varepsilon$, $|y - \eta| > \delta$, prove that $h(t, y) \in \Gamma L(\Delta)$.

2.[ss] If $\square$ and Δ are as in Problem 1, and the vector function $f(t, y)$ is of class $\text{Lip}[\square; \kappa_1(t)]$ in y, prove that the vector function $h(t, y)$ defined in Problem 1 is of class $\text{Lip}[\Delta; \kappa_1(t)]$ in y.

3.[ss] Suppose that the vector function $f(t, y)$ is locally of class ΓL in an open region $\mathfrak{R}$ of (t, y)-space, and that $\phi(t)$, $t \in [a, b]$, is a continuous vector function with graph in $\mathfrak{R}$. Prove that there exists a $\zeta > 0$ such that:

(i) the set $S_\zeta = \{(t, y) \mid t \in [a, b],\ |y - \phi(t)| \leq \zeta\}$ lies in $\mathfrak{R}$, and there exists a real-valued function $k(t)$ which is integrable on $[a, b]$ and such that $|f(t, y)| \leq k(t)$ for $(t, y) \in S_\zeta$;

(ii) the function $h(t, y)$ defined on $\Delta = [a, b] \times \mathbf{C}_n$ as $h(t, y) = f(t, y)$ on S_ζ, $h(t, y) = f(t, \phi(t) + \zeta[y - \phi(t)]/|y - \phi(t)|)$ if $t \in [a, b]$, $|y - \phi(t)| > \zeta$, is of class $\Gamma(\Delta)$ and $|h(t, y)| \leq k(t)$ for $(t, y) \in \Delta$.

4. If $\Delta_1 = [\tau, b] \times \mathbf{C}_n$ and $f(t, y)$ is of class $\Gamma(\Delta_1)$, show that the result corresponding to that of Theorem I.5.3 holds when there exists a non-negative integrable function $\lambda(t)$ on $[\tau, b]$ such that

$$\mathfrak{Re}\{[z^* - y^*][f(t, z) - f(t, y)]\} \leq \lambda(t)\, |z - y|^2$$

for arbitrary (t, y), (t, z) on Δ_1. Formulate and prove, the corresponding analogue of the result of Theorem I.5.4.

5.[ss] For (a_0, b_0) an open interval on the real line, and $\mathfrak{M}(n, r)$ the class of $n \times r$ matrices U with complex elements, suppose that the $n \times n$ matrix function $C(t, U)$ is such that:

(i) $C(t, U)$ is locally of class ΓL on $(a_0, b_0) \times \mathfrak{M}(n, r)$;

(ii) the solutions of the $n \times r$ matrix differential equation

$$U' = C(t, U)U \tag{3.8}$$

are locally unique on $(a_0, b_0) \times \mathfrak{M}(n, r)$;

(iii) there exists a non-negative real-valued function $\kappa(t)$ which is integrable on compact subintervals of (a_0, b_0), and such that

$$\nu[C(t, U) + \{C(t, U)\}^*] \leq 2\kappa(t) \quad \textit{for} \quad (t, U) \in (a_0, b_0) \times \mathfrak{M}(n, r).$$

If $\tau \in (a_0, b_0)$, and $N \in \mathfrak{M}(n, r)$, prove that the solution $U = U(t; \tau, N)$ of (3.8) satisfying the initial condition $U(\tau) = N$ has for its maximal interval of existence the interval (a_0, b_0), and that

$$\nu[U(t; \tau, N)] \leq \nu[N]\exp\left\{\left| \int_\tau^t \kappa(s)\, ds \right|\right\} \quad \textit{for} \quad t \in (a_0, b_0).$$

6.[s] Suppose that $Q(t; \Phi, \Psi)$ is an $n \times n$ matrix function for $(t, \Phi, \Psi) \in (a_0, b_0) \times \mathfrak{M}_{nn} \times \mathfrak{M}_{nn}$ which is hermitian, locally of class ΓL, and such that the solutions of the differential system

$$\Phi' = Q(t; \Phi, \Psi)\Psi, \qquad \Psi' = -Q(t; \Phi, \Psi)\Phi,$$

are locally unique. In particular, these conditions hold for $Q(t; \Phi, \Psi) = Q_0(t) + \Psi(t)B(t)\Psi^*(t) + \Psi(t)A(t)\Phi^*(t) + \Phi(t)A^*(t)\Psi^*(t) - \Phi(t)C(t)\Phi^*(t)$, where $Q_0(t)$, $A(t)$, $B(t)$, $C(t)$ are $n \times n$ matrix functions of class $\mathfrak{L}_{nn}$ for arbitrary compact subintervals $[a, b]$ of (a_0, b_0), with $Q_0(t)$, $B(t)$, $C(t)$ hermitian. With the aid of the preceding problem, prove that:

(i) If $\tau \in (a_0, b_0)$, and Φ_0, Ψ_0 are given $n \times n$ matrices, then there is a unique solution $\Phi = \Phi(t; \tau, \Phi_0, \Psi_0), \Psi = \Psi(t; \tau, \Phi_0, \Psi_0)$ such that $\Phi = \Phi_0$, $\Psi = \Psi_0$ for $t = \tau$, and the maximal interval of existence of this solution is the given interval (a_0, b_0); moreover, the $n \times n$ matrix functions $\Phi^*\Phi + \Psi^*\Psi$ and $\Phi^*\Psi - \Psi^*\Phi$ are constant on (a_0, b_0). In particular, if $\Phi^*\Phi + \Psi^*\Psi = E$ and $\Phi^*\Psi - \Psi^*\Phi = 0$ for $t \in (a_0, b_0)$, then on this interval the $2n \times 2n$ matrix

$$M(t; \Phi, \Psi) = \begin{bmatrix} \Phi(t) & \Psi(t) \\ \Psi(t) & -\Phi(t) \end{bmatrix}$$

is unitary, and also $\Phi\Phi^* + \Psi\Psi^* = E$, $\Phi\Psi^* - \Psi\Phi^* = 0$ for $t \in (a_0, b_0)$.

(ii) If the solution Φ, Ψ is such that $\Phi(\tau)\,\Phi^*(\tau) + \Psi(\tau)\,\Psi^*(\tau) = E$, $\Psi(\tau)\,\Phi^*(\tau) - \Phi(\tau)\,\Psi^*(\tau) = 0$, then $\Phi\Phi^* + \Psi\Psi^* \equiv E$, $\Psi\Phi^* - \Phi\Psi^* \equiv 0$ on (a_0, b_0); moreover, $M(t; \Phi, \Psi)$ is unitary and also $\Phi^*\Phi + \Psi^*\Psi \equiv E$, $\Phi^*\Psi - \Psi^*\Phi \equiv 0$ on (a_0, b_0).

Hint. (ii) Show that $G(t) = \Phi(t)\,\Phi^*(t) + \Psi(t)\,\Psi^*(t)$, $H(t) = \Psi(t)\,\Phi^*(t) - \Phi(t)\,\Psi^*(t)$ are solutions of the differential system

$$G' = QH - HQ, \qquad H' = GQ - QG, \qquad G(\tau) = E, \qquad H(\tau) = 0,$$

and that $G(t) \equiv E$, $H(t) \equiv 0$ is the unique solution of this system.

7. In the spirit of the above Section 3, formulate and prove a theorem analogous to Theorem I.4.8.

8.[ss] If $A(t)$, $B(t)$, $C(t)$, $D(t)$ are matrix functions of class $\mathfrak{L}_{nn}(I)$, where I is a nondegenerate interval on the real line, prove that the first order matrix differential system

$$\begin{aligned} L_1[U, V] &\equiv -V' + C(t)U - D(t)V = 0, \\ L_2[U, V] &\equiv U' - A(t)U - B(t)V = 0, \end{aligned} \tag{3.9}$$

is satisfied by a pair of $n \times n$ matrix functions $(U(t); V(t))$ with $U(t)$ nonsingular on I if and only if there is on I a solution $W(t)$ of the *Riccati matrix differential equation*

$$\mathfrak{R}[W] \equiv W' + WA(t) + D(t)W + WB(t)W - C(t) = 0 \tag{3.10}$$

such that $W(\tau) = V(\tau)\,U^{-1}(\tau)$ for some, (and consequently all), $\tau \in I$.

Hint. Establish the identity

$$U^*\mathfrak{R}[VU^{-1}]U = -U^*(L_1[U, V] + VU^{-1}\,L_2[U, V]).$$

9.[ss] Suppose that the matrix coefficient functions are as in Problem 8, and $W = W_0(t)$ is a solution of (3.10) on a nondegenerate interval I. If $s \in I$, and the matrix functions $G(t) = G(t, s \mid W_0)$, $H(t) = H(t, s \mid W_0)$ are the solutions of the linear differential systems

$$G' + (D + W_0 B)G = 0, \qquad G(s) = E, \tag{3.11}$$

$$H' + H(A + BW_0) = 0, \qquad H(s) = E, \tag{3.12}$$

and

$$Z(t, s \mid W_0) = \int_s^t H(r, s \mid W_0)\, B(r)\, G(r, s \mid W_0)\, dr, \tag{3.13}$$

show that $W(t)$ is a solution of (3.10) on I if and only if the constant matrix $\Gamma = W(s) - W_0(s)$ is such that $E + Z(t, s \mid W_0)\Gamma$ is nonsingular on I, and

(3.14)
$$W(t) = W_0(t) + G(t, s \mid W_0)\Gamma[E + Z(t, s \mid W_0)\Gamma]^{-1} H(t, s \mid W_0).$$

The solution $W(t)$ of (3.14) is also given by

(3.14′)
$$W(t) = W_0(t) + G(t, s \mid W_0)[E + \Gamma Z(t, s \mid W_0)]^{-1}\, \Gamma H(t, s \mid W_0).$$

Conclude from (3.14) or (3.14′) that if $W(t)$ and $W_0(t)$ are solutions of (3.10) on I then $W(t) - W_0(t)$ is of constant rank on this interval.

Hints. For $W_0(t)$ a particular solution of (3.10) on I, and G, H and Z defined by (3.11), (3.12), (3.13), respectively, show that $W(t)$ is a solution of (3.10) on I if and only if the matrix function F defined by $W - W_0 = GFH$ is such that $F_1 = F[E + Z\Gamma] - \Gamma$ is a solution of the linear homogeneous system $F_1' + FHBGF_1 = 0$, $F_1(s) = 0$, and hence $F_1 \equiv 0$. For arbitrary $n \times n$ matrices Z, Γ show that the identity $(E + \Gamma Z)\Gamma = \Gamma(E + Z\Gamma)$ implies that $E + \Gamma Z$ is nonsingular if and only if $E + Z\Gamma$ is nonsingular, in which case $\Gamma[E + Z\Gamma]^{-1} = [E + \Gamma Z]^{-1}\Gamma$.

10.[ss] For arbitrary $n \times n$ matrices M_β, $(\beta = 1, 2, 3, 4)$ with $M_3 - M_2$ and $M_4 - M_1$ nonsingular, let $\{M_1, M_2, M_3\} = (M_3 - M_1)(M_3 - M_2)^{-1}$ and $\{M_1, M_2, M_3, M_4\} = \{M_1, M_2, M_3\}\{M_2, M_1, M_4\}$; clearly $\{M_1, M_2, M_3, M_4\}$ is a direct generalization of the scalar cross, (or anharmonic), ratio. If $W_0(t)$, $W_j(t)$, $(j = 1, 2, 3, 4)$, are solutions of (3.10) on a nondegenerate interval I with $W_3 - W_2$ and $W_4 - W_1$ nonsingular on this interval show that for $(t, s) \in I \times I$ one has the identity

$$\{W_1(t), W_2(t), W_3(t), W_4(t)\}$$
$$= Q(t, s \mid W_0, W_1)\{W_1(s), W_2(s), W_3(s), W_4(s)\}\, Q^{-1}(t, s \mid W_0, W_1),$$

where $Q(t, s \mid W_0, W_1) = G(t, s \mid W_0)[E + \Gamma_1 Z(t, s \mid W_0)]^{-1}$ and $\Gamma_1 = W_1(s) - W_0(s)$.

Hint. If $\Gamma_\alpha = W_\alpha(s) - W_0(s)$, show that

$$W_\alpha(t) - W_\beta(t) = G(t, s \mid W_0) S^{(\alpha,\beta)}(t, s \mid W_0)\, H(t, s \mid W_0),$$

where

$$S^{(\alpha,\beta)}(t, s \mid W_0) = [E + \Gamma_\beta Z(t, s \mid W_0)]^{-1}(\Gamma_\alpha - \Gamma_\beta)[E + Z(t, s \mid W_0)\Gamma_\alpha]^{-1}.$$

11.[ss] If $W(t)$ and $W_0(t)$ are solutions of (3.10) on a nondegenerate interval I, $s \in I$, and $\Gamma = W(s) - W_0(s)$, prove that

$$\begin{aligned} G(t, s \mid W) &= G(t, s \mid W_0)[E + \Gamma Z(t, s \mid W_0)]^{-1}, \\ H(t, s \mid W) &= [E + Z(t, s \mid W_0)\Gamma]^{-1} H(t, s \mid W_0), \\ Z(t, s \mid W) &= [E + Z(t, s \mid W_0)\Gamma]^{-1} Z(t, s \mid W_0), \\ &= Z(t, s \mid W_0)[E + \Gamma Z(t, s \mid W_0)]^{-1}. \end{aligned} \tag{3.15}$$

In particular, if $Z(t, s \mid W_0)$ is nonsingular for a fixed s, and t on a subinterval I_0 of I, then for the same fixed s the matrix $Z(t, s \mid W)$ is also nonsingular for t on I_0, and

$$Z^{-1}(t, s \mid W) = Z^{-1}(t, s \mid W_0) + \Gamma.$$

Hint. Let $\mathcal{A}(t)$ denote the $2n \times 2n$ matrix

$$\mathcal{A}(t) = \begin{bmatrix} A(t) & 0 \\ 0 & 0 \end{bmatrix},$$

where 0 is the $n \times n$ zero matrix, with similar definitions for $\mathcal{B}(t)$, $\mathcal{C}(t)$, $\mathcal{D}(t)$ in terms of the corresponding $B(t)$, $C(t)$, $D(t)$. Verify directly that a $2n \times 2n$ matrix function $\mathcal{W}(t)$ is a solution of the Riccati matrix differential equation

$$\mathcal{W}' + \mathcal{W}\mathcal{A}(t) + \mathcal{D}(t)\mathcal{W} + \mathcal{W}\mathcal{B}(t)\mathcal{W} - \mathcal{C}(t) = 0 \tag{3.16}$$

on an interval I if and only if

$$\mathcal{W}(t) = \begin{bmatrix} W(t) & G(t) \\ H(t) & -Z(t) \end{bmatrix}, \tag{3.17}$$

where $W(t)$, $G(t)$, $H(t)$, $Z(t)$ are $n \times n$ matrix functions which satisfy on this interval the Riccati system

$$\begin{aligned} W' + WA(t) + D(t)W + WB(t)W - C(t) &= 0, \\ G' + [D(t) + WB(t)]G &= 0, \\ H' + H[A(t) + D(t)W] &= 0, \\ Z' - HB(t)G &= 0. \end{aligned} \tag{3.18}$$

In particular, if $W_0(t)$ is a solution of (3.10) on I, and $G(t, s \mid W_0)$, $H(t, s \mid W_0)$, $Z(t, s \mid W_0)$ are defined respectively by (3.11), (3.12), (3.13), show that the solution $\mathfrak{W} = \mathfrak{W}_0(t)$ of (3.16) satisfying the initial condition

$$\mathfrak{W}_0(s) = \begin{bmatrix} W_0(s) & E \\ E & 0 \end{bmatrix}$$

is given by

$$\mathfrak{W}_0(t) = \begin{bmatrix} W_0(t) & G(t, s \mid W_0) \\ H(t, s \mid W_0) & -Z(t, s \mid W_0) \end{bmatrix}.$$

For such a solution $\mathfrak{W}_0(t)$ of (3.16), and a second solution $\mathfrak{W}(t)$ of this equation related in a similar fashion to a solution $W(t)$ of (3.10), consider the equation involving $\mathfrak{W}_0(t)$, $\mathfrak{W}(t)$ corresponding to (3.14).

4. DIFFERENTIABILITY OF SOLUTIONS

For $q = 1, 2, \ldots$ a real-valued vector function $g(t, u) = (g_\alpha(t, u_1, \ldots, u_n))$, $(\alpha = 1, \ldots, n)$, is said to be *locally of class* $\Gamma^{(q)}L$ in u on an open region $\mathfrak{R}$ of real (t, u)-space if in this region g has partial derivatives with respect to the u_β, $(\beta = 1, \ldots, n)$ of orders not exceeding q, and $g(t, u)$ and each such partial derivative vector function is locally of class ΓL on $\mathfrak{R}$. In particular, if $g(t, u)$ is locally of class $\Gamma^{(1)}L$ in $\mathfrak{R}$ then by the mean value theorem $g(t, u)$ is locally of class Lip in u on $\mathfrak{R}$, so that the solutions of

$$u' = g(t, u) \tag{4.1}$$

are locally unique in $\mathfrak{R}$. Corresponding to Theorem I.10.1, one has the following result.

Theorem 4.1. *Suppose that the real-valued vector function* $g(t, u) = (g_\alpha(t, u_1, \ldots, u_n))$, $(\alpha = 1, \ldots, n)$, *is locally of class* $\Gamma^{(1)}L$ *in* u *on an open region* $\mathfrak{R}$ *of real* (t, u)*-space and that* $u = u(t, h)$ *is a one-parameter family of solutions of* (4.1) *with graphs in* $\mathfrak{R}$ *for* $t \in [a, b]$, $h \in (-\delta, \delta)$. *If for a fixed* τ *on* $[a, b]$ *the vector function* $\eta(h) = u(\tau, h)$, $h \in (-\delta, \delta)$, *is continuous and has a finite derivative at* $h = 0$, *then on* $[a, b]$ *the derivative vector function* $z(t) = u_h(t, 0) = (u_{\alpha h}(t, 0))$ *exists and is a solution of the linear vector differential equation*

$$z' = g_u(t, u(t, 0))z, \tag{4.2}$$

where $g_u(t, u(t, 0))$ *is the* $n \times n$ *matrix function* $[g_{\alpha u_\beta}(t, u(t, 0))]$.

Theorem 4.1 may be established by essentially the same argument as that employed for the proof of Theorem I.10.1. In particular, for the matrix $A(t, h)$ occurring in the proof of Theorem I.10.1, some of the results of

Problem 1 below are useful in proving the desired continuity property of this matrix under the weakened hypotheses of the present chapter.

For $q = 1, 2, \ldots$, an n-dimensional real-valued vector function $g(t, u; c) = (g_\alpha(t, u_1, \ldots, u_n; c_1, \ldots, c_k))$ will be said to be "locally of class $\Gamma^{(q)}L$ in u on an open region $\mathfrak{D}$ of real (t, u, c)-space" if on this region g has partial derivatives with respect to the u_β, $(\beta = 1, \ldots, n)$, of orders not exceeding q, and $g(t, u; c)$ and each such partial derivative vector function is locally of class ΓL on $\mathfrak{D}$, in accordance with the definition given immediately following the statement of Theorem 3.8.

If $g(t, u; c)$ is locally of class $\Gamma^{(q)}L$, $(q \geq 1)$, in u on an open region $\mathfrak{D}$ of real (t, u, c)-space, and $\mathfrak{D}_c$ denotes the section consisting of all (t, u) such that $(t, u, c) \in \mathfrak{D}$, then on any nonvacuous section $\mathfrak{D}_c$ the solutions of the vector differential equation

$$u' = g(t, u; c) \tag{4.3}$$

are locally unique, and for $(\tau, \eta, c) \in \mathfrak{D}$ the solution $u = \phi(t; \tau, \eta, c)$ of (4.3) passing through the initial point (τ, η) in $\mathfrak{D}_c$ has a maximal interval of existence $I(\tau, \eta, c) = \{t \mid a(\tau, \eta, c) < t < b(\tau, \eta, c)\}$, and $\phi(t; \tau, \eta, c)$ is continuous in (t, τ, η, c) on

$$\{(t, \tau, \eta, c) \mid (\tau, \eta, c) \in \mathfrak{D}, t \in I(\tau, \eta, c)\}. \tag{4.4}$$

Corresponding to Theorem I.10.2, one has for the differential equation (4.3) the following result.

Theorem 4.2. *If $g(t, u, c)$ is locally of class $\Gamma^{(q)}L$, $(q \geq 1)$, in u on an open region $\mathfrak{D}$ of real (t, u, c)-space, then for the solutions $u = \phi(t; \tau, \eta, c)$ of* (4.3) *the partial derivatives*

$$\phi_{\eta_\gamma}(t; \tau, \eta, c) = (\phi_{\alpha\eta_\gamma}(t; \tau, \eta, c)), \qquad (\gamma = 1, \ldots, n), \tag{4.5}$$

exist and are of class $\mathfrak{C}^{(q-1)}$ in η on the open region (4.4) *of (t, τ, η, c)-space. The vector functions $z(t) = \phi_{\eta_\gamma}(t; \tau, \eta, c) \equiv (\phi_{\alpha\eta_\gamma}(t; \tau, \eta, c))$ are solutions of the linear vector homogeneous differential equation*

$$z' = g_u(t, \phi(t; \tau, \eta, c); c)z \tag{4.6}$$

on the corresponding interval $I(\tau, \eta, c)$, satisfying the initial conditions

$$\phi_{\alpha\eta_\gamma}(\tau; \tau, \eta, c) = \delta_{\alpha\gamma}, \qquad (\alpha, \gamma = 1, \ldots, n). \tag{4.7}$$

In view of the weakened hypothesis on $g(t, u; c)$ as a function of t, clearly it is no longer true that the vector functions ϕ_{η_γ} of (4.5) possess partial derivatives ϕ'_{η_γ} with respect to t which are of class $\mathfrak{C}^{(q-1)}$ in η on the region (4.4), as was the case in Theorem I.10.3. It is true, however, that each of the vector functions (4.5), and the partial derivatives of these functions with respect to the η_β of orders not exceeding $q - 1$, are absolutely continuous functions of t which individually are solutions of respective differential systems. For example, if $q \geq 2$, then $w(t) = \phi_{\eta_\gamma \eta_\nu}(t; \tau, \eta, c)$, $(\gamma, \nu = 1, \ldots, n)$, is the solution of the vector differential equation

$$w' = g_u(t, \phi(t; \tau, \eta, c); c)w + h^{(\gamma,\nu)}(t; \tau, \eta, c), \qquad w(\tau) = 0,$$

where $h^{(\gamma,\nu)}(t; \tau, \eta, c) = \sum_{\alpha,\beta=1}^{n} g_{u_\alpha u_\beta}(t, \phi; c)\phi_{\alpha\eta_\gamma}\phi_{\beta\eta_\nu}$, and the arguments of ϕ, $\phi_{\alpha\eta_\gamma}$, $\phi_{\beta\eta_\nu}$ are $(t; \tau, \eta, c)$.

In comparison with the result of Theorem I.10.1, under the weakened hypothesis that $g(t, u; c)$ is locally of class $\Gamma^{(q)}L$, $(q \geq 1)$, in u on an open region $\mathfrak{D}$ of real (t, u, c)-space, it is not true in general that the partial derivative vector function $\phi_\tau(t; \tau, \eta, c) = (\phi_{\alpha\tau}(t; \tau, \eta, c))$ exists for all values (t, τ, η, c) in the region (4.4). By the method used to prove Theorem I.10.1, however, one may establish the following result.

Theorem 4.3. *Suppose that the real-valued vector function $g(t, u; c)$ is locally of class $\Gamma^{(q)}L$, $(q \geq 1)$, in u on an open region $\mathfrak{D}$ of real (t, u, c)-space, and for a fixed set of values η, c we have that $(\tau^0, \eta, c) \in \mathfrak{D}$, and*

$$\lim_{h \to 0} \frac{1}{h} \int_{\tau^0}^{\tau^0 + h} g(t, \eta, c)\, dt = g(\tau^0, \eta, c). \tag{4.8}$$

Then for the fixed values η, c the vector function $\phi(t; \tau, \eta, c)$ has at τ^0 a partial derivative $\phi_\tau(t; \tau^0, \eta, c)$ with respect to τ, and $\phi_\tau(t; \tau^0, \eta, c)$ is equal to $-\sum_{\gamma=1}^{n} \phi_{\eta_\gamma}(t; \tau^0, \eta, c)\, g_\gamma(\tau^0, \eta, c)$, the solution $z(t)$ of the corresponding equation of variation

$$z' = g_u(t, \phi(t; \tau^0, \eta, c); c)z \tag{4.9}$$

on the interval $I(\tau^0, \eta, c)$, and satisfying the initial condition $z(\tau^0) = -g(\tau^0, \eta, c)$. If τ^0 is a value such that (4.8) *holds for all (η, c) such that $(\tau^0, \eta, c) \in \mathfrak{D}$, then as a function of (t, η, c) the vector function $\phi_\tau(t; \tau^0, \eta, c)$ has partial derivatives with respect to the η_β of the first $q - 1$ orders, each such partial derivative vector function is continuous in (t, η, c) on*

$$\{(t, \eta, c) \mid (\tau^0, \eta, c) \in \mathfrak{D},\, t \in I(\tau^0, \eta, c)\},$$

and on $I(\tau^0, \eta, c)$ is an absolutely continuous function of t.

Finally, suppose that the real-valued vector function $g(t, u; c)$ is locally of class $\Gamma^{(q)}L$, $(q \geq 1)$, in (u, c) on an open region $\mathfrak{D}$ of (t, u, c)-space, in the sense that on this region g has partial derivatives with respect to the u_β, c_ν, $(\beta = 1, \ldots, n; \nu = 1, \ldots, k)$, of orders not exceeding q, and $g(t, u; c)$ and each such partial derivative vector function is locally of class ΓL on $\mathfrak{D}$. Under this condition, corresponding to Theorem I.10.4 one has the result that the partial derivatives (4.5) and also the partial derivatives $\phi_{\alpha c_\nu}(t; \tau, \eta, c)$,

$(\alpha = 1, \ldots, n; \nu = 1, \ldots, k)$ exist; moreover, each such partial derivative vector function is continuous in (t, η, c) on (4.4), and on $I(\tau, \eta, c)$ is an absolutely continuous function of t.

Problems II.4

1.[ss] Suppose that the real-valued vector function $g(t, u) \equiv (g_\alpha(t, u_1, \ldots, u_n))$, $(\alpha = 1, \ldots, n)$, is locally of class $\Gamma^{(1)}L$ in u on an open region $\mathfrak{R}$ of real (t, u)-space, and that $u^0(t) = (u_\alpha^{\ 0}(t))$, $t \in [a, b]$, is a continuous vector function with graph in $\mathfrak{R}$. If $\zeta > 0$ is such that $S_\zeta = \{(t, u) \mid t \in [a, b], |u - u^0(t)| \leq \zeta\}$ lies in $\mathfrak{R}$, and on $Q_\zeta = \{(t, v) \mid t \in [a, b], |v| \leq \zeta\}$ the $n \times n$ matrix function $F(t, v) = F(t, v_1, \ldots, v_n)$ is defined as

$$F(t, v) = \int_0^1 g_u(t, u^0(t) + \theta v)\ d\theta,$$

show that $F(t, v)$ is of class $\Gamma L[Q_\zeta]$. Moreover,

(i) there exists an integrable function $\kappa(t)$ on $[a, b]$ such that $\nu[F(t, v)] \leq \kappa(t)$ for $(t, v) \in Q_\zeta$, and if $w(t)$ is an n-dimensional vector function which on $[a, b]$ is measurable and satisfies $|w(t)| \leq \zeta$, then $F(t, w(t))$ is integrable on $[a, b]$ and

$$\nu\left[\int_\tau^\sigma F(t, w(t))\ dt\right] \leq \int_\tau^\sigma \kappa(t)\ dt, \quad \text{for} \quad a \leq \tau \leq \sigma \leq b;$$

(ii) if on $\Delta_1 = [a, b] \times (-\delta, \delta)$ the n-dimensional vector function $v(t, h)$ is continuous and satisfies $|v(t, h)| \leq \zeta$, then $B(t, h) = F(t, v(t, h))$ is an $n \times n$ matrix function of class $\Gamma L[\Delta_1]$;

(iii) if for $j = 1, 2, \ldots$ the n-dimensional vector function $w^{(j)}(t)$, $t \in [a, b]$, is continuous, has graph in S_ζ, and $\{w^{(j)}(t)\} \to 0$ as $j \to \infty$ for $t \in [a, b]$, then $\{\int_a^b \nu[F(t, w^{(j)}(t)) - F(t, 0)]\ dt\} \to 0$ and $\{\int_a^b F(t, w^{(j)}(t))\ dt\} \to \int_a^b F(t, 0)\ dt$ as $j \to \infty$.

2. Suppose that $A(t)$, $B(t)$, $C(t)$, $D(t)$ are real-valued matrix functions of class $\mathfrak{L}_{nn}(I)$ on arbitrary compact subintervals I of $(-\infty, \infty)$, and that $W = W(t; \tau, \Omega)$ is the solution of the Riccati matrix differential equation (3.10) satisfying the real initial condition $W(\tau) = \Omega$, and with maximal interval of existence $a(\tau, \Omega) < t < b(\tau, \Omega)$. If $\Omega = [\omega_{ij}]$, $(i, j = 1, \ldots, n)$, prove that the partial derivative matrix function $W_{\omega_{ij}}(t; \tau, \Omega)$ is given by

$$W_{\omega_{ij}}(t; \tau, \Omega) = G(t, \tau \mid W(\quad; \tau, \Omega))\ \Delta^{ij}\ H(t, \tau \mid W(\quad; \tau, \Omega)),$$

where $\Delta^{ij} = (\Delta_{\alpha\beta}^{\ \ ij})$, $(\alpha, \beta = 1, \ldots, n)$, with $\Delta_{\alpha\beta}^{\ \ ij}$ equal to 1 if $(\alpha, \beta) = (i, j)$, and equal to 0 if $(\alpha, \beta) \neq (i, j)$. Moreover, if τ^0 and Ω are such that

$F(t, W) = -WA(t) - D(t)W - WB(t)W + C(t)$ satisfies

$$\lim_{h\to 0} \frac{1}{h} \int_{\tau^0}^{\tau^0+h} F(t, \Omega)\, dt = F(\tau^0, \Omega),$$

then the partial derivative matrix function $W_\tau(t; \tau^0, \Omega)$ exists and

$$W_\tau(t; \tau^0, \Omega) = -G(t, \tau^0 \mid W(\quad; \tau^0, \Omega))\, F(\tau^0, \Omega)\, H(t, \tau^0 \mid W(\quad; \tau^0, \Omega))$$

for $a(\tau^0, \Omega) < t < b(\tau^0, \Omega)$.

5. NOTES AND REMARKS

Additional references for the material of this chapter are Carathéodory [1, Ch. XI], McShane [2, Ch. IX], Sansone [1-II, Ch. VIII], and Coddington and Levinson [2, Ch. II]. For a discussion of local existence and uniqueness theorems under hypotheses of Carathéodory type, and including problems in which the prescribed initial conditions are not confined to a single point, the reader is referred to Whyburn [3].

In view of the manner in which the material of Chapter I has been presented, many basic results for differential equations in the setting of this chapter have been merely listed with statements to the effect that these results may be established by methods that are identical with, or closely related to, the methods used in Chapter I to prove corresponding results under stronger hypotheses. Accordingly, the student is urged to carry through the pertinent steps of the proofs of individual results to verify the validity of such statements. In this chapter, and also in Chapter VII, the student is supposed to be familiar with the theory of the Lebesgue integral in one dimension.

No attempt is made in this work to introduce the reader to the concept of a differential equation in a more abstract space setting. For an introduction to this area, it is suggested that one consult Bourbaki [1], Hille [4, Ch. 6], or Massera and Schäffer [1].

RELATED REFERENCES AND COMMENTS FOR SPECIFIC PROBLEMS

II.3:4 Giuliano [1,2].

II.3:6 The result of this problem is intimately related to results of Barrett [3], Reid [18], Etgen [1].

II.3:8 Radon [1]; Reid [10,23].

II.3:9 Reid [23]. For additional results on Riccati matrix differential equations, see also Redheffer [1,2,3,4] and Reid [19,21,26].

II.3:10 Sandor [1]; Levin [1]; Reid [23].

II.3:11 Reid [23].

III

Linear Differential Systems

1. INTRODUCTION

We shall proceed to a more thorough study of n-dimensional linear vector differential equations, which were previously considered in Sections I.9 and II.3. Most of the specific discussion of this chapter will be concerned with vector differential operators which formally are of the type

$$L[y](t) = A_1(t)[A_2(t)y]' + A_0(t)y, \tag{1.1}$$

where the coefficient matrices satisfy the following hypothesis:

($\mathfrak{H}$) *the* $A_j(t)$, $(j = 0, 1, 2)$, *are continuous* $n \times n$ *matrix functions on a given compact interval* $[a, b]$, *with* $A_1(t)$, $A_2(t)$ *nonsingular on this interval.*

As usual, the linear space of n-dimensional vector functions which are continuous and have continuous derivatives of the first m orders on $[a, b]$ will be denoted by $\mathfrak{C}_n{}^m[a, b]$, with the understanding that $\mathfrak{C}_n{}^0[a, b]$ signifies the set $\mathfrak{C}_n[a, b]$ of continuous n-dimensional vector functions on $[a, b]$. Moreover, for $m \geq 0$ the set $\{y \mid y \in \mathfrak{C}_n{}^m[a, b],\ y^{(j-1)}(a) = y^{(j-1)}(b) = 0,\ j = 1, \ldots, m\}$ will be denoted by $\mathfrak{C}_{n0}{}^m[a, b]$.

Also, for brevity, we write $\mathfrak{C}^m[a, b]$ and $\mathfrak{C}_0{}^m[a, b]$ for $\mathfrak{C}_1{}^m[a, b]$ and $\mathfrak{C}_{10}{}^m[a, b]$, respectively.

For the formal differential operator L defined by (1.1), let $\mathfrak{D}(L)$ denote the linear space of vector functions y in $\mathfrak{C}_n[a, b]$ such that $y = A_2^{-1}u_y$, with $u_y \in \mathfrak{C}_n{}^1[a, b]$. In particular, if $y \in \mathfrak{D}(L)$ then

$$u_y = A_2 y \tag{1.2}$$

is such that $L[y](t) = A_1(t)u_y'(t) + A_0(t)A_2^{-1}(t)u_y(t)$ is a vector function in $\mathfrak{C}_n[a, b]$. Moreover, if $g \in \mathfrak{C}_n[a, b]$ then $y(t)$ is a *solution* of

$$L[y](t) = g(t), \qquad t \in [a, b], \tag{1.3}$$

in the sense that $y \in \mathfrak{D}(L)$ and $L[y](t) = g(t)$ on $[a, b]$ if and only if $u = u_y(t)$ defined by (1.2) is a solution of

$$u' = A(t)u + h(t), \qquad t \in [a, b], \tag{1.4}$$

where

$$A = -A_1^{-1}A_0A_2^{-1}, h = A_1^{-1}g. \tag{1.5}$$

As in the earlier discussions in Chapters I and II, for the consideration of solutions of (1.3) the matrix differential equation

$$L[Y](t) \equiv A_1(t)[A_2(t)Y(t)]' + A_0(t)Y(t) = 0, \qquad t \in [a, b], \tag{1.6}$$

is of basic importance. In particular, a fundamental matrix for

$$L[y](t) = 0, \qquad t \in [a, b], \tag{1.3$_\circ$}$$

is by definition an $n \times n$ matrix $Y(t)$ whose column vectors are linearly independent solutions of $(1.3_\circ)$.

A particular advantage of the general form (1.1) for the differential operator is that when the involved matrix functions $A_j(t)$, $(j = 0, 1, 2)$, are assumed to be merely continuous the corresponding adjoint operator, which will be defined in the following section, is of the same form as the given differential operator. Moreover, when these matrix functions are allowed to satisfy still weaker conditions, and "solution" is taken in the Carathéodory sense, this general formulation permits the automatic incorporation of certain types of interface conditions, as shown by Problems 8, 9 of III.2.

2. ADJOINT VECTOR DIFFERENTIAL EQUATIONS

Suppose that the vector function $\phi(t) = (\phi_\alpha(t))$, $(\alpha = 1, \ldots, n)$, is such that

$$(y, \phi(t)) \equiv \phi^*(t)y \equiv \sum_{\alpha=1}^{n} \bar{\phi}_\alpha(t)\, y_\alpha(t) \tag{2.1}$$

is a *first integral* for the homogeneous linear vector differential equation $(1.3_\circ)$; that is, for each solution $y = y(t)$ of $(1.3_\circ)$ the function $(y(t), \phi(t))$ is constant on $[a, b]$. Since $A_1(t)$, $A_2(t)$ are nonsingular matrices, without loss of generality we may write $\phi(t) = A_2^*(t)A_1^*(t)z(t)$. Then for $Y(t)$ a fundamental matrix for $(1.3_\circ)$ there is an n-dimensional constant vector $\gamma = (\gamma_\alpha)$ such that $\gamma^* = z^*(t)A_1(t)A_2(t)Y(t) = z^*(t)A_1(t)U_Y(t)$ on $[a, b]$ and consequently $A_1^*(t)z(t) = [U_Y(t)]^{*-1}\gamma$ on this interval. Conversely, if $\phi(t) = A_2^*(t)A_1^*(t)z(t)$, with $A_1^*(t)z(t)$ of the form $[U_Y(t)]^{*-1}\gamma$ on $[a, b]$, then $\phi^*(t)Y(t) \equiv \gamma^*$, and (2.1) is a first integral for $(1.3_\circ)$ since each solution

$y(t)$ of (1.3_o) is of the form $y(t) = Y(t)\xi$, where ξ is an n-dimensional constant vector. Moreover, since the identity $E \equiv U_Y^{-1}U_Y$ implies $0 = (U_Y^{-1})'U_Y + (U_Y^{-1})U_Y' = [(U_Y^{-1})' - U_Y^{-1}A_1^{-1}A_0A_2^{-1}]U_Y$, it follows that $Z(t) = A_1^{*-1}(t)\, U_Y^{*-1}(t)$ is a fundamental matrix for the differential equation

$$(2.2) \qquad L^\star[z](t) = 0, \qquad t \in [a, b],$$

where $L^\star[z]$ denotes the linear homogeneous vector differential expression

$$(2.3) \qquad L^\star[z](t) = -A_2^*(t)[A_1^*(t)z(t)]' + A_0^*(t)z(t).$$

It is to be noted that the expression $L^\star$ is obtained from the expression L by substituting A_0^*, $-A_2^*$, A_1^* for the respective coefficient matrices A_0, A_1, A_2. Moreover, condition ($\mathfrak{H}$) for L implies the corresponding condition for $L^\star$. Now for $L^\star$ the set $\mathfrak{D}(L^\star)$ consists of those vector functions z in $\mathfrak{C}_n[a, b]$ such that $z = A_1^{*-1}v_z$, with $v_z \in \mathfrak{C}_n^{\,1}[a, b]$, and

$$(2.4) \qquad L^\star[z](t) = -A_2^*(t)v_z'(t) + A_0^*(t)A_1^{*-1}(t)v_z(t).$$

Corresponding to (1.6), we have the matrix differential equation

$$(2.5) \qquad \begin{aligned} L^\star[Z](t) &= -A_2^*(t)[A_1^*(t)Z(t)]' + A_0^*(t)Z(t) = 0, \\ &= -A_2^*(t)V_Z'(t) + A_0^*(t)A_1^{*-1}(t)V_Z(t) = 0, \qquad t \in [a, b], \end{aligned}$$

where $V_Z(t) = A_1^*(t)Z(t)$.

The vector differential equation (2.2) is called the *adjoint* of (1.3_o). Clearly the property of adjointness is reciprocal; that is, the adjoint of (2.2) is (1.3_o). Moreover, in view of these comments the results of the following theorem are immediate.

Theorem 2.1 (a) *If $y(t)$ and $z(t)$ are solutions of (1.3_o) and (2.2), respectively, then $v_z^*(t)u_y(t)$ is constant on $[a, b]$*; (b) *if $Y(t)$ and $Z(t)$ are solutions of the respective matrix differential equations* (1.6) *and* (2.5), *then there is a constant matrix C such that $C = V_Z^*(t)U_Y(t) = Z^*(t)A_1(t)A_2(t)Y(t)$ on $[a, b]$*; (c) *if $Y(t)$ is a fundamental matrix for (1.3_o) and $Z(t)$ is defined by $V_Z^*(t)U_Y(t) = C$, where C is a constant matrix, then $Z(t)$ is a solution of* (2.5); *moreover, $Z(t)$ is a fundamental matrix for* (2.2) *if and only if C is nonsingular.*

It is to be noted that we have the identity

$$(2.6) \quad (L[y], z) - (y, L^\star[z]) = (v_z^* u_y)', \quad \text{for} \quad y \in \mathfrak{D}(L), \quad z \in \mathfrak{D}(L^\star).$$

In turn, (2.6) implies that

$$(2.7) \quad \int_a^b (L[y], z)\, dt - \int_a^b (y, L^\star[z])\, dt = v_z^* u_y \Big|_a^b, \quad \text{for} \quad y \in \mathfrak{D}(L), \quad z \in \mathfrak{D}(L^\star).$$

For brevity let $\mathfrak{D}_0(L)$ denote the set of all $y \in \mathfrak{D}(L)$ such that $u_y(a) = 0 = u_y(b)$ and, correspondingly, let $\mathfrak{D}_0(L^\star)$ denote the set of all $z \in \mathfrak{D}(L^\star)$ such that $v_z(a) = 0 = v_z(b)$. Clearly the condition $u_y(a) = 0 = u_y(b)$ holds if and only if $y(a) = 0 = y(b)$, and, similarly, $v_z(a) = 0 = v_z(b)$ if and only if $z(a) = 0 = z(b)$. It seems preferable to state these conditions in terms of the end-values of u_y and v_z, however, in view of the later statements for more general boundary conditions, and also for comparison with the conditions occurring when the hypothesis $(\mathfrak{H})$ is relaxed and the concept of a solution is generalized, in the sense discussed in Problem 8 of the following problem set III.2.

In particular, from (2.7) it follows that

$$(2.8) \quad \int_a^b (L[y], z)\, dt - \int_a^b (y, L^\star[z])\, dt = 0, \quad \textit{if} \quad y \in \mathfrak{D}_0(L), \quad z \in \mathfrak{D}(L^\star).$$

Moreover, if $z(t) \in \mathfrak{C}_n[a, b]$ and there exists a vector function $f = f_z(t)$ in $\mathfrak{C}_n[a, b]$ such that

$$(2.9) \qquad \int_a^b (L[y], z)\, dt - \int_a^b (y, f_z)\, dt = 0 \quad \textit{for} \quad y \in \mathfrak{D}_0(L),$$

then since (2.9) may be written as

$$(2.10) \quad \int_a^b (u', A_1^* z)\, dt + \int_a^b (u, A_2^{*-1}[A_0^* z - f_z])\, dt = 0 \quad \textit{for} \quad u \in \mathfrak{C}_{n0}^{\ 1}[a, b],$$

it follows from Problem 1 below that $v_z = A_1^* z \in \mathfrak{C}_n^{\ 1}[a, b]$, and $f_z(t) = L^\star[z](t)$. Consequently, one has the following alternative determination of the linear manifold $\mathfrak{D}(L^\star)$ and the adjoint differential operator $L^\star(z)$.

Theorem 2.2. *The class $\mathfrak{D}(L^\star)$ is characterized as the set of vector functions $z \in \mathfrak{C}_n[a, b]$ such that there exists a corresponding $f_z \in \mathfrak{C}_n[a, b]$ for which* (2.9) *holds, and for $z \in \mathfrak{D}(L^\star)$ the corresponding f_z is uniquely determined as $f_z = L^\star[z]$.*

Problems III.2

1.[ss] (*Fundamental lemma of the calculus of variations*). (i) If $g(t) \in \mathfrak{C}_n[a, b]$, prove that

$$(2.11) \qquad \int_a^b (u', g)\, dt = 0, \quad \textit{for arbitrary} \quad u \in \mathfrak{C}_{n0}^{\ 1}[a, b],$$

if and only if there exists a constant vector γ such that $g(t) = \gamma$ for $t \in [a, b]$. (ii) If $h(t)$ and $k(t)$ are vector functions in $\mathfrak{C}_n[a, b]$, prove that

$$(2.12) \quad \int_a^b [(u', h) + (u, k)]\, dt = 0, \quad \textit{for arbitrary} \quad u \in \mathfrak{C}_{n0}^{\ 1}[a, b]$$

if and only if there exists a constant vector γ such that $h(t) - \int_a^t k(s)\,ds = \gamma$ for $t \in [a, b]$; in particular, $h \in \mathfrak{C}_n^1[a, b]$ and $h'(t) = k(t)$ on $[a, b]$.

Hints. (i) Note that if γ is an arbitrary n-dimensional vector then

$$\int_a^b (u', \gamma)\,dt = 0 \quad \text{for} \quad u \in \mathfrak{C}_{n0}^1[a, b].$$

Conclude that if g satisfies (2.11) then also $\int_a^b (u', g - \gamma)\,dt = 0$ for arbitrary $u \in \mathfrak{C}_{n0}^1[a, b]$. In particular, for $\gamma = (b - a)^{-1} \int_a^b g(t)\,dt$ the vector function $u(t) = \int_a^t [g(s) - \gamma]\,ds$ belongs to $\mathfrak{C}_{n0}^1[a, b]$ and $u'(t) = g(t) - \gamma$, so that $0 = \int_a^b (u', g - \gamma)\,dt = \int_a^b |g - \gamma|^2\,dt$, and $g(t) = \gamma$ for $t \in [a, b]$.

(ii) Integrate by parts, to reduce (2.12) to (2.11) with

$$g(t) = h(t) - \int_a^t k(s)\,ds.$$

2.[ss] If $g(t) \in \mathfrak{C}_n[a, b]$, and m is a non-negative integer, prove that

$$\int_a^b (u^{[m]}, g)\,dt = 0, \quad \textit{for arbitrary} \quad u \in \mathfrak{C}_{n0}^m[a, b], \tag{2.13}$$

if and only if there exists an n-dimensional vector function which is a polynomial $P_{m-1}(t)$ in t of degree at most $m - 1$, and such that $g(t) = P_{m-1}(t)$ for $t \in [a, b]$, where it is understood that a polynomial vector function of degree at most -1 is the identically zero vector function.

Hint. For $m = 0$, and $a \leq t_1 < t_2 \leq b$, consider functions $u(t)$ of the form $u(t) = (t - t_1)(t_2 - t)\eta$ for $t \in [t_1, t_2]$, $u(t) = 0$ for $t \in [a, b] - [t_1, t_2]$, where η is a constant n-dimensional vector, and show that the real and pure imaginary parts of $(\eta, g(t))$ individually vanish at least once on $[t_1, t_2]$. From the arbitrariness of t_1, t_2, conclude that $(\eta, g(t)) \equiv 0$ on $[a, b]$ for arbitrary η, and hence that $g(t) \equiv 0$ on this interval. For $m > 0$, note that the identity

$$(u^{[m]}, v) = \{(u^{[m-1]}, v) - (u^{[m-2]}, v') + \cdots + (-1)^{m-1}(u, v^{[m-1]})\}' + (-1)^m(u, v^{[m]})$$

holds for arbitrary u, v in $\mathfrak{C}_n^m[a, b]$. Deduce that if $P_{m-1}(t)$ is a polynomial vector function in t of degree at most $m - 1$ then

$$\int_a^b (u^{[m]}, P_{m-1})\,dt = 0, \quad \textit{for arbitrary} \quad u \in \mathfrak{C}_{n0}^m[a, b], \tag{2.14}$$

and conclude that if g satisfies (2.13) then also

$$\int_a^b (u^{[m]}, g - P_{m-1})\,dt = 0, \quad \textit{for arbitrary} \quad u \in \mathfrak{C}_{n0}^m[a, b].$$

With the aid of (2.14) show that if $P_{m-1}(t)$ is a polynomial vector function of degree at most $m-1$ such that there exists a $u \in \mathfrak{C}_{n0}{}^m[a, b]$ satisfying $u^{[m]}(t) = P_{m-1}(t)$ on $[a, b]$ then $P_{m-1}(t) \equiv 0$, and in turn conclude that if g satisfies (2.13) then there is a unique $P_{m-1}(t)$ for which there is a $u \in \mathfrak{C}_{n0}{}^m[a, b]$ satisfying $u^{[m]}(t) = g(t) - P_{m-1}(t)$ on $[a, b]$.

3.[ss] If $f(t) \in \mathfrak{C}_n[a, b]$, let $S_0(t;f) = f(t)$, and $S_{k+1}(t;f) = \int_c^t S_k(s;f)\,ds$ for $t \in [a, b]$ and $k = 0, 1, 2, \ldots$, where c is some fixed value on (a, b). For $r_j(t) \in \mathfrak{C}_n[a, b]$, $(j = 0, 1, \ldots, m \geq 0)$, and $k = m, m+1, \ldots$, let H_k denote the condition:

$$I[u] \equiv \int_a^b \sum_{j=0}^m (u^{[j]}(t), r_j(t))\,dt = 0, \quad \textit{for} \quad u \in \mathfrak{C}_{n0}{}^k[a, b]. \tag{2.15}$$

Show that for a given $k \geq m$ the condition H_k holds if and only if there exists a polynomial vector function $P_{m-1}(t)$ of degree at most $m-1$ such that $\sum_{j=0}^m (-1)^j S_j(t; r_{m-j}) = P_{m-1}(t)$ on $[a, b]$; in particular, H_k is equivalent to H_j for $k \geq m$, $j \geq m$. Moreover, if H_k holds for some $k \geq m$ and $r_i(t) \equiv 0$ on $[a, b]$ for $m - q < i < m$, $q \geq 1$, then $r_m(t) \in \mathfrak{C}_n{}^q[a, b]$; in particular, in all cases $r_m(t) \in \mathfrak{C}_n{}^1[a, b]$.

Hints. For $m = 0$, modify appropriately the function $u(t)$ in the corresponding Hint of Problem 2. If $k \geq m \geq 1$, and $u \in \mathfrak{C}_{n0}{}^k[a, b]$, note that integration by parts yields

$$I[u] = \int_a^b (u^{[m]}(t), r(t))\,dt, \quad \textit{where} \quad r(t) = \sum_{j=0}^m (-1)^j S_j(t; r_{m-j});$$

moreover, since $\mathfrak{C}_{n0}^{k+1}[a, b] \subset \mathfrak{C}_{n0}{}^k[a, b]$,

$$I[u] = (-1)^{k-m+1} \int_a^b (u^{[k+1]}(t), S_{k-m+1}(t; r))\,dt, \quad \textit{for} \quad u \in \mathfrak{C}_{n0}^{k+1}[a, b].$$

Apply the result of Problem 2 to establish the existence of a polynomial vector function $P_k(t)$ of degree at most k such that $S_{k-m+1}(t; r) = P_k(t)$ on $[a, b]$, and, therefore, $r(t) = P_k^{[k-m+1]}(t) = P_{m-1}(t)$, a polynomial vector function of degree at most $m-1$. If $r_i(t) \equiv 0$ on $[a, b]$ for $m - q < i < m$, then

$$r(t) = P_{m-1}(t) - \sum_{j=q}^m (-1)^j S_j(t; r_{m-j}) \quad \text{and} \quad r(t) \in \mathfrak{C}_n{}^q[a, b].$$

4.[ss] Under the hypotheses of Problem 3, show that the condition H_k of that problem holds if and only if there exist vector functions $w_i(t)$, $(i = 1, \ldots, m)$, of class $\mathfrak{C}_n{}^1[a, b]$ such that

$$w_1 = r_m, \quad w_j' + w_{j+1} = r_{m-j}, \qquad (j = 1, \ldots, m-1),\ w_m' = r_0. \tag{2.16}$$

Hints. If $w_i \in \mathfrak{C}_n^1[a, b]$, $(i = 1, \ldots, m)$, and (2.16) holds, show that the integrand of (2.15) is equal to the derivative of $\sum_{j=0}^{m-1} (u^{[j]}, w_{m-j})$, and conclude that H_k holds for $k \geq m$. Conversely, if H_k holds for some integer $k \geq m$, and $P_{m-1}(t)$ is of degree at most $m - 1$ and satisfies the condition $\sum_{j=0}^{m} (-1)^j S_j(t; r_{m-j}) = P_{m-1}(t)$ of Problem 3, show that

$$w_i(t) = (-1)^{i-1} P_{m-1}^{[i-1]}(t) + \sum_{j=0}^{m-i} (-1)^j S_{j+1}(t; r_{m-i-j}), \qquad (i = 1, \ldots, m),$$

are vector functions of class $\mathfrak{C}_n^1[a, b]$ satisfying (2.16).

5.[s] Let $\mathfrak{C}_n^m[a, b; \theta]$ denote the class of vector functions $y \in \mathfrak{C}_n^m[a, b]$ for which $y(t) \equiv 0$ outside some compact subinterval $[a_1, b_1]$ of (a, b), where the values a_1, b_1 are dependent upon the particular y, and denote by H_k^θ the condition that the integral $I[u]$ of (2.15) is zero for $u \in \mathfrak{C}_n^k[a, b; \theta]$. Show that if $k \geq m$ then condition H_k^θ holds if and only if condition H_m holds.

Hints. As $\mathfrak{C}_n^k[a, b; \theta] \subset \mathfrak{C}_{n0}^k[a, b]$, clearly H_k implies H_k^θ. On the other hand, if H_k^θ holds consider a nondegenerate subinterval $[a_1, b_1]$ of (a, b), and for $u \in \mathfrak{C}_{n0}^{k+1}[a_1, b_1]$ set $u_\theta(t) = u(t)$ for $t \in [a_1, b_1]$, $u_\theta(t) = 0$ for $t \in [a, b] - [a_1, b_1]$. From the relation $0 = I[u_\theta] = \int_{a_1}^{b_1} \sum_{j=0}^{m} (u^{[j]}(t), r_j(t))\, dt$, establish with the aid of the result of Problem 3 that there exists a polynomial vector function $P_{m-1} = P_{m-1}(t; a_1, b_1)$ of degree at most $m - 1$ such that $\sum_{j=0}^{m} (-1)^j S_j(t; r_{m-j}) = P_{m-1}(t; a_1, b_1)$ for $t \in [a_1, b_1]$. Conclude that this polynomial vector function is independent of the subinterval $[a_1, b_1]$, and deduce from Problem 3 that condition H_m holds.

6.[ss] Suppose that for $j = 0, 1, \ldots, m$ and $i = 0, 1, \ldots, h$ the $n \times n$ matrix functions $R_j(t)$, $S_i(t)$ are continuous on $[a, b]$, and for $u \in \mathfrak{C}_n^m[a, b]$, $v \in \mathfrak{C}_n^h[a, b]$ let $L[u] = \sum_{j=0}^{m} R_j(t)u^{[j]}$, $M[v] = \sum_{i=0}^{h} S_i(t)v^{[i]}$. For $k \geq \max\{m, h\}$, let $(L, M; k)$ denote the condition that the functional

$$\{u, v\} \equiv \int_a^b (L[u], v)\, dt - \int_a^b (u, M[v])\, dt \tag{2.17}$$

is zero for $v \in \mathfrak{C}_n^k[a, b]$, $u \in \mathfrak{C}_{n0}^k[a, b]$. Correspondingly, let $(L, M; k)^0$ denote the condition that (2.17) is zero for $v \in \mathfrak{C}_{n0}^k[a, b]$, $u \in \mathfrak{C}_{n0}^k[a, b]$, and let $(L, M; k)^\theta$ denote the condition that (2.17) is zero for $v \in \mathfrak{C}_n^k[a, b; \theta]$, $u \in \mathfrak{C}_n^k[a, b; \theta]$. If $k_0 = \max\{m, h\}$, show that for $k \geq k_0$ the conditions $(L, M; k)$, $(L, M; k)^0$ and $(L, M; k)^\theta$ are equivalent to $(L, M; k_0)$.

Hints. Conclude from Problem 3 that for $k \geq k_0$ the condition $(L, M; k)$ is equivalent to $(L, M; k_0)$, and the problem is reduced to showing that for each $k \geq k_0$ the conditions $(L, M; k)$, $(L, M; k)^0$ and $(L, M; k)^\theta$ are equivalent. As $\mathfrak{C}_n^k[a, b; \theta] \subset \mathfrak{C}_{n0}^k[a, b] \subset \mathfrak{C}_n^k[a, b]$, clearly $(L, M; k) \to (L, M; k)^0 \to (L, M; k)^\theta$, and one need only show that $(L, M; k)^\theta \to (L, M; k)$. For a fixed integer $k \geq \max\{m, h\}$, and $0 < \varepsilon < (b - a)/2$, denote by $\phi_\varepsilon(t)$ a function of class $\mathfrak{C}^k[a, b]$ such that $\phi_\varepsilon(t) = 1$ for $t \in [a + \varepsilon, b - \varepsilon]$, while $\phi_\varepsilon(t) = 0$ for $t \in [a, a + \varepsilon/2]$ and $t \in [b - \varepsilon/2, b]$. If $v \in \mathfrak{C}_n^{\,k}[a, b]$ and $u \in \mathfrak{C}_n^{\,k}[a, b; \theta]$, then for ε so small that $u(t) = 0$ on $[a, a + \varepsilon]$ and $[b - \varepsilon, b]$, it follows that $\{u; v\} = \{u; \phi_\varepsilon v\} = 0$ in case $(L, M; k)^\theta$ holds. Note that for $v \in \mathfrak{C}_n^{\,k}[a, b]$ the functional $\{u; v\}$ is of the form (2.15) and from Problem 5 it follows that for fixed $v \in \mathfrak{C}_n^{\,k}[a, b]$ the condition $\{u; v\} = 0$ for arbitrary $u \in \mathfrak{C}_n^{\,k}[a, b; \theta]$ is equivalent to the condition that $\{u; v\} = 0$ for arbitrary $u \in \mathfrak{C}_{n0}^{\,k}[a, b]$.

7.[s] Establish the following generalization of Problem 3: If $r_j(t) \in \mathfrak{L}_n[a, b]$, $(j = 0, 1, \ldots, m \geq 0)$, and $k = m, m + 1, \ldots,$ as in Problem 3 let H_k denote the condition (2.15), where now the integral is in the sense of Lebesgue. Prove that for a given $k \geq m$ the condition H_k holds if and only if there exists a polynomial vector function $P_{m-1}(t)$ of degree at most $m - 1$ such that $\sum_{j=0}^{m} (-1)^j S_j(t; r_{m-j}) = P_{m-1}(t)$ a.e. on $[a, b]$; in particular, H_k is equivalent to H_l for $k \geq m, l \geq m$. Moreover, if H_k holds for some $k \geq m$ and $r_i(t) \equiv 0$ on $[a, b]$ for $m - q < i < m, q \geq 1$, then there exists a vector function $\rho_m(t) \in \mathfrak{C}_n^{q-1}[a, b]$ with $\rho_m^{(q-1)}(t)$ absolutely continuous on $[a, b]$ and such that $\rho_m(t) = r_m(t)$ a.e.; in particular, in all cases there exists an absolutely continuous function $\rho_m(t)$ such that $\rho_m(t) = r_m(t)$ a.e. on $[a, b]$.

Hint. Note that if $k \geq m$ and H_k holds, then integration by parts yields the result

$$I[u] = \int_a^b \sum_{j=1}^{m+1} (u^{[j]}(t), - S_1(t; r_{j-1}))\, dt = 0 \quad for \quad u \in \mathfrak{C}_{n0}^{k+1}[a, b].$$

8.[ss] Suppose that the $n \times n$ matrices $A_j(t)$, $(j = 0, 1, 2)$, are such that $A_1(t)$, $A_2(t)$ are nonsingular on $[a, b]$, the matrices $A_1(t)$, $A_2(t)$, $A_1^{-1}(t)$, $A_2^{-1}(t)$ all belong to $\mathfrak{L}_{nn}^{\infty}[a, b]$, and $A_0(t) \in \mathfrak{L}_{nn}[a, b]$. Let $\hat{\mathfrak{D}}(L)$ denote the linear space of vector functions y in $\mathfrak{L}_n[a, b]$ such that $y = A_2^{-1}u_y$ with u_y an absolutely continuous vector function on $[a, b]$, and $\hat{\mathfrak{D}}_0(L) = \{y \mid y \in \hat{\mathfrak{D}}(L), u_y(a) = 0 = u_y(b)\}$. If $L[y](t) = A_1(t)[A_2(t)y]' + A_0(t)y$, then $y(t)$ is said to be a solution of $L[y] = 0$ on $[a, b]$ if $y \in \hat{\mathfrak{D}}(L)$ and $u = u_y$ is a solution of $u' + A_1^{-1}(t)A_0(t)A_2^{-1}(t)u = 0$, $t \in [a, b]$, in the sense of Chapter II. Correspondingly, let $\hat{\mathfrak{D}}(L^\star)$ be the linear space of vector functions z in $\mathfrak{L}_n[a, b]$ such that $z = A_1^{*-1}v_z$, with v_z absolutely continuous on $[a, b]$, and for $z \in \hat{\mathfrak{D}}(L^\star)$ let

$$\begin{aligned} L^\star[z](t) &= -A_2^*(t)[A_1^*(t)z(t)]' + A_0^*(t)z(t) \\ &= -A_2^*(t)v_z'(t) + A_0^*(t)A_1^{*-1}(t)v_z(t). \end{aligned}$$

Show that the class $\hat{\mathfrak{D}}(L^\star)$ is characterized as the set of vector functions

$z \in \mathfrak{L}_n[a, b]$ such that there exists a corresponding $f_z \in \mathfrak{L}_n[a, b]$ for which

$$\int_a^b (L[y], z)\, dt - \int_a^b (y, f_z)\, dt = 0, \quad for \quad y \in \hat{\mathfrak{D}}_0(L),$$

and for $z \in \hat{\mathfrak{D}}(L^\star)$ the corresponding f_z is uniquely determined as $f_z = L^\star[z]$.

Hint. Imitate the steps in the proof of Theorem 2.2, now using the result of the above Problem 7.

9. With solution defined as in the preceding problem determine the solution of the scalar differential equation $A_1(t)[A_2(t)y]' + A_0(t)y = 0$ on $(-\infty, \infty)$ which satisfies the initial condition $y(-1) = 1$, when:

(i) $A_0(t) \equiv A_2(t) \equiv 1,\ A_1(t) = \dfrac{|t|}{t}$ for $t \neq 0$;

(ii) $A_0(t) \equiv A_1(t) \equiv 1,\ A_2(t) = \dfrac{|t|}{t}$ for $t \neq 0$.

Ans. (i) $y(t) = e^{t+1}$ for $t \leq 0$, $y(t) = e^{1-t}$ for $t \geq 0$;
(ii) $y(t) = e^{t+1}$ for $t < 0$, $y(t) = -e^{1-t}$ for $t > 0$.

3. ADJOINT *n*-th ORDER DIFFERENTIAL EQUATIONS

The results of the preceding sections may be applied to an n-th order linear homogeneous differential equation

$$(3.1) \qquad l_n[u] \equiv p_n(t)u^{[n]} + \cdots + p_1(t)u' + p_0(t)u = 0, \qquad t \in [a, b],$$

where the functions $p_0(t), \ldots, p_n(t)$ are continuous on $[a, b]$ and $p_n(t) \neq 0$ on this interval. As shown in Section I.9, under the substitution $y_\alpha = u^{[\alpha-1]}$, $(\alpha = 1, \ldots, n)$, the differential equation (3.1) is equivalent to a vector differential equation of the first order in the n-dimensional vector function $y(t) = (y_\alpha(t))$. In line with the form studied in the preceding section, we now write this system as $L[y] \equiv (L_\alpha[y]) = 0$, with

$$(3.2) \qquad \begin{aligned} L_1[y] &\equiv p_n(t)y_n' + p_0(t)y_1 + \cdots + p_{n-1}(t)y_n = 0, \\ L_\beta[y] &\equiv y_{n-\beta+1}' - y_{n-\beta+2} = 0, \qquad (\beta = 2, \ldots, n), \end{aligned}$$

where in the form (1.1) the $n \times n$ matrices $A_0(t)$, $A_1(t)$, $A_2(t)$ are given by
(3.3)

$$A_2(t) = E,\ A_1(t) = \begin{bmatrix} 0 & \cdots 0 & p_n(t) \\ 0 & \cdots 1 & 0 \\ \cdot & \cdots\cdot & \cdot \\ 1 & 0 \cdots & 0 \end{bmatrix},\ A_0(t) = \begin{bmatrix} p_0(t) & \cdots & p_{n-1}(t) \\ 0 & \cdot\cdot 0 & -1 \\ \cdot & \cdots & \cdot \\ 0 & -1 \cdot\cdot & 0 \end{bmatrix},$$

In particular, $u(t)$ is a solution of the nonhomogeneous differential equation

$$l_n[u](t) = r(t), \qquad t \in [a, b], \tag{3.4}$$

if and only if $y(t) = (u^{[\alpha-1]}(t))$ is a solution of the corresponding nonhomogeneous vector differential system (1.3) with $g(t)$ the n-dimensional vector function $(g_\alpha(t))$ with $g_1(t) = r(t)$, $g_\beta(t) \equiv 0$, $\beta = 2, \ldots, n$.

For (3.2) the associated adjoint vector differential system (2.5) is then $L^\star[z] \equiv (L_\alpha^\star[z]) = 0$, where

$$\begin{aligned} L_1^\star[z] &\equiv -z_n' + \bar{p}_0(t)z_1 = 0, \\ L_\gamma^\star[z] &\equiv -z_{n-\gamma+1}' + \bar{p}_{\gamma-1}(t)z_1 - z_{n-\gamma+2} = 0, \qquad (\gamma = 2, \ldots, n-1), \\ L_n^\star[z] &\equiv -[\bar{p}_n(t)z_1]' + \bar{p}_{n-1}(t)z_1 - z_2 = 0. \end{aligned} \tag{3.5}$$

In general, the system (3.5) is not equivalent to an n-th order differential equation of the form (3.1). For $p_j(t)$, $(j = 0, 1, \ldots, n)$, continuous on $[a, b]$, *and independent of the condition that* $p_n(t) \neq 0$ *on this interval*, let $\mathfrak{D}(l_n^\star)$ denote the set of scalar functions $v(t)$ for which $\bar{p}_n(t)v \in \mathfrak{C}^1[a, b]$, and the scalar functions $z_\beta(t) = z_{\beta;v}(t)$, $(\beta = 2, \ldots, n)$, defined recursively by the last $n - 1$ equations of (3.5) with $z_1(t) = v(t)$, are all of class $\mathfrak{C}^1[a, b]$. For $v \in \mathfrak{D}(l_n^\star)$ let

$$l_n^\star[v](t) = \bar{p}_0(t)v(t) - z_{n;v}'(t), \quad t \in [a, b]. \tag{3.6}$$

In particular, for $v \in \mathfrak{D}(l_n^\star)$ we have that $l_n^\star[v] \in \mathfrak{C}[a, b]$. An expression (3.6) thus defined, is also written sometimes as

$$l_n^\star[v] = \bar{p}_0 v - (\bar{p}_1 v - [\bar{p}_2 v - \cdots - \{\bar{p}_{n-1}v - (\bar{p}_n v)'\}' \cdots]')', \tag{3.6$'$}$$

and has been designated a "quasi-differential" expression by Bôcher.

Corresponding to relation (2.6), one now has the identity

$$\bar{v}l_n[u] - \overline{l_n^\star[v]}u = \left\{\bar{v}p_n u^{[n-1]} + \sum_{j=0}^{n-2} \bar{z}_{n-j;v}u^{[j]}\right\}', \tag{3.7}$$
$$\textit{for} \quad v \in \mathfrak{D}(l_n^\star), \quad u \in \mathfrak{C}^n[a, b].$$

Moreover, the following result, corresponding to that of Theorem 2.2, may be established with the aid of Problem III.2:4.

Theorem 3.1. *The class $\mathfrak{D}(l_n^\star)$ is characterized as the set of scalar functions $v \in \mathfrak{C}[a, b]$ such that there exists a corresponding scalar function $f_v \in \mathfrak{C}[a, b]$ for which*

$$\int_a^b \bar{v}l_n[u]\,dt - \int_a^b f_v u\,dt = 0, \quad \textit{for} \quad u \in \mathfrak{C}_0{}^n[a, b], \tag{3.8}$$

and for $v \in \mathfrak{D}(l_n^\star)$ *the corresponding* f_v *is uniquely determined as* $f_v(t) = l_n^\star[v](t)$ *for* $t \in [a, b]$.

Corollary. *If* $v \in \mathfrak{C}[a, b]$, *then*

$$\int_a^b \bar{v} l_n[u]\, dt = 0 \quad \textit{for} \quad u \in \mathfrak{C}_0{}^n[a, b], \tag{3.9}$$

if and only if $v \in \mathfrak{D}(l_n^\star)$ *and* $l_n^\star[v] = 0$.

In general, for $v \in \mathfrak{D}(l_n^\star)$ the functional $l_n^\star[v]$ of (3.6) is called the *adjoint* of $l_n[u]$, and the equation

$$l_n^\star[v](t) = 0, \qquad t \in [a, b], \tag{3.10}$$

is called the *adjoint equation* to (3.1). In particular, if $p_n(t) \neq 0$ for $t \in [a, b]$, $\tau \in [a, b]$, $r \in \mathfrak{C}[a, b]$, and $\zeta_1, \ldots, \zeta_n$ are given constant values, then from the theory of vector differential systems of the first order it follows that there is a unique solution of the system

$$\begin{aligned} &l_n^\star[v](t) = r(t), \qquad t \in [a, b], \\ &v(\tau) = \zeta_1,\, z_{\beta;v}(\tau) = \zeta_\beta, \qquad (\beta = 2, \ldots, n). \end{aligned} \tag{3.11}$$

In case the coefficient functions $p_j(t)$, $(j = 0, 1, \ldots, n)$, are respectively of class $\mathfrak{C}^j[a, b]$ then $\mathfrak{C}^n[a, b] \subset \mathfrak{D}(l_n^\star)$; that is, if $v \in \mathfrak{C}^n[a, b]$ then the scalar functions $z_\beta = z_{\beta;v}$, $(\beta = 2, \ldots, n)$, defined recursively by the last $n - 1$ equations of (3.5) with $z_1 = v$ are all of class $\mathfrak{C}^1[a, b]$. Indeed, for such a function v,

(3.12)

$$z_{\gamma+1} = (\bar{p}_{n-\gamma} v) - (\bar{p}_{n-\gamma+1} v)' + \cdots + (-1)^\gamma (\bar{p}_n v)^\gamma, \qquad (\gamma = 1, \ldots, n - 1),$$

and

$$l_n^\star[v] = \sum_{j=0}^{n} (-1)^j (\bar{p}_j(t) v)^{[j]}. \tag{3.13}$$

Moreover,

$$l_n^\star[v] = q_n(t) v^{[n]} + \cdots + q_1(t) v' + q_0(t) v, \tag{3.13'}$$

where

$$q_j(t) = \sum_{i=j}^{n} (-1)^i {}_iC_j \bar{p}_i^{[i-j]}(t), \qquad (j = 0, 1, \ldots, n), \tag{3.14}$$

and ${}_iC_j$ is the binomial coefficient $i!/[j!\,(i - j)!]$. In particular, $q_j \in \mathfrak{C}^j[a, b]$ for $j = 0, 1, \ldots, n$, and $q_n(t) = (-1)^n \bar{p}_n(t)$ so that if $p_n(t) \neq 0$ for $t \in [a, b]$ then also $q_n(t) \neq 0$ for $t \in [a, b]$.

In view of the identity (3.7), we have the following result.

Theorem 3.2. *If $p_j(t) \in \mathfrak{C}^j[a, b]$ for $j = 0, 1, \ldots, n$, then for $u \in \mathfrak{C}^n[a, b]$, $v \in \mathfrak{C}^n[a, b]$ we have*

$$\bar{v}l_n[u] - \overline{l_n^\star[v]}u = [K\{u; v\}]', \tag{3.15}$$

where

$$\begin{aligned} K\{u; v\}(t) &= \sum_{\beta=1}^{n}\left\{\sum_{\gamma=1}^{n+1-\beta}(-1)^{\gamma-1}[p_{\beta+\gamma-1}(t)\,\bar{v}(t)]^{[\gamma-1]}\right\}u^{[\beta-1]}(t), \\ &= \sum_{j=2}^{n+1}\sum_{\alpha+\beta=j}(-1)^{\alpha-1}[p_{j-1}(t)\,\bar{v}(t)]^{[\alpha-1]}\,u^{[\beta-1]}(t). \end{aligned} \tag{3.16}$$

The expression $K\{u; v\}$, which is called the *bilinear concomitant of u and v*, is seen to be of the form

$$K\{u; v\}(t) = \sum_{\alpha,\beta=1}^{n} \bar{v}^{[\alpha-1]}(t)\,K_{\alpha\beta}(t)\,u^{[\beta-1]}(t), \tag{3.17}$$

where $K_{\alpha\beta}(t) \equiv 0$, for $\alpha + \beta > n + 1$, $K_{\alpha\beta}(t) = (-1)^{\alpha-1}p_n(t)$ for $\alpha + \beta = n + 1$. In particular, since $p_n(t) \neq 0$ for $t \in [a, b]$, the matrix $[K_{\alpha\beta}(t)]$ is non-singular on this interval.

Whereas the condition that $p_j(t) \in \mathfrak{C}^j[a, b]$ for $j = 0, 1, \ldots, n$ is a sufficient condition for $\mathfrak{C}^n[a, b] \subset \mathfrak{D}(l_n^\star)$ and $l_n^\star[v]$ to be of the form (3.13′) with continuous coefficients for $v \in \mathfrak{C}^n[a, b]$, it is to be emphasized that this condition is not necessary for $l_n^\star[v]$ to have the form (3.13′) for $v \in \mathfrak{C}^n[a, b]$. For example, if $p_2(t)$ is merely of class $\mathfrak{C}^1[a, b]$ with $p_2(t) \neq 0$ for $t \in [a, b]$, and $l_2[u] = p_2(t)u'' + p_2'(t)u' = [p_2(t)u']'$, then from (3.5) it follows that $\mathfrak{D}(l_2^\star)$ consists of those scalar functions $v \in \mathfrak{C}^1[a, b]$ for which there exists a $z_2 \in \mathfrak{C}^1[a, b]$ such that $z_2 = \bar{p}_2'v - [\bar{p}_2v]' = -\bar{p}_2v'$. It then follows that $v \in \mathfrak{D}(l_2^\star)$ if and only if $v \in \mathfrak{C}^2[a, b]$, and $l_2^\star[v] = -z_2' = (\bar{p}_2v')' = \bar{p}_2v'' + \bar{p}_2'v'$, which is of the form (3.13′) with continuous coefficients.

Theorem 3.3. *If $l_n[u] = \sum_{j=0}^{n} p_j(t)u^{[j]}$ and $l_n^+[v] = \sum_{j=0}^{n} q_j(t)v^{[j]}$, where the coefficient functions $p_j(t)$, $q_j(t)$, $(j = 0, 1, \ldots, n)$ are continuous on $[a, b]$, and*

$$\int_a^b \bar{v}l_n[u]\,dt - \int_a^b \overline{l_n^+[v]}u\,dt = 0, \quad \text{for} \quad v \in \mathfrak{C}^n[a, b], \qquad u \in \mathfrak{C}_0{}^n[a, b], \tag{3.18}$$

then

$$\mathfrak{C}^n[a, b] \subset \mathfrak{D}(l_n^\star), \quad \text{and} \quad l_n^\star[v] = l_n^+[v] \quad \text{for} \quad v \in \mathfrak{C}^n[a, b]. \tag{3.19}$$

Moreover, if $p_n(t) \neq 0$ and $q_n(t) \neq 0$ for $t \in [a, b]$, then

(i) *a function $v \in \mathfrak{C}[a, b]$ is such that*

$$\int_a^b \bar{v}l_n[u]\,dt = 0, \quad \text{for} \quad u \in \mathfrak{C}_0{}^n[a, b], \tag{3.20}$$

if and only if $v \in \mathfrak{C}^n[a, b]$ *and* $l_n^+[v] = 0$;

(ii) $\mathfrak{D}(l_n^\star) = \mathfrak{C}^n[a, b]$, *and* $l_n^\star[v] = l_n^+[v]$ *for* $v \in \mathfrak{D}(l_n^\star)$.

The conclusion that condition (3.18) implies (3.19) is an immediate consequence of Theorem 3.1. Moreover, from (3.18) it follows directly that (3.20) holds for a function $v \in \mathfrak{C}^n[a, b]$ which is a solution of $l_n^+[v] = 0$. On the other hand, if $w(t)$ is a nonidentically vanishing solution of $l_n^+[w] = 0$ then the solution of $l_n[u] = w$ satisfying the initial conditions $u^{[\alpha-1]}(a) = 0$, $(\alpha = 1, \ldots, n)$, cannot be such that $u^{[\alpha-1]}(b) = 0$, $(\alpha = 1, \ldots, n)$, for if these end-conditions at $t = b$ held then (3.18) would imply $0 = \int_a^b \bar{w} l_n[u]\, dt = \int_a^b |w|^2\, dt$, and hence the contradictory result $w(t) \equiv 0$ for $t \in [a, b]$. Consequently, if $v \in \mathfrak{C}[a, b]$ and (3.20) holds, then for $w_1(t), \ldots, w_n(t)$ linearly independent solutions of $l_n^+[w] = 0$ there exist unique constants $c_1, \ldots, c_n$ such that if $v_1(t) = v(t) - \sum_{\alpha=1}^n c_\alpha w_\alpha(t)$ then the solution $u_1(t)$ of $l_n[u_1] = v_1(t)$, $u_1^{[\alpha-1]}(a) = 0$, $(\alpha = 1, \ldots, n)$, satisfies also the conditions $u_1^{[\alpha-1]}(b) = 0$, $(\alpha = 1, \ldots, n)$, and consequently by (3.20) we have $\int_a^b \bar{v} l_n[u_1]\, dt = 0$. Since also $\int_a^b \bar{w}_\alpha l_n[u_1]\, dt = 0$, $(\alpha = 1, \ldots, n)$, it then follows that

$$0 = \int_a^b \bar{v}_1\, l_n[u_1]\, dt = \int_a^b |v_1|^2\, dt,$$

and hence on $[a, b]$ we have $v_1(t) \equiv 0$ and $v(t) \equiv \sum_{\alpha=1}^n c_\alpha\, w_\alpha(t)$, so that $v \in \mathfrak{C}^n[a, b]$ and $l_n^+[v] = 0$.

In order to prove conclusion (ii), suppose that $v \in \mathfrak{D}(l_n^\star)$ and let $w_1(t), \ldots, w_n(t)$ be linearly independent solutions of $l_n^+[w] = 0$, which for simplicity of detail we will choose to be orthonormal on $[a, b]$ in the sense that

$$\int_a^b \bar{w}_\alpha\, w_\beta\, dt = \delta_{\alpha\beta}, (\alpha, \beta = 1, \ldots, n).$$

If $w_0(t)$ is a particular solution of $l_n^+[w_0] = l_n^\star[v]$, then

$$w_1(t) = w_0(t) + \sum_{\alpha=1}^n \left[\int_a^b \bar{w}_\alpha(v - w_0)\, ds \right] w_\alpha(t)$$

is the solution of $l_n^+[w_1] = l_n^\star[v]$ such that $\int_a^b \bar{w}(v - w_1)\, dt = 0$ for arbitrary solutions w of $l_n^+[w] = 0$. Then for $u \in \mathfrak{C}_0{}^n[a, b]$ we have

$$\begin{aligned}\int_a^b \bar{v} l_n[u]\, dt &= \int_a^b \overline{l_n^\star[v]} u\, dt,\\ &= \int_a^b \overline{l_n^+[w_1]} u\, dt,\\ &= \int_a^b \bar{w}_1 l_n[u]\, dt.\end{aligned}$$

That is, $\int_a^b (\bar{v} - \bar{w}_1)\, l_n[u]\, dt = 0$ for $u \in \mathfrak{C}_0{}^n[a, b]$, and from conclusion (i) it follows that $v - w_1 \in \mathfrak{C}^n[a, b]$ and $l_n^+[v - w_1] = 0$. Moreover, since $\int_a^b \bar{w}(v - w_1)\, dt = 0$ for arbitrary solutions w of $l_n^+[w] = 0$, it follows that $0 = \int_a^b |v - w_1|^2\, dt$ and hence $v(t) = w_1(t)$ for $t \in [a, b]$. Consequently, for $v \in \mathfrak{D}(l_n^\star)$ we have that $v \in \mathfrak{C}^n[a, b]$ and $l_n^\star[v] = l_n^+[v]$.

A differential expression $l_n[u]$ given by (3.1) with continuous coefficients is said to be *symmetric* if

$$\int_a^b \bar{v} l_n[u]\, dt - \int_a^b \overline{l_n[v]} u\, dt = 0 \tag{3.21}$$

for arbitrary $u \in \mathfrak{C}_0{}^n[a, b]$, $v \in \mathfrak{C}_0{}^n[a, b]$. In view of Problem III.2:6 this condition is equivalent to the relation (3.21) holding for arbitrary $v \in \mathfrak{C}^n[a, b]$, $u \in \mathfrak{C}_0{}^n[a, b]$, and consequently $l_n[u]$ is symmetric if and only if $\mathfrak{C}^n[a, b] \subset \mathfrak{D}(l_n^\star)$ and $l_n^\star[v] = l_n[v]$ for $v \in \mathfrak{C}^n[a, b]$. If $p_j(t) \in \mathfrak{C}^j[a, b]$, $(j = 0, 1, \ldots, n)$, then in view of Theorem 3.2, Theorem 3.3, and the result of Problem III.2:3 for $m = 0$, it follows that $l_n[u]$ is symmetric if and only if

$$l_n^\star[v] - l_n[v] = 0, \quad \text{for} \quad v \in \mathfrak{C}^n[a, b],$$

and this condition holds if and only if each coefficient function $p_j(t)$ in $l_n[u]$ is equal to the respective coefficient function $q_j(t)$ of $l_n^\star[u]$ when written in the form (3.13′). In case $p_j(t) \in \mathfrak{C}^j[a, b]$, $(j = 0, 1, \ldots, n)$, and $p_j(t) = p_j{}^1(t) + i\, p_j{}^2(t)$, $(j = 0, 1, \ldots, n)$, with the functions $p_j{}^1$, $p_j{}^2$ real-valued on $[a, b]$, then $l_n[u] = l_n{}^1[u] + i l_n{}^2[u]$ with $l_n{}^\sigma[u] = \sum_{j=0}^n p_j{}^\sigma(t) u^{[j]}$, $(\sigma = 1, 2)$. Correspondingly, $l_n^\star[v] = l_n^{1\star}[v] - i l_n^{2\star}[v]$, where

$$l_n^{\sigma\star}[v] = \sum_{j=0}^n (-1)^j [p_j{}^\sigma(t) v]^{[j]},$$

and $l_n[u]$ is symmetric if and only if $l_n{}^1[u] = l_n^{1\star}[u]$, $l_n{}^2[u] = -l_n^{2\star}[u]$ for arbitrary $u \in \mathfrak{C}^n[a, b]$. Finally, from Theorem 3.3 it follows that if $l_n[u]$ is symmetric, and $p_n(t) \neq 0$ for $t \in [a, b]$, then $\mathfrak{D}(l_n^\star) = \mathfrak{C}^n[a, b]$ and $l_n^\star[v] = l_n[v]$ for all $v \in \mathfrak{D}(l_n^\star)$.

Problems III.3

1. If $p_j(t)$, $(j = 0, 1, 2)$, are of class $\mathfrak{C}^j[a, b]$ and real-valued on $[a, b]$, show that $l_2[u] = p_2(t)u'' + p_1(t)u' + p_0(t)u$ is symmetric if and only if $p_2'(t) = p_1(t)$ for $t \in [a, b]$. For any such $l_2[u]$ with $p_2(t) \neq 0$ on $[a, b]$, show that there is a multiplier $\mu(t)$ for which $\lambda[u] = \mu(t)\, l_2[u]$ is symmetric.

2. If $s_j(t)$, $(j = 0, 1, 2, 3)$, are of class $\mathfrak{C}^j[a, b]$ and real-valued on $[a, b]$, show that $l_3[u] = i[s_3(t)u''' + s_2(t)u'' + s_1(t)u' + s_0(t)u]$ is symmetric if and only if $l_3[u] = i[(r_1(t)u')'' + (r_1(t)u'')' + (r_0(t)u)' + r_0(t)u']$ for suitable choices of $r_0(t)$, $r_1(t)$.

3. If $p_j(t) \in \mathfrak{C}^j[a, b]$, $(j = 0, 1, \ldots, n)$, and $l_n[u] = \sum_{j=0}^n p_j(t)u^{[j]}$ is symmetric, show that the matrix $K(t) = [K_{\alpha\beta}(t)]$ of (3.17) is skew-Hermitian; that is, $K_{\alpha\beta}(t) = -\bar{K}_{\beta\alpha}(t)$ for $t \in [a, b]$, $\alpha, \beta = 1, \ldots, n$.

Hint. For $u \in \mathfrak{C}^n[a, b]$, $v \in \mathfrak{C}^n[a, b]$, show that (3.15) implies that the function

$$\sum_{\alpha,\beta=1}^{n} \bar{v}^{[\alpha-1]}(t)[K_{\alpha\beta}(t) + \bar{K}_{\beta\alpha}(t)]\, u^{[\beta-1]}(t)$$

is constant on $[a, b]$, and hence if $u^{[\alpha-1]}(a) = 0 = v^{[\alpha-1]}(a)$ then the value of this function is zero.

4. For $m = 0, 1, 2, \ldots$, define $\Lambda_m(u; \pi)$ as follows:

$$\text{(3.22)}\qquad \begin{aligned} \Lambda_0(u; \pi) &= \pi(t)u, \ \Lambda_{2r}(u; \pi) = (\pi(t)u^{[r]})^{[r]}, \\ \Lambda_{2r-1}(u; \pi) &= \tfrac{1}{2}\{(\pi(t)u^{[r-1]})^{[r]} + (\pi(t)u^{[r]})^{[r-1]}\}, \end{aligned}$$

where in the definition of $\Lambda_m(u; \pi)$ it is supposed that $\pi(t) \in \mathfrak{C}^{k(m)}[a, b]$, and $k(m)$ is used to denote $m/2$ or $(m + 1)/2$ according as m is even or m is odd. If $l_n[u] = \sum_{m=0}^n c_m \Lambda_m(u; \pi_m)$, where the c_m's are constants, show that $l_n[u]$ is of the form (3.1) with continuous coefficients, $\mathfrak{C}^n[a, b] \subset \mathfrak{D}(l_n^\star)$, and $l_n^\star[v] = \sum_{m=0}^n \bar{c}_m \Lambda_m(v; (-1)^m \bar{\pi}_m)$ for $v \in \mathfrak{C}^n[a, b]$. Moreover, if $l_n[u]$ is of the form (3.1) with $p_j(t) \in \mathfrak{C}^j[a, b]$, show that there exist unique functions $\pi_m(t) \in \mathfrak{C}^m[a, b]$, $(m = 0, 1, \ldots)$, such that $l_n[u] = \sum_{m=0}^n \Lambda_m(u; \pi_m)$ for $u \in \mathfrak{C}^n[a, b]$.

Hints. Show that if $m \geq 1$ then

$$\bar{v}\Lambda_m(u; \pi) - \overline{\Lambda_m(v; (-1)^m \bar{\pi})}u = [K(u; v \mid \Lambda_m)]',$$

where

$$K(u; v \mid \Lambda_{2r}) = \sum_{\alpha=1}^{r} (-1)^{\alpha-1}\{\bar{v}^{[\alpha-1]}(\pi u^{[r]})^{[r-\alpha]} - (\bar{v}^{[r]}\pi)^{[r-\alpha]}u^{[\alpha-1]}\},$$

$$\begin{aligned} K(u; v \mid \Lambda_{2r-1}) = {} & \tfrac{1}{2} \sum_{\alpha=1}^{r} (-1)^{\alpha-1}\{\bar{v}^{[\alpha-1]}(\pi u^{[r-1]})^{[r-\alpha]} + (\bar{v}^{[r-1]}\pi)^{[r-\alpha]}u^{[\alpha-1]}\} \\ & + \tfrac{1}{2} \sum_{\alpha=1}^{r-1} (-1)^{\alpha-1}\{\bar{v}^{[\alpha-1]}(\pi u^{[r]})^{[r-\alpha]} + (\bar{v}^{[r]}\pi)^{[r-\alpha]}u^{[\alpha-1]}\}. \end{aligned}$$

To establish the last conclusion, note that $l_n[u] - \Lambda_n(u; p_n)$ is a differential expression of the form $\sum_{j=0}^{n-1} r_j(t)u^{[j]}$ with $r_j(t) \in \mathfrak{C}^j[a, b]$ for $j = 0, 1, \ldots, n - 1$.

5. As in Problem 4, let the differential expression $\Lambda_m(u; \pi)$, $(m = 0, 1, \ldots)$, be defined by (3.22), and denote by $k(m)$ the number $m/2$ or $(m + 1)/2$, according as m is even or odd. Moreover, let $g_0(t) \equiv 1$, $g_r(t) = t^r/r!$, $r = 1, 2, \ldots$. Suppose that $l_n[u] = \sum_{j=0}^n p_j(t)u^{[j]}$, where the coefficient functions $p_j(t)$ are all continuous on $[a, b]$.

(i) If $g_r(t) \in \mathfrak{D}(l_n^\star)$ for $r = 0, 1, \ldots, k(n) - 1$, prove that there exist functions $\pi_m(t)$, $(m = 0, 1, \ldots, n)$, such that $\pi_m(t) \in \mathfrak{C}^{k(m)}[a, b]$, $(m = 0, 1, \ldots, n)$, and

$$l_n[u] = \sum_{m=0}^{n} \Lambda_m(u; \pi_m), \quad for \quad u \in \mathfrak{C}^n[a, b]; \tag{3.23}$$

also, $\mathfrak{C}^n[a, b] \subset \mathfrak{D}(l_n^\star)$ and

$$l_n^\star[v] = \sum_{m=0}^{n} \Lambda_m(v; (-1)^m \bar{\pi}_m), \quad for \quad v \in \mathfrak{C}^n[a, b]. \tag{3.24}$$

(ii) If $l_n[u]$ satisfies the above specified conditions and $p_n(t) \not\equiv 0$ on $[a, b]$, while $l_h^+[v] = \sum_{i=0}^{h} q_i(t) v^{[i]}$ with continuous coefficients $q_i(t)$ and $q_h(t) \not\equiv 0$ on $[a, b]$, and such that

$$\int_a^b \bar{v}\, l_n[u]\, dt - \int_a^b \overline{l_h^+[v]} u\, dt = 0, \quad for \quad v \in \mathfrak{C}^h[a, b],\ u \in \mathfrak{C}_0^n[a, b], \tag{3.25}$$

then $h = n$, and $l_h^+[v] = \sum_{m=0}^{n} \Lambda_m(v; (-1)^m \bar{\pi}_m)$.

Hints. (i) Prove by induction on n that under the stated conditions $l_n[u]$ has the form (3.23), noting that the result is immediate for $n = 0$. As an indication of the basic ingredient of proof, although logically unnecessary, note that for $n = 1$ the fact that $g_0(t) \equiv 1$ is a member of $\mathfrak{D}(l_1^\star)$ implies that

$$\int_a^b (p_1(t)u' + [p_0(t) - \bar{q}_0(t)]u)\, dt = 0, \quad for \quad u \in \mathfrak{C}_0^{\,1}[a, b],$$

where $q_0(t) = l_1^\star[g_0](t)$, and from Problem III.2:1 it follows that $p_1(t) \in \mathfrak{C}^1[a, b]$, and hence $l_1[u] = \Lambda_1(u; \pi_1) + \Lambda_0(u; \pi_0)$, where $\pi_1(t) = p_1(t)$, $\pi_0(t) = p_0(t) - \pi_1'(t)/2$. Suppose that the conclusion (3.23) as to the form of $l_n[u]$ holds for $n = 0, 1, \ldots, s - 1$, and let $l_s[u] = \sum_{j=0}^{s} p_j(t) u^{[j]}$ with continuous coefficient functions $p_j(t)$ be such that $g_r(t) \in \mathfrak{D}(l_s^\star)$, $r = 0, 1, \ldots, k(s) - 1$. If z and w belong to $\mathfrak{C}^s[a, b]$, verify that

$$wz^{[i]} = \sum_{j=0}^{i} (-1)^j\, {}_iC_j [w^{[j]}z]^{[i-j]}, \qquad (i = 0, 1, \ldots, s),$$

where ${}_iC_j = i!/[j!\,(i-j)!]$. In particular, for u and v in $\mathfrak{C}^s[a, b]$,

$$\bar{v} l_s[u] = \sum_{\sigma=0}^{s} (-1)^\sigma \lambda_\sigma[\bar{v}^{[\sigma]} u],$$

with

$$\lambda_\sigma[y] = \sum_{i=\sigma}^{s} {}_iC_\sigma\, p_i(t) y^{[i-\sigma]}, \qquad (\sigma = 0, 1, \ldots, s).$$

Note that the condition that $g_r(t) \in \mathfrak{D}(l_s^\star)$, for $r = 0, 1, \ldots, k(s) - 1$, implies that for $f_r(t) = l_s^\star[g_r](t)$ we have

$$(3.26) \quad \int_a^b \bar{f}_r u \, dt = \int_a^b g_r \, l_s[u] \, dt = \sum_{\sigma=0}^{r} (-1)^\sigma \int_a^b \lambda_\sigma[g_{r-\sigma} u] \, dt,$$

$$for \quad u \in \mathfrak{C}_0^s[a, b], \qquad r = 0, 1, \ldots, k(s) - 1.$$

If the continuous functions $q_r(t)$ are defined as

$$q_0(t) = f_0(t), \qquad q_r(t) = f_r(t) - \sum_{j=0}^{r-1} g_{r-j}(t) \, q_j(t),$$

for $r = 0, 1, \ldots, k(s) - 1$, in view of the fact that $g_r u \in \mathfrak{C}_0{}^s[a, b]$ whenever $u \in \mathfrak{C}_0{}^s[a, b]$, conclude by induction that (3.26) implies

$$I_\sigma[u] \equiv \int_a^b \{\lambda_\sigma[u] - (-1)^\sigma \bar{q}_\sigma u\} \, dt = 0,$$

$$for \quad u \in \mathfrak{C}_0{}^s[a, b], \qquad (\sigma = 0, 1, \ldots, k(s) - 1).$$

As $\lambda_\sigma[u]$ is of order at most $s - \sigma$, deduce from Problem III.2:3 that

$$I_\sigma[y] = 0 \quad for \quad y \in \mathfrak{C}_0^{s-\sigma}[a, b], \qquad (\sigma = 0, 1, \ldots, k(s) - 1);$$

moreover, since $u^{[\sigma]} \in \mathfrak{C}_0^{s-\sigma}[a, b]$ when $u \in \mathfrak{C}_0{}^s[a, b]$, conclude that

$$I_\sigma[u^{[\sigma]}] = 0 \quad for \quad u \in \mathfrak{C}_0{}^s[a, b], \qquad (\sigma = 0, 1, \ldots, k(s) - 1).$$

Note that for $j = \sigma, \ldots, s$ the coefficient of $u^{[j]}$ in $\lambda_\sigma[u^{[\sigma]}]$ is ${}_jC_\sigma p_j(t)$. Show that for $0 \le 2\tau \le s$ the $(\tau + 1) \times (\tau + 1)$ matrix $[{}_{s-\beta}C_\alpha]$, $(\alpha, \beta = 0, 1, \ldots, \tau)$, is nonsingular; indeed, if $\Delta(s, \tau)$ denotes the determinant of this matrix, then $\Delta(s, 0) = 1$, $(s = 0, 1, \ldots)$, and the readily derivable relation $\Delta(s, \tau) = (-1)^\tau \Delta(s - 1, \tau - 1)$, imply that $\Delta(s, \tau) = (-1)^{\tau(\tau+1)/2}$. Conclude that for $\tau = k(s) - 1$ there exist unique constants e_α such that $\sum_{\alpha=0}^{\tau} {}_{s-\beta}C_\alpha e_\alpha$ is equal to 1 if $\beta = 0$, and equal to 0 for $\beta = 1, \ldots, \tau$, so that

$$\sum_{\alpha=0}^{\tau} e_\alpha\{\lambda_\alpha[u^{[\alpha]}] - (-1)^\alpha \bar{q}_\alpha u^{[\alpha]}\} = \sum_{j=0}^{s} r_j(t) u^{[j]}$$

with $r_s(t) = p_s(t)$ and $r_j(t) = 0$ for $s - k(s) < j < s$. From Problem III.2:3 conclude that $p_s(t) \in \mathfrak{C}^{k(s)}[a, b]$, and consequently for $\pi_s(t) = p_s(t)$ we have that $l_s[u] - \Lambda_s(u; \pi_s)$ is of the form $l_{s-1}[u] = \sum_{j=0}^{s-1} p_j(t) u^{[j]}$ with continuous coefficients, and with $g_r(t) \in \mathfrak{D}(l_{s-1}^\star)$ for $r = 0, 1, \ldots, k(s - 1) - 1$, so that $l_s[u]$ has the form (3.23) in view of the inductive hypothesis. In order to obtain (3.24), use results of Problem 4 above and Theorem 3.3.

(ii) Note that the stated hypotheses imply that $g_r(t) \in \mathfrak{D}(l_n^\star)$, $(r = 0, 1, \ldots)$, and with the aid of the results of the preceding parts of the problem conclude that $l_h^+[g_r] = \sum_{m=0}^{n} \Lambda_m(g_r; (-1)^m \bar{\pi}_m)$, for $r = 0, 1, \ldots$.

6. With the aid of the result of Problem III.2:7, formulate and establish a result corresponding to that of Problem 5 above for $l_n[u] = \sum_{j=0}^{n} p_j(t) u^{[j]}$, where now it is assumed that each of the coefficient functions is merely Lebesgue integrable on $[a, b]$.

4. HOMOGENEOUS DIFFERENTIAL SYSTEMS INVOLVING TWO-POINT BOUNDARY CONDITIONS

Attention is now redirected to differential operators of the form (1.1) introduced in Section 1. It is supposed throughout that hypothesis $(\mathfrak{H})$ holds so that the $n \times n$ coefficient matrices $A_j(t)$, $(j = 0, 1, 2)$ are continuous on $[a, b]$ and $A_1(t)$, $A_2(t)$ are nonsingular on this interval. Moreover, it is to be recalled that $\mathfrak{D}(L)$ denotes the linear space of vector functions $y(t) \in \mathfrak{C}_n[a, b]$ for which $y = A_2^{-1}u_y$ with $u_y \in \mathfrak{C}_n^{1}[a, b]$ and $\mathfrak{D}_0(L) = \{y \mid y \in \mathfrak{D}(L), u_y(a) = 0 = u_y(b)\}$.

We now consider a linear differential operator L with domain $D(L)$, where $D(L)$ is a linear manifold in $\mathfrak{C}_n[a, b]$ that satisfies

$$\mathfrak{D}_0(L) \subset D(L) \subset \mathfrak{D}(L). \tag{4.1}$$

The object of immediate attention is the *null space* of L; that is, the set of vector functions $y(t)$ which satisfy the homogeneous differential system

(4.2)

$$L[y](t) \equiv A_1(t)[A_2(t)\,y(t)]' + A_0(t)\,y(t) = 0, \qquad t \in [a, b], \qquad y \in D(L).$$

Clearly the totality of solutions of (4.2) is a vector space. If there are non-trivial solutions of (4.2), then, in view of Theorem I.9.1, there is a uniquely determined integer k, $(1 \leq k \leq n)$ such that there are k linearly independent solutions $y^{(1)}(t), \ldots, y^{(k)}(t)$ of $L[y] = 0$ which form a basis for the set of solutions of (4.2); that is, the totality of solutions of this system is given by $y(t) = y^{(1)}(t)\xi_1 + \cdots + y^{(k)}(t)\xi_k$, with $\xi_1, \ldots, \xi_k$ arbitrary constants. In this case the system (4.2) is said to be *compatible* and to have *index* (of compatibility) equal to k. If (4.2) has only the trivial solution $y(t) \equiv 0$, this system is said to be *incompatible* or to have index equal to zero.

If $u(t)$ is an n-dimensional vector function defined on $[a, b]$, the $2n$-dimensional vector $\omega = (\omega_\sigma)$, $(\sigma = 1, \ldots, 2n)$, with $\omega_\alpha = u_\alpha(a)$, $\omega_{n+\alpha} = u_\alpha(b)$, $(\alpha = 1, \ldots, n)$, will be denoted by $\hat{u}$. Correspondingly, if $U(t)$ is an $n \times n$ matrix function defined on $[a, b]$ the $2n \times r$ matrix $\Omega = [\Omega_{\sigma\beta}]$, $(\sigma = 1, \ldots, 2n;\ \beta = 1, \ldots, r)$, with $\Omega_{\alpha\beta} = U_{\alpha\beta}(a)$, $\Omega_{n+\alpha,\beta} = U_{\alpha\beta}(b)$, $(\alpha = 1, \ldots, n)$, will be denoted by $\hat{U}$. If $D(L; a, b)$ denotes the set of $2n$-dimensional vectors ω such that there exists a $y \in D(L)$ with $\hat{u}_y = \omega$, then $D(L; a, b)$ is a linear subspace in $\mathbf{C}_{2n}$, and since $\mathfrak{D}_0(L) \subset D(L)$ it follows that

$$D(L) = \{y \mid y \in \mathfrak{D}(L),\ \hat{u}_y \in D(L; a, b)\}; \tag{4.3}$$

that is, the domain $D(L)$ of the differential operator $L[y]$ is specified by the condition that y is a member of the above defined class $\mathfrak{D}(L)$, together with the condition that the associated vector u_y of class $\mathfrak{C}^1[a, b]$ satisfies the two-point linear homogeneous boundary conditions expressing the fact that the end-value vector $\hat{u}_y$ belongs to the linear subspace $D(L; a, b)$ in $\mathbf{C}_{2n}$. An algebraic formulation of the boundary conditions of (4.2) is obtained by the following consideration of the possible cases for the dimension of $D(L; a, b)$.

CASE 1°. $\dim D(L; a, b) = 2n$. In this case $D(L; a, b) = \mathbf{C}_{2n}$, the end-values $y(a)$, $y(b)$ are not subjected to additional restrictions, so that $D(L) = \mathfrak{D}(L)$ and (4.2) is merely

$$(4.2_\circ) \qquad L[y](t) = 0, \qquad t \in [a, b], \qquad y \in \mathfrak{D}(L);$$

the index of $(4.2_\circ)$ is clearly equal to n.

CASE 2°. $\dim D(L; a, b) = 0$. In this instance $D(L; a, b)$ consists of only the null vector, $D(L) = \mathfrak{D}_0(L)$, (4.2) becomes

$$(4.2_{2n}) \qquad L[y](t) = 0, \qquad t \in [a, b], \qquad y \in \mathfrak{D}_0(L),$$

and the index of this system is equal to zero.

CASE 3°. $\dim D(L; a, b) = 2n - m$, $1 \leq m \leq 2n - 1$. In this case there exists a $2n \times (2n - m)$ matrix P of rank $2n - m$ such that the column vectors of P form a basis for $D(L; a, b)$; that is, $D(L; a, b)$ consists of all $\hat{u}_y$ of the form

$$(4.4) \qquad \hat{u}_y = P\zeta,$$

where ζ is a $(2n - m)$-dimensional vector. If M is an $m \times 2n$ matrix of rank m, and such that

$$(4.5) \qquad MP = 0,$$

an equivalent form of the boundary conditions is the vector equation

$$(4.6) \qquad M\hat{u}_y = 0.$$

In this case we shall introduce $\mu[y]$ for the vector $M\hat{u}_y$ determined by the end-values of the vector function $u_y(t)$, and write (4.2) as

$$(4.2_m) \qquad L[y](t) = 0, \qquad t \in [a, b]; \qquad y \in \mathfrak{D}(L), \qquad \mu[y] = 0.$$

By Theorem I.9.1 the general solution of the differential equation of (4.2_m) is $y(t) = Y(t)\xi$, where $Y(t)$ is a fundamental matrix for this equation and $\xi \equiv (\xi_\alpha)$ is a constant n-dimensional vector; moreover, $y^{(\rho)}(t) = Y(t)\xi^{(\rho)}$, $(\rho = 1, \ldots, k)$, are linearly independent solutions if and only if $\xi^{(\rho)}$, $(\rho = 1, \ldots, k)$, are linearly independent vectors of $\mathbf{C}_n$. Now for

$y(t) = Y(t)\xi$ the vector boundary condition $\mu[y] = 0$ of (4.2_m) becomes

$$D\xi = 0, \tag{4.7}$$

where D is the $m \times n$ constant matrix

$$D = M\hat{U}_Y, \tag{4.8}$$

and hence (4.2_m) has index equal to k if and only if the matrix D has rank $n - k$. It is to be remarked that if M^a and M^b are $m \times n$ matrices such that $M = [M^a \ M^b]$, then $D = M^a A_2(a) Y(a) + M^b A_2(b) Y(b)$. Correspondingly, if $y(t) = Y(t)\xi$ satisfies (4.4) then

$$\hat{U}_Y \xi - P\zeta = 0,$$

and as the $2n \times (2n - m)$ matrix P has rank $2n - m$ it follows that (4.2_m) has index equal to k if and only if the $2n \times (3n - m)$ matrix

$$[\hat{U}_Y \quad P] \tag{4.9}$$

has rank $3n - m - k$.

The above derived results on the index of (4.2) are summarized in the following theorem.

Theorem 4.1. *The differential system* (4.2_0) *has index equal to* n, *and the system* (4.2_{2n}) *has index equal to zero. If* $1 \leq m \leq 2n - 1$, *and* (4.2_m) *has index equal to* k, *then the rank of the* $m \times n$ *matrix* (4.8) *is equal to* $n - k$, *and the rank of the* $2n \times (3n - m)$ *matrix* (4.9) *is equal to* $3n - m - k$.

Problems III.4

1.[s] Determine the index of compatibility of each of the following differential systems:

(i) The differential equations $y_1' = y_2, y_2' = 0$, with boundary conditions
  (a) $y_1(0) = 0 = y_1(1)$;
  (b) $y_1(0) + y_1(1) = 0 = y_2(0) - y_2(1)$.

(ii) The differential equations $y_1' = y_2, y_2' = -y_1$, with boundary conditions
  (a) $y_1(0) - y_1(2\pi) = 0$;
  (b) $y_1(0) = 0 = y_1(2\pi) = y_2(0) - y_2(2\pi)$.

(iii) With $y = (y_\alpha)$, $(\alpha = 1, 2)$, $a = -1$, $b = 1$, the system

(a) $$\begin{bmatrix} 1 & 0 \\ 0 & 5 - t^2 \end{bmatrix} y' = \begin{bmatrix} 0 & 1 \\ -12 & 0 \end{bmatrix} y, \qquad \begin{bmatrix} 1 & 0 & 0 & 0 \\ 0 & 0 & 1 & 0 \end{bmatrix} \hat{u}_y = 0;$$

(b) $$\begin{bmatrix} 0 & 1 \\ -1 & 0 \end{bmatrix} \left(\begin{bmatrix} 1 + t^2 & 0 \\ 0 & (1 + t^2)^{-1} \end{bmatrix} y \right)' = \begin{bmatrix} 1 + t^2 & 0 \\ 0 & -(1 + t^2)^{-1} \end{bmatrix} y,$$

$$\begin{bmatrix} 1 & 0 & -1 & 0 \\ 0 & 1 & 0 & 1 \end{bmatrix} \hat{u}_y = 0.$$

Hint. (iii-a) Show that the scalar differential equation $(5 - t^2)u'' + 12u = 0$ has a solution of the form $u(t) = at^4 + bt^2 + c$.

2.[s] Suppose that the $n \times n$ matrix coefficients $A_j(t)$, $(j = 0, 1, 2)$, satisfy hypothesis $(\mathfrak{H})$ on $(-\infty, \infty)$, and are all periodic with positive period T; that is, $A_j(t + T) \equiv A_j(t)$. Prove that $L[y] \equiv A_1(t)[A_2(t)y]' + A_0(t)y = 0$ has a nonidentically vanishing solution which is periodic with period T if and only if the differential system

$$(4.10) \qquad L[y] = 0, \qquad u_y(0) - u_y(T) = 0,$$

is compatible; moreover, if $Y(t)$ is a fundamental matrix for $L[y] = 0$ such that $Y(0) = E$ then $y(t)$ is a nonidentically vanishing solution of (4.10) if and only if $y(t) = Y(t)\eta$, where η is a nonzero vector satisfying the equation $[E - Y(T)]\eta = 0$.

Hint. Show that $Y(t + T) = Y(t)Y(T)$.

3. Suppose that $L[y]$ is as in Problem 2, and that the differential system

$$(4.11) \qquad L[y] = 0, \qquad u_y(0) + u_y(T) = 0,$$

is compatible. Prove that the equation $L[y] = 0$ has a nonidentically vanishing solution which is periodic with period $2T$.

4. If $L[y]$ is as in Problem 2, and there exists a positive integer k such that all solutions of $L[y] = 0$ are periodic with period kT, prove that the matrix $A(t) = A_1^{-1}(t)A_0(t)A_2^{-1}(t)$ is such that the integral of its trace over $[0, kT]$ is equal to an integral multiple of $2\pi i$; in particular, if $A(t)$ is real-valued then $\int_0^{kT} \sum_{\alpha=1}^{n} A_{\alpha\alpha}(t)\,dt = 0$.

5. NONHOMOGENEOUS DIFFERENTIAL SYSTEMS INVOLVING TWO-POINT BOUNDARY CONDITIONS

If $D(L)$ is a manifold of n-dimensional vector functions on $[a, b]$ that satisfies the conditions specified at the beginning of Section 4, then, corresponding to the homogeneous differential system (4.2), we have the nonhomogeneous system

$$(5.1) \qquad L[y](t) = g(t), \qquad t \in [a, b], \qquad y(t) - w(t) \in D(L),$$

where $w(t)$ and $g(t)$ are given n-dimensional vector functions with $w(t) \in \mathfrak{D}(L)$ and $g(t) \in \mathfrak{C}_n[a, b]$. Now for an arbitrary $2n$-dimensional vector $\omega = (\omega_\sigma)$, $(\sigma = 1, \ldots, 2n)$ there exists a corresponding n-dimensional vector function $w(t) \in \mathfrak{D}(L)$ that satisfies the boundary condition $\hat{u}_w = \omega$; for example, $w(t) = A_2^{-1}(t)\,x(t)$, where $x(t) = (x_\alpha(t))$, $(\alpha = 1, \ldots, n)$, with $x_\alpha(t) = (b - a)^{-1}[(b - t)\omega_\alpha + (t - a)\omega_{n+\alpha}]$. Consequently the boundary restriction of (5.1) is equivalent to the condition that for a given

$2n$-dimensional vector ω the vector function $y(t)$ belongs to $\mathfrak{D}(L)$ and $\hat{u}_y - \omega \in D(L; a, b)$. If the nonhomogeneous system (5.1) has a particular solution $y = y^p(t)$, clearly the general solution of (5.1) is the sum of $y^p(t)$ and the general solution of the corresponding homogeneous system (4.2).

For (5.1) the following cases correspond to the respective cases of the preceding section for the homogeneous system (4.2).

CASE 1°. $\dim D(L; a, b) = 2n$. Here $y(a) - w(a)$ and $y(b) - w(b)$ are not subject to additional restrictions so that (5.1) is

$$(5.1_{\circ}) \qquad L[y](t) = g(t), \qquad t \in [a, b], \qquad y \in \mathfrak{D}(L).$$

Clearly the general solution of $(5.1_{\circ})$ is

$$(5.2) \qquad y(t) = y^0(t) + Y(t)\xi,$$

where $y = y^0(t)$ is a particular solution of $(5.1_{\circ})$, $Y(t)$ is a fundamental matrix for the corresponding homogeneous vector differential equation $(4.2_{\circ})$, and ξ is an arbitrary n-dimensional constant vector.

CASE 2°. $\dim D(L; a, b) = 0$. In this instance system (5.1) becomes

$$(5.1_{2n}) \qquad L[y](t) = g(t), \qquad t \in [a, b], \qquad y(t) - w(t) \in \mathfrak{D}_0(L).$$

As the general solution of the vector differential equation of (5.1_{2n}) is of the form (5.2), the boundary restrictions of this system become the existence of an n-dimensional vector ξ satisfying

$$\hat{U}_Y \xi = \hat{u}_w - \hat{u}_{y^0}.$$

Clearly this vector equation has a solution if and only if

$$(5.3) \qquad -U_Y^{-1}(a)[u_w(a) - u_{y^0}(a)] + U_Y^{-1}(b)[u_w(b) - u_{y^0}(b)] = 0;$$

an equivalent condition is that the $2n \times (n + 1)$ matrix

$$(5.4) \qquad [\hat{U}_Y \quad \hat{u}_w - \hat{u}_{y^0}]$$

has rank n. If a solution exists for the system (5.1_{2n}) it is unique, since the corresponding homogeneous system (4.2_{2n}) has only the trivial solution.

CASE 3°. $\dim D(L; a, b) = 2n - m$, $1 \leq m \leq 2n - 1$. If P is a $2n \times (2n - m)$ matrix as in the corresponding discussion of the preceding section, the system (5.1) may be written with boundary conditions in parametric form as

$$(5.1_m)^{\circ} \qquad L[y](t) = g(t), \qquad t \in [a, b], \qquad \hat{u}_y = \hat{u}_w + P\zeta.$$

Since the involved nonhomogeneous vector differential equation has its general solution of the form (5.2), the system $(5.1_m)^{\circ}$ has a solution if and only if there is an n-dimensional vector ξ and a $(2n - m)$-dimensional vector ζ such that

$$\hat{U}_Y \xi - P\zeta = \hat{u}_w - \hat{u}_{y^0}.$$

Consequently $(5.1_m)^\circ$ has a solution if and only if the matrices

$$(5.5)^\circ \qquad [\hat{U}_Y \quad P], \qquad [\hat{U}_Y \quad P \quad \hat{u}_w - \hat{u}_{y^0}]$$

have the same rank. An alternate form for (5.1) is

$$(5.1_m) \qquad L[y](t) = g(t), \qquad t \in [a, b], \qquad \mu[y] \equiv M\hat{u}_y = \eta,$$

where M is an $m \times 2n$ matrix related to P, as in the preceding section, and the m-dimensional vector η is related to $\hat{u}_w$ of $(5.1_m)^\circ$ by $\eta = \mu[w] \equiv M\hat{u}_w$. It is to be noted that for a given m-dimensional vector η the $2n$-dimensional vector $\hat{u}_w$ is such that the nonparametric boundary conditions of (5.1_m) are equivalent to the parametric boundary conditions of $(5.1_m)^\circ$ if and only if there is a $(2n - m)$-dimensional vector ζ^0 for which $\hat{u}_w = P^0\eta + P\zeta^0$, where P^0 is a $2n \times m$ matrix such that $MP^0 = E_m$. Now the nonparametric vector boundary condition $\mu[y] = \eta$ has a solution if and only if there is an n-dimensional vector ξ such that $D\xi = \eta - \mu[y^0]$, where D is the $m \times n$ matrix (4.8); hence (5.1_m) has a solution if and only if the matrices

$$(5.5) \qquad D = M\hat{U}_Y, \qquad [D \quad \eta - \mu[y^0]]$$

have the same rank.

For clarity of future reference the above results are stated as the following theorem.

Theorem 5.1. *If $y^0(t)$ is a particular solution of the nonhomogeneous vector differential equation* (5.1) *and $Y(t)$ is a fundamental matrix for the corresponding homogeneous vector differential equation, then*

(i) *the system* (5.1_0) *has an n-parameter family of solutions* (5.2) *involving an arbitrary n-dimensional vector ξ;*

(ii) *the system* (5.1_{2n}) *has a solution if and only if the matrix* (5.4) *is of rank n, in which case the solution is unique;*

(iii) *for $1 \leq m \leq 2n - 1$ the system* (5.1_m) *has a solution if and only if the matrices* $(5.5)^\circ$ *have the same rank or equivalently the matrices* (5.5) *have the same rank; whenever this solvability condition holds, the general solution of* (5.1_m) *is $y = y^p(t) + y^{(1)}(t)\xi_1 + \cdots + y^{(k)}(t)\xi_k$, where $y = y^p(t)$ is a particular solution of this system, $y^{(1)}(t), \ldots, y^{(k)}(t)$ is a basis for the set of solutions of the corresponding homogeneous system* (4.2_m), *and $\xi_1, \ldots, \xi_k$ are arbitrary constants.*

Problems III.5

1. Show that for arbitrary continuous functions $g_1(t), g_2(t)$, and constants η_1, η_2, the nonhomogeneous differential system

$$\begin{gathered} y_1' = y_2 + g_1(t), \\ y_2' = -y_1 + g_2(t), \\ y_1(0) = \eta_1, \qquad y_2(2\pi) = \eta_2, \end{gathered}$$

has a unique solution.

2. If $g_1(t), g_2(t)$ are continuous on $[0, 2\pi]$, what conditions must be satisfied by $g_1(t), g_2(t)$, and η in order that the nonhomogeneous differential system

$$\begin{gathered} y_1' = y_2 + g_1(t), \\ y_2' = -y_1 + g_2(t), \\ y_1(0) - y_1(2\pi) = \eta, \end{gathered}$$

possess a solution?

3. What condition must the continuous function $g(t)$ satisfy in order that the differential system

$$\begin{gathered} y_1' = y_2, \\ y_2' = g(t), \\ y_1(0) + y_1(1) = 0, \qquad y_2(0) - y_2(1) = 0, \end{gathered}$$

possess a solution?

4. Suppose that the coefficient matrices $A_j(t)$, $(j = 0, 1, 2)$, satisfy the conditions specified in Problem 2 of Section 4, and that $g(t)$ is a continuous n-dimensional vector function on $(-\infty, \infty)$ which is also periodic of period T. Show that if the system (4.10) is incompatible, then the differential equation $L[y](t) = g(t)$, $t \in (-\infty, \infty)$, possesses a unique solution which is periodic with period T.

6. ADJOINT DIFFERENTIAL SYSTEMS

If $D(L)$ is a manifold of n-dimensional vector functions on $[a, b]$ satisfying the conditions specified at the beginning of Section 4, the symbol $D(L^\star)$ will denote the manifold of all n-dimensional vector functions $z(t)$ continuous on $[a, b]$, and for which there is a corresponding continuous n-dimensional vector function $f(t) = f_z(t)$ such that

$$\int_a^b (L[y], z)\, dt - \int_a^b (y, f_z)\, dt = 0, \quad \textit{for} \quad y \in D(L). \tag{6.1}$$

Since $\mathfrak{D}_0(L) \subset D(L)$, it follows from Theorem 2.2 that $D(L^\star) \subset \mathfrak{D}(L^\star)$, and

$f_z(t) = L^\star[z](t)$, $t \in [a, b]$, for each $z \in D(L^\star)$. That is, $D(L^\star)$ is the set of vector functions $z \in \mathfrak{D}(L^\star)$ such that

$$(6.2) \qquad \int_a^b (L[y], z)\,dt - \int_a^b (y, L^\star[z])\,dt = 0, \quad \textit{for} \quad y \in D(L).$$

Now from (2.7) it follows that

$$(6.3) \quad \int_a^b (L[y], z)\,dt - \int_a^b (y, L^\star[z])\,dt = \hat{v}_z^* Q \hat{u}_y, \quad \textit{for} \quad y \in \mathfrak{D}(L), z \in \mathfrak{D}(L^\star),$$

where, for brevity, Q denotes the $2n \times 2n$ matrix

$$(6.4) \qquad Q = \begin{bmatrix} -E_n & 0 \\ 0 & E_n \end{bmatrix}.$$

Thus $D(L^\star)$ has the characterization

$$(6.5) \qquad D(L^\star) = \{z \mid z \in \mathfrak{D}(L^\star), \hat{v}_z^* Q \hat{u}_y = 0 \textit{ for } y \in D(L)\}.$$

In particular, if $z \in \mathfrak{D}_0(L^\star) = \{z \mid z \in \mathfrak{D}(L^\star), \hat{v}_z = 0\}$ then relation (6.2) holds and $z \in D(L^\star)$, so that as a set of vector functions associated with $L^\star[z]$ the set $D(L^\star)$ satisfies all the conditions specified for the associated $D(L)$ for $L[y]$ at the beginning of Section 4. Corresponding to the notation of Section 4, the symbol $D(L^\star; a, b)$ denotes the set of $2n$-dimensional vectors ω such that there exists a $z \in D(L^\star)$ with $\hat{v}_z = \omega$, so that also

$$(6.5') \qquad D(L^\star) = \{z \mid z \in \mathfrak{D}(L^\star), \hat{v}_z \in D(L^\star; a, b)\}.$$

Consequently, in view of (6.5), $D(L^\star; a, b)$ is the linear subspace in $\mathbf{C}_{2n}$ which is the orthogonal complement of the linear subspace consisting of vectors ω of the form $Q\hat{u}_y$, where $y \in D(L)$.

The homogeneous differential system

$$(6.6) \qquad L^\star[z](t) = 0, \qquad t \in [a, b], \qquad z \in D(L^\star),$$

is said to be the *adjoint* of the system (4.2). In particular, if $\dim D(L; a, b) = 2n$ then (6.5) implies that $\dim D(L^\star; a, b) = 0$, so that the system adjoint to (4.2_0) is

$$(6.6_{2n}) \qquad L^\star[z](t) = 0, \qquad t \in [a, b], \qquad z \in \mathfrak{D}_0(L^\star).$$

Correspondingly, if $\dim D(L; a, b) = 0$ then condition (6.2) holds for arbitrary $v_z(a)$, $v_z(b)$, so that $\dim D(L^\star; a, b) = 2n$, and the adjoint of (4.2_{2n}) is

$$(6.6_0) \qquad L^\star[z](t) = 0, \qquad t \in [a, b], \qquad z \in \mathfrak{D}(L^\star).$$

Finally, if $\dim D(L; a, b) = 2n - m$, $1 \leq m \leq 2n - 1$, and the $2n \times (2n - m)$ matrix P is as in Section 4, then from (4.4) and (6.5) it follows that

$D(L^\star; a, b)$ consists of all $\hat{v}_z$ belonging to $z(t) \in \mathfrak{D}(L^\star)$, and such that $\hat{v}_z^* QP\zeta = 0$ for arbitrary $(2n - m)$-dimensional vectors ζ. Consequently the differential system adjoint to (4.2_m) may be written as

(6.6_{2n-m})

$$L^\star[z](t) = 0, \qquad t \in [a, b], \qquad z \in \mathfrak{D}(L^\star), \qquad P^* Q\hat{v}_z = 0.$$

For M an $m \times 2n$ matrix of rank m and satisfying (4.5), the matrix MQ is of rank m and $P^*Q[MQ]^* = P^*M^* = 0$, so that an equivalent form of this adjoint system with the boundary conditions expressed parametrically is given by

$(6.6_{2n-m})^\circ$

$$L^\star[z](t) = 0, \qquad t \in [a, b], \qquad z \in \mathfrak{D}(L^\star), \qquad \hat{v}_z = QM^*\chi,$$

where χ is an m-dimensional vector.

Theorem 6.1. *If k is the index of the homogeneous differential system* (4.2), *and k^* is the index of the corresponding adjoint system* (6.6), *then $n + k^* = m + k$.*

For the system $(4.2_\circ)$ we have $m = 0$ and $k = n$, and for the corresponding adjoint system (6.6_{2n}) we have $k^* = 0$, so that the result of Theorem 6.1 is true for this case. Correspondingly, for (4.2_{2n}) we have $k = 0$, $m = 2n$, and $k^* = n$ for the corresponding adjoint $(6.6_\circ)$, so that Theorem 6.1 holds for this case also. Finally, if $1 \le m \le 2n - 1$ and k is the index of the differential system (4.2_m), it follows from Theorem 4.1 that the $2n \times (3n - m)$ matrix (4.9) has rank $3n - m - k$. Therefore, for $Z(t)$ the fundamental matrix solution of (2.5) such that $V_Z^*(t)U_Y(t) \equiv E$ it follows that the matrix

$$\begin{bmatrix} V_Z^*(a) & V_Z^*(b) \\ 0 & V_Z^*(b) \end{bmatrix} \cdot Q \cdot [\hat{U}_Y \quad P]$$

also has rank $3n - m - k$. As this latter matrix is of the form

$$\begin{bmatrix} 0 & T^* \\ E & N \end{bmatrix},$$

where T is the $(2n - m) \times n$ matrix $P^*Q\hat{V}_Z$, and E is the $n \times n$ identity matrix, it follows that T has rank $2n - m - k$. On the other hand, application of Theorem 4.1 to the adjoint differential system yields the result that $P^*Q\hat{V}_Z$ has rank $n - k^*$, where k^* is the index of (6.6_{2n-m}). Consequently $2n - m - k = n - k^*$ and $n + k^* = m + k$.

Theorem 6.2. *The nonhomogeneous system* (5.1) *has a solution if and only if the relation*

$$\int_a^b z^*(t)\, g(t)\, dt = \hat{v}_z^* Q \hat{u}_w \tag{6.7}$$

holds for all solutions $z(t)$ of the homogeneous adjoint system (6.6).

In the proof of this theorem we consider the separate cases already encountered in the preceding sections.

Case 1°. dim $D(L; a, b) = 2n$. In this case (5.1) has solutions for arbitrary $g(t)$, $w(t)$, as shown in Section 5; moreover, the corresponding adjoint system (6.6_{2n}) has only the trivial solution $z(t) \equiv 0$, and (6.7) imposes no restriction on $g(t)$, $w(t)$.

Case 2°. dim $D(L; a, b) = 0$. In this instance (5.1) has a solution if and only if the $2n \times (n + 1)$ matrix (5.4) has rank n, or equivalently that the vector equation (5.3) holds. The corresponding homogeneous adjoint system is (6.6_0), with general solution $z(t) = Z(t)\xi$, where ξ is an n-dimensional constant vector and $Z(t)$ is a fundamental matrix solution of (2.5) such that $V_Z^*(t) U_Y(t) \equiv E$. Consequently (5.3) is equivalent to the condition that

$$\hat{v}_z^*\, Q[\hat{u}_w - \hat{u}_{y^0}] = 0 \tag{6.8}$$

holds for arbitrary solutions $z(t)$ of (6.6_0), where $y = y^0(t)$ is a particular solution of the nonhomogeneous vector differential equation of (5.1). As $L[y^0] = g(t)$ and $L^\star[z] = 0$, it follows from (6.3) that

$$\int_a^b z^*(t)\, g(t)\, dt = \hat{v}_z^* Q \hat{u}_{y^0}, \tag{6.9}$$

and hence (6.8) reduces to condition (6.7).

Case 3°. dim $D(L; a, b) = 2n - m$, $1 \le m \le 2n - 1$. In this case it follows from Section 5 that the nonhomogeneous system (5.1_m) has a solution if and only if the matrices $(5.5)^\circ$ have the same rank. Moreover, from Theorems 4.1 and 6.1 it follows that if k and k^* denote the indices of (5.1_m) and the corresponding adjoint system (6.6_{2n-m}), respectively, then the $2n \times (3n - m)$ matrix

$$[\hat{U}_Y \quad P], \tag{6.10}$$

which is the first matrix of $(5.5)^\circ$, has rank $3n - m - k = 2n - k^*$. Now if $k^* = 0$, the two matrices of $(5.5)^\circ$ have common rank equal to $2n$, and (5.1_m) has a solution for arbitrary $g(t)$, $w(t)$; moreover, the corresponding adjoint system (6.6_{2n-m}) has only the trivial solution $z(t) \equiv 0$, and (6.7) places no restriction on $g(t)$, $w(t)$. In case $k^* > 0$, let $z^1(t), \ldots, z^{k^*}(t)$ be k^*

linearly independent solutions of (6.6_{2n-m}). From the boundary conditions of (6.6_{2n-m}) it follows that $P^*Q\hat{v}_{z^\rho} = 0$, $(\rho = 1, \ldots, k^*)$. Moreover, as the column vectors of $Y(t)$ are solutions of $L[y] = 0$, and $L^\star[z^\rho] = 0$, it follows from (6.1) that $\hat{U}_Y^* Q\hat{v}_{z^\rho} = 0$, $(\rho = 1, \ldots, k^*)$. Then the $2n$-dimensional vectors $\omega = Q\hat{v}_{z^\rho}$, $(\rho = 1, \ldots, k^*)$, are k^* linearly independent vectors satisfying $\omega^*\hat{U}_Y = 0$, $\omega^* P = 0$, and as (6.10) is of rank $2n - k^*$ it follows that the matrix $[\hat{U}_Y \quad P \quad \hat{u}_w - \hat{u}_{y0}]$ has the same rank as (6.10) if and only if relation (6.8) holds for arbitrary solutions $z(t)$ of (6.6_{2n-m}).

As in the consideration of Case 2°, it follows that (6.9) holds and (6.8) reduces to (6.7), thus completing the proof of Theorem 6.2.

Theorem 6.2.′ *If* $\dim D(L; a, b) = 2n - m$, $1 \leq m \leq 2n - 1$, *and* P^0 *is a* $2n \times m$ *matrix which satisfies with the matrix* M *of* (5.1_m) *the matrix equation* $MP^0 = E_m$, *then the nonhomogeneous differential system* (5.1_m) *has a solution if and only if the relation*

$$\int_a^b z^*(t)g(t)\,dt = \hat{v}_z^* Q P^0 \eta \tag{6.11}$$

holds for arbitrary solutions $z(t)$ *of the corresponding homogeneous adjoint system* (6.3_{2n-m}).

As noted in Section 5, the differential system (5.1_m) is equivalent to $(5.1_m)^\circ$ with $\hat{u}_w = P^0\eta$, where P^0 is a $2n \times m$ matrix satisfying $MP^0 = E_m$, and consequently the result of Theorem 6.2′ is an immediate corollary of Theorem 6.2.

It is to be commented that if $\dim D(L; a, b) = 0$ then the corresponding solvability criterion (6.7) may equally well be given the form (6.11). Indeed, the boundary conditions of (5.1_{2n}) are of the form $M\hat{u}_y = \eta$, where M is the $2n \times 2n$ identity matrix, so that correspondingly P^0 is the $2n \times 2n$ identity matrix and $\hat{v}_z^* QP^0\eta = \hat{v}_z^* Q\hat{u}_w$.

For L a differential operator with domain $D(L)$ as discussed above, let $\mathfrak{N}(L)$ denote the *null space* of L, that is, the set of vector functions $y(t)$ which satisfy the differential equation (4.2). Correspondingly, let $\mathfrak{R}(L)$ denote the *range* of L, that is, the set of vector functions $g \in \mathfrak{C}_n[a, b]$ for which there is an associated $y \in D(L)$ such that $L[y](t) = g(t)$ for $t \in [a, b]$. Clearly $\mathfrak{N}(L)$ and $\mathfrak{R}(L)$ are individually vector spaces of vector functions belonging to $\mathfrak{C}_n[a, b]$; also, the dimension k of $\mathfrak{N}(L)$ does not exceed n. Similarly, let $\mathfrak{N}(L^\star)$ and $\mathfrak{R}(L^\star)$ denote the null space and range of the adjoint differential operator $L^\star$. Moreover, for S an arbitrary set of vector functions $f \in \mathfrak{C}_n[a, b]$, let $S^\perp$ denote the orthogonal complement of S in $\mathfrak{C}_n[a, b]$, consisting of those vector functions $g \in \mathfrak{C}_n[a, b]$ such that

$$0 = \int_a^b (g(t), f(t))\,dt = \int_a^b f^*(t)g(t)\,dt, \quad \textit{for all} \quad f \in S.$$

The result of Theorem 6.1, and that of Theorem 6.2 for the special case of $w(t) \equiv 0$, together with the corresponding result when the roles of L and $L^\star$ are interchanged, may be stated as follows.

Theorem 6.3. $\mathfrak{R}(L) = [\mathfrak{N}(L^\star)]^\perp$, *and* $\mathfrak{R}(L^\star) = [\mathfrak{N}(L)]^\perp$. *Moreover,* $n + \dim \mathfrak{N}(L^\star) = m + \dim \mathfrak{N}(L)$; *in particular, if* $n = m$, *so that*

$$\dim D(L; a, b) = \dim D(L^\star; a, b),$$

then $\dim \mathfrak{N}(L) = \dim \mathfrak{N}(L^\star)$.

Problems III.6

1. Determine the system adjoint to each of the differential systems listed in Problem III.4:1 and compute the index of each adjoint system.

2. Show that the differential system $y_1' = 2y_2$, $y_2' = -2y_1 + g(t)$, $y_2(0) = 0$, $y_2(\pi) = k$, does not have a solution for arbitrary continuous functions $g(t)$ and constants k, and determine a necessary and sufficient condition for this system to have a solution.

3.[ss] Suppose that $1 \leq m \leq 2n - 1$, Q is the $2n \times 2n$ matrix defined by (6.4), and M is an $m \times 2n$ matrix of rank m. Show that:

(i) there exists a $2n \times (2n - m)$ matrix P, together with a $2n \times m$ matrix P^0 and a $(2n - m) \times 2n$ matrix M^0 such that

(6.12)
$$\mu[y] = M\hat{u}_y, \qquad \mu^0[y] = M^0\hat{u}_y, \qquad \nu[z] = P^*Q\hat{v}_z, \qquad \nu^0[z] = P^{0*}Q\hat{v}_z$$

satisfy for arbitrary $\hat{u}_y$, $\hat{v}_z$ the identity

$$(6.13) \qquad (\nu^0[z])^* \mu[y] + (\nu[z])^* \mu^0[y] = \hat{v}_z^* Q\hat{u}_y;$$

(ii) if $\mu[y] \equiv M\hat{u}_y = 0$ specifies the boundary conditions for a system (4.2_m), while P, P^0, M are any matrices of the dimensions specified in (i) and such that the linear forms (6.12) satisfy the identity (6.13), then P is of rank $2n - m$ and $MP = 0$, so that the boundary conditions of the corresponding adjoint system (6.6_{2n-m}) are specified by $\nu[z] \equiv P^*Q\hat{v}_z = 0$.

Hints. Choose M^0 so that the $2n \times 2n$ matrix

$$(6.14) \qquad \begin{bmatrix} M \\ M_0 \end{bmatrix}$$

is nonsingular, and determine the $2n \times (2n - m)$ matrix P and the $2n \times m$ matrix P^0 so that $[P^0 \quad P]$ is the inverse of (6.14). In particular, note that

$$Q \cdot [P^0 \quad P] \cdot \begin{bmatrix} M \\ M^0 \end{bmatrix} = Q.$$

4. If the coefficient matrices $A_j(t)$, $(j = 0, 1, 2)$, satisfy the conditions specified in Problem 2 of Section 4, show that the differential equation $L[y] = A_1(t)[A_2(t)y]' + A_0(t)y = 0$ has k linearly independent solutions which are periodic with period T if and only if the corresponding adjoint equation

$$L^\star[z] \equiv -A_2^*(t)[A_1^*(t)z]' + A_0^*(t)z = 0 \tag{6.15}$$

has k linearly independent solutions which are periodic with period T. Moreover, if $g(t)$ is a continuous n-dimensional vector function on $(-\infty, \infty)$ which is also of period T, then the equation $L[y](t) = g(t)$ possesses a solution that is periodic with period T if and only if

$$\int_0^T z^*(t)g(t)dt = 0$$

for arbitrary solutions $z(t)$ of (6.15) which are periodic with period T.

7. THE GREEN'S MATRIX

When $m = n$ it follows from Theorem 6.1 that the index of the homogeneous system (4.2_n) is equal to the index of the corresponding adjoint system (6.6_n); in particular, if one of these systems is incompatible so is the other also. Moreover, if (4.2_n) is incompatible then by Theorem 5.1 the nonhomogeneous system (5.1_n) has a unique solution for given vector η, and continuous vector function $g(t)$. Indeed, if $y^0(t)$ is a particular solution of the differential equation of (5.1_n), and $Y(t)$ is a fundamental matrix for the corresponding homogeneous differential equation, then $y(t)$ is a solution of (5.1_n) if and only if $y(t) = y^0(t) + Y(t)\xi$, where the n-dimensional constant vector ξ satisfies with the $n \times n$ matrix $D = M\hat{U}_Y$ the algebraic condition $D\xi = \eta - \mu[y^0]$. The system (4.2_n) is incompatible if and only if D is nonsingular, and thus in case this system is incompatible the unique solution of the nonhomogeneous system (5.1_n) is given by

$$y(t) = Y(t)D^{-1}\eta + y^0(t) - Y(t)D^{-1}\mu[y^0]. \tag{7.1}$$

Relation (7.1) displays the individual contributions of the vector function $g(t)$ and the vector η. Specifically, $y(t) = Y(t)D^{-1}\eta$ is the solution of

$$L[y](t) = 0, \qquad t \in [a, b], \qquad \mu[y] \equiv M\hat{u}_y = \eta, \tag{7.2}$$

and $y(t) = y^0(t) - Y(t)D^{-1}\mu[y^0]$ is the solution of

$$L[y](t) = g(t), \qquad t \in [a, b], \qquad \mu[y] \equiv M\hat{u}_y = 0. \tag{7.3}$$

That is, the function with domain $\mathfrak{C}_n[a, b]$ and range in $D(L)$, defined as the solution of (7.3), is the *inverse of the differential operator* $L[y]$, $y \in D(L)$. For simplification in references, we list specifically the corresponding homogeneous system

$$(7.4) \qquad L[y](t) = 0, \qquad t \in [a, b], \qquad \mu[y] = 0.$$

Throughout the discussion of this section it will be understood that M is an $n \times 2n$ matrix of rank n. The associated homogeneous adjoint system is then

$$(7.5) \qquad L^\star[z](t) = 0, \qquad t \in [a, b], \qquad \nu[z] \equiv P^*Q\hat{v}_z = 0,$$

where P is a $2n \times n$ matrix of rank n, and such that $MP = 0$.

Theorem 7.1. *If* (7.4) *is incompatible then the unique solution of* (7.3) *is given by*

$$(7.6) \qquad y(t) = \int_a^b G(t, s)g(s)\, ds, \qquad t \in [a, b],$$

with $G(t, s)$ defined for $(t, s) \in [a, b] \times [a, b]$, $t \neq s$, by

$$(7.7) \qquad G(t, s) = \tfrac{1}{2} Y(t)[E \operatorname{sgn}(t - s) + D^{-1}\Delta]Z^*(s),$$

where $Y(t)$ and $Z(t)$ are fundamental matrix solutions of $L[Y] = 0$ and $L^\star[Z] = 0$ such that $V_Z^*(t)U_Y(t) \equiv E$, $D = M\hat{U}_Y$, $\Delta = -MQ\hat{U}_Y$, with Q the $2n \times 2n$ matrix defined by (6.4).

For the differential equation $L[y](t) = g(t)$ the method of variation of parameters applied to $A_1(t)u_y' + A_0(t)A_2^{-1}(t)u_y = g(t)$ yields $u_y(t) = U_Y(t)w(t)$ with $A_1(t)U_Y(t)w'(t) = g(t)$. Consequently, the condition $V_Z^*(t)U_Y(t) = Z^*(t)A_1(t)U_Y(t) = E$ for $t \in [a, b]$ implies that $w'(t) = Z^*(t)\,g(t)$, and hence a particular solution $y^0(t)$ of the differential equation of (7.3) is given by

$$y^0(t) = \tfrac{1}{2}Y(t)\left[\int_a^t Z^*(s)g(s)\, ds - \int_t^b Z^*(s)g(s)\, ds\right].$$

Moreover, for such a choice of $y^0(t)$ it follows that

$$y^0(a) = -\tfrac{1}{2}Y(a)\int_a^b Z^*(s)g(s)\, ds, \qquad y^0(b) = \tfrac{1}{2}Y(b)\int_a^b Z^*(s)g(s)\, ds,$$

$$\hat{u}_y = \tfrac{1}{2}Q\hat{U}_Y\int_a^b Z^*(s)g(s)\, ds, \qquad \mu[y^0] = M\hat{u}_{y^0} = -\tfrac{1}{2}\Delta\int_a^b Z^*(s)g(s)\, ds,$$

where Q is the $2n \times 2n$ matrix (6.4) and $\Delta = -MQ\hat{U}_Y$. Relation (7.6) then follows from the above noted fact that the solution $y(t)$ of (7.3) is given by $y(t) = y^0(t) - Y(t)D^{-1}\mu[y^0]$, and the observation that $y^0(t)$ may be written as

$$y^0(t) = \tfrac{1}{2}Y(t)\int_a^b sgn(t - s)\, Z^*(s)g(s)\, ds, \qquad t \in [a, b].$$

The kernel matrix $G(t, s)$ in the integral transformation of (7.6) which defines the inverse of the differential operator $L[y]$, $y \in D(L)$, is called the *Green's matrix* for the incompatible homogeneous system (7.4).

Theorem 7.2. *For an incompatible system* (7.4) *the Green's matrix* $G(t, s) \equiv [G_{\alpha\beta}(t, s)]$, $(\alpha, \beta = 1, \ldots, n)$, *is characterized by the condition that the matrix* $\mathbf{G}(t, s) = A_2(t)G(t, s)A_1(s)$ *possesses the following properties:*

(i) *on each of the triangular regions* $\Delta_1 = \{(t, s) \mid (t, s) \in [a, b] \times [a, b], s < t\}$ *and* $\Delta_2 = \{(t, s) \mid (t, s) \in [a, b] \times [a, b], t < s\}$ *the elements of* $\mathbf{G}(t, s)$ *are continuous in* (t, s)*;*

(ii) *if* $s \in (a, b)$ *then:*

(a) *on each of the half-open intervals* $a \leq t < s$ *and* $s < t \leq b$ *the partial derivatives* $(\partial/\partial t)\mathbf{G}_{\alpha\beta}(t, s) = \mathbf{G}^{[1,0]}_{\alpha\beta}(t, s)$ *exist, are continuous functions of* t, *and also* $\mathbf{G}(t, s) = [\mathbf{G}_{\alpha\beta}(t, s)]$ *satisfies the differential equation*

$$A_1(t)\mathbf{G}^{[1,0]}(t, s) + A_0(t)A_2^{-1}(t)\,\mathbf{G}(t, s) = 0,$$

(b) $M\mathbf{G}(\ \ , s) = 0$,

(c) *as functions of* t, *the elements* $\mathbf{G}_{\alpha\beta}(t, s)$ *have at* $t = s$ *finite right-hand limits* $\mathbf{G}_{\alpha\beta}(s^+, s)$, *and finite left-hand limits* $\mathbf{G}_{\alpha\beta}(s^-, s)$*; moreover,* $\mathbf{G}(s^+, s) = [\mathbf{G}_{\alpha\beta}(s^+, s)]$ *and* $\mathbf{G}(s^-, s) = [\mathbf{G}_{\alpha\beta}(s^-, s)]$ *satisfy the relation*

$$\mathbf{G}(s^+, s) - \mathbf{G}(s^-, s) = E. \tag{7.8}$$

It may be verified directly that if $G(t, s)$ is defined by (7.7) then $\mathbf{G}(t, s) = A_2(t)G(t, s)A_1(s)$ possesses the properties (i), (ii-a, b, c). Conversely, a matrix which satisfies these properties is uniquely determined on the square region $[a, b] \times [a, b]$, except along the diagonal $t = s$. Indeed, if $G_j(t, s)$, $(j = 1, 2)$, are such that the corresponding matrices

$$\mathbf{G}_j(t, s) = A_2(t)G_j(t, s)A_1(s)$$

satisfy conditions (i), (ii-a, b, c), for $s \in [a, b]$ let $U(t, s) = \mathbf{G}_1(t, s) - \mathbf{G}_2(t, s)$ for $t \in [a, s)$ and $t \in (s, b]$, and $U(s, s) = \mathbf{G}_1(s^+, s) - \mathbf{G}_2(s^+, s)$. From properties (i), (ii-a, b, c) for the individual $\mathbf{G}_j(t, s)$ it follows that if $s \in (a, b)$ then as a function of t the matrix $U(t, s)$ has the following properties: (1°) $U(t, s)$ is continuous in t on $[a, b]$; (2°) $A_1(t)\,U_t(t, s) + A_0(t)A_2^{-1}(t)\,U(t, s) = 0$ for $t \in [a, s)$ and $t \in (s, b]$; (3°) $M\hat{U}(\ \ , s) = 0$. From 1° and 2° it follows that

$$U(t, s) = U(a, s) - \int_a^t A_1^{-1}(r)A_0(r)A_2^{-1}(r)U(r, s)\,dr, \quad \text{for} \quad t \in [a, b].$$

Hence for fixed $s \in (a, b)$, and ξ an n-dimensional constant vector, the vector function $u(t) = U(t, s)\xi$ is a solution of the differential system

$$A_1(t)u'(t) + A_0(t)A_2^{-1}(t)u(t) = 0, \qquad t \in [a, b], \qquad M\hat{u} = 0.$$

In view of the assumption that (7.4) is incompatible, it follows that $U(t, s)\xi = 0$ for $t \in [a, b]$, $s \in (a, b)$, and arbitrary ξ, so that $U(t, s) \equiv 0$ on this set of values for (t, s). Consequently $\mathbf{G}_1(t, s) \equiv \mathbf{G}_2(t, s)$ on $\{(t, s) \mid s \in (a, b), t \in [a, s)\}$ and $\{(t, s) \mid s \in (a, b), t \in (s, b]\}$, and by continuity it follows that $\mathbf{G}_1(t, s) \equiv \mathbf{G}_2(t, s)$ on Δ_1 and on Δ_2.

Although Green's matrix for an incompatible system (7.4) is characterized by the conditions of Theorem 7.2, it is to be emphasized that this matrix function possesses continuity and differentiability properties beyond those specifically stated in these conditions. Indeed, from (7.7) it follows that if $\tau \in [a, b]$ then for $j = 1, 2$ the matrix $\mathbf{G}(t, s)$ tends to the limit

(7.9) $$\mathbf{G}^j(\tau) = \tfrac{1}{2}[(-1)^{j+1}E + U_Y(\tau)D^{-1}\Delta V_Z^*(\tau)]$$

as (t, s) approaches (τ, τ), with (t, s) restricted to the region Δ_j, $(j = 1, 2)$. This limit relation is indicated by the notation

(7.10$_1$) $$\mathbf{G}(t, s) \to G^j(\tau) \quad \text{as} \quad (t, s)\,\Delta_j \to (\tau, \tau), \qquad (j = 1, 2).$$

Moreover, since $U_Y(t) = A_2(t)Y(t)$ is a solution of $U_Y' = AU_Y$, where $A = -A_1^{-1}A_0A_2^{-1}$, and $(U_Y^{-1})' = -U_Y^{-1}A$, from the expression (7.7) and the fact that U_Y and V_Z^* are inverses it follows that $\mathbf{G}^{[1,0]}(t, s) = (\partial/\partial t)\,\mathbf{G}(t, s)$ and $\mathbf{G}^{[0,1]}(t, s) = (\partial/\partial s)\,\mathbf{G}(t, s)$ satisfy the following conditions

(7.10$_2$)
$$\mathbf{G}^{[1,0]}(t, s) \to A(\tau)\,\mathbf{G}^j(\tau) \quad \text{as} \quad (t, s)\,\Delta_j \to (\tau, \tau), \qquad (j = 1, 2),$$

(7.10$_3$)
$$\mathbf{G}^{[0,1]}(t, s) \to -\mathbf{G}^j(\tau)\,A(\tau) \quad \text{as} \quad (t, s)\,\Delta_j \to (\tau, \tau), \qquad (j = 1, 2).$$

Since the hypotheses of Theorem 7.1 require the system (7.4) to be of the form (4.2$_n$) and incompatible, as noted above the corresponding adjoint system (7.5) is of the form (6.6$_n$) and also incompatible. Let $H(t, s)$ be the corresponding Green's matrix for (7.5), so that for an arbitrary vector function $h(t) \in \mathfrak{C}_n[a, b]$ the solution $z(t)$ of the system

(7.11) $$L^\star[z](t) = h(t), \qquad t \in [a, b], \quad \nu[z] = 0,$$

is given by

(7.12) $$z(t) = \int_a^b H(t, s)h(s)\,ds.$$

Now for $y(t)$ and $z(t)$ solutions of the respective systems (7.3) and (7.11), it follows from the definition of the boundary conditions $\mu[y] = 0$, $\nu[z] = 0$ that

$$\int_a^b (L[y], z)\,dt - \int_a^b (y, L^\star[z])\,dt = 0.$$

In terms of the expressions (7.6) and (7.12) for $y(t)$ and $z(t)$ it then follows that

$$\int_a^b \int_a^b h^*(t)[H^*(s, t) - G(t, s)]g(s)\, ds\, dt = 0.$$

Consequently, in view of the arbitrariness of the vector functions $h(t)$ and $g(t)$ one has that $G(t, s) = H^*(s, t)$ for $(t, s) \in \Delta_j$, $(j = 1, 2)$, and the following result has been established.

Theorem 7.3. *If $G(t, s)$ is the Green's matrix for an incompatible system* (7.4), *then $H(t, s) = G^*(s, t)$ is the Green's matrix for the incompatible adjoint system* (7.5).

Problems III.7

1. Show that each of the following differential systems is incompatible, and determine the corresponding Green's matrix:

(i) $y_1' = y_2$, $y_2' = 0$, with boundary conditions:
(a) $y_1(0) = 0 = y_1(1)$;
(b) $y_1(0) + y_1(1) = 0 = y_2(0) + y_2(1)$.

(ii) $y_1' = ky_2$, $y_2' = -ky_1$, $y_1(0) = 0 = y_1(\pi)$, where k is a real number distinct from $0, \pm 1, \pm 2, \ldots$.

(iii) $y_1' = y_2$, $y_2' = -\lambda y_1$, $y_1(0) - y_1(2\pi) = 0 = y_2(0) - y_2(2\pi)$, where λ is a complex number distinct from $0, 1, 2^2, 3^2, \ldots$.

(iv) $y_1' = y_2$, $y_2' = -y_1$, $y_1(0) - y_2(0) = 0 = y_1(\pi)$.

(v) $y_1' = ky_2$, $y_2' = ky_1$, where k is non-zero and real, with boundary conditions:
(a) $y_1(0) = 0 = y_1(1)$;
(b) $y_2(0) = 0 = y_2(1)$.

(vi) $y = (y_\alpha)$, $(\alpha = 1, 2)$, $a = -1$, $b = 1$, and

$$\begin{bmatrix} 0 & 1 \\ -1 & 0 \end{bmatrix}\left(\begin{bmatrix} (1 + t^2) & 0 \\ 0 & (1 + t^2)^{-1} \end{bmatrix} y\right)' = \begin{bmatrix} 1 + t^2 & 0 \\ 0 & -(1 + t^2)^{-1} \end{bmatrix} y,$$

$$\begin{bmatrix} 1 & 0 & 0 & 0 \\ 0 & 0 & 1 & 0 \end{bmatrix} \hat{u}_y = 0.$$

2. For the case of an incompatible system (7.4) express the Green's matrix for the corresponding incompatible adjoint system (7.5) in the form (7.7), and establish the result of Theorem 7.3 by direct comparison of the expressions for the Green's matrices for (7.4) and (7.5).

8. DIFFERENTIAL SYSTEMS INVOLVING A SINGLE *n*-th ORDER LINEAR DIFFERENTIAL EQUATION

We shall consider now differential systems involving an n-th order linear homogeneous differential expression

$$l_n[u] \equiv p_n(t)u^{[n]} + \cdots + p_1(t)u' + p_0(t)u,$$

where the scalar functions $p_0(t), \ldots, p_n(t)$ are continuous on $[a, b]$, and $p_n(t) \neq 0$ on this interval. As noted in Section 3, under the substitution $y_\alpha = u^{[\alpha-1]}$, $(\alpha = 1, \ldots, n)$, the differential equation

$$l_n[u] = 0, \qquad t \in [a, b], \tag{8.1}$$

is equivalent to the system of first order differential equations

$$\begin{aligned} L_1[y] &\equiv p_n(t)y_n' + p_0(t)y_1 + \cdots + p_{n-1}(t)y_n = 0, \\ L_\beta[y] &\equiv y_{n-\beta+1}' - y_{n-\beta+2} = 0, \qquad (\beta = 2, \ldots, n). \end{aligned} \tag{8.2}$$

Similarly, if $r(t)$ is a continuous scalar function on $[a, b]$, then under this substitution the nonhomogeneous differential equation

$$l_n[u](t) = r(t), \qquad t \in [a, b], \tag{8.3}$$

is equivalent to the nonhomogeneous first order system

$$\begin{aligned} L_1[y] &\equiv p_n(t)y_n' + p_0(t)y_1 + \cdots + p_{n-1}(t)y_n = r(t), \\ L_\beta[y] &\equiv y_{n-\beta+1}' - y_{n-\beta+2} = 0, \qquad (\beta = 2, \ldots, n). \end{aligned} \tag{8.4}$$

In view of these remarks, the preceding sections clearly provide results for differential systems involving the differential equation (8.1) or (8.3), and linear boundary conditions in the end-values of $u, u', \ldots, u^{[n-1]}$ at $t = a$ and $t = b$. If a homogeneous differential system involving (8.1) and homogeneous linear boundary conditions has nontrivial solutions then the index (of compatibility) of this system is defined as the largest integer k such that there is a set of k linearly independent solutions of this system; if such a differential system has only the trivial solution $u(t) \equiv 0$ then its index is defined as zero, and it is termed incompatible. Clearly an equivalent procedure is to define the index of such a differential system as equal to the index of the corresponding differential system involving the first order differential equations (8.2).

Corresponding to Case 1° of Section 4 we have

$$l_n[u](t) = 0, \qquad t \in [a, b], \tag{8.5$_\circ$}$$

with index equal to n. Similarly, corresponding to Case 2° of Section 4, we have the system

$$(8.5_{2n})\qquad \begin{aligned} &l_n[u](t) = 0, && t \in [a, b], \\ &u^{[\alpha-1]}(a) = 0, && u^{[\alpha-1]}(b) = 0, \qquad (\alpha = 1, \ldots, n), \end{aligned}$$

with index equal to zero. Finally, corresponding to Case 3° of Section 4, there is the system

$$(8.5_m)^\circ\qquad \begin{aligned} &l_n[u](t) = 0, \qquad t \in [a, b], \\ &u^{[\alpha-1]}(a) = \sum_{\tau=1}^{2n-m} P_{\alpha\tau}\zeta_\tau, \qquad u^{[\alpha-1]}(b) = \sum_{\tau=1}^{2n-m} P_{n+\alpha,\tau}\zeta_\tau, \end{aligned}$$

with the boundary conditions expressed parametrically in terms of $\zeta_1, \ldots, \zeta_{2n-m}$, where P is a $2n \times (2n - m)$ matrix of rank $2n - m$. An equivalent system with boundary conditions in nonparametric form is

$$(8.5_m)\qquad \begin{aligned} &l_n[u](t) = 0, \qquad t \in [a, b], \\ &\mu_\sigma\{u\} \equiv \sum_{\beta=1}^{n} [M_{\sigma\beta}u^{[\beta-1]}(a) + M_{\sigma,n+\beta}u^{[\beta-1]}(b)] = 0, \quad (\sigma = 1, \ldots, m), \end{aligned}$$

where, as in Section 4, M is an $m \times 2n$ matrix of rank m, and such that $MP = 0$. For $1 \leq m \leq 2n - 1$ it follows from Theorem 4.1 that if (8.5_m) has index k then the $m \times n$ matrix (4.8) has rank $n - k$, and the $2n \times (3n - m)$ matrix (4.9) has rank $3n - m - k$, where in each case $Y(t)$ is now the Wronskian matrix $[u_\beta^{[\alpha-1]}(t)]$, $(\alpha, \beta = 1, \ldots, n)$, for linearly independent solutions $u_1(t), \ldots, u_n(t)$ of (8.1).

Corresponding to Case 1° of Section 5, we now have the nonhomogeneous differential equation

$$(8.6_\circ)\qquad l_n[u](t) = r(t), \qquad t \in [a, b],$$

with $r(t) \in \mathfrak{C}[a, b]$. The general solution of $(8.6_\circ)$ is

$$(8.7)\qquad u(t) = u_0(t) + u_1(t)\xi_1 + \cdots + u_n(t)\xi_n,$$

where $u = u_0(t)$ is a particular solution of $(8.6_\circ)$, $u_1(t), \ldots, u_n(t)$ are linearly independent solutions of (8.1), and $\xi_1, \ldots, \xi_n$ are arbitrary constants. Corresponding to Case 2° of Section 5, we have the system

$$(8.6_{2n})\qquad \begin{aligned} &l_n[u](t) = r(t), \qquad t \in [a, b], \\ &u^{[\alpha-1]}(a) = w_\alpha^{\ a}, \quad u^{[\alpha-1]}(b) = w_\alpha^{\ b}, \quad (\alpha = 1, \ldots, n), \end{aligned}$$

where $w^a = (w_\alpha^{\ a})$ and $w^b = (w_\alpha^{\ b})$ are given n-dimensional vectors. Finally, corresponding to Case 3° of Section 5, there is the system

$$(8.6_m)^\circ\qquad \begin{aligned} &l_n[u](t) = r(t), \qquad t \in [a, b], \\ &u^{[\alpha-1]}(a) = w_\alpha^{\ a} + \sum_{\tau=1}^{2n-m} P_{\alpha\tau}\zeta_\tau, \qquad u^{[\alpha-1]}(b) = w_\alpha^{\ b} + \sum_{\tau=1}^{2n-m} P_{n+\alpha,\tau}\zeta_\tau, \\ &\qquad (\alpha = 1, \ldots, n), \end{aligned}$$

with boundary conditions in parametric form, or an equivalent system

$$(8.6_m)\qquad \begin{aligned} l_n[u](t) &= r(t), \qquad t \in [a, b],\\ \mu_\sigma\{u\} &= \eta_\sigma, \qquad (\sigma = 1, \ldots, m), \end{aligned}$$

with boundary conditions in nonparametric form, where the linear forms $\mu_\sigma\{u\}$ in the end-values $u^{[\alpha-1]}(a)$, $u^{[\alpha-1]}(b)$ are as in (8.5_m), and the matrices P and M satisfy the same conditions indicated above in the formulation of $(8.5_m)^\circ$ and (8.5_m). Theorem 5.1 then yields the following result on the solvability of these nonhomogeneous differential systems, in terms of a particular solution $u = u_0(t)$ of $(8.6_\circ)$ and n linearly independent solutions $u_1(t), \ldots, u_n(t)$ of $(8.5_\circ)$.

Theorem 8.1. *The system* $(8.6_\circ)$ *has an n-parameter family of solutions given by* (8.7). *The system* (8.6_{2n}) *has a solution if and only if the* $2n \times (n + 1)$ *matrix*

$$\begin{bmatrix} u_\beta^{[\alpha-1]}(a) & w_\alpha{}^a - u_0^{[\alpha-1]}(a) \\ u_\beta^{[\alpha-1]}(b) & w_\alpha{}^b - u_0^{[\alpha-1]}(b) \end{bmatrix}$$

is of rank n, in which case the solution is unique. For $1 \leq m \leq 2n - 1$ *the system* (8.6_m) *has a solution if and only if the matrices*

$$\begin{bmatrix} u_\beta^{[\alpha-1]}(a) & P_{\alpha\tau} \\ u_\beta^{[\alpha-1]}(b) & P_{n+\alpha,\tau} \end{bmatrix}, \qquad \begin{bmatrix} u_\beta^{[\alpha-1]}(a) & P_{\alpha\tau} & w_\alpha{}^a - u_0^{[\alpha-1]}(a) \\ u_\beta^{[\alpha-1]}(b) & P_{n+\alpha,\tau} & w_\alpha{}^b - u_0^{[\alpha-1]}(b) \end{bmatrix}$$

have the same rank, or, equivalently, if the matrices

$$[D_{\sigma\beta}] \equiv \left[\sum_{\alpha=1}^{n} \{M_{\sigma\alpha} u_\beta^{[\alpha-1]}(a) + M_{\sigma,n+\alpha} u_\beta^{[\alpha-1]}(b)\} \right], \quad [D_{\sigma\beta} \qquad \eta_\sigma - \mu_\sigma\{u_0\}],$$
$$(\sigma = 1, \ldots, m\,;\, \beta = 1, \ldots, n),$$

have the same rank; whenever this solvability condition is satisfied the general solution of (8.6_m) *is* $u = u^p(t) + u_1(t)\xi_1 + \cdots + u_k(t)\xi_k$, *where* $u = u^p(t)$ *is a particular solution of this system,* $u_1(t), \ldots, u_k(t)$ *form a basis for the set of solutions of the corresponding homogeneous system* (8.5_m), *and* $\xi_1, \ldots, \xi_k$ *are arbitrary constants.*

If $m = n$, and the system (8.5_n) is incompatible, then for the non-homogeneous system

$$(8.8)\qquad \begin{aligned} l_n[u](t) &= r(t),\\ \mu_\alpha\{u\} \equiv \sum_{\beta=1}^{n} \{M_{\alpha\beta}\, u^{[\beta-1]}(a) &+ M_{\alpha,n+\beta} u^{[\beta-1]}(b)\} = 0, \qquad (\alpha = 1, \ldots, n), \end{aligned}$$

a solvability theorem may be obtained by applying the results of Section 7 to the equivalent first order differential system

$$
\begin{aligned}
&L_1[y] \equiv p_n(t)y_n' + p_0(t)y_1 + \cdots + p_{n-1}(t)y_n = r(t),\\
(8.9)\qquad &L_\beta[y] \equiv y_{n-\beta+1}' - y_{n-\beta+2} = 0, \qquad (\beta = 2, \ldots, n),\\
&\sum_{\beta=1}^{n} \{M_{\alpha\beta}y_\beta(a) + M_{\alpha,n+\beta}y_\beta(b)\} = 0, \qquad (\alpha = 1, \ldots, n).
\end{aligned}
$$

Indeed, the following result is an immediate consequence of Theorem 7.1.

Theorem 8.2. *If* (8.5_n) *is incompatible then the unique solution of* (8.8) *is given by*

$$(8.10)\qquad u(t) = \int_a^b g(t, s)\, r(s)\, ds, \qquad t \in [a, b],$$

where $g(t, s) = G_{11}(t, s)$, *and* $G(t, s) \equiv [G_{\alpha\beta}(t, s)]$, $(\alpha, \beta = 1, \ldots, n)$ *is the Green's matrix for the first order homogeneous differential system given by* (8.9) *with* $r(t) \equiv 0$.

The formula (7.7) does not specify values for the elements of the matrix $G(t, s)$ along the line $t = s$. In view of the values (3.3) of the coefficient matrices $A_2(t)$, $A_1(t)$ for the system (8.9), however, it follows that the associated matrix $\mathbf{G}(t, s) = A_2(t)G(t, s)A_1(s)$ occurring in Theorem 7.2 has elements

$$
\begin{aligned}
\mathbf{G}_{\alpha\beta}(t, s) &= G_{\alpha,n-\beta+1}(t, s), \qquad (\alpha = 1, \ldots, n;\ \beta = 1, \ldots, n-1),\\
\mathbf{G}_{\alpha n}(t, s) &= G_{\alpha 1}(t, s)\, p_n(s), \qquad (\alpha = 1, \ldots, n).
\end{aligned}
$$

Moreover, from relations (7.9), (7.10_1) it follows that the elements $\mathbf{G}_{\alpha\beta}(t, s)$ approach finite limits $\mathbf{G}_{\alpha\beta}{}^j(\tau)$ as (t, s) tends to (τ, τ) on the set Δ_j, $(j = 1, 2)$, and that $\mathbf{G}_{\alpha\beta}{}^1(\tau) = \mathbf{G}_{\alpha\beta}{}^2(\tau)$ for $\alpha \neq \beta$. That is, if $n > 1$ and the nondiagonal elements $\mathbf{G}_{\alpha\beta}(t, s)$, $(\alpha \neq \beta)$, of $\mathbf{G}(t, s)$ are defined as equal to $\mathbf{G}_{\alpha\beta}{}^1(\tau)$ at a point (τ, τ), $\tau \in [a, b]$, then these non-diagonal elements are continuous functions of (t, s) on $\square = [a, b] \times [a, b]$.

In particular, with the non-diagonal elements of the matrix $\mathbf{G}(t, s)$ for (8.9) chosen as continuous on $\square$, it follows that the function $g(t, s)$ of (8.10) is continuous in (t, s) on $\square$, and *throughout the following discussion it will be supposed that* $g(t, s)$ *has this property*. The function $g(t; s)$ thus determined is called the *Green's function* for the incompatible system (8.5_n).

The following result is a ready consequence of the result of Theorem 7.2 applied to the system (8.9), and the above remarks.

Theorem 8.3. *If $n > 1$, and the homogeneous system (8.5_n) is incompatible, then the Green's function is characterized by the following properties:*
(i) *the function $g(t, s)$, and the partial derivatives $g^{[j,0]}(t, s) \equiv (\partial^j/\partial t^j)\, g(t, s)$, $(j = 1, \ldots, n-2)$, are continuous in (t, s) on $\Box = [a, b] \times [a, b]$; on each of the triangular regions $\Delta_1 = \{(t, s) \mid (t, s) \in \Box, s < t\}$ and $\Delta_2 = \{t, s \mid (t, s) \in \Box, t < s\}$ the partial derivative $g^{[n-1,0]}(t, s)$ is continuous in (t, s);*

(ii) *if $s \in (a, b)$, then*

(a) *on each of the half-open intervals $a \le t < s$ and $s < t \le b$ the partial derivative $g^{[n,0]}(t, s)$ exists, is continuous in t, and*

$$p_n(t) g^{[n,0]}(t, s) + \cdots + p_1(t)\, g^{[1,0]}(t, s) + p_0(t)\, g(t, s) = 0,$$

(b) $$\sum_{\beta=1}^{n} \{M_{\alpha\beta} g^{[\beta-1,0]}(a, s) + M_{\alpha,n+\beta} g^{[\beta-1,0]}(b, s)\} = 0, \qquad (\alpha = 1, \ldots, n),$$

where it is understood that $g^{[0,0]}(t, s) \equiv g(t, s)$,

(c) *$g^{[n-1,0]}(t, s)$ has at $t = s$ a finite right-hand limit $g^{[n+1,0]}(s^+, s)$, and a finite left-hand limit $g^{[n-1,0]}(s^-, s)$; moreover*

$$g^{[n-1,0]}(s^+, s) - g^{[n-1,0]}(s^-, s) = \frac{1}{p_n(s)}\,. \tag{8.11}$$

In view of the above properties one may show that if (8.5_n) is incompatible and $u(t)$ is the solution of (8.8) then

$$u^{[\alpha-1]}(t) = \int_a^b g^{[\alpha-1,0]}(t, s)\, r(s)\, ds, \qquad t \in [a, b], \qquad (\alpha = 1, \ldots, n). \tag{8.10'}$$

As an example of the use of the conditions of Theorem 8.3 in the calculation of a Green's function, consider the incompatible system

$$u'' = 0, \qquad u(a) = 0 = u(b). \tag{8.12}$$

From properties (i), (ii-a, b) it follows that there are functions $k_1(s)$, $k_2(s)$ such that

$$g(t, s) = (t - a)\, k_2(s) \quad \text{for} \quad t \in [a, s],$$
$$g(t, s) = (b - t)\, k_1(s) \quad \text{for} \quad t \in [s, b].$$

The conditions that $g(t, s)$ is continuous in (t, s), and $g^{[1,0]}(s^+, s) - g^{[1,0]}(s^-, s) = 1$, then impose on $k_1(s)$, $k_2(s)$ the relations

$$(b - s)\, k_1(s) - (s - a)\, k_2(s) = 0, \qquad -k_1(s) - k_2(s) = 1.$$

Consequently $k_1(s) = (s - a)/(a - b)$, $k_2(s) = (b - s)/(a - b)$, and

$$g(t, s) = \frac{(t - a)(b - s)}{(a - b)} \quad \text{for} \quad a \le t \le s, \quad a \le s \le b,$$

$$= \frac{(s - a)(b - t)}{(a - b)} \quad \text{for} \quad s \le t \le b, \quad a \le s \le b.$$

For a homogeneous differential system of the form (8.5_j) and corresponding non-homogeneous differential system (8.6_j), $(j = 0, 1, \ldots, 2n)$, results are obtained by applying the theorems of Section 6 to the corresponding equivalent differential systems involving (8.2) and (8.4). These results are phrased in terms of the solutions of the corresponding homogeneous adjoint differential system which involves the differential equations (3.5), and, as pointed out in Section 3, the set of differential equations (3.5) is not equivalent to an n-th order differential equation unless the coefficients possess certain differentiability properties.

In the remainder of this section it will be supposed that the coefficients $p_j(t)$ *are of class* $\mathfrak{C}^j[a, b]$, *and* $p_n(t) \neq 0$ *on this interval.* In this case, as established in Section 3, the system of differential equations (3.5) has a solution $(z_\alpha(t))$, $(\alpha = 1, \ldots, n)$, if and only if $v(t) = z_1(t)$ is a solution of $l_n^\star[v] = 0$, where the adjoint differential expression $l_n^\star[v]$ is given by (3.13), and $\mathfrak{D}(l_n^\star) = \mathfrak{C}^n[a, b]$. Moreover, the components of a solution of (3.3) are given in terms of $v(t)$ and its derivatives by (3.12), which in turn is equivalent to

$$(8.13)\quad z_1(t) = v(t), \quad z_\gamma(t) = \sum_{\beta=1}^{n} \bar{K}_{\beta, n-\gamma+1}(t)\, v^{[\beta-1]}(t), \qquad (\gamma = 2, \ldots, n),$$

where $K(t) = [K_{\alpha\beta}(t)]$ is the coefficient matrix of the bilinear concomitant (3.17). Relations (8.13) are equivalent to the condition that the associated n-dimensional vector function $v_z(t) = A_1^*(t)z(t)$ is given by

$$(8.13')\qquad v_z(t) = \left(\sum_{\beta=1}^{n} \bar{K}_{\beta\alpha}(t)\, v^{[\beta-1]}\right), \qquad (\alpha = 1, \ldots, n).$$

From the fact that the matrix $K(t)$ is nonsingular it follows that under the transformation (8.13) the differential system (6.6_{2n}) is equivalent to

$$(8.14_{2n})\qquad \begin{aligned} &l_n^\star[v](t) = 0, \qquad t \in [a, b], \\ &v^{[\alpha-1]}(a) = 0, \qquad v^{[\alpha-1]}(b) = 0, \qquad (\alpha = 1, \ldots, n), \end{aligned}$$

while the differential system $(6.6_\circ)$ is equivalent to

$$(8.14_\circ)\qquad l_n^\star[v](t) = 0, \qquad t \in [a, b].$$

Finally, for $1 \leq m \leq 2n - 1$ the differential system (6.6_{2n-m}) is equivalent to the system

$$(8.14_{2n-m})\qquad \begin{aligned} &l_n^\star[v](t) = 0, \qquad t \in [a, b], \\ &v_\tau\{v\} \equiv \sum_{\alpha,\beta=1}^{n} \{-\bar{P}_{\alpha\tau}\bar{K}_{\beta\alpha}(a)\, v^{[\beta-1]}(a) + \bar{P}_{n+\alpha,\tau}\bar{K}_{\beta\alpha}(b)\, v^{[\beta-1]}(b)\} = 0, \\ &\qquad\qquad (\tau = 1, \ldots, 2n - m), \end{aligned}$$

with boundary conditions in nonparametric form. An equivalent system with boundary conditions in parametric form is

$$
(8.14_{2n-m})^{\circ} \quad
\begin{aligned}
&l_n^{\star}[v](t) = 0, \qquad t \in [a, b], \\
&\sum_{\beta=1}^{n} \bar{K}_{\beta\alpha}(a)\, v^{[\beta-1]}(a) = -\sum_{\sigma=1}^{m} \bar{M}_{\sigma\alpha}\chi_{\sigma}, \\
&\sum_{\beta=1}^{n} \bar{K}_{\beta\alpha}(b)\, v^{[\beta-1]}(b) = \sum_{\sigma=1}^{m} \bar{M}_{\sigma,n+\alpha}\chi_{\sigma}, \qquad (\alpha = 1, \ldots, n).
\end{aligned}
$$

For $j = 0, 1, \ldots, 2n$ let $D_j(l_n)$ denote the class of functions $u \in \mathfrak{C}^n[a, b]$ satisfying the boundary conditions of the system (8.5_j), and let $D_j(l_n^{\star})$ denote the class of functions $v \in \mathfrak{C}^n[a, b]$ satisfying the boundary conditions of the system (8.14_j). It follows readily from the corresponding result for linear vector differential operators of the first order that for $j = 0, 1, \ldots, 2n$ we have

$$
(8.15) \qquad \int_a^b \bar{v} l_n[u]\, dt - \int_a^b \overline{l_n^{\star}[v]}\, u\, dt = 0, \quad \text{for} \quad u \in D_j(l_n), \quad v \in D_{2n-j}(l_n^{\star}).
$$

The following theorems are ready consequences of Theorems 6.1, 6.2 and 6.2′, respectively.

Theorem 8.4. *If k is the index of the homogeneous differential system (8.5_m), $(m = 0, 1, \ldots, 2n)$, and k^* is the index of the corresponding adjoint system (8.14_{2n-m}), then $n + k^* = m + k$.*

Theorem 8.5. *For $m = 0, 1, \ldots, 2n$ the nonhomogeneous differential system (8.6_m) has a solution if and only if the relation*

$$
(8.16) \qquad \int_a^b \bar{v}(t)\, r(t)\, dt = \sum_{\alpha,\beta=1}^{n} \{-\bar{v}^{[\beta-1]}(a)\, K_{\beta\alpha}(a) w_{\alpha}^{a} + \bar{v}^{[\beta-1]}(b)\, K_{\beta\alpha}(b) w_{\alpha}^{b}\}.
$$

holds for all solutions $v(t)$ of the homogeneous adjoint system (8.14_{2n-m}).

Theorem 8.5.′ *For $m = 1, 2, \ldots, 2n - 1$ the nonhomogeneous differential system (8.6_m) has a solution if and only if the relation*

$$
(8.16') \qquad \int_a^b \bar{v}(t)\, r(t)\, dt = \sum_{\sigma=1}^{m} \sum_{\alpha,\beta=1}^{n} \{-\bar{v}^{[\beta-1]}(a)\, K_{\beta\alpha}(a) P_{\alpha\sigma}^{\circ} + \bar{v}^{[\beta-1]}(b)\, K_{\beta\alpha}(b) P_{n+\alpha,\sigma}^{\circ}\}\eta_{\sigma}
$$

holds for all solutions $v(t)$ of the homogeneous adjoint system (8.14_{2n-m}), where P° is a $2n \times m$ matrix which satisfies with the matrix M of (8.6_m) the equation $MP^{\circ} = E_m$.

If a differential system (8.5_n) is incompatible, then by Theorem 8.4 the corresponding adjoint system (8.14_n) is also incompatible, and there exists a Green's function $k(t, s)$ of (8.14_n) such that for $q(t) \in \mathfrak{C}[a, b]$ the unique solution of the system

$$(8.17) \qquad l_n^{\star}[v](t) = q(t),\ t \in [a, b],\ \nu_\tau\{v\} = 0, \qquad (\tau = 1, \ldots, n),$$

is given by

$$(8.18) \qquad v(t) = \int_a^b h(t, s)\, q(s)\, ds, \qquad t \in [a, b].$$

With the aid of (8.15) one establishes readily the following result, which corresponds to Theorem 7.3.

Theorem 8.6. *If $g(t, s)$ is the Green's function of an incompatible differential system (8.5_n), then the corresponding adjoint system (8.14_n) is incompatible and has Green's function $h(t, s) = \overline{g(s,t)}$.*

Problems III.8

1. What condition must the continuous function $r(t)$ satisfy in order that the differential system

$$u'' + u = r(t), \qquad u(0) = 0 = u(\pi),$$

possess a solution?

2. Determine the Green's function for each of the following differential systems:

(a) $u'' = 0,\ u(0) = 0,\ u'(1) = 0$;

(b) $u'' - u = 0,\ u'(-1) = 0,\ u'(1) = 0$;

(c) $u'' + k^2 u = 0,\ \ u(0) - u(2\pi) = 0,\ \ u'(0) - u'(2\pi) = 0,\ \ k \neq 0,\ \pm 1, \pm 2, \ldots$;

(d) $u^{[iv]} = 0,\ u(0) = u'(0) = u(b) = u'(b) = 0$, where $b > 0$;

(e) $u^{[iv]} = 0,\ u(0) = u'(0) = u''(b) = u'''(b) = 0$, where $b > 0$.

3. If a differential system (8.5_n) is incompatible, show that in terms of linearly independent solutions $u_1(t), \ldots, u_n(t)$ of $l_n[u] = 0$ the Green's function $g(t, s)$ of (8.5_n) may be written for $t \neq s$ as

$$g(t, s) = \frac{J(t, s)}{J p_n(s)}, \quad \text{where} \quad J = \det[\mu_\alpha\{u_\beta\}], \qquad (\alpha, \beta = 1, \ldots, n),$$

$$J(t, s) = \det\begin{bmatrix} g_0(t, s) & u_\beta(s) \\ \mu_\alpha\{g_0(\ , s)\} & \mu_\alpha\{u_\beta\} \end{bmatrix},$$

with $g_0(t, s)$ defined for $t \neq s$ by $g_0(t, s) = \frac{1}{2}[sgn(t - s)/W(s)]\, K(t, s)$, where

$$K(t, s) = \det \begin{bmatrix} u_\beta(s) \\ \cdot \\ \cdot \\ \cdot \\ u_\beta^{[n-2]}(s) \\ u_\beta(t) \end{bmatrix}, \quad W(s) = \det[u_\beta^{[\alpha-1]}(s)], \quad (\alpha, \beta = 1, \ldots, n),$$

and the symbol $\mu_\alpha\{g_0(\ , s)\}$ signifies the boundary form $\mu_\alpha\{u\}$ of (8.5_n) computed for $u(t) = g_0(t, s)$, with s as parameter.

Hint. Show that $g(t, s)$ defined above for $t \neq s$ may be defined along the line $t = s$ in such a manner as to be continuous on $\square = [a, b] \times [a, b]$, and that the resulting function satisfies the conditions of Theorem 8.3.

4. A differential operator l_n is said to be self-adjoint if $D(l_n) = D(l_n^\star)$ and $l_n[u] \equiv l_n^\star[u]$ for $u \in D(l_n)$. If $l_n[u](t) = p_n(t)u^{[n]}(t) + \cdots + p_0(t)u(t)$, where $p_j \in \mathfrak{C}^j[a, b]$, $(j = 0, 1, \ldots, n)$, and $D(l_n)$ and $D(l_n^\star)$ are specified by the boundary conditions $\mu_\sigma\{u\} = 0$, $(\sigma = 1, \ldots, m)$, and $\nu_\tau\{v\} = 0$, $(\tau = 1, \ldots, 2n - m)$, of (8.5_m) and (8.14_{2n-m}), respectively, prove that if l_n is self-adjoint then $m = n$. Moreover, if the differential system (8.5_n) is incompatible, then the involved differential operator $l_n[u]$ is self-adjoint if and only if its Green's functions $g(t, s)$ is hermitian in the sense that $g(t, s) = \overline{g(s, t)}$.

5. Suppose that $u = u_\alpha(t)$, $(\alpha = 1, 2)$ is the solution of the differential system

$$u'' = r_\alpha(t), \qquad u(0) = 0 = u(1),$$

where $r_1(t)$ and $r_2(t)$ are continuous functions on $[0, 1]$. If $|r_1(t) - r_2(t)| \leq \varepsilon$ for $t \in [0, 1]$, show that $|u_1(t) - u_2(t)| \leq \varepsilon/8$ for $t \in [0, 1]$. Correspondingly, if $\int_0^1 |r_1(t) - r_2(t)|^2\, dt \leq \varepsilon^2$ show that $\int_0^1 |u_1(t) - u_2(t)|^2\, dt \leq \varepsilon^2/90$.

6. Suppose that $p_0(t)$ is a real-valued continuous function on $[a, b]$, and that $p_0(t) \leq k$ on this interval. If $u(t)$ is a nonidentically vanishing real-valued solution of the differential equation

$$u'' + p_0(t)u = 0, \qquad t \in [a, b],$$

and $u(t_1) = 0 = u(t_2)$, where $a \leq t_1 < t_2 \leq b$, prove that $(t_2 - t_1)^2 \geq 8/k$.

Hint. Note that attention may be restricted to the case where t_1 and t_2 are consecutive zeros of $u(t)$, and that without loss of generality it may be assumed that $u(t) > 0$ for $t \in (t_1, t_2)$. For $g(t, s; t_1, t_2)$ the Green's function for the differential system $u'' = 0$, $u(t_1) = 0 = u(t_2)$, from the relation

$$u(t) = -\int_{t_1}^{t_2} g(t, s; t_1, t_2)\, p_0(s)\, u(s)\, ds, \qquad t \in [t_1, t_2],$$

and the specific value of $g(t, s; t_1, t_2)$, estimate $u(t_0)$, where $t_0 \in (t_1, t_2)$ and $u(t_0)$ is equal to the maximum of $u(t)$ on $[t_1, t_2]$.

9. DIFFERENTIAL SYSTEMS INVOLVING AN *n*-th ORDER LINEAR VECTOR DIFFERENTIAL EQUATION

The results of Sections 3 and 8 on n-th order linear scalar differential equations may be extended readily to differential systems involving an n-th order linear vector differential expression of the form

$$\mathcal{L}_n[u] \equiv P_n(t)u^{[n]} + \cdots + P_1(t)u' + P_0(t)u, \tag{9.1}$$

where $P_j(t) \equiv [P_{\sigma\rho j}(t)]$, $(\sigma, \rho = 1, \ldots, h;\ j = 0, 1, \ldots, n)$, are $h \times h$ matrices with elements continuous on $[a, b]$, the matrix $P_n(t)$ is nonsingular on this interval, and $u \equiv (u_\sigma(t))$, $\sigma = 1, \ldots, h)$, is an h-dimensional vector function.

If the h-dimensional vector functions $y_\beta(t) \equiv (y_{\sigma\beta}(t))$, $(\sigma = 1, \ldots, h;$ $\beta = 1, \ldots, n)$, are defined as

$$y_\beta(t) \equiv u^{[\beta-1]}(t) \equiv (u_\sigma^{[\beta-1]}(t)), \qquad (\sigma = 1, \ldots, h\ ;\ \beta = 1, \ldots, n), \tag{9.2}$$

then the homogeneous vector differential equation

$$\mathcal{L}_n[u](t) = 0, \qquad t \in [a, b], \tag{9.3}$$

is equivalent to the system of vector equations

$$\begin{aligned} &P_n(t)y_n' + P_0(t)y_1 + \cdots + P_{n-1}(t)y_n = 0, \\ &y_{n-\beta+1}' - y_{n-\beta+2} = 0, \qquad (\beta = 2, \ldots, n), \end{aligned} \tag{9.4}$$

which may be written as a hn-dimensional vector differential equation in $\mathbf{y}(t) \equiv (\mathbf{y}_\nu(t))$, $(\nu = 1, \ldots, hn)$, with $\mathbf{y}_{(\beta-1)h+\sigma}(t) = y_{\sigma\beta}(t) = u_\sigma^{[\beta-1]}(t)$, $(\sigma = 1, \ldots, h;\ \beta = 1, \ldots, n)$. System (9.4) in the h-dimensional vector functions $y_1, \ldots, y_n$ is an obvious extension of (8.2), with relatively minor adjustments in notation due to the fact that the coefficients of (9.1) are matrices.

All the results of Section 8 have direct analogues for vector differential systems involving $\mathcal{L}_n[u]$ and two-point boundary conditions. For brevity, specific attention will be limited to the direct analogues of the scalar systems (8.5_m) and (8.6_m) for $1 \le m \le 2n - 1$, as the analogues of the cases $m = 0$

and $m = 2n$ are immediate. For such a system, analogous to (8.5_m) one has

$$(9.6_m)\qquad \begin{gathered} \mathfrak{L}_n[u](t) = 0, \qquad t \in [a, b], \\ \sum_{\beta=1}^{n} \{M_{\theta\beta} u^{[\beta-1]}(a) + M_{\theta,n+\beta} u^{[\beta-1]}(b)\} = 0, \qquad (\theta = 1, \ldots, m), \end{gathered}$$

where $M_{\theta\gamma} = [M_{\sigma\rho;\theta\gamma}]$, $(\sigma, \rho = 1, \ldots, h)$, are $h \times h$ constant matrices for $\theta = 1, \ldots, m$, $\gamma = 1, \ldots, 2n$, such that the $hm \times 2hn$ matrix

$$(9.7)\qquad \mathcal{M} = \begin{bmatrix} M_{\sigma\rho;11} & \cdots & M_{\sigma\rho;1,2n} \\ \cdot & & \\ \cdot & & \\ M_{\sigma\rho;m1} & \cdots & M_{\sigma\rho;m,2n} \end{bmatrix}$$

is of rank hm. In particular, if $m = n$ and the system (9.6_n) is incompatible, then for arbitrary h-dimensional vector functions $r(t) \equiv (r_\sigma(t))$ continuous on $[a, b]$ the nonhomogeneous system

$$(9.8_n)\qquad \begin{gathered} \mathfrak{L}_n[u](t) = r(t), \\ \sum_{\beta=1}^{n} \{M_{\theta\beta} u^{[\beta-1]}(a) + M_{\theta,n+\beta} u^{[\beta-1]}(b)\} = 0, \qquad (\theta = 1, \ldots, n), \end{gathered}$$

has a unique solution, which is given by

$$(9.9)\qquad u(t) = \int_a^b \mathcal{G}(t, s) r(s)\, ds, \qquad t \in [a, b],$$

where $\mathcal{G}(t, s)$ is an $h \times h$ Green's matrix for the incompatible system (9.6_n). Indeed, the Green's matrix for an incompatible system (9.6_n) is characterized by conditions that are direct matrix generalizations of the conditions of Theorem 8.3.

If on $[a, b]$ the coefficient matrices $P_j(t)$ are of class $\mathfrak{C}^j[a, b]$, $(j = 0, 1, \ldots, n)$, then for (9.1) the corresponding formal adjoint differential expression is

$$(9.10)\qquad \mathfrak{L}_n^\star[v] = \sum_{j=0}^{n} (-1)^j (P_j^*(t) v)^{[j]}.$$

Corresponding to the result of Theorem 3.2 one has that if $u \in \mathfrak{C}_h{}^n[a, b]$, $v \in \mathfrak{C}_h{}^n[a, b]$, then

$$(9.11)\qquad v^*\mathfrak{L}_n[u] - (\mathfrak{L}_n^\star[v])^* u = \left[\sum_{\alpha,\beta=1}^{n} v^{*[\alpha-1]}(t)\, K_{\alpha\beta}(t)\, u^{[\beta-1]}(t)\right]',$$

where $K_{\alpha\beta}(t)$, $(\alpha, \beta = 1, \ldots, n)$, is an $h \times h$ matrix whose elements are of class $\mathfrak{C}^1[a, b]$. Moreover, $K_{\alpha\beta}(t)$ is the zero matrix for $\alpha + \beta > n + 1$, and $K_{\alpha\beta}(t) = (-1)^{\alpha-1} P_n(t)$ for $\alpha + \beta = n + 1$.

Now for $\mathcal{M}$ an $hm \times 2hn$ matrix of the form (9.7), and of rank hm, let

$\mathcal{T}$ be a $2hn \times (2hn - hm)$ matrix of rank $2hn - hm$ and such that $\mathcal{M}\mathcal{T} = 0$; moreover, let $P_{\gamma\tau} = [P_{\sigma\rho;\gamma\tau}]$, $(\gamma = 1, \ldots, 2n;\quad \tau = 1, \ldots, 2n - m)$, be $h \times h$ matrices such that

$$\mathcal{T} = \begin{bmatrix} P_{\sigma\rho;11} & \cdots P_{\sigma\rho;1,2n-m} \\ \cdot & \\ \cdot & \\ P_{\sigma\rho;2n,1} & \cdots P_{\sigma\rho;2n,2n-m} \end{bmatrix}. \tag{9.12}$$

Corresponding to (8.14_{2n-m}), the system adjoint to (9.6_m) is

$$\mathfrak{L}_n^\star[v](t) = 0, \qquad t \in [a, b],$$

$$\sum_{\alpha,\beta=1}^{n} \{-P^*_{\alpha\tau}\, K^*_{\beta\alpha}(a)\, v^{[\beta-1]}(a) + P^*_{n+\alpha,\tau}\, K^*_{\beta\alpha}(b)\, v^{[\beta-1]}(b)\} = 0, \tag{9.13_{2n-m}}$$

$$(\tau = 1, \ldots, 2n - m).$$

Corresponding to Theorem 8.4, one has that if k is the index of the homogeneous system (9.6_m), and k^* is the index of the corresponding adjoint system (9.13_{2n-m}), then $hn + k^* = hm + k$. For the non-homogeneous system

$$\mathfrak{L}_n[u](t) = r(t), \qquad t \in [a, b],$$

$$\mu_\theta\{u\} = \sum_{\beta=1}^{n} \{M_{\theta\beta}\, u^{[\beta-1]}(a) + M_{\theta,n+\beta}\, u^{[\beta-1]}(b)\} = \eta_\theta, \tag{9.14_m}$$

$$(\theta = 1, \ldots, m),$$

where now $r(t) \in \mathfrak{C}_h[a, b]$, and η_θ, $(\theta = 1, \ldots, m)$, are h-dimensional vectors, one has solvability criteria corresponding to Theorems 8.5 and 8.5′. In particular, let $\mathcal{T}^0$ be a $2hn \times hm$ matrix such that $\mathcal{M}\mathcal{T}^0$ is the $hm \times hm$ identity matrix, and $P_{\gamma\theta}{}^0 = [P^0_{\sigma\rho;\gamma\theta}]$, $(\gamma = 1, \ldots, 2n;\ \theta = 1, \ldots, m)$ are $h \times h$ matrices such that

$$\mathcal{T}^0 = \begin{bmatrix} P^0_{\sigma\rho;\,11} & \cdots & P^0_{\sigma\rho;\,1m} \\ P^0_{\sigma\rho;\,2n,1} & \cdots & P^0_{\sigma\rho;\,2n,m} \end{bmatrix}. \tag{9.15}$$

Then the analogue of Theorem 8.5′ is the conclusion that the non-homogeneous system (9.14_m) has a solution if and only if the relation

$$\int_a^b v^*(t)\, r(t)\, dt = \sum_{\theta=1}^{m} \sum_{\alpha,\beta=1}^{n} \{-v^{*[\alpha-1]}(a)\, K_{\alpha\beta}(a) P^0_{\beta\theta} + v^{*[\alpha-1]}(b)\, K_{\alpha\beta}(b) P^0_{n+\beta,\theta}\}\eta_\theta \tag{9.16}$$

holds for all solutions $v(t)$ of the homogeneous adjoint system (9.13_{2n-m}).

Also, corresponding to Theorem 8.6, we have that if $m = n$ and the system (9.6_n) is incompatible with Green's matrix $\mathcal{G}(t, s)$, the corresponding adjoint system (9.13_n) is incompatible and has Green's matrix $\mathcal{H}(t, s) = \mathcal{G}^*(s, t)$.

Problems III.9

1. Show that the differential system

$$u_1'' + u_1' + u_2 = 0,$$
$$u_2'' = 0, \qquad u_1(0) = u_2(0) = u_1(1) = u_2(1) = 0,$$

is incompatible.

2. Determine the Green's matrix for the differential system of Problem 1.

3. Show that the system consisting of the differential equations

$$u_1'' - u_2 = 0,$$
$$u_2'' + 4u_1 = 0,$$

and each of the following sets of boundary conditions is incompatible:

(i) $u_1 = u_1' = 0$ at $t = 0$ and $t = \pi$;

(ii) $u_1 = u_2 = 0$ at $t = 0$ and $t = \pi$.

4. Determine the differential system adjoint to each of the systems of Problem 3.

5. If $P_j(t)$, $(j = 0, 1, 2)$, are $h \times h$ matrices with elements continuous on $[a, b]$, and $P_2(t)$ is nonsingular on this interval, show that there exists a non-singular $h \times h$ matrix $W(t)$ such that

$$W(t)[P_2(t)u'' + P_1(t)u' + P_0(t)u] \equiv (R_2(t)u')' + R_0(t)u,$$

where $R_2(t)$ is a non-singular matrix of class $\mathfrak{C}^1[a, b]$, and $R_0(t)$ is continuous on this interval.

6. Suppose that $R_0(t)$, $R_2(t)$ are $h \times h$ matrices with $R_2(t)$ nonsingular and of class $\mathfrak{C}^1[a, b]$, while $R_0(t)$ is continuous on this interval. Show that:

(i) The adjoint of

$$\text{(A)} \qquad [R_2(t)u']' + R_0(t)u = 0, \qquad t \in [a, b],$$

is

$$\text{(A*)} \qquad [R_2^*(t)v']' + R_0^*(t)v = 0, \qquad t \in [a, b].$$

(ii) If $u(t)$ and $v(t)$ are h-dimensional vector functions which are solutions of (A) and (A*), respectively, then $v^*R_2u' - v^{*\prime}R_2u \equiv$ const. on $[a, b]$.

(iii) The system adjoint to

$$\text{(B)} \qquad [R_2(t)u']' + R_0(t)u = 0, \qquad t \in [a, b], \qquad u(a) = 0 = u(b),$$

is

$$\text{(B*)} \qquad [R_2^*(t)v']' + R_0^*(t)v = 0, \qquad t \in [a, b], \qquad v(a) = 0 = v(b).$$

(iv) If (B) is incompatible then the Green's matrix for this system is

$$\begin{aligned} \mathcal{G}(t, s) &= U_1(t)V_2^*(s), \qquad a \le t \le s, \qquad a \le s \le b, \\ &= U_2(t)V_1^*(s), \qquad s \le t \le b, \qquad a \le s \le b, \end{aligned}$$

where $U_1(t)$, $U_2(t)$ are $h \times h$ matrices whose column vectors are solutions of (A) with $U_1(a) = 0 = U_2(b)$, and $V_1(t)$, $V_2(t)$ are $h \times h$ matrices whose columns are solutions of (A*) with $V_1(a) = 0 = V_2(b)$ and

$$\begin{aligned} &V_1^* R_2 U_1' - V_1^{*\prime} R_2 U_1 \equiv 0 \equiv V_2^* R_2 U_2' - V_2^{*\prime} R_2 U_2, \\ &V_1^* R_2 U_2' - V_1^{*\prime} R_2 U_2 \equiv E_h, \qquad V_2^* R_2 U_1' - V_2^{*\prime} R_2 U_1 \equiv -E_h. \end{aligned}$$

10. THE GENERALIZED GREEN'S MATRIX

In this section we shall return to the study of a vector differential operator L of the form

$$L[y](t) = A_1(t)[A_2(t)y(t)]' + A_0(t)y(t), \tag{10.1}$$

where the coefficient matrices satisfy hypothesis ($\mathfrak{H}$) of Section 1; that is, each $A_j(t)$ is a continuous $n \times n$ matrix function on a given compact interval $[a, b]$, with $A_1(t)$ and $A_2(t)$ nonsingular on this interval. As in Sections 4, 5, 6, and 7, we shall consider a linear vector differential operator whose domain $D(L)$ is a linear manifold in $\mathfrak{C}_n[a, b]$ satisfying $\mathfrak{D}_0(L) \subset D(L) \subset \mathfrak{D}(L)$, and functional value (10.1).

A matrix $G(t, s)$ is called a *generalized Green's matrix* for a differential operator L, if as a function of (t, s) it is bounded on $\square = [a, b] \times [a, b]$, continuous on each of the triangular domains $\Delta_1 = \{(t, s) \mid (t, s) \in \square, s < t\}$, $\Delta_2 = \{(t, s) \mid (t, s) \in \square, t < s\}$, and

$$y(t) = \int_a^b G(t, s)g(s)\, ds, \qquad t \in [a, b], \tag{10.2}$$

provides a linear mapping of $\mathfrak{R}(L)$ into $D(L)$; that is, if $g \in \mathfrak{C}_n[a, b]$ and is such that the differential system

$$L[y](t) = g(t), \qquad t \in [a, b], \qquad y \in D(L), \tag{10.3}$$

has a solution, then a particular solution of (10.3) is given by (10.2). If $m = n$ and the dimension k of $\mathfrak{N}(L)$ is equal to zero, then the dimension k^* of $\mathfrak{N}(L^*)$ is also zero, and by Theorem 7.1 the generalized Green's matrix is unique and given by (7.7). As a consequence of the relation $n + k^* = m + k$ of Theorem 6.1, however, if $m \neq n$ then either $k \neq 0$ or $k^* \neq 0$.

If $k \neq 0$ we shall denote by $Y_0(t)$ an $n \times k$ matrix function whose column vectors form a basis for $\mathfrak{N}(L)$; correspondingly, if $k^* \neq 0$ we shall denote by $Z_0(t)$ an $n \times k^*$ matrix function whose column vectors form a basis for

$\mathfrak{N}(L^\star)$. As in Section 7, $Y(t)$ will denote a fundamental matrix solution of $L[Y] = 0$, and $Z(t)$ a fundamental matrix solution of $L^\star[Z] = 0$ such that $V_Z{}^*(t)U_Y(t) \equiv E_n$.

Now if $\tilde{G}(t, s)$ denotes the matrix function

$$\tilde{G}(t, s) = \tfrac{1}{2}\, sgn(t - s)\; Y(t)Z^*(s), \qquad (t, s) \in [a, b] \times [a, b],$$

and for $g(t) \in \mathfrak{C}_n[a, b]$ the vector function $y^0(t)$ is defined as

$$y^0(t) = \int_a^b \tilde{G}(t, s)g(s)\, ds, \qquad t \in [a, b], \tag{10.4}$$

then, as in the proof of Theorem 7.1, it follows that $y = y^0(t)$ is a particular solution of the differential equation $L[y] = g$ on $[a, b]$. Consequently, if $\tilde{L}$ denotes the differential operator with domain $D(\tilde{L}) = \mathfrak{D}(L)$ and functional value $\tilde{L}[y](t) = L[y](t)$, $t \in [a, b]$, then $m = 0$, $k = n$, $k^* = 0$, and $G(t, s) = \tilde{G}(t, s)$ is a particular generalized Green's matrix for $\tilde{L}$. In this case it is to be noted that if $H(t)$ is an arbitrary continuous $n \times n$ matrix function on $[a, b]$ then $G(t, s) = \tilde{G}(t, s) + Y(t)H^*(s)$ is also a generalized Green's matrix for $\tilde{L}$.

Now suppose that L is a differential operator with functional value given by (10.1), and domain

$$D(L) = \{y \mid y \in \mathfrak{D}(L), \qquad \mu[y] \equiv M\hat{u}_y = 0\}, \tag{10.5}$$

where M is an $m \times 2n$ matrix of rank m, and $1 \leq m \leq 2n$. In particular, for $m = 2n$ we have $D(L) = \mathfrak{D}_0(L)$. As in the proof of Theorem 7.1, the general solution of the differential equation $L[y](t) = g(t)$ is

$$y(t) = y^0(t) + Y(t)\xi, \tag{10.6}$$

where $y^0(t)$ is given by (10.4), and ξ is an arbitrary n-dimensional vector. Consequently, a vector function $g(t) \in \mathfrak{C}_n[a, b]$ belongs to $\mathfrak{R}(L)$ if and only if there exists a solution ξ of the algebraic vector equation

$$[M\hat{U}_Y]\xi + \mu[y^0] = 0, \tag{10.7}$$

and, if $g(t)$ is such that (10.7) has a solution, then a particular solution of the differential system (10.3) is given by

$$y(t) = y^0(t) - Y(t)D^{\#}\mu[y^0],$$

where $D^{\#}$ is a generalized inverse, (see Appendix B), of the $m \times n$ matrix $D = M\hat{U}_Y = \mu[Y]$. As in the proof of Theorem 7.1, it follows that

$$\mu[y^0] = M\hat{u}_{y^0} = -\tfrac{1}{2}\Delta \int_a^b Z^*(s)g(s)\, ds,$$

where $\Delta = -MQ\hat{U}_Y$, and consequently we have the following result.

Theorem 10.1. *For a vector differential operator L of the form* (10.1) *a generalized Green's matrix is given by*

(10.8)
$$G^{\#}(t, s) = \tfrac{1}{2} Y(t)\{E_n \operatorname{sgn}(t - s) + \Gamma\} Z^*(s), \qquad (t, s) \in [a, b] \times [a, b],$$

where $Y(t)$ and $Z(t)$ are fundamental matrix solutions of $L[Y] = 0$ and $L^\star[Z] = 0$ such that $V_Z{}^*(t)\, U_Y(t) \equiv E_n$, and: (i) *if $D(L) = \mathfrak{D}(L)$ then $\Gamma = 0$;* (*ii*) *if $D(L)$ is given by* (10.5), *where M is an $m \times 2n$ matrix of rank m, then $\Gamma = D^{\#}\Delta$ with $D^{\#}$ any generalized inverse of $D = M\hat{U}_Y$ and $\Delta = -MQ\hat{U}_Y$, where Q is the $2n \times 2n$ matrix defined by* (6.4).

The most general form of a generalized Green's matrix is presented in the following theorem, where it is to be understood that the stated relations involving Y_0 and Θ are to be deleted if $k \equiv \dim \mathfrak{N}(L) = 0$, and the stated relations involving Z_0 and Ψ are to be deleted if $k^* \equiv \dim \mathfrak{N}(L^\star) = 0$.

Theorem 10.2. *If the column vectors of $Y_0(t)$ and $Z_0(t)$ form a basis for $\mathfrak{N}(L)$ and $\mathfrak{N}(L^\star)$, respectively, then the most general generalized Green's matrix for the differential operator L is of the form*

$$G(t, s) = G_0(t, s) + Y_0(t)H^*(s) + K(t)Z_0^*(s), \tag{10.9}$$

for $t \neq s$, where $G_0(t, s)$ is any particular generalized Green's matrix, while $H(t)$ and $K(t)$ are continuous matrix functions on $[a, b]$ of respective dimensions $n \times k$ and $n \times k^*$.

From the fact that $\mathfrak{R}(L) = [\mathfrak{N}(L^\star)]^\perp$ it follows immediately that if $G_0(t, s)$ is a particular generalized Green's matrix for L then a matrix function $G(t, s)$ of the form (10.9) is also a generalized Green's matrix for L.

On the other hand, if $G_0(t, s)$ and $G(t, s)$ are individually generalized Green's matrices for L, then $K(t, s) = G(t, s) - G_0(t, s)$ is a matrix function on $[a, b] \times [a, b]$ which is bounded, continuous on each of the triangular domains Δ_1, Δ_2, and such that if $g(t) \in \mathfrak{R}(L)$ then there exists a k-dimensional vector $v = v[g]$ such that

$$\int_a^b K(t, s)g(s)\, ds = Y_0(t)v[g], \qquad t \in [a, b]. \tag{10.10}$$

Now let $\Theta(t)$ and $\Psi(t)$ be continuous matrix functions on $[a, b]$, of respective dimensions $n \times k$ and $n \times k^*$, and such that the matrices

$$T = \int_a^b \Theta^*(t) Y_0(t)\, dt, \qquad S = \int_a^b Z_0^*(t)\Psi(t)\, dt, \tag{10.11}$$

are non-singular. The vector $v[g]$ of (10.10) is then given by

$$v[g] = \int_a^b H^*(s)g(s)\,ds, \quad \text{where} \quad H^*(s) = T^{-1}\int_a^b \Theta^*(t)K(t,s)\,dt,$$

and

$$\int_a^b [K(t,s) - Y_0(t)H^*(s)]\,g(s)\,ds = 0, \qquad t \in [a,b], \quad \text{for} \quad g(t) \in \Re(L).$$

For $h(t) \in \mathfrak{C}_n[a,b]$ the vector function

$$g_0(t) = h(t) - \Psi(t)S^{-1}\int_a^b Z_0^*(s)h(s)\,ds \tag{10.12}$$

is such that $\int_a^b Z_0^*(t)g_0(t)\,dt = 0$, and hence $g_0(t) \in \Re(L)$. Consequently,

$$\begin{aligned} 0 &= \int_a^b [K(t,s) - Y_0(t)H^*(s)]g_0(s)\,ds, \\ &= \int_a^b [K(t,s) - Y_0(t)H^*(s) - K(t)Z_0^*(s)]h(s)\,ds, \end{aligned} \tag{10.13}$$

where the matrices $H(t)$ and $K(t)$ are defined as

$$H(t) = \left[\int_a^b K^*(\tau,t)\Theta(\tau)\,d\tau\right]T^{*-1},$$

$$K(t) = \left(\int_a^b [K(t,\sigma) - Y_0(t)H^*(\sigma)]\Psi(\sigma)\,d\sigma\right)S^{-1},$$

and in view of the arbitrariness of h in (10.13) it follows that

$$K(t,s) = Y_0(t)H^*(s) + K(t)Z_0^*(s) \quad \text{on} \quad \Delta_1 \cup \Delta_2.$$

Theorem 10.3. *If $\Theta(t)$ and $\Psi(t)$ are continuous matrix functions on $[a,b]$, of respective dimensions $n \times k$ and $n \times k^*$, and such that the matrices T and S of (10.11) are nonsingular, then there exists a unique generalized Green's matrix $G(t,s) = G_{\Theta,\Psi}(t,s:L)$ for L such that*

$$\int_a^b \Theta^*(t)G(t,s)\,dt = 0, \quad \text{for} \quad s \in [a,b], \tag{10.14}$$

$$\int_a^b G(t,s)\Psi(s)\,ds = 0, \quad \text{for} \quad t \in [a,b]. \tag{10.15}$$

If $G_0(t,s)$ is a particular generalized Green's matrix for L, then it follows directly that the generalized Green's matrix defined by (10.9) with

$$H(t) = -\left(\int_a^b G_0^*(\tau,t)\Theta(\tau)\,d\tau\right)T^{*-1},$$

$$K(t) = -\left\{\int_a^b G_0(t,\sigma)\Psi(\sigma)\,d\sigma + Y_0(t)\int_a^b H^*(\sigma)\Psi(\sigma)\,d\sigma\right\}S^{-1},$$

is a generalized Green's matrix satisfying (10.14) and (10.15).

On the other hand, if $G_1(t, s)$ and $G_2(t, s)$ are individually generalized Green's matrices for L satisfying (10.14) and (10.15), then there exist continuous matrix functions $H(t)$ and $K(t)$ on $[a, b]$, of respective dimensions $n \times k$ and $n \times k^*$, and such that $F(t, s) = G_1(t, s) - G_2(t, s)$ satisfies for $t \neq s$ the relation

$$F(t, s) = Y_0(t)H^*(s) + K(t)Z_0^*(s);$$

moreover, $F(t, s)$ satisfies the conditions

$$\text{(10.16)} \qquad \int_a^b \Theta^*(t)F(t, s)\,dt = 0, \quad \text{for} \quad s \in [a, b],$$

$$\text{(10.17)} \qquad \int_a^b F(t, s)\Psi(s)\,ds = 0, \quad \text{for} \quad t \in [a, b].$$

Now the conditions (10.16), (10.17) may be written respectively as

$$\text{(10.16')} \quad TH^*(s) + T_1Z_0^*(s) = 0, \quad s \in [a, b], \quad \text{where} \quad T_1 = \int_a^b \Theta^*(t)K(t)\,dt,$$

$$\text{(10.17')} \quad Y_0(t)S_1 + K(t)S = 0, \quad t \in [a, b], \quad \text{where} \quad S_1 = \int_a^b H^*(s)\Psi(s)\,ds.$$

Then (10.16') implies that $H^*(s) = -T^{-1}T_1Z_0^*(s)$, and hence

$$S_1 = \textstyle\int_a^b H^*(s)\Psi(s)\,ds = -T^{-1}T_1 \int_a^b Z_0^*(s)\Psi(s)\,ds = -T^{-1}T_1S.$$

In turn, relation (10.17') implies that $[-Y_0(t)T^{-1}T_1 + K(t)]S = 0$, and consequently $K(t) = Y_0(t)T^{-1}T_1$. Finally, from (10.16') it follows that $0 = Y_0(t)\,T^{-1}[TH^*(s) + T_1Z_0^*(s)] = Y_0(t)\,H^*(s) + Y_0(t)T^{-1}T_1\,Z_0^*(s) = Y_0(t)H^*(s) + K(t)\,Z_0^*(s) = F(t, s)$, and hence $G_1(t, s) = G_2(t, s)$ on $\Delta_1 \cup \Delta_2$.

Problems III.10

1. For $a = -\pi$, $b = \pi$, let L denote the two-dimensional vector differential operator $L[y](t) = y'(t) + A_0(t)\,y(t)$, with

$$A_0(t) = \begin{bmatrix} 0 & -1 \\ 1 & 0 \end{bmatrix}, \qquad D(L) = \{y \mid y \in \mathfrak{C}_2{}^1[a, b]\,, \ \ M\hat{y} = 0\}.$$

(i) Determine a generalized Green's matrix for L of the form (10.8), in each of the following cases:

(a) $M = [1 \quad 0 \quad -1 \quad 0]$,

(b) $M = \begin{bmatrix} 1 & 0 & 0 & 0 \\ 0 & 0 & 1 & 0 \end{bmatrix}$.

(ii) For $M = \begin{bmatrix} 1 & 0 & -1 & 0 \\ 0 & 1 & 0 & 1 \end{bmatrix}$, determine the generalized Green's matrix $G_{\Theta,\Psi}(t, s{:}L)$ as specified in Theorem 10.3, in each of the following cases:

(a) $\Theta(t) = \Psi(t) = \dfrac{1}{2\pi}\begin{bmatrix} \cos t \\ -\sin t \end{bmatrix}$;

(b) $\Theta(t) = \dfrac{1}{\pi}\begin{bmatrix} 0 \\ -\sin t \end{bmatrix}$, $\quad \Psi(t) = \dfrac{1}{\pi}\begin{bmatrix} \cos t \\ 0 \end{bmatrix}$.

2. If L is a vector differential operator with functional value (10.1), and domain $D(L) = \mathfrak{D}_0(L) = \{y \mid y \in \mathfrak{D}(L),\ \hat{u}_y = 0\}$, show that the matrix function $\tilde{G}(t, s)$ appearing in (10.4) is a generalized Green's matrix for L.

3. With $\Theta(t)$ and $\Psi(t)$ continuous matrix functions such that the matrices (10.11) are non-singular, consider the vector function $\eta(t)$ of dimension $N = n + k^* + k = m + 2k$, where $\eta_\alpha = y_\alpha$, $(\alpha = 1, \ldots, n)$, $\eta_{n+\beta} = \rho_\beta$, $(\beta = 1, \ldots, k^*)$, $\eta_{n+k^*+\gamma} = \mu_\gamma$, $(\gamma = 1, \ldots, k)$, with the understanding that if either k or k^* is zero then the corresponding components are nonexistent. Let $\mathfrak{L}[\eta] \equiv (\mathfrak{L}_\sigma[\eta])$, $(\sigma = 1, \ldots, N)$, denote the corresponding differential operator with domain

$$D(\mathfrak{L}) = \{(y, \rho, \mu) \mid y \in D(L),\ \rho \in \mathfrak{C}_{k^*}^{1}[a, b],\ \mu \in \mathfrak{C}_k^{1}[a, b],\ \mu(a) = 0 = \mu(b)\},$$

and functional value

$$\begin{aligned} (\mathfrak{L}_\alpha[\eta]) &= L[y] + \Psi(t)\rho, \\ (\mathfrak{L}_{n+\beta}[\eta] &= \rho', \\ (\mathfrak{L}_{n+k^*+\gamma}[\eta]) &= \mu' + \Theta^*(t)y. \end{aligned}$$

Show that the null-space $\mathfrak{N}(\mathfrak{L})$ is zero-dimensional; that is, the differential system

$$\text{(10.18)} \qquad \mathfrak{L}[\eta](t) = 0, \qquad t \in [a,b] \qquad \eta \in D(\mathfrak{L}),$$

has only the trivial solution $\eta(t) \equiv 0$. Consequently, for $\phi \in \mathfrak{C}_N[a, b]$ there is a unique solution of the differential system

$$\text{(10.19)} \qquad \mathfrak{L}[\eta](t) = \phi(t), \qquad t \in [a,b] \qquad \eta \in D(\mathfrak{L}),$$

and this solution is given by

$$\text{(10.20)} \qquad \eta(t) = \int_a^b \mathfrak{G}(t, s)\,\phi(s)\,ds, \qquad t \in [a, b],$$

where $\mathfrak{G}(t, s)$ is the ordinary Green's matrix for (10.18). In particular, if we set $\mathfrak{G}(t, s) = [\mathfrak{G}_{\alpha\beta}(t, s)]$, $(\alpha, \beta = 1, 2, 3)$, where $\mathfrak{G}_{\alpha\beta}(t, s)$ is an $r_\alpha \times r_\beta$ matrix with $r_1 = n$, $r_2 = k^*$, $r_3 = k$, and consider (10.19) for $\phi_\alpha(t) = f_\alpha(t)$, $(\alpha = 1, \ldots, n)$, $\phi_\tau(t) \equiv 0$ for $\tau = n+1, \ldots, N$, then the component vector function $y(t)$ of the solution vector $\eta(t)$ in (10.20) is given by

$$(10.21) \qquad y(t) = \int_a^b \mathcal{G}_{11}(t, s) f(s)\, ds, \qquad t \in [a, b],$$

and the transformation defined by the integral of (10.21) is a generalized inverse of the original differential operator L, for which the kernel matrix $\mathcal{G}_{11}(t, s)$ is the generalized Green's matrix $G_{\Theta,\Psi}(t, s:L)$ of Theorem 10.3.

4. If in Theorem 10.3 the matrix $\Theta(t)$ is chosen as equal to $Y_0(t)$, show that the corresponding generalized Green's matrix $G_{Y_0,\Psi}(t, s:L)$ is such that if $g \in \mathfrak{R}(L)$ then

$$y(t) = \int_a^b \mathcal{G}_{Y_0,\Psi}(t, s:L) g(s)\, ds, \qquad t \in [a, b],$$

is the vector function $y(t)$ which minimizes

$$N[y] = \left(\int_a^b |y(t)|^2\, dt\right)^{1/2}$$

in the class of vector functions $\{y \mid y \in D(L),\ L[y](t) = g(t),\ t \in [a, b]\}$.

Hint. Note that one may choose the column vectors $y^{(j)}(t)$, $(j = 1, \ldots, k)$, of $Y_0(t)$ such that

$$\int_a^b [y^{(j)}(t)]^* y^{(i)}(t)\, dt = \delta_{ij}, \qquad (i, j = 1, \ldots, k).$$

5. If the conditions of Theorem 10.3 hold, show that $G_{\Psi,\Theta}(t, s:L^\star)$ also exists, and $G_{\Psi,\Theta}(t, s:L^\star) = (G_{\Theta,\Psi}(s, t:L))^*$.

Hint. Proceed in a manner similar to that used in the proof of Theorem 7.3, noting that if $g(t)$ and $h(t)$ belong to $\mathfrak{C}_n[a, b]$ then in view of Theorem 6.3 there exist constant vectors ξ and ζ of respective dimensions k and k^* such that $g(t) - \Psi(t)\zeta \in \mathfrak{R}(L)$ and $h(t) - \Theta(t)\xi \in \mathfrak{R}(L^\star)$.

6. Suppose that $m_\sigma(t) \equiv (m_{\alpha\sigma}(t))$, $(\sigma = 1, \ldots, N)$, are n-dimensional vector functions of bounded variation on $[a, b]$, and consider a vector differential operator L with functional value $L[y](t) = A_1(t)[A_2(t)\, y(t)]' + A_0(t)\, y(t)$, $t \in [a, b]$, where the $n \times n$ coefficient matrices $A_j(t)$, $(j = 0, 1, 2)$, satisfy hypothesis $(\mathfrak{H})$, and the domain $D(L)$ of L consists of those n-dimensional vector functions in $\mathfrak{D}(L)$ satisfying the conditions

$$(10.22) \quad \mu_\sigma\langle y\rangle \equiv \int_a^b [dm_\sigma^*(s)]\, u_y(s) \equiv \int_a^b \sum_{j=1}^n (u_y(s))_j\, d\bar{m}_{j\sigma}(s) = 0, \qquad (\sigma = 1, \ldots, N),$$

where the integrals are in the Riemann-Stieltjes sense. Show that:

(i) Boundary conditions of the form $M\hat{u}_y = 0$, where M is an $N \times 2n$ matrix, may be written in the form (10.22), with

$$\bar{m}_{j\sigma}(t) \equiv 0 \quad \text{for} \quad t \in (a, b), \qquad \bar{m}_{j\sigma}(a) = -M_{\sigma j}, \qquad \bar{m}_{j\sigma}(b) = M_{\sigma, n+j}.$$

(ii) If $\mathfrak{R}(L)$ denotes the range of L, then there exists a kernel matrix, (generalized Green's matrix), $\hat{G}(t, s)$ such that if $g \in \mathfrak{R}(L)$ then a solution of

$$\text{(10.23)} \quad L[y](t) = g(t), \quad t \in [a, b], \quad \mu_\sigma\langle y\rangle = 0, \qquad \sigma = 1, \ldots, N,$$

is given by

$$\text{(10.24)} \qquad y(t) = \int_a^b \hat{G}(t, s)g(s)\, ds, \qquad t \in [a, b].$$

In particular, if $N = n$ and the column vectors $y^{(j)}(t)$, $(j = 1, \ldots, n)$, of a fundamental matrix $Y(t)$ of $L[y](t) = 0$, $t \in [a, b]$, are such that the $n \times n$ matrix $[\mu_\sigma\langle y^{(j)}\rangle]$, $(\sigma, j = 1, \ldots, n)$, is nonsingular, then $\hat{G}(t, s)$ is uniquely determined on $\{(t, s) \mid (t, s) \in [a, b] \times [a, b], t \neq s\}$.

Hints. (ii) Proceed as in the proof of Theorem 10.1, and use appropriately integration by parts to evaluate $\mu_\sigma\langle y^0\rangle$, where $y^0(t)$ is given by (10.4).

7. Show that for arbitrary continuous functions $r(t)$ on $[-1, 1]$ there is a unique solution of the differential system

$$u'' = r(t), \qquad \int_{-1}^1 u(s)\, ds = 0, \qquad \int_{-1}^1 su(s)\, ds = 0,$$

and determine a "Green's function" $g(t, s)$, $(t, s) \in [-1, 1] \times [-1, 1]$, such that the solution of this system is given by

$$u(t) = \int_{-1}^1 g(t, s)\, r(s)\, ds, \qquad t \in [-1, 1].$$

8. For $a = a_0 < a_1 < \cdots < a_k = b$, let $A_j(t)$, $(j = 0, 1, 2)$, be $n \times n$ matrix functions continuous on each open subinterval (a_{q-1}, a_q), and possessing finite right-hand limits $A_j(a_{q-1}^+)$ and finite left-hand limits $A_j(a_q^-)$, for $j = 0, 1, 2$ and $q = 1, \ldots, k$; moreover, for $i = 1, 2$ and $q = 1, \ldots, k$ the matrices $A_i(t)$, $t \in (a_{q-1}, a_q)$, and $A_i(a_{q-1}^+)$, $A_i(a_q^-)$ are non-singular. Let $\hat{\mathfrak{D}}$ denote the n-dimensional vector functions $y(t)$ such that $u_y(t) = A_2(t)\, y(t)$ is continuously differentiable on each subinterval (a_{q-1}, a_q), and has finite right- and left-hand derivatives $u_y'(a_{q-1}^+)$ and $u_y'(a_q^-)$ for $q = 1, \ldots, k$. Let L denote the differential operator with domain

$$D(L) = \left\{y \mid y \in \hat{\mathfrak{D}}, \sum_{q=1}^k [M_q u_y(a_{q-1}^+) + N_q u_y(a_q^-)] = 0\right\},$$

where M_q, N_q, $(q = 1, \ldots, k)$, are $r \times n$ matrices with $1 \leq r \leq 2kn$ such that the $r \times 2kn$ matrix $[M_1 \cdots M_k \quad N_1 \cdots N_k]$ is of rank r, and functional value

$$L[y](t) = A_1(t)[A_2(t)y(t)]' + A_0(t)y(t),$$
$$= A_1(t)u_y'(t) + A_0(t)A_2^{-1}(t)u_y(t), \quad \text{for} \quad y \in D(L).$$

Discuss the solvability of the homogeneous differential system

(10.25) $$L[y](t) = 0, \quad t \in [a, b], \qquad y \in D(L),$$

and the corresponding adjoint differential system

(10.26) $$L^\star[z](t) = -A_2^*(t)[A_1^*(t)z(t)]' + A_0^*(t)z(t) = 0, \qquad z \in D(L^\star),$$

where $D(L^\star)$ consists of those n-dimensional vector functions $z(t)$ which satisfy with the corresponding $v_z(t) = A_1^*(t)z(t)$ conditions similar to those described above for $y(t)$, $u_y(t)$, and are such that

$$\int_a^b z^*L[y]\,dt - \int_a^b (L^\star[z])^*y\,dt = 0, \quad \text{for} \quad y \in D(L).$$

Show that there exists for L a "generalized Green's matrix" $G(t, s)$, such that if the non-homogeneous differential system

(10.27) $$L[y](t) = g(t), \qquad y \in D(L),$$

has a solution, then a particular solution of this system is given by

(10.28) $$y(t) = \int_a^b G(t, s)\, g(s)\, ds.$$

Hint. For $q = 1, \ldots, k$, let $t_q(s) = a_{q-1} + s[a_q - a_{q-1}]$, $s \in [0, 1]$, and define the $nk \times nk$ matrix functions $\mathcal{A}_0(s)$, $\mathcal{A}_1(s)$, $\mathcal{A}_2(s)$ as

$$\mathcal{A}_\beta(s) = \operatorname{diag}\{A_\beta(t_1[s]), \ldots, A_\beta(t_k[s])\}, \qquad \beta = 1, 2;$$
$$\mathcal{A}_0(s) = \operatorname{diag}\{(a_1 - a_0)\, A_0(t_1[s]), \ldots, (a_k - a_{k-1})\, A_0(t_k[s])\}.$$

For $\eta(s) = (\eta_\nu(s))$, $(\nu = 1, \ldots, kn)$, the kn-dimensional vector function defined as $\eta_{(q-1)n+\alpha}(s) = y_\alpha(t_q[s])$, $(\alpha = 1, \ldots, n; q = 1, \ldots, k)$, show that the system (10.25) is equivalent to

(10.25′)
$$\mathfrak{L}[\eta](s) \equiv \mathcal{A}_1(s)[\mathcal{A}_2(s)\eta(s)]' + \mathcal{A}_0(s)\eta(s) = 0, \qquad s \in [0, 1],$$
$$\mathcal{M}u_\eta(0) + \mathcal{N}u_\eta(1) = 0,$$

where $u_\eta(s) = \mathcal{A}_2(s)\eta(s)$, $\mathcal{M} = [M_1 \cdots M_k]$, $\mathcal{N} = [N_1 \cdots N_k]$, and (10.27) is equivalent to a corresponding nonhomogeneous system

(10.27′) $$\mathfrak{L}[\eta](s) = \phi(s), \qquad s \in [0, 1], \qquad \mathcal{M}u_\eta(0) + \mathcal{N}u_\eta(1) = 0.$$

9. Let $\check{\mathfrak{D}}$ denote the class of scalar valued functions $u(t)$ of class $\mathfrak{C}^2[a, b]$ with $u(a) = 0 = u(b) = u(c)$, where c is a given value on (a, b), and $l_2[u](t) = u''(t)$, $t \in [a, b]$, for $u \in \check{\mathfrak{D}}$. Apply the method of the preceding problem to a related two-dimensional vector differential system in $y(t) = (u^{[\alpha-1]}(t))$, $(\alpha = 1, 2)$, to solve the following problems.

(a) Show that the operator $l_2^\star$ adjoint to l_2 has domain $\check{\mathfrak{D}}^\star$ consisting of scalar valued functions $v(t)$ of class $\mathfrak{C}^2$ on the individual intervals $[a, c)$ and $(c, b]$, continuous on $[a, b]$, with finite right- and left-hand first order derivatives at $t = c$, satisfying the conditions $v(a) = 0 = v(b)$, and with functional value $l_2^\star[v](t) = v''(t)$ for $t \in [a, c) \cup (c, b]$.

(b) Determine the conditions that $r(t)$, $t \in [a, b]$, must satisfy in order that the differential system

$$(10.29) \qquad l_2[u](t) = r(t), \qquad u \in \check{\mathfrak{D}},$$

possess a solution.

(c) Determine a function $g(t, s)$, $(t, s) \in [a, b] \times [a, b]$, which is such that if (10.29) has a solution then a solution of this system is given by

$$u(t) = \int_a^b g(t, s)\, r(s)\, ds, \qquad t \in [a, b],$$

Is such a function $g(t, s)$ unique?

11. EQUIVALENT DIFFERENTIAL SYSTEMS

In addition to a differential expression $L[y]$ of the form (1.1), and with coefficient matrix functions satisfying hypothesis $(\mathfrak{H})$, consider a second similar differential expression

$$L_0[z](t) = A_1{}^0(t)[A_2{}^0(t)z(t)]' + A_0{}^0(t)z(t),$$

with coefficient matrices which also satisfy hypothesis $(\mathfrak{H})$; that is, $A_j{}^0(t)$, $(j = 0, 1, 2)$, are continuous $n \times n$ matrix functions on $[a, b]$, with $A_1{}^0(t)$ and $A_2{}^0(t)$ nonsingular on this interval. Correspondingly, we set $\mathfrak{D}(L_0) = \{z \mid z \in \mathfrak{C}_n[a, b],\, u_z{}^0 \equiv A_2{}^0 z \in \mathfrak{C}_n{}^1[a, b]\}$, and $\mathfrak{D}_0(L_0) = \{z \mid z \in \mathfrak{D}(L_0),\, \hat{u}_z{}^0 = 0\}$.

The differential equation

$$(11.1) \qquad L[y](t) \equiv A_1(t)[A_2(t)y(t)]' + A_0(t)y(t) = 0, \qquad t \in [a, b],$$

is said to be equivalent to the differential equation

$$(11.2) \qquad L_0[z](t) \equiv A_1{}^0(t)[A_2{}^0(t)z(t)]' + A_0{}^0(t)z(t) = 0, \qquad t \in [a, b],$$

under the transformation

$$(11.3) \qquad z(t) = T(t)y(t), \qquad t \in [a, b],$$

provided that $T(t)$ is a nonsingular matrix function on $[a, b]$ with the property that (11.3) defines a one-to-one mapping of $\mathfrak{D}(L)$ onto $\mathfrak{D}(L_0)$ such that $L[y](t) = 0$, $t \in [a, b]$, if and only if $L_0[z](t) = 0$, $t \in [a, b]$.

Theorem 11.1. *The differential equation* (11.1) *is equivalent to the differential equation* (11.2) *under the transformation* (11.3) *if and only if*

$$T(t) = [A_2{}^0(t)]^{-1}T_1(t)A_2(t), \qquad t \in [a, b], \tag{11.4}$$

where $T_1(t)$ is a nonsingular, continuously differentiable matrix function which satisfies the matrix differential equation

$$L[T_1] \equiv T_1' + A^0(t)T_1 - T_1A(t) = 0, \qquad t \in [a, b], \tag{11.5}$$

with the matrix functions $A(t)$, $A^0(t)$ defined as

$$A = A_1{}^{-1}A_0A_2{}^{-1}, \qquad A^0 = (A_1{}^0)^{-1}A_0{}^0(A_2{}^0)^{-1}. \tag{11.6}$$

It is to be noted that the transformation (11.3) is equivalent to

$$u_z{}^0(t) = T_1(t)u_y(t), \qquad t \in [a, b], \tag{11.3'}$$

where $T_1(t)$ is the nonsingular $n \times n$ matrix function related to $T(t)$ by (11.4). Consequently, if (11.1) is equivalent to (11.2) under the transformation (11.3), then $U(t)$ is a nonsingular matrix solution of $A_1U' + A_0A_2{}^{-1}U = 0$ on $[a, b]$ if and only if $U^0(t) = T_1(t)U(t)$ is a nonsingular matrix solution of $A_1{}^0U^{0\prime} + A_0{}^0(A_2{}^0)^{-1}U^0 = 0$ on $[a, b]$; in particular, $T_1(t) = U^0(t)U^{-1}(t)$ is a nonsingular matrix function which is continuously differentiable on $[a, b]$. Moreover, as $U' = -AU$ and $U^{0\prime} = -A^0U^0$, where A and A^0 are defined by (11.6), it follows directly that $T_1(t)$ is a solution of the matrix differential equation (11.5).

Conversely if $T_1(t)$ is a nonsingular, continuously differentiable matrix function satisfying (11.5), and $T(t)$ is defined by (11.4), then (11.3) is seen to define a one-to-one mapping of $\mathfrak{D}(L)$ onto $\mathfrak{D}(L_0)$, and if $z(t) = T(t)y(t)$ then

$$L_0[z](t) = A_1{}^0(t)T_1(t)A_1{}^{-1}(t)L[y](t), \qquad t \in [a, b], \tag{11.7}$$

so that $L[y](t) = 0$, $t \in [a, b]$, if and only if $L_0[z](t) = 0$, $t \in [a, b]$.

The matrix differential equation (11.5) may be considered as a linear vector differential equation in the n^2 elements of $T_1(t)$. The condition that $T_1(t)$ is nonsingular on $[a, b]$ is a property of an individual solution, however, and it is pertinent to determine the character of the general solution of this equation. If $U(t)$ is a fundamental matrix solution of $A_1U' + A_0A_2{}^{-1}U = 0$, and $U^0(t)$ is a fundamental matrix solution of $A_1{}^0U^{0\prime} + A_0{}^0(A_2{}^0)^{-1}U^0 = 0$, then the matrix function $F(t) = [U^0(t)]^{-1}T_1(t)U(t)$ is such that

$$L[T_1](t) = U^0(t)F'(t)U^{-1}(t),$$

and hence $T_1(t)$ is a solution of (11.5) if and only if $F(t)$ is a constant matrix function on $[a, b]$. Consequently, we have the following result.

Theorem 11.2. *The general solution of the matrix differential equation* (11.5) *is* $T_1(t) = U^0(t)CU^{-1}(t)$, *where* $U(t)$ *is a fundamental matrix solution of* $A_1U' + A_0A_2^{-1}U = 0$, $U^0(t)$ *is a fundamental matrix solution of* $A_1^0U^{0\prime} + A_0^0(A_2^0)^{-1}U^0 = 0$, *and* C *is a constant* $n \times n$ *matrix. An individual solution* $T_1(t)$ *of* (11.5) *is of constant rank on* $[a, b]$; *in particular, if* $T_1(t)$ *is nonsingular at one point on* $[a, b]$ *then* $T_1(t)$ *is nonsingular throughout this interval.*

The following corollary follows from the linear homogeneous nature of the differential equation (11.5) and the transformation (11.4).

Corollary. *If the differential equation* (11.1) *is equivalent to the differential equation* (11.2) *under each of the transformations* $z(t) = T^{(\alpha)}(t)y(t)$, $\alpha = 1, 2$, *and* c_1, c_2 *are constants such that* $T(t) = c_1T^{(1)}(t) + c_2T^{(2)}(t)$ *is nonsingular at some point on* $[a, b]$, *then* $T(t)$ *is nonsingular for* $t \in [a, b]$ *and* (11.1) *is equivalent to* (11.2) *under the transformation* $z(t) = T(t)y(t)$. *Moreover, if* (11.1) *is equivalent to* (11.2) *under each of the transformations*

$$z(t) = T^{(\gamma)}(t)y(t), \ (\gamma = 1, 2, 3),$$

then (11.1) *is equivalent to* (11.2) *under the transformation*

$$z(t) = T^{(3)}(t)[T^{(2)}(t)]^{-1}T^{(1)}(t)\,y(t).$$

As in the preceding sections, let $D(L)$ be a linear subspace in $\mathfrak{C}_n[a, b]$ satisfying $\mathfrak{D}_0(L) \subset D(L) \subset \mathfrak{D}(L)$, and $D(L; a, b)$ the linear subspace in $\mathbf{C}_{2n}$ such that $D(L) = \{y \mid y \in \mathfrak{D}(L),\ \hat{u}_y \in D(L; a, b)\}$. Similarly, let $D(L_0)$ be a linear subspace in $\mathfrak{C}_n[a, b]$ satisfying $\mathfrak{D}_0(L_0) \subset D(L_0) \subset \mathfrak{D}(L_0)$, and $D(L_0; a, b)$ the linear subspace in $\mathbf{C}_{2n}$ such that $D(L_0) = \{z \mid z \in \mathfrak{D}(L_0),\ \hat{u}_z^0 \in D(L_0; a, b)\}$. The differential system

$$L[y](t) = 0, \qquad t \in [a, b], \qquad y \in D(L), \tag{11.8}$$

is said to be equivalent to the differential system

$$L_0[z](t) = 0, \qquad t \in [a, b], \qquad z \in D(L_0), \tag{11.9}$$

under the transformation (11.3) if the differential equation $L[y] = 0$ is equivalent to $L_0[z] = 0$ under this transformation, and $y \in D(L)$ if and only if $z = Ty \in D(L_0)$. Since (11.3) implies that (11.3′) holds with $T_1 = A_2^0TA_2^{-1}$, and hence $\hat{u}_z^0 = [\mathrm{diag}\{T_1(a), T_1(b)\}]\hat{u}_y$, the following result is immediate.

Theorem 11.3. *The differential system* (11.8) *is equivalent to the differential system* (11.9) *under the transformation* (11.3) *if and only if* $T(t)$ *is of the form* (11.4), *where* $T_1(t)$ *is a nonsingular continuously differentiable matrix function* *which satisfies the differential equation* (11.5), *and the following condition holds*:

(i) *if* $\dim D(L; a, b)$ *is either* 0 *or* $2n$, *then* $\dim D(L_0; a, b) = \dim D(L; a, b)$;

(ii) *if* $\dim D(L; a, b) = 2n - m$, $1 \leq m \leq 2n - 1$, *and* (11.8) *is written as*

$$(11.8') \qquad L[y](t) = 0, \qquad t \in [a, b], \qquad M\hat{u}_y = 0,$$

where M *is an* $m \times 2n$ *matrix of rank* m, *then* $\dim D(L_0; a, b) = \dim D(L; a, b)$, *and* (11.9) *may be written as*

$$(11.9') \qquad L_0[z](t) = 0, \qquad t \in [a, b], \qquad M_0\hat{u}_z^{\,0} = 0,$$

where M_0 *is an* $m \times 2n$ *matrix of rank* m *which satisfies the equation*

$$(11.10) \qquad M_0[\operatorname{diag}\{T_1(a), T_1(b)\}]P = 0,$$

with P *a* $2n \times (2n - m)$ *matrix of rank* $2n - m$ *such that* $MP = 0$.

Theorem 11.4. *If the differential system* (11.8) *is equivalent to the differential system* (11.9) *under the transformation* (11.3), *then:*

(i) *the systems* (11.8) *and* (11.9) *have the same index of compatibility*;

(ii) *if* $\dim D(L; a, b) = n$, *and* (11.8) *is incompatible, then* (11.9) *is incompatible also, and if* $G(t, s)$ *and* $G^0(t, s)$ *are the Green's matrices for* (11.8) *and* (11.9), *respectively, then for* $(t, s) \in \Delta_1 \cup \Delta_2$ *we have*

$$(11.11) \qquad T(t)G(t, s)A_1(s)A_2(s) = G^0(t, s)A_1^{\,0}(s)A_2^{\,0}(s)T(s).$$

Conclusion (i) is a ready consequence of the fact that $y = y^{(j)}(t)$, $(j = 1, \ldots, k)$, are solutions of (11.8) which are linearly independent on $[a, b]$ if and only if $z = z^{(j)}(t) = T(t)y^{(j)}(t)$, $(j = 1, \ldots, k)$, are solutions of (11.9) which are linearly independent on this interval. Now if $\dim D(L; a, b) = n$, and (11.8) is incompatible, the differential system

$$(11.12) \qquad L[y](t) = g(t), \qquad t \in [a, b], \qquad y \in D(L),$$

has for arbitrary $g(t) \in \mathfrak{C}_n[a, b]$ a unique solution given by

$$y(t) = \int_a^b G(t, s)g(s)\, ds, \qquad t \in [a, b].$$

In view of the identity (11.7) for $z(t) = T(t)y(t)$ and $y \in D(L)$, it then follows that

$$z(t) = \int_a^b T(t)G(t, s)g(s)\, ds, \qquad t \in [a, b],$$

is the unique solution of the differential system

$$L_0[z](t) = A_1^{\,0}(t)T_1(t)A_1^{-1}(t)g(t), \qquad t \in [a, b], \qquad z \in D(L_0),$$

and hence

$$z(t) = \int_a^b G^0(t, s)A_1^{\,0}(s)T_1(s)A_1^{-1}(s)g(s)\, ds, \qquad t \in [a, b],$$

That is, for arbitrary $g(t) \in \mathfrak{C}_n[a, b]$ we have

$$\int_a^b [T(t)G(t, s) - G^0(t, s)A_1{}^0(s)T_1(s)A_1{}^{-1}(s)]\, g(s)\, ds = 0, \qquad t \in [a, b],$$

and in view of the continuity of $G(t, s)$ and $G^0(t, s)$ on the individual triangular domains Δ_1 and Δ_2 it follows that

$$T(t)G(t, s) = G^0(t, s)A_1{}^0(s)T_1(s)A_1{}^{-1}(s), \quad \textit{for} \quad (t, s) \in \Delta_1 \cup \Delta_2,$$

and (11.11) follows since $T_1(s) = A_2{}^0(s)T(s)A_2{}^{-1}(s)$.

Of particular interest is the case in which the system (11.9) is the system

$$\text{(11.13)} \quad L^\star[z](t) \equiv -A_2^*(t)[A_1^*(t)z]' + A_0^*(t)z = 0, \qquad t \in [a, b],\ z \in D(L^\star),$$

adjoint to (11.8). In particular, if (11.8) is equivalent to (11.13) under the transformation (11.3) then $\dim D(L; a, b) = \dim D(L^\star; a, b) = n$ by conclusion (i) of Theorem 11.3, and hence (11.13) may be written as

$$\text{(11.13')} \qquad L^\star[z](t) = 0, \qquad t \in [a, b], \qquad P^*Q\hat{v}_z = 0,$$

where P is an $n \times 2n$ matrix of rank n and such that $MP = 0$, where M is the coefficient matrix of the boundary conditions of (11.8′) and Q is the $2n \times 2n$ matrix defined by (6.4). Now (11.13) is of the form (11.9) with $A_1{}^0(t) = -A_2^*(t)$, $A_2{}^0(t) = A_1^*(t)$, and $A_0{}^0(t) = A_0^*(t)$, and the following theorem is a consequence of the results of Theorems 11.3 and 11.4 for (11.8) and its adjoint (11.13).

Theorem 11.5. *The system* (11.8) *is equivalent to its adjoint* (11.13) *under the transformation* (11.3) *if and only if* $\dim D(L; a, b) = n$, *and* $T(t)$ *is of the form*

$$\text{(11.14)} \qquad T(t) = A_1^{*-1}(t)T_1(t)A_2(t), \qquad t \in [a, b],$$

where $T_1(t)$ *is a nonsingular continuously differentiable matrix function which satisfies the matrix differential equation*

$$\text{(11.15)} \qquad T_1' - A^*(t)T_1 - T_1A(t) = 0, \qquad t \in [a, b],$$

with $A(t) = A_1{}^{-1}(t)A_0(t)A_2{}^{-1}(t)$, *and*

$$\text{(11.16)} \qquad P^*[\operatorname{diag}\{-T_1(a), T_1(b)\}]P = 0,$$

where P *is a* $2n \times n$ *matrix of rank* n *which satisfies with the coefficient matrix* M *of* (11.8′) *the equation* $MP = 0$. *Moreover, if* (11.8) *is an incompatible system that is equivalent to its adjoint* (11.13) *under the transformation* (11.3), *then the Green's matrix* $G(t, s)$ *for* (11.8) *satisfies with the* $T_1(t)$ *of* (11.14) *the equation*

$$\text{(11.17)} \qquad T_1(t)\mathbf{G}(t, s) + \mathbf{G}^*(s, t)T_1(s) = 0$$

for $(t, s) \in \Delta_1 \cup \Delta_2$, *where, as in Theorem* 7.2, $\mathbf{G}(t, s) = A_2(t)G(t, s)A_1(s)$.

Now if $T_1 = S(t)$ is a nonsingular, continuously differentiable matrix function which satisfies (11.15), (11.16) or (11.17), then clearly $T_1 = S^*(t)$ is also a nonsingular, continuously differentiable matrix function which satisfies the same condition. In view of the Corollary to Theorem 11.2, therefore, we have the following result.

Corollary. *If the differential system* (11.8) *is equivalent to its adjoint* (11.13) *under the transformation* (11.3), *then* (11.8) *is also equivalent to its adjoint under the transformation*

$$z(t) = \hat{T}(t)y(t), \qquad t \in [a, b], \tag{11.18}$$

where $\hat{T}(t)$ *is any matrix of the form*

$$\hat{T}(t) = c_1 T(t) + c_2 A_1^{*-1}(t) A_2^{*-1}(t) T^*(t) A_1(t) A_2(t), \tag{11.19}$$

with c_1, c_2 *constants such that* $\hat{T}(t)$ *is nonsingular for some* $t_0 \in [a, b]$.

Problems III.11

1.[ss] Suppose that $A(t)$, $B(t)$, $C(t)$ are $n \times n$ continuous matrix functions on $[a, b]$, with $B(t)$ and $C(t)$ hermitian on this interval, and that κ_1, μ_1, κ_2, μ_2 are $2n \times n$ constant matrices such that the $2n \times 4n$ matrix

$$[\kappa_1 \quad -\mu_1 \quad \kappa_2 \quad \mu_2]$$

is of rank $2n$, and the $2n \times 2n$ matrix $\kappa_1\mu_1^* + \kappa_2\mu_2^*$ is hermitian. Show that the differential system

$$y'(t) + \begin{bmatrix} -A(t) & -B(t) \\ -C(t) & A^*(t) \end{bmatrix} y(t) = 0,$$

$$[\kappa \quad -\mu_1]\, y(a) + [\kappa_2 \quad \mu_2]\, y(b) = 0,$$

in the $2n$-dimensional vector function $y(t)$ is equivalent to its adjoint under the transformation $z(t) = T(t)y(t)$, with $T(t)$ the $2n \times 2n$ constant matrix

$$T = \begin{bmatrix} 0 & E_n \\ -E_n & 0 \end{bmatrix}.$$

2. Suppose that the coefficients of $l_n[u] \equiv p_n(t)u^{[n]} + \cdots + p_0(t)u$ are such that $p_j(t) \in \mathfrak{C}^j[a, b]$, $(j = 0, 1, \ldots, n)$, with $p_n(t) \neq 0$ on this interval, and that the differential system

$$l_n[u](t) = 0, \qquad t \in [a, b],$$

$$\mu_\sigma\{u\} \equiv \sum_{\beta=1}^{n} [M_{\sigma\beta}u^{[\beta-1]}(a) + M_{\sigma,n+\beta}u^{[\beta-1]}(b)] = 0, \quad (\sigma = 1, \ldots, n), \tag{11.20}$$

is self-adjoint, (see Problem III.8:4). Show that:

(a) The corresponding first order differential system

$$L_1[y] \equiv p_n(t)y_n' + p_0(t)y_1 + \cdots + p_{n-1}(t)y_n = 0,$$
$$L_\beta[y] \equiv y_{n-\beta+1}' - y_{n-\beta+2} = 0, \qquad (\beta = 2, \ldots, n),$$
$$\sum_{\beta=1}^{n} [M_{\sigma\beta}y_\beta(a) + M_{\sigma,n+\beta}y_\beta(b)] = 0, \qquad (\sigma = 1, \ldots, n),$$

is equivalent to its adjoint under the transformation $z_\alpha(t) = \sum_{\beta=1}^{n} T_{\alpha\beta}(t)\, y_\beta(t)$, $(\alpha = 1, \ldots, n)$, where $T(t) \equiv [T_{\alpha\beta}(t)]$ is such that the corresponding matrix $T_1(t) \equiv [T_{1\alpha\beta}(t)]$ of (11.4) is given by $T_{1\alpha\beta}(t) = \overline{K_{\beta\alpha}(t)}$, $(a, \beta = 1, \ldots, n)$, and $K(t) \equiv [K_{\alpha\beta}(t)]$ is the matrix of the bilinear concomitant (3.17).

(b) If (11.20) is incompatible, then conclusion (a), together with the last result of Theorem 11.5, provide an alternate proof of the result that the Green's function $g(t, s)$ for (11.20) is such that $g(t, s) = \overline{g(s, t)}$.

Hints. (a) Use relations (8.13) and (8.13′).
(b) Use (11.17), and the result of Problem III.3:3.

3. Show that the differential system $u'' + iu' + u = 0$, $u(-1) = 0$, $u(1) = 0$ is self-adjoint. Determine the corresponding matrix $K(t) = [K_{\alpha\beta}(t)]$, $(\alpha, \beta = 1, 2)$, and for this system verify directly conclusion (a) of Problem 2.

4. For a matrix differential system of the form (9.6_n) state and prove a result corresponding to that of the above Problem 2.

5. If a differential system (11.8) is such that $A_1(t)A_2(t) \equiv E$, and this system is equivalent to its adjoint (11.13) under a transformation $z(t) = T(t)y(t)$, show that there is a $\hat{T}(t)$ which is skew-hermitian, $[\hat{T}^*(t) \equiv -\hat{T}(t)]$, and such that (11.8) is equivalent to (11.13) under the transformation $z(t) = \hat{T}(t)y(t)$.

Hint. Use the Corollary to Theorem 11.5 to show that the stated result holds for

$$\hat{T}(t) = \cos\theta[T(t) - T^*(t)] + i\sin\theta[T(t) + T^*(t)],$$

where θ is a real value such that the $n \times n$ matrix $T^*(t)T^{-1}(t) - e^{2i\theta}E$ is nonsingular for some value on $[a, b]$.

6. Show that the differential system

$$\begin{aligned}
y_1' &= 0, & y_1(0) + y_2(2\pi) &= 0,\\
y_2' &= 0, & y_2(0) + y_1(2\pi) &= 0,\\
y_3' &= y_4, & y_3(0) - y_3(2\pi) &= 0,\\
y_4' &= -y_3, & y_4(0) - y_4(2\pi) &= 0,
\end{aligned}$$

is equivalent to its adjoint under the transformation $z(t) = T(t)y(t)$ where $T(t) \equiv [T_{\alpha\beta}(t)]$ is the skew-hermitian matrix

$$T = \begin{bmatrix} i & 0 & 0 & 0 \\ 0 & i & 0 & 0 \\ 0 & 0 & 0 & 1 \\ 0 & 0 & -1 & 0 \end{bmatrix},$$

but that there is no real-valued skew-hermitian admissible matrix $T(t)$ such that this differential system is equivalent to its adjoint under the transformation $z(t) = T(t)y(t)$.

12. NOTES AND REMARKS

The general subject matter of this chapter is presented to some degree in most books on boundary problems for ordinary differential equations. In particular, the reader is referred to Kamke [5, *B*-§2], Collatz [1, Chs. 2, 3], Sansone [1-I, Ch. IV], and Coddington and Levinson [2, Chs. 7, 11]. Specific papers that have contributed to the development of the particular presentation of this chapter include, Birkhoff [1,2], Bounitzky [1], Tamarkin [1,2], Birkhoff and Langer [1], Bliss [3,4], and Reid [9,17,29]. The presented treatment is in the general formulation of Reid [29], which possesses the advantages listed at the end of Section 1.

It is to be mentioned that historically the concept of the adjoint of an n-th order scalar differential equation was introduced by Lagrange in about 1765, and the notion of a system adjoint to a first order differential system was given by Jacobi in about 1837. The actual designation of such associated equations as "adjoint" is due to Fuchs in 1873.

The kernel function $g(t, s)$ of (8.10) for the special differential system $u'' = 0$, $u(a) = 0 = u(b)$ was first introduced by Picard [1] in 1890. In 1894 the case of a differential system involving the equation $u'' = 0$ and more general two-point boundary conditions was considered by Burkhardt [1], who introduced the term "Green's function," since for the differential systems studied this function assumed a role similar to that played by a function introduced much earlier by Green [1] for certain boundary problems involving partial differential equations. Extensions of the concept of a Green's function for differential systems involving a scalar differential system of order n, as well as the introduction of a Green's matrix for incompatible first order systems of the sort studied in Section 7, were attained in about the first decade of the twentieth century, largely through the works of Bôcher [2,3,4,5], G. D. Birkhoff [1,2], and Bounitzky [1].

The concept of a generalized Green's function dates from the work of

Hilbert [1], and the existence of generalized Green's functions for certain special types of compatible differential systems are to be found in several Göttingen theses written under the direction of Hilbert. Specific papers of the Bibliography that deal with this concept are Bounitzky [1], Elliott [1,2], Reid [2,29], Greub and Rheinboldt [1], Wyler [1,2], Bradley [1,2], and Loud [1]. In the consideration of boundary problems associated with the Jacobi equation for a variational problem, the generalized Green's matrix has been used specifically by the author [3,6] and by E. Hölder [1,2,3]. The reader is referred to Reid [31] for a general survey of this topic, the historical development of generalized inverses for differential and integral operators, and relations with the E. H. Moore [1,2] generalized inverse of an algebraic matrix.

The treatment of equivalent differential systems in Section 11 had its genesis in the transformation theory used by Bliss [3,4] in his definition of "definitely self-adjoint differential systems." The specific discussion of this section follows closely that of Reid [17,29], and, in particular, uses a result of Reid [8].

For the discussion of differential systems involving more general boundary conditions than those treated in the present chapter, the reader is referred to Tamarkin [2], Whyburn [1,4,5], Mansfield [1] and Jones [1]. For a general survey of boundary problems, and references to recent literature, see Conti [4].

RELATED REFERENCES AND COMMENTS FOR SPECIFIC PROBLEMS

III.2:1	Bliss [5].
III.2:3,4,5,6,7,8	Reid [14].
III.3:4,5,6	Reid [14].
III.6:3	Bliss [3].
III.10:3	Reid [29].
III.10:8	Mansfield [1]; Reid [29].
III.11:5,6	Reid [17].

IV

Differential Systems Involving Linearly a Parameter

1. FORMULATION OF THE PROBLEM

As in the preceding chapter, let L be a vector differential operator of the form

$$(1.1) \qquad L[y](t) = A_1(t)[A_2(t)y(t)]' + A_0(t)y(t), \qquad t \in [a, b],$$

where the $A_j(t)$, $(j = 0, 1, 2)$, are continuous $n \times n$ matrix functions on $[a, b]$ with $A_1(t)$, $A_2(t)$ nonsingular on this interval. In this chapter we shall be concerned with two-point boundary problems involving such a differential operator with domain $D(L)$ having the related subspace $D(L; a, b)$ of $\mathbf{C}_{2n}$ of dimension n, and in which the characteristic parameter enters linearly. Specifically, the system to be considered may be written

$$(1.2) \qquad \begin{aligned} &\text{(a)} \ L[y](t) = \lambda B(t)y(t), \qquad t \in [a, b], \\ &\text{(b)} \ \mu[y] \equiv M\hat{u}_y = 0, \end{aligned}$$

where M is an $n \times 2n$ matrix of rank n, and $\hat{u}_y$ is the $2n$-dimensional boundary vector for the related vector function $u_y(t) = A_2(t)\,y(t)$. It will be assumed that on $[a, b]$ the $n \times n$ matrix function $B(t)$ is continuous and non-identically zero.

For each complex number λ the symbol $k(\lambda)$ will denote the index of compatibility of (1.2). A value λ for which (1.2) is compatible, (i.e., $k(\lambda) > 0$), is called a *proper value* (*eigenvalue*, or *characteristic value*) of (1.1) of index $k(\lambda)$, and a corresponding nontrivial solution $y(t)$ is called a *proper* (*vector*) *function*, (*eigenfunction*, or *characteristic function*) of this differential system. For a differential system involving a parameter the problem of determining properties of proper values and proper functions is frequently identified as a *proper value problem*, an *eigenvalue problem*, or a *boundary value problem*.

Suppose that $l_n[u] \equiv p_n(t)u^{[n]} + \cdots + p_0(t)u$, $m_k(u) \equiv q_k(t)u^{[k]} + \cdots q_0(t)u$, are scalar differential expressions with coefficients continuous on $[a, b]$, $0 \leq k < n$, and $p_n(t) \neq 0$ on $[a, b]$. The differential system

$$(1.3)\quad \begin{aligned} &l_n[u] = \lambda m_k[u], \qquad t \in [a, b], \\ &\sum_{\beta=1}^{n} [M_{\alpha\beta}\, u^{[\beta-1]}(a) + M_{\alpha,n+\beta}\, u^{[\beta-1]}(b)] = 0, \qquad (\alpha = 1, \ldots, n), \end{aligned}$$

with $M \equiv [M_{\alpha\tau}]$, $(\alpha = 1, \ldots, n; \tau = 1, \ldots, 2n)$, of rank n, is equivalent to a system (1.2) under the transformation $y_\alpha(t) = u^{[\alpha-1]}(t)$, $(\alpha = 1, \ldots, n)$. For differential systems analogous to (1.3) and involving linear vector differential expressions $\mathfrak{L}_n[u]$ and $\mathcal{M}_k[u]$, $0 \leq k < n$, in $u \equiv (u_\sigma(t))$, $(\sigma = 1, \ldots, h)$, the results of Section III.9 imply a similar reduction to an equivalent boundary problem of the form (1.2).

If P is a $2n \times n$ matrix of rank n, and such that $MP = 0$, then

$$(1.4)\quad \begin{aligned} &(a)\ L^\star[z](t) = \lambda B^*(t)z(t), \qquad t \in [a, b], \\ &(b)\ v[z] \equiv P^*Q\hat{v}_z = 0, \end{aligned}$$

is called the *boundary problem adjoint to* (1.2), where, as in the preceding chapter,

$$L^\star[z](t) = -A_2^*(t)[A_1^*(t)z(t)]' + A_0^*(t)z(t), \qquad t \in [a, b],$$

$\hat{v}_z$ is the $2n$-dimensional boundary vector for the related vector function $v_z(t) = A_1^*(t)z(t)$, and $Q = \text{diag}\{-E_n, E_n\}$. It is to be emphasized that for a given complex value λ the differential system (1.4) is not the differential system adjoint to (1.2) for this same λ; rather, for a given $\lambda = \lambda_0$ the differential system adjoint to (1.2) is the system (1.4) with $\lambda = \bar{\lambda}_0$.

2. ELEMENTARY PROPERTIES OF BOUNDARY VALUE PROBLEMS

If $Y(t; \lambda)$ is a fundamental matrix solution of the matrix differential equation $L[Y](t) = \lambda B(t)Y(t)$, we write $U_Y(t; \lambda)$ for the corresponding $A_2(t)Y(t; \lambda)$, and denote by $D(\lambda)$ the $n \times n$ matrix

$$(2.1)\qquad D(\lambda) = \mu[Y(\ ;\lambda)] = M\hat{U}_Y(\ ;\lambda).$$

The results of the following four lemmas are ready consequences of the respective Theorems III.4.1, III.6.1, III.7.1 and III.7.3.

***Lemma* 2.1.** *If* $r(\lambda)$ *is the rank of* $D(\lambda)$ *for a given* λ, *then* $r(\lambda) = n - k(\lambda)$.

***Lemma* 2.2** *If* $k^*(\lambda)$ *is the index of compatibility of* (1.4) *for a value* λ, *then* $k^*(\lambda) = k(\bar{\lambda})$.

Lemma 2.3. *If $\lambda = \lambda_0$ is not a proper value of* (1.2), *and $G(t, s; \lambda_0)$ is the Green's matrix of this differential system with $\lambda = \lambda_0$, then $y(t)$ is a solution of* (1.2) *for a value λ if and only if $y(t)$ is a solution of the vector integral equation*

$$y(t) = (\lambda - \lambda_0)\int_a^b G(t, s; \lambda_0)B(s)y(s)\, ds, \qquad t \in [a, b]. \tag{2.2}$$

Lemma 2.4. *If $\lambda = \lambda_0$ is not a proper value of* (1.2), *and $G(t, s; \lambda_0)$ is the Green's matrix of this system for $\lambda = \lambda_0$, then $G^*(s, t; \lambda_0)$ is the Green's matrix of* (1.4) *for $\lambda = \bar{\lambda}_0$.*

The following result is a direct consequence of formula (III.6.3), and the fact that $\hat{v}_z^* Q \hat{u}_y = 0$ whenever $\mu[y] = 0$, $\nu[z] = 0$.

Lemma 2.5. *If $y(t)$ is a solution of* (1.2) *for $\lambda = \lambda_1$, and $z(t)$ is a solution of* (1.4) *for $\lambda = \lambda_2$ then*

$$\int_a^b z^*(t)B(t)y(t)\, dt = 0$$

whenever $\lambda_2 \neq \bar{\lambda}_1$.

If $\nabla(\lambda)$ denotes the determinant of the matrix $D(\lambda)$ of (2.1) corresponding to a fundamental matrix $Y(t; \lambda)$ of the vector differential equation (1.2-a), then clearly $\lambda = \lambda_0$ is a proper value of (1.2) if and only if $\nabla(\lambda_0) = 0$. Now a fundamental matrix $Y(t; \lambda)$ of (1.2-a) may be chosen so that its elements are entire functions of the complex variable λ for arbitrary values of t on $[a, b]$; in particular, in view of Problem I.10:4, such is assured if for some fixed value $t = t_0$ of $[a,b]$ the elements of $Y(t_0; \lambda)$ or $U_Y(t_0; \lambda)$ are entire functions of λ, for example, $Y(t_0; \lambda) = E$ or $U_Y(t_0; \lambda) = E$. Moreover, it is to be noted that if $Y(t; \lambda)$ is a fundamental matrix for (1.2-a) with elements entire functions of λ, and for a given $\lambda = \lambda_0$ the matrix $Y = Y_0(t)$ is a fundamental matrix for (1.2-a) with $\lambda = \lambda_0$, then for arbitrary $t_0' \in [a, b]$ the matrix

$$Y(t; \lambda)\, Y^{-1}(t_0'; \lambda_0)\, Y_0(t_0')$$

is a fundamental matrix of this differential equation with elements entire functions of λ and reducing to $Y_0(t)$ for $\lambda = \lambda_0$.

If $Y(t; \lambda)$ is a fundamental matrix of (1.2-a) with elements entire functions of λ, then $\nabla(\lambda) \equiv \det D(\lambda)$ is an entire function of λ, and for each λ_0 the function $\nabla(\lambda)$ is given throughout the complex λ-plane by its Taylor's expansion $\sum_{j=0}^{\infty} [\nabla^{(j)}(\lambda_0)/j!](\lambda - \lambda_0)^j$ about $\lambda = \lambda_0$. In particular, either $\nabla(\lambda) \equiv 0$ or each zero $\lambda = \lambda_0$ of $\nabla(\lambda)$ has finite *multiplicity* $m = m(\lambda_0) \geq 1$, that is, $\nabla^{(j)}(\lambda_0) = 0$, $(j = 0, 1, \ldots, m-1)$, and $\nabla^{(m)}(\lambda_0) \neq 0$; moreover, if $\nabla(\lambda) \not\equiv 0$ then the set of zeros of $\nabla(\lambda)$ has no finite limit point.

In view of these remarks it follows that for a boundary value problem (1.2) each of the following situations might occur: (a) all values of λ are proper values; (b) no value of λ is a proper value; (c) the set of proper values is finite; (d) there exists an infinite sequence of proper values, which is necessarily unbounded. The first three possibilities may be illustrated by boundary value problems involving the differential equations

$$y_1' = \lambda y_2, \qquad y_2' = 0. \tag{2.3}$$

Indeed, a fundamental matrix solution of (2.3) is given by

$$Y(t; \lambda) = \begin{bmatrix} 1 & \lambda t \\ 0 & 1 \end{bmatrix}, \tag{2.4}$$

and it may be verified readily that the possibilities (a), (b), (c) are illustrated by the system consisting of (2.3) and the following respective boundary conditions

$$\text{(a)} \begin{array}{l} y_1(0) - y_1(1) = 0, \\ y_2(0) - y_2(1) = 0; \end{array} \qquad \text{(b)} \begin{array}{l} y_1(1) = 0, \\ y_2(0) = 0; \end{array} \qquad \text{(c)} \begin{array}{l} y_1(0) + y_2(0) = 0, \\ y_1(1) - y_2(1) = 0. \end{array}$$

From the specific form of (2.4) it follows that for arbitrary end-points $t = a$, $t = b$, and 2×4 matrices M, the determinant $\nabla(\lambda)$ of $D(\lambda) = M\hat{U}_Y(\ ;\lambda)$ is a linear function of λ, and hence the system consisting of (2.3) and the boundary conditions $M\hat{u}_y = 0$ has either all values of λ as proper values or at most one proper value. In particular, possibility (d) cannot be illustrated by a system of the form (1.2) involving the differential equations (2.3). This possibility is illustrated, however, by the differential system

$$\begin{array}{ll} y_1' = y_2, & y_1(0) = 0, \\ y_2' = -\lambda y_1, & y_1(1) = 0, \end{array} \tag{2.5}$$

which has proper values $\lambda = \lambda_j = j^2\pi^2$, $(j = 1, 2, \ldots)$, with corresponding proper vector functions $y_1 = y_1^{(j)}(t) \equiv \sin j\pi t$, $y_2 = y_2^{(j)}(t) \equiv j\pi \cos j\pi t$.

Theorem 2.1. *The index $k = k(\lambda_0)$ of a proper value λ_0 of* (1.2) *does not exceed its multiplicity $m = m(\lambda_0)$; moreover, $k(\lambda_0) < m(\lambda_0)$ if and only if there is a proper vector function $y = y^0(t)$ of* (1.2) *for $\lambda = \lambda_0$ such that*

$$\int_a^b z^*(t)B(t)y^0(t)\,dt = 0 \tag{2.6}$$

for all solutions $z(t)$ of the adjoint boundary problem (1.4) *for $\lambda = \bar{\lambda}_0$.*

In view of the above remarks on fundamental matrices with elements entire functions of λ, it follows that if $k = k(\lambda_0)$ is the index of $\lambda = \lambda_0$ as a proper value of (1.2) then there exists a fundamental matrix $Y(t; \lambda)$ of the differential equation (1.2-a) with elements entire functions of λ, and such that the first k

column vectors of $Y(t; \lambda_0)$ are proper vector functions of (1.2) for $\lambda = \lambda_0$. For such a choice of $Y(t; \lambda)$ the matrix $D(\lambda)$ has elements that are entire functions of λ, while $D(\lambda_0)$ is of rank $n - k$ and has the elements in its first k columns all zero. If the derivatives of $\nabla(\lambda) = \det D(\lambda)$ are computed as sums of determinants of matrices obtained from $D(\lambda)$ by column-wise differentiation, it is clear that $\nabla^{(j)}(\lambda_0) = 0$, $(j = 0, 1, \ldots, k - 1)$, and $k(\lambda_0) \leq m(\lambda_0)$, thus establishing the first result of the theorem.

Now if $y^{(\beta)}(t; \lambda)$ is the β-th column vector of $Y(t; \lambda)$ and $y_\lambda^{(\beta)}(t; \lambda) \equiv (y_{\alpha\lambda}^{(\beta)}(t; \lambda)) \equiv (\partial/\partial\lambda) Y_{\alpha\beta}(t; \lambda)$, then for $y^{(\beta,0)}(t) = y^{(\beta)}(t; \lambda_0)$, $y_\lambda^{(\beta,0)}(t) = y_\lambda^{(\beta)}(t; \lambda_0)$ we have

$$\text{(2.7)}\quad \nabla^{(k)}(\lambda_0) = k! \det[\mu[y_\lambda^{(1,0)}] \cdots \mu[y_\lambda^{(k,0)}] \qquad \mu[y^{(k+1,0)}] \cdots \mu[y^{(n,0)}]].$$

In particular, the condition $k(\lambda_0) < m(\lambda_0)$ is equivalent to $\nabla^{(k)}(\lambda_0) = 0$, and from (2.7) it follows that this condition holds if and only if there are constants $d_1, \ldots, d_k, c_{k+1}, \ldots, c_n$ not all zero and such that for $u^{(1)} = u^{(1)}(t; d_1, \ldots, d_k)$, $u^{(2)} = u^{(2)}(t; c_{k+1}, \ldots, c_n)$ defined as

$$\text{(2.8)}\quad u^{(1)} = y_\lambda^{(1,0)} d_1 + \cdots + y_\lambda^{(k,0)} d_k, \quad u^{(2)} = y^{(k+1,0)} c_{k+1} + \cdots + y^{(n,0)} c_n,$$

we have $\mu[u^{(1)} + u^{(2)}] = 0$; moreover, not all the constants $d_1, \ldots, d_k$ may be zero, as this condition would require $y = u^{(2)}$ to be a proper solution of (1.2) for $\lambda = \lambda_0$ and consequently $D(\lambda_0)$ of rank less than $n - k(\lambda_0)$, contrary to Lemma 2.1. That is, $k(\lambda_0) < m(\lambda_0)$ if and only if there are constants $d_1, \ldots, d_k$ not all zero with which there are associated constants $c_{k+1}, \ldots, c_n$ such that the functions $u^{(1)}$, $u^{(2)}$ of (2.8) satisfy $\mu[u^{(1)} + u^{(2)}] = 0$. Now identically in λ,

$$L[y_\lambda^{(\beta)}](t) - \lambda B(t) y_\lambda^{(\beta)}(t) = B(t) y^{(\beta)}(t), \quad t \in [a, b], \quad \beta = 1, \ldots, n,$$

and hence for $y^0(t) = y^{(0)}(t; d_1, \ldots, d_k) = y^{(1,0)} d_1 + \cdots + y^{(k,0)} d_k$ the vector differential equation

$$\text{(2.9)}\qquad L[y](t) - \lambda_0 B(t) y(t) = B(t) y^0(t), \qquad t \in [a, b],$$

has as particular solution the vector $u^{(1)} = y_\lambda^{(1,0)} d_1 + \cdots + y_\lambda^{(k,0)} d_k$ defined by (2.8). Consequently the general solution of the differential equation (2.9) is $y = u^{(1)} + y^{(1,0)} c_1 + \cdots + y^{(n,0)} c_n$, and as $\mu[y^{(1,0)}] = 0 = \cdots = \mu[y^{(k,0)}]$, the condition that there exist constants $c_{k+1}, \ldots, c_n$ such that the functions $u^{(1)}$, $u^{(2)}$ of (2.8) satisfy $\mu[\mu^{(1)} + u^{(2)}] = 0$ is equivalent to the condition that (2.9) has a solution y satisfying $\mu[y] = 0$. As the most general proper function of (1.2) for $\lambda = \lambda_0$ is of the form $y^0(t) = y^{(1,0)} d_1 + \cdots + y^{(k,0)} d_k$ with $d_1, \ldots, d_k$ not all zero, from Theorem III.6.3 it follows that $k(\lambda_0) < m(\lambda_0)$ if and only if there is a proper function $y^0(t)$ of (1.2) for $\lambda = \lambda_0$ such that (2.6) holds for all solutions $z(t)$ of (1.4) for $\lambda = \bar{\lambda}_0$.

The result of the following corollary is an immediate consequence of Theorem 2.1.

Corollary. *Suppose that $\lambda = \lambda_0$ is a proper value of (1.2) of index $k = k(\lambda_0)$, with $y = y^{(\tau)}(t)$, $(\tau = 1, \ldots, k)$, linearly independent proper functions of (1.2) for $\lambda = \lambda_0$, and $z = z^{(\sigma)}(t)$, $(\sigma = 1, \ldots, k)$, linearly independent proper functions of (1.4) for $\lambda = \bar{\lambda}_0$. Then $k(\lambda_0) = m(\lambda_0)$ if and only if the $k \times k$ matrix*

$$\left[\int_a^b z^{(\sigma)*}(t)B(t)y^{(\tau)}(t)\,dt\right], \qquad (\sigma, \tau = 1, \ldots, k),$$

is nonsingular.

An illustration of a boundary value problem in which $k(\lambda_0) < m(\lambda_0)$ for certain proper values is afforded by the system

$$u'' + \lambda u = 0, \qquad u'(0) = 0, \qquad u(0) - u(2) + u'(2) = 0, \tag{2.10}$$

for which $\lambda = 0$ is a proper value with corresponding proper function $u(t) \equiv 1$. Under the transformation $y_1 = u$, $y_2 = u'$ the system (2.10) is equivalent to

$$\begin{aligned} y_1' &= y_2, & y_2(0) &= 0,\\ y_2' &= -\lambda y_1, & y_1(0) - y_1(2) + y_2(2) &= 0, \end{aligned} \tag{2.10$'$}$$

which is a system of the form (1.2) with

$$A_0(t) \equiv \begin{bmatrix} 0 & -1 \\ 0 & 0 \end{bmatrix}, \qquad A_1(t) = A_2(t) \equiv E_2, \qquad B(t) \equiv \begin{bmatrix} 0 & 0 \\ -1 & 0 \end{bmatrix},$$

$$M = \begin{bmatrix} 0 & 1 & 0 & 0 \\ 1 & 0 & -1 & 1 \end{bmatrix}.$$

Now if $\lambda \neq 0$, and we set $\lambda = \omega^2$, then the fundamental matrix $Y(t; \lambda)$ of the set of differential equations of (2.10′) which satisfies the initial condition $Y(0; \lambda) = E_2$ is given by

$$Y(t; \lambda) = \begin{bmatrix} \cos \omega t & \dfrac{\sin \omega t}{\omega} \\ -\omega \sin \omega t & \cos \omega t \end{bmatrix}.$$

It follows readily that for the example (2.10′) we have

$$D(\lambda) = \begin{bmatrix} 0 & 1 \\ 1 - \cos 2\omega - \omega \sin 2\omega & \cos 2\omega - \dfrac{\sin 2\omega}{\omega} \end{bmatrix},$$

and

$$\nabla(\lambda) = -1 + \cos 2\omega + \omega \sin 2\omega = -\tfrac{2}{3}\lambda^2 + \tfrac{8}{15}\lambda^3 + \cdots,$$

so that $\lambda = 0$ is a proper value of multiplicity two, while its index is one.

If the differential system (1.2) has proper values, and $\nabla(\lambda) \not\equiv 0$, it is desirable to label the proper values and corresponding proper functions of this system as a simple sequence

$$\lambda_j, y^{(j)}(t), \qquad (j = 1, 2, \ldots), \tag{2.11}$$

where each proper value of (1.2) is repeated a number of times equal to its index of compatibility, and the corresponding proper functions are selected as linearly independent solutions of (1.2) for this proper value. It is to be emphasized that the sequence in (2.11) may be finite or infinite. In the important case where each proper value of (1.2) has index equal to multiplicity it follows from Lemma 2.5 and the above Corollary that for each integer h there is a solution $z^{(h)}$ of (1.4) for $\lambda = \bar{\lambda}_h$ such that

$$\int_a^b z^{(h)*}(t)B(t)y^{(j)}(t)\,dt = \delta_{hj}, \qquad (h, j = 1, 2, \ldots). \tag{2.12}$$

With such a choice of proper functions $y^{(j)}(t)$, $z^{(j)}(t)$, $(j = 1, 2, \ldots)$ of (1.2) and (1.4) it follows that if $c_1, c_2, \ldots$ are constants such that the series

$$y^{(1)}(t)c_1 + y^{(2)}(t)c_2 + \cdots \tag{2.13}$$

converges uniformly on $[a, b]$ to a vector function $f(t)$, then

$$c_h = \int_a^b z^{(h)*}(t)B(t)f(t)\,dt, \qquad (h = 1, 2, \ldots). \tag{2.14}$$

If $f(t)$ is any vector function integrable on $[a, b]$ the integrals (2.14) exist, and the series (2.13) with coefficients (2.14) is well defined, quite independent of any properties of convergence of this series. For such a vector function $f(t)$ the constants c_h of (2.14) are called the *(generalized) Fourier coefficients of $f(t)$ relative to the sequence* $\{y^{(j)}(t)\}$ of proper functions of (1.2), and the corresponding series (2.13) is referred to as the *(generalized) Fourier series of $f(t)$ relative to* $\{y^{(j)}(t)\}$.

For a system (1.2) with $\nabla(\lambda) \not\equiv 0$ we shall consider now the behavior of the solutions of the nonhomogeneous system

$$L[y](t) = \lambda B(t)y(t) + g(t), \qquad t \in [a, b], \qquad \mu[y] = 0, \tag{2.15}$$

near a proper value.

Theorem 2.2. *If $\nabla(\lambda) \not\equiv 0$ for a system* (1.2), *and $\lambda = \lambda_0$ is a proper value of this system of index $k = k(\lambda_0)$ and multiplicity $m = m(\lambda_0)$, then for a given vector function $g(t)$ the solution of* (2.15) *is either regular at $\lambda = \lambda_0$ or at this value has a pole of order not exceeding $m - k + 1$. Moreover, if* (2.15) *has a solution for $\lambda = \lambda_0$ then at this value the solution of* (2.15) *is either regular or has a pole of order not exceeding $m - k$.*

Let $y^{(0)}(t; \lambda)$ be a solution of the vector differential equation of (2.15) that is an entire function of λ; such a solution is obtained by selecting a particular solution $y = y^0(t)$ of this equation for $\lambda = \lambda_0$ and choosing the solution $y = y^{(0)}(t; \lambda)$ satisfying $y^{(0)}(a; \lambda) = y^0(a)$, (see, Problem I.10:5). Now let $Y(t; \lambda)$ be a fundamental matrix for $L[y](t) = \lambda B(t)y(t)$, $t \in [a, b]$, whose elements are entire functions of λ, and denote the column vectors of $Y(t; \lambda)$ by $y^{(\beta)}(t; \lambda)$, $(\beta = 1, \ldots, n)$. In a deleted neighborhood of $\lambda = \lambda_0$ containing no other zero of $\nabla(\lambda)$ the system (2.15) has a unique solution $y = y(t; \lambda)$ with γ-th component given by

(2.16)

$$y_\gamma(t; \lambda) = \frac{(-1)^n}{\nabla(\lambda)} \det \begin{bmatrix} \mu_\alpha[y^{(0)}(\ ;\lambda)] & \mu_\alpha[y^{(1)}(\ ;\lambda)] \cdots \mu_\alpha[y^{(n)}(\ ;\lambda)] \\ y_\gamma^{(0)}(t;\lambda) & y_\gamma^{(1)}(t;\lambda) \quad \cdots \quad y_\gamma^{(n)}(t;\lambda) \end{bmatrix}.$$

For $t \in [a, b]$ and $\gamma = 1, \ldots, n$ the determinant in (2.16) is an entire function of λ. As $\lambda = \lambda_0$ is a zero of $\nabla(\lambda)$ of order m, if for all t on $[a, b]$ and $\gamma = 1, \ldots, n$ the determinant in (2.16) has $\lambda = \lambda_0$ as a zero of order not less than m then each component of $y(t; \lambda)$ is regular at $\lambda = \lambda_0$.

However, if for some γ and some $t \in [a, b]$ the determinant in (2.16) does not have a zero of order at least m at $\lambda = \lambda_0$, then some of the components of $y(t; \lambda)$ will have a pole at this value.

Now $D(\lambda) \equiv M\hat{U}_y(\ ; \lambda)$ is such that $D(\lambda_0)$ has rank $n - k$, and the $n \times (n + 1)$ matrix

$$[\mu_\alpha[y^{(0)}(\ ;\lambda_0)] \quad \mu_\alpha[y^{(1)}(\ ;\lambda_0)] \cdots \mu_\alpha[y^{(n)}(\ ;\lambda_0)]]$$

has rank at most $n - k + 1$. Without loss of generality we may assume that for $\lambda = \lambda_0$ the elements of the first k rows of $D(\lambda_0)$ are all zero and at most one of the components $\mu_1[y^{(0)}(\ ; \lambda_0)], \ldots, \mu_k[y^{(0)}(\ ; \lambda_0)]$ is different from zero. Indeed, such is obtained upon replacing M by CM, with C a suitable non-singular $n \times n$ matrix. With the coefficient matrix M so adjusted the matrix whose determinant is displayed in (2.16) is such that in at least $k - 1$ of the first k rows all elements vanish for $\lambda = \lambda_0$. Consequently the expansion of this determinant about $\lambda = \lambda_0$ has $(\lambda - \lambda_0)^{k-1}$ as a factor, and a component $y_\gamma(t; \lambda)$, $t \in [a, b]$, $(\gamma = 1, \ldots, n)$, that is not regular at $\lambda = \lambda_0$ has at this value a pole of order not exceeding $m - (k - 1)$. If (2.15) has a solution $y = y^0(t)$ for $\lambda = \lambda_0$ then $y^{(0)}(t; \lambda)$ may be chosen to satisfy $y^{(0)}(t; \lambda_0) = y^0(t)$ and consequently $\mu[y^{(0)}(\ ; \lambda_0)] = 0$. In this case the matrix whose determinant is displayed in (2.16) has only zero elements in the first k rows for $\lambda = \lambda_0$, so that the expansion of this determinant about $\lambda = \lambda_0$ then has the factor $(\lambda - \lambda_0)^k$, and a given component $y_\gamma(t; \lambda)$, $t \in [a, b]$, $(\gamma = 1, \ldots, n)$, is either regular at $\lambda = \lambda_0$ or has at this value a pole of order not exceeding $m - k$.

Corollary. *If each proper value of* (1.2) *has index equal to its multiplicity, and* $g(t)$ *is a continuous vector function such that*

$$\int_a^b z^*(t)g(t)\,dt = 0 \tag{2.17}$$

for all proper vector functions $z(t)$ *of* (1.4) *corresponding to proper values* λ *satisfying* $|\lambda| < R$, *then on* $|\lambda| < R$ *the system* (2.15) *has a solution* $y = y(t; \lambda)$ *with Maclaurin expansion*

$$y(t; \lambda) = y^{(0)}(t) + \lambda y^{(1)}(t) + \cdots + \lambda^j y^{(j)}(t) + \cdots, \qquad t \in [a, b], \qquad |\lambda| < R. \tag{2.18}$$

Moreover, if $\lambda = 0$ *is not a proper value of* (1.2) *then* $y^{(0)}(t)$, $y^{(j)}(t)$, ($j = 1, 2, \ldots$), *are the unique solutions of the differential systems*

$$\begin{aligned} &L[y^{(0)}](t) = g(t), \qquad t \in [a, b], \qquad \mu[y^{(0)}] = 0, \\ &L[y^{(j)}](t) = B(t)y^{(j-1)}(t), \qquad t \in [a, b], \qquad \mu[y^{(j)}] = 0, \ (j = 1, 2, \ldots). \end{aligned} \tag{2.19}$$

In particular, if (2.17) *holds for all proper vector functions of* (1.4) *then the solution* (2.18) *of* (1.2) *is an entire function of* λ.

It is to be remarked that the condition that each proper value of (1.2) has index equal to multiplicity implies that $\nabla(\lambda) \not\equiv 0$. If λ_0 is a proper value with $|\lambda_0| < R$, then the condition (2.17) for all solutions of (1.4) for $\lambda = \bar{\lambda}_0$ implies that (2.15) has a solution for $\lambda = \lambda_0$. Since $m(\lambda_0) = k(\lambda_0)$ by the hypotheses of the Corollary, from the above theorem it follows that the solution of (2.15) given by (2.16) in a deleted neighborhood of $\lambda = \lambda_0$ is regular at $\lambda = \lambda_0$, and $y(t; \lambda_0) = \lim_{\lambda \to \lambda_0} y(t; \lambda)$ is a solution of (2.15) for $\lambda = \lambda_0$. As this is true for each proper value λ of (1.2) satisfying $|\lambda| < R$, there is a solution $y(t; \lambda)$ which is regular on this circular domain, and hence has a Maclaurin expansion of the form (2.18). The remainder of the conclusions of the Corollary are immediate.

Problems IV.2

1. Show that the proper values of $u'' + \lambda u = 0$, $u(0) = 0 = u(\pi)$, are $\lambda = \lambda_j = j^2$, ($j = 1, 2, \ldots$), with corresponding proper functions $u = A_j \sin jt$, and that $k(\lambda_j) = m(\lambda_j) = 1$, ($j = 1, 2, \ldots$).

2. Show that the proper values of

$$u'' + \lambda u = 0, \qquad u(0) - u(2\pi) = 0 = u'(0) - u'(2\pi),$$

are $\lambda = \lambda_j = (j - 1)^2$, ($j = 1, 2, \ldots$), with $k(\lambda_1) = 1$, $k(\lambda_j) = 2$ if $j > 1$, and $k(\lambda_j) = m(\lambda_j)$. ($j = 1, 2, \ldots$); show that for $\lambda = \lambda_1 = 0$ the proper function of the system is $u \equiv B_0 \neq 0$, while for $j = 2, 3, \ldots$ the proper functions corresponding to $\lambda = \lambda_j$ are of the form $u = A_j \sin(j-1)t + B_j \cos(j-1)t$, where A_j, B_j are constants.

3. Show that $u'' + \lambda u = 0$, $u(-1) + u(1) = 0 = u'(-1) - u'(1)$, is compatible for all values of λ.

4. Show that $u^{(iv)} + \lambda u'' = 0$, $u(-1) = u'(-1) = u''(1) = u'''(1) = 0$ has no proper values.

5. Show that

$$\begin{aligned} y_1' &= y_2 + \lambda(t - \tfrac{1}{2})y_1, & y_1(0) - y_1(1) &= 0, \\ y_2' &= \lambda y_1, & y_2(0) + y_2(1) &= 0, \end{aligned}$$

is compatible for all values of λ.

Hint. Show that the system adjoint to the given system has for all values of λ the proper vector functions $z_1 = 1$, $z_2 = \frac{1}{2} - t$.

6. Prove that the proper values of (2.10) distinct from $\lambda = 0$ are given by $\lambda = \omega^2$, where either $\omega = j\pi$, $(j = 1, 2, \ldots)$, or ω is a positive root of the transcendental equation $\tan \omega = \omega$; show that each non-zero proper value is of multiplicity one.

7. Show that the proper values of

$$\begin{aligned} y_1' &= \lambda y_2, & 2y_1(0) + y_1(1) &= 0, \\ y_2' &= -\lambda y_1, & y_2(0) - y_2(1) &= 0, \end{aligned}$$

are given by $\lambda = 2j\pi$, $(j = 0, \pm 1, \pm 2, \ldots)$, and that each of these proper values is of index one and multiplicity two.

3. PROPERTIES OF THE GREEN'S MATRIX

Throughout this section we shall consider a system (1.2) for which not all values of λ are proper values. If λ is not a proper value of (1.2), then by Theorem III.7.1 there exists a Green's matrix $G(t; s; \lambda)$ for this system, and on $\Delta_1 \cup \Delta_2$ we have

$$G(t, s; \lambda) = \tfrac{1}{2} Y(t; \lambda)[E_n \operatorname{sgn}(t - s) + D^{-1}(\lambda)\, \Delta(\lambda)][Z(s; \bar{\lambda})]^*, \tag{3.1}$$

where $Y(t; \lambda)$, $Z(t; \lambda)$ are fundamental matrices for the respective differential systems (1.2-a), (1.4-a) with $U_Y(a; \lambda) \equiv A_2(a) Y(a; \lambda) \equiv E_n$, $V_Z(a; \lambda) \equiv A_1^*(a) Z(a, \lambda) \equiv E_n$, $D(\lambda) = M\hat{U}_Y(\ ; \lambda)$, $\Delta(\lambda) = -MQ\hat{U}_Y(\ ; \lambda)$, and $Q = \operatorname{diag}\{-E_n, E_n\}$. As pointed out above, the matrix functions $Y(t; \lambda)$ and $U_Y(t; \lambda)$ are entire functions of λ, and these matrix functions are nonsingular for $t \in [a, b]$ and arbitrary complex numbers λ. Correspondingly, $Z(t; \lambda)$ and $V_Z(t; \lambda)$ are nonsingular entire matrix functions of λ. Moreover, since $[A_1^*(t)Z(t; \bar{\lambda})]^*[A_2(t)Y(t; \lambda)] = [Z(t; \bar{\lambda})]^* A_1(t) A_2(t) Y(t; \lambda)$ is a constant matrix on $[a, b]$ for each value of λ, it follows that this matrix function is the identity matrix for arbitrary (t, λ). In particular, $[Z(t; \bar{\lambda})]^* A_1(t)$ is the inverse of the nonsingular matrix function $A_2(t) Y(t; \lambda) = U_Y(t; \lambda)$, and hence $[Z(t; \bar{\lambda})]^*$ and $[Z(t; \bar{\lambda})]^* A_1(t)$ are entire functions of λ. As an initial step in the study of the character of $G(t, s; \lambda)$ near a proper value, the following lemma will be established.

***Lemma* 3.1.** *If λ_0 is a proper value of* (1.2) *of multiplicity $m = m(\lambda_0)$, then at $\lambda = \lambda_0$ the matrix $D^{-1}(\lambda)\, \Delta(\lambda)$ has a pole of order r with $1 \leq r \leq m$; that is, in a deleted circular neighborhood of $\lambda = \lambda_0$ containing no proper value of* (1.2) *the matrix $D^{-1}(\lambda)\, \Delta(\lambda)$ has a Laurent expansion*

$$D^{-1}(\lambda)\, \Delta(\lambda) = \sum_{j=-r}^{\infty} (\lambda - \lambda_0)^j C_j, \tag{3.2}$$

where r is a positive integer satisfying $1 \leq r \leq m$ and C_{-r} is not the zero matrix.

As the elements of $D(\lambda)$ and $\Delta(\lambda)$ are entire functions of λ, and $\nabla(\lambda) = \det D(\lambda)$ has a zero of multiplicity m at $\lambda = \lambda_0$, it follows that in a deleted circular neighborhood of $\lambda = \lambda_0$ containing no proper value of (1.1) the matrix $D^{-1}(\lambda)\, \Delta(\lambda)$ has a Laurent expansion of the form $\sum_{j=-m}^{\infty} (\lambda - \lambda_0)^j C_j$. Consequently the conclusion of the lemma is equivalent to the condition that not all the matrices $C_{-m}, \ldots, C_{-1}$ are the zero matrix. If each of these matrices were the zero matrix then the elements of $D^{-1}(\lambda)\, \Delta(\lambda)$ would be bounded in a neighborhood of $\lambda = \lambda_0$, and from (3.1) it would follow that $(\lambda - \lambda_0)\, G(t, s; \lambda) \to 0$ as $\lambda \to \lambda_0$, uniformly in (t, s) on $\Delta_1 \cup \Delta_2$. Now if $y(t)$ is a solution of (1.2) for $\lambda = \lambda_0$, then

$$L[y](t) = \lambda B(t) y(t) - (\lambda - \lambda_0)\, B(t) y(t), \qquad t \in [a, b], \qquad \mu[y] = 0,$$

for arbitrary λ. Hence for λ not a proper value of (1.2) we have

$$y(t) = -(\lambda - \lambda_0) \int_a^b G(t, s; \lambda) B(s) y(s)\, ds, \tag{3.3}$$

and if $0 = C_{-m} = \cdots = C_{-1}$ then the right-hand member of (3.3) would tend to the zero vector as $\lambda \to \lambda_0$, and for $\lambda = \lambda_0$ the system (1.2) would have only the solution $y(t) \equiv 0$, contrary to the hypothesis of the lemma that λ_0 is a proper value. Hence at least one of the matrices $C_{-m}, \ldots, C_{-1}$ is not the zero matrix, and (3.2) holds with $1 \leq r \leq m$ and $C_{-r} \neq 0$.

In view of the remarks preceding Lemma 3.1, for $(t, s) \in \square = [a, b] \times [a, b]$ the $n \times n$ matrix function $K(t, s; \lambda) = \frac{1}{2} Y(t; \lambda)[Z(s; \bar{\lambda})]^*$ is an entire function of λ, and thus for a given complex number λ_0 this matrix function has as Taylor's series expansion about $\lambda = \lambda_0$ a permanently convergent power series

$$K(t, s; \lambda) = \sum_{j=0}^{\infty} (\lambda - \lambda_0)^j K_j(t, s). \tag{3.4}$$

In view of the differential equations satisfied by $Y(t; \lambda)$ and $Z(t, \lambda)$ it may be verified readily that on $\square$ the coefficient matrix functions $K_j(t, s)$ are continuous, and the associated matrix functions $\mathbf{K}_j(t, s) = A_2(t)K_j(t, s)A_1(s)$ are continuous and have continuous partial derivatives of the first order with respect to t and s.

If $F(t, s)$, $(t, s) \in \square$, is an $n \times n$ matrix function with

$$\mathbf{F}(t, s) = A_2(t)F(t, s)A_1(s)$$

continuous and having continuous partial derivatives with respect to t and s, for a complex value ζ let $\mathfrak{L}_t[\mathbf{F} \mid \zeta]$ and $\tilde{\mathfrak{L}}_s[\mathbf{F} \mid \zeta]$ denote the partial derivative expressions

$$\begin{aligned} \mathfrak{L}_t[\mathbf{F} \mid \zeta] &= A_1(t)\mathbf{F}_t(t, s) + [A_0(t) - \zeta B(t)]A_2^{-1}(t)\mathbf{F}(t, s), \\ \tilde{\mathfrak{L}}_s[\mathbf{F} \mid \zeta] &= -\mathbf{F}_s(t, s)A_2(s) + \mathbf{F}(t, s)A_1^{-1}(s)[A_0(s) - \zeta B(s)]. \end{aligned} \tag{3.5}$$

Since $\mathbf{K}(t, s; \lambda) = A_2(t)K(t, s; \lambda)A_1(s)$ satisfies the conditions

$$\begin{aligned} &\text{(a)} \quad \mathfrak{L}_t[\mathbf{K}(\ ,\ ; \lambda) \mid \lambda](t, s) = 0, \qquad \text{for} \quad (t, s) \in \square, \\ &\text{(b)} \quad \tilde{\mathfrak{L}}_s[\mathbf{K}(\ ,\ ; \lambda) \mid \lambda](t, s) = 0, \qquad \text{for} \quad (t, s) \in \square, \\ &\text{(c)} \qquad\qquad \mathbf{K}(s, s; \lambda) = \tfrac{1}{2}E_n, \quad \text{for} \quad s \in [a, b], \end{aligned} \tag{3.6}$$

it follows that the coefficient matrix functions $K_j(t, s)$ of (3.4) are such that the corresponding matrix functions $\mathbf{K}_j(t, s) = A_2(t)K_j(t, s)A_1(s)$ are characterized by the following differential systems:

(3.7)

$$\begin{aligned} &\mathfrak{L}_t[\mathbf{K}_0 \mid \lambda_0](t, s) = 0,\ \tilde{\mathfrak{L}}_s[\mathbf{K}_0 \mid \lambda_0](t, s) = 0, \qquad \mathbf{K}_0(s, s) = \tfrac{1}{2}E_n; \\ &\mathfrak{L}_t[\mathbf{K}_j \mid \lambda_0](t, s) = B(t)A_2^{-1}(t)\mathbf{K}_{j-1}(t, s), \\ &\tilde{\mathfrak{L}}_s[\mathbf{K}_j \mid \lambda_0](t, s) = \mathbf{K}_{j-1}(t, s)A_1^{-1}(s)B(s), \qquad \mathbf{K}_j(s, s) = 0, \qquad j = 1, 2, . \end{aligned}$$

Moreover, if $\lambda = \lambda_0$ is a proper value of (1.2) and the integer r is determined as in Lemma 3.1, then in a deleted circular neighborhood of $\lambda = \lambda_0$ containing no proper value of (1.2) the matrix $\frac{1}{2}Y(t; \lambda)D^{-1}(\lambda)\Delta(\lambda)[Z(s; \bar{\lambda})]^*$ has a Laurent expansion of the form

$$\tfrac{1}{2}Y(t; \lambda)D^{-1}(\lambda)\Delta(\lambda)[Z(s; \bar{\lambda})]^* = \sum_{j=-r}^{\infty} (\lambda - \lambda_0)^j W_j(t, s), \tag{3.8}$$

where as functions of (t, s) on $\square$ the corresponding matrix functions

$$\mathbf{W}_j(t, s) = A_2(t)W_j(t, s)A_1(s)$$

are continuous and have continuous partial derivatives with respect to t and s. Moreover, these functions satisfy the following conditions:

$$
\begin{aligned}
\mathfrak{L}_t[\mathbf{W}_{-r} \mid \lambda_0](t, s) &= 0, \\
\tilde{\mathfrak{L}}_s[\mathbf{W}_{-r} \mid \lambda_0](t, s) &= 0, \\
\mathbf{W}_{-r}(t, s) &= \tfrac{1}{2} U_Y(t; \lambda_0)\, C_{-r}[V_Z(s; \bar{\lambda}_0)]^* \\
\mathfrak{L}_t[\mathbf{W}_j \mid \lambda_0](t, s) &= B(t) A_2^{-1}(t) \mathbf{W}_{j-1}(t, s), \\
\tilde{\mathfrak{L}}_s[\mathbf{W}_j \mid \lambda_0](t, s) &= \mathbf{W}_{j-1}(t, s) A_1^{-1}(s) B(s), \\
j &= -r + 1, -r + 2, \ldots .
\end{aligned}
\tag{3.9}
$$

Now, if λ is not a proper value of (1.2), it follows from Theorems 7.2 and 7.3 that $\mathbf{G}(t, s; \lambda) = A_2(t) G(t, s; \lambda) A_1(s)$ satisfies the boundary conditions

$$
\begin{aligned}
&\text{(a)}\ M \cdot \begin{bmatrix} \mathbf{G}(a, s; \lambda) \\ \mathbf{G}(b, s; \lambda) \end{bmatrix} = 0, \quad \textit{for} \quad s \in (a, b), \\
&\text{(b)}\ [\mathbf{G}(t, a; \lambda) \qquad \mathbf{G}(t, b; \lambda)] QP = 0, \quad \textit{for} \quad t \in (a, b).
\end{aligned}
\tag{3.10}
$$

Consequently, if λ_0 is a proper value of (1.2), then for $(t, s) \in \Delta_1 \cup \Delta_2$ and λ in a deleted circular neighborhood of λ_0 containing no proper value of this system, the Green's matrix $G(t, s; \lambda)$ has the Laurent expansion

$$
G(t, s; \lambda) = \sum_{j=-r}^{\infty} (\lambda - \lambda_0)^j G_j(t, s),
\tag{3.11}
$$

where the coefficient matrix functions $G_j(t, s)$ have the following properties:

P_1: (a) $G_j(t, s) = W_j(t, s), \quad (j = -r, \ldots, -1; \quad (t, s) \in \square)$,

(b) $G_j(t, s) = \operatorname{sgn}(t - s)\, K_j(t, s) + W_j(t, s), \quad (j = 0, 1, \ldots;$

$(t, s) \in \Delta_1 \cup \Delta_2)$;

P_2: (a) *for* $j \neq 0, 1$ *the matrix functions* $\mathbf{G}_j(t, s) = A_2(t) G_j(t, s) A_1(s)$ *are continuous and have continuous partial derivatives with respect to* t *and* s *on* $\square$;

(b) $\mathbf{G}_0(t, s)$ *is continuous and has continuous partial derivatives with respect to* t *and* s *on* $\Delta_1 \cup \Delta_2$, *while* $\mathbf{G}_0(s^+, s) - \mathbf{G}_0(s^-, s) = E_n$ *for* $s \in (a, b)$, *and* $\mathbf{G}_0(t, t^-) - \mathbf{G}_0(t, t^+) = E_n$ *for* $t \in (a, b)$;

(c) $\mathbf{G}_1(t, s)$ *is continuous on* $\square$, *and has continuous partial derivatives with respect to* t *and* s *on* $\Delta_1 \cup \Delta_2$;

P_3: (a) $\mathfrak{L}_t[\mathbf{G}_{-r} \mid \lambda_0](t, s) = 0$, $\tilde{\mathfrak{L}}_s[\mathbf{G}_{-r} \mid \lambda_0](t, s) = 0$, $\mathbf{G}_{-r}(t, s) \not\equiv 0$ *on* $\square$;
(b) *for* $j \neq -r, 0, 1$ *the matrix function* $\mathbf{G}_j(t, s)$ *satisfies the differential equations*

$$\mathfrak{L}_t[\mathbf{G}_j \mid \lambda_0](t, s) = B(t)A_2^{-1}(t)\mathbf{G}_{j-1}(t, s),$$

$$\tilde{\mathfrak{L}}_s[\mathbf{G}_j \mid \lambda_0](t, s) = \mathbf{G}_{j-1}(t, s)A_1^{-1}(s)B(s),$$

for $(t, s) \in \square$, *whereas for* $j = 0, 1$ *these equations hold for* $(t, s) \in \Delta_1 \cup \Delta_2$;

P_4: *for* $j \neq 0$ *the boundary conditions*

$$M \cdot \begin{bmatrix} \mathbf{G}_j(a, s) \\ \mathbf{G}_j(b, s) \end{bmatrix} = 0, \qquad [\mathbf{G}_j(t, a) \quad \mathbf{G}_j(t, b)]QP = 0,$$

hold for $s \in [a, b]$ *and* $t \in [a, b]$, *respectively, whereas for* $j = 0$ *these conditions hold for* $s \in (a, b)$ *and* $t \in (a, b)$, *respectively.*

The above properties of $G(t, s; \lambda)$ are summarized in the following theorem.

Theorem 3.1. *If* λ_0 *is a proper value of* (1.2) *of multiplicity* $m = m(\lambda_0)$, *then there is an integer* r *satisfying* $1 \leq r \leq m$ *such that in a deleted circular neighborhood of* $\lambda = \lambda_0$ *containing no proper value of this system the Green's matrix has for* $(t, s) \in \Delta_1 \cup \Delta_2$ *the Laurent expansion* (3.11) *with the matrices* $\mathbf{G}_j(t, s) = A_2(t)G_j(t, s)A_1(s)$ *possessing the properties* $P_1 - P_4$; *in particular, if* λ_0 *is of index* $k = k(\lambda_0)$ *then*

$$G_{-r}(t, s) = \sum_{\sigma,\tau=1}^{k} y^{(\sigma)}(t)C_{\sigma\tau}\, z^{(\tau)*}(s), \tag{3.12}$$

where $y^{(\sigma)}(t)$, $(\sigma = 1, \ldots, k)$, *are linearly independent solutions of* (1.2) *for* $\lambda = \lambda_0$, $z^{(\tau)}(t)$, $(\tau = 1, \ldots, k)$, *are linearly independent solutions of* (1.4) *for* $\lambda = \bar{\lambda}_0$, *and* $[C_{\sigma\tau}]$ *is a nonzero* $k \times k$ *matrix.*

A further characterization of the Green's matrix is given by the following theorem.

Theorem 3.2. *If* λ_0 *is a proper value of* (1.2) *the order* r *of the pole of* $G(t, s; \lambda)$ *at* $\lambda = \lambda_0$ *is characterized by the property that* r *is the largest integer* q *such that there exist vector functions* $u_1(t) \not\equiv 0$, $u_2(t), \ldots, u_q(t)$ *satisfying the differential system*

$$\begin{aligned} &L[u_1](t) = \lambda_0\, B(t)u_1(t), \qquad t \in [a, b], \qquad \mu[u_1] = 0, \\ &L[u_j](t) = \lambda_0\, B(t)u_j(t) + B(t)u_{j-1}(t), \qquad t \in [a, b], \qquad \mu[u_j] = 0, \\ &\qquad\qquad (j = 2, \ldots, q), \end{aligned} \tag{3.13}$$

As $G_{-r}(t, s) \not\equiv 0$ on $\square$, there exists a value $s_0 \in [a, b]$, and a constant

n-dimensional vector ξ such that $G_{-r}(t, s_0)\xi$ is not the null vector function on $[a, b]$. From properties P_2, P_3, P_4 it follows that (3.13) holds for $q = r$ and $u_j(t) = G_{-r+j-1}(t, s_0)\xi$, $(j = 1, \ldots, r)$. On the other hand, if (3.13) were valid for an integer $q > r$ and vector functions $u_1(t) \not\equiv 0, u_2(t), \ldots, u_q(t)$, then

$$u = u(t; \lambda) = u_1(t) + (\lambda - \lambda_0)\, u_2(t) + \cdots + (\lambda - \lambda_0)^{q-1}\, u_q(t), \qquad t \in [a, b],$$

would satisfy the differential system

$$L[u](t) = \lambda B(t)u(t) - (\lambda - \lambda_0)^q\, B(t)u_q(t), \qquad t \in [a, b], \qquad \mu[u] = 0,$$

and hence for λ not a proper value of (1.2) we would have

$$\text{(3.14)} \quad u(t; \lambda) = -(\lambda - \lambda_0)^q \textstyle\int_a^b G(t, s; \lambda)B(s)u_q(s)\, ds, \qquad t \in [a, b].$$

From the assumption that $q > r$ it would then follow that the right-hand member of (3.14) tends to the null vector as $\lambda \to \lambda_0$ and $u_1(t) \equiv 0$, contrary to the assumption that $u_1(t) \not\equiv 0$. Thus r is the largest integer q for which the system (3.13) possesses a solution set $u_1(t), \ldots, u_q(t)$ with $u_1(t) \not\equiv 0$, and the theorem is proved.

Theorem 3.3. *If λ_0 is a proper value of* (1.2) *of index $k = k(\lambda_0)$ and multiplicity $m = m(\lambda_0)$, then $k(\lambda_0) = m(\lambda_0)$ if and only if $G(t, s; \lambda)$ has at $\lambda = \lambda_0$ a simple pole. If $k(\lambda_0) = m(\lambda_0)$ and $y = y^{(\sigma)}(t)$, $(\sigma = 1, \ldots, k)$, are linearly independent solutions of* (1.2) *for $\lambda = \lambda_0$, then for $z = z^{(\sigma)}(t)$, $(\sigma = 1, \ldots, k)$, solutions of* (1.4) *for $\lambda = \bar{\lambda}_0$ such that*

$$\text{(3.15)} \qquad \int_a^b z^{(\sigma)*}(t)B(t)y^{(\tau)}\, dt = \delta_{\sigma\tau}, \qquad (\sigma, \tau = 1, \ldots, k),$$

the residue of $G(t, s; \lambda)$ at $\lambda = \lambda_0$ is given by

$$\text{(3.16)} \qquad G_{-1}(t, s) = -\sum_{\sigma=1}^{k} y^{(\sigma)}(t)\, z^{(\sigma)*}(s).$$

In view of the solvability criterion of Theorem III.6.3, it follows from the above Theorem 3.2 that the order of the pole of $G(t, s; \lambda)$ at $\lambda = \lambda_0$ exceeds one if and only if there is a proper vector function $y = y^0(t) \equiv u_1(t)$ of (1.2) for $\lambda = \lambda_0$ satisfying condition (2.6) with all solutions $z(t)$ of (1.4) for $\lambda = \bar{\lambda}_0$, and in view of Theorem 2.1 this latter condition is equivalent to $k(\lambda_0) < m(\lambda_0)$. Now if $k(\lambda_0) = m(\lambda_0)$ it follows from the Corollary to Theorem 2.1 that if $y^{(1)}(t), \ldots, y^{(k)}(t)$ are linearly independent solutions of (1.2) for $\lambda = \lambda_0$ then there are solutions $z^{(1)}(t), \ldots, z^{(k)}(t)$ of (1.4) for $\lambda = \bar{\lambda}_0$ which satisfy (3.15). As $k(\lambda_0) = m(\lambda_0)$ implies $r = 1$ in Theorem 3.1, it follows that there are constants $C_{\sigma\tau}$ such that (3.12) holds for $r = 1$. Since equation (3.3) is satisfied by $y(t) = y^{(\rho)}(t)$, $(\rho = 1, \ldots, k)$, it then results that

$$y^{(\rho)}(t) = -\int_a^b G_{-1}(t, s)B(s)y^{(\rho)}(s)\,ds = -\sum_{\sigma=1}^{k} y^{(\sigma)}(t)C_{\sigma\rho}, \qquad (\rho = 1, \ldots, k)$$

so that $G_{-1}(t, s)$ is of the form (3.16).

The result of the following theorem is an immediate consequence of Theorem 3.3 and the Cauchy Residue Theorem.

Theorem 3.4. *Suppose that for a system* (1.2) *each proper value has index equal to its multiplicity, and that* (2.11) *denotes the proper values and corresponding proper vector functions ordered as a simple sequence, where each proper value is repeated a number of times equal to its index, and* $y = y^{(j)}(t)$, $z = z^{(j)}(t)$ *are proper functions of* (1.2), (1.4) *for* $\lambda = \lambda_j$ *and* $\lambda = \bar{\lambda}_j$, *respectively, which satisfy* (2.12). *If* C *is a simple closed regular curve in the* λ*-plane passing through no proper value of* (1.2), *and* $\mathfrak{D}_C$ *denotes the interior of* C, *then for* $f(s)$ *integrable on* $[a, b]$ *we have*

$$(3.17) \qquad -\frac{1}{2\pi i}\int_C d\lambda \int_a^b G(t, s; \lambda)B(s)f(s)\,ds = \sum_{\lambda_j \in \mathfrak{D}_C} y^{(j)}(t)\, c_j[f],$$

where $c_j[f]$ *is the generalized Fourier coefficient given by* (2.14).

When the matrix $B(t)$ satisfies certain conditions one may determine the asymptotic character of solutions of the vector differential equation (1.2-a) for large values of $|\lambda|$, and from the attendant asymptotic behavior of the Green's matrix for (1.2) determine the limit of the integral in (3.17) as the curve C in the λ-plane tends to infinity in a suitable manner. Although elements of this procedure occurred as early as the work of Liouville [1], a systematic treatment of boundary value problems by this method dates from the considerations of G. D. Birkhoff [1,2]. In our present consideration this method will not be pursued further, and we shall turn our attention to the important case of self-adjoint systems for which expansion theorems may be obtained by essentially real variable methods.

Problems IV.3

1. Show that if λ_1 and λ_2 are not proper values of (1.2) then:

(a) $G(t, s; \lambda_2) - G(t, s; \lambda_1) = (\lambda_2 - \lambda_1)\int_a^b G(t, r; \lambda_1)B(r)G(r, s; \lambda_2)\,dr,$

(b) $\int_a^b G(t, r; \lambda_1)B(r)G(r, s; \lambda_2)\,dr = \int_a^b G(t, r; \lambda_2)B(r)G(r, s; \lambda_1)\,dr.$

Hints. (a) Show that for $s \in (a, b)$ there is a matrix $F(t, s) = F(t, s; \lambda_1, \lambda_2)$ continuous in t on $[a, b]$ and such that

$$F(t, s) = G(t, s; \lambda_2) - G(t, s; \lambda_1), \quad \text{for} \quad (t, s) \in \Delta_1 \cup \Delta_2, \; M\hat{U}_F(\;, s) = 0,$$

$$L[F(\;, s)](t) = \lambda_1 B(t)F(t, s) + (\lambda_2 - \lambda_1)\, B(t)G(t, s; \lambda_2), \quad t \in [a, s) \cup (s, b].$$

(b) Use (a), together with the relation obtained by interchanging λ_1 and λ_2.

2. Show that if λ is not a proper value of (1.2) then

$$G_\lambda(t, s; \lambda) = \int_a^b G(t, r; \lambda)B(r)G(r, s; \lambda)\, dr.$$

Hint. Use the result (a) of Problem 1.

4. BOUNDARY PROBLEMS INVOLVING AN *n*-th ORDER LINEAR VECTOR DIFFERENTIAL EQUATION

The results of the preceding section will be applied now to a boundary problem of the form

$$\begin{aligned} &\text{(a)} \quad \mathcal{L}[u] = \lambda\, \mathcal{M}[u], \qquad t \in [a, b], \\ (4.1) \quad &\text{(b)} \quad \mu_\alpha[u] \equiv \sum_{\beta=1}^{n} \{M_{\alpha\beta}\, u^{[\beta-1]}(a) + M_{\alpha, n+\beta}\, u^{[\beta-1]}(b)\} = 0, \qquad (\alpha = 1, \ldots, n), \end{aligned}$$

in an h-dimensional vector function $u(t) \equiv (u_\sigma(t))$, $(\sigma = 1, \ldots, h)$. The vector differential expressions $\mathcal{L}$ and $\mathcal{M}$ in (4.1-a) are given by

(4.2)

$$\mathcal{L}[u] = P_n(t)u^{(n)} + \cdots + P_0(t)u, \qquad \mathcal{M}[u] = Q_r(t)u^{(r)} + \cdots + Q_0(t)u,$$

with $0 \leq r < n$ and the $h \times h$ matrix functions $P_0(t), \ldots, P_n(t), Q_0(t), \ldots, Q_r(t)$ continuous on $[a, b]$, such that $Q_r(t) \not\equiv 0$ and $P_n(t)$ is nonsingular on this interval. Moreover, $M_{\alpha\gamma} = [M_{\sigma\rho;\alpha\gamma}]$, $(\sigma, \rho = 1, \ldots, h)$, are $h \times h$ constant matrices for $\alpha = 1, \ldots, n$, $\gamma = 1, \ldots, 2n$ such that the $hn \times 2hn$ matrix

$$(4.3) \qquad \tilde{\mathcal{M}} = [M_{\alpha\gamma}] = \begin{bmatrix} M_{\sigma\rho;11} & \cdots & M_{\sigma\rho;1,2n} \\ \cdot & \cdots & \cdot \\ \cdot & \cdots & \cdot \\ M_{\sigma\rho;n1} & \cdots & M_{\sigma\rho;n,2n} \end{bmatrix}$$

is of rank hn.

Corresponding to the discussion of Section III.9, in terms of the h-dimensional vector functions $y_\beta(t) = y_{\sigma\beta}(t))$ defined as

$$y_\beta(t) = u^{[\beta-1]}(t) = (u_\sigma^{[\beta-1]}(t)), \qquad (\sigma = 1, \ldots, h; \beta = 1, \ldots, n),$$

the boundary problem (4.1) may be written

$$
\begin{aligned}
& P_n(t)y'_n + \sum_{j=1}^{n} P_{j-1}(t)y_j = \lambda \sum_{k=1}^{m+1} Q_{j-1}(t)y_j, \\
(4.4) \qquad & y'_{n-\gamma+1} - y_{n-\gamma+2} = 0, \qquad (\gamma = 2, \ldots, n), \\
& \sum_{\beta=1}^{n} \{M_{\alpha\beta}\, y_\beta(a) + M_{\alpha,n+\beta}\, y_\beta(b)\} = 0, \qquad (\alpha = 1, \ldots, n),
\end{aligned}
$$

which is of the form (1.2) in the hn-dimensional vector function $\eta(t) = (\eta_\nu(t))$, $(\nu = 1, \ldots, hn)$, with $\eta_{(\beta-1)h+\sigma}(t) = y_{\sigma\beta}(t) = u_\sigma^{[\beta-1]}(t)$, $(\sigma = 1, \ldots, h;\ \beta = 1, \ldots, n)$.

Let $\tilde{\mathfrak{T}}$ be a $2hn \times hn$ matrix of the form

$$
(4.5) \qquad \tilde{\mathfrak{T}} = [P_{\gamma\beta}] = \begin{bmatrix} P_{\sigma\rho;11} & \cdots & P_{\sigma\rho;1n} \\ \cdot & \cdots & \cdot \\ \cdot & \cdots & \cdot \\ P_{\sigma\beta;2n,1} & \cdots & P_{\sigma\rho;2n,n} \end{bmatrix},
$$

where $P_{\gamma\beta} = [P_{\sigma\rho;\gamma\beta}]$, $(\sigma, \rho = 1, \ldots, h)$, are $h \times h$ matrices for $\gamma = 1, \ldots, 2n$, $\beta = 1, \ldots, n$, such that $\tilde{\mathfrak{T}}$ is of rank hn and $\tilde{\mathcal{M}}\tilde{\mathfrak{T}} = 0$. For $z_\beta(t) = (z_{\sigma\beta}(t))$, $(\beta = 1, \ldots, n)$, h-dimensional vector functions, let $\zeta = (\zeta_\nu(t))$ denote the hn-dimensional vector function with $\zeta_{(\beta-1)h+\sigma}(t) = z_{\sigma\beta}(t)$, $(\sigma = 1, \ldots, h;\ \beta = 1, \ldots, n)$. Moreover, for conciseness of notation, let $Q_{r+1}(t), \ldots, Q_n(t)$ be defined as the zero $h \times h$ matrix function. Then the boundary problem adjoint to (4.4) may be written as

$$
\begin{aligned}
& -z'_n + P_0^*(t)z_1 = \lambda Q_0^*(t)z_1, \\
& -z'_{n-\gamma} + P_\gamma^*(t)z_1 - z_{n-\gamma+1} = \lambda Q_\gamma^*(t)z_1, \qquad (\gamma = 1, \ldots, n-2), \\
(4.6) \qquad & -[P_n^*(t)z_1]' + P_{n-1}^*(t)z_1 - z_2 = \lambda Q_{n-1}^*(t)z_1, \\
& \tilde{\mathfrak{T}}^*[\operatorname{diag}\{-E_{hn}, E_{hn}\}]\hat{v}_\zeta = 0.
\end{aligned}
$$

For the remainder of this section it will be supposed that for $j = 0, 1, \ldots, n$ *the matrix functions* $P_j(t)$, $Q_j(t)$ *are of class* $\mathfrak{C}^j[a, b]$. It then follows readily that $z_1(t), \ldots, z_n(t)$ is a solution of the differential equations of (4.6) if and only if $z_2(t), \ldots, z_n(t)$ are defined by the last $n - 1$ equations in terms of $z_1(t) = v(t)$, and the h-dimensional vector function $v(t)$ is a solution of the n-th order vector differential equation

$$
\mathfrak{L}^\star[v] = \lambda \mathcal{M}^\star[v], \qquad t \in [a, b],
$$

where $\mathfrak{L}^\star$ and $\mathcal{M}^\star$ are the adjoint differential expressions

$$
\mathfrak{L}^\star[v] = \sum_{j=0}^{n} (-1)^j [P_j^*(t)v]^{[j]}, \quad \mathcal{M}^\star[v] = \sum_{j=0}^{n} (-1)^j [Q_j^*(t)v]^{[j]}.
$$

If $K_{\alpha\beta}(t)$, $H_{\alpha\beta}(t)$, $(\alpha, \beta = 1, \ldots, n)$, are the $h \times h$ matrices such that the identities

$$(4.7)\quad \begin{aligned} (\mathfrak{L}[u], v) - (u, \mathfrak{L}^\star[v]) &= \left\{\sum_{\alpha,\beta=1}^{n} v^{*[\alpha-1]}(t)\, K_{\alpha\beta}(t) u^{[\beta-1]}(t)\right\}', \\ (\mathcal{M}[u], v) - (u, \mathcal{M}^\star[v]) &= \left\{\sum_{\alpha,\beta=1}^{n} v^{*[\alpha-1]}\, H_{\alpha\beta}(t)\, u^{[\beta-1]}(t)\right\}' \end{aligned}$$

hold for arbitrary vector functions u, v of class $\mathfrak{C}_h{}^n[a, b]$, then the elements of $K_{\alpha\beta}(t)$, $H_{\alpha\beta}(t)$ are continuously differentiable on $[a, b]$, $K_{\alpha\beta}(t) \equiv 0$ if $\alpha + \beta > n + 1$, $H_{\alpha\beta}(t) \equiv 0$ if $\alpha + \beta > r + 1$, and $K_{\alpha\beta}(t) = (-1)^{\alpha-1} P_n(t)$ if $\alpha + \beta = n + 1$. Moreover, in terms of v the adjoint system (4.6) becomes

$$(4.8)\quad \begin{aligned} &\mathfrak{L}^\star[v] = \lambda \mathcal{M}^\star[v], \qquad t \in [a, b], \\ &\sum_{\alpha,\beta=1}^{n} \{-P^*_{\alpha\gamma}[K^*_{\beta\alpha}(a) - \lambda H^*_{\beta\alpha}(a)]\, v^{[\beta-1]}(a) \\ &\qquad + P^*_{n+\alpha,\gamma}[K^*_{\beta\alpha}(b) - \lambda H^*_{\beta\alpha}(b)]\, v^{[\beta-1]}(b)\} = 0, \qquad (\gamma = 1, \ldots, n). \end{aligned}$$

By definition a value λ_0 is a proper value of (4.1) of index $k = k(\lambda_0)$ and multiplicity $m = m(\lambda_0)$ if λ_0 is a proper value of (4.4) of index k and multiplicity m; clearly λ_0 is a proper value of (4.1) of index k if and only if k is the maximum number of linearly independent solutions $u(t)$ of (4.1) for $\lambda = \lambda_0$. Corresponding definitions and properties hold for (4.8) and the system (4.6).

Lemmas 2.2 and 2.5 provide the following results for the system (4.1).

***Lemma* 4.1.** *If $k^*(\lambda)$ is the index of compatibility of* (4.8) *for a value λ then $k^*(\lambda) = k(\bar{\lambda})$.*

***Lemma* 4.2.** *If $u(t)$ is a solution of* (4.1) *for $\lambda = \lambda_1$, and $v(t)$ is a solution of* (4.8) *for $\lambda = \lambda_2$, then $\int_a^b (\mathcal{M}[u], v)\, dt = 0$ whenever $\lambda_2 \neq \bar{\lambda}_1$.*

Correspondingly, Theorem 2.1 and its corollary imply the following result for the boundary problem (4.1).

Theorem 4.1. *The index $k(\lambda_0)$ of a proper value λ_0 of* (4.1) *does not exceed its multiplicity $m(\lambda_0)$. Moreover, $k(\lambda_0) < m(\lambda_0)$ if and only if there is a proper vector function $u = u^0(t)$ of* (4.1) *for $\lambda = \lambda_0$ such that*

$$\int_a^b (\mathcal{M}[u^0], v)\, dt = 0$$

for all solutions $v(t)$ of (4.8) *for $\lambda = \bar{\lambda}_0$; an equivalent condition is that $u = u^0(t)$ is a proper vector function of* (4.1) *for $\lambda = \lambda_0$ such that the system*

$$\mathfrak{L}[u] = \lambda_0 \mathcal{M}[u] + \mathcal{M}[u^0], \qquad \mu_\alpha[u] = 0, \qquad (\alpha = 1, \ldots, n),$$

is compatible.

The following result is an immediate consequence of Theorem 2.2.

Theorem 4.2. *Suppose that not all values of λ are proper values of* (4.1), *and that λ_0 is a proper value of this system of index $k = k(\lambda_0)$ and multiplicity $m = m(\lambda_0)$. For a given continuous h-dimensional vector function $r(t)$ the solution of*

$$\mathfrak{L}[u] = \lambda \mathcal{M}[u] + r(t), \qquad \mu_\alpha[u] = 0, \qquad (\alpha = 1, \ldots, n). \tag{4.9}$$

is either regular at $\lambda = \lambda_0$, or at this value has a pole of order not exceeding $m - k + 1$. Moreover, if $r(t)$ is such that (4.9) *has a solution for $\lambda = \lambda_0$ then at this value the solution of* (4.9) *is either regular or has a pole of order not exceeding $m - k$.*

Finally, as a corollary to the above theorem one has the following result, corresponding to the Corollary to Theorem 2.2.

Corollary. *Suppose that each proper value of* (4.1) *has index equal to its multiplicity. If $r(t)$ is a continuous h-dimensional vector function such that*

$$\int_a^b (r, v)\, dt = 0 \tag{4.10}$$

for all proper vector functions $v(t)$ of (4.8) *corresponding to proper values λ satisfying $|\lambda| < R$, then on $|\lambda| < R$ the system* (4.9) *has a solution $u = u(t; \lambda)$ with Maclaurin expansion*

$$u(t; \lambda) = u^{(0)}(t) + \lambda u^{(1)}(t) + \cdots + \lambda^j u^{(j)}(t) + \cdots. \tag{4.11}$$

Moreover, if $\lambda = 0$ is not a proper value of (4.1), *then $u^{(0)}(t)$, $u^{(j)}(t)$, $(j = 1, 2, \ldots)$, are the unique solutions of the differential systems*

$$\begin{aligned} \mathfrak{L}[u^{(0)}] &= r(t), \qquad \mu_\alpha[u^{(0)}] = 0, \qquad (\alpha = 1, \ldots, n), \\ \mathfrak{L}[u^{(j)}] &= \mathcal{M}[u^{(j-1)}], \qquad \mu_\alpha[u^{(j)}] = 0, \qquad (\alpha = 1, \ldots, n). \end{aligned} \tag{4.12}$$

In particular, if (4.10) *holds for all proper vector functions of* (4.8) *then the solution* (4.11) *of* (4.8) *is an entire function of λ.*

It is to be emphasized that in general the boundary conditions of (4.8) involve specifically the parameter λ, and hence in form this adjoint boundary problem expressed in terms of v differs from the original problem (4.1) in the vector function u. For example, for the boundary value problem

$$u^{(\text{iv})} = \lambda u'', \qquad u(a) = u'(a) = u''(b) = u'''(b) = 0, \tag{4.13}$$

the adjoint boundary value problem is

$$v^{(\text{iv})} = \lambda v'', \qquad v(a) = v'(a) = v'''(b) - \lambda v'(b) = v''(b) - \lambda v(b) = 0. \tag{4.14}$$

Of particular importance is the case in which the restrictions imposed on $v^{[\beta-1]}(a)$, $v^{[\beta-1]}(b)$, $(\beta = 1, \ldots, n)$, by the boundary conditions of (4.8) are independent of λ. It is to be emphasized that this property does not require the boundary conditions, as written in (4.8), to be independent of λ. Indeed, if $M^a_{\alpha\gamma} \equiv [M^a_{\sigma\rho;\alpha\gamma}]$, $(\sigma, \rho = 1, \ldots, h)$, are $h \times h$ matrices for $\alpha = 1, \ldots, n$, $\gamma = 1, \ldots, 2n$ such that the $hn \times 2hn$ matrix $\mathcal{M}^a = [M^a_{\alpha\gamma}]$ is of rank hn, then the boundary conditions of (4.7) are equivalent to the conditions

$$\text{(4.15)} \qquad \mu^a_\alpha[v] = \sum_{\beta=1}^{n} \{M^a_{\alpha\beta} v^{[\beta-1]}(a) + M^a_{\alpha,n+\beta} v^{[\beta-1]}(b)\} = 0, \qquad (\alpha = 1, \ldots, n),$$

if and only if there are $h \times h$ matrices $C_{\alpha\beta} \equiv [C_{\sigma\rho;\alpha\beta}]$, $C^1_{\alpha\beta} \equiv [C^1_{\sigma\rho;\alpha,\beta}]$ $(\alpha, \beta = 1, \ldots, n)$, such that

$$\text{(4.16)} \qquad \begin{aligned} -\sum_{\alpha=1}^{n} P^*_{\alpha\gamma} K^*_{\beta\alpha}(a) &= \sum_{\alpha=1}^{n} C_{\gamma\alpha} M^a_{\alpha\beta}, \quad \sum_{\alpha=1}^{n} P^*_{n+\alpha,\gamma} K^*_{\beta\alpha}(b) = \sum_{\alpha=1}^{n} C_{\gamma\alpha} M^a_{\alpha,n+\beta}, \\ -\sum_{\alpha=1}^{n} P^*_{\alpha\gamma} H^*_{\beta\alpha}(a) &= \sum_{\alpha=1}^{n} C^1_{\gamma\alpha} M^a_{\alpha\beta}, \quad \sum_{\alpha=1}^{n} P^*_{n+\alpha,\gamma} H^*_{\beta\alpha}(b) = \sum_{\alpha=1}^{n} C^1_{\gamma\alpha} M^a_{\alpha,n+\beta}, \end{aligned}$$

for $\beta, \gamma = 1, \ldots, n$, where the $hn \times hn$ matrices $C = [C_{\alpha\beta}]$, $C^1 = [C^1_{\alpha\beta}]$ are such that $C - \lambda C^1$ is non-singular for all values of λ. In view of the definition of the matrix $\mathfrak{T} = [\mathfrak{T}_{\gamma\beta}]$ and the identities (4.7), the existence of such matrices C, C^1 is equivalent to the condition that

$$\text{(4.17)} \qquad \int_a^b (\mathfrak{L}[u], v)\, dt = \int_a^b (u, \mathfrak{L}^\star[v])\, dt, \quad \int_a^b (\mathcal{M}[u], v)\, dt = \int_a^b (u, \mathcal{M}^\star[v])\, dt,$$

for arbitrary vector functions $u(t)$, $v(t)$ of class $\mathfrak{C}_h{}^n[a, b]$ satisfying the end-conditions

$$\text{(4.18)} \qquad \begin{aligned} \mu_\alpha[u] &\equiv \sum_{\beta=1}^{n} \{M_{\alpha\beta} u^{[\beta-1]}(a) + M_{\alpha,n+\beta} u^{[\beta-1]}(b)\} = 0, \qquad (\alpha = 1, \ldots, n), \\ \hat{\mu}^a_\alpha[v] &\equiv \sum_{\beta=1}^{n} \{M^a_{\alpha\beta} v^{[\beta-1]}(a) + M^a_{\alpha,n+\beta}\, v^{[\beta-1]}(b)\} = 0, \qquad (\alpha = 1, \ldots, n). \end{aligned}$$

For example, for the boundary problem

$$\text{(4.19)} \qquad u^{(\text{iv})} = \lambda u'', \; 2u^{[\alpha-1]}(a) - u^{[\alpha-1]}(b) = 0, \qquad (\alpha = 1, 2, 3, 4),$$

the boundary conditions of the corresponding system (4.8) are

$$\begin{gathered} v(a) - 2v(b) = 0, \qquad -v'(a) + 2v'(b) = 0, \\ v''(a) - \lambda v(a) - 2[v''(b) - \lambda v(b)] = 0, \\ -[v'''(a) - \lambda v'(a)] + 2[v'''(b) - \lambda v'(b)] = 0, \end{gathered}$$

so that the adjoint boundary problem may be written in the form

$$\text{(4.20)} \qquad v^{(\text{iv})} = \lambda v'', \; v^{[\alpha-1]}(a) - 2v^{[\alpha-1]}(b) = 0, \qquad (\alpha = 1, 2, 3, 4),$$

with the boundary conditions independent of λ.

Now consider (4.1) and a second boundary problem

$$(4.21)\quad \begin{aligned} &\text{(a) } \mathfrak{L}^a[v] = \lambda\, \mathfrak{M}^a[v], \qquad t \in [a, b],\\ &\text{(b) } \mu_\alpha^a[v] \equiv \sum_{\beta=1}^{n} \{M_{\alpha\beta}^a\, v^{[\beta-1]}(a) + M_{\alpha,n+\beta}^a\, v^{[\beta-1]}(b)\} = 0,\\ &\hspace{20em} (\alpha = 1, \ldots, n), \end{aligned}$$

in an h-dimensional vector function $v(t) = (v_\sigma(t))$, $(\sigma = 1, \ldots, h)$, where the $h \times h$ coefficient matrices of

$$\mathfrak{L}^a[v] = P_n^a(t)v^{[n]} + \cdots + P_0^a(t)v,\ \mathfrak{M}^a[v] = Q_r^a(t)v^{[r]} + \cdots + Q_0^a(t)v,$$

and the $h \times h$ constant matrices $M_{\alpha\gamma}^a$, $(\alpha = 1, \ldots, n;\ \gamma = 1, \ldots, 2n)$ satisfy conditions analogous to those specified for the coefficients of the system (4.1). The systems (4.1) and (4.21) will be said to satisfy *condition* (S) if

$$(4.22)\quad \begin{aligned} &\text{(a) } \int_a^b (\mathfrak{L}[u], v)\, dt = \int_a^b (u, \mathfrak{L}^a[v])\, dt,\\ &\text{(b) } \int_a^b (\mathfrak{M}[u], v)\, dt = \int_a^b (u, \mathfrak{M}^a[v])\, dt, \end{aligned}$$

for arbitrary vector functions u, v of class $\mathfrak{C}_h{}^n[a, b]$ satisfying $\mu_\alpha[u] = 0 = \mu_\alpha^a[v]$, $(\alpha = 1, \ldots, n)$.

For a complex value λ let $\mathfrak{L}[u; \lambda] = \mathfrak{L}[u] - \lambda\mathfrak{M}[u]$, $\mathfrak{L}^a[v; \lambda] = \mathfrak{L}^a[v] - \lambda\mathfrak{M}^a[v]$. Condition (S) for systems (4.1), (4.21) then implies that

$$(4.23)\quad \int_a^b (\mathfrak{L}[u; \lambda], v)\, dt = \int_a^b (u, \mathfrak{L}^a[v; \bar{\lambda}])\, dt$$

for all complex values λ and arbitrary vector functions u, v of class $\mathfrak{C}_h{}^n[a, b]$ satisfying $\mu_\alpha[u] = 0 = \mu_\alpha[v]$, $(\alpha = 1, \ldots, n)$.

With the aid of the result of Problem III.2:6, together with a result for linear matrix differential operators corresponding to that of Theorem III.3.3 for scalar differential operators, it follows that for arbitrary complex values λ the differential system

$$(4.24)\quad \mathfrak{L}^a[v; \bar{\lambda}] = 0, \qquad \mu_\alpha^a[v] = 0, \quad (\alpha = 1, \ldots, n),$$

is the adjoint of the system

$$(4.25)\quad \mathfrak{L}[u; \lambda] = 0, \qquad \mu_\alpha[u] = 0, \quad (\alpha = 1, \ldots, n).$$

In particular, $v = v(t)$ is a solution of (4.21) for a value λ if and only if for this λ there is a solution $(z_\beta(t))$, $(\beta = 1, \ldots, n)$, of (4.6) with $z_1(t) = v(t)$. Consequently, we have the following result.

Theorem 4.3. *Suppose that* (4.1) *and* (4.21) *are differential systems whose coefficient matrices satisfy the conditions specified in the first paragraph of this section, and that these systems satisfy Condition* (*S*). *Then*

(i) *if* $u = u(t)$ *is a solution of* (4.1) *for* $\lambda = \lambda_1$, *and* $v = v(t)$ *is a solution of* (4.21) *for* $\lambda = \lambda_2$, *then in case* $\lambda_2 \neq \bar{\lambda}_1$ *we have*

$$(4.26) \qquad \int_a^b (\mathcal{M}[u], v)\, dt = 0, \qquad \int_a^b (u, \mathcal{M}^a[v])\, dt = 0;$$

(ii) *the index* $k(\lambda_0)$ *of* λ_0 *as a proper value of* (4.1) *is equal to the index* $k^a(\bar{\lambda}_0)$ *of* $\bar{\lambda}_0$ *as a proper value of* (4.21);

(iii) *for a continuous h-dimensional vector function* $r(t)$ *the differential system*

$$(4.27) \qquad \mathcal{L}[u] = \lambda_0 \mathcal{M}[u] + r(t), \qquad \mu_\alpha[u] = 0, \quad (\alpha = 1, \ldots, n),$$

has a solution if and only if $\int_a^b (g, v)\, dt = 0$ *for all solutions* $v(t)$ *of* (4.21) *for* $\lambda = \bar{\lambda}_0$;

(iv) *if* $\mathcal{G}(t, s; \lambda)$ *and* $\mathcal{G}^a(t, s, \lambda)$ *are Green's matrices of* (4.1) *and* (4.21) *for values of* λ *that are not proper values of these respective systems, then* $\mathcal{G}(t, s; \lambda) \equiv [\mathcal{G}^a(s, t; \bar{\lambda})]^*$;

(v) *the results of Theorem* 4.1 *and the corollary to Theorem* 4.2 *remain valid when in each instance* "(4.8)" *is replaced by* (4.21)".

5. SELF-ADJOINT BOUNDARY PROBLEMS

Attention will now be re-directed to boundary problems of the form

$$(5.1) \qquad \begin{aligned} &\text{(a)}\ \ L[y](t) = \lambda B(t)y(t), \qquad t \in [a, b], \\ &\text{(b)}\ \ \mu[y] \equiv M\hat{u}_y = 0, \end{aligned}$$

where, as in Section 1, $L[y](t) = A_1(t)[A_2(t)y(t)]' + A_0(t)y(t)$, and the $n \times n$ matrix functions $B(t)$, $A_j(t)$, $(j = 0, 1, 2)$, are continuous on $[a, b]$, with $B(t) \not\equiv 0$ and $A_1(t)$, $A_2(t)$ nonsingular on this interval, while M is an $n \times 2n$ matrix of rank n. Moreover, as before, $\mathfrak{D}(L)$ will denote the linear vector space of $y(t) \in \mathfrak{C}_n[a, b]$ such that $y = A_2^{-1}u_y$ with $u_y \in \mathfrak{C}_n^{\,1}[a, b]$, and $\mathfrak{D}_0(L) = \{y \mid y \in \mathfrak{D}(L),\ \hat{u}_y = 0\}$, while $D(L)$ is a linear manifold in $\mathfrak{C}_n[a, b]$ satisfying $\mathfrak{D}_0(L) \subset D(L) \subset \mathfrak{D}(L)$, and determined by the end-conditions $D(L) = \{y \mid y \in \mathfrak{D}(L),\ M\hat{u}_y = 0\}$.

For typographical simplicity, in the following discussion we shall employ the symbol $((x, y)) = ((x, y))_n$ for the integral

$$(5.2) \qquad ((x, y)) = \int_a^b (x(t), y(t))\, dt = \int_a^b y^*(t)x(t)\, dt,$$

where $x(t)$ and $y(t)$ are n-dimensional vector functions on $[a, b]$ such that the integral of (5.2) exists. Indeed, in most of the discussion the involved vector functions are continuous. In connection with this notation, and related properties of integrals, the reader is referred to Appendix F.

A boundary problem (5.1) is said to be *self-adjoint* if the following conditions hold:

$$(5.3)\qquad \begin{aligned} &\text{(a)}\ \ B(t) \textit{ is hermitian for } t \in [a, b];\\ &\text{(b)}\ \ ((L[y], z)) = ((y, L[z])), \quad \textit{for} \quad y \in D(L),\ z \in D(L). \end{aligned}$$

In particular, by an argument similar to that suggested for Problem III.2:6, it follows that condition (5.3-b) holds if and only if $\mathfrak{D}(L) = \mathfrak{D}(L^\star)$ and $L[y](t) = L^\star[y](t)$ for $y \in \mathfrak{D}(L) = \mathfrak{D}(L^\star)$, while $u_y = A_2 y$ and $v_z = A_1^* z$ satisfy the boundary condition

$$(5.2')\qquad \hat{v}_z^*[\operatorname{diag}\{-E_n, E_n\}]\hat{u}_y = 0 \quad \text{if} \quad M\hat{u}_y = 0, \qquad P^*Q\hat{v}_z = 0,$$

where $Q = \operatorname{diag}\{-E_n, E_n\}$ and P is a $2n \times n$ matrix of rank n and such that $MP = 0$.

The following results are immediate consequences of Lemmas 2.2, 2.4, and 2.5.

Lemma 5.1. *A self-adjoint boundary problem* (5.1) *possesses the following properties*:

(i) *if* $\lambda = \lambda_0$ *is a proper value of index k for* (5.1), *then* $\lambda = \bar{\lambda}_0$ *is also a proper value of index k for this problem;*

(ii) *if* $\lambda = \lambda_0$ *is not a proper value of* (5.1), *and* $G(t, s; \lambda_0)$ *is Green's matrix of this system for* $\lambda = \lambda_0$, *then* $\lambda = \bar{\lambda}_0$ *is not a proper value of* (5.1) *and* $G(t, s; \bar{\lambda}_0) = [G(s, t; \lambda_0)]^*$; *in particular, if* λ_0 *is real then* $G(t, s; \lambda_0) = [G(s, t; \lambda_0)]^*$;

(iii) *if* $y = y_1(t)$ *and* $y = y_2(t)$ *are solutions of* (5.1) *for respective values* $\lambda = \lambda_1$ *and* $\lambda = \lambda_2$, *with* $\lambda_2 \neq \bar{\lambda}_1$, *then*

$$(5.4)\qquad \begin{aligned} &\text{(a)}\ \ ((By_1, y_2)) = 0,\\ &\text{(b)}\ \ ((L[y_1], y_2)) = 0. \end{aligned}$$

It is to be remarked that the result of conclusion (i) of the above lemma is essentially the only restriction placed upon the proper values of a system (5.1) by the condition of self-adjointness. This fact is illustrated by Problem 1 below.

A boundary problem (5.1) is said to be *fully self-adjoint* if it is self-adjoint and $((By, y))$ is nonzero for each proper solution $y(t)$ of this problem. It is to be emphasized that this property may hold vacuously, as a self-adjoint boundary problem without proper values is fully self-adjoint; an example of such a boundary problem is given in Problem 2 below. The property that we have called full self-adjointness, and other closely related properties, have appeared repeatedly in the theory of boundary problems and integral equations, and various terminologies have been used in their designation. Certain important implications of this property are given in the following theorem.

Theorem 5.1. *If* (5.1) *is fully self-adjoint, then*

(a) *all proper values are real, and the set of proper values is at most denumerably infinite, with no finite limit point;*

(b) *the index of each proper value is equal to its multiplicity;*

(c) *if* λ *is a proper value of index* k, *then there exist corresponding proper solutions* $y^{(1)}(t), \ldots, y^{(k)}(t)$ *such that*

$$((By^{(\tau)}, y^{(\sigma)})) = \varepsilon(\lambda)\delta_{\sigma\tau}, \qquad (\sigma, \tau = 1, \ldots, k), \tag{5.5}$$

where either $\varepsilon(\lambda) = +1$ or $\varepsilon(\lambda) = -1$;

(d) *each proper value is a simple pole of the Green's matrix* $G(t, s; \lambda)$ *of* (5.1), *and at a proper value the residue of the Green's matrix is equal to*

$$-\varepsilon(\lambda)\sum_{\sigma=1}^{k} y^{(\sigma)}(t)[y^{(\sigma)}(s)]^*, \tag{5.6}$$

where $\varepsilon(\lambda)$ *and* $y^{(1)}(t), \ldots, y^{(k)}(t)$ *are determined as in* (c) *to satisfy* (5.5).

From conclusion (iii) of the above Lemma 5.1 it follows that all proper values of a fully self-adjoint system (5.1) are real. Now from Section IV.2 it follows that the proper values of a system (5.1) are the zeros of an entire function of λ, and the reality of all of the zeros of this entire function implies that the totality of proper values forms a set which is at most denumerably infinite, and possesses no finite limit point. Conclusion (b) for a fully self-adjoint boundary problem (5.1) is an immediate consequence of Theorem 2.1. Now if λ is a proper value of index k for a fully self-adjoint boundary problem (5.1), and $u^{(1)}(t), \ldots, u^{(k)}(t)$ are corresponding linearly independent solutions of this system, then the condition of full self-adjointness implies that for $\varepsilon(\lambda) = +1$ or $\varepsilon(\lambda) = -1$ the hermitian form $\varepsilon(\lambda)\sum_{\sigma,\tau=1}^{k} ((Bu^{(\sigma)}, u^{(\tau)}))\bar{\pi}_\tau\pi_\sigma$ is positive definite. Consequently, a set of proper solutions $y^{(1)}(t), \ldots, y^{(k)}(t)$ of the desired type is provided by the Gram-Schmidt orthonormalization process

$$y^{(1)}(t) = \frac{u^{(1)}(t)}{|((Bu^{(1)}, u^{(1)}))|^{1/2}},$$

$$y^{(\sigma)}(t) = \frac{u^{(\sigma)}(t) - \varepsilon(\lambda)\sum_{\tau=1}^{\sigma-1}((Bu^{(\sigma)}, y^{(\tau)}))y^{(\tau)}(t)}{[|((Bu^{(\sigma)}, u^{(\sigma)}))| - \sum_{\tau=1}^{\sigma-1}|((Bu^{(\sigma)}, y^{(\tau)}))|^2]^{1/2}}, \qquad (\sigma = 2, \ldots, k).$$

Finally, conclusion (d) is a direct consequence of Theorem 3.3.

Problems IV.5

1. For an arbitrary boundary problem (1.2) and its adjoint (1.4), let $w(t) = (w_\sigma(t))$, $(\sigma = 1, \ldots, 2n)$, with $w_\alpha(t) = y_\alpha(t)$, $w_{n+\alpha}(t) = z_\alpha(t)$, $(\alpha = 1, \ldots, n)$, and consider the corresponding boundary problem in $w(t)$ defined by

$$(5.7) \qquad \begin{aligned} L^\star[z](t) &= \lambda\, B^*(t)z(t), \qquad & \nu[z] &\equiv P^*[\operatorname{diag}\{-E, E\}]\hat{v}_z = 0, \\ L[y](t) &= \lambda\, B(t)y(t), & \mu[y] &\equiv M\hat{u}_y = 0. \end{aligned}$$

Show that (5.7) is of the form

$$(5.7') \qquad \mathcal{A}_1(t)[\mathcal{A}_2(t)w(t)]' + \mathcal{A}_0(t)w(t) = \lambda\, \mathcal{B}(t)w(t), \qquad \mathcal{M}\hat{u}_w = 0,$$

where the $2n \times 2n$ matrix functions $\mathcal{B}(t)$, $\mathcal{A}_j(t)$, $(j = 0, 1, 2)$, are given by $\mathcal{A}_1 = \operatorname{diag}\{A_2^*, A_1\}$, $\mathcal{A}_2 = J_0 \operatorname{diag}\{A_2, -A_1^*\}$, $\mathcal{A}_0 = J_0 \operatorname{diag}\{A_0, A_0^*\}$, $\mathcal{B} = J_0 \operatorname{diag}\{B, B^*\}$, with

$$J_0 = \begin{bmatrix} 0 & E_n \\ E_n & 0 \end{bmatrix};$$

moreover, if M_1, M_2, P_1, P_2 are $n \times n$ matrices such that $M = [M_1 \quad M_2]$, $P^* = [P_1^* \quad P_2^*]$, then

$$\mathcal{M} = \begin{bmatrix} -P_1^* & 0 & P_2^* & 0 \\ 0 & M_1 & 0 & M_2 \end{bmatrix}.$$

Prove that (5.7′) is self-adjoint. Also, show that λ is a proper value of (5.7′) if and only if λ is a proper value for either (1.2) or (1.4), and its index as a proper value of (5.7′) is equal to the sum of its indices as a proper value of (1.2) and (1.4).

2. If $f(t)$ is a real-valued scalar function on $[a, b]$, show that

$$\begin{aligned} -y_2' + \lambda f(t)y_1 &= 0, \qquad & y_1(a) - y_2(a) &= 0, \\ y_1' &= 0, & y_1(b) + y_2(b) &= 0, \end{aligned}$$

is a self-adjoint boundary value problem which has no proper value, or a single proper value, according as $\int_a^b f(t)\, dt$ is zero or different from zero.

3.[s] Suppose that $C(t)$, $F(t)$, $G(t)$, $H(t)$ are $h \times h$ matrices whose elements are continuous on $[a, b]$, with $C(t)$, $G(t)$ and $H(t)$ hermitian on this interval, and that κ^1, μ^1, κ^2, μ^2 are $2h \times h$ constant matrices such that the $2h \times 4h$ matrix $[\kappa^1 \quad -\mu^1 \quad \kappa^2 \quad \mu^2]$ is of rank $2h$, and the $2h \times 2h$ matrix $\kappa^1\mu^{1*} + \kappa^2\mu^{2*}$ is hermitian. Show that the boundary problem in the $2h$-dimensional vector function $y(t) = (y_\alpha(t))$, with $y_\sigma(t) = \eta_\sigma(t)$, $y_{h+\sigma}(t) = \zeta_\sigma(t)$, $(\sigma = 1, \ldots, h)$, defined by

$$\begin{aligned} -\zeta'(t) + C(t)\eta(t) - F^*(t)\zeta(t) &= \lambda\, H(t)\eta(t), \\ \eta'(t) - F(t)\eta(t) - G(t)\zeta(t) &= 0, \qquad t \in [a, b], \\ \kappa^1\eta(a) - \mu^1\zeta(a) + \kappa^2\eta(b) + \mu^2\zeta(b) &= 0, \end{aligned}$$

is self-adjoint.

Hint. Show that the general solution of the end-conditions of this system is given by

$$\eta(a) = \mu^{1*}\xi, \quad \zeta(a) = \kappa^{1*}\xi, \quad \eta(b) = \mu^{2*}\xi, \quad \zeta(b) = -\kappa^{2*}\xi,$$

where ξ is an arbitrary $2h$-dimensional vector.

6. DEFINITE BOUNDARY PROBLEMS

For boundary problems (5.1) there are certain linear vector spaces of vector functions that are of basic importance. For such a problem we shall denote by $\mathcal{F}_0$ the set of all n-dimensional vector functions $y(t)$ which are of class $\mathfrak{D}(L)$, and satisfy the boundary conditions $\mu[y] = 0$ of (5.1). For $j = 1, 2, \ldots$ the symbol $\mathcal{F}_j$ will denote the set of all vector functions $y(t)$ of class $\mathfrak{D}(L)$ for which there is an associated $z(t) \in \mathcal{F}_{j-1}$ such that

$$(6.1) \qquad L[y](t) = B(t)z(t), \qquad t \in [a, b], \quad \mu[y] = 0;$$

it is to be recalled that $\mathfrak{D}(L)$ is the class of $y(t) \in \mathfrak{C}_n[a, b]$ for which $y = A_2^{-1}u_y$ with $u_y \in \mathfrak{C}_n^1[a, b]$. Finally, the symbol $\mathcal{F}_\infty$ will designate the set of all vector functions common to $\mathcal{F}_j$, $(j = 0, 1, \ldots)$. Clearly, for each $j = 0, 1, \ldots, \infty$ the set $\mathcal{F}_j$ is a linear vector space of n-dimensional vector functions on $[a, b]$, $\mathcal{F}_{j+1}$ is contained in $\mathcal{F}_j$, and all proper solutions of (5.1) belong to $\mathcal{F}_\infty$. Moreover, if $z \in \mathcal{F}_\infty$, and there is a $y(t) \in \mathfrak{D}(L)$ satisfying (6.1) then also $y \in \mathcal{F}_\infty$.

A boundary problem (5.1) will be called *$\mathcal{F}_j$-definite*, $(j = 0, 1, \ldots)$, provided

1°. *the problem is self-adjoint*;

2°. *the only vector function $y(t) \in \mathcal{F}_0$ satisfying $L[y](t) = 0$, and $B(t)y(t) = 0$ for $t \in [a, b]$, is $y(t) \equiv 0$ on $[a, b]$*;

3°. *there exist real constants c_1, c_2 not both zero and such that*

$$(6.2) \qquad ((c_1 L[y] + c_2 By, y)) \geq 0,$$

for arbitrary $y \in \mathcal{F}_j$, and if the equality sign in (6.2) *holds for a $y \in \mathcal{F}_j$ then $B(t)y(t) \equiv 0$ on $[a, b]$.*

If a problem (5.1) is $\mathcal{F}_j$-definite, then clearly this problem is $\mathcal{F}_k$-definite for $k > j$. A significant property of definite problems is given in the following theorem.

Theorem 6.1. *If there is a value* $j = 0, 1, \ldots, \infty$ *such that* (5.1) *is* $\mathcal{F}_j$*-definite, then this system is fully self-adjoint.*

If $y(t)$ is a solution of (5.1) for a value λ and $((By, y)) = 0$, then $((L[y], y)) = \lambda((By, y)) = 0$, and for this vector function $y(t)$ the equality sign holds in (6.2). Since such a $y(t)$ is contained in each $\mathcal{F}_j$, it results from condition 3° that $B(t)y(t) \equiv 0$, and from (5.1) and condition 2° it follows that $y(t) \equiv 0$ on $[a, b]$. That is, if $y(t)$ is a proper solution of (5.1) then $((By, y)) \neq 0$, and hence this system is fully self-adjoint.

If $L^{(1)}[y](t) = d_1 L[y](t) + d_2 B(t)y(t)$, $B^{(1)}(t) = d_3 B(t)$, where d_1, d_2, d_3 are constants with $d_1 \neq 0$, $d_3 \neq 0$, then $y(t)$ is a solution of the boundary problem

$$(6.3) \qquad L^{(1)}[y](t) = \lambda\, B^{(1)}(t)y(t), \qquad t \in [a, b], \quad \mu[y] = 0,$$

for $\lambda = \lambda_0$ if and only if $y(t)$ is a solution of (5.1) for $\lambda = (\lambda_0 d_3 - d_2)/d_1$. Now for (6.3) the above defined classes $\mathcal{F}_j$ are identical with the corresponding classes $\mathcal{F}_j$ for (5.1), since $\mathcal{F}_0$ is clearly identical for the two problems, and $L^{(1)}[y](t) = B^{(1)}(t)\, z(t)$ if and only if $L[y] = B(t)[(d_3 z(t) - d_2 y(t))/d_1]$. Moreover, if d_1, d_2, d_3 are real-valued constants then (5.1) is $\mathcal{F}_j$-definite if and only if (6.3) is $\mathcal{F}_j$-definite. In view of these remarks it follows that if (5.1) is $\mathcal{F}_j$-definite then by suitable choice of real constants $d_1 \neq 0$, d_2, $d_3 \neq 0$ the boundary problem (6.3) is $\mathcal{F}_j$-definite, and for this problem condition 3° above holds with either $c_1 = 0$, $c_2 = 1$ or $c_1 = 1$, $c_2 = 0$. That is, for the consideration of an $\mathcal{F}_j$-definite system (5.1) it may be assumed without loss of generality that condition 3° holds either with $c_1 = 0$, $c_2 = 1$ or $c_1 = 1$, $c_2 = 0$; in these cases the system will be termed $B\mathcal{F}_j$-definite and $L\mathcal{F}_j$-definite, respectively. It is to be remarked that for $B\mathcal{F}_0$-definite problems the last statement of condition 3° is extraneous, since the condition that $((By, y))$ be non-negative for all $y \in \mathcal{F}_0$ implies that the hermitian matrix $B(t)$ is non-negative definite for $t \in [a, b]$, and consequently if $y \in \mathcal{F}_0$ and $((By, y)) = 0$ then $B(t)y(t) \equiv 0$ on $[a, b]$.

It is to be noted also that if (5.1) is $L\mathcal{F}_j$-definite than $\lambda = 0$ is not a proper value for this problem. Moreover, in view of the above comments on the equivalence of the systems (5.1) and (6.3), for a $B\mathcal{F}_j$-definite system it may be assumed without loss of generality that $\lambda = 0$ is not a proper value, since this condition holds for a system (6.3) with $d_1 = 1 = d_3$, and d_2 a real constant such that $\lambda = -d_2$ is not a proper value of the given system.

The following elementary result is presented here as a lemma, in order to facilitate future references.

***Lemma* 6.1.** *Suppose that* (5.1) *is* $\mathcal{F}_j$*-definite, and that* $y = u(t)$ *is a proper solution of this system for a proper value* λ*. If* $y(t)$ *is a vector of* $\mathcal{F}_1$

satisfying (6.1) *with a vector function* $z(t)$ *of* $\mathfrak{F}_0$, *then*

$$((Bu, z)) = \lambda((Bu, y)); \tag{6.4}$$

in particular, if $\lambda \neq 0$ *then* $((Bu, z)) = 0$ *if and only if* $((Bu, y)) = 0$.

The result of this lemma follows from the relations

$$((Bu, z)) = ((u, Bz)) = ((u, L[y])) = ((L[u], y)) = \lambda((Bu, y)).$$

In particular, for a $B\mathfrak{F}_j$-definite system the condition that $((By, y)) \geq 0$ for all $y \in \mathfrak{F}_j$ implies the Cauchy-Bunyakovsky-Schwarz inequality

$$|((By_1, y_2))|^2 \leq ((By_1, y_1))((By_2, y_2)), \quad \text{for} \quad y_1 \in \mathfrak{F}_j,\ y_2 \in \mathfrak{F}_j. \tag{6.5}$$

The result of the following lemma will be used frequently in the subsequent discussion.

Lemma 6.2. *If* (5.1) *is* $B\mathfrak{F}_j$*-definite, and* $y \in \mathfrak{F}_{j+1}$, *then*:
(a) *if* $((By, y)) = 0$, *then* $((L[y], y)) = 0$;
(b) *if* $((By, y)) > 0$ *and* $w(t)$ *is a vector function of* $\mathfrak{F}_{j+2}$ *such that*

$$L[w](t) = B(t)y(t), \qquad t \in [a, b], \quad \mu[w] = 0, \tag{6.6}$$

then $((L[u], u)) \neq 0$ *and* $((Bu, u)) > 0$ *for* u *equal to at least one of the vector functions* y, w *and* $y + w$ *of* $\mathfrak{F}_{j+1}$.

Conclusion (a) is an immediate consequence of (6.5) and the fact that if $z(t)$ is a vector function of $\mathfrak{F}_j$ which satisfies with $y(t)$ the system (6.1), then $|((L[y], y))|^2 = |((Bz, y))|^2 \leq ((Bz, z))((By, y)) = 0$. If $((By, y)) > 0$, and $w(t)$ is a vector function of $\mathfrak{F}_{j+2}$ satisfying (6.6), then

$$((L[w], y)) = ((By, y)) = ((y, By)) = ((y, L[w])) = ((L[y], w))$$

and

$$((L[y + w], y + w)) = ((L[y], y)) + 2((By, y)) + ((L[w], w)),$$

so that $((L[u], u)) \neq 0$ for u equal to at least one of the vector functions y, w, $y + w$ of $\mathfrak{F}_{j+1}$. For such u the positiveness of $((Bu, u))$ is a consequence of conclusion (a).

As noted above, for a $B\mathfrak{F}_j$-definite problem it may be assumed without loss of generality that $\lambda = 0$ is not a proper value. This property will be used in the following considerations.

Lemma 6.3. *Suppose that* (5.1) *is* $B\mathfrak{F}_j$*-definite, and* $\lambda = 0$ *is not a proper value for this problem. If* $y^{(0)}(t) \in \mathfrak{F}_j$, *and* $y^{(p)}(t)$, $(p = 1, 2, \ldots)$, *are vector functions in* $\mathfrak{F}_{j+p}$ *defined by*

$$L[y^{(p)}](t) = B(t)y^{(p-1)}(t), \qquad t \in [a, b], \qquad \mu[y^{(p)}] = 0, \tag{6.7}$$

then the (Schwarz) constants

(6.8) $$W_p = ((By^{(0)}, y^{(p)})), \qquad (p = 0, 1, \ldots),$$

are real, and possess the following properties:

(6.9) $$W_{q+p} = ((By^{(q)}, y^{(p)})), \qquad (p, q = 0, 1, \ldots);$$

(6.10) $$W_{2p-1}^2 \le W_{2p}W_{2p-2}, \qquad W_{2p}^2 \le W_{2p-2}W_{2p+2}, \qquad (p = 1, 2, \ldots);$$

(6.11) *if* $W_2 \neq 0$, *then* $W_{2p} > 0$ *and* $W_{2p} \ge W_0\left(\dfrac{W_2}{W_0}\right)^p$, $(p = 0, 1, \ldots);$

(6.12) *if* $W_1 \neq 0$, *then* $W_{2p} > 0, (p = 0, 1, \ldots)$, *and* $\dfrac{W_0}{W_2} \le \dfrac{W_0{}^2}{W_1{}^2}$.

Conclusion (6.9) follows from the definition (6.8) and the relations

$$\begin{aligned}((By^{(q)}, y^{(p)})) &= ((y^{(q)}, By^{(p)})) = ((y^{(q)}, L[y^{(p+1)}]))\\ &= ((L[y^{(q)}], y^{(p+1)})) = ((By^{(q-1)}, y^{(p+1)})),\end{aligned}$$

for $q = 1, 2, \ldots$ and $p = 0, 1, \ldots$. In particular, from (6.9) we have the relations

(6.13) $$\begin{aligned}&\text{(a)}\ W_{2p} = ((By^{(p)}, y^{(p)})),\\ &\text{(b)}\ W_{2p-1} = ((By^{(p-1)}, y^{(p)})) = ((L[y^{(p)}], y^{(p)})),\end{aligned}$$

which imply that these constants are all real. The first inequality of (6.10) is obtained by applying the Schwarz inequality (6.5) to the first expression for W_{2p-1} in (6.13-b). Similarly, the second inequality of (6.10) follows from (6.5) and the relations $W_{2p} = ((By^{(p-1)}, y^{(p+1)}))$. If $W_2 \neq 0$, then the condition of $B\mathfrak{F}_j$-definiteness implies that $W_2 > 0$, and by induction it follows from the second inequality of (6.10) that $W_{2p} > 0$, $p = 0, 1, \ldots$, and $W_{2p}/W_{2p-2} \le W_{2p+2}/W_{2p}$ for $p = 1, 2, \ldots$; in particular, this last inequality implies that $W_{2p} \ge W_0(W_2/W_0)^p$. Finally, if $W_1 \neq 0$ then the first inequality of (6.10) implies that W_2 is nonzero, and hence $W_{2p} > 0$, $(p = 0, 1, \ldots)$, by (6.11); in view of the positiveness of W_0 and W_2, the last inequality of (6.12) is equivalent to the first inequality of (6.10) for $p = 1$.

Theorem 6.2. *Suppose that* (5.1) *is* $B\mathfrak{F}_j$*-definite, and* $\lambda = 0$ *is not a proper value of this problem. Then either* $((By, y)) = 0$ *for all vector functions* $y(t) \in \mathfrak{F}_{j+1}$ *and* (5.1) *has no proper values, or there is a proper value* λ_1 *of this problem such that for arbitrary* $y(t) \in \mathfrak{F}_{j+1}$ *and associated* $z(t) \in \mathfrak{F}_j$ *satisfying* (6.1) *we have*

(6.14) $$|((L[y], y))| \le |\lambda_1|^{-1}((Bz, z)).$$

If (5.1) *has proper values* $\lambda_1, \ldots, \lambda_r$ *and* $\Gamma\{\lambda_1, \ldots, \lambda_r\}$ *denotes the set of all* $y(t) \in \mathfrak{F}_{j+1}$ *satisfying* $((By, u)) = 0$ *for all proper vector solutions* $u(t)$ *of* (5.1) *corresponding to a proper value of the set* $\lambda_1, \ldots, \lambda_r$, *then either* $((By, y)) = 0$ *for all* $y(t) \in \Gamma\{\lambda_1, \ldots, \lambda_r\}$ *and* (5.1) *has no other proper value, or there is a proper value* λ_{r+1} *distinct from* $\lambda_1, \ldots, \lambda_r$ *and such that for arbitrary* $y(t) \in \Gamma\{\lambda_1, \ldots, \lambda_r\}$ *and associated* $z(t) \in \mathcal{F}_j$ *satisfying* (6.1) *we have*

$$|((L[y], y))| \leq |\lambda_{r+1}|^{-1}((Bz, z)). \tag{6.15}$$

If $((By, y)) = 0$ for all $y(t) \in \mathcal{F}_{j+1}$ then, in view of Theorem 6.1, it follows that the boundary problem (5.1) has no proper value. On the other hand, if $((By, y))$ does not vanish for all $y(t) \in \mathcal{F}_{j+1}$ then by Lemma 6.2 there is a $y(t) \in \mathcal{F}_{j+1}$ for which $((L[y], y)) \neq 0$ and $((By, y)) > 0$. For such a vector function $y(t)$, and $z(t)$ an associated vector function of $\mathcal{F}_j$ satisfying (6.1), let $y^{(0)}(t) = z(t)$ and define the $y^{(p)}(t)$, $(p = 1, 2, \ldots)$, recursively by (6.7); in particular, $y^{(1)}(t) = y(t)$. Now from the Corollary to Theorem 2.2 it follows that

$$y(t; \lambda) = y^{(1)}(t) + \lambda\, y^{(2)}(t) + \cdots + \lambda^{p-1}\, y^{(p)}(t) + \cdots \tag{6.16}$$

is the Maclaurin series expansion for the solution of the differential system

$$L[y](t) = \lambda\, B(t)y(t) + B(t)y^{(0)}(t), \qquad t \in [a, b], \qquad \mu[y] = 0. \tag{6.17}$$

Moreover, if ρ is a positive constant such that (5.1) has no proper value λ satisfying $|\lambda| < \rho$, then the series (6.16) converges uniformly in (t, λ) on each set of the form $t \in [a, b]$, $|\lambda| \leq \rho_1 < \rho$. In particular, if $w = w(t; \lambda) = y^{(0)}(t) + \lambda\, y(t; \lambda)$, then for $|\lambda| < \rho$ termwise integration of the series $(Bw, y^{(0)})$ may be employed to show that for such values of λ the infinite series

$$W_0 + \lambda W_1 + \cdots + \lambda^p W_p + \cdots \tag{6.18}$$

converges to the value $((Bw, y^{(0)}))$. In turn, the convergence of (6.18) implies that the power series

$$W_0 + \lambda^2 W_2 + \cdots + \lambda^{2p} W_{2p} + \cdots \tag{6.18$'$}$$

converges for all λ satisfying $|\lambda| < \rho$, where ρ is a positive constant such that (5.1) has no proper value on the circular disk $\{\lambda \mid |\lambda| < \rho\}$.

Now $y(t) = y^{(1)}(t)$ has been chosen so that $0 \neq ((L[y^{(1)}], y^{(1)})) = ((By^{(0)}, y^{(1)})) = W_1$. In view of the conclusion (6.12) of Lemma 6.3 it follows that $W_{2p} > 0$, $(p = 0, 1, \ldots)$, and the infinite series (6.18′) does not converge for $\lambda = W_0/|W_1|$, and consequently $\rho \leq W_0/|W_1|$. That is, whenever $((By, y))$ does not vanish for all $y(t) \in \mathcal{F}_{j+1}$ there exists a proper value λ_1 of (5.1) such that the inequality

$$|\lambda_1| \leq \frac{((Bz, z))}{|((L[y], y))|} \tag{6.14$'$}$$

holds for all $y(t) \in \mathcal{F}_{j+1}$ with $((L[y], y)) \neq 0$, and where $z(t)$ is an associated vector function of $\mathcal{F}_j$ with which $y(t)$ satisfies (6.1). Clearly inequality (6.14′) for the indicated class of vector functions $y(t)$ is equivalent to (6.14) for arbitrary $y(t) \in \mathcal{F}_{j+1}$ and $z(t)$ an associated vector function of $\mathcal{F}_j$ which satisfies with $y(t)$ the system (6.1).

The proof of the second part of the theorem is quite similar to that of the first part. If $((By, y)) = 0$ for all $y(t) \in \Gamma\{\lambda_1, \ldots, \lambda_r\}$, then with the aid of conclusion (iii) of Lemma 5.1 and Theorem 6.1 it follows that (5.1) has no proper value distinct from $\lambda_1, \ldots, \lambda_r$. Now if $y(t) \in \Gamma\{\lambda_1, \ldots, \lambda_r\}$, and $w(t)$ satisfies with $y(t)$ the system (6.6), then Lemma 6.1 implies that $w(t) \in \Gamma\{\lambda_1, \ldots, \lambda_r\}$. Consequently, from Lemma 6.2 it follows that if $((By, y))$ does not vanish for all $y(t) \in \Gamma\{\lambda_1, \ldots, \lambda_r\}$ then there is a vector function $y(t)$ of this set for which $((L[y], y)) \neq 0$ and $((By, y)) > 0$. For $y(t)$ such a vector function, and $z(t)$ an associated vector function of $\mathcal{F}_j$ that satisfies with $y(t)$ the system (6.1), let $y^{(0)}(t) = z(t)$, and define the $y^{(p)}(t)$, $(p = 1, 2, \ldots,)$, recursively by (6.7). In particular, $y^{(1)}(t) = y(t)$ and from Lemma 6.1 it follows that $((By^{(0)}, u)) = 0$ for all solutions $u(t)$ of (5.1) corresponding to proper values $\lambda = \lambda_j$, $(j = 1, \ldots, r)$. Then the results of Theorem 5.1 and the Corollary to Theorem 2.2 imply that the solution (6.16) of (6.17) is regular at each of the proper values $\lambda_1, \ldots, \lambda_r$. Hence for ρ a positive constant such that the disk $\{\lambda \mid |\lambda| < \rho\}$ contains no proper value of (5.1) distinct from $\lambda_1, \ldots, \lambda_r$, the series (6.16) converges uniformly in (t, λ) on each set of the form $t \in [a, b]$, $|\lambda| \leq \rho_1 < \rho$. By argument identical with that used in the proof of the first part of the theorem, one is led to the conclusion that if $((By, y))$ does not vanish for all $y(t) \in \Gamma\{\lambda_1, \ldots, \lambda_r\}$ then there is a proper value λ_{r+1} of (5.1) distinct from $\lambda_1, \ldots, \lambda_r$, and such that if $y(t) \in \Gamma\{\lambda_1, \ldots, \lambda_r\}$ with $((L[y], y)) \neq 0$, and $z(t)$ is an associated vector function of $\mathcal{F}_j$ which satisfies with $y(t)$ the system (6.1), then

$$(6.15') \qquad |\lambda_{r+1}| \leq \frac{((Bz, z))}{|((L[y], y))|},$$

which is equivalent to (6.15).

If a $B\mathcal{F}_j$-definite problem (5.1) possesses proper values, we shall consider the proper values and corresponding proper solutions labelled as a simple sequence

$$(6.19) \qquad \lambda_p, \; u^{(p)}(t), \qquad (p = 1, 2, \ldots),$$

where it is understood that each proper value is repeated a number of times equal to its index of compatibility, and the corresponding proper solutions are selected to be B-orthonormal in the sense that

$$(6.20) \qquad ((Bu^{(p)}, u^{(q)})) = \delta_{pq}, \qquad (p, q = 1, 2, \ldots).$$

We also write

$$c_p[y] = ((By, u^{(p)})), \qquad (p = 1, 2, \ldots), \tag{6.21}$$

for the corresponding Fourier coefficients of an n-dimensional vector function $y(t)$ relative to the B-orthonormal set of proper solutions $\{u^{(p)}(t)\}$ of (5.1). It is to be emphasized that the sequence (6.19) may be vacuous, finite, or denumerably infinite.

Theorem 6.3. *If* (5.1) *is $B\mathcal{F}_j$-definite, and $y(t) \in \mathcal{F}_{j+1}$, then*

$$((L[y], y)) = \sum_p \lambda_p |c_p[y]|^2, \tag{6.22}$$

$$((By, y)) = \sum_p |c_p[y]|^2. \tag{6.23}$$

Clearly the results of this theorem hold for a boundary problem (5.1) if and only if they hold for an equivalent boundary problem (6.3). Consequently, under the assumption that (5.1) is $B\mathcal{F}_j$-definite it may be assumed without loss of generality that $\lambda = 0$ is not a proper value. Now if $y(t) \in \mathcal{F}_{j+1}$, and we set $w^{(q)}(t) = y(t) - \sum_{p=1}^{q} c_p[y]\, u^{(p)}(t)$, $(q = 1, 2, \ldots)$, then one may verify readily the following relations:

$$((Bw^{(q)}, u^{(p)})) = 0, \qquad (p = 1, \ldots, q); \tag{6.24}$$

$$((Bw^{(q)}, w^{(q)})) = ((By, y)) - \sum_{p=1}^{q} |c_p[y]|^2; \tag{6.25}$$

$$((L[w^{(q)}], w^{(q)})) = ((L[y], y)) - \sum_{p=1}^{q} \lambda_p |c_p[y]|^2. \tag{6.26}$$

If the sequence (6.19) contains only a finite number of proper values $\lambda_1, \ldots, \lambda_q$, then from Theorem 6.2 and Lemma 6.2 it follows that $((Bw^{(q)}, w^{(q)})) = 0$ and $((L[w^{(q)}], w^{(q)})) = 0$, so that relations (6.22) and (6.23) follow from (6.25) and (6.26), respectively. On the other hand, if the sequence (6.19) is infinite, let $w^{(q)}(t)$ be as defined above, and set

$$v^{(q)}(t) = z(t) - \sum_{p=1}^{q} c_p[z]u^{(p)}(t),$$

where $z(t)$ is an element of $\mathcal{F}_j$ satisfying (6.1) with $y(t)$. Then $v^{(q)}(t) \in \mathcal{F}_j$, $w^{(q)}(t) \in \mathcal{F}_{j+1}$, and $L[w^{(q)}](t) = B(t)v^{(q)}(t)$, $\mu[w^{(q)}] = 0$, and Theorem 6.2 implies that

$$|((L[w^{(q)}], w^{(q)}))| \leq ((Bv^{(q)}, v^{(q)})) \max_{p>q} \frac{1}{|\lambda_p|}. \tag{6.27}$$

Now an inequality corresponding to (6.25) implies that $((Bv^{(q)}, v^{(q)})) \leq ((Bz, z))$, and as $|\lambda_p| \to \infty$ as $p \to \infty$, it follows from the inequality (6.27) that $((L[w^{(q)}], w^{(q)})) \to 0$ as $q \to \infty$. Equation (6.26) then implies that the infinite series of (6.22) converges to the sum $((L[y], y))$. Indeed, since a rearrangement of the terms in the series of (6.22) is equivalent to a re-labelling of the proper values and corresponding proper solutions in (6.19), we have that a series obtained by any rearrangement of terms in the series of (6.22) converges, and hence the series of (6.22) converges absolutely.

Corresponding to (6.22) we have the sesquilinear relation

$$((L[w], y)) = \sum_p \lambda_p c_p[w]\,\overline{c_p[y]} \tag{6.28}$$

for arbitrary vector functions $y(t)$ and $w(t)$ of $\mathcal{F}_{j+1}$. Indeed, if $y(t)$ and $w(t)$ belong to $\mathcal{F}_{j+1}$, then for $k = 1, 2, 3, 4$ the vector functions $y(t) + i^k w(t)$ belong to $\mathcal{F}_{j+1}$, and

$$\begin{aligned} 4((L[w], y)) &= \sum_{k=1}^{4} i^k((L[w + i^k y], w + i^k y)), \\ 4c_p[w]\,\overline{c_p[y]} &= \sum_{k=1}^{4} i^k\,|c_p[w + i^k y]|^2, \end{aligned} \tag{6.29}$$

so that (6.28) is a ready consequence of (6.22) for the vector functions $w(t) + i^k y(t)$.

Now if $y(t) \in \mathcal{F}_{j+1}$, and $w(t)$ is a vector function satisfying (6.6) with $y(t)$, then $c_p[y] = \lambda_p\, c_p[w]$ by Lemma 6.1, and with the aid of (6.28) we obtain the result

$$((By, y)) = ((L[w], y)) = \sum_p \lambda_p\, c_p[w]\,\overline{c_p[y]} = \sum_p |c_p[y]|^2,$$

thus establishing (6.23).

In view of (6.26), the condition that (6.23) holds for vector functions $y(t) \in \mathcal{F}_{j+1}$ is equivalent to the condition that for such $y(t)$ the associated vector function $w^{(q)}(t) = y(t) - \sum_{p=1}^{q} c_p[y]\, u^{(p)}(t)$ satisfies the condition that $((Bw^{(q)}, w^{(q)})) \to 0$ as $q \to \infty$. Now if $z(t) \in \mathcal{F}_j$ it may be verified readily that

$$((Bw^{(q)}, z)) = ((By, z)) - \sum_{p=1}^{q} c_p[y]\,\overline{c_p[z]},$$

and with the aid of the Schwarz inequality it follows that $|((Bw^{(q)}, z))| \leq ((Bw^{(q)}, w^{(q)}))((Bz, z))$. Consequently, $((Bw^{(q)}, z)) \to 0$ as $q \to \infty$, thus establishing the following result.

***Corollary* 1.** *If (5.1) is $B\mathcal{F}_j$-definite, then*

$$((By, z)) = \sum_p c_p[y]\,\overline{c_p[z]}, \quad \textit{for} \quad y(t) \in \mathcal{F}_{j+1} \quad \textit{and} \quad z(t) \in \mathcal{F}_j. \tag{6.30}$$

If a $B\mathcal{F}_j$-definite boundary problem (5.1) has no proper values, it follows from Theorem 6.2 and Lemma 6.2 that $((By, y)) = 0$ and $((L[y], y)) = 0$ for arbitrary $y(t) \in \mathcal{F}_{j+1}$. Moreover, in this case the Schwarz inequality (6.5) implies that $((By, z)) = 0$ for arbitrary $y(t) \in \mathcal{F}_{j+1}$ and $z(t) \in \mathcal{F}_j$. Now for a $B\mathcal{F}_j$-definite problem (5.1) with no proper values the system (6.19) is vacuous, and the series in (6.22), (6.23) are undefined. In view of the above remarks, however, it is clear that in each of these cases the relation remains valid if the indicated series of the right-hand member is replaced by zero. Consequently we shall adopt this convention, and write the indicated relations even though the boundary problem may have no proper values. Since for a $B\mathfrak{F}_j$-definite boundary problem a vector function $y(t) \in \mathfrak{F}_j$ satisfies $((By, y)) = 0$ if and only if $B(t)y(t) \equiv 0$ on $[a, b]$, relation (6.23) implies the following result.

Corollary 2. *If* (5.1) *is $B\mathfrak{F}_j$-definite, then a vector function $y(t) \in \mathfrak{F}_{j+1}$ is such that $B(t)y(t) \equiv 0$ on $[a, b]$ if and only if $((By, u)) = 0$ for all proper solutions $u(t)$ of* (5.1).

Theorem 6.4. *Suppose that for a $B\mathfrak{F}_j$-definite problem* (5.1) *there is a real constant λ_0 such that this system has no proper value less than λ_0, and let the set* (6.19) *be ordered so that $\lambda_1 \leq \lambda_2 \leq \cdots$. If Γ_1 denotes the totality of vector functions $y(t) \in \mathfrak{F}_{j+1}$ satisfying $((By, y)) = 1$, and Γ_1 is non-vacuous, then λ_1 exists and is the minimum of $((L[y], y))$ on Γ_1; moreover, this minimum is attained by a $y(t)$ of Γ_1 if and only if $y(t) = u(t) + \eta(t)$, where $u(t)$ is a proper solution of* (5.1) *for $\lambda = \lambda_1$ and $\eta(t) \in \mathfrak{F}_{j+1}$ with $B(t)\eta(t) \equiv 0$ on $[a, b]$. In general, if proper values $\lambda_1, \ldots, \lambda_r$ exist, and Γ_{r+1} denotes the totality of vector functions $y(t) \in \mathfrak{F}_{j+1}$ satisfying $((By, y)) = 1$, $c_p[y] = ((By, u^{(p)})) = 0$, $(p = 1, \ldots, r)$, then whenever Γ_{r+1} is nonvacuous the proper value λ_{r+1} exists and is the minimum of $((L[y], y))$ on Γ_{r+1}; moreover, this minimum is attained by a $y(t)$ of Γ_{r+1} if and only if $y(t) = u(t) + \eta(t)$, where $u(t)$ is a proper solution of* (5.1) *for $\lambda = \lambda_{r+1}$ and $\eta(t) \in \mathfrak{F}_{j+1}$ with $B(t)\eta(t) \equiv 0$ on $[a, b]$.*

Whenever Γ_1 is nonvacuous the relations (6.22), (6.23) imply that λ_1 exists and $((L[y], y)) \geq \lambda_1$ on Γ_1. Furthermore, if $\lambda_1 = \lambda_2 = \cdots = \lambda_q < \lambda_{q+1}$, then $((L[y], y)) = \lambda_1$ for a $y(t) \in \Gamma_1$ if and only of $c_p[y] = 0$ for $p > q$. For such a $y(t)$ the vector function $u(t) = c_1[y]\, u^{(1)}(t) + \cdots + c_q[y]\, u^{(q)}(t)$ is a proper solution of (5.1) for $\lambda = \lambda_1$, while $\eta(t) = y(t) - u(t)$ is a vector function of $\mathfrak{F}_{j+1}$ satisfying $c_p[\eta] = 0$, $(p = 1, 2, \ldots)$, so that $B(t)\eta(t) \equiv 0$ on $[a, b]$ by Corollary 2 to Theorem 6.3. In general, if $\lambda_1, \ldots, \lambda_r$ exist, and the class Γ_{r+1} is nonvacuous, the relations (6.22), (6.23) imply that λ_{r+1} exists and $((L[y], y)) \geq \lambda_{r+1}$ on this class. Moreover, if $\lambda_{r+1} = \lambda_{r+2} = \cdots = \lambda_{r+q} < \lambda_{r+q+1}$, then $((L[y], y)) = \lambda_{r+1}$ for a $y(t) \in \Gamma_{r+1}$ if and only if $c_p[y] = 0$ for $p > r + q$. For such a $y(t)$ the vector function $u(t) = c_{r+1}[y]\, u^{(r+1)}(t) + \cdots + c_{r+q}[y]\, u^{(r+q)}(t)$ is a proper solution of (5.1) for $\lambda = \lambda_{r+1}$, while $\eta(t) = y(t) - u(t)$ is a vector function of $\mathfrak{F}_{j+1}$ satisfying $c_p[\eta] = 0$, $(p = 1, 2, \ldots)$, so that again Corollary 2 to Theorem 6.3 implies that $B(t)\eta(t) \equiv 0$ on $[a, b]$.

Boundary problems (5.1) that are $L\mathfrak{F}_j$-definite possess properties similar to those established above for $B\mathcal{F}_j$-definite problems. Moreover, although the specific results for these two types of problems are appreciably different, the methods of proof are so similar that details of argument will be omitted.

In the first place, for a $L\mathcal{F}_j$-definite problem the condition that $((L[y], y))$ is non-negative for $y(t) \in \mathcal{F}_j$ implies the Schwarz inequality

$$(6.31)\quad |((L[y], z))|^2 \leq ((L[y], y))((L[z], z)), \quad \textit{for} \quad y \in \mathcal{F}_j, \qquad z \in \mathcal{F}_j.$$

Also, as noted above, $\lambda = 0$ is not a proper value for a $L\mathcal{F}_j$-definite boundary problem (5.1).

Corresponding to Lemmas 6.2, 6.3, and Theorem 6.2, one has for $L\mathcal{F}_j$-definite problems the following results.

***Lemma* 6.4.** *If* (5.1) *is* $L\mathcal{F}_j$*-definite, then:*

(a) *if* $y(t) \in \mathcal{F}_j$ *and* $((L[y], y)) = 0$, *then* $((By, y)) = 0$;

(b) *if* $y(t) \in \mathcal{F}_{j+1}$ *with* $((L[y], y)) > 0$, *and* $z(t)$ *is a vector function of* $\mathcal{F}_j$ *satisfying* (6.1) *with* $y(t)$, *then* $((Bu, u)) \neq 0$ *and* $((L[u], u)) > 0$ *for* u *equal to at least one of the vector functions* y, z *and* $y + z$ *of* $\mathcal{F}_j$.

***Lemma* 6.5.** *If* (5.1) *is* $L\mathcal{F}_j$*-definite, and for a given* $y^{(o)}(t) \in \mathcal{F}_j$ *the vector functions* $y^{(p)}(t) \in \mathcal{F}_{j+p}$, $(p = 1, 2, \ldots)$, *are defined by* (6.7), *then the (Schwarz) constants*

$$(6.32)\qquad V_p = ((L[y^{(o)}]\, y^{(p)})), \qquad (p = 0, 1, \ldots),$$

are real, and possess the following properties:

$$(6.33)\qquad V_{q+p} = ((L[y^{(q)}], y^{(p)})), \qquad (p, q = 0, 1, \ldots);$$

$$(6.34)\qquad V_{2p-1}^2 \leq V_{2p}V_{2p-2}, \qquad V_{2p}^2 \leq V_{2p-2}V_{2p+2}, \qquad (p = 1, 2, \ldots);$$

$$(6.35)\quad \textit{if} \quad V_2 \neq 0, \quad \textit{then} \quad V_{2p} > 0 \quad \textit{and} \quad V_{2p} \geq V_0\left(\frac{V_2}{V_0}\right)^p, \qquad (p = 0, 1, \ldots);$$

$$(6.36)\quad \textit{if} \quad V_1 \neq 0, \quad \textit{then} \quad V_{2p} > 0, \quad (p = 0, 1, \ldots), \quad \textit{and} \quad \frac{V_0}{V_2} \leq \frac{V_0^2}{V_1^2}.$$

Theorem 6.5. *Suppose that the boundary problem* (5.1) *is* $L\mathcal{F}_j$*-definite. Then either* $((By, y)) = 0$ *for all* $y(t) \in \mathcal{F}_j$ *and* (5.1) *has no proper values, or there is a proper value* λ_1 *of this problem such that*

$$(6.37)\qquad |((By, y))| \leq |\lambda_1|^{-1}((L[y], y))$$

for arbitrary $y(t) \in \mathcal{F}_j$. *If* (5.1) *has proper values* $\lambda_1, \ldots, \lambda_r$, *and* $\Gamma_0\{\lambda_1, \ldots, \lambda_r\}$ *denotes the set of all* $y(t) \in \mathcal{F}_j$ *satisfying* $((By, u)) = 0$ *for all proper vector solutions* $u(t)$ *of* (5.1) *corresponding to a proper value of the set* $\Gamma_0\{\lambda_1, \ldots, \lambda_r\}$, *then either* $((By, y)) = 0$ *for all* $y(t) \in \Gamma_0\{\lambda_1, \ldots, \lambda_r\}$, *and*

(5.1) *has no other proper value, or there is a proper value* λ_{r+1} *distinct from* $\lambda_1, \ldots, \lambda_r$ *and such that*

$$|((By, y))| \le |\lambda_{r+1}|^{-1}((L[y], y))$$

for arbitrary $y(t) \in \Gamma_0\{\lambda_1, \ldots, \lambda_r\}$.

If a $L\mathcal{F}_j$-definite problem (5.1) possesses proper values we shall consider the proper values and corresponding proper solutions labelled as a simple sequence

$$\lambda_p, \; u^{(p)}(t), \qquad (p = 1, 2, \ldots), \tag{6.38}$$

where again it is understood that each proper value is repeated a number of times equal to its index of compatibility, and the corresponding proper solutions are selected as linearly independent solutions for this proper value. As $0 < ((L[y], y)) = \lambda((By, y))$ for a proper solution $y(t)$ corresponding to a proper value λ of a $L\mathcal{F}_j$-definite problem, the proper solutions of the sequence (6.38) may be selected to satisfy the conditions

$$((Bu^{(p)}, u^{(q)})) = \frac{|\lambda_p|}{\lambda_p}\,\delta_{pq}, \qquad (p, q = 1, 2, \ldots), \tag{6.39}$$

and *throughout the following discussion it will be assumed that such a choice has been made*. Also, we shall denote by $d_p[y]$ the corresponding Fourier coefficients

$$d_p[y] = \frac{|\lambda_p|}{\lambda_p}((By, u^{(p)})), \qquad (p = 1, 2, \ldots), \tag{6.40}$$

of an n-dimensional vector function $y(t)$ relative to the above described sequence $\{u^{(p)}(t)\}$ of proper solutions of (5.1). Again, it is to be emphasized that the sequence (6.38) may be vacuous, finite, or infinite.

Corresponding to Theorem 6.3 and its Corollaries, we now have the following results.

Theorem 6.6. *If* (5.1) *is* $L\mathcal{F}_j$*-definite, and* $y(t) \in \mathcal{F}_j$, *then*

$$((L[y], y)) \ge \sum_p |\lambda_p|\, |d_p[y]|^2, \tag{6.41}$$

$$((By, y)) = \sum_p (|\lambda_p|/\lambda_p)\, |d_p[y]|^2. \tag{6.42}$$

***Corollary* 1.** *If* (5.1) *is* $L\mathcal{F}_j$*-definite, then*

$$((L[y], w)) = \sum_p |\lambda_p|\, d_p[y]\, \overline{d_p[w]} \tag{6.43}$$

for arbitrary $y(t) \in \mathcal{F}_{j+1}$ *and* $w(t) \in \mathcal{F}_j$; *in particular, the equality sign holds in* (6.41) *if* $y(t) \in \mathcal{F}_{j+1}$.

***Corollary* 2.** *If (5.1) is $L\mathcal{F}_j$-definite, then a vector $y(t) \in \mathcal{F}_{j+1}$ is such that $B(t)y(t) \equiv 0$ on $[a, b]$ if and only if $((By, u)) = 0$ for all proper solutions $u(t)$ of (5.1).*

Problems IV.6

1. Show that if (5.1) is both $B\mathcal{F}_j$-definite and $L\mathcal{F}_j$-definite then all proper values of this boundary problem are positive.

2. Show that the boundary problem

$$-y_2' = \lambda y_1, \qquad y_1' = 0, \qquad t \in [a, b],$$
$$y_1(a) = 0, \qquad y_2(b) = 0,$$

is $B\mathcal{F}_0$-definite and has no proper value, while the boundary problem involving the same differential equations and the boundary conditions

$$y_1(a) - y_2(a) = 0, \qquad y_1(b) + y_2(b) = 0,$$

is $B\mathcal{F}_0$-definite and has a single proper value $\lambda = 2/(b - a)$ of index one.

3. Show that the boundary problem of Problem IV.5:3 is $B\mathcal{F}_0$-definite whenever the matrix $H(t)$ is positive definite, and the only solution of $\zeta'(t) + F^*(t)\zeta(t) = 0$ satisfying $G(t)\zeta(t) \equiv 0$ on $[a, b]$ is $\zeta(t) \equiv 0$; in particular, such a system is $B\mathcal{F}_0$-definite if $H(t)$ is positive definite and $G(t)$ is non-singular on $[a, b]$.

4. Show that the boundary problem

$$-y_2' = \lambda y_1,$$
$$y_1' - y_2 = 0,$$
$$y_1(-1) = 0, \qquad y_1(1) = 0,$$

is both $B\mathcal{F}_0$-definite and $L\mathcal{F}_1$-definite. For $y^{(0)}(t) = (y_\alpha^{(0)}(t))$, $(\alpha = 1, 2)$, with $y_1^{(0)}(t) = 1 - t^2$, $y_2^{(0)}(t) = -2t$, compute the solution $y^{(1)}(t)$ of (6.7) for $p = 1$. Use the result of the first part of Theorem 6.2 to show that the least proper value of this boundary problem is such that $0 < \lambda_1 < 42/17$. Use the first part of Theorem 6.5 to conclude that $0 < \lambda_1 < 153/62$. Compare these estimates with the true value $\lambda_1 = \pi^2/4$, and indicate why these estimates are quite close to the true value.

5. With the understanding that each proper value is counted a number of times equal to its index of compatibility, show that a $B\mathcal{F}_j$-definite problem has at least r proper values if and only if $((By, y))$ is positive definite on a linear manifold in $\mathcal{F}_{j+1}$ of dimension r; that is, there exist vector functions $\eta^{(1)}(t), \ldots, \eta^{(r)}(t)$ of $\mathcal{F}_{j+1}$ such that the hermitian form

$$\sum_{\sigma,\tau=1}^{r} ((B\eta^{(\sigma)}, \eta^{(\tau)}))\bar{\pi}_\tau \pi_\sigma$$

is positive definite. Show that for a given constant λ_0 such a boundary problem has at least r proper values greater, {less}, then λ_0 if and only if the functional $((L[y] - \lambda_0 By, y))$ is positive, {negative}, definite on a linear manifold in $\mathcal{F}_{j+1}$ of dimension r.

Hint. Use the result of Theorem 6.3.

6. If (5.1) is $B\mathcal{F}_j$-definite show that $((Bz, z)) \geq \sum_p |c_p[z]|^2$ if $z \in \mathcal{F}_j$; moreover, if $y(t) \in \mathcal{F}_{j+1}$ then $\sum_p \lambda_p^2 |c_p[y]|^2$ converges and $((Bz, z)) \geq \sum_p \lambda_p^2 |c_p[y]|^2$, where $z(t)$ is a vector function of $\mathcal{F}_j$ which satisfies with $y(t)$ condition (6.1).

7. For a $B\mathcal{F}_j$-definite problem with $\lambda = 0$ not a proper value, label the sequence (6.19) as

$$\lambda_\beta, u^{(\beta)}(t), \qquad (\beta = \pm 1, \pm 2, \ldots),$$

with $\cdots \leq \lambda_{-2} \leq \lambda_{-1} < 0 < \lambda_1 \leq \lambda_2 \leq \cdots$, it being understood that the individual sequences $\lambda_1, \lambda_2, \ldots$ and $\lambda_{-1}, \lambda_{-2}, \ldots$ may be either vacuous or finite. If $y(t) \in \mathcal{F}_{j+1}$, and $z(t)$ is a vector function of $\mathcal{F}_j$ satisfying $L[y](t) = B(t)z(t)$, $t \in [a, b]$, show that the conditions $c_\beta[y] \equiv ((By, u^{(\beta)})) = 0$ for $-s \leq \beta \leq r$ imply

$$\frac{1}{\lambda_{-s-1}}((Bz, z)) \leq ((L[y], y)) \leq \frac{1}{\lambda_{r+1}}((Bz, z)),$$

with suitable interpretation of this inequality whenever there are only a finite number of positive proper values, or only a finite number of negative proper values.

8.[s] Show that the following results hold for a boundary problem (5.1) that is $B\mathcal{F}_0$-definite, and for which the matrix function $A_2(t)$ is of class $\mathfrak{C}^1$ on $[a, b]$:

1°. $B(t) \geq 0$ for $t \in [a, b]$.

2°. If $\mathcal{F}_0^0 = \mathfrak{C}_n[a, b]$, and for $j = 1, 2, \ldots$ the class $\mathcal{F}_j^0$ is defined as the set of n-dimensional vector functions $y(t)$ for which there is a corresponding $z(t) \in \mathcal{F}_{j-1}^0$ satisfying with $y(t)$ the differential system (6.1), then the results of Lemma 6.3 and Theorems 6.2, 6.3 hold for $j = 0$ and $\mathcal{F}_j$ replaced by $\mathcal{F}_j^0$.

3°. The expansion (6.23) holds for any $y(t) \in \mathfrak{L}_n^2[a, b]$ which is such that for an arbitrary $\varepsilon > 0$ there is a $y_\varepsilon(t) \in \mathcal{F}_1^0$ satisfying the inequality $(B([y - y_\varepsilon], y - y_\varepsilon)) < \varepsilon$.

4°. If $B(t)$ is non-singular for $t \in [a, b]$, then the expansion (6.23) holds for arbitrary $y(t) \in \mathfrak{L}_n^2[a, b]$; in particular, $c_p[y] = 0$, $(p = 1, 2, \ldots)$, for such a $y(t)$ if and only if $y(t) = 0$ for t a.e. on $[a, b]$. Moreover,

$$((L[y], w)) = \sum_p \lambda_p c_p[y] \overline{c_p[w]} \tag{6.44}$$

for $w(t) \in \mathfrak{L}_n^2[a, b]$, and $y(t)$ a solution of a system (6.1) with $z(t) \in \mathfrak{L}_n^2[a, b]$.

5°. If λ is not a proper value of (5.1), then the infinite series

$$\sum_p |\lambda_p - \lambda|^{-2} |u^{(p)}(t)|^2 \tag{6.45}$$

converges for $t \in [a, b]$, and does not exceed $\int_a^b \mathrm{Tr}\{G(t, s; \lambda)B(s)G(s, t; \bar{\lambda})\}\, ds$, so that for each such λ there is a value $\kappa(\lambda)$ such that the sum of this series does not exceed $\kappa(\lambda)$.

6°. If $y(t)$ is a solution of a system (6.1) with $z(t) \in \mathfrak{L}_n^2[a, b]$, then the infinite series

$$\sum_p c_p[y]\, u^{(p)}(t) \tag{6.46}$$

converges uniformly for t on the interval $[a, b]$; moreover, if $B(t)$ is non-singular for $t \in [a, b]$, then the series (6.46) converges to $y(t)$.

Hints. 1°. If $\pi \in \mathbf{C}_n$, and $\theta(t)$ is an arbitrary real-valued scalar function of class $\mathfrak{C}^1[a, b]$ with $\theta(a) = 0 = \theta(b)$, note that $y_0(t) = \theta(t)\pi \in \mathcal{F}_0$ and hence

$$0 \le ((By_0, y_0)) = \int_a^b [\pi^* B(t)\pi]\, \theta^2(t)\, dt.$$

2°. Note that the sequence of relations presented in the text to establish (6.9) does not require $y = y^{(q-1)}(t)$ and $y = y^{(p)}(t)$ to satisfy the boundary conditions $\mu[y] = 0$.

3°. If $r_\varepsilon^{(q)}(t) = \sum_{q=1}^{p} c_p[y_\varepsilon]\, u^{(p)}(t)$, show first that the non-negative definiteness of the matrix function $B(t)$ implies the inequality

$$\begin{aligned} ((B[y - r_\varepsilon^{(q)}], y - r_\varepsilon^{(q)})) &\le 2((B[y - y_\varepsilon], y - y_\varepsilon)) \\ &\quad + 2((B[y_\varepsilon - r_\varepsilon^{(q)}], y_\varepsilon - r_\varepsilon^{(q)})). \end{aligned}$$

For $w^{(q)}(t) = y(t) - \sum_{p=1}^{q} c_p[y]u^{(p)}(t)$, note that relation (6.25) remains valid under the weaker assumption that $y(t) \in \mathfrak{L}_n^2[a, b]$, and use the Bessel inequality for the B-orthonormal set $\{u^{(p)}(t)\}$, (see Section 5 of Appendix A), to conclude that $((Bw^{(q)}, w^{(q)})) \le ((B[y - r_\varepsilon^{(q)}], y - r_\varepsilon^{(q)}))$.

4°. If $u(t) \in \mathfrak{C}_n^1[a, b]$, note that since $A_2(t)$ is of class $\mathfrak{C}^1$ and $B(t)$ is non-singular there is a continuous $v(t)$ such that $L[u](t) = B(t)v(t)$ for $t \in [a,b]$; moreover, if $u(a) = 0 = u(b)$ then $u(t) \in \mathcal{F}_1^0$. Proceed to show that if $y(t) \in \mathfrak{L}_n^2[a, b]$ then for a given $\varepsilon > 0$ there exists a $y_\varepsilon(t) \in \mathfrak{C}_n^1[a, b]$ with $y_\varepsilon(a) = 0 = y_\varepsilon(b)$ and $((B[y - y_\varepsilon], y - y_\varepsilon)) < \varepsilon$, and apply the result of 3° above. In turn, note that the validity of (6.23) for $y(t) \in \mathfrak{L}_n^2[a, b]$ implies the corresponding expansion

$$((Bz, w)) = \sum_p c_p[z]\, \overline{c_p[w]}, \quad \text{for} \quad z \in \mathfrak{L}_n^2[a, b], \qquad w \in \mathfrak{L}_n^2[a, b].$$

If $z \in \mathfrak{L}_n^2[a, b]$, and $y(t)$ is an associated solution of (6.1), then (6.44) is a consequence of the relations $((L[y], w)) = ((Bz, w))$, $c_p[z] = \lambda_p c_p[y]$.

5°. If λ is not a proper value of (5.1), and $G(t, s; \lambda)$ is the Green's matrix for the differential system

$$L[y](t) - \lambda\, B(t)y(t) = 0, \qquad \mu[y] = 0,$$

note that

$$u^{(p)}(t) = (\lambda_p - \lambda)\int_a^b G(t, s; \lambda)B(s)u^{(p)}(s)\,ds, \qquad (p = 1, 2, \ldots).$$

As $G(t, s; \lambda) = [G(s, t; \bar{\lambda})]^*$, conclude that if $\xi \in \mathbf{C}_n$ then for fixed $t \in [a, b]$ the vector function $g(s) = G(s, t; \bar{\lambda})\xi$, $s \in [a, b]$, belongs to $\mathfrak{L}_n{}^2[a, b]$ with

$$(\lambda_p - \lambda)^{-1}\, \xi^* u^{(p)}(t) = \overline{c_p[g]}, \qquad (p = 1, 2, \ldots).$$

Since $\sum_{p=1}^{q} |c_p[g]|^2 \leq ((Bg, g))$, $(q = 1, 2, \ldots)$, conclude that the series $\sum_p |\lambda_p - \lambda|^{-2}\, |\xi^* u^{(p)}(t)|^2$ converges, and that for $t \in [a, b]$ the sum of this series does not exceed

$$\xi^*\left[\int_a^b G(t, s; \lambda)B(s)G(s, t; \bar{\lambda})\,ds\right]\xi.$$

6°. Note that without loss of generality it may be assumed that $\lambda = 0$ is not a proper value of (5.1). For $z(t) \in \mathfrak{L}_n{}^2[a, b]$, and $y(t)$ the solution of the corresponding system (6.1), use the relations $\lambda_p\, c_p[y] = c_p[z]$, $(p = 1, 2, \ldots)$, and $\sum_p |c_p[z]|^2 \leq ((Bz, z))$, together with the inequality

$$\left|\sum_{p=q}^{r} c_p[y]\, u^{(p)}(t)\right|^2 \leq \left(\sum_{p=q}^{r} |c_p[y]|\, |u^{(p)}(t)|\right)^2$$
$$\leq \left(\sum_{p=q}^{r} \lambda_p{}^2\, |c_p[y]|^2\right)\left(\sum_{p=q}^{r} \lambda_p{}^{-2}\, |u^{(p)}(t)|^2\right),$$

and the result of 5°.

9. Utilize certain results of the above Problem 8 for appropriate boundary problems involving the differential system $-y_2' = \lambda y_1$, $y_1' - y_2 = 0$, to establish the following results:

(a) If $\eta(t)$ is a real-valued function of class $\mathfrak{C}^1[0, \pi]$ which possesses a second derivative $\eta''(t)$ belonging to $\mathfrak{L}^2[0, \pi]$, and $\eta(0) = 0 = \eta(\pi)$, then

$$\int_0^\pi [\eta'(t)]^2\,dt \geq \int_0^\pi \eta^2(t)\,dt,$$

and the equality sign holds if and only if there exists a constant C such that $\eta(t) = C \sin t$ for $t \in [0, \pi]$.

(b) If $\eta(t)$ is a real-valued function of class $\mathfrak{C}^1[-\pi, \pi]$ which possesses a second derivative $\eta''(t)$ belonging to $\mathfrak{L}^2[-\pi, \pi]$, and

$$\eta(-\pi) - \eta(\pi) = 0,$$
$$\eta'(-\pi) - \eta'(\pi) = 0,$$
$$\int_{-\pi}^{\pi} \eta(t)\,dt = 0,$$

then

$$\int_{-\pi}^{\pi} [\eta'(t)]^2 \, dt \geq \int_{-\pi}^{\pi} \eta^2(t) \, dt,$$

and the equality sign holds if and only if there exist constants A, B such that $\eta(t) = A \cos t + B \sin t$ for $t \in [-\pi, \pi]$.

Furthermore, with the aid of certain approximation theorems, establish the validity of each of the above functional inequalities for a larger class of functions $\eta(t)$ than that prescribed in the corresponding statement (a) or (b).

10.[s] A boundary problem of the form (4.1) which satisfies the conditions specified in the first paragraph of Section 4 is said to be self-adjoint provided $((\mathcal{M}[u], v)) = ((u, \mathcal{M}[v]))$ and $((\mathfrak{L}[u], v)) = ((u, \mathfrak{L}[v]))$ for arbitrary h-dimensional vector functions $u(t)$, $v(t)$ of class $\mathfrak{C}_h{}^n[a, b]$ satisfying the boundary conditions $\mu_\alpha[u] = 0 = \mu_\alpha[v]$, $(\alpha = 1, \ldots, n)$, defined by (4.1-b). Correspondingly, for such a problem let the symbol $\mathbf{F}_0$ denote the class of vector functions $u(t) \in \mathfrak{C}_h{}^n[a, b]$ satisfying the boundary conditions (4.1-b), and for $j = 1, 2, \ldots$ let $\mathbf{F}_j$ denote the class of vector functions $u(t) \in \mathfrak{C}_h{}^n[a, b]$ for which there is a corresponding $v(t) \in \mathbf{F}_{j-1}$ such that

$$\mathfrak{L}[u](t) = \mathcal{M}[v](t), \qquad t \in [a, b], \qquad \mu_\alpha[u] = 0, \qquad \alpha = 1, \ldots, n.$$

Such a boundary problem (4.1) will be called $\mathbf{F}_j$-definite, $(j = 0, 1, \ldots)$, provided:

1°. the problem is self-adjoint;

2°. the only vector function $u(t) \in \mathbf{F}_0$ which satisfies $\mathfrak{L}[u](t) = 0$ and $\mathcal{M}[u](t) = 0$ for $t \in [a, b]$ is $u(t) \equiv 0$ on $[a, b]$;

3°. there exist real constants c_1, c_2 not both zero and such that

$$((c_1 \mathfrak{L}[u] + c_2 \mathcal{M}[u], u)) \geq 0, \tag{6.47}$$

for arbitrary $u(t) \in \mathbf{F}_j$, and if the equality sign in (6.47) holds for a $u(t) \in \mathbf{F}_j$ then $\mathcal{M}[u](t) \equiv 0$ on $[a, b]$.

For such self-adjoint and $\mathbf{F}_j$-definite boundary problems formulate and establish results analogous to those of Sections 5 and 6 for systems (5.1).

11. If $s(t)$ is a real-valued scalar function of class $\mathfrak{C}^1[a, b]$, show that in the terminology of Problem 10 the boundary problem

$$\begin{aligned} &u^{(\text{iv})}(t) = \lambda[s(t)\, u'(t)]', \qquad && t \in [a, b], \\ &u(a) = u'(a) = u(b) = u'(b) = 0, && \end{aligned} \tag{6.48}$$

is self-adjoint, whereas the boundary problem

$$\begin{aligned} &u^{(\text{iv})}(t) = \lambda[s(t)\, u'(t)]', \qquad && t \in [a, b], \\ &u(a) = u'(a) = u''(b) = u'''(b) = 0, && \end{aligned} \tag{6.49}$$

is self-adjoint if and only if $s(b) = 0$.

12.[s] If $k(t)$, $r_j(t)$, $(j = 0, 1, \ldots, n)$ are real-valued scalar functions such that on $[a, b]$ the function $k(t)$ is continuous, $r_j(t) \in \mathfrak{C}^j[a, b]$, and $r_n(t) > 0$, prove that the functional

$$J[\eta] = \int_a^b \sum_{j=0}^n r_j(t)\{\eta^{[j]}(t)\}^2\,dt \tag{6.50}$$

is non-negative on the class $\mathfrak{D}$ of real-valued functions $\eta(t) \in \mathfrak{C}^{2n}[a, b]$ satisfying the boundary conditions

$$\eta^{[\alpha-1]}(a) = 0, \qquad \eta^{[\alpha-1]}(b) = 0, \qquad (\alpha = 1, \ldots, n), \tag{6.51}$$

if and only if all proper values of the boundary problem

$$\begin{aligned} &\text{(a)}\ l[u] \equiv \sum_{j=0}^n (-1)^j [r_j(t)\, u^{[j]}(t)]^{[j]} = \lambda\, k(t)\, u(t), \qquad t \in [a, b], \\ &\text{(b)}\ u^{[\alpha-1]}(a) = 0, \qquad u^{[\alpha-1]}(b) = 0, \qquad (\alpha = 1, \ldots, n), \end{aligned} \tag{6.52}$$

are non-negative. Moreover, when this condition holds there exists a non-identically vanishing function $\eta(t) \in \mathfrak{D}$ such that $J[\eta] = 0$ if and only if $\lambda = 0$ is a proper value of (6.52) and $u(t) = \eta(t)$ is a corresponding proper solution of this boundary problem.

Hint. Prove that (6.52) is an $\mathbf{F}_0$-definite boundary problem in the sense of Problem 10, with corresponding $c_1 = 0$, $c_2 = 1$ in the inequality (6.47), and that if $\eta(t) = \eta_1(t) + i\eta_2(t)$ with $\eta_\alpha(t) \in \mathfrak{D}$, $(\alpha = 1, 2)$, then

$$\int_a^b \overline{\eta(t)}\, l[\eta](t)\,dt = J[\eta_1] + J[\eta_2].$$

Deduce the stated conclusions from the results for the boundary problem (6.52) which correspond to the results of Theorems 6.2 and 6.4.

13. In each of the following cases determine the largest constant κ such that the stated functional inequality holds for an arbitrary real-valued scalar function $\eta(t)$ which satisfies the given end-conditions and possesses suitable differentiability conditions.

(i) $\displaystyle\int_0^1 \{\eta'(t)\}^2\,dt \geq \kappa \int_0^1 \eta^2(t)\,dt$, if $\eta(0) = 0 = \eta'(1)$;

(ii) $\displaystyle\int_1^b \{\eta'(t)\}^2\,dt \geq \kappa \int_1^b t^{-2}\eta^2(t)\,dt$, if $\eta(1) = 0 = \eta(b)$, $\quad b > 1$;

(iii) $\displaystyle\int_0^1 \{\eta''(t)\}^2\,dt \geq \kappa \int_0^1 \eta^2(t)\,dt$, if $\eta(0) = \eta'(0) = \eta(1) = \eta'(1) = 0$;

(iv) $\displaystyle\int_0^1 \{\eta''(t)\}^2\,dt \geq \kappa \int_0^1 \eta^2(t)\,dt$, if $\eta(0) = \eta''(0) = \eta(1) = \eta''(1) = 0$.

7. NOTES AND REMARKS

The treatment of this chapter is a continuation of Chapter III, and in the same general setting. The references that are most pertinent as collateral reading are those of Birkhoff [1,2], Tamarkin [1,2], Birkhoff and Langer [1], Ince [1, Ch. XI], Bliss [3,4], Collatz [1, Chs. 3, 4], Coddington and Levinson [2, Ch. 12], and Reid [9,17,29]. The influence of Bliss [3,4] is evident throughout. In particular, the use of formula (3.17), coupled with the asymptotic behavior of proper values and proper solutions, is a central feature of the proof of the expansion theorems established by Birkhoff [1,2], Tamarkin [1,2], Birkhoff and Langer [1], and Cole [1,2].

The class of self-adjoint problems considered in Sections 5 and 6 is a generalization of the definitely self-adjoint problems introduced in Bliss [3,4], and of the class of problems considered in Reid [9]. In particular, the class of problems considered here includes those treated by Krein [1], Kamke [4], and Schubert [1]. In turn, important instances of the considered classes of problems are special cases of symmetrizable completely continuous operators, as considered by Zaanen [1,2] and Reid [11]. The basic idea of the treatment in Sections 5, 6 is that employed by Bliss [3]. It is to be remarked that essentially the same idea was also used by Schmidt [1] in the treatment of linear integral equations with real symmetric kernel, and goes back to Schwarz. In particular, the "Schwarz constants" appearing in Lemmas 6.2 and 6.5 are the analogues of constants appearing in the work of Schwarz [1] on the solution of certain boundary problems involving the partial differential equation $\partial^2 u/\partial x^2 + \partial^2 u/\partial y^2 + p(x, y)u = 0$.

RELATED REFERENCES AND COMMENTS FOR SPECIFIC PROBLEMS

IV.5:3 The boundary problem occurring in this exercise includes as a most important instance the canonical form of the accessory differential system for a variational problem of Lagrange or Bolza type, (see, for example, Morse [2, Ch. II], and Bliss [5, Ch. VIII-Secs. 81, 91]). For a survey of boundary problems associated with the calculus of variations, the reader is referred to Reid [5].

IV.6:9(b) The inequality of this problem is known as Wirtinger's inequality. For various similar and related inequalities, see Coles [1,2], and Diaz and Metcalf [1].

V

Second Order Linear Differential Equations

1. INTRODUCTION

In the present chapter we shall be concerned with a linear homogeneous second-order differential equation

$$(1.1) \qquad l[u] \equiv [r(t)u' + q(t)u]' - [q(t)u' + p(t)u] = 0, \qquad t \in I,$$

where on a given interval I of the real line the coefficient functions r, q, p satisfy the hypothesis:

(H) $\quad$ *r, q, p are real-valued, continuous, and $r(t) \neq 0$ for $t \in I$.*

By a solution of (1.1) is meant a continuous function u which has a continuous derivative u', and for which there is an associated function v which is also continuously differentiable, and satisfies the equations

$$r(t)\,u'(t) + q(t)\,u(t) = v(t), \qquad v'(t) = q(t)\,u'(t) + p(t)\,u(t), \qquad t \in I.$$

That is, by definition u is a solution of (1.1) on I if and only if there is an associated function v such that (u, v) is a solution of the first order system

$$(1.2) \qquad \begin{aligned} L_1[u, v] &\equiv -v' + c(t)u - a(t)v = 0, \\ L_2[u, v] &\equiv u' - a(t)u - b(t)v = 0, \end{aligned} \qquad t \in I,$$

in the "canonical variables" u, $v = ru' + qu$, where the coefficient functions in (1.2) are given by

$$(1.3) \qquad a = -\frac{q}{c}, \qquad b = \frac{1}{r}, \qquad c = p - \frac{q^2}{r};$$

in particular, the functions a, b and c are real-valued, continuous, and $b(t) \neq 0$ for $t \in I$.

In terms of the two-dimensional vector function $y = (y_\alpha)$, $(\alpha = 1, 2)$, with $y_1 = u$, $y_2 = v$, the system (1.2) may be written as

$$\mathbf{J}y'(t) + A(t)y(t) = 0, \qquad t \in I, \tag{1.2'}$$

where $\mathbf{J}$ is the real skew matrix

$$\mathbf{J} = \begin{bmatrix} 0 & -1 \\ 1 & 0 \end{bmatrix}, \tag{1.4}$$

and $A(t)$ is the real-valued symmetric matrix function

$$A(t) = \begin{bmatrix} c(t) & -a(t) \\ -a(t) & -b(t) \end{bmatrix}. \tag{1.5}$$

If in (1.1) the coefficient functions r, q are continuously differentiable on I, then this equation may be written as

$$r(t)u'' + r'(t)u' + [q'(t) - p(t)]u = 0,$$

with continuous coefficients. Conversely, if $p_j(t)$, $(j = 0, 1, 2)$, are continuous and $p_2(t) \neq 0$ for $t \in I$, then

$$\mu(t) = \frac{1}{p_2(t)} \exp\left\{\int_{t_0}^{t} \frac{p_1(s)}{p_2(s)}\, ds\right\}$$

is such that

$$\mu[p_2 u'' + p_1 u' + p_0 u] = [ru']' - pu,$$

where

$$r(t) = \exp\left[\int_{t_0}^{t} \frac{p_1(s)}{p_2(s)}\, ds\right], \qquad p(t) = -\frac{p_0(t)}{p_2(t)}\, r(t), \quad \text{for} \quad t \in I, \tag{1.6}$$

and hence the differential equation

$$p_2(t)u'' + p_1(t)u' + p_0(t)u = 0, \qquad t \in I, \tag{1.7}$$

may be written in the form (1.1) with r, p given by (1.6), and $q = 0$.

The basic existence and uniqueness theorem for linear differential equations implies that if $t_0 \in I$, and u_0, v_0 are given real numbers, then there exists a unique solution $u = \phi(t; t_0, u_0, v_0)$, $v = \psi(t; t_0, u_0, v_0)$ of (1.2) on I satisfying the initial conditions $u(t_0) = u_0$, $v(t_0) = v_0$, and this solution is real-valued on I.

It is to be remarked that the study of a differential equation of the form (1.1) is reducible readily to the consideration of an equation of the same form with $q \equiv 0$ on I. Indeed, if g is the solution of the first order differential system

$$r(t)g' + q(t)g = 0, \qquad g(t_0) = 1, \tag{1.8}$$

that is, if $g(t) = \exp\{-\int_{t_0}^{t} [q(s)/r(s)]\, ds\}$, then for $\tilde{u} = (1/g)u$ we have that

$gl[u] = \tilde{l}[\tilde{u}]$, where

$$\tilde{l}[\tilde{u}](t) = [\tilde{r}(t)\, \tilde{u}'(t)]' - \tilde{p}(t)\, \tilde{u}(t), \tag{1.9}$$

with

$$\tilde{r} = rg^2, \qquad \tilde{p} = \left(p - \frac{q^2}{r}\right)g^2 = cg^2.$$

Correspondingly, under the transformation $\tilde{u} = (1/g)u$, $\tilde{v} = gv$, we have $gL_1[u, v] = \tilde{L}_1[\tilde{u}, \tilde{v}]$, $(1/g)L_2[u, v] = \tilde{L}_2[\tilde{u}, \tilde{v}]$, where

$$\begin{aligned} \tilde{L}_1[\tilde{u}, \tilde{v}] &= -\tilde{v}' + \tilde{c}(t)\tilde{u}, \\ \tilde{L}_2[\tilde{u}, \tilde{v}] &= \tilde{u}' - \tilde{b}(t)\tilde{v}, \end{aligned} \tag{1.10}$$

with

$$\tilde{b} = \frac{b}{g^2}, \qquad \tilde{c} = cg^2 = \tilde{p}.$$

In particular, since b is nonzero on I the function $\tilde{b}$ is also non-zero on I.

Under the trigonometric substitution

$$u(t) = \rho(t)\sin\theta(t), \qquad v(t) = \rho(t)\cos\theta(t), \tag{1.11}$$

the system (1.2) is equivalent to

$$\begin{aligned} \theta' &= a(t)\sin 2\theta + b(t)\cos^2\theta - c(t)\sin^2\theta, \\ \rho' &= \{\tfrac{1}{2}[b(t) + c(t)]\sin 2\theta - a(t)\cos 2\theta\}\rho. \end{aligned} \tag{1.12}$$

Although the first equation of (1.12) is non-linear, for t on a compact sub-interval $[t_1, t_2]$ of I, and $-\infty < \theta < \infty$, the right-hand member is bounded and satisfies a Lipschitz condition in θ. Consequently, for $t_0 \in I$ and θ_0 a given real value, there is a unique solution $\theta(t) = \theta(t; t_0, \theta_0)$ of this equation satisfying the initial condition $\theta(t_0) = \theta_0$, and in view of Theorem I.5.1 the maximal interval of existence of this solution is I. With the function $\theta(t)$ thus determined, the second equation of (1.12) has the unique solution $\rho(t) = \rho_0 \exp\{\int_{t_0}^{t} f(s)\, ds\}$, where

$$f(t) = \tfrac{1}{2}[b(t) + c(t)]\sin 2\theta(t) - a(t)\cos 2\theta(t), \quad \text{and} \quad \rho_0 = \rho(t_0).$$

The first published use of the substitution (1.11) in the derivation of certain results of the Sturmian theory for a real linear homogeneous differential equation appears to have been by Prüfer [1], and consequently this substitution is frequently termed the "Prüfer substitution", and the system (1.12) is called the "Prüfer transform of (1.2)". Other authors, (see, in particular, Whyburn [2] and statements in Reid [32]), have also used it extensively, and we shall refer to it as a "polar coordinate transformation".

Problems V.1

1. For the differential equation $u'' - u = 0$ determine solutions (θ, ρ) of the associated system (1.12) for each of the following particular solutions: (i) $u(t) = \sinh t$; (ii) $u(t) = e^t$.

Ans. (i) $\theta(t) = \arctan(\tanh t)$, $\rho(t) = \sqrt{\cosh 2t}$;
(ii) $\theta(t) \equiv \pi/4 \pmod{2\pi}$, $\rho(t) = \sqrt{2}e^t$.

2. If $[a, b] \subset I$, and (u, v) is the solution of (1.2) satisfying the initial conditions $u(a) = 0$, $v(a) \neq 0$, show that u has at most m zeros on $(a, b]$ if

$$\int_a^t [|a(s)| + |b(s) + c(s)| - c(s)]\, ds \leq m\pi \quad \textit{for} \quad t \in [a, b].$$

Hint. Use the system (1.12), noting that if $t_1 \in [a, b]$ and $u(t_1) = 0$, then $\theta'(t_1) > 0$.

3. Consider the differential equation

$$[r(t)u']' + [\lambda k(t) - p(t)]u = 0, \qquad t \in [a, b], \tag{1.13}$$

where on $[a, b]$ the functions r, k, p are real-valued continuous functions such that $r(t) > 0$, $k(t) > 0$, $r(t)$ has a continuous derivative, and $r(t)\,k(t)$ has continuous derivatives of the first two orders. Show that under the substitution

$$x = \frac{1}{K}\int_a^t \left[\frac{k(s)}{r(s)}\right]^{1/2} ds, \quad \text{where} \quad K = \frac{1}{\pi}\int_a^b \left[\frac{k(s)}{r(s)}\right]^{1/2} ds,$$

$$U = (kr)^{1/4}\, u, \quad \text{and} \quad \mu = K^2\lambda,$$

(1.13) is transformed into

$$\frac{d^2U}{dx^2} + [\mu - P(x)]U = 0, \qquad 0 \leq x \leq \pi,$$

where $P = [d^2f/dx^2]/f - K^2g$, and f, g are respectively $(kr)^{1/4}$ and p/k expressed as functions of x.

2. PRELIMINARY PROPERTIES OF SOLUTIONS OF (1.1)

Suppose that u is a nonidentically vanishing solution of (1.1), and let corresponding functions u, v and $\tilde{u}$, $\tilde{v}$ be determined as in Section 1 above. If t_1 and t_2 are zeros of u on I with $t_1 < t_2$, then $\tilde{u}(t_1) = 0 = \tilde{u}(t_2)$ and

$$0 = \tilde{u}(t_2) - \tilde{u}(t_1) = \int_{t_1}^{t_2} \tilde{b}(s)\, \tilde{v}(s)\, ds.$$

Moreover, $u(t) = u'(t_1)\, u_0(t)$, where $u_0(t)$ is the real-valued solution of (1.1) satisfying $u_0(t_1) = 0$, $u_0'(t_1) = 1$, and hence for the consideration of zeros of $u(t)$ we may without loss of generality assume that $u(t)$ is real-valued. In this instance the corresponding $\tilde{u}(t)$ is also real-valued, and as $\tilde{b}(t)$ is of constant sign the above equation implies that there exists a value $\tau \in (t_1, t_2)$ such that $\tilde{v}(\tau) = 0$, and hence also $v(\tau) = 0$. In particular, if there were an infinite sequence of distinct values $t_j \in I$ with $u(t_j) = 0$, $(j = 1, 2, \ldots)$, and the sequence $\{t_j\}$ had a finite accumulation point t_0, then it would follow that there exists an infinite sequence of values $\tau_j \in I$ such that $v(\tau_j) = 0$, $(j = 1, 2, \ldots)$, and with t_0 as an accumulation point of $\{\tau_j\}$. By continuity it would then follow that $u(t_0) = 0$, $v(t_0) = 0$, and the uniqueness theorem for (1.2) would imply that $u(t) \equiv 0$, $v(t) \equiv 0$ on I, contrary to the assumption that $u(t)$ is a non-identically vanishing solution of (1.1). Thus we have established the following result.

Theorem 2.1. *If u is a nonidentically vanishing solution of* (1.1) *then the zeros of u are isolated; moreover, between any two zeros of u on I there is a zero of the associated function $v = ru' + qu$.*

If u_1 and u_2 are solutions of (1.1), we shall denote by $\{u_1, u_2\}$ the expression

$$\{u_1, u_2\}(t) = v_2(t)\, u_1(t) - u_2(t)\, v_1(t), \tag{2.1}$$

where for $\alpha = 1, 2$ it is understood that v_α is the corresponding function such that $(u, v) = (u_\alpha, v_\alpha)$ is a solution of (1.2). It is to be noted that the abbreviated notation $\{u_1, u_2\}$ is unambiguous since for a given solution u of (1.1) the associated $v = ru' + qu$ is uniquely determined. If $y^{(\alpha)}(t)$ is the corresponding solution of (1.2′) with $y_1^{(\alpha)}(t) = u_\alpha(t)$, $y_2^{(\alpha)}(t) = v_\alpha(t)$, $(\alpha = 1, 2)$, then (2.1) may be written as

$$\{u_1, u_2\}(t) = y^{(2)*}(t)\, \mathbf{J} y^{(1)}(t). \tag{2.1′}$$

In particular, since $\mathbf{J}[y^{(\alpha)}(t)]' + A(t) y^{(\alpha)}(t) = 0$, and the matrix $A(t)$ is symmetric, it follows readily that $\{u_1, u_2\}(t)$ is constant on I for any pair of solutions $u_1(t)$, $u_2(t)$ of (1.1). Finally if u_1, u_2 are solutions of (1.1), and u_1 is not identically zero, then for $t_1 \in I$ the two-dimensional vector $(u_1(t_1), v_1(t_1))$ is non-null. Consequently, $\{u_1, u_2\}(t_1) = 0$ if and only if there exists a constant k such that $u_2(t_1) = k u_1(t_1)$, $v_2(t_1) = k v_1(t_1)$, and in turn these initial conditions hold if and only if $u_2(t) = k u_1(t)$, $v_2(t) = k v_1(t)$ for $t \in I$. For reference, these results are presented as the following theorem.

Theorem 2.2. *If u_1 and u_2 are solutions of* (1.1) *then $\{u_1, u_2\}$ is constant on I, and $\{u, u\} = 0$ for an arbitrary solution u of* (1.1). *Moreover, if u_1 and u_2 are solutions of* (1.1), *and u_1 is not identically zero on I, then $\{u_1, u_2\} = 0$ if and only if there is a constant k such that $u_2 = k u_1$ on I.*

Theorem 2.3. *If u_1 is a solution of* (1.1) *such that $u_1(t) \neq 0$ for $t \in I$, then u is a solution of* (1.1) *if and only if*

$$(2.2) \qquad u(t) = u_1(t)\, h(t), \quad \text{with} \quad h(t) = k_0 + k_1 \int_{t_1}^{t} \frac{b(s)}{u_1^2(s)}\, ds, \\ = k_0 + k_1 \int_{t_1}^{t} \frac{ds}{r(s)\, u_1^2(s)}$$

where $t_1 \in I$ and k_0, k_1 are constants; in particular, the corresponding function v is given by

$$(2.3) \qquad v(t) = v_1(t)\, h(t) + \frac{k_1}{u_1(t)},$$

and

$$k_0 = \frac{u(t_1)}{u_1(t_1)}, \qquad k_1 = \{u_1, u\}.$$

If u_1 is a solution of (1.1) with $u_1(t) \neq 0$ for $t \in I$, then upon direct substitution it follows that

$$(2.4) \qquad u(t) = u_1(t)\, h(t), \qquad v(t) = v_1(t)\, h(t) + w(t),$$

is a solution of (1.2) if and only if $h(t)$ and $w(t)$ are such that $u_1 h' = bw$, $v_1 h' + w' = -aw$. These two equations are equivalent to the equations $u_1 w' = -(au_1 + bv_1)w$, $h' = bw/u_1$, and, therefore, $(u_1 w)' = 0$. Consequently, there is a constant k_1 such that $w(t) = k_1/u_1(t)$ for $t \in I$, and hence $h' = k_1 b/u_1^2$, so that h is of the form given in (2.2) and v is given by (2.3). With u, v thus expressed it follows immediately that $k_0 = u(t_1)/u_1(t_1)$ and $k_1 = \{u_1, u\}$.

Theorem 2.4. *If u_1 and u_2 are linearly independent real-valued solutions of* (1.1) *on I, then the zeros of these functions separate each other; that is, between successive zeros of u_α there is exactly one zero of u_β, ($\beta \neq \alpha$; $\alpha, \beta = 1, 2$).*

Suppose that t_1 and t_2, $(t_1 < t_2)$, are zeros of u_2 on I. The condition that u_1 and u_2 are linearly independent solutions of (1.1) implies that $u_1(t_1) \neq 0$ and $u_1(t_2) \neq 0$. Now if $u_1(t)$ has no zero between t_1 and t_2, then $u_1(t) \neq 0$ for $t \in [t_1, t_2]$, and in this case it follows from Theorem 2.3 that there exist constants k_0, k_1 such that

$$u_2(t) = u_1(t)\left[k_0 + k_1 \int_{t_1}^{t} \frac{b(s)}{u_1^2(s)}\, ds\right] \quad \text{for} \quad t \in I.$$

From the conditions $u_2(t_1) = 0 = u_2(t_2)$ it follows that $k_0 = 0 = k_1$, and hence $u_2(t) \equiv 0$ on I, contrary to the assumption that u_1 and u_2 are linearly independent on this interval. Thus between any two zeros of u_2 there is at least one zero of u_1, and interchanging the roles of u_1 and u_2 in the above discussion, between any two zeros of u_1 there is at least one zero of u_2. Consequently, for $\alpha, \beta = 1, 2$ and $\alpha \neq \beta$, it follows that between successive zeros of u_α there is exactly one zero of u_β.

3. AN ASSOCIATED FUNCTIONAL

If I_0 is a subinterval of I let $D' = D'(I_0)$ denote the class of real-valued functions η which are continuous and have a piecewise continuous derivative on I_0. If $\eta \in D'(I_0)$ then $\zeta = r\eta' + q\eta$ is piecewise continuous on this interval, and to indicate its association with η we write $\eta \in D'(I_0):\zeta$. The subclass of functions η in $D'(I_0)$ such that η is continuously differentiable and there is a function $\zeta \in D'(I_0)$ satisfying $\zeta = r\eta' + q\eta$ will be denoted by $D''(I_0)$, and the notation $\eta \in D''(I_0):\zeta$ is used to indicate this association. If I_0 is a compact subinterval $[a, b]$ of I, the abbreviated notations $D'[a, b]$, $D''[a, b]$ are used for the respective precise symbols $D'([a, b])$, $D''([a, b])$. The subclass of functions η of classes $D'[a, b]$ and $D''[a, b]$ for which $\eta(a) = 0 = \eta(b)$ will be designated by $D_0'[a, b]$ and $D_0''[a, b]$, respectively.

Now for $[a, b] \subset I$ and $\eta_\alpha \in D'[a, b]:\zeta_\alpha$, $(\alpha = 1. 2)$, let

$$(3.1') \quad J[\eta_1, \eta_2; a, b] = \int_a^b \{\eta_2'[r(t)\eta_1' + q(t)\eta_1] + \eta_2[q(t)\eta_1' + p(t)\eta_1]\}\, dt,$$

and, for brevity, set $J[\eta_1; a, b] = J[\eta_1, \eta_1; a, b]$. It follows readily that also

$$(3.1'') \qquad J[\eta_1, \eta_2; a, b] = \int_a^b \{\zeta_2 b(t)\zeta_1 + \eta_2\, c(t)\eta_1\}\, dt,$$

where the functions $b(t)$, $c(t)$ are defined by (1.3).

***Lemma* 3.1.** *If $[a, b] \subset I$, then:*

(i) *$J[\eta_1, \eta_2; a, b]$ is a real-valued symmetric functional on $D'[a, b] \times D'[a, b]$; that is, if $\eta_\alpha \in D'[a, b]$, $(\alpha = 1, 2, 3)$, then*

$$(3.2) \qquad \begin{aligned} &(a)\ J[\eta_1, \eta_2; a, b] = J[\eta_2, \eta_1; a, b], \\ &(b)\ J[c\eta_1, \eta_2; a, b] = cJ[\eta_1, \eta_2; a, b] \textit{ for } c \textit{ a real constant}, \\ &(c)\ J[\eta_1 + \eta_2, \eta_3; a, b] = J[\eta_1, \eta_3; a, b] + J[\eta_2, \eta_3; a, b]. \end{aligned}$$

(ii) *If $\eta_1 \in D''[a, b]:\zeta_1$, and $\eta_2 \in D'[a, b]:\zeta_2$, then*

$$(3.3') \qquad \begin{aligned} J[\eta_1, \eta_2; a, b] &= \eta_2\,\zeta_1\Big|_a^b - \int_a^b \eta_2\, l[\eta_1]\, dt, \\ &= \eta_2\zeta_1\Big|_a^b + \int_a^b \eta_2\, L_1[\eta_1, \zeta_1]\, dt, \end{aligned}$$

$$(3.3'') \qquad J[\eta_1; a, b] = \eta_1\zeta_1\Big|_a^b + \int_a^b \eta_1\, L_1[\eta_1, \zeta_1]\, dt.$$

Moreover, if $\eta_\alpha \in D''[a, b]:\zeta_\alpha$, $(\alpha = 1, 2)$, *then*

$$\text{(3.4)} \qquad \int_a^b \eta_1\, l[\eta_2]\, dt - \int_a^b \eta_2\, l[\eta_1]\, dt = (\zeta_2\eta_1 - \eta_2\zeta_1)\Big|_a^b = \{\eta_1, \eta_2\}\Big|_a^b.$$

Conclusion (i) is obvious from the definition (3.1′) of the functional $J[\eta_1, \eta_2; a, b]$, and (3.3′) is an immediate result of an integration by parts; relation (3.3″) is (3.3′) with $\eta_2 = \eta_1$. Finally, (3.4) is a direct consequence of (3.2-a) and (3.3′).

For $[a, b] \subset I$, let $\hat{D}'[a, b]$ be a linear manifold in $D'[a, b]$; that is, $\hat{D}'[a, b] \subset D'[a, b]$, and if $\eta_\alpha \in \hat{D}'[a, b]$, $(\alpha = 1, 2)$, then $c_1\eta_1 + c_2\eta_2 \in \hat{D}'[a, b]$ for arbitrary real constants c_1, c_2. The functional $J[\eta; a, b]$ is said to be *non-negative definite on* $\hat{D}'[a, b]$ if $J[\eta; a, b] \geq 0$ for all $\eta \in \hat{D}'[a, b]$. The functional $J[\eta; a, b]$ is said to be *positive definite* on $\hat{D}'[a, b]$ if it is non-negative definite on this manifold, and $J[\eta; a, b] = 0$ for an $\eta \in \hat{D}'[a, b]$ only if $\eta(t) \equiv 0$ on $[a, b]$.

Theorem 3.1. *If* $[a, b] \subset I$ *then u is a solution of* (1.1) *on* $[a, b]$ *if and only if* $u \in D'[a, b]$ *and*

$$\text{(3.5)} \qquad J[u, \eta; a, b] = 0 \quad \textit{for} \quad \eta \in D_0'[a, b].$$

A function u is a solution of (1.1) *on I if and only if for arbitrary* $[a, b] \subset I$ *we have that* $u \in D'[a, b]$ *and* (3.5) *holds.*

If u is a solution of (1.1) on a subinterval $[a, b]$ of I, then $u \in D'[a, b]$, and (3.3′) with $\eta_1 = u$, $\eta_2 = \eta$ implies (3.5). Conversely, if $u \in D'[a, b]$ and (3.5) holds, then upon suitable integration by parts one obtains that the function

$$g(t) = r(t)\, u'(t) + q(t)\, u(t) - \int_a^t [q(s)\, u'(s) + p(s)\, u(s)]\, ds$$

is such that $\int_a^b \eta'(t)\, g(t)\, dt = 0$ for $\eta \in D_0'[a, b]$. From Problem III.2:1 it then follows that there is a constant c such that $g(t) = c$ for $t \in [a, b]$; consequently $u(t)$, $v(t) = c + \int_a^t [q(s)\, u'(s) + p(s)\, u(s)]\, ds$ is a solution of (1.2) on $[a, b]$, and u is a solution of (1.1) on this interval. The last statement of the theorem follows from the fact that u is a solution of (1.1) on I if and only if it is a solution of this equation on arbitrary subintervals $[a, b]$ of I.

***Corollary* 1.** *If* $[a, b] \subset I$ *and* $J[\eta; a, b]$ *is non-negative definite on* $D_0'[a, b]$, *and u is an element of* $D_0'[a, b]$ *such that* $J[u; a, b] = 0$, *then u is a solution of* (1.1).

For $u \in D_0'[a, b]$, $\eta \in D_0'[a, b]$, and σ a real constant, the function $\eta_\sigma = u + \sigma\eta$ belongs to $D_0'[a, b]$, and hence

$$\begin{aligned} 0 \leq J[\eta_\sigma; a, b] &= J[u; a, b] + 2\sigma\, J[u, \eta; a, b] + \sigma^2\, J[\eta; a, b], \\ &= 2\sigma\, J[u, \eta; a, b] + \sigma^2\, J[\eta; a, b]. \end{aligned}$$

Consequently, $J[u, \eta; a, b] = 0$ for $\eta \in D_0'[a, b]$, and from Theorem 3.1 it follows that u is a solution of (1.1).

If u is a solution of (1.1) on a subinterval $[a, b]$ of I, and $y \in D'[a, b]$, then

$$J[y; a, b] = J[u; a, b] + 2J[u, y - u; a, b] + J[y - u; a, b].$$

In particular, if $y(a) = u(a)$ and $y(b) = u(b)$, then $y - u \in D_0'[a, b]$, so that $J[u, y - u; a, b] = 0$ by Theorem 3.1, and we have the following result.

***Corollary* 2.** *Suppose that $[a, b] \subset I$, and $J[\eta; a, b]$ is non-negative definite on $D_0'[a, b]$. If u is a solution of* (1.1) *on I, and y is a function of $D'[a, b]$ such that $y(a) = u(a)$, $y(b) = u(b)$, then $J[y; a, b] \geq J[u; a, b]$; moreover, if the functional $J[\eta; a, b]$ is positive definite on $D_0'[a, b]$, then $J[y; a, b] \geq J[u; a, b]$ with the equality sign holding only if $y(t) = u(t)$ for $t \in [a, b]$.*

Theorem 3.2. (i) *If $[a, b] \subset I$, $u \in D''[a, b]:v$, and $h \in D'[a, b]$, then $\eta = uh$ is such that $\eta \in D'[a, b]$ and*

$$(3.6) \quad \eta'[r\eta' + q\eta] + \eta[q\eta' + p\eta] = r[uh']^2 + [\eta vh]' + h^2 uL_1[u, v];$$

(ii) *if u, v is a solution of* (1.2) *on a subinterval $[a, b]$ of I and $h \in D'[a, b]$, then $\eta = uh$ is such that $\eta \in D'[a, b]$ and*

$$(3.7) \qquad J[\eta; a, b] = \eta vh\Big|_a^b + \int_a^b r(t)[\eta'(t) - u'(t)h(t)]^2\, dt;$$

in particular, if $u(t) \neq 0$ for $t \in [a, b]$, then (3.7) *holds with $h = \eta/u$.*

The identity (3.6) may be verified directly. If u, v is a solution of (1.2), then (3.7) is a ready consequence of (3.6) in view of the relations $L_1[u, v] = 0$ and $\eta' = u'h + uh'$.

Theorem 3.3. *If $[a, b] \subset I$ and $J[\eta; a, b]$ is non-negative definite on $D_0'[a, b]$, then*:

(i) *$r(t) \geq 0$ for $t \in [a, b]$;*

(ii) *if $u_a(t)$, $v_a(t)$ is a solution of* (1.2) *satisfying $u_a(a) = 0$, $v_a(a) \neq 0$, then $u_a(t) \neq 0$ for $t \in (a, b)$;*

(iii) *if $u_b(t)$, $v_b(t)$ is a solution of* (1.2) *satisfying $u_b(b) = 0$, $v_b(b) \neq 0$, then $u_b(t) \neq 0$ or $t \in (a, b)$;*

(iv) *if the $u_a(t)$, $v_a(t)$ of* (ii) *is such that $u_a(t) \neq 0$ for $t \in (a, b]$, then there exists a real-valued solution u of* (1.1) *with $u(t) \neq 0$ for $t \in [a, b]$.*

For $s \in (a, b)$, and $0 < \varepsilon < \min\{s - a, b - s\}$, let

$$\eta_\varepsilon(t) = (\varepsilon - |t - s|)/\sqrt{\varepsilon} \quad \text{for} \quad t \in [s - \varepsilon, s + \varepsilon],$$

and ${}_\varepsilon'\eta(t) = 0$ for $t \in [a, s - \varepsilon]$ and $t \in [s + \varepsilon, b]$. As $0 \leq \eta_\varepsilon(t) \leq \sqrt{\varepsilon}$, $|\eta_\varepsilon'(t)| \leq 1/\sqrt{\varepsilon}$, while q and p are bounded on $[a, b]$, there exists a constant k

such that

$$J[\eta_\varepsilon; a, b] = \frac{1}{\varepsilon}\int_{s-\varepsilon}^{s+\varepsilon} r(t)\,dt + \rho(\varepsilon), \quad and \quad |\rho(\varepsilon)| \leq k\varepsilon.$$

Since the function $r(t)$ is continuous it follows that $J[\eta_\varepsilon; a, b]$ has limit $2r(s)$ as $\varepsilon \to 0$, and as $J[\eta_\varepsilon; a, b] \geq 0$ we have that $r(s) \geq 0$ at each point s on the open interval (a, b). Finally, by continuity it follows that r is also non-negative at the endpoints $t = a$ and $t = b$.

In order to establish conclusion (ii), suppose that there exists a $c \in (a, b)$ such that $u_a(c) = 0$. Then u defined as $u(t) = u_a(t)$ for $t \in [a, c]$, $u(t) = 0$ for $t \in (c, b]$, is such that $u \in D_0'[a, b]$, with

$$J[u; a, b] = J[u_a; a, c] = u_a v_a\Big|_a^c = 0,$$

and from Corollary 1 to Theorem 3.1 it follows that u is a solution of (1.1) on $[a, b]$. If v is the function such that u, v is a solution of (1.2) on $[a, b]$, then $v(t) = 0$ for $t \in (c, b]$, and hence also $v(c) = 0$ by continuity. As $u(c) = 0$, $v(c) = 0$, it then follows that $u(t) \equiv 0$ on $[a, b]$, whereas $u(t) = u_a(t)$ for $t \in [a, c]$ and $u_a(a) = 0$, $v_a(a) \neq 0$ leads to the contradictory result that $u(t) \not\equiv 0$ on $[a, c]$. In a similar fashion conclusion (iii) may be established.

If $t_1 \in I$, and u_0, v_0 is the solution of (1.2) satisfying the initial conditions $u_0(t_1) = 0$, $v_0(t_1) = 1$, then any solution u, v of (1.2) for which $u(t_1) = 0$ is of the form $u(t) = ku_0(t)$, $v(t) = kv_0(t)$, with $k = v(t_1)$. Consequently, in conclusions (ii) and (iii) there is no loss of generality in restricting u_a, v_a and u_b, v_b to be real-valued solutions of (1.2), and we shall make this restriction for the following argument establishing conclusion (iv). If $u_a(b) \neq 0$, then the constant function $\{u_a, u_b\} = v_b u_a - u_b v_a$ is equal to the nonzero value $v_b(b)\, u_a(b)$, and equal to $-u_b(a)\, v_b(a)$ also, so that $u_b(a) \neq 0$. Moreover, by a suitable choice of $v_a(a)$ or $v_b(b)$, one may attain the normalization

$$\{u_a, u_b\} = -1. \tag{3.8}$$

With this choice, set $u = u_a + u_b$, $v = v_a + v_b$. Then u is a real-valued solution of (1.1) on $[a, b]$, and it will be established that $u(t) \neq 0$ for $t \in [a, b]$. It is to be noted that $u(a) = u_b(a)$ and $u(b) = u_a(b)$ are nonzero. Now if $c \in (a, b)$ and $u(c) = 0$, define $\eta(t) = u_a(t)$ for $t \in [a, c]$, $\eta(t) = -u_b(t)$ for $t \in (c, b]$. Then $\eta \in D_0'[a, b]$, and (3.3′) for the individual subintervals $[a, c]$ and $[c, b]$, with $(\eta_1, \zeta_1) = (u_a, v_a)$ and $(\eta_1, \zeta_1) = (u_b, v_b)$, respectively, yields the contradictory relation

$$\begin{aligned} 0 \leq J[\eta; a, b] &= J[u_a; a, c] + J[u_b; c, b], \\ &= u_a(c)v_a(c) - v_b(c)u_b(c), \\ &= -u_b(c)v_a(c) + v_b(c)u_a(c), \\ &= \{u_a, u_b\}(c) = -1. \end{aligned}$$

Thus $u(t) \neq 0$ for $t \in (a, b)$ also, and hence $u(t) \neq 0$ for $t \in [a, b]$.

A partial converse of Theorem 3.3 is given by the following theorem.

Theorem 3.4. *If $[a, b] \subset I$, $r(t) > 0$ for $t \in [a, b]$, and there exists a real-valued solution u of* (1.1) *such that $u(t) \neq 0$ on $[a, b]$, then $J[\eta; a, b]$ is positive definite on $D_0'[a, b]$.*

If $\eta \in D_0'[a, b]$, and $h(t) = \eta(t)/u(t)$, then from Theorem 3.2 it follows that

$$J[\eta; a, b] = \int_a^b r(t)[\eta'(t) - u'(t)h(t)]^2\,dt. \tag{3.9}$$

As $r(t) > 0$ for $t \in [a, b]$, by hypothesis, it follows that $J[\eta; a, b] \geq 0$, and the equality sign holds if and only if $0 = \eta' - u'h = uh'$ on $[a, b]$. As $u(t) \neq 0$ for $t \in [a, b]$, this condition holds if and only if $h'(t) = 0$ and $h(t) = h(a) = 0$, $\eta(t) = u(t)\,h(t) = 0$ for $t \in [a, b]$.

Corollary. *If $[a, b] \subset I$, $r(t) > 0$ for $t \in [a, b]$, and there exists a solution u_0 of* (1.1) *such that $u_0(a) = 0$ and $u_0(t) \neq 0$ for $t \in (a, b)$, then*

$$J[\eta; a, b] \geq 0 \quad \textit{for} \quad \eta \in D_0'[a, b]. \tag{3.9'}$$

If $u_0(b) \neq 0$, then the equality sign holds in (3.9′) *if and only if $\eta(t) = 0$ for $t \in [a, b]$, whereas, if $u_0(b) = 0$ then the equality sign holds in* (3.9′) *if and only if there is a constant k such that $\eta(t) = ku_0(t)$ for $t \in [a, b]$.*

For $\eta \in D_0'[a, b]$, and $0 \leq \varepsilon \leq \varepsilon_0 < (b - a)/2$, let

$$T_\varepsilon(t) = [(b - a)t - (b + a)\varepsilon]/[(b - a) - 2\varepsilon],$$

and define

$$\begin{aligned} \eta_\varepsilon(t) &= \eta(T_\varepsilon(t)), \quad \textit{for} \quad t \in [a + \varepsilon, b - \varepsilon], \\ &= 0, \quad \textit{for} \quad t \in [a, a + \varepsilon], \quad \textit{and} \quad t \in [b - \varepsilon, b]. \end{aligned}$$

Then $\eta_\varepsilon(t) \in D_0'[a + \varepsilon, b - \varepsilon]$ and $\eta_\varepsilon(t) \in D_0'[a, b]$, while $\eta_0 = \eta$. Moreover, $|\eta_\varepsilon(t)|$ and $|\eta_\varepsilon'(t)|$ are uniformly bounded for $t \in [a, b]$, $0 \leq \varepsilon \leq \varepsilon_0$, and on any closed subinterval of $[a, b]$ not containing a point of discontinuity of $\eta'(t)$ the functions $\eta_\varepsilon(t)$, $\eta_\varepsilon'(t)$ tend uniformly to the respective functions $\eta(t)$, $\eta'(t)$ as $\varepsilon \to 0$. Consequently, we have the limit relation

$$J[\eta_\varepsilon; a + \varepsilon, b - \varepsilon] = J[\eta_\varepsilon; a, b] \to J[\eta; a, b] \quad \textit{as} \quad \varepsilon \to 0. \tag{3.10}$$

Now, by hypothesis, u_0 is a solution of (1.1) such that $u_0(a) = 0$ and $u_0(t) \neq 0$ for $t \in (a, b)$. As noted above, there is no loss of generality in supposing that u_0 is real-valued, and this restriction will be assumed in the following argument. In view of Theorem 3.4 it then follows that $J[\eta_\varepsilon; a + \varepsilon, b - \varepsilon] \geq 0$ for $0 < \varepsilon \leq \varepsilon_0$, and hence (3.10) implies that $J[\eta; a, b] \geq 0$. If $u_0(b) \neq 0$, from (iv) of Theorem 3.3 and Theorem 3.4 it follows that equality holds in (3.9′) if and only if $\eta(t) = 0$ for $t \in [a, b]$. Moreover, if $u_0(b) = 0$ then since

u_0 is a solution of (1.1) satisfying the conditions $u_0(a) = 0 = u_0(b)$ it follows from (3.3″) that $J[u_0; a, b] = 0$, and in view of Corollary 1 to Theorem 3.1 a function $\eta \in D_0'[a, b]$ is such that $J[\eta; a, b] = 0$ if and only if η is a constant multiple of u_0.

As $u_0(t) = \sin[\pi(t - a)/(b - a)]$ is a solution of the differential equation $u'' + [\pi/(b - a)]^2 u = 0$, application of the above Corollary yields the following result.

***Example* 3.1.** *If $\eta \in D_0'[a, b]$, then*

$$\int_a^b \eta'^2\, dt \geq \frac{\pi^2}{(b - a)^2} \int_a^b \eta^2\, dt, \tag{3.11}$$

and the equality sign holds if and only if there is a constant k such that $\eta(t) = k \sin[\pi(t - a)/(b - a)]$.

It is to be emphasized that in Theorem 3.4 the hypothesis that the non-vanishing solution u_0 be real-valued is essential. For example, $u(t) = \exp\{it\}$ is a non-real solution of $u'' + u = 0$ that is nonzero for all t, while for $(b - a) > \pi$ the function $\eta = \sin[\pi(t - a)/(b - a)]$ belongs to $D_0'[a, b]$ and, in view of the above example, the corresponding functional $J[\eta; a, b] = \int_a^b (\eta'^2 - \eta^2)\, dt$ has the negative value $-[(b - a)^2 - \pi^2]/[2(b - a)]$.

Problems V.3

1. If $[a, b]$ is a given interval on the real line, prove that there exists a real-valued solution u of the differential equation $[(1 + e^t)u']' - e^t u = 0$ such that $u(t) \neq 0$ for $t \in [a, b]$.

2. If $b > 0$, prove that

$$\int_0^b e^{mt}[\eta'^2 - \eta^2]\, dt \geq 0 \quad \textit{for} \quad \eta \in D_0'[0, b], \tag{3.12}$$

if m is a real constant such that $m^2 \geq 4$, whereas if $-2 < m < 2$ then (3.12) holds only if $b \leq 2\pi/\sqrt{4 - m^2}$.

3. If $0 < a < b$, show that $\int_a^b (\eta'^2 - \{1/(4t^2)\}\eta^2)\, dt > 0$ for $\eta \in D_0'[a, b]$ and $\eta(t) \not\equiv 0$ on $[a, b]$.

4. If $[a, b] \subset I$, and $r(t) > 0$ for $t \in [a, b]$, prove that there exist positive constants k_0, k_1 such that

$$J[\eta; a, b] \geq k_0 \int_a^b \eta'^2\, dt - k_1 \int_a^b \eta^2\, dt \quad \textit{for} \quad \eta \in D'[a, b]. \tag{3.13}$$

Hint. For $\mu_1 = \min r(t)$, $\mu_2 = \max |p(t)|$, $\mu_3 = \max |q(t)|$ for $t \in [a, b]$, and $\varepsilon > 0$, note that $|2q\eta\eta'| \leq \varepsilon\eta'^2 + (1/\varepsilon)\mu_3{}^2\eta^2$, so that

$$\eta'[r\eta' + q\eta] + \eta[q\eta' + p\eta] \geq [\mu_1 - \varepsilon]\eta'^2 - [\mu_2 + (1/\varepsilon)\mu_3{}^2]\eta^2,$$

and for $\varepsilon = \mu_1/2$ conclude that (3.13) holds with $k_0 = \mu_1/2$, $k_1 \geq \mu_2 + 2\mu_3^2/\mu_1$.

5. If $[a, b] \subset I$, $r(t) > 0$ for $t \in [a, b]$, and there exists a real-valued solution u of (1.1) such that $u(t) \neq 0$ for $t \in [a, b]$, prove that there exist positive constants κ and κ_1 such that

$$J[\eta; a, b] \geq \kappa \int_a^b \eta^2 \, dt \quad \textit{for} \quad \eta \in D_0'[a, b], \tag{3.14}$$

$$J[\eta; a, b] \geq \kappa_1 \int_a^b (\eta'^2 + \eta^2) \, dt \quad \textit{for} \quad \eta \in D_0'[a, b]. \tag{3.15}$$

Hints. Note that if $\eta \in D_0'[a, b]$, and $\eta = uh$, then $h \in D_0'[a, b]$; with the aid of (3.9), and the result of Example 3.1 for the function h, conclude that (3.14) holds with

$$\kappa = \frac{\pi^2}{(b-a)^2} \frac{\min\{r(t)u^2(t) \mid t \in [a, b]\}}{\max\{u^2(t) \mid t \in [a, b]\}}.$$

In order to establish (3.15), write $J[\eta; a, b] = (1 - \theta)J[\eta; a, b] + \theta J[\eta; a, b]$ for $0 < \theta < 1$, and with the aid of (3.13) and (3.14) conclude that

$$J[\eta; a, b] \geq (1 - \theta)k_0 \int_a^b \eta'^2 \, dt + [\theta\kappa - (1 - \theta)k_1] \int_a^b \eta^2 \, dt,$$

which for $\theta = (k_0 + k_1)/(\kappa + k_0 + k_1)$ implies (3.15) with

$$\kappa_1 = \kappa k_0/(\kappa + k_0 + k_1).$$

6. If $\eta \in D_0'[a, b]$, and $c \in (a, b)$, prove the inequality

$$\int_a^b \eta'^2 \, dt \geq \frac{4}{(b-a)} \eta^2(c). \tag{3.16}$$

Moreover, if $\eta \not\equiv 0$ on $[a, b]$, prove that the equality sign in (3.16) holds only if $c = (a + b)/2$ and $\eta(t) = \eta(c)[1 - 2\,|t - c|/(b - a)]$ for $t \in [a, b]$. In particular, if $\eta \not\equiv 0$ and η has a continuous derivative on (a, b), then the inequality sign holds in (3.16).

Hint. Establish the inequality

$$\int_a^b \eta'^2 \, dt \geq \eta^2(c)\left[\frac{1}{c-a} + \frac{1}{b-c}\right], \quad \textit{for} \quad c \in (a, b),$$

by applying to the functionals $\int_a^c \eta'^2 \, dt$ and $\int_c^b \eta'^2 \, dt$ the result of Corollary 2 to Theorem 3.1.

4. THE ASSOCIATED RICCATI DIFFERENTIAL EQUATION

If I_0 is a subinterval of I, and $u \in D''(I_0):v$ with $u(t) \neq 0$ for $t \in I_0$, then $w(t) = v(t)/u(t)$ is of class $D'(I_0)$ and

$$ul[u] = u^2\mathfrak{k}[w], \tag{4.1}$$

where $\mathfrak{k}[w]$ is the *Riccati* formal differential operator

$$\mathfrak{k}[w](t) = w'(t) + 2a(t)w(t) + b(t)w^2(t) - c(t), \qquad t \in I_0. \tag{4.2}$$

Moreover, if w is a solution of the first order differential equation

$$\mathfrak{k}[w] = 0, \qquad t \in I_0, \tag{4.3}$$

and $u(t)$ is a nonidentically vanishing solution of the first order linear homogeneous differential equation

$$u' = [a(t) + b(t)w(t)]u, \qquad t \in I_0, \tag{4.4}$$

then $u(t)$, $v(t) = w(t)\,u(t)$ is a solution of (1.2) with $u(t) \neq 0$ for $t \in I_0$. Thus we have established the following result.

Theorem 4.1. *The Riccati differential equation* (4.3) *has a solution w on a subinterval I_0 of I if and only if there is a solution u, v of* (1.2) *such that $u(t) \neq 0$ and $w(t) = v(t)/u(t)$ for $t \in I_0$.*

Theorem 4.2. *If w_0 is a solution of* (4.3) *on a subinterval I_0 of I, and for $\tau \in I_0$ the functions $g = g(t, \tau \mid w_0)$, $\theta = \theta(t, \tau \mid w_0)$ are defined as*

$$g(t, \tau \mid w_0) = \exp\left\{-\int_\tau^t [a(s) + b(s)w_0(s)]\,ds\right\}, \tag{4.5}$$

$$\theta(t, \tau \mid w_0) = \int_\tau^t b(s)g^2(s, \tau \mid w_0)\,ds, \tag{4.6}$$

then w is a solution of (4.3) *on I_0 if and only if the constant $\gamma = w(\tau) - w_0(\tau)$ is such that $1 + \gamma\theta(t, \tau \mid w_0) \neq 0$ for $t \in I_0$, and*

$$w(t) = w_0(t) + \frac{g^2(t, \tau \mid w_0)\gamma}{1 + \gamma\theta(t, \tau \mid w_0)}, \qquad t \in I_0. \tag{4.7}$$

If $\mathfrak{k}[w_0] = 0$ on I_0, and for an arbitrary $w \in D'(I_0)$ we set $\psi = w - w_0$ then, since g of (4.5) is characterized as the solution of the differential equation $g' + [a + bw_0]g = 0$ satisfying the initial condition $g(\tau) = 1$, it follows immediately that w is a solution of (4.3) on I_0 if and only if on this subinterval the function $f = \psi/g^2$ is a solution of the special Riccati differential equation

$$f' + bg^2f^2 = 0, \quad \textit{with} \quad f(\tau) = \gamma = w(\tau) - w_0(\tau). \tag{4.8}$$

Moreover, the function θ of (4.6) is the solution of $\theta' = bg^2$ satisfying the initial condition $\theta(\tau) = 0$, so that if f is a solution of (4.8) on I_0 then $f_1 = f[1 + \gamma\theta] - \gamma$ satisfies the linear homogeneous system

$$f_1' + fbg^2 f_1 = 0, \qquad f_1(\tau) = 0,$$

and hence on I_0 the function f_1 is identically zero and $f = \gamma/[1 + \gamma\theta]$.

If m_1, m_2, m_3, m_4 are four numbers, and we introduce the notations

$$\{m_1, m_2, m_3\} = \frac{(m_3 - m_1)}{(m_3 - m_2)},$$

$$\{m_1, m_2, m_3, m_4\} = \{m_1, m_2, m_3\}\{m_2, m_1, m_4\},$$

then $\{m_1, m_2, m_3, m_4\}$ is called the *anharmonic ratio*, or *cross ratio*, of m_1, m_2, m_3, m_4. One of the most important properties of solutions of the Riccati differential equation is presented in the following theorem.

Theorem 4.3. *If* $w_\alpha(t)$, $(\alpha = 1, 2, 3, 4)$, *are solutions of* (4.3) *on a subinterval* I_0 *of* I, *with* $w_3 - w_2$ *and* $w_4 - w_1$ *nonzero on this subinterval, then* $\{w_1(t), w_2(t), w_3(t), w_4(t)\}$ *is constant on* I_0.

Indeed, if w_0 and w_α, $(\alpha = 1, 2, 3, 4)$, are solutions of (4.3) on I_0, and $\tau \in I_0$, $\gamma_\alpha = w_\alpha(\tau) - w_0(\tau)$, then from Theorem 4.2 it follows that

$$w_\alpha(t) - w_0(t) = \frac{g^2(t, \tau \mid w_0)\gamma_\alpha}{1 + \gamma_\alpha\theta(t, \tau \mid w_0)}, \qquad (\alpha = 1, 2, 3, 4).$$

Consequently, for $\alpha, \beta = 1, 2, 3, 4$,

$$w_\alpha(t) - w_\beta(t) = \frac{(\gamma_\alpha - \gamma_\beta)g^2}{(1 + \gamma_\alpha\theta)(1 + \gamma_\beta\theta)},$$

with $g = g(t, \tau \mid w_0)$, $\theta = \theta(t, \tau \mid w_0)$, and one obtains readily that

$$\{w_1(t), w_2(t), w_3(t), w_4(t)\} = \{\gamma_1, \gamma_2, \gamma_3, \gamma_4\} \quad \textit{for} \quad t \in I_0.$$

It is to be remarked that the results of Theorems 4.1, 4.2 and 4.3 are particular instances of the results of the respective Problems 8, 9, and 10 of Section II.3 for matrix differential equations. In view of the widespread occurrence of Riccati differential equations in the study of linear differential equations of the second order, however, it seems fitting to present here these special results as theorems.

Problems V.4

1. If $[a, b] \subset I$ and $r(t) > 0$ for $t \in [a, b]$, prove that $J[\eta; a, b]$ of Section 3 above is positive definite on $D_0'[a, b]$ if and only if there exists on $[a, b]$ a real-valued solution of the corresponding Riccati differential equation (4.3).

Hint. Use results of Theorems 3.3 and 3.4.

2. If $h(t)$ is a real-valued non-negative continuous function on $[0, \pi]$ which is not identically zero on this interval, prove that there is no real-valued solution of the Riccati differential equation

$$w' + w^2 + 1 + h(t) = 0$$

whose interval of existence contains the open interval $(0, \pi)$.

5. OSCILLATION CRITERIA

If t_1 and t_2 are distinct points on I then t_2 is said to be *conjugate* to t_1, [with respect to (1.1) or (1.2)], if there exists a solution u of (1.1) such that $u(t_1) = 0 = u(t_2)$ and $u(t) \not\equiv 0$ for $t \in (t_1, t_2)$. The equation (1.1), or the system (1.2), is called *disconjugate on a subinterval* I_0 *of* I in case no two distinct points of I_0 are conjugate. Such an equation is said to be *oscillatory* on I if one, (and hence every), nonidentically vanishing real-valued solution has infinitely many zeros on I, and *nonoscillatory* on I in case one, (and hence every), nonidentically vanishing real-valued solution has only a finite number of zeros on this interval.

As a simple application of the conclusion of Theorem 2.4 one has the following preliminary result.

***Lemma* 5.1.** *If* $[a, b] \subset I$ *and either* $t_1 = a$ *or* $t_1 = b$, *then* (1.1) *is disconjugate on the open interval* (a, b) *if and only if there is no value* t_2 *on* (a, b) *which is conjugate to* t_1.

The results of Theorems 3.3 and 3.4, the Corollary to Theorem 3.4, and the above Lemma 5.1 lead to the following result on disconjugacy.

Theorem 5.1. *If* $[a, b] \subset I$, *and* $r(t) > 0$ *for* $t \in [a, b]$, *then* (1.1) *is disconjugate on the open interval* (a, b) *if and only if* $J[\eta; a, b]$ *is nonnegative definite on* $D_0'[a, b]$, *and* (1.1) *is disconjugate on* $[a, b]$ *if and only if* $J[\eta; a, b]$ *is positive definite on* $D_0'[a, b]$.

In considering the behavior of solutions of equations (1.1) for which condition (H) holds there is clearly no loss of generality in assuming $r(t) > 0$ for $t \in I$. Consequently, in view of the results of Theorem 3.4, its Corollary, Theorem 4.2 and the above Theorem 5.1, one has the following result.

Corollary. *If* $[a, b] \subset I$, *then the condition that* (1.1) *is disconjugate on* $[a, b]$ *is equivalent to each of the following conditions:*

(i) *there exists a real-valued solution* $u(t)$ *of* (1.1) *such that* $u(t) \neq 0$ *for* $t \in [a, b]$;

(ii) *if* u_a, v_a *is a solution of* (1.2) *such that* $u_a(a) = 0$, $v_a(a) \neq 0$, *then* $u_a(t) \neq 0$ *for* $t \in (a, b]$;

(iii) *there exists on* $[a, b]$ *a real-valued solution* w *of the Riccati differential equation* (4.3).

Suppose that r_1, p_1, q_1 and r_2, p_2, q_2 are real-valued continuous functions on I with $r_1(t) \neq 0$ and $r_2(t) \neq 0$ on this interval. Corresponding to the differential equations

(5.1)
$$l_\alpha[u] \equiv [r_\alpha(t)u' + q_\alpha(t)u]' - [q_\alpha(t)u' + p_\alpha(t)u] = 0, \qquad (\alpha = 1, 2),$$

there are the associated functionals

$$(5.2)\quad J_\alpha[\eta; a, b] = \int_a^b \{\eta'[r_\alpha(t)\eta' + q_\alpha(t)\eta] + \eta[q_\alpha(t)\eta' + p_\alpha(t)\eta]\}\, dt, \qquad (\alpha = 1, 2),$$

for $[a, b] \subset I$. In particular, the *difference functional*

$$(5.3)\qquad J_{1,2}[\eta; a, b] = J_1[\eta; a, b] - J_2[\eta; a, b]$$

is a quadratic functional whose integrand quadratic form has coefficients $r_1 - r_2, p_1 - p_2, q_1 - q_2$.

Theorem 5.2. *Suppose that* $r_\alpha, p_\alpha, q_\alpha$, $(\alpha = 1, 2)$, *are real-valued continuous functions with* $r_2(t) > 0$ *for* $t \in I$, *and* $J_{1,2}[\eta; a, b]$ *is non-negative definite on* $D_0'[a, b]$ *for arbitrary* $[a, b] \subset I$. *If* $[a_0, b_0] \subset I$ *and* $l_2[u] = 0$ *is disconjugate on* $[a_0, b_0]$, *then* $l_1[u] = 0$ *is also disconjugate on* $[a_0, b_0]$. *Moreover, if* $J_{1,2}[\eta; a, b]$ *is positive definite on* $D_0'[a, b]$ *for arbitrary* $[a, b] \subset I$ *then the solutions of* $l_2[u] = 0$ *oscillate more rapidly than the solutions of* $l_1[u] = 0$; *that is, if* u_1 *is a nonidentically vanishing real-valued solution of* $l_1[u] = 0$ *with successive zeros at* t_1 *and* t_2, *then any real-valued solution* u_2 *of* $l_2[u] = 0$ *has at least one zero on* (t_1, t_2).

As in the proof of (i) of Theorem 3.3, the condition that $J_{1,2}[\eta; a, b]$ is non-negative definite on $D_0'[a, b]$ implies that $r_1(t) - r_2(t) \geq 0$ for $t \in [a, b]$, and thus the hypothesis of the theorem implies that $r_1(t) \geq r_2(t) > 0$ for $t \in I$. Now if $l_2[u] = 0$ is disconjugate on $[a_0, b_0]$ then $J_2[\eta; a_0, b_0]$ is positive definite on $D_0'[a_0, b_0]$ by Theorem 5.1, and the non-negativeness of $J_{1,2}[\eta; a_0, b_0]$ on $D_0'[a_0, b_0]$ implies that $J_1[\eta; a_0, b_0]$ is positive definite on $D_0'[a_0, b_0]$, so that in turn Theorem 5.1 implies that $l_1[u] = 0$ is disconjugate on $[a_0, b_0]$.

Now suppose that u_1 is a nonidentically vanishing real-valued solution of $l_1[u] = 0$ with successive zeros t_1 and t_2. Then $u_1(t) \not\equiv 0$ on $[t_1, t_2]$ and $J_1[u_1; t_1, t_2] = 0$, so that if $J_{1,2}[\eta; a, b]$ is positive definite on $D_0'[a, b]$ for arbitrary $[a, b] \subset I$ then $J_2[u_1; t_1, t_2] < 0$. If u_3 is a nonidentically vanishing real-valued, solution of $l_2[u_3] = 0$ satisfying the initial condition $u_3(t_1) = 0$, then the result of the Corollary to Theorem 3.4 applied to $l_2[u] = 0$ implies that there exists a t_3 such that $t_3 \in (t_1, t_2)$ and $u_3(t_3) = 0$. Now if u_2 is any real-valued solution of $l_2[u] = 0$ then either u_2 is a constant multiple of u_3, in which case $u_2(t_3) = 0$, or u_2, u_3 are linearly independent solutions of $l_2[u] = 0$, and by Theorem 2.4 there is a value $t_4 \in (t_1, t_3) \subset (t_1, t_2)$ such that $u_2(t_4) = 0$.

Associated with the functional $J_{1,2}[\eta; a, b]$ is the differential equation

(5.4)

$$l_{1,2}[u] \equiv [(r_1 - r_2)u' + (q_1 - q_2)u]' - [(q_1 - q_2)u' + (p_1 - p_2)u] = 0,$$

where in general one is not assured that the coefficient function $r_1 - r_2$ is non-zero on $[a, b]$. In view of Theorems 5.1 and 5.2, however, one has the following result.

Corollary 1. *If $[a, b] \subset I$ and $r_1(t) > r_2(t) > 0$ for $t \in [a, b]$, while $l_{1,2}[u] = 0$ is disconjugate on (a, b), then whenever $l_2[u] = 0$ is disconjugate on $[a, b]$ the equation $l_1[u] = 0$ is also disconjugate on $[a, b]$.*

Corollary 2. *If $[a, b] \subset I$ and $r(t) > 0$ for $t \in [a, b]$, then (1.1) is disconjugate on $[a, b]$ if and only if one of the following conditions holds:*

(i) *there exists a real-valued function $u \in D''[a, b]:v$ with $v \in \mathfrak{C}^1[a, b]$, such that $u(t) \neq 0$ and $u(t)l[u](t) \leq 0$ for $t \in [a, b]$;*

(ii) *there exists a real-valued function w of class $\mathfrak{C}^1[a, b]$ such that $\mathfrak{k}[w](t) \leq 0$ for $t \in [a, b]$, where $\mathfrak{k}[w]$ is the associated Riccati differential expression (4.2).*

If (1.1) is disconjugate on $[a, b]$ then in view of Theorems 5.1 and 3.3 it follows that there is a real-valued solution $u(t)$ of (1.1) with $u(t) \neq 0$ for $t \in [a, b]$, so that conclusion (*i*) holds for such a function u, and by Theorem 4.1 conclusion (ii) holds for w equal to the corresponding function v/u. On the other hand, if u is a real-valued function such that $u \in D''[a, b]:v$ with $v \in \mathfrak{C}^1[a, b]$, while $u(t) \neq 0$ and $u(t)\, l[u](t) \leq 0$ for $t \in [a, b]$, then $\phi(t) = u(t)\, l[u](t)$ is a non-positive continuous real-valued function on $[a, b]$, and for

$$(5.5) \qquad r_2 = r, \qquad q_2 = q, \qquad p_2 = p + \frac{\phi}{u^2},$$

it follows that u is a real-valued solution of $l_2[u] = 0$ such that $u(t) \neq 0$ for $t \in [a, b]$. Hence, by (i) of the Corollary to Theorem 5.1, the equation $l_2[u] = 0$ is disconjugate on $[a, b]$, and as $p_2(t) \leq p(t)$ for $t \in [a, b]$ it follows from Theorem 5.2 that $l[u] = 0$ is also disconjugate on $[a, b]$. In turn, if w is a real-valued function of class $\mathfrak{C}^1[a, b]$ such that $\mathfrak{k}[w](t) \leq 0$ for $t \in [a, b]$ then $\psi(t) = \mathfrak{k}[w](t)$ is a non-positive continuous real-valued function on $[a, b]$, and if u is the function $u(t) = \exp\{\int_a^t [a(s) + b(s)w(s)]\, ds\}$, which is the solution of the system

$$u' = [a(t) + b(t)w(t)]u, \qquad u(a) = 1,$$

then $u(t)$ is a real-valued function such that $v = wu \in \mathfrak{C}^1[a, b]$, $u \in D''[a, b]:v$, and $u(t)l[u(t)] = u^2(t)\psi(t) \leq 0$ for $t \in [a, b]$, so that $l[u] = 0$ is disconjugate on $[a, b]$ by the preceding result.

Problems V.5

1. If $r_\alpha, p_\alpha, q_\alpha$, $(\alpha = 1, 2)$, are real-valued continuous functions with $r_2(t) > 0$ for $t \in I$, show that the results of Theorem 5.2 for the differential equations $l_\alpha[u] = 0$, $(\alpha = 1, 2)$, remain valid when the functional $J_{1,2}[\eta; a, b]$ of that theorem is replaced by $k_1 J_1[\eta; a, b] - k_2 J_2[\eta; a, b]$, where k_1, k_2 are given positive constants.

2. If $0 < a < b < \infty$, show that $\int_a^b [\eta'^2 - \{h(t)/(4t^2)\}\eta^2]\, dt$ is positive definite on $D_0'[a, b]$, when $h(t)$ is real-valued, continuous, and $h(t) < 1$ on $(0, \infty)$.

3. Show that on $D_0'[-1, 1]$ the functional $\int_{-1}^{1} [\eta'^2 - \{12/(5 - t^2)\}\eta^2]\, dt$ is non-negative definite, but is not positive definite.

Hint. Determine a solution of $u'' + \{12/(5 - t^2)\}u = 0$ of the form $u = a + bt^2 + ct^4$.

4. If $g(t)$ is a real-valued continuous function such that $g(t) > k^2 > 0$ on I, show that on any subinterval $[a, b]$ of I of length π/k an arbitrary real-valued solution of

$$u'' + g(t)u = 0, \tag{5.6}$$

has at least one zero on (a, b). Moreover, if there exists a positive integer m such that $g(t) > m^2\pi^2/(b - a)^2$ on $[a, b]$, then an arbitrary real-valued solution of (5.6) has at least m zeros on (a, b).

Hint. Apply results of Theorem 5.2 to (5.6) and $u'' + k^2u = 0$.

5.[ss] If r, p are real-valued continuous functions on $[a, b]$ and $r(t) > 0$ for $t \in [a, b]$, show that:

(a) the differential equation

$$[r(t)u']' - p(t)u = 0, \tag{5.7}$$

is disconjugate on $[a, b]$ if

$$\frac{\min\{p(t) \mid t \in [a, b]\}}{\min\{r(t) \mid t \in [a, b]\}} > -\frac{\pi^2}{(b - a)^2};$$

(b) an arbitrary real-valued solution of (5.7) has at least m zeros on (a, b) if

$$\frac{\max\{p(t) \mid t \in [a, b]\}}{\max\{r(t) \mid t \in [a, b]\}} < -\frac{m^2\pi^2}{(b-a)^2}.$$

Hint. To establish conclusion (a), apply results of Theorem 5.2 to (5.7) and $r_0 u'' - p_0 u = 0$, where r_0, p_0 are constants such that

$$\min\{p(t) \mid t \in [a, b]\} > p_0, \qquad \min\{r(t) \mid t \in [a, b]\} > r_0 > 0,$$

and

$$p_0/r_0 > -\pi^2(b-a)^2.$$

Use a similar comparison technique to establish conclusion (b).

6. If $g(t)$ is a real-valued continuous function on an interval I, then each of the following conditions is sufficient for the differential equation (5.6) to be disconjugate on I:

(a) there exists a $\tau \in I$ such that

$$4\left[\int_t^\tau g(s)\,ds\right]^2 \leq g(t) \quad \textit{for} \quad t \in I;$$

(b) $I \subset (0, \infty)$, and there exists a $\tau \in I$ such that

$$-3 \leq 4t\int_t^\tau g(s)\,ds \leq 1 \quad \textit{for} \quad t \in I.$$

If there exists a t_0 such that $(t_0, \infty) \subset I$ and for $t_1 \in (t_0, \infty)$ the integral $\int_{t_1}^\infty g(s)\,ds = \lim_{t\to\infty} \int_{t_1}^t g(s)\,ds$ exists and is finite, then either (a) or (b), with $\tau = \infty$, is a sufficient condition for disconjugacy on I.

Hint. Compute $\mathfrak{k}[w]$, for $w(t) = 2\int_t^\tau g(s)\,ds$ in (a), and $w(t) = \int_t^\tau g(s)\,ds + 1/(4t)$ in (b).

7. Show that an arbitrary solution u of $l[u] \equiv 4u'' + (19 - 2t^2)u = 0$ has at least one zero on each of the intervals $(0, \sqrt{3})$ and $(-\sqrt{3}, 0)$.

Hint. Note that $l_1[u] = u'' + (\frac{7}{2} - \frac{1}{4}t^2)u = 0$ has the solution $u_1(t) = (t^3 - 3t)e^{-t^2/4}$. Show that $8l_1[u] - l[u] = 0$ is disconjugate on $[0, \sqrt{3}]$ and $[-\sqrt{3}, 0]$, and use the result of Corollary 1 to Theorem 5.2.

8. Suppose that r and p are real-valued continuous functions with $r(t) > 0$ on $[a, \infty)$, and that r and p are nonincreasing functions on this interval. If u is a non-identically vanishing real-valued solution of the differential equation (5.7) which has two zeros on $[a, \infty)$, show that u has infinitely many zeros on $[a, \infty)$, and that if $\{t_j\}$ denotes these zeros in increasing order then $t_{j+2} - t_{j+1} \leq t_{j+1} - t_j$, $(j = 1, 2, \ldots)$.

9. Suppose that r and p are real-valued continuous functions on $[a, b]$ with $r(t) > 0$ on this interval, and let $p^+(t) = \frac{1}{2}[|p(t)| + p(t)], p^-(t)]$ $\frac{1}{2}[|p(t)| - p(t)]$. Show that if $[r(t)u']' - p(t)u = 0$ is disconjugate on $[a, b]$ then $[r(t)u']' - p^+(t)u = 0$ is also disconjugate on $[a, b]$.

10. Suppose that r_α and p_α, $(\alpha = 1, 2)$, are real-valued continuous functions with $r_1(t) \geq r_2(t) > 0$ and $p_1(t) > p_2(t)$ for $t \in [a, b]$. Use the trigonometric transformation (1.11) to prove that on $[a, b]$ the solutions of $l_2[u] \equiv [r_2(t)u']' - p_2(t)u = 0$ oscillate more rapidly than the solutions of $l_1[u] \equiv [r_1(t)u']' - p_1(t)u = 0$.

Hint. If u_α is a nonidentically vanishing real-valued solution of $l_\alpha[u_\alpha] = 0$, $(\alpha = 1, 2)$, and θ_α, ρ_α the corresponding polar coordinate functions, then $\theta_\alpha' = F_\alpha(t, \theta_\alpha)$, where $F_\alpha(t, \theta) = [1/r_\alpha(t)]\cos^2\theta - p_\alpha(t)\sin^2\theta$, and the hypothesis of the problem implies that $F_1(t, \theta) \leq F_2(t, \theta)$ for $(t, \theta) \in [a, b] \times (-\infty, \infty)$. If t_1 and t_2 are successive zeros of u_1, with $a \leq t_1 < t_2 \leq b$, note that without loss of generality one may assume $u_2(t_1) \geq 0$, so that one may choose $\theta_1(t_1) = 0$, $0 \leq \theta_2(t_1) < \pi$. Proceed, with the aid of Problem 1, to show that the hypothesis of the problem implies $\theta_2(t) > \theta_1(t)$ for $t > t_1$.

6. COMPARISON THEOREMS

For a given subinterval $[a, b]$ of I we shall denote by $D'_{*0}[a, b]$ the set of functions η of class $D'[a, b]$ with $\eta(b) = 0$; similarly, $D'_{0*}[a, b]$ will denote the class $\{\eta \mid \eta \in D'[a, b], \eta(a) = 0\}$. In particular, we have $D'_0[a, b] = D'_{*0}[a, b] \cap D'_{0*}[a, b]$. Attention will now be devoted to functionals of the form

$$
\begin{aligned}
(6.1)\quad \hat{J}[\eta_1, \eta_2; a, b] &= \gamma\eta_2(a)\,\eta_1(a) \\
&\quad + \int_a^b \{\eta_2'[r(t)\eta_1' + q(t)\eta_1] + \eta_2[q(t)\eta_1' + p(t)\eta_1]\}\,dt, \\
&= \gamma\eta_2(a)\,\eta_1(a) + J[\eta_1, \eta_2; a, b],
\end{aligned}
$$

where the functions r, p, q satisfy hypothesis **(H)** and γ is a real constant; corresponding to the previous notation, $\hat{J}[\eta; a, b]$ will denote $\hat{J}[\eta, \eta; a, b]$. Clearly $\hat{J}[\eta_1, \eta_2; a, b]$ is a real-valued symmetric quadratic functional on $D'[a, b] \times D'[a, b]$; that is, it possesses properties analogous to properties (3.2) for $J[\eta_1, \eta_2; a, b]$. Moreover, corresponding to the results of Theorem 3.1 and its Corollary, one has the following results for $\hat{J}[\eta_1, \eta_2; a, b]$.

Theorem 6.1. *If $[a, b] \subset I$ then u is a solution of* (1.1) *on $[a, b]$ which satisfies with its associated function v the initial condition*

$$
(6.2)\qquad \gamma u(a) - v(a) = 0
$$

if and only if $u \in D'[a, b]$ and $\hat{J}[u, \eta; a, b] = 0$ for $\eta \in D'_{*0}[a, b]$.

Corollary. *If $[a, b] \subset I$ and $\hat{J}[\eta; a, b]$ is non-negative definite on $D'_{*0}[a, b]$, and u is an element of $D'_{*0}[a, b]$ such that $\hat{J}[u; a, b] = 0$, then u is a solution of* (1.1) *which satisfies with its associated function v the boundary conditions*

$$\gamma u(a) - v(a) = 0, \qquad u(b) = 0. \tag{6.3}$$

Indeed, if $u \in D'[a, b]$ and $\hat{J}[u, \eta; a, b] = 0$ for $\eta \in D'_{*0}[a, b]$, then since $D'_0[a, b] \subset D'_{*0}[a, b]$ it follows from Theorem 3.1 that u is a solution of (1.1) on $[a, b]$. Moreover, if v is the associated function such that u, v is a solution of (1.2), then with the aid of identity (3.3′) it follows that

$$0 = \hat{J}[u, \eta; a, b] = [\gamma u(a) - v(a)]\eta(a) \quad \textit{for} \quad \eta \in D'_{*0}[a, b],$$

and (6.2) holds as there exists an $\eta \in D'_{*0}[a, b]$ for which $\eta(a)$ assumes an arbitrary value; for example, $\eta(t) = \eta(a)[1 - (t - a)/(b - a)]$. Conversely, if u, v is a solution of (1.2) satisfying the initial condition (6.2), with the aid of (3.3′) it follows readily that $\hat{J}[u, \eta; a, b] = 0$ for arbitrary $\eta \in D'_{*0}[a, b]$.

All details of proof of the above Corollary 1 will be omitted, as they parallel those of the proof of Corollary 1 to Theorem 3.1.

Theorem 6.2. *$\hat{J}[\eta; a, b]$ is non-negative definite on $D'_{*0}[a, b]$ if and only if $r(t) \geq 0$ on $[a, b]$, and the solution u, v of* (1.2) *determined by the initial conditions*

$$u(a) = 1, \qquad v(a) = \gamma, \tag{6.4}$$

is such that $u(t) \neq 0$ for $t \in [a, b)$; if $u(b) \neq 0$, then $\hat{J}[\eta; a, b]$ is positive definite on $D'_{*0}[a, b]$, whereas, if $u(b) = 0$ then $\hat{J}[\eta; a, b] = 0$ for an $\eta \in D'_{*0}[a, b]$ if and only if there is a constant k such that $\eta(t) = ku(t)$ for $t \in [a, b]$.

The conclusion that $r(t) \geq 0$ on $[a, b]$ whenever $\hat{J}[\eta; a, b]$ is non-negative on $D'_{*0}[a, b]$ follows from (i) of Theorem 3.3, and the fact that $D'_0[a, b] \subset D'_{*0}[a, b]$. If u, v is the solution of (1.2) determined by conditions (6.4), and there exists a value c on $(a, b]$ such that $u(c) = 0$, then $u_0(t) = u(t)$ for $t \in [a, c]$, $u_0(t) = 0$ for $t \in (c, b]$, is such that $u_0 \in D'_{*0}[a, b]$ and

$$\hat{J}[u_0; a, b] = \hat{J}[u; a, c] = \gamma u^2(a) + uv\Big|_a^c = u(a)[\gamma u(a) - v(a)] = 0. \tag{6.5}$$

Consequently, if $\hat{J}[\eta; a, b] \geq 0$ for $\eta \in D'_{*0}[a, b]$ it follows from the above Corollary to Theorem 6.1 that u_0 is a solution of (1.1) on $[a, b]$. As $u_0(a) = 1$, the function u_0 is nonidentically vanishing on $[a, b]$ and thus by Theorem 2.1 its zeros on this interval are isolated. In particular, u_0 cannot be zero throughout a nondegenerate interval, so that c cannot lie in the open interval (a, b) whenever $\hat{J}[\eta; a, b]$ is non-negative definite on $D'_{*0}[a, b]$. Moreover, if $\hat{J}[\eta; a, b]$ is positive definite on $D'_{*0}[a, b]$, then the above equation (6.5) shows that $c = b$ is also impossible, so that in this case $u(t) \neq 0$ for $t \in (a, b]$.

Now if $r(t) > 0$ for $t \in [a, b]$, and the solution u, v of (1.2) determined by the initial conditions (6.4) is such that $u(t) \neq 0$ for $t \in [a, b_1]$, $a < b_1 \leq b$, then for $\eta \in D'_{*0}[a, b_1]$ and $h(t) = \eta(t)/u(t)$ it follows with the aid of Theorem 3.2 that

$$\hat{J}[\eta; a, b_1] = \int_a^{b_1} r(t)[uh']^2\, dt.$$

Consequently, $\hat{J}[\eta; a, b_1] \geq 0$, with equality sign holding only if $0 = u(t)\, h'(t)$ and $h(t) = h(b_1) = 0$ for $t \in [a, b_1]$. In particular, if $b_1 = b$ then $\hat{J}[\eta; a, b]$ is positive definite on $D'_{*0}[a, b]$. Also, if $u(t) \neq 0$ for $t \in [a, b)$ and $u(b) = 0$, an argument similar to that employed in the proof of the Corollary to Theorem 3.4 shows that if $\eta \in D'_{*0}[a, b]$ then $\hat{J}[\eta; a, b] \geq 0$, with the equality sign holding only if there is a constant k such that $\eta(t) = ku(t)$ for $t \in [a, b]$.

***Example* 6.1.** $\displaystyle\int_a^b \eta'^2\, dt \geq \frac{\pi^2}{4(b-a)^2}\int_a^b \eta^2\, dt$ *for* $\eta \in D'_{*0}[a, b]$, *and the equality sign holds if and only if there is a constant* k *such that* $\eta(t) = k\cos[\pi(t-a)/2(b-a)]$.

Relative to the functional (6.1), or relative to the differential equation (1.1) with initial condition (6.2), a value $\tau \in I$ is a *right-hand*, {*left-hand*} *focal point* to $t = a$ if $\tau > a$, $\{\tau < a\}$, and there is a nonidentically vanishing solution u of (1.1) which with its associated function v satisfies the initial condition (6.2) and $u(\tau) = 0$.

Theorem 6.3. *Suppose that* $r_\alpha, p_\alpha, q_\alpha$, $(\alpha = 1, 2)$, *are real-valued continuous functions with* $r_2(t) > 0$ *for* $t \in I$, *and that* γ_α, $(\alpha = 1, 2)$, *are real constants such that*

$$\begin{aligned}\hat{J}_{1,2}[\eta; a, b] &= \hat{J}_1[\eta; a, b] - \hat{J}_2[\eta; a, b],\\ &= (\gamma_1 - \gamma_2)\,\eta^2(a) + J_{1,2}[\eta; a, b],\end{aligned}$$

is non-negative definite on $D'_{*0}[a, b]$. *If relative to* $\hat{J}_2[\eta; a, b]$ *there is no right-hand focal point to* $t = a$ *on* $(a, b]$, *then relative to* $\hat{J}_1[\eta; a, b]$ *there is also no right-hand focal point to* $t = a$ *on* $(a, b]$. *Moreover, if* $\hat{J}_{1,2}[\eta; a, b]$ *is positive definite on* $D'_{*0}[a, b]$, *and relative to* $\hat{J}_1[\eta; a, b]$ *there are* m *right-hand focal points* t_j^1 *to* $t = a$ *on* $(a, b]$ *with* $a < t_1^1 < t_2^1 < \cdots < t_m^1 \leq b$, *then relative to* $\hat{J}_2[\eta; a, b]$ *there are* r *focal points* t_k^2, $(k = 1, \ldots, r)$, *to* $t = a$ *on* $(a, b]$ *with* $r \geq m$, *and if* $a < t_1^2 < t_2^2 < \cdots < t_r^2 \leq b$ *then* $t_j^2 < t_j^1$ *for* $j = 1, \ldots, m$.

If relative to $\hat{J}_2[\eta; a, b]$ there is no right-hand focal point to $t = a$ on $(a, b]$, then by Theorem 6.2 the functional $\hat{J}_2[\eta; a, b]$ is positive definite on $D'_{*0}[a, b]$. In this case, the condition that $\hat{J}_{1,2}[\eta; a, b]$ is non-negative on $D'_{*0}[a, b]$ implies that $\hat{J}_1[\eta; a, b]$ is also positive definite on $D'_{*0}[a, b]$, and from Theorem 6.2 it follows that relative to $\hat{J}_1[\eta; a, b]$ there is also no right-hand focal point to $t = a$ on $(a, b]$.

Now if $[a_0, b_0] \subset [a, b]$ and $\eta_0 \in D'_0[a_0, b_0]$, then $\eta(t) = \eta_0(t)$ for $t \in [a_0, b_0]$, $\eta(t) = 0$ for $t \in [a, a_0)$ and $t \in (b_0, b]$, defines an η which belongs to $D'_0[a, b] \subset D'_{*0}[a, b]$ and $\hat{J}_{1,2}[\eta; a, b] = J_{1,2}[\eta_0; a_0, b_0]$. Consequently, the condition that $\hat{J}_{1,2}[\eta; a, b]$ is positive definite on $D'_{*0}[a, b]$ implies that $\hat{J}_{1,2}[\eta; a_0, b_0]$ is positive definite on $D'_0[a_0, b_0]$ for arbitrary $[a_0, b_0] \subset [a, b]$, and hence by Theorem 5.2 the solutions of $l_2[u] = 0$ oscillate more rapidly than the solutions of $l_1[u] = 0$. Thus, in this case, if u_α, v_α, $(\alpha = 1, 2)$, is the solution of the corresponding system (1.3) satisfying the initial conditions $u_\alpha(a) = 1$, $v_\alpha(a) = \gamma_\alpha$, it follows that between consecutive zeros of u_1 on $[a, b]$ there is at least one zero of u_2. Moreover, for the first zero $t_1{}^1$ of u_1 on $(a, b]$ that is greater than $t = a$ one has that $\hat{J}_1[u_1; a, t_1{}^1] = 0$ in view of Theorem 6.1. Now the condition that $\hat{J}_{1,2}[\eta; a, b]$ is positive definite on $D'_{*0}[a, b]$ implies that $\hat{J}_{1,2}[\eta; a, b_0]$ is positive definite on $D'_{*0}[a, b_0]$ for arbitrary b_0 on $(a, b]$. Thus the positive definiteness of $\hat{J}_{1,2}[\eta; a, b]$ on $D'_{*0}[a, b]$ implies that $\hat{J}_2[u_1; a, t_1{}^1] < 0$, and from Theorem 6.2 it follows that relative to $\hat{J}_2[\eta; a, b]$ there is a right-hand focal point to $t = a$ on the open interval $(a, t_1{}^1)$.

Theorem 6.4. *Suppose that $r_\alpha, p_\alpha, q_\alpha$, $(\alpha = 1, 2)$, are real-valued continuous functions with $r_\alpha(t) > 0$ for $t \in [a, b]$, while u_α is a nonidentically vanishing real-valued solution of the corresponding differential equation $l_\alpha[u_\alpha] = 0$, such that*

$$\text{(6.6)} \qquad \textit{if } u_1(a) \neq 0, \textit{ then } u_2(a) \neq 0 \textit{ and } \frac{v_1(a)}{u_1(a)} \geq \frac{v_2(a)}{u_2(a)}.$$

(i) *If $J_{1,2}[\eta; a, b_0]$ is positive definite on $D'_{*0}[a, b_0]$ for arbitrary $[a, b_0] \subset [a, b_0]$, and u_1 has exactly m zeros $t = t_j{}^1$, $(j = 1, \ldots, m)$, with*

$$a < t_1{}^1 < t_2{}^1 < \cdots < t_m{}^1 \leq b,$$

then u_2 has r zeros $t_k{}^2$, $(k = 1, \ldots, r)$, with

$$a < t_1{}^2 < t_2{}^2 < \cdots < t_r{}^2 \leq b, \qquad r \geq m,$$

and $t_j{}^2 < t_j{}^1$, $(j = 1, \ldots, m)$.

(ii) *If $J_{1,2}[\eta; a, b_0]$ is positive definite on $D'[a, b_0]$ for arbitrary $[a, b_0] \subset [a, b]$, and there exists a $c \in (a, b]$ such that $u_1(c) \neq 0$, $u_2(c) \neq 0$, while u_1 and u_2 have the same number of zeros on (a, c), then*

$$\text{(6.7)} \qquad \frac{v_1(c)}{u_1(c)} > \frac{v_2(c)}{u_2(c)}.$$

Conclusions (i) and (ii) of Theorem 6.4 are, respectively, the *first* and *second comparison theorems of Sturm*, where the differential equations have been written in slightly more general form, and the hypotheses are somewhat weaker than those employed by Sturm in his famous paper of 1836.

As the condition that $J_{1,2}[\eta; a, b_0]$ is positive definite on $D'_{*0}[a, b_0]$ implies that $J_{1,2}[\eta; a_0, b_0]$ is positive definite on $D'_0[a_0, b_0]$ for $[a_0, b_0] \subset [a, b_0]$, in view of Theorem 5.2 the solutions of $l_2[u] = 0$ oscillate more rapidly than the solutions of $l_1[u] = 0$. Thus, in case $u_1(a) = 0$, conclusion (i) is a consequence of Theorem 5.2. On the other hand, if $u_1(a) \neq 0$ then $u_2(a) \neq 0$ and for $\gamma_\alpha = v_\alpha(a)/u_\alpha(a)$, $(\alpha = 1, 2)$, we have $\gamma_1 \geq \gamma_2$ by (6.6). In this case, the respective functionals $\hat{J}_\alpha[\eta; a, b]$ are such that $\hat{J}_{1,2}[\eta; a, b]$ is positive definite on $D'_{*0}[a, b]$, and conclusion (i) is a consequence of Theorem 6.3.

Now suppose that $J_{1,2}[\eta; a, b_0]$ is positive definite on $D'[a, b_0]$ for arbitrary $[a, b_0] \subset [a, b]$, and there exists a $c \in (a, b]$ such that $u_1(c) \neq 0$, $u_2(c) \neq 0$, while u_1 and u_2 each has exactly m zeros on (a, c). If $m \geq 1$ and t_j^{α}, $(j = 1, \ldots, m)$, are the zeros of u_α on this interval with $t_1^{\alpha} < t_2^{\alpha} < \cdots t_m^{\alpha} < c$, then since $D'_{*0}[a, b_0] \subset D'[a, b_0]$, it follows as in (i) that $t_j^2 < t_j^1$, $(j = 1, \ldots, m)$. In particular, $t_m^2 < t_m^1$ and $u_2(t) \neq 0$ for $t \in [t_m^1, c]$. Now $J_1[u_1; t_m^1, c] = u_1(c)\, v_1(c)$ in view of relation (3.3″), whereas conclusion (ii) of Theorem 3.2 implies that

$$J_2[u_1; t_m^1, c) = \frac{u_1^2 v_2}{u_2}\bigg|_{t_m^1}^{c} + \int_{t_m^1}^{c} r_2\left[u_1' - u_2' \frac{u_1}{u_2}\right]^2 dt,$$

$$> \frac{u_1^2(c)\, v_2(c)}{u_2(c)},$$

and, therefore

$$u_1^2(c)\left[\frac{v_1(c)}{u_1(c)} - \frac{v_2(c)}{u_2(c)}\right] > J_{1,2}[u_1; t_m^1, c]. \tag{6.8}$$

Now if $s \in [a, c)$ and $\eta \in D'_{0*}[s, c]$, then $\eta_0(t) = \eta(t)$ for $t \in [s, c]$, $\eta_0(t) = 0$ for $t \in [a, s)$, defines a function η_0 such that $\eta_0 \in D'[a, c]$ and $J_{1,2}[\eta_0; a, c] = J_{1,2}[\eta; s, c]$. Consequently, the hypothesis of (ii) implies that $J_{1,2}[u_1; t_m^1, c] \geq 0$, and (6.7) holds in this case.

If u_1 and u_2 both have no zeros on (a, c), and $u_2(a) \neq 0$, $u_1(a) = 0$, then the above argument applied to the interval $[a, c]$ implies that the left-hand member of (6.8) exceeds $J_{1,2}[u_1; a, c]$, and in turn, that (6.7) holds. Correspondingly, if $u_1(a) \neq 0$, then in view of (6.6) we have $u_2(a) \neq 0$, so that a similar argument yields

$$u_1^2(c)\left[\frac{v_1(c)}{u_1(c)} - \frac{v_2(c)}{u_2(c)}\right] \geq u_1^2(a)\left[\frac{v_1(a)}{u_1(a)} - \frac{v_2(a)}{u_2(a)}\right] + J_{1,2}[u_1; a, c], \tag{6.9}$$

and (6.7) follows from (6.6) and the positive definiteness of $J_{1,2}[\eta; a, c]$ on $D'[a, c]$. Finally, if $u_1(a) = 0$ and $u_2(a) = 0$, for $\varepsilon > 0$ let $u_{1\varepsilon}(t)$ be the solution of $l_1[u] = 0$ satisfying the initial conditions $u_{1\varepsilon}(a + \varepsilon) = 0$, $v_{1\varepsilon}(a + \varepsilon) = v_1(a)$, where $0 < \varepsilon < c - a$. As $u_1(t) \neq 0$ for $t \in (a, c]$, it follows from

Theorem 2.4 that $u_{1\varepsilon}(t) \neq 0$ for $t \in (a + \varepsilon, c]$. Moreover, as $u_2(t) \neq 0$ for $t \in (a, c]$, we have $u_2(t) \neq 0$ for $t \in [a + \varepsilon, c]$, and application of a preceding result to $u_{1\varepsilon}$ and u_2 implies that

$$u_{1\varepsilon}^2(c)\left[\frac{v_{1\varepsilon}(c)}{u_{1\varepsilon}(c)} - \frac{v_2(c)}{u_2(c)}\right] > J_{1,2}[u_{1\varepsilon}; a + \varepsilon, c].$$

Now it may be shown readily that as $\varepsilon \to 0$ the functions $u_{1\varepsilon}(t)$ and $v_{1\varepsilon}(t)$ tend uniformly on $[a, c]$ to the respective limit functions $u_1(t)$, $v_1(t)$; hence in this case

$$u_1^2(c)\left[\frac{v_1(c)}{u_1(c)} - \frac{v_2(c)}{u_2(c)}\right] \geq J_{1,2}[u_1; a, c] > 0,$$

thus completing the proof of conclusion (ii).

Problems V.6

1. Formulate, and prove, results on left-hand focal points corresponding to the results of Theorems 6.3 and 6.4.

2. If r and p are real-valued continuous functions with $r(t) > 0$ for $t \in [a, b]$, show that relative to the functional

$$J[\eta; a, b] = \int_a^b [r(t)\eta'^2 + p(t)\eta^2]\, dt$$

each of the following conditions is necessary and sufficient for $(a, b]$ to contain no focal point to $t = a$:

(a) if u_0 is the solution of $l[u_0] \equiv [r(t)u_0']' - p(t)u_0 = 0$ satisfying the initial conditions $u_0(a) = 1$, $v_0(a) = 0$, then $u_0(t) \neq 0$ for $t \in [a, b]$;

(b) there exists a function $u \in D''[a, b]:v$ with $v \in \mathfrak{C}^1[a, b]$ satisfying $u(a)\, v(a) \leq 0$, $u(t) \neq 0$ on $[a, b]$, and $u(t)\, l[u](t) \leq 0$ for $t \in [a, b]$;

(c) there exists a function $w \in \mathfrak{C}^1[a, b]$ such that $w(a) \leq 0$, and $w'(t) + [1/r(t)]\, w^2(t) - p(t) \leq 0$ for $t \in [a, b]$.

3. Suppose that r_α and p_α are real-valued functions with $r_\alpha \in \mathfrak{C}^1[a, b]$, $p_\alpha \in \mathfrak{C}[a, b]$, $r_\alpha(t) > 0$ for $t \in I$, and let $l_\alpha[u] = [r_\alpha(t)u']' - p_\alpha(t)u$, $\alpha = 1, 2$. Show that:

(a) if $(r_2/r_1)' \geq 0$, $p_1 \leq 0$, and $(p_1/r_1) - (p_2/r_2) \leq 0$ on I, while relative to

$$l_2[u] = 0, \qquad v(t_0) = 0, \tag{6.10}$$

there exists on I a right-hand focal point t^2 of $t = t_0$, then there exists on I a right-hand focal point of $t = t_0$ relative to

$$l_1[u] = 0, \qquad v(t_0) = 0, \tag{6.11}$$

and the first such right-hand focal point t^1 satisfies the inequality $t^1 \leq t^2$;

(b) correspondingly, if $(r_2/r_1)' \geq 0$, $p_2 \leq 0$ and $(p_2/r_2) - (p_1/r_1) \leq 0$ on I, while relative to (6.11) there exists on I a left-hand focal point ${}^1t$ of $t = t_0$, then relative to (6.10) there exists on I a left-hand focal point of $t = t_0$ and the first such left-hand focal point ${}^2t$ satisfies the inequality ${}^1t \leq {}^2t$.

Hint. Note that if the conclusion of (a) were not true then the solution u_1 of $l_1[u_1] = 0$ satisfying the initial conditions $u_1(t_0) = 1$, $v_1(t_0) = 0$ would be such that $u_1(t) > 0$ on $[t_0, t^2]$ and $v_1(t) < 0$ on $(t_0, t^2]$. Proceed to show that $u_1 l_2[u_1] = u_1^2 v_1 (r_2/r_1)' + r_2 u_1^2[(p_1/r_1) - (p_2/r_2)] \leq 0$ on $[t_0, t^2]$, and from (b) of Problem 2 above obtain a contradiction to the condition that t^2 is a right-hand focal point to $t = t_0$, relative to (6.10). Establish conclusion (b) by a similar argument.

4. If r_α and p_α, $(\alpha = 1, 2)$, are real-valued continuous functions with $r_1(t) \geq r_2(t) > 0$ and $p_1(t) > p_2(t)$ for $t \in [a, b]$, use the trigonometric transformation (1.11) to establish the conclusion of Theorem 6.4 for the differential equations $l_\alpha[u] \equiv [r_\alpha(t)u']' - p_\alpha(t)u = 0$, $(\alpha = 1,2)$.

7. DIFFERENTIAL SYSTEMS INVOLVING A REAL PARAMETER

Suppose that for intervals $I = (a^0, b^0)$ and $\Delta = \{\lambda \mid \Lambda_1 < \lambda < \Lambda_2\}$ on the real line the functions $r(t, \lambda)$, $p(t, \lambda)$, $\alpha_a(\lambda)$, $\beta_a(\lambda)$ satisfy the following hypotheses:

$(\mathbf{H}_0)$ *$r(t, \lambda)$ and $p(t, \lambda)$ are real-valued continuous functions with $r(t, \lambda) > 0$ for $(t, \lambda) \in I \times \Delta$;*

$(\mathbf{H}_a)$ *$\alpha_a(\lambda)$ and $\beta_a(\lambda)$ are real-valued continuous functions such that $\alpha_a^2(\lambda) + \beta_a^2(\lambda) > 0$ for $\lambda \in \Delta$, and either $\alpha_a(\lambda) \equiv 0$ on Δ, or $\alpha_a(\lambda) \neq 0$ for all $\lambda \in \Delta$.*

Moreover, let $u = u(t, \lambda)$ be the solution of

$$l[u:\lambda] \equiv [r(t, \lambda)u']' - p(t, \lambda)u = 0, \qquad t \in I, \tag{7.1}$$

determined by the initial conditions

$$u(a, \lambda) = \alpha_a(\lambda), \qquad v(a, \lambda) = \beta_a(\lambda), \qquad \lambda \in \Delta, \tag{7.2}$$

where $v(t, \lambda)$ signifies the canonical variable $r(t, \lambda)\, u'(t, \lambda)$, and denote by $t_j(\lambda)$ the j-th zero of $u(t, \lambda)$ on (a, b^0), with $a < t_1(\lambda) < t_2(\lambda) < \cdots < b^0$, whenever this zero exists.

The functions $u(t, \lambda)$, $v(t, \lambda)$ are continuous in (t, λ) on $I \times \Delta$ by Theorem I.5.8, and by an elementary argument one may establish the following preliminary result.

***Lemma* 7.1.** *If hypotheses $(\mathbf{H}_0)$ and $(\mathbf{H}_a)$ hold, and for $\lambda = \lambda_0$ the j-th zero $t_j(\lambda_0)$ of $u(t, \lambda_0)$ exists, then there is a neighborhood $N(\lambda_0) = \{\lambda \mid \lambda \in \Delta, |\lambda - \lambda_0| < \delta\}$ such that $t_j(\lambda)$ exists and is a continuous function for $\lambda \in N(\lambda_0)$.*

For the differential equation (7.1) there is the associated functional

$$J[\eta;\lambda;c,d]=\int_c^d [r(t,\lambda)\eta'^2+p(t,\lambda)\eta^2]\,dt \tag{7.3}$$

for $[c, d] \subset I$ and $\eta \in D'[c, d]$. It will now be supposed that in addition to the above hypotheses the following conditions hold.

($\mathbf{H}_a^+$) *The functions $\alpha_a(\lambda)$, $\beta_a(\lambda)$ satisfy the conditions of* ($\mathbf{H}_a$); *moreover, if $\alpha_a(\lambda) \neq 0$ for $\lambda \in \Delta$ then $\beta_a(\lambda)/\alpha_a(\lambda)$ is a monotone non-increasing function on Δ.*

($\mathbf{H}_1$) *If $[a, b] \subset I$ and $\Lambda_1 < \lambda' < \lambda'' < \Lambda_2$, then*

$$J[\eta;\lambda';a,b]-J[\eta;\lambda'';a,b]$$

is positive definite on $D'_{*0}[a, b]$.

The following result is a direct consequence of Theorem 6.3 and Lemma 7.1.

Theorem 7.1. *If hypotheses* ($\mathbf{H}_0$), ($\mathbf{H}_a^+$), *and* ($\mathbf{H}_1$) *are satisfied, and for $\lambda = \lambda_0$ the j-th zero $t_j(\lambda_0)$ of $u(t, \lambda_0)$ on (a, b^0) exists, then $t_j(\lambda)$ exists for $\lambda \in [\lambda_0, \Lambda_2)$, and is a monotone decreasing continuous function on this interval.*

The fundamental result of this section concerns the following differential system

$$\begin{aligned} l[u:\lambda] &\equiv [r(t,\lambda)u']'-p(t,\lambda)u=0,\\ \beta_a(\lambda)u(a)&-\alpha_a(\lambda)v(a)=0,\\ \beta_b(\lambda)u(b)&+\alpha_b(\lambda)v(b)=0, \end{aligned} \tag{7.4}$$

involving boundary conditions at the end-points of a compact subinterval $[a, b]$ of I. Corresponding to the terminology of Chapter IV, a value λ on (Λ_1, Λ_2) for which (7.4) has a nonidentically vanishing solution is called a *proper value*, (*characteristic value*, or *eigenvalue*), of this system, and a corresponding nonidentically vanishing solution is called a *proper function*, (*characteristic function*, or *eigenfunction*). For the system (7.4) *it is supposed that the coefficients $r(t, \lambda)$, $p(t, \lambda)$, $\alpha_a(\lambda)$, $\beta_a(\lambda)$ are as specified above, and $\alpha_b(\lambda)$, $\beta_b(\lambda)$ satisfy the following condition.*

($\mathbf{H}_b^+$) *$\alpha_b(\lambda)$ and $\beta_b(\lambda)$ are real-valued continuous functions such that $\alpha_b^2(\lambda) + \beta_b^2(\lambda) > 0$ for $\lambda \in \Delta$, and either $\alpha_b(\lambda) \equiv 0$ on Δ or $\alpha_b(\lambda) \neq 0$ for all $\lambda \in \Delta$; moreover, if $\alpha_b(\lambda) \neq 0$ for $\lambda \in \Delta$ then $\beta_b(\lambda)/\alpha_b(\lambda)$ is a monotone nonincreasing function on Δ.*

It will be supposed also that the coefficient functions r, p satisfy the following condition:

($\mathbf{H}_2$) *There exist a subinterval $[a_0, b_0]$ of $[a, b]$ such that*

$$\frac{\max\{p(t,\lambda)\mid t\in[a_0,b_0]\}}{\max\{r(t,\lambda)\mid t\in[a_0,b_0]\}}\to-\infty \quad as \quad \lambda\to\Lambda_2.$$

For $\lambda_\alpha \in \Delta$, $(\alpha = 1, 2)$, $[c, d] \subset I$, and $\eta \in D'[c, d]$, the difference functional $J[\eta; \lambda_1; \lambda_2; c, d]$ is defined to be

$$J[\eta; \lambda_1, \lambda_2; c, d] = J[\eta; \lambda_1; c, d] - J[\eta; \lambda_2; c, d]. \tag{7.5}$$

Theorem 7.2. *Suppose that hypotheses* $(\mathbf{H}_0)$, $(\mathbf{H}_2)$, $(\mathbf{H}_a^+)$, *and* $(\mathbf{H}_b^+)$ *hold for a given subinterval* $[a, b]$ *of* I, *and for* $\Lambda_1 < \lambda' < \lambda'' < \Lambda_2$ *we have*:

(i) *if* $\alpha_a(\lambda) \equiv 0$, $\alpha_b(\lambda) \equiv 0$, *then* $J[\eta; \lambda', \lambda''; a, b]$ *is positive definite on* $D_0'[a, b]$;

(ii) *if* $\alpha_a(\lambda) \not\equiv 0$, $\alpha_b(\lambda) \equiv 0$, *then* $J[\eta; \lambda', \lambda''; a, b]$ *is positive definite on* $D_{*0}'[a, b]$;

(iii) *if* $\alpha_a(\lambda) \equiv 0$, $\alpha_b(\lambda) \not\equiv 0$, *then* $J[\eta; \lambda', \lambda''; a, b]$ *is positive definite on* $D_{0*}'[a, b]$;

(iv) *if* $\alpha_a(\lambda) \not\equiv 0$, $\alpha_b(\lambda) \not\equiv 0$, *then* $J[\eta; \lambda', \lambda''; a, b_0]$ *is positive definite on* $D'[a, b_0]$ *for arbitrary* $[a, b_0] \subset [a, b]$.

Then there exists a positive integer m *such that the proper values of* (7.4) *on* Δ *may be written as an increasing sequence* $\lambda_m < \lambda_{m+1} < \cdots$, *and for* $j \geq m$ *a proper function* $u_j(t)$ *of* (7.4) *for* $\lambda = \lambda_j$ *has exactly* $j - 1$ *zeros on* (a, b); *moreover*, $\{\lambda_j\} \to \Lambda_2$ *as* $j \to \infty$.

In cases (i), (ii) and (iv), consider the solution $u(t, \lambda)$ of (7.1) which with the associated $v(t, \lambda) = r(t, \lambda)u'(t, \lambda)$ satisfies the initial conditions (7.2). In view of conclusion (b) of Problem V.5:5, and the hypothesis $(\mathbf{H}_2)$, it follows that for a given positive integer j the j-th zero $t_j(\lambda)$ of $u(t, \lambda)$ on $[a, b^0)$ exists and satisfies $t_j(\lambda) < b$ provided λ is sufficiently large. Now in view of Lemma 7.1 the function $t_j(\lambda)$ is continuous on its domain of existence. Moreover, with the aid of Theorem 5.2 in case (i), and Theorem 6.4 in cases (ii) and (iv), the positive definiteness of the involved difference functional implies that if $t_j(\lambda_0) \in [a, b]$ then $t_j(\lambda) \in [a, b]$ for $\lambda \in [\lambda_0, \Lambda_2)$, and $t_j(\lambda)$ is a monotone decreasing function on $[\lambda_0, \Lambda_2)$. Let k be the smallest positive integer such that the condition $t_k(\lambda) < b$ is not satisfied for λ in a suitably small neighborhood of Λ_1, and define $\lambda = \mu_j$ as the solution of $t_j(\lambda) = b$, $(j = k, k + 1, \ldots)$. In cases (i), (ii) where $\alpha_b(\lambda) \equiv 0$, the conclusion of the theorem holds for $m = k$ and $\lambda_j = \mu_j$, $(j = k, k + 1, \ldots)$. In case (iv), conclusion (ii) of Theorem 6.4 implies that for $j \geq k$ the function $v(b, \lambda)/u(b, \lambda)$ is monotone decreasing on $\mu_j < \lambda < \mu_{j+1}$, and tends to ∞ and $-\infty$ as $\lambda \to \mu_j^+$ and $\lambda \to u_{j+1}^-$, respectively. Consequently, in view of hypothesis $(\mathbf{H}_b^+)$, in this case for $j \geq k$ there exists a unique value $\lambda = \lambda_{j+1}$ on (μ_j, μ_{j+1}) such that the second boundary condition of (7.4) holds for $\lambda = \lambda_{j+1}$, $u = u(t, \lambda_{j+1})$, $v = v(t, \lambda_{j+1})$, and $u = u(t, \lambda_{j+1})$ has j zeros on (a, b). Now for $\lambda \in (\Lambda_1, \mu_k)$ the function $u(t, \lambda)$ has $k - 1$ zeros on $(a, b]$, and by conclusion (ii) of Theorem 6.4 the function $v(b, \lambda)/u(b, \lambda)$ is monotone decreasing on (Λ_1, μ_k), and tends to $-\infty$ as $\lambda \to \mu_k^-$. As the function $-\beta_b(\lambda)/\alpha_b(\lambda)$ is monotone nondecreasing on (Λ_1, μ_k), it follows that if

$$\lim_{\lambda \to \Lambda_1} \frac{-\beta_b(\lambda)}{\alpha_b(\lambda)} < \lim_{\lambda \to \Lambda_1} \frac{v(b, \lambda)}{u(b, \lambda)}, \tag{7.6}$$

then (7.4) has a single proper value λ_k on (Λ_1, μ_k), the corresponding proper function $u(t, \lambda_k)$ has $k - 1$ zeros on (a, b), and the conclusion of the theorem holds with $m = k$. On the other hand, if (7.6) does not hold then there is no proper value of (7.4) on (Λ_1, μ_k), and the conclusion of the theorem holds for $m = k + 1$.

Now in view of Theorem 2.1 any proper function of (7.4) has only a finite number of zeros on the compact interval $[a, b]$, and, in view of the monotoneity of the zeros $t_j(\lambda)$ as defined above, it follows that $\{\mu_j\} \to \Lambda_2$ and hence $\{\lambda_j\} \to \Lambda_2$ as $j \to \infty$.

Finally, the proof of the stated result in case (iii) is dual to the proof in case (ii), with the roles of $t = a$ and $t = b$ interchanged. Moreover, a corresponding interchange of the roles of $t = a$ and $t = b$ in the proof of case (iv) establishes the conclusion of the theorem under the assumption that hypotheses $(\mathbf{H}_0)$, $(\mathbf{H}_2)$, $(\mathbf{H}_a^+)$, and $(\mathbf{H}_b^+)$ hold and for $\Lambda_1 < \lambda' < \lambda'' < \Lambda_2$ we have:

(iv′) *if* $\alpha_a(\lambda) \neq 0$, $\alpha_b(\lambda) \neq 0$, *then* $J[\eta; \lambda', \lambda''; a_0, b]$ *is positive definite on* $D'[a_0, b]$ *for arbitrary* $[a_0, b] \subset [a, b]$.

Theorem 7.3. *If the hypotheses of Theorem 7.2 hold, and for* $\lambda \in \Delta$ *the functions* $\tau(\lambda)$, $\rho(\lambda)$ *are defined as*

$$\tau(\lambda) = \min\{p(t, \lambda) \mid t \in [a, b]\}, \qquad \rho(\lambda) = \min\{r(t, \lambda) \mid t \in [a, b]\},$$

then the positive integer m in the conclusion of Theorem 7.2 is equal to 1 if:

(a) *case* (i) *holds, and there is a value* $\hat{\lambda} \in \Delta$ *such that* $\tau(\hat{\lambda})/\rho(\hat{\lambda}) > -\pi^2/(b - a)^2$;

(b) *case* (ii) *holds, and there is a value* $\lambda' \in \Delta$ *such that*

$$\tau(\lambda') > 0, \sqrt{\tau(\lambda')\,\rho(\lambda')} + \frac{\beta_a(\lambda')}{\alpha_a(\lambda')} > 0; \tag{7.7′}$$

(c) *case* (iii) *holds, and there is a value* $\lambda'' \in \Delta$ *such that*

$$\tau(\lambda'') > 0, \sqrt{\tau(\lambda'')\,\rho(\lambda'')} + \frac{\beta_b(\lambda'')}{\alpha_b(\lambda'')} > 0; \tag{7.7″}$$

(d) *any one of the cases* (i)–(iv) *holds, and* $\tau(\lambda)/\rho(\lambda) \to \infty$ *as* $\lambda \to \Lambda_1$.

Under condition (a), it follows from (a) of Problem V.5:5 that the differential equation $l[u:\hat{\lambda}] = 0$ is disconjugate on $[a, b]$. Moreover, in this case the hypotheses of Theorem 7.2 imply that if $\lambda \in (\Lambda_1, \hat{\lambda})$ then $J[\eta; \lambda, \hat{\lambda}; a, b]$ is positive definite on $D_0'[a, b]$, and from Theorem 5.2 it follows that

$l[u:\lambda] = 0$ is disconjugate on $[a, b]$. Hence in this case the integer k in the above proof of Theorem 7.2 is equal to 1, and correspondingly $m = 1$.

For the consideration of condition (b), let

$$\hat{J}[\eta; \lambda; a, b] = \frac{\beta_a(\lambda)}{\alpha_a(\lambda)} \eta^2(a) + J[\eta; \lambda; a, b], \quad \text{for} \quad \lambda \in \Delta, \tag{7.8}$$

and

$$\hat{J}_2[\eta; a, b] = \frac{\beta_a(\lambda')}{\alpha_a(\lambda')} \eta^2(a) + \int_a^b [\rho(\lambda')\eta'^2 + \tau(\lambda')\eta^2]\, dt, \tag{7.8'}$$

where λ' is a value for which conditions (7.7′) hold. The solution $u = u_2(t)$ of the differential system

$$\hat{l}_2[u] \equiv \rho(\lambda')u'' - \tau(\lambda')u = 0, \quad u(a) = 1, \quad v(a) \equiv \rho(\lambda')u'(a) = \frac{\beta_a(\lambda')}{\alpha_a(\lambda')}$$

has the specific value

$$u_2(t) = \cosh k(t - a) + \frac{\beta_a(\lambda')/\alpha_a(\lambda')}{\sqrt{\tau(\lambda')\,\rho(\lambda')}} \sinh k(t - a),$$

where $k = \sqrt{\tau(\lambda')/\rho(\lambda')}$, and from (7.7′) it follows readily that $u_2(t) > 0$ for $t > a$. Thus relative to $\hat{J}_2[\eta; a, b]$ there is no right-hand focal point to $t = a$ on $(a, b]$, and, as $\hat{J}[\eta; \lambda'; a, b] - \hat{J}_2[\eta; a, b]$ is non-negative on $D'_{*0}(a, b]$, Theorem 6.3 implies that relative to $\hat{J}[\eta; \lambda'; a, b]$ there is no right-hand focal point to $t = a$ on $(a, b]$. Moreover, since for $\lambda \in (\Lambda_1, \lambda')$ we have $\beta_a(\lambda)/\alpha_a(\lambda) \geq \beta_a(\lambda')/\alpha_a(\lambda')$ and $J[\eta; \lambda, \lambda'; a, b]$ is positive definite on $D'_{*0}[a, b]$, from Theorem 6.3 it follows that if $\lambda \in (\Lambda_1, \lambda')$ then relative to $\hat{J}[\eta; \lambda; a, b]$ there is no right-hand focal point to $t = a$ on $(a, b]$, so that in this case $k = 1$ and $m = 1$.

The consideration of condition (*c*) is dual to that of (*b*), with (7.8), (7.8′) replaced by the respective functionals

$$\frac{\beta_b(\lambda)}{\alpha_b(\lambda)} \eta^2(b) + J[\eta; \lambda; a, b], \tag{7.9}$$

$$\frac{\beta_b(\lambda'')}{\alpha_b(\lambda'')} \eta^2(b) + \int_a^b [\rho(\lambda'')\eta'^2 + \tau(\lambda'')\eta^2]\, dt, \tag{7.9''}$$

where λ'' is a value for which conditions (7.7″) hold.

Finally, suppose that $\tau(\lambda)/\rho(\lambda) \to \infty$ as $\lambda \to \Lambda_1$. In particular, this condition implies the existence of a $\hat{\lambda}$ such that $\tau(\hat{\lambda})/\rho(\hat{\lambda}) > -\pi^2/(b - a)^2$, so that if case (i) holds then $m = 1$ in view of the above conclusion (a).

It is to be noted that the positive definiteness of the difference functional in any one of the cases (i)–(iv) implies that if $\Lambda_1 < \lambda' < \lambda'' < \Lambda_2$, and $[a_0, b_0] \subset [a, b]$, then $J[\eta; \lambda', \lambda''; a_0, b_0]$ is positive definite on $D'_0[a_0, b_0]$.

Conclusion (i) of Theorem 3.3 then implies that $r(t, \lambda') \geq r(t, \lambda'')$, and hence $\rho(\lambda)$ is a nonincreasing function on Δ. Consequently,

$$\sqrt{\tau(\lambda)\rho(\lambda)} = \rho(\lambda)\sqrt{\tau(\lambda)/\rho(\lambda)}$$

also tends to ∞ as $\lambda \to \Lambda_1$, whenever $\tau(\lambda)/\rho(\lambda) \to \infty$ as $\lambda \to \Lambda_1$. Therefore, hypothesis $(\mathbf{H}_a^+)$ and the condition that $\tau(\lambda)/\rho(\lambda) \to \infty$ as $\lambda \to \Lambda_1$ imply the existence of a λ' satisfying conditions (7.7'), so that $m = 1$ in view of the above conclusion (b).

By a similar argument, if case (iii) holds, and $\tau(\lambda)/\rho(\lambda) \to \infty$ as $\lambda \to \Lambda_1$, then there exists a λ'' satisfying conditions (7.7''), and $m = 1$ in view of the above conclusion (c).

Now suppose that case (iv) holds, and $\tau(\lambda)/\rho(\tau) \to \infty$ as $\lambda \to \Lambda_1$. Since also $\sqrt{\tau(\lambda)\rho(\lambda)} \to \infty$ as $\lambda \to \Lambda_1$, for a given c satisfying $0 < c < 1$ there exists a value $\lambda_0 \in \Delta$ such that

$$\text{(7.10)} \qquad \tau(\lambda) > 0, \qquad c\sqrt{\tau(\lambda)\rho(\lambda)} + \frac{\beta_a(\lambda_0)}{\alpha_a(\lambda_0)} > 0, \quad \textit{for} \quad \lambda \in (\Lambda_1, \lambda_0].$$

Let $u_2 = u_2(t, \lambda)$ be the solution of the differential system

$$c\rho(\lambda)u_2'' - c\tau(\lambda)u_2 = 0, \qquad u_2(a,\lambda) = 1,$$
$$v_2(a,\lambda) \equiv c\rho(a)u_2'(a,\lambda) = \beta_a(\lambda_0)/\alpha_a(\lambda_0).$$

Then

$$\text{(7.11)} \quad u_2(t, \lambda) = \cosh[k(\lambda)(t - a)] + \frac{\beta_a(\lambda_0)/\alpha_a(\lambda_0)}{c\sqrt{\tau(\lambda)\,\rho(\lambda)}} \sinh[k(\lambda)(t - a)],$$

where $k(\lambda) = \sqrt{\tau(\lambda)/\rho(\lambda)}$, and the following properties are ready consequences of (7.10), (7.11):

$$\text{(7.12)} \qquad u_2(t, \lambda) > 0, \quad \textit{for} \quad t > a, \qquad \lambda \in (\Lambda_1, \lambda_0];$$

$$\text{(7.13)} \qquad \frac{v_2(b, \lambda)/u_2(b, \lambda)}{c\sqrt{\tau(\lambda)\rho(\lambda)}} \to 1 \quad \textit{as} \quad \lambda \to \Lambda_1.$$

In particular, (7.13) implies that $v_2(b, \lambda)/u_2(b, \lambda) \to \infty$ as $\lambda \to \Lambda_1$. Now consider the functional

$$\hat{J}_c[\eta; \lambda; a, b_0] = \frac{\beta_a(\lambda_0)}{\alpha_a(\lambda_0)}\eta^2(a) + \int_a^{b_0} [c\rho(\lambda)\eta'^2 + c\tau(\lambda)\eta^2]\,dt,$$

for $\lambda \in (\Lambda_1, \lambda_0]$, $\eta \in D'[a, b_0]$, and $[a, b_0] \subset [a, b]$. As $\tau(\lambda) > 0$, and $\beta_a(\lambda)/\alpha_a(\lambda) \geq \beta_a(\lambda_0)/\alpha_a(\lambda_0)$ for $\lambda \in (\Lambda_1, \lambda_0]$, it follows that if $\lambda \in (\Lambda_1, \lambda_0]$ then $\hat{J}[\eta; \lambda; a, b_0] - \hat{J}_c[\eta; \lambda; a, b_0]$ is positive definite on $D'[a, b_0]$ for arbitrary $[a, b_0] \subset [a, b]$. If $u(t, \lambda)$ is the solution of (7.1) satisfying the initial conditions (7.2), it then follows from (7.12) and conclusion (i) of Theorem 6.4 that if $\lambda \in (\Lambda_1, \lambda_0]$ then $u(t, \lambda) \neq 0$ for $t \in [a, b]$, so that the integer k in the proof of Theorem 7.2 is equal to 1. In turn, conclusion (ii) of Theorem 6.4 implies that $v(b, \lambda)/u(b, \lambda) > v_2(b, \lambda)/u_2(b, \lambda)$ for $\lambda \in (\Lambda_0, \lambda_0]$; as $v_2(b, \lambda)/u_2(b, \lambda) \to \infty$ as $\lambda \to \Lambda_1$ it then follows that $v(b, \lambda)/u(b, \lambda) \to \infty$ as $\lambda \to \Lambda_1$. In particular, this latter relation implies that inequality (7.6) holds, which, as noted in the proof of Theorem 7.2, implies $m = k$ and thus $m = 1$.

Of particular interest is a differential system of the form

$$\begin{aligned} &[r(t)u']' - [p(t) - \lambda k(t)]u = 0, \\ &\beta_a u(a) - \alpha_a r(a)u'(a) = 0, \qquad (7.14) \\ &\beta_b u(b) + \alpha_b r(b)u'(b) = 0, \end{aligned}$$

in which λ is a complex parameter, and the following hypothesis is satisfied.

($\mathbf{H}^0$) *On the compact interval* $[a, b]$ *the functions* r, p, k *are real-valued, continuous, with* $r(t) > 0$, $k(t) \not\equiv 0$, *while* $\alpha_a, \beta_a, \alpha_b, \beta_b$ *are real constants such that* $\alpha_a^2 + \beta_a^2 > 0$, $\alpha_b^2 + \beta_b^2 > 0$.

A system (7.14) for which hypothesis ($\mathbf{H}^0$) holds is called a *Sturm-Liouville system.*

Now if $u = u_0(t)$ is a proper solution of (7.14) for a proper value $\lambda = \lambda_0$, then

$$\begin{aligned} \lambda \int_a^b k(t)\,|u_0(t)|^2\,dt &= \int_a^b \overline{u_0}\{-[ru_0']' + pu_0\}\,dt, \\ &= -\overline{u_0}ru_0'\Big|_a^b + \int_a^b \{r\,|u_0'|^2 + p\,|u_0|^2\}\,dt, \qquad (7.15) \\ &= \gamma_a\,|u_0(a)|^2 + \gamma_b\,|u_0(b)|^2 + \int_a^b \{r\,|u_0'|^2 + p\,|u_0|^2\}\,dt, \end{aligned}$$

where

$$\gamma_a = 0 \quad \textit{if} \quad \alpha_a = 0, \qquad \gamma_a = \frac{\beta_a}{\alpha_a} \quad \textit{if} \quad \alpha_a \neq 0;$$

$$\gamma_b = 0 \quad \textit{if} \quad \alpha_b = 0, \qquad \gamma_b = \frac{\beta_b}{\alpha_b} \quad \textit{if} \quad \alpha_b \neq 0.$$

Consequently, in view of (7.15) one has the following result.

***Lemma* 7.2.** *For a Sturm-Liouville system* (7.14) *all proper values are real in each of the following cases:*

(i) $k(t)$ *is of constant sign on* $[a, b]$;

(ii) $k(t)$ *changes sign on* $[a, b]$, $p(t) > 0$ *for* $t \in [a, b]$, *and* $\alpha_a\beta_a \geq 0$, $\alpha_b\beta_b \geq 0$.

In general, if the coefficient functions r, p, k satisfy the conditions specified in ($\mathbf{H}^0$), then there exist real-valued extensions of these functions on an interval (a^0, b^0) containing $[a, b]$ with $r(t) > 0$ on (a^0, b^0), so that the above analysis used in treating system (7.4) is applicable to (7.14). In particular, if

$r(t, \lambda) = r(t)$, $p(t, \lambda) = p(t) - \lambda k(t)$, where $k(t) > 0$ for $t \in [a, b]$, and $-\infty < \lambda' < \lambda'' < \infty$, then

$$J[\eta; \lambda', \lambda''; a_0, b_0] = (\lambda'' - \lambda') \int_{a_0}^{b_0} k(t)\, \eta^2(t)\, dt$$

is positive definite on $D'[a_0, b_0]$ for arbitrary $[a_0, b_0] \subset [a, b]$. Moreover, if $\rho(\lambda)$, $\tau(\lambda)$ denote the respective minima of $r(t, \lambda)$, $p(t, \lambda)$ on $[a, b]$, then $\tau(\lambda)/\rho(\lambda) \to \infty$ as $\lambda \to -\infty$, and the results of Theorems 7.2, 7.3 applied to (7.14) for $\lambda \in (-\infty, \infty)$ yield the following result.

Theorem 7.4. *Suppose that* (7.14) *satisfies hypothesis* ($\mathbf{H}^0$), *and* $k(t) > 0$ *for* $t \in [a, b]$. *Then all proper values of this system are real, the totality of proper values may be written as a sequence* $\{\lambda_j\}$, *where* $\lambda_1 < \lambda_2 < \cdots$, $\{\lambda_j\} \to \infty$ *as* $j \to \infty$, *and a proper function* $u = u_j(t)$ *of* (7.14) *for* $\lambda = \lambda_j$ *has exactly* $j - 1$ *zeros on* (a, b).

If $k(t)$ changes sign on $[a, b]$ then the results of Theorems 7.2, 7.3 cannot be applied directly to (7.14). However, as initially pointed out by M. Bôcher, there is a simple modification of the form of this system to which one may apply results of the type presented in the above theorems.

Theorem 7.5. *Suppose that* (7.14) *satisfies hypothesis* ($\mathbf{H}^0$), $k(t)$ *changes sign on* $[a, b]$, $p(t) > 0$ *for* $t \in [a, b]$, *and* $\alpha_a\beta_a \geq 0$, $\alpha_b\beta_b \geq 0$. *Then all proper values of* (7.14) *are real, the totality of proper values may be written as two sequences* $\{\lambda_j^+\}$, $\{\lambda_j^-\}$, *with* $0 < \lambda_1^+ < \lambda_2^+ < \cdots$, $0 > \lambda_1^- > \lambda_2^- > \cdots$, $\{\lambda_j^+\} \to \infty$ *and* $\{\lambda_j^-\} \to -\infty$ *as* $j \to \infty$, *and a proper function* $u = u_j^+(t)$ *or* $u = u_j^-(t)$ *for respectively* $\lambda = \lambda_j^+$ *or* $\lambda = \lambda_j^-$ *has exactly* $j - 1$ *zeros on* (a, b).

Consider the system (7.14) for $\lambda \in (0, \infty)$ in the form

$$\begin{aligned} \left[\frac{r(t)}{\lambda} u'\right]' - \left[\frac{p(t)}{\lambda} - k(t)\right] u &= 0, \\ \frac{1}{\lambda} \beta_a u(a) - \alpha_a v(a) &= 0, \\ \frac{1}{\lambda} \beta_b u(b) + \alpha_b v(b) &= 0, \end{aligned} \tag{7.16}$$

where now $v(t) = v(t, \lambda)$ is given by $v(t, \lambda) = (1/\lambda)\, r(t)\, u'(t, \lambda)$. The system (7.16) is clearly of the form (7.4), with $r(t, \lambda) = r(t)/\lambda$, $p(t, \lambda) = [p(t)/\lambda] - k(t)$, $\alpha_a(\lambda) = \alpha_a$, $\beta_a(\lambda) = \beta_a/\lambda$, $\alpha_b(\lambda) = \alpha_b$, $\beta_b(\lambda) = \beta_b/\lambda$. It will also be supposed that r, p, k are extended to be continuous, and with $r(t) > 0$, for t on an open interval (a^0, b^0) containing $[a, b]$. In view of the hypothesis that $k(t)$ changes sign on $[a, b]$ and $\alpha_a\beta_a \geq 0$, $\alpha_b\beta_b \geq 0$, it then follows that for $(t, \lambda) \in (a^0, b^0) \times (0, \infty)$ the system (7.16) satisfies hypotheses ($\mathbf{H}_0$), ($\mathbf{H}_2$),

(H_a^+) and (H_b^+). Moreover, for such a system the functional (7.5) is given by

$$J[\eta; \lambda_1, \lambda_2; c, d] = (\lambda_1^{-1} - \lambda_2^{-1}) \int_c^d [r(t)\eta'^2 + p(t)\eta^2]\, dt,$$

so that if $0 < \lambda' < \lambda'' < \infty$ then $J[\eta; \lambda', \lambda''; a_0, b_0]$ is positive definite on $D'[a_0, b_0]$ for arbitrary $[a_0, b_0] \subset [a, b]$. Consequently, from Theorem 7.2 it follows that there exists an integer m such that the proper values of (7.16) on $(0, \infty)$ may be written as a sequence $\lambda_m^+ < \lambda_{m+1}^+ < \cdots$, and for $j \geq m$ a proper function $y = u_j^+(t)$ of (7.16) has exactly $j - 1$ zeros on (a, b); moreover, $\{\lambda_j^+\} \to \infty$ as $j \to \infty$.

Now to prove that $m = 1$, consider the solution $u(t, \lambda)$ of the differential equation of (7.14) such that $u(a, \lambda) = \alpha_a$, $r(a)\, u'(a, \lambda) = \beta_a$. The functions $u(t, \lambda)$ and $u'(t, \lambda)$ are continuous in (t, λ) on $[a, b] \times [0, \infty)$, and $u(t, \lambda) \to u(t, 0)$, $u'(t, \lambda) \to u'(t, 0)$ uniformly on $[a, b]$ as $\lambda \to 0^+$. As conclusion (ii) of Lemma 3.1 implies that

$$u(t, \lambda)r(t)u'(t, \lambda) = \alpha_a\beta_a + \int_a^t \{r(s)u'^2(s, \lambda) + [p(s) - \lambda k(s)]u^2(s, \lambda)\}\, ds,$$

and the hypotheses of the theorem require $\alpha_a\beta_a \geq 0$, $p(t) > 0$ for $t \in [a, b]$, it follows that if $\lambda_0 > 0$ is such that $p(t) - \lambda k(t) > 0$ for $(t, \lambda) \in [a, b] \times [0, \lambda_0]$, then $u(t, \lambda)r(t)u'(t, \lambda) > 0$ for $(t, \lambda) \in (a, b] \times [0, \lambda_0]$. In particular, for $\lambda \in (0, \lambda_0]$ we have $u(t, \lambda) \neq 0$ for $t \in (a, b]$, so that the integer k in the proof of Theorem 7.2 is equal to 1, and also $m = 1$ if $\alpha_b = 0$. On the other hand, if $\alpha_b \neq 0$ then the function $v(b, \lambda)/u(b, \lambda) = (1/\lambda)r(b)u'(b, \lambda)/u(b, \lambda)$ is such that $v(b, \lambda)/u(b, \lambda) \to \infty$ as $\lambda \to 0^+$, since $r(b)u'(b, \lambda)/u(b, \lambda) \to r(b)u'(b, 0)/u(b, 0) = u(b, 0)r(b)u'(b, 0)/u^2(b, 0) > 0$. In view of the condition $\alpha_b\beta_b \geq 0$ it then follows that

$$\lim_{\lambda \to 0^+} \frac{-\beta_b}{\lambda\alpha_b} \leq 0 < \lim_{\lambda \to 0^+} \frac{v(b, \lambda)}{u(b, \lambda)},$$

so that inequality (7.6) holds, and $m = 1$ in this case also.

Finally, the results of Theorem 7.5 concerning negative proper values follow from the preceding case of positive proper values for the related system obtained from (7.14) upon replacing $k(t)$ by $-k(t)$.

Problems V.7

1. Determine the sequence of proper values $\{\lambda_j\}$, $\lambda_1 < \lambda_2 < \cdots$, of the boundary value problem

$$u'' + \lambda u = 0, \qquad u(0) = 0, \qquad u(\pi) + u'(\pi) = 0.$$

Prove that $\lambda_j/(j - \frac{1}{2})^2 \to 1$ as $j \to \infty$.

2. For a system (7.14) satisfying hypothesis ($\mathbf{H}^0$), and with $k(t) > 0$, use the trigonometric transformation to establish the result of Theorem 7.4.

3. Verify the result of Theorem 7.3 for the differential system

$$\begin{aligned} u'' + \lambda u &= 0, \\ u(0) &= 0, \\ (-1 - \lambda)\, u(1) + u'(1) &= 0. \end{aligned}$$

4. Suppose that $g(t)$ is a real-valued continuous function on $[a, b]$, and that $\phi(t)$ is a non-trivial real-valued solution of the differential system

$$\phi'' + g(t)\phi = 0, \qquad \phi'(a) = 0 = \phi'(b).$$

(For example, one might have $a = 0$, $b = \pi$, $g(t) \equiv 1$, $\phi(t) = \cos t$). Show that the boundary value problem

$$u'' + \frac{g(t)\, \phi^2(t)}{1 + \phi^2(t)}\, u + \lambda\, \frac{g(t)\, \phi(t)}{1 + \phi^2(t)}\, u = 0, \qquad u'(a) = 0 = u'(b),$$

has the proper value $\lambda = i$ with corresponding proper function $u(t) = \phi(t) - i$. Wherein do the hypotheses of Theorems 7.4 and 7.5 fail to be satisfied in this example?

8. FUNDAMENTAL QUADRATIC FORMS FOR CONJUGATE AND FOCAL POINTS

In this section we shall continue to consider the differential equation (1.1) with coefficients satisfying hypothesis (**H**), and $r(t) > 0$ for $t \in I$. If I_0 is compact subinterval of I, then there exists a $\delta > 0$ such that (1.1) is disconjugate on any subinterval of I_0 of length not exceeding δ. Indeed, if u_1 and u_2 are real-valued solutions of (1.1) with the constant function $\{u_1, u_2\} = v_2 u_1 - u_2 v_1$ equal to 1, and for $s \in I$ the functions $u(t, s)$, $v(t, s)$ are defined as

$$u(t, s) = \det\begin{bmatrix} u_1(t) & u_2(t) \\ v_1(s) & v_2(s) \end{bmatrix}, \qquad v(t, s) = \det\begin{bmatrix} v_1(t) & v_2(t) \\ v_1(s) & v_2(s) \end{bmatrix},$$

then, as function of t, $u = u(t, s)$ is a real-valued solution of (1.1) with canonical variable $v = v(t, s)$ such that $u(s) = 1$, $v(s) = 0$. For a compact subinterval I_0 of I the function $u(t, s)$ is uniformly continuous in (t, s) on $I_0 \times I_0$, and in view of Theorems 3.4 and 5.1 a possible choice of δ is a value such that $|u(t, s) - u(s, s)| < 1$ for $t \in I_0$, $s \in I_0$, $|t - s| < \delta/2$.

For a compact subinterval $[a, b]$ of I a partition

$$\Pi : a = t_0 < t_1 < \cdots t_m < t_{m+1} = b$$

will be called a *fundamental partition of* $[a, b]$, (*relative to* (1.1), *or* (1.2)), *provided* (1.1) *is disconjugate on each of the associated subintervals* $[t_{j-1}, t_j]$, $(j = 1, \ldots, m + 1)$. The result of the preceding paragraph shows that fundamental partitions of $[a, b]$ always exist, and provides an estimate for the number of points occurring in such a partition. It is to be emphasized, however, that there is no finite upper bound to the number of points in a fundamental partition. In particular, if Π is a fundamental partition of $[a, b]$ then any refinement of Π, that is, a partition obtained by adding further partition points to those of Π, is also a fundamental partition of $[a, b]$.

Now for a fundamental partition $\Pi : a = t_0 < t_1 < \cdots t_m < t_{m+1} = b$, let $\mathcal{S}(\Pi)$ denote the totality of real sequences $\xi = (\xi_0, \xi_1, \ldots, \xi_m, \xi_{m+1})$. As (1.1) is disconjugate on $[t_{j-1}, t_j]$, $(j = 1, \ldots, m + 1)$, there exists a unique solution $u = u^{(j)}(t)$ of (1.1) such that

$$(8.1) \qquad u^{(j)}(t_{j-1}) = \xi_{j-1}, \qquad u^{(j)}(t_j) = \xi_j, \qquad (j = 1, \ldots, m + 1).$$

Indeed, if u_1 and u_2 are real-valued linearly independent solutions of (1.1), and for c and d distinct values on $[a, b]$ we set

$$D[c, d] = \det\begin{bmatrix} u_1(d) & u_2(d) \\ u_1(c) & u_2(c) \end{bmatrix},$$

then $D[c, d] \neq 0$ if both c and d belong to a subinterval $[t_{j-1}, t_j]$, and

$$(8.2) \qquad u^{(j)}(t) = \frac{1}{D[t_{j-1}, t_j]} \det\begin{bmatrix} u_1(t) & u_2(t) & 0 \\ u_1(t_{j-1}) & u_2(t_{j-1}) & \xi_{j-1} \\ u_1(t_j) & u_2(t_j) & \xi_j \end{bmatrix},$$

with corresponding canonical variable

$$(8.3) \qquad v^{(j)}(t) = \frac{1}{D[t_{j-1}, t_j]} \det\begin{bmatrix} v_1(t) & v_2(t) & 0 \\ u_1(t_{j-1}) & u_2(t_{j-1}) & \xi_{j-1} \\ u_1(t_j) & u_2(t_j) & \xi_j \end{bmatrix}.$$

For $\xi \in \mathcal{S}(\Pi)$ the function $u_\xi(t)$ defined by

$$(8.4) \qquad u_\xi(t) = u^{(j)}(t), \qquad t \in [t_{j-1}, t_j], \qquad (j = 1, \ldots, m + 1),$$

is such that $u_\xi \in D'[a, b] : v_\xi$, where $v_\xi(t)$ is the piecewise continuous function such that

$$(8.5) \qquad v_\xi(t) = v^{(j)}(t), \qquad t \in (t_{j-1}, t_j), \qquad (j = 1, \ldots, m + 1).$$

The subclass of elements $\xi = (\xi_0, \xi_1, \ldots, \xi_{m+1})$ of $\mathcal{S}(\Pi)$ for which $\xi_{m+1} = 0$ will be denoted by $\mathcal{S}_{*0}(\Pi)$. Similarly, $\mathcal{S}_{0*}(\Pi)$ will denote the subclass of elements ξ for which $\xi_0 = 0$, and $\mathcal{S}_0(\Pi) = \mathcal{S}_{*0}(\Pi) \cap \mathcal{S}_{0*}(\Pi)$, the subclass of elements ξ for which $\xi_0 = 0 = \xi_{m+1}$. Consequently, the function u_ξ belongs to $D_0'[a, b]$, $D_{*0}'[a, b]$, or $D_{0*}'[a, b]$, according as ξ is a member of the respective class $\mathcal{S}_0(\Pi)$, $\mathcal{S}_{*0}(\Pi)$, or $\mathcal{S}_{0*}(\Pi)$.

Now for Π a fundamental partition of $[a, b]$, consider the functional

$$(8.6)\quad Q^0[\xi, \zeta \mid \Pi] = J[u_\xi, u_\zeta; a, b], \quad \textit{for} \quad \xi \in \mathcal{S}_0(\Pi), \qquad \zeta \in \mathcal{S}_0(\Pi).$$

Since u_ξ and u_ζ belong to $D_0'[a, b]$, and $J[u_\xi, u_\zeta; a, b] = J[u_\zeta, u_\xi; a, b]$, it follows that

$$(8.7)\quad \begin{aligned} Q^0[\xi, \zeta \mid \Pi] &= \sum_{\alpha,\beta=1}^{m} Q^0_{\alpha\beta}[\Pi]\zeta_\alpha\xi_\beta, \\ Q^0_{\alpha\beta}[\Pi] &= Q^0_{\beta\alpha}[\Pi], \qquad (\alpha, \beta = 1, \ldots, m). \end{aligned}$$

As with other quadratic functionals, we write $Q^0[\xi \mid \Pi]$ for $Q^0[\xi, \xi \mid \Pi]$.

Since for $\xi \in \mathcal{S}_0(\Pi)$ the function u_ξ is a solution of (1.1) on each component subinterval $[t_{j-1}, t_j]$ with corresponding canonical variable v_ξ, with the aid of (ii) of Lemma 3.1 it follows that

$$(8.8)\quad Q^0[\xi, \zeta \mid \Pi) = \sum_{j=1}^{m} \zeta_j[v_\xi(t_j^-) - v_\xi(t_j^+)], \quad \textit{for} \quad \xi \in \mathcal{S}_0(\Pi), \qquad \zeta \in \mathcal{S}_0(\Pi).$$

Therefore, in view of the continuity properties of the functions (8.2), (8.3), one has the following preliminary result.

***Lemma* 8.1.** *If for a fixed positive integer m the symbol T_m denotes the set of points $(t_0, t_1, \ldots, t_m, t_{m+1})$ belonging to fundamental partitions $a = t_0 < t_1 < \cdots < t_m < t_{m+1} = b$ of compact subintervals $[a, b]$ of I, then the coefficients $Q^0_{\alpha\beta}[\Pi]$ are continuous functions of $(t_0, t_1, \ldots, t_m, t_{m+1})$ on T_m.*

In general, a real quadratic form $Q[z] \equiv \sum_{i,j=1}^N Q_{ij}z_iz_j$, with $Q_{ij} = Q_{ji}$, is said to be *singular* if the matrix $[Q_{ij}]$ is singular; an equivalent condition is that $\lambda = 0$ is a proper value of $Q[z]$, i.e.,

$$(8.9)\quad \Delta_Q(\lambda) \equiv \det[\lambda E - Q] = 0,$$

or that there exists a nonzero N-tuple $z^0 \equiv (z_i^0)$ such that $Q[z^0; z] \equiv \sum_{i,j=1}^N Q_{ij}z_j^0 z_i$ is zero for arbitrary real N-tuples z. The order of $\lambda = 0$ as a root of (8.9), or equivalently, the dimension of the null space

$$\left\{z \;\middle|\; z = (z_i),\ \sum_{j=1}^{N} Q_{ij}z_j = 0, \qquad (i = 1, \ldots, N)\right\},$$

is called the *nullity* of $Q[z]$. Such a quadratic form $Q[z]$ is said to have (*negative*) *index* equal to r, if r is the largest non-negative integer such that there is a subspace of $\mathbf{R}_N$ of dimension r on which $Q[z]$ is negative definite; in particular, $Q[z]$ has index equal to zero if and only if $Q[z] \geq 0$ for all $z \in \mathbf{R}_N$. From the elementary theory of quadratic forms, (see Sections 1 and 2 of Appendix E), one has that $Q[z]$ has index equal to r if and only if $Q[z]$ has r

negative proper values, where each proper value is counted according to its multiplicity. Correspondingly, the sum of the nullity and index of $Q[z]$ is equal to s if and only if s is the largest non-negative integer such that there is a subspace of $\mathbf{R}_N$ of dimension s on which $Q[z]$ is nonpositive, or, equivalently, if s is the number of nonpositive roots of (8.9), where each root is counted according to its multiplicity. Moreover, if the Q_{ij} are continuous functions of a parameter $\tau \equiv (\tau_1, \ldots, \tau_q)$ on a set $\mathfrak{D}$ in q-dimensional Euclidean space, then the roots of (8.9) are continuous functions of τ on $\mathfrak{D}$.

Lemma 8.2. *If $\Pi: a = t_0 < t_1 < \cdots < t_m < t_{m+1} = b$ is a fundamental partition of $[a, b]$, then $Q^0[\xi \mid \Pi]$ is singular if and only if $t = b$ is conjugate to $t = a$. Moreover, if $Q^0[\xi \mid \Pi]$ is singular then its nullity is equal to one.*

If $t = b$ is conjugate to $t = a$, and $u(t)$ is a real-valued nonidentically vanishing solution of (1.1) with $u(a) = 0 = u(b)$, then for $\xi_i[u] = u(t_i)$, $(i = 0, 1, \ldots, m + 1)$, we have $u_{\xi[u]}(t) \equiv u(t)$, and, in particular, $\xi[u]$ is a non-null element of $\mathcal{S}_0(\Pi)$. Moreover, for arbitrary $\zeta \in \mathcal{S}_0(\Pi)$ we have $u_\zeta \in D_0'[a, b]$ and $Q[\xi[u], \zeta \mid \Pi] = J[u, u_\zeta; a, b] = 0$, where the latter equality follows from Theorem 3.1. Consequently $\sum_{j=1}^m Q_{ij}^0 [\Pi]\, \xi_j[u] = 0$, $(i = 1, \ldots, m)$, and the quadratic form $Q^0[\xi \mid \Pi]$ is singular

On the other hand, suppose $Q^0[\xi \mid \Pi]$ is singular. Then there exists a non-null $\xi \in \mathcal{S}_0(\Pi)$ such that $Q[\xi; \zeta \mid \Pi] = 0$ for arbitrary $\zeta \in \mathcal{S}_0(\Pi)$, and from (8.8) it follows that $v_\xi(t_j^-) = v_\xi(t_j^+)$ for $j = 1, \ldots, m$. Consequently, $u = u_\xi(t)$ is a solution of (1.1) on the entire interval $[a, b]$, with corresponding canonical variable $v_\xi(t)$, and as $u_\xi(a) = \xi_0 = 0$, $u_\xi(b) = \xi_{m+1} = 0$, it follows that $u_\xi(t)$ is a nonidentically vanishing solution of (1.1) which vanishes at $t = a$ and $t = b$, so that $t = b$ is conjugate to $t = a$.

The final statement of the Lemma is a consequence of the fact that if $t = a$ is conjugate to $t = b$, then an arbitrary solution $u(t)$ of (1.1) that vanishes at both $t = a$ and $t = b$ is a constant multiple of any one such solution that is nonidentically vanishing.

Now if $\Pi: a = t_0 < t_1 < \cdots < t_m < t_{m+1} = b$ is a fundamental partition of $[a, b]$, and Π' is a partition obtained by inserting an additional division point t' which belongs to (t_{k-1}, t_k), then clearly Π' is also a fundamental partition of $[a, b]$. Moreover, if $\xi' \in \mathcal{S}_0(\Pi')$, and $\xi \in \mathcal{S}_0(\Pi)$ with $\xi_i = \xi_i'$, $(i = 1, \ldots, k - 1)$, $\xi_i = \xi_{i+1}'$, $(i = k, k + 1, \ldots, m + 1)$, then $J[u_{\xi'}; t_{j-1}, t_j] = J[u_\xi; t_{j-1}, t_j]$, $j = 1, \ldots, m + 1, j \neq k$, while in view of Corollary 2 to Theorem 3.1 we have $J[u_{\xi'}; t_{k-1}, t_k] \geq J[u_\xi; t_{k-1}, t_k]$, with the equality sign holding only if $\xi_k' = u_\xi(t')$. Therefore, for ξ thus specified in terms of a given $\xi' \in \mathcal{S}_0(\Pi')$ it follows that $Q^0[\xi \mid \Pi] \leq Q^0[\xi' \mid \Pi']$, and consequently the index of $Q^0[\xi' \mid \Pi']$ does not exceed the index of $Q^0[\xi \mid \Pi]$. On the other hand, if $\xi \in \mathcal{S}_0(\Pi)$, and $\xi' \in \mathcal{S}_0(\Pi')$ is defined as $\xi_i' = \xi_i$, $(i = 1, \ldots, k - 1)$, $\xi_k' = u_\xi(t')$, $\xi_i' = \xi_{i-1}$, $(i = k + 1, \ldots, m + 2)$, then

$Q^0[\xi' \mid \Pi'] = Q^0[\xi \mid \Pi]$, and consequently the index of $Q^0[\xi \mid \Pi]$ does not exceed the index of $Q^0[\xi' \mid \Pi']$. Hence it has been established that the quadratic form $Q^0[\xi' \mid \Pi']$ has the same index as $Q^0[\xi \mid \Pi]$, whenever Π is a fundamental partition of $[a, b]$, and Π' is a refinement of Π obtained by inserting a single additional division point. Now any partition Π^2 that is a refinement of Π is the result of a finite number of successive refinements, each of which involves the insertion of a single additional division point, and thus the index of the corresponding form $Q^0[\xi \mid \Pi^2]$ is equal to the index of $Q^0[\xi \mid \Pi]$. Finally, if Π^1 and Π^2 are two fundamental partitions of $[a, b]$, and Π'' denotes the partition whose division points consist of those points belonging either to Π^1 or Π^2, then Π'' is a refinement of each of the fundamental partitions Π^1 and Π^2. In view of this result and Lemma 8.2, we then have the following theorem.

Theorem 8.1. *If Π^1 and Π^2 are both fundamental partitions of a compact subinterval $[a, b]$ of I, then the two quadratic forms $Q^0[\xi \mid \Pi^1]$ and $Q^0[\xi \mid \Pi^2]$ have the same index, and the same nullity.*

Lemma 8.3. *Suppose that $[a, b_1] \subset [a, b_2] \subset I$, and Π^1, Π^2 are fundamental partitions of $[a, b_1]$ and $[a, b_2]$, respectively. If i_1, n_1 denote the index and nullity of $Q^0[\xi \mid \Pi^1]$, and i_2, n_2 denote the index and nullity of $Q_0[\xi \mid \Pi^2]$, then $i_1 \leq i_2$ and $i_1 + n_1 \leq i_2 + n_2$.*

In view of Theorem 8.1 it may be assumed that $\Pi^1 : a = t_0 < t_1 < \cdots < t_{m+1} = b_1$, and $\Pi^2 : a = t_0 < t_1 < \cdots < t_{m+1} < \cdots < t_q = b_2$, and the result of the Lemma is a direct consequence of the fact that if $\xi \in \mathcal{S}_0(\Pi^1)$, then $\zeta_i = \xi_i$, $(i = 1, \ldots, m + 1)$, $\zeta_i = 0$, $(i = m + 2, \ldots, q)$, defines an element $\zeta \in \mathcal{S}_0(\Pi^2)$ such that $Q^0[\zeta \mid \Pi^2] = Q^0[\xi \mid \Pi^1]$.

The basic result relating the quadratic form $Q^0[\xi \mid \Pi]$ to the existence of conjugate points is the following theorem.

Theorem 8.2. *If Π is a fundamental partition of a compact subinterval $[a, b]$ of I, then the index of $Q^0[\xi \mid \Pi]$ is equal to the number of points on the open interval (a, b) which are conjugate to $t = a$.*

If Π is a fundamental partition $a = t_0 < t_1 < \cdots < t_m < t_{m+1} = b$ of $[a, b]$, and $\eta \in D_0'[a, b]$, then for $\xi_i[\eta] = \eta(t_i)$ it follows with the aid of Corollary 2 to Theorem 3.1 that $J[\eta; a, b] \geq Q^0[\xi[\eta] \mid \Pi]$, with the equality sign holding if and only if $\eta(t) = u_{\xi[\eta]}(t)$ for $t \in [a, b]$. Consequently, $J[\eta; a, b]$ is non-negative or positive definite on $D_0'[a, b]$ if and only if $Q^0[\xi \mid \Pi]$ is correspondingly non-negative or positive definite on $\mathcal{S}_0[\Pi]$. From Theorem 5.1 it then follows that the index of $Q^0[\xi \mid \Pi]$ is zero if and only if there are no points on (a, b) which are conjugate to $t = a$.

Now suppose that $[a, b] \subset I$, and denote by $c_1, \ldots, c_k$ the points on (a, b) which are conjugate to $t = a$, with $a < c_1 < c_2 < \cdots c_k < b$. Let m

be a positive integer such that $(b - a)/(m + 1) < \delta$, where δ is a positive constant such that (1.1) is disconjugate on any subinterval of $[a, b]$ with length not exceeding δ. If $\Pi^\theta : a = t_0{}^\theta < t_1{}^\theta < \cdots < t^\theta_{m+1}$, where $t_j{}^\theta = a + j\theta(b - a)/(m + 1)$, then Π^θ is a fundamental partition of $[a, t^\theta_{m+1}]$ whenever $0 < \theta \leq 1$. For θ sufficiently small we have $a < t^\theta_{m+1} < c_1$ and $Q^0[\xi \mid \Pi^\theta]$ is positive definite in view of the above result. Moreover, if i_θ and n_θ denote the respective index and nullity of $Q^0[\xi \mid \Pi^\theta]$, then, in view of Lemma 8.3, i_θ and $i_\theta + n_\theta$ are non-decreasing functions of θ on $(0, 1]$. Now i_θ and $i_\theta + n_\theta$ denote, respectively, the number of negative and non-positive proper values of $Q^0[\xi \mid \Pi^\theta]$. Also, if $\lambda_1(\theta) \leq \lambda_2(\theta) \leq \cdots \leq \lambda_m(\theta)$ denote the proper values of $Q^0[\xi \mid \Pi^\theta]$ in non-decreasing order, then as a consequence of Lemma 8.1 each $\lambda_\alpha(\theta)$ is a continuous function of θ on $(0, 1]$. Therefore, if $0 < \theta_0 < 1$ and $\lambda_\alpha(\theta_0) = 0$, from the nondecreasing nature of $i_\theta + n_\theta$ it follows that $\lambda_\alpha(\theta) \leq 0$ for $\theta \in (\theta_0, 1]$. Indeed, since the points conjugate to $t = a$ are isolated, and $\lambda_\alpha(\theta) = 0$ implies that t^θ_{m+1} is conjugate to $t = a$, it follows that if $\lambda_\alpha(\theta^0) = 0$ then $\lambda_\alpha(\theta) < 0$ for $\theta \in (\theta^0, 1]$. Hence, the index of $Q^0[\xi \mid \Pi^\theta]$ remains constant as θ ranges over the interval specified by the condition $a + \theta(b - a) \in (c_{j-1}, c_j]$, where $c_0 = a$, and the index of this form increases by 1 as θ passes through the value determined by $a + \theta(b - a) = c_j$, $(j = 1, \ldots, k)$. Consequently, for $\theta = 1$ the index of $Q^0[\xi \mid \Pi^1]$ is equal to k, the number of points on (a, b) conjugate to $t = a$.

Now for $[a, b]$ a compact subinterval of I, consider the quadratic functional

$$\begin{aligned} \hat{J}[\eta; a, b] &= \gamma\, \eta^2(a) + \int_a^b [r(t)\eta'^2 + 2q(t)\eta'\eta + p(t)\eta^2]\, dt, \\ &= \gamma\, \eta^2(a) + J[\eta; a, b], \end{aligned} \tag{8.10}$$

as defined by (6.1), where γ is a real constant. For $\Pi : a = t_0 < t_1 < \cdots < t_{m+1} = b$, define

(8.11)

$$Q^{*0}[\xi, \zeta \mid \Pi] = \hat{J}[u_\xi, u_\zeta; a, b], \quad \textit{for} \quad \xi \in \mathcal{S}_{*0}(\Pi), \qquad \zeta \in \mathcal{S}_{*0}(\Pi),$$

and, as usual, write $Q^{*0}[\xi \mid \Pi]$ for $Q^{*0}[\xi, \xi \mid \Pi]$. Corresponding to (8.7) and (8.8), we now have the relations

$$\begin{aligned} Q^{*0}[\xi, \zeta \mid \Pi] &= \sum_{\alpha,\beta=0}^{m} Q^{*0}_{\alpha\beta}[\Pi]\zeta_\alpha\xi_\beta, \\ Q^{*0}_{\alpha\beta}[\Pi] &= Q^{*0}_{\beta\alpha}[\Pi], \qquad (\alpha, \beta = 0, 1, \ldots, m), \\ Q^{*0}[\xi, \zeta \mid \Pi] &= \zeta_0[\gamma\xi_0 - v_\xi(a)] + \sum_{j=1}^{m} \zeta_j[v_\xi(t_j^-) - v_\xi(t_j^+)], \\ &\qquad \textit{for} \quad \xi \in \mathcal{S}_{*0}(\Pi), \qquad \zeta \in \mathcal{S}_{*0}(\Pi). \end{aligned} \tag{8.12}$$

By an argument similar to that employed in the proof of Lemma 8.2, and with the aid of results of Theorem 6.1, its Corollary, and Theorem 6.2, one may establish the following result relating this quadratic form to the existence of right-hand focal points to $t = a$ relative to the functional (8.10), or relative to the differential equation (1.1) with initial condition $\gamma u(a) - v(a) = 0$.

***Lemma* 8.4.** *If* Π *is a fundamental partition of* $[a, b]$, *then* $Q^{*0}[\xi \mid \Pi]$ *is singular if and only if* $t = b$ *is a right-hand focal point to* $t = a$ *relative to the functional* $\hat{J}[\eta; a, b]$ *of* (8.10). *Moreover, if* $Q^{*0}[\xi \mid \Pi]$ *is singular then its nullity is equal to one.*

Results corresponding to those of Theorems 8.1, Lemma 8.3, and Theorem 8.2 hold for the form $Q^{*0}[\xi \mid \Pi]$; in particular, the analogue of Theorem 8.2 is the following basic result.

Theorem 8.3. *If* Π *is a fundamental partition of* $[a, b]$, *then the index of* $Q^{*0}[\xi \mid \Pi]$ *is equal to the number of points on* (a, b) *which are right-hand focal points to* $t = a$, *relative to the functional* $\hat{J}[\eta; a, b]$ *of* (8.10).

To establish corresponding results for left-hand focal points to $t = b$, relative to the functional

$$\check{J}[\eta; a, b] = -\gamma\eta^2(b) + J[\eta; a, b], \tag{8.13}$$

or relative to the differential equation (1.1) with initial condition

$$-\gamma u(b) + v(b) = 0,$$

consider the form

$$Q^{0*}[\xi, \zeta \mid \Pi] = \check{J}[u_\xi, u_\zeta; a, b], \quad for \quad \xi \in \mathcal{S}_{0*}(\Pi), \qquad \zeta \in \mathcal{S}_{0*}(\Pi). \tag{8.14}$$

Corresponding to Theorem 8.3, one has the following result.

Theorem 8.4. *If* Π *is a fundamental partition of* $[a, b]$, *then the index of* $Q^{0*}[\xi \mid \Pi]$ *is equal to the number of points on* (a, b) *which are left-hand focal points to* $t = b$, *relative to the functional* $\check{J}[\eta; a, b]$ *of* (8.13).

In regard to the chosen forms for $\hat{J}$ and $\check{J}$ defined respectively by (8.10) and (8.13), it is to be noted that if $u(t)$ is any nonidentically vanishing solution of (1.1) on I, and $t = c$ is a point on (a, b) where $u(c) \neq 0$, then the zeros of $u(t)$ on $(c, b]$ are the right-hand focal points to $t = c$ relative to the functional $\hat{J}[\eta; c, b]$, and the zeros of $u(t)$ on $[a, c)$ are the left-hand focal points to $t = c$ relative to the functional $\check{J}[\eta; a, c]$, where *in each case* $\gamma = v(c)/u(c)$.

The introduction of the algebraic quadratic forms Q^0, Q^{*0}, Q^{0*}, and their systematic use, was due to Marston Morse, who employed them as basic tools in extending the results of the Sturm-Liouville theory for scalar real second-order linear differential equations to self-adjoint systems of the second order.

Problems V.8

1. Show that the index of $Q^0[\xi \mid \Pi]$ is equal to r if and only if r is the largest integer such that $J[\eta; a, b]$ is negative definite on an r-dimensional subspace of $D_0'[a, b]$, i.e., there exist r functions $\eta^\sigma(t) \in D_0'[a, b]$ for $\sigma = 1, 2, \ldots, r$, and $J[\sum_{\sigma=1}^n c_\sigma \eta^\sigma; a, b] < 0$ if $(c_\sigma) \neq 0$.

2. State and prove a result on the sum of the index and nullity of $Q^0[\xi \mid \Pi]$, corresponding to that of Problem 1 for the index.

3. For each of the quadratic forms $Q^{*0}[\xi \mid \Pi]$ and $Q^{0*}[\xi \mid \Pi]$ state and prove results corresponding to those of Problems 1 and 2.

4. If $\Pi: a = t_0 < t_1 < \cdots < t_{m+1} = b$ is a fundamental partition of $[a, b]$ relative to (1.1), for $k = 0, 1, \ldots, m+1$ let $\xi^k = (\xi_0^k, \ldots, \xi_{m+1}^k)$, with $\xi_\alpha^k = 0$ for $\alpha \neq k$, $\xi_k^k = 1$, and set $\hat{u}_k(t) = u_{\xi^k}(t)$, $(k = 0, 1, \ldots, m+1)$. Show that if $\xi \in \mathcal{S}(\Pi)$ and $\zeta \in \mathcal{S}(\Pi)$ then $J[u_\xi, u_\zeta; a, b] = \sum_{h,k=0}^{m+1} Q_{hk}[\Pi]\zeta_h\xi_k$ with $Q_{hk}[\Pi] = J[\hat{u}_k, \hat{u}_h; a, b]$; moreover $Q_{hk}(\Pi) = Q_{kh}(\Pi)$, and $Q_{hk}(\Pi) = 0$ if $h, k = 0, 1, \ldots, m+1$ and $|h - k| > 1$.

5. For $a = 0$, and $m\pi/2 < b \leq (m+1)\pi/2$, $m = 1, 2, \ldots$, consider the quadratic functional

$$J[\eta; 0, b] = \int_0^b \{\eta'^2 - \eta^2\}\, dt, \tag{8.15}$$

and the fundamental partition

$$\Pi: 0 = t_0 < t_1 < \cdots < t_m < t_{m+1} = b, \tag{8.16}$$

of $[0, b]$, with $t_j = j\pi/2$, $(j = 1, \ldots, m)$.

(i) Compute the corresponding quadratic form $Q^0[\xi \mid \Pi] = Q^0[\xi, \xi \mid \Pi]$ as defined by (8.6), and verify directly the result of Theorem 8.2 for this particular functional and fundamental partition (8.16).

(ii) For $\hat{J}[\eta; 0, b] = J[\eta; 0, b]$, as defined by (8.15), compute the corresponding quadratic form $Q^{*0}[\xi \mid \Pi] = Q^{*0}[\xi, \xi \mid \Pi]$ as defined by (8.11), and verify directly the result of Theorem 8.3 for this particular functional and fundamental partition (8.16).

9. NOTES AND REMARKS

The central results of this chapter deal with properties of solutions of a real scalar linear homogeneous differential equation of the second order, notably those results on oscillation and comparison that form the basis of the famous theory of Sturm [1] for such equations. Related references are Bôcher [4; 6], Ince [1, Ch. X], Sansone [1-I, Ch. IV], Morse [1, 2, Ch. IV], Birkhoff and

Rota [1, Chs. II, X], Hartman [7, Ch. XI], and Hille [4, Ch. 8]. The treatment presented herein is consistently related to the behavior of associated quadratic functionals, and thus may be considered to be essentially variational in nature, although no previous knowledge of the calculus of variations is needed for a complete understanding of the material.

For additional discussion of scalar Riccati differential equations, the reader is referred to Ince [1, Ch. II] and Sansone [1-I, Ch. II]. The appearance of Riccati inequality conditions, as in (ii) of Corollary 2 to Theorem 5.2, dates from Wintner [4].

For further specific results dealing with criteria of disconjugacy and oscillation for scalar second-order differential equations, the reader is referred to A. Kneser [1], Richardson [1], Hille [1], Wintner [4, 5, 6], Hartman [2, 3, 4], Hartman and Wintner [1, 2, 3, 4], Putnam [1], Leighton [3, 4], Taam [1, 2, 3], Potter [1], Barrett [1, 2], and Reid [13].

The proofs of Theorems 6.4, 7.2, 7.3, and 7.4 are in the spirit of the classical Sturmian theory as presented by Bôcher [5] and Ince [1, Ch. X], while the proof of Theorem 7.5 follows the ingenious simplification due to Bôcher [4, p. 173]. The methods of Bôcher have also been used by Ettlinger [1, 2] to establish more sophisticated comparison and oscillation theorems for a differential system involving a self-adjoint scalar differential equation of the second order.

The material of Section 8 is introduced here, to afford an easy introduction to the work of Morse [1, 2], which has played such a basic role in the extension of the classical Sturmian theory to the self-adjoint differential systems to be discussed in Chapter VII.

RELATED REFERENCES AND COMMENTS FOR SPECIFIC PROBLEMS

V.1:3 The transformation of this problem is known as "Liouville's transformation." For a more detailed discussion, the reader is referred to Birkhoff and Rota [1, Ch. X-Sec. 9].

V.3:2 Reid [13].

V.3:6 The result of this problem dates from Liapunov [1].

V.5:1 Leighton [2].

V.5:6 Wintner [4].

V.5:10 Prüfer [1]; Kamke [3].

V.6:3 Reid [13].

V.6:4 Prüfer [1]; Kamke [3].

VI

Self-Adjoint Boundary Problems Associated with Second-Order Linear Differential Equations

1. CANONICAL FORMS FOR SELF-ADJOINT BOUNDARY PROBLEMS

We now consider a differential system of the form

$$(B)\quad \begin{aligned} &\text{(a)}\quad l[u] \equiv [r(t)u' + q(t)u]' - [q(t)u' + p(t)u] = 0,\\ &\text{(b)}\quad s_\alpha[u, v] \equiv M_{\alpha 1}u(a) + M_{\alpha 2}v(a) + M_{\alpha 3}u(b) + M_{\alpha 4}v(b) = 0,\\ &\qquad\qquad\qquad\qquad\qquad\qquad\qquad\qquad\qquad\qquad (\alpha = 1, 2), \end{aligned}$$

where the coefficient functions r, q, p satisfy the hypothesis (**H**) *of Chapter* V, *$v(t)$ denotes the canonical variable $v(t) = r(t)u'(t) + q(t)u(t)$, and $M_{\alpha\tau}$, $(\alpha = 1, 2; \tau = 1, 2, 3, 4)$, are real constants such that the 2×4 matrix $M = [M_{\alpha\tau}]$ is of rank* 2. Now if $u_\alpha \in D''[a, b]{:}v_\alpha$, $(\alpha = 1, 2)$, then

$$(1.1)\qquad \int_a^b \bar{u}_2 l[u_1]\,dt - \int_a^b \overline{l[u_2]}u_1\,dt = (\bar{u}_2 v_1 - \bar{v}_2 u_1)\Big|_a^b,$$

and consequently (B) is self-adjoint if and only if

$$(1.2)\quad (\bar{u}_2 v_1 - \bar{v}_2 u_1)\big|_a^b = 0 \; \textit{for arbitrary}\; u_\alpha(a), v_\alpha(a), u_\alpha(b), v_\alpha(b),\\ (\alpha = 1, 2),\; \textit{satisfying}\; (B\text{-b}).$$

As in the general discussion of Section III.5, let P be a 4×2 real matrix of rank 2 and such that $MP = 0$. Then $u(a)$, $v(a)$, $u(b)$, $v(b)$ are values that satisfy the boundary conditions (B-b) if and only if there exists a two-dimensional vector ξ such that

$$u(a) = \sum_{\beta=1}^{2} P_{1\beta}\xi_\beta, \qquad u(b) = \sum_{\beta=1}^{2} P_{3\beta}\xi_\beta,$$
$$v(a) = \sum_{\beta=1}^{2} P_{2\beta}\xi_\beta, \qquad v(b) = \sum_{\beta=1}^{2} P_{4\beta}\xi_\beta,$$

and consequently condition (1.2) is equivalent to the matrix equation

$$(1.3') \qquad P^*[\operatorname{diag}\{-\mathbf{J}, \mathbf{J}\}]P = 0,$$

where $\mathbf{J}$ is the real skew matrix (V.1.4). In view of the definitive properties of P, (1.3′) is equivalent to the matrix equation

$$(1.3'') \qquad M[\operatorname{diag}\{-\mathbf{J}, \mathbf{J}\}]M^* = 0.$$

Moreover, if for brevity we set

$$(1.4) \qquad S^{\sigma\tau} = \begin{bmatrix} M_{1\sigma} & M_{1\tau} \\ M_{2\sigma} & M_{2\tau} \end{bmatrix}, \qquad D^{\sigma\tau} = \det S^{\sigma\tau}, \quad (\sigma, \tau = 1, 2, 3, 4),$$

it may be verified directly that

$$(1.5) \qquad M[\operatorname{diag}\{-\mathbf{J}, \mathbf{J}\}]M^* = (D^{34} - D^{12})\mathbf{J},$$

and hence one has the following result.

Lemma 1.1. *A system* (B) *is self-adjoint if and only if*

$$(1.6) \qquad D^{12} = D^{34}.$$

Now as a special instance of the general discussion of Section III.4, the two-point boundary conditions (B-b) are equally well specified by similar equations in which the coefficient matrix M is replaced by KM, where K is an arbitrary nonsingular real 2×2 matrix. The replacement of M by KM replaces $S^{\sigma\tau}$ by $KS^{\sigma\tau}$, and hence $D^{\sigma\tau}$ is replaced by $D^{\sigma\tau} \det K$; in particular, (1.6) holds for (B-b) if and only if this relation holds for the modified form of these boundary conditions. We shall proceed to show that by proper choices of K the following canonical forms for the boundary conditions are obtained.

***Lemma* 1.2.** *For a self-adjoint system* (B) *the boundary conditions may be written in one of the following forms, where the involved coefficients* $\gamma_{\alpha\beta}$,

$(\alpha, \beta = 1, 2)$, ψ_1, ψ_2 *are real constants:*

$$
\begin{aligned}
&\text{I.} \quad u(a) = 0, \qquad u(b) = 0;\\
&\text{II}_1. \quad u(a) = 0, \qquad \gamma_{22}u(b) + v(b) = 0;\\
&\text{II}_2. \quad u(b) = 0, \qquad \gamma_{11}u(a) - v(a) = 0;
\end{aligned}
$$

(1.7)

$$
\begin{aligned}
&\text{II}_3. \qquad \psi_1 u(a) + \psi_2 u(b) = 0,\\
&-\psi_2[\gamma_{11}u(a) - v(a)] + \psi_1[\gamma_{22}u(b) + v(b)] = 0; \qquad \psi_1^2 + \psi_2^2 > 0;\\
&\text{III.} \quad \gamma_{11}u(a) + \gamma_{12}u(b) - v(a) = 0,\\
&\qquad \gamma_{12}u(a) + \gamma_{22}u(b) + v(b) = 0.
\end{aligned}
$$

The classification (1.7) is based upon the rank of the matrix S^{24}. If S^{24} is of rank 0, then the matrix of coefficients of $u(a)$, $u(b)$ in (B-b) is non-singular, and these boundary conditions are equivalent to (1.7-I).

Now if S^{24} is of rank 1, then each of the 2×3 matrices

$$
S_1 = \begin{bmatrix} M_{12} & M_{13} & M_{14} \\ M_{22} & M_{23} & M_{24} \end{bmatrix} \qquad S_2 = \begin{bmatrix} M_{11} & M_{12} & M_{14} \\ M_{21} & M_{22} & M_{24} \end{bmatrix}
$$

has rank at least 1; moreover, since M is of rank 2 at least one of these matrices is of rank 2. Therefore, when S^{24} is of rank 1 the boundary conditions (B-b) are reducible to the form

$$
\begin{aligned}
&\psi_1 u(a) + \psi_2 u(b) = 0,\\
&M_{21}^0 u(a) + M_{22}^0 v(a) + M_{23}^0 u(b) + M_{24}^0 v(b) = 0,
\end{aligned}
\tag{1.8}
$$

where ψ_1, ψ_2 are real constants such that $\psi_1^2 + \psi_2^2 > 0$, and $M_{2\tau}^0$, $(\tau = 1, 2, 3, 4)$, are real constants with M_{22}^0 and M_{24}^0 not both zero. Now if S^{24} and S_1 have rank 1 and S_2 has rank 2, then $\psi_2 = 0$, $\psi_1 \neq 0$, condition (1.6) implies $M_{22}^0 = 0$, and hence $M_{24}^0 \neq 0$. In this case, the boundary conditions (B-b) are reducible to the form (1.7-II$_1$), with $\gamma_{22} = M_{23}^0/M_{24}^0$. Similarly, if S^{24} and S_2 have rank 1, and S_1 has rank 2, then $\psi_1 = 0$, $\psi_2 \neq 0$, condition (1.6) implies $M_{24}^0 = 0$, so that $M_{22}^0 \neq 0$, and the boundary conditions (B-b) are reducible to the form (1.7-II$_2$), with $\gamma_{11} = -M_{21}^0/M_{22}^0$. If S^{24} has rank 1, while both S_1 and S_2 are of rank 2, then $\psi_1 \neq 0$, $\psi_2 \neq 0$, condition (1.6) implies the existence of a nonzero d such that $M_{22}^0 = \psi_2 d$, $M_{24}^0 = \psi_1 d$, and in this case the boundary conditions (B-b) are reducible to the form (1.7-II$_3$), with $\gamma_{11} = -M_{21}^0/M_{22}^0$, $\gamma_{22} = M_{23}^0/M_{24}^0$.

Finally, if S^{24} is nonsingular, then the boundary conditions (B-b) are reducible to the form (1.7-III), with

$$\gamma_{11} = -\frac{D^{14}}{D^{24}}, \qquad \gamma_{22} = \frac{D^{23}}{D^{24}},$$
$$\gamma_{12} = -\frac{D^{12}}{D^{24}} = -\frac{D^{34}}{D^{24}}.$$

It is to be remarked that in cases II_1, II_2, II_3 an alternate form of the boundary conditions (1.7) is

$$\begin{aligned} \psi_1 u(a) + \psi_2 u(b) &= 0, \\ \gamma_{11} u(a) - v(a) + \psi_1 \nu &= 0, \\ \gamma_{22} u(b) + v(b) + \psi_2 \nu &= 0, \end{aligned} \tag{1.7-II$'$}$$

where ψ_1, ψ_2 are real constants with $\psi_1^2 + \psi_2^2 > 0$, and ν is a real parameter.

For each of the canonical forms (1.7) of the boundary conditions (B-b), let $\hat{s}[u \mid B] \equiv (\hat{s}_\rho[u \mid B],\ (\rho = 1, \ldots, h_B)$, denote the boundary conditions independent of $v(a)$, $v(b)$. Moreover, for each such set of boundary conditions, let

$$\begin{aligned} Q[u, y] &= [y(a)\gamma_{11} + y(b)\gamma_{12}]\, u(a) + [y(a)\gamma_{12} + y(b)\gamma_{22}]\, u(b), \\ Q[u] &= Q[u, u], \end{aligned}$$

where γ_{11}, γ_{12}, γ_{22} are the constants in the boundary conditions (1.7) if they appear explicitly, and zero otherwise. Specifically, the individual cases are presented in the following table.

CASE	h_B	$\hat{s}[\eta \mid B] = 0$	$Q[\eta]$
I	2	$\eta(a) = 0,\ \eta(b) = 0$	—
II_1	1	$\eta(a) = 0$	$\gamma_{22}\eta^2(b)$
II_2	1	$\eta(b) = 0$	$\gamma_{11}\eta^2(a)$
II_3	1	$\psi_1\eta(a) + \psi_2\eta(b) = 0,$ $\psi_1^2 + \psi_2^2 > 0$	$\gamma_{11}\eta^2(a) + \gamma_{22}\eta^2(b)$
III	0	—	$\gamma_{11}\eta^2(a) + 2\gamma_{12}\eta(a)\,\eta(b) +$ $\gamma_{22}\eta^2(b)$

For a general real quadratic form

$$Q[\eta] = \gamma_{11}\eta^2(a) + 2\gamma_{12}\eta(a)\,\eta(b) + \gamma_{22}\eta^2(b), \tag{1.9}$$

in the end-values $\eta(a)$, $\eta(b)$, let $Q_1[\eta]$ and $Q_2[\eta]$ denote the "partial derivative" linear forms

$$Q_1[\eta] = \gamma_{11}\eta(a) + \gamma_{12}\eta(b), \qquad Q_2[\eta] = \gamma_{12}\eta(a) + \gamma_{22}\eta(b),$$

so that $Q[u, \eta] = Q_1[\eta]\, u(a) + Q_2[\eta]\, u(b) = Q_1[u]\, \eta(a) + Q_2[u]\, \eta(b)$. Self-adjoint differential systems (B) then admit the following characterization, which is of basic significance in the treatment of associated boundary value problems.

Theorem 1.1. *A differential system (B) is self-adjoint if and only if there exists a real quadratic form $Q = Q[\eta \mid B]$ of the form (1.9), and a linear subspace $\mathcal{S} = \mathcal{S}[B]$ of real two-dimensional Euclidean space $\mathbf{R}_2$, such that end-values $u(a)$, $v(a)$, $u(b)$, $v(b)$ satisfy the boundary conditions (B-b) if and only if*

$$(1.10)\quad (u(a), u(b)) \in \mathcal{S}[B], \qquad (Q_1[u] - v(a), Q_2[u] + v(b)) \in \mathcal{S}^{\perp}[B],$$

where $\mathcal{S}^{\perp}[B]$ denotes the orthogonal complement of $\mathcal{S}[B]$ in $\mathbf{R}_2$.

If (B) is self-adjoint, then the boundary conditions (1.7) are seen to be of the form (1.10), where $\mathcal{S}[B]$ is of dimension 0, 1 and 2 in the respective cases I, II and III. Conversely, if the end-values $u(a)$, $v(a)$, $u(b)$, $v(b)$ are required to satisfy (1.10), then whenever dim $\mathcal{S}[B]$ is either 0 or 2 the requirement is clearly equivalent to the respective conditions (1.7-I), (1.7-III), and the system (B) is self-adjoint. If dim $\mathcal{S}[B] = 1$, then there exist real constants ψ_1, ψ_2 such that $\psi_1^2 + \psi_2^2 > 0$, and (1.10) is equivalent to (1.7-II′) or

$$(1.11)\quad \begin{gathered} \psi_1 u(a) + \psi_2 u(b) = 0, \\ -\psi_2\{Q_1[u] - v(a)\} + \psi_1\{Q_2[u] + v(b)\} = 0. \end{gathered}$$

In this case the self-adjointness of (B) follows from Lemma 1.1, and the fact that (1.11) is of the form (B-b) with

$$M = \begin{bmatrix} \psi_1 & 0 & \psi_2 & 0 \\ M_{21} & \psi_2 & M_{23} & \psi_1 \end{bmatrix},$$

so that $D^{12} = D^{24} = \psi_1\psi_2$.

For a given boundary problem (B), let $D[B]$ denote the set of functions $\eta \in D'[a, b]$ with end-values $\eta(a)$, $\eta(b)$ such that $(\eta(a), \eta(b)) \in \mathcal{S}[B]$. In particular, in the cases I, II_1, II_2, II_3 and III the set $D[B]$ is respectively equal to $D_0'[a, b]$, $D_{0*}'[a, b]$, $D_{*0}'[a, b]$, $\{\eta \mid \eta \in D'[a, b], \psi_1\eta(a) + \psi_2\eta(b) = 0\}$, and $D'[a, b]$. As a consequence of the definitive property of $\mathcal{S}[B]$, one has the following result.

Corollary. *If (B) is self-adjoint, and $\mathcal{S}[B]$ is determined as in Theorem 1.1, then a set of end-values $u(a)$, $v(a)$, $u(b)$, $v(b)$ satisfies the boundary conditions (B-b) if and only if $(u(a), u(b)) \in \mathcal{S}[B]$ and*

$$(1.12)\quad Q[u, \eta] + \eta(t)v(t)\Big|_a^b = 0, \quad \textit{for} \quad \eta \in D[B].$$

Corresponding to the notation of Section V.3, for a problem (B) with associated end-forms $Q = Q[\eta \mid B]$, and $\eta_\alpha \in D'[a, b] : \zeta_\alpha$, let

$$(1.13)\quad \begin{gathered} J[\eta_1, \eta_2 \mid B] = Q[\eta_1, \eta_2 \mid B] + \int_a^b \{\eta_2'[r\eta_1' + q\eta_1] + \eta_2[q\eta_1' + p\eta_1]\}\, dt, \\ J[\eta_1 \mid B] = J[\eta_1, \eta_1 \mid B]. \end{gathered}$$

With the aid of conclusion (ii) of Lemma V.3.1 and the above relation (1.12), one may establish the following result.

***Lemma* 1.3.** *If (B) is self-adjoint, and u is a solution of the differential system*

$$(1.14)\quad l[u](t) + f(t) = 0, \qquad t \in [a, b], \qquad s_\alpha[u, v] = 0, \qquad (\alpha = 1, 2),$$

where $f \in \mathfrak{C}[a, b]$, *then*

$$(1.15)\qquad J[u, \eta \mid B] = \int_a^b \eta(t) f(t)\, dt \quad for \quad \eta \in D[B];$$

in particular,

$$(1.16)\qquad J[u \mid B] = \int_a^b u(t) f(t)\, dt.$$

Problems VI.1

1.[ss] If (B) is self-adjoint, prove that u is a solution of this system if and only if $u \in D[B]$, and $J[u, \eta \mid B] = 0$, for $\eta \in D[B]$.

2. Let (B_1) and (B_2) denote the boundary problems involving the differential equation $u'' - p(t)u = 0$, where $p(t)$ is a real-valued continuous function on $[a, b]$, and the respective set of boundary conditions:

$$(1)\qquad \begin{aligned} u(a) + 2u(b) + u'(b) &= 0, \\ 2u(a) - u(b) - u'(a) - u'(b) &= 0; \end{aligned}$$

$$(2)\qquad \begin{aligned} u(a) + u'(b) &= 0, \\ u(b) + u'(a) &= 0. \end{aligned}$$

In each case, determine whether or not the boundary problem is self-adjoint.

3. For a boundary problem (B) involving the differential equation $(B\text{-a})$, and boundary conditions of the form

$$s_\alpha{}^0[u, u'] \equiv M_{\alpha 1}{}^0 u(a) + M_{\alpha 2}{}^0 u'(a) + M_{\alpha 3}{}^0 u(b) + M_{\alpha 4}{}^0 u'(b) = 0,$$

where $M_{\alpha\tau}{}^0$, $(\alpha = 1, 2;\ \tau = 1, 2, 3, 4)$ are real constants such that the 2×4 matrix $M^0 = [M_{\alpha\tau}{}^0]$ is of rank 2, determine the condition for (B) to be self-adjoint corresponding to the result of Lemma 1.1.

2. EXTREMUM PROPERTIES FOR SELF-ADJOINT SYSTEMS (B)

For a self-adjoint system (B) with $D[B]$, $Q[\eta \mid B]$ and $J[\eta \mid B]$ determined as in the preceding section, we shall denote by $D_N[B]$ the set of functions $\eta \in D[B]$ satisfying the norming condition

$$\int_a^b \eta^2(t)\, dt = 1.$$

Theorem 2.1. *Suppose that* (B) *is self-adjoint, and that* $J[\eta \mid B]$ *is non-negative definite on* $D[B]$. *Then the infimum of* $J[\eta \mid B]$ *on* $D_N[B]$ *is zero if and only if* (B) *has a nonidentically vanishing solution; moreover, if* $u \in D[B]$ *and* $J[u \mid B] = 0$, *then* u *is a solution of* (B).

If u is a nonidentically vanishing solution of (B), then either the real part of u, or the pure imaginary part of u, is a nonidentically vanishing real-valued solution u_1 of (B). Moreover, $u_0(t) = u_1(t)[\int_a^b u_1^2(s)\,ds]^{-1/2}$ is an element of $D_N[B]$ with $J[u_0 \mid B] = 0$, in view of relation (1.16) of Lemma 1.3. On the other hand, if (B) possesses no nonidentically vanishing solution, then from the general results of Section III.7 it follows that for arbitrary $\eta \in D[B]$ the nonhomogeneous system (1.14) with $f = \eta$ has a unique solution u. In this case, relations (1.15) and (1.16) of Lemma 1.3 imply

$$(2.1)\qquad J[u, \eta \mid B] = \int_a^b \eta^2(t)\,dt, \qquad J[u \mid B] = \int_a^b \eta(t)\,u(t)\,dt.$$

Moreover, if $g(t, s)$ denotes the Green's function for the incompatible system (B), then

$$(2.2)\qquad u(t) = -\int_a^b g(t, s)\eta(s)\,ds,$$

and with the aid of Schwarz' inequality it follows that

$$(2.3)\qquad \int_a^b u^2(t)\,dt \le \kappa^2 \int_a^b \eta^2(t)\,dt, \quad \textit{where} \quad \kappa^2 = \int_a^b \int_a^b g^2(t, s)\,dt\,ds.$$

In turn, Schwarz' inequality also implies

$$J[u \mid B] = \left|\int_a^b \eta u\,dt\right| \le \left(\int_a^b \eta^2\,dt\right)^{1/2} \left(\int_a^b u^2\,dt\right)^{1/2} \le \kappa \int_a^b \eta^2\,dt.$$

Since

$$0 \le J\left[\eta - \frac{1}{\kappa} u \mid B\right] = J[\eta \mid B] - \frac{2}{\kappa} J[u, \eta \mid B] + \frac{1}{\kappa^2} J[u \mid B],$$

$$\le J[\eta \mid B] - \frac{1}{\kappa} \int_a^b \eta^2\,dt,$$

it then follows that the infimum of $J[\eta \mid B]$ on $D_N[B]$ is not less then $1/\kappa$.

Now if $J[\eta \mid B]$ is non-negative definite on $D[B]$, and u is an element of $D[B]$ such that $J[u \mid B] = 0$, then an argument similar to that used in the proof of Corollary 1 to Theorem V.3.1 implies that $J[u, \eta \mid B] = 0$ for $\eta \in D[B]$, and from Problem VI.1:1 it follows that u is a solution of (B).

For a finite set $F = \{f_1, \ldots, f_k\}$ of functions f_i, $(i = 1, \ldots, k)$, which are real-valued and continuous on $[a, b]$, and for which the $k \times k$ matrix

$$\left[\int_a^b f_i(t)f_j(t)\,dt\right], \qquad (i, j = 1, \dots, k), \tag{2.4}$$

is nonsingular, consider the differential system

$$(B/F)\quad \begin{aligned} &\text{(a)} \quad l[u](t) + \sum_{j=1}^{k} e_j f_j(t) = 0, && t \in [a, b],\\ &\text{(b)} \quad s_\alpha[u, v] = 0, && (\alpha = 1, 2),\\ &\text{(c)} \quad \int_a^b f_i(t)u(t)\,dt = 0, && (i = 1, \dots, k), \end{aligned}$$

where the boundary conditions $s_\alpha[u, v]$ are as in (B-b), and $e_1, \dots, e_k$ are parameters. For brevity, the class of functions η in $D[B]$ and $D_N[B]$, respectively, which satisfy (B/F-c) will be denoted by $D[B/F]$ and $D_N[B/F]$.

Theorem 2.2. *Suppose that* (B) *is self-adjoint, and that* $J[\eta \mid B]$ *is nonnegative definite on* $D[B/F]$. *Then the infimum of* $J[\eta \mid B]$ *on* $D_N[B/F]$ *is zero if and only if there exist real constants* $e_1, \dots, e_k$ *such that* (B/F) *has a real-valued nonidentically vanishing solution; moreover, if* $u_0 \in D[B/F]$ *and* $J[u_0 \mid B] = 0$, *then* u_0 *is a solution of* (B/F) *for suitable real constants* $e_1, \dots, e_k$. *In particular, if the functions* f_i *of* F *are such that there exist constants* γ_i *and functions* u_i *satisfying the differential system*

$$\begin{aligned} l[u_i](t) + \gamma_i f_i(t) &= 0, && t \in [a, b],\\ s_\alpha[u_i, v_i] &= 0, && (\alpha = 1, 2;\ i = 1, \dots, k), \end{aligned} \tag{2.5}$$

and the $k \times k$ *matrix* $[\int_a^b u_i(t)f_j(t)\,dt]$, $(i, j = 1, \dots, k)$, *is nonsingular, then whenever* u *is a solution of* (B/F) *with constants* $e_1, \dots, e_k$, *one must have* $e_j = 0$, $(j = 1, \dots, k)$.

If u is a nonidentically vanishing function on $[a, b]$ which satisfies (B/F) with constants $e_1, \dots, e_k$, then either the real part of u or the pure imaginary part of u is a nonidentically vanishing real-valued function u_1 on $[a, b]$ which satisfies (B/F) with real constants $e_1^1, \dots, e_k^1$, and $u_0(t) = \mu u_1(t)$, with $\mu = [\int_a^b u_1^2(t)\,dt]^{-1/2}$, is an element of $D_N[B/F]$ which satisfies (B/F) with the real constants $\mu e_1^1, \dots, \mu e_k^1$, and

$$J[u_0 \mid B] = \mu \sum_{j=1}^{k} e_j^1 \int_a^b f_j u_0\,dt = 0.$$

On the other hand, suppose that there do not exist real constants $e_1, \dots, e_k$ with associated nonidentically vanishing real-valued solution of (B/F). Let $u_1(t)$, $u_2(t)$ be linearly independent real-valued solutions of $l[u] = 0$, and denote by $u = u_{2+j}(t)$ the solution of the differential system

$$l[u](t) + f_j(t) = 0, \qquad u(a) = 0 = v(a), \qquad (j = 1, \ldots, k);$$

necessarily, $u_{2+j}(t)$ is real-valued on $[a, b]$. Then the general solution u of the differential equation $(B/F\text{-a})$ with associated constants $e_j = c_{2+j}$, $(j = 1, \ldots, k)$, is

$$u(t) = u_1(t)c_1 + u_2(t)c_2 + \sum_{j=1}^{k} u_{2+j}(t)c_{2+j}. \tag{2.6}$$

In view of the assumption that the matrix (2.4) is nonsingular, it then follows that $u_1(t), \ldots, u_{2+j}(t)$ are linearly independent functions on $[a, b]$, and $u(t) \equiv 0$ on $[a, b]$ if and only if $c_\sigma = 0$, $(\sigma = 1, \ldots, 2 + k)$; moreover, $u(t)$ is real-valued on $[a, b]$ if and only if all the coefficients c_σ are real-valued. Consequently, if there do not exist real constants $e_1, \ldots, e_k$ with an associated nonidentically vanishing real-valued solution $u(t)$ of (B/F), then the $(2 + k) \times (2 + k)$ matrix

$$\begin{bmatrix} s_\alpha[u_\sigma, v_\sigma] \\ \int_a^b f_i(t)u_\sigma(t)\,dt \end{bmatrix}$$

is non-singular. As in the proof of the existence of a Green's function for an incompatible system (B), it follows that if f is a continuous real-valued function on $[a, b]$ then there exist unique constants e_j and function u, both necessarily real-valued, such that

$$\begin{aligned} l[u](t) + \sum_{j=1}^{k} e_j f_j(t) + f(t) &= 0, \qquad t \in [a, b], \\ s_\alpha[u, v] &= 0, \qquad (\alpha = 1, 2), \\ \int_a^b f_i(t)u(t)\,dt &= 0, \qquad (i = 1, \ldots, k). \end{aligned} \tag{2.7}$$

Moreover, the function $u(t)$ belonging to such a solution is given by an integral transform

$$u(t) = -\int_a^b g_1(t, s)f(s)\,ds, \qquad t \in [a, b], \tag{2.8}$$

where $g_1(t, s)$ is continuous in (t, s) on $[a, b] \times [a, b]$.

If $\eta \in D[B/F]$, and u denotes the solution of (2.7) with $f = \eta$, then it follows readily that the functions η and u again satisfy relations (2.1). By steps analogous to those in the proof of Theorem 2.1 it then follows that $0 \leq J[\eta \mid B] - (1/\kappa_1) \int_a^b \eta^2 dt$, where now $\kappa_1^2 = \int_a^b \int_a^b g_1^2(t, s)\,dt\,ds$, so that the infimum of $J[\eta \mid B]$ on $D_N[B/F]$ is not less than $1/\kappa_1$.

Now suppose that there exists a $u_0 \in D[B/F]$ such that $J[u_0 \mid B] = 0$. If $u_0 \equiv 0$ on $[a, b]$, then u_0 is a solution of (B/F) with the real constants $e_j = 0$, $(j = 1, \ldots, k)$, so that we need consider only the case $u_0 \not\equiv 0$ on $[a, b]$. It then follows from the result established above that there exist real constants $e_1, \ldots, e_k$ such that there is an associated nonidentically vanishing real-valued solution u of (B/F). As any such u is of the form (2.6), it then follows that there are h, $0 < h \leq k + 2$, such solutions $u^{(1)}(t), \ldots, u^{(h)}(t)$ which are linearly independent on $[a, b]$, and such that each u which satisfies (B/F) with associated constants is a linear combination of $u^{(1)}(t), \ldots, u^{(h)}(t)$. Now let F_1 denote the set of $k + h$ functions $\{f_1(t), \ldots, f_k(t), u^{(1)}(t), \ldots, u^{(h)}(t)\}$. If $u(t)$ is a solution of the corresponding system (B/F_1) with associated constants $e_1, \ldots, e_k, e_{k+1}, \ldots, e_{k+h}$, then from relations (1.1), (1.2) for $u_1 = u$ and $u_2 = u^{(\rho)}$ it follows that $e_{k+\rho} = 0$, $(\rho = 1, \ldots, h)$; that is, $u(t)$ is also a solution of (B/F), so that it is of the form $u(t) = d_1 u^{(1)}(t) + \cdots + d_h u^{(h)}(t)$. From the conditions $\int_a^b u^{(\sigma)} u\, dt = 0$, $(\sigma = 1, \ldots, h)$, it then follows that $d_\sigma = 0$, $(\sigma = 1, \ldots, h)$, so that $u(t) \equiv 0$ on $[a, b]$. The first result of the theorem then implies that the infimum of $J[\eta \mid B]$ on $D_N[B/F_1]$ is positive. Now for the given u_0 such that $u_0 \in D[B/F]$ and $J[u_0 \mid B] = 0$, there exist constants $c_1, \ldots, c_h$ such that $\eta = u_0(t) - \sum_{\rho=1}^h c_\rho u^{(\rho)}(t)$ is a member of $D[B/F_1]$. Since $u^{(\rho)}(t)$ is a solution of (B/F) with associated constants, it then follows as in the proof of the first part of the theorem that $J[u^{(\rho)} \mid B] = 0$ and $J[u_0, u^{(\rho)} \mid B] = 0$ for $\rho = 1, \ldots, h$. Consequently, η is an element of $D[B/F_1]$ such that $J[\eta \mid B] = J[u_0 \mid B] = 0$, and hence $\eta(t) \equiv 0$ and $u_0(t) \equiv \sum_{\rho=1}^h c_\rho u^{(\rho)}(t)$ for $t \in [a, b]$.

The final conclusion of the theorem follows from the fact that if u is a solution of (B/F) with constants e_j, and there exist constants γ_i and functions u_i satisfying the differential system (2.5), and such that the matrix $[\int_a^b u_i(t) f_j(t)\, dt]$ is nonsingular, then for $i = 1, \ldots, k$ we have that

$$\sum_{j=1}^k e_j \int_a^b u_i(t) f_j(t)\, dt = -\int_a^b u_i l[u]\, dt = -\int_a^b u l[u_i]\, dt,$$

where the last equality is a consequence of relations (1.1) and (1.2). In turn, the differential equation of (2.5), with the boundary condition $(B/F\text{-c})$, implies that

$$-\int_a^b u l[u_i]\, dt = \gamma_i \int_a^b f_i(t) u(t)\, dt = 0, \qquad (i = 1, \ldots, k),$$

and, in view of the nonsingularity of the matrix $[\int_a^b u_i(t) f_j(t)\, dt]$ each of the constants e_j is zero.

3. EXISTENCE OF PROPER VALUES

Attention will be directed now to the existence of proper values for a self-adjoint boundary value problem of the form

$$(3.1)\quad \begin{aligned} &\text{(a)}\ l[u;\lambda] \equiv l[u] + \lambda\, k(t)u = 0, \qquad t \in [a,b],\\ &\text{(b)}\ s_\alpha[u,v] \equiv M_{\alpha 1}u(a) + M_{\alpha 2}v(a) + M_{\alpha 3}u(b) + M_{\alpha 4}v(b) = 0,\\ &\qquad\qquad\qquad\qquad\qquad\qquad\qquad\qquad\qquad\qquad (\alpha = 1, 2), \end{aligned}$$

where

$$(3.2)\quad l[u] \equiv [r(t)u' + q(t)u]' - [q(t)u' + p(t)u], \qquad v = ru' + qu,$$

and the following hypothesis is satisfied.

($\hat{\text{H}}$) *r, q, p, k are real-valued continuous functions on $[a, b]$, with $r(t) > 0$ and $k(t) > 0$ on this interval, and the real coefficients $M_{\alpha\tau}$, ($\alpha = 1, 2$; $\tau =$ 1, 2, 3, 4, in* (3.1-b) *are such that the matrix M is of rank two and the self-adjointness condition* (1.2) *holds.*

In particular, when hypothesis ($\hat{\text{H}}$) is satisfied, as in Section 1 above one has the quadratic end-form

$$(3.3)\qquad Q[\eta \mid B] = \gamma_{11}\eta^2(a) + 2\gamma_{12}\eta(a)\,\eta(b) + \gamma_{22}\eta^2(b),$$

the linear subspace $\mathcal{S}[B]$ of real end-values $(\eta(a), \eta(b))$, the associated function space

$$(3.4)\qquad D[B] = \{\eta \mid \eta \in D'[a,b],\quad (\eta(a), \eta(b)) \in \mathcal{S}[B]\},$$

and the quadratic functional

$$(3.5)\qquad J[\eta \mid B] = Q[\eta \mid B] + \int_a^b \{r\eta'^2 + 2q\eta'\eta + p\eta^2\}\, dt.$$

Also, for brevity, we set

$$(3.6)\qquad K[\eta_1, \eta_2] = \int_a^b \eta_2(t)k(t)\eta_1(t)\, dt, \qquad K[\eta] = K[\eta, \eta].$$

Now let c be a constant such that $|Q[\eta \mid B]| \leq c[\eta^2(a) + \eta^2(b)]$ for arbitrary real end-values $\eta(a)$, $\eta(b)$; for example, one might choose $c = |\gamma_{12}| + \max\{|\gamma_{11}|, |\gamma_{22}|\}$. Then, independent of the particular set $\mathcal{S}[B]$, one has the inequality

$$Q[\eta \mid B] \geq -c[\eta^2(a) + \eta^2(b)] = \int_a^b c\{2\theta\eta'\eta + \theta'\eta^2\}\, dt, \quad \textit{for} \quad \eta \in D'[a,b],$$

where $\theta(t) = (a + b - 2t)/(b - a)$, and hence

$$J[\eta \mid B] \geq \int_a^b \{r\eta'^2 + 2(q + c\theta)\eta'\eta + (p + c\theta')\eta^2\}\, dt, \quad \textit{for} \quad \eta \in D'[a,b].$$

Consequently, the result of Problem V.3:4 implies that there exist constants $k_0 > 0$, $k_1 > 0$ such that

$$(3.7) \qquad J[\eta \mid B] \geq k_0 \int_a^b \eta'^2\,dt - k_1 \int_a^b \eta^2\,dt, \quad \textit{for} \quad \eta \in D'[a, b].$$

Moreover, since in view of hypothesis ($\hat{\mathbf{H}}$) there exists a positive constant κ_0 such that $k(t) \geq \kappa_0$ for $t \in [a, b]$, we have that

$$K[\eta] \geq \kappa_0 \int_a^b \eta^2\,dt,$$

and consequently the functional

$$(3.8) \qquad J[\eta; \hat{\lambda} \mid B] \equiv J[\eta \mid B] - \hat{\lambda}\,K[\eta]$$

is positive definite on $D'[a, b]$ if $\hat{\lambda} < -k_1/\kappa_0$.

If $\lambda = \lambda_0$ is a proper value of (3.1), and $u_0 = u_1 + iu_2$ is a corresponding proper solution, where u_1 and u_2 are real-valued, then

$$(3.9) \quad \lambda\,K[u_0, \bar{u}_0] = -\int_a^b \bar{u}_0 l[u_0]\,dt,$$

$$= -\bar{u}_0 v_0\Big|_a^b + \int_a^b \{\bar{u}_0'[ru_0' + qu_0] + \bar{u}_0[qu_0' + pu_0]\}\,dt,$$

$$= J[u_1 \mid B] + J[u_2 \mid B],$$

where the last equality follows from the preceding in view of the fact that the end-values of u, v satisfy the real boundary conditions (3.1-b) if and only if the end-values of $\bar{u}$, $\bar{v}$ satisfy these boundary conditions, and hence, with the aid of the Corollary to Theorem 1.1, it follows that

$$-\bar{u}_0 v_0\Big|_a^b = Q[u_1 \mid B] + Q[u_2 \mid B].$$

As the integral $K[u_0, \bar{u}_0]$ is positive for u_0 a proper solution of (3.1) corresponding to the proper value λ_0, relation (3.9) implies the following result.

***Lemma* 3.1.** *Under hypothesis* ($\hat{\mathbf{H}}$) *all proper values of* (3.1) *are real and greater than* $\hat{\lambda}$, *where* $\hat{\lambda}$ *is a real value such that the functional* $J[\eta; \hat{\lambda} \mid B]$ *of* (3.8) *is positive definite on* $D'[a, b]$; *moreover, the proper functions corresponding to a proper value may be chosen to be real-valued.*

For the boundary problem (3.1) the basic existence theorem is the following result.

Theorem 3.1. *If hypothesis* ($\hat{\mathbf{H}}$) *holds then for the boundary problem* (3.1) *there exists an infinite sequence of real proper values* $\lambda_1 \leq \lambda_2 \leq \cdots$, *with* *corresponding real proper functions* $u = u_j(t)$ *for* $\lambda = \lambda_j$ *such that:*

(a) $K[u_i, u_j] = \delta_{ij}$, $(i, j = 1, 2, \ldots)$;

(b) $\lambda_1 = J[u_1 \mid B]$ *is the minimum of* $J[\eta \mid B]$ *on the class*

$$D_N[B \mid K] = \{\eta \mid \eta \in D[B],\ K[\eta] = 1\}; \tag{3.10}$$

(c) *for* $j = 2, 3, \ldots$, *the class*

$$D_{Nj}[B \mid K] = \{\eta \mid \eta \in D_N[B \mid K],\ K[\eta, u_i] = 0,\ i = 1, \ldots, j-1\} \tag{3.11}$$

is non-empty, and $\lambda_j = J[u_j \mid B]$ *is the minimum of* $J[\eta \mid B]$ *on* $D_{Nj}[B \mid K]$.

Since $k(t) > 0$ for $t \in [a, b]$, if $h \in D_0'[a, b]$ and $h \not\equiv 0$ on $[a, b]$, then $\kappa_1 = K[h] > 0$ and $\eta = (1/\kappa_1^{1/2})h$ is an element of $D_N[B \mid K]$. Therefore $D_N[B \mid K]$ is non-empty, and if λ_1 is defined as the infimum of $J[\eta \mid B]$ on $D_N[B \mid K]$, then $\lambda_1 \geq \hat{\lambda}$, where $\hat{\lambda}$ is such that the functional $J[\eta; \hat{\lambda} \mid B]$ of (3.8) is positive definite on $D'[a, b]$. Consequently, $J[\eta; \lambda_1 \mid B] = J[\eta \mid B] - \lambda_1 K[\eta]$ is non-negative for $\eta \in D[B]$, and if μ is the infimum of $J[\eta; \lambda_1 \mid B]$ on the class

$$D_N[B] = \left\{\eta \mid \eta \in D[B],\ \int_a^b \eta^2(t)\, dt = 1\right\},$$

then application of Theorem 2.1 to the functional $J[\eta; \lambda_1 \mid B]$ implies that either $\mu > 0$ or $\mu = 0$, and in the latter case there is a real-valued non-identically vanishing solution of (3.1) for $\lambda = \lambda_1$. Now if

$$\kappa = \max\{k(t) \mid t \in [a, b]\}$$

then

$$K[\eta] \leq \kappa \int_a^b \eta^2(t)\, dt \quad \textit{for} \quad \eta \in D'[a, b],$$

and hence for $\eta \in D_N[B \mid K]$ we have $J[\eta; \lambda_1 \mid B] \geq (\mu/\kappa)\, K[\eta]$ and $J[\eta \mid B] \geq [\lambda_1 + (\mu/\kappa)]\, K[\eta]$, which contradicts the definition of λ_1 in case $\mu > 0$. Consequently $\mu = 0$, the boundary problem (3.1) has a real-valued non-identically vanishing solution u_1 for $\lambda = \lambda_1$, and this solution may be chosen normalized so that $K[u_1] = 1$.

Now suppose that proper values λ_j and corresponding real-valued proper solutions $u_j(t)$, $(j = 1, 2, \ldots, m-1)$, have been determined to satisfy the corresponding conditions (a), (b), (c) of (3.7) for $i, j = 1, \ldots, m-1$. Let $h_1(t), \ldots, h_m(t)$ be functions of class $D_0'[a, b]$ which are linearly independent on $[a, b]$; for example,

$$h_j(t) = (t-a)^j (b-t)^j$$

or

$$h_j(t) = \sin[j\pi(t-a)/(b-a)],$$

for $j = 1, \ldots, m$. Then there exist real constants $c_1, \ldots, c_m$ not all zero and such that $h(t) = c_1h_1(t) + \cdots + c_mh_m(t)$ satisfies the $m - 1$ linear homogeneous algebraic equations which specify that h belongs to the class

$$D_m[B \mid K] = \{\eta \mid \eta \in D[B],\ K[\eta, u_j] = 0,\ (j = 1, \ldots, m-1)\}.$$

Then $\kappa_m = K[h]$ is positive, $\eta = (1/\kappa_m^{1/2})h$ is an element of the class

$$D_{Nm}[B \mid K] = \{\eta \mid \eta \in D_N[B \mid K],\ K[\eta, u_j] = 0,\ (j = 1, \ldots, m-1)\},$$

and, therefore, we may define λ_m as the infimum of $J[\eta \mid B]$ on $D_{Nm}[B \mid K]$. As $D_{Nm}[B \mid K] \subset D_{Nj}[B \mid K]$ for $j = 1, \ldots, m-1$, in view of the induction hypothesis that $\lambda_j = J[u_j \mid B]$ is the minimum of $J[\eta \mid B]$ on $D_{Nj}[B \mid K]$ for $j = 1, \ldots, m-1$, it follows that $\lambda_m \geq \lambda_{m-1} \geq \cdots \geq \lambda_1$. Consequently, $J[\eta; \lambda_m \mid B] \geq 0$ for $\eta \in D_m[B \mid K]$, and if $F = \{f_1, \ldots, f_{m-1}\} = \{ku_1, \ldots, ku_{m-1}\}$, then in the notation of Theorem 2.2 we have

$$D_N[B/F] = \{\eta \mid \eta \in D[B],\ K[\eta, u_j] = 0 \ \ for \ \ j = 1, \ldots, m-1,\ \int_a^b \eta^2\,dt = 1\},$$

and $J[\eta; \lambda_m \mid B] \geq 0$ for $\eta \in D_N[B/F]$. Moreover, since $u = u_j$ is a solution of (3.1) for $\lambda = \lambda_j$, $(j = 1, \ldots, m-1)$, we have that

$$\begin{aligned} l[u_j; \lambda_m](t) + (\lambda_j - \lambda_m)f_j(t) &= 0, \qquad t \in [a, b], \\ s_\alpha[u_j, v_j] &= 0, \qquad (\alpha = 1, 2), \end{aligned}$$

and the matrix $[\int_a^b u_i f_j\,dt] = [K[u_i, u_j]] = [\delta_{ij}]$ is nonsingular. Therefore, if μ_m is defined to be the infimum of $J[\eta; \lambda_m \mid B]$ on $D_N[B/F]$, the result of Theorem 2.2 applied to $J[\eta; \lambda_m \mid B]$ implies that either $\mu_m > 0$ or $\mu_m = 0$, and in this latter case there is a real-valued nonidentically vanishing solution u of the differential system

$$\begin{aligned} l[u; \lambda_m](t) &= 0, \\ s_\alpha[u, v] &= 0, \qquad (\alpha = 1, 2), \\ K[u, u_j] &= 0, \qquad (j = 1, \ldots, m-1), \end{aligned}$$

so that $\lambda = \lambda_m$ is a proper value of (3.1). If $\kappa = \max\{k(t) \mid t \in [a, b]\}$, then it follows that $J[\eta \mid B] \geq [\lambda_m + (\mu_m/\kappa)]\,K[\eta]$ for $\eta \in D_{Nm}[B \mid K]$, which contradicts the definition of λ_m in case $\mu_m > 0$. Hence $\mu_m = 0$, $\lambda = \lambda_m$ is a proper value of (3.1) with corresponding proper function u_m belonging to $D_m[B \mid K]$, and this proper solution may be normalized so that $K[u_m] = 1$. Finally, it is to be noted that in view of Lemma 1.3 we have

$$J[u_m \mid B] = \lambda_m\,K[u_m] = \lambda_m,$$

thus completing the inductive argument for the proof of Theorem 3.1.

Theorem 3.2. *Suppose that hypothesis* ($\hat{\mathbf{H}}$) *holds, and* $\{\lambda_j, u_j\}$, $(j = 1, 2, \ldots)$, *is a sequence of proper values and corresponding proper functions as determined in Theorem* 3.1. *If* λ_0 *is a real value satisfying* $\lambda_0 < \lambda_1$, *then the infinite series*

$$\sum_{j=1}^{\infty}(\lambda_j - \lambda_0)^{-2}\, u_j^2(t), \quad \sum_{j=1}^{\infty}(\lambda_j - \lambda_0)^{-2}\, v_j^2(t), \quad t \in [a, b], \tag{3.12}$$

converge and the sums of these series do not exceed the respective values

$$\int_a^b k(s) g^2(t, s; \lambda_0)\, ds, \quad \int_a^b k(s) g_1^2(t, s; \lambda_0)\, ds, \tag{3.13}$$

where $g(t, s; \lambda_0)$ *is the Green's function for the incompatible differential system*

$$\begin{aligned} l[u; \lambda_0](t) &= 0, \qquad t \in [a, b], \\ s_\alpha[u, v] &= 0, \qquad (\alpha = 1, 2), \end{aligned} \tag{3.14}$$

and

$$g_1(t, s; \lambda_0) = r(t) g_t(t, s; \lambda_0) + q(t) g(t, s; \lambda_0). \tag{3.15}$$

Moreover, the infinite series $\sum_{j=1}^{\infty} (\lambda_j - \lambda_0)^{-2}$ *converges, and*

$$\sum_{j=1}^{\infty}(\lambda_j - \lambda_0)^{-2} \leq \int_a^b \int_a^b k(t) k(s) g^2(t, s; \lambda_0)\, dt\, ds. \tag{3.16}$$

If $h(t)$ is real-valued and piecewise continuous on $[a, b]$, and $c_j[h] = K[h, u_j]$, $(j = 1, 2, \ldots)$, then in view of the orthogonality relations $K[u_i, u_j] = \delta_{ij}$ it follows that

$$0 \leq K\left[h - \sum_{j=1}^{m} c_j[h] u_j\right] = K[h] - \sum_{j=1}^{m} c_j^2[h], \qquad m = 1, 2, \ldots. \tag{3.17}$$

Consequently, the infinite series $\sum_{j=1}^{\infty} c_j^2[h]$ converges, and

$$\sum_{j=1}^{\infty} c_j^2[h] \leq K[h]. \tag{3.18}$$

Now if $\lambda_0 < \lambda_1$, then $u_j(t)$ is a solution of

$$\begin{aligned} l[u_j; \lambda_0](t) + (\lambda_j - \lambda_0) k(t) u_j(t) &= 0, \qquad t \in [a, b], \\ s_\alpha[u_j, v_j] &= 0, \qquad (\alpha = 1, 2), \end{aligned}$$

and, therefore, for $j = 1, 2, \ldots$ we have

$$\begin{aligned} u_j(t) &= -(\lambda_j - \lambda_0) \int_a^b g(t, s; \lambda_0) k(s) u_j(s)\, ds, \\ v_j(t) &= -(\lambda_j - \lambda_0) \int_a^b g_1(t, s; \lambda_0) k(s) u_j(s)\, ds, \end{aligned} \tag{3.19}$$

where $g(t, s; \lambda_0)$ is the Green's function for (3.14), and $g_1(t, s; \lambda_0)$ is defined by (3.15). Consequently, if for fixed $t \in [a, b]$ the inequalities (3.17), (3.18) are applied to $h(s) = g(t, s; \lambda_0)$ and $h(s) = g_1(t, s; \lambda_0)$, it results that for fixed $t \in [a, b]$ the infinite series (3.12) converge and have sums not exceeding the respective values in (3.13). Moreover, since $k(t) \geq 0$ for $t \in [a, b]$, we have the inequality

$$\text{(3.20)} \quad k(t) \sum_{j=1}^{m} (\lambda_j - \lambda_0)^{-2} u_j^2(t) \leq k(t) \int_a^b k(s) g^2(t, s; \lambda_0)\, ds, \quad (m = 1, 2, \ldots).$$

Since $K[u_j] = 1$, $(j = 1, \ldots, m)$, upon integration of the members of inequality (3.20) it follows that

$$\sum_{j=1}^{m} (\lambda_j - \lambda_0)^{-2} \leq \int_a^b \int_a^b k(t) k(s) g^2(t, s; \lambda_0)\, dt\, ds, \qquad (m = 1, 2, \ldots),$$

so that the infinite series $\sum_{j=1}^{\infty} (\lambda_j - \lambda_0)^{-2}$ converges and satisfies the inequality (3.16).

In view of the convergence of the infinite series in (3.16), the following Corollary is immediate.

Corollary. *$\{\lambda_j\} \to +\infty$ as $j \to \infty$.*

It is to be remarked that the result of this Corollary also follows from the discussion of Section IV.2. Indeed, the proper values of (3.1) form the set of zeros of an entire function of λ, and as this system has only real proper values it then follows that the set of proper values has no finite accumulation point, so that $\{\lambda_j\} \to +\infty$ as $j \to \infty$. However, *the above proof is entirely by real variable methods.* It is to be noted also that since $\lambda K[u] = J[u \mid B]$ for a proper solution u of (3.1) corresponding to a proper value λ, the result of this Corollary implies that each proper value of (3.1) occurs in the sequence $\{\lambda_j\}$ of Theorem 3.1, and occurs a number of times equal to its index.

Theorem 3.3. *Suppose that hypothesis* ($\hat{\mathbf{H}}$) *holds and $\{\lambda_j, u_j\}$ is a sequence of proper values and corresponding proper functions as determined in Theorem* 3.1.

(a) *If r is a positive integer and $d_1, \ldots, d_r$ are real constants such that $d_1^2 + \cdots + d_r^2 = 1$, then $\eta(t) = d_1 u_1(t) + \cdots + d_r u_r(t)$ belongs to $D[B]$ and $J[\eta \mid B] \leq \lambda_r$.*

(b) *(Courant-Hilbert max-min property) If $F = \{f_1, \ldots, f_k\}$ is a set of real-valued continuous functions on $[a, b]$, and $\lambda[F]$ denotes the minimum of $J[\eta \mid B]$ on*

$$D_N[B/F \mid K] = \left\{\eta \mid \eta \in D_N[B \mid K], \quad \int_a^b f_i \eta\, dt = 0, \quad (i = 1, \ldots, k)\right\},$$

then λ_{k+1} is the maximum of $\lambda[F]$.

Clearly $\eta = d_1u_1 + \cdots + d_ru_r$ belongs to $D[B]$, since each u_i is an element of $D[B]$. Moreover, as Lemma 1.3 implies that $J[u_i, u_j \mid B] = \lambda_j K[u_i, u_j] = \lambda_j\delta_{ij}$, it follows that

$$J[\eta \mid B] = \sum_{i,j=1}^{r} d_i d_j J[u_i, u_j \mid B] = \sum_{j=1}^{r} \lambda_j d_j^{\,2} \leq \lambda_r \sum_{j=1}^{r} d_j^{\,2} = \lambda_r.$$

For a given set $F = \{f_1, \ldots, f_k\}$, there exist sets of real constants $d_1, \ldots, d_{k+1}$, not all zero, and satisfying the k linear homogeneous algebraic equations which specify that the function $\eta(t) = d_1u_1(t) + \cdots + d_{k+1}u_{k+1}(t)$ is such that $\int_a^b f_i\eta\, dt = 0$, $(i = 1, \ldots, k)$. Such a set may then be normalized so that $d_1^2 + \cdots + d_{k+1}^2 = 1$, and the resulting function η is such that $\eta \in D_N[B/F \mid K]$, and hence $\lambda[F] \leq J[\eta \mid B] \leq \lambda_{k+1}$, where the last inequality is a consequence of conclusion (a). On the other hand, for $f_j(t) = k(t)\, u_j(t)$, $(j = 1, \ldots, k)$, the set $D_N[B/F \mid K]$ is the set $D_{N,k+1}[B \mid K]$ of Theorem 3.1, and for this particular set F we have $\lambda[F] = \lambda_{k+1}$.

Problems VI.3

1. If hypothesis ($\hat{\mathbf{H}}$) holds for a boundary problem (3.1), prove that $J[\eta \mid B]$ is positive, {non-negative}, definite on $D[B]$ if and only if all proper values of (3.1) are positive, {non-negative}.

2. If hypothesis ($\hat{\mathbf{H}}$) holds for a boundary problem involving the differential equation (3.1-a) and the boundary conditions $u(a) = 0 = u(b)$, and l is a real number which is not a proper value of this boundary problem, prove that the corresponding Green's function $g(t, s; l)$ is of constant sign on $[a, b] \times [a, b]$ if and only if $l < \lambda_1$, where λ_1 is the smallest proper value of this problem.

3. Suppose that hypothesis ($\hat{\mathbf{H}}$) holds, and $\{\lambda_j, u_j\}$ is a sequence of proper values and corresponding proper functions of (3.1) as determined in Theorem 3.1. If $h_\alpha(t)$, $(\alpha = 1, \ldots, m)$ are given functions of class $D[B]$ satisfying $K[h_\alpha, h_\beta] = \delta_{\alpha\beta}$, $(\alpha, \beta = 1, \ldots, m)$, and $\mu_1 \leq \mu_2 \leq \cdots \leq \mu_m$ denote the proper values of the real quadratic form $Q[\xi] \equiv \sum_{\alpha,\beta=1}^{m} J[h_\alpha, h_\beta \mid B]\xi_\alpha\xi_\beta$, prove that $\mu_j \geq \lambda_j$ for $j = 1, \ldots, m$.

Hint. Use appropriate results of Theorems 3.1, 3.3, and the result of Theorem 2.2 of Appendix E.

4. For the boundary value problem

$$u'' + \lambda u = 0, \qquad u(0) = 0, \qquad u'(1) = 0,$$

and $h_1(t) = \sqrt{3}t$, $h_2(t) = \sqrt{\tfrac{7}{4}}(3t - 5t^3)$, use Problem 3 to obtain upper bounds for the proper values λ_1, λ_2, and compare with the exact values of λ_1, λ_2.

5. If the coefficients of the differential equation (3.1-a) satisfy the conditions specified in ($\hat{\mathbf{H}}$), and λ_0 is a given real value, prove that the number of points on the open interval (a, b) which are conjugate to $t = a$, relative to the equation $l[u; \lambda_0] = 0$, is equal to the number of proper values of the boundary problem

$$(3.21) \qquad l[u; \lambda](t) = 0, \qquad u(a) = 0 = u(b),$$

which are less than λ_0.

Hint. Use results of Theorem V.8.2, Problem V.8:1, and the extremizing properties of the proper values of (3.21).

4. COMPARISON THEOREMS

A boundary problem B of the type (3.1) depends upon the real quadratic integrand form

$$2\omega(t, \eta, \zeta) = 2\omega(t, \eta, \zeta \mid B) = r(t \mid B)\zeta^2 + 2q(t \mid B)\zeta\eta + p(t \mid B)\eta^2,$$

the real quadratic end-form

$$Q[\eta \mid B] = \gamma_{11}{}^B \eta^2(a) + 2\gamma_{12}{}^B \eta(a)\eta(b) + \gamma_{22}{}^B \eta^2(b),$$

the end-space $\mathcal{S}[B]$ in $\mathbf{R}_2$, the set

$$D[B] = \{\eta \mid \eta \in D'[a, b],\ (\eta(a), \eta(b)) \in \mathcal{S}[B]\},$$

and also the coefficient function $k(t) = k(t \mid B)$. In this section we shall establish some comparison theorems for a problem B and a second problem $\tilde{B}$ involving corresponding $\omega(t, \eta, \zeta \mid \tilde{B})$, $Q[\eta \mid \tilde{B}]$, $\mathcal{S}[\tilde{B}]$ and $k(t \mid \tilde{B})$. In particular, if $\mathcal{S}[B] = \mathcal{S}[\tilde{B}]$ then $D[B] = D[\tilde{B}]$. *In all cases it will be supposed that each of the problems B, $\tilde{B}$ satisfies the hypothesis* ($\hat{\mathbf{H}}$) *of Section* 3. Moreover, $\{\lambda_i, u_i\}$ and $\{\tilde{\lambda}_j, \tilde{u}_j\}$ will denote for these respective problems a set of proper values and corresponding proper functions which satisfy the conditions of Theorem 3.1.

Theorem 4.1. *Suppose that B and $\tilde{B}$ are such that $\mathcal{S}[\tilde{B}] = \mathcal{S}[B] = \mathcal{S}$ and $k(t \mid \tilde{B}) \equiv k(t \mid B) \equiv k(t)$. If $\Delta J[\eta \mid B, \tilde{B}] \equiv J[\eta \mid \tilde{B}] - J[\eta \mid B]$ is non-negative definite on $D[B] = D[\tilde{B}]$, then $\tilde{\lambda}_j \geq \lambda_j$, $(j = 1, 2, \ldots)$; moreover, if $\Delta J[\eta \mid B, \tilde{B}]$ is positive definite on $D[B] = D[\tilde{B}]$ then $\tilde{\lambda}_j > \lambda_j$, $(j = 1, 2, \ldots)$.*

Corresponding to a positive integer j, let $d_1, \ldots, d_j$ be real constants such that $d_1^2 + \cdots + d_j^2 = 1$ and $\eta(t) = d_1 \tilde{u}_1(t) + \cdots + d_j \tilde{u}_j(t)$ satisfies the $j - 1$ relations $K[\eta, u_i] = 0$, $1 \leq i < j$. Then $\eta \in D[B] = D[\tilde{B}]$, and $K[\eta] = d_1^2 + \cdots + d_j^2 = 1$, so that in view of conclusion (a) of Theorem 3.3, and the minimizing property of λ_j, we have the inequalities

$$(4.1) \qquad \tilde{\lambda}_j \geq J[\eta \mid \tilde{B}] \geq J[\eta \mid B] \geq \lambda_j.$$

Moreover, if $\Delta J[\eta \mid B, \tilde{B}]$ is positive definite on $D[B] = D[\tilde{B}]$, then $J[\eta \mid \tilde{B}] > J[\eta \mid B]$ in (4.1), so that $\tilde{\lambda}_j > \lambda_j$.

If for two problems B and $\tilde{B}$ we have $\mathcal{S}[B] = \mathcal{S}[\tilde{B}] = \mathcal{S}$, and $k(t \mid B) \equiv k(t \mid \tilde{B}) \equiv k(t)$, then the "difference problem" involving $\mathcal{S}$, $k(t)$, and

$$\begin{aligned}\Delta J[\eta \mid B, \tilde{B}] &\equiv J[\eta \mid \tilde{B}] - J[\eta \mid B],\\ &= \gamma_{11}{}^{\Delta}\eta^2(a) + 2\gamma_{12}{}^{\Delta}\eta(a)\,\eta(b) + \gamma_{22}{}^{\Delta}\eta^2(b)\\ &\quad + \int_a^b \{(\tilde{r} - r)\eta'^2 + 2(\tilde{q} - q)\eta'\eta + (\tilde{p} - p)\eta^2\}\,dt,\end{aligned}$$

where we write r, q, p and $\tilde{r}, \tilde{q}, \tilde{p}$ in place of $r(t \mid B)$, $r(t \mid \tilde{B})$, etc., and $\gamma_{\alpha\beta}{}^{\Delta} = \gamma_{\alpha\beta}{}^{\tilde{B}} - \gamma_{\alpha\beta}{}^{B}$, $(\alpha, \beta = 1, 2;\ \alpha \leq \beta)$, will be denoted by $\Delta(B, \tilde{B})$. The conditions of hypothesis $(\hat{\mathrm{H}})$ are satisfied by $\Delta(B, \tilde{B})$, with the possible exception of the requirement that $\tilde{r}(t) - r(t)$ be positive for $t \in [a, b]$. If this further condition is satisfied, however, the corresponding difference boundary problem

$$\begin{gathered}[(\tilde{r} - r)u' + (\tilde{q} - q)u]' - [(\tilde{q} - q)u' + (\tilde{p} - p)u] + \lambda k u = 0,\\ (u(a), u(b)) \in \mathcal{S},\\ (\gamma_{11}{}^{\Delta}u(a) + \gamma_{12}{}^{\Delta}u(b) - w(a),\ \gamma_{12}{}^{\Delta}u(a) + \gamma_{22}{}^{\Delta}u(b) + w(b)) \in \mathcal{S}^{\perp},\end{gathered}$$

where $w = (\tilde{r} - r)u' + (\tilde{q} - q)u$, has an infinite sequence $\{\lambda_j{}^{\Delta}, u_j{}^{\Delta}\}$ of proper values and corresponding proper functions, determined as in Theorem 3.1.

Theorem 4.2. *Suppose that $\mathcal{S}[B] = \mathcal{S}[\tilde{B}] = \mathcal{S}$ and $k(t \mid B) \equiv k(t \mid \tilde{B}) \equiv k(t)$ for two problems B and $\tilde{B}$, while hypothesis $(\hat{\mathrm{H}})$ is satisfied by each of the problems B, $\tilde{B}$, $\Delta(B, \tilde{B})$. If $\{\lambda_j{}^{\Delta}, u_j{}^{\Delta}\}$ denotes a sequence of proper values and corresponding proper functions for $\Delta(B, \tilde{B})$ determined as in Theorem 3.1, then $\tilde{\lambda}_{j+k-1} \geq \lambda_j + \lambda_k{}^{\Delta}$, $(j, k = 1, 2, \ldots)$.*

Let $d_1, \ldots, d_{j+k-1}$ be real constants such that the function $\eta(t) = d_1\tilde{u}_1(t) + \cdots + d_{j+k-1}\tilde{u}_{j+k-1}(t)$ satisfies the $j + k - 2$ relations

$$K[\eta, u_i] = 0, \quad \textit{for} \quad 1 \leq i < j;\ K[\eta, u_\sigma{}^{\Delta}] = 0, \quad \textit{for} \quad 1 \leq \sigma < k,$$

and $1 = d_1{}^2 + \cdots + d_{j+k-1}^2 = K[\eta]$. In view of conclusion (a) of Theorem 3.3, and the respective minimizing properties of λ_j and $\lambda_k{}^{\Delta}$, we then have the inequalities

$$\tilde{\lambda}_{j+k-1} \geq J[\eta \mid \tilde{B}] = J[\eta \mid B] + \Delta J[\eta \mid B, \tilde{B}] \geq \lambda_j + \lambda_k{}^{\Delta}.$$

Corollary. *Under the hypotheses of Theorem 4.2, if g_j denotes the number of proper values of $\Delta(B, \tilde{B})$ less than λ_j, then $\tilde{\lambda}_{j+g_j} \geq 2\lambda_j$.*

Now consider two problems B and $\tilde{B}$ which differ only in the end-forms

$Q[\eta \mid B]$ and $Q[\eta \mid \tilde{B}]$; that is $\omega(t, \eta, \zeta \mid B) \equiv \omega(t, \eta, \zeta \mid \tilde{B}), k(t \mid B) \equiv k(t \mid \tilde{B}) \equiv k(t)$, $\mathcal{S}[B] = \mathcal{S}[\tilde{B}] = \mathcal{S}$, so that $D[B] = D[\tilde{B}]$. Let

$$Q^{\Delta}[\eta] = Q[\eta \mid \tilde{B}] - Q[\eta \mid B] = \gamma_{11}{}^{\Delta}\, \eta^2(a) + 2\gamma_{12}{}^{\Delta}\, \eta(a)\, \eta(b) + \gamma_{22}{}^{\Delta}\, \eta^2(b),$$

and consider the polynomial $D(\rho \mid B, \tilde{B})$ defined as follows: If $\dim \mathcal{S} = 0$, then $D(\rho \mid B, \tilde{B}) \equiv 1$; if $\dim \mathcal{S} = 1, 2$, then $D(\rho \mid B, \tilde{B}) = \det \Gamma(\rho \mid B, \tilde{B})$ where $\Gamma(\rho \mid B, \tilde{B})$ is the matrix defined as

$$\text{(i)}\quad \Gamma(\rho \mid B, \tilde{B}) = \begin{bmatrix} \rho - \gamma_{11}{}^{\Delta} & -\gamma_{12}{}^{\Delta} & \psi_1 \\ -\gamma_{12}{}^{\Delta} & \rho - \gamma_{22}{}^{\Delta} & \psi_2 \\ \psi_1 & \psi_2 & 0 \end{bmatrix},$$

if $\dim \mathcal{S} = 1$, *and* $\mathcal{S} = \{(\eta(a), \eta(b)) \mid \psi_1\eta(a) + \psi_2\eta(b) = 0\}$;

$$\text{(ii)}\quad \Gamma(\rho \mid B, \tilde{B}) = \begin{bmatrix} \rho - \gamma_{11}{}^{\Delta} & -\gamma_{12}{}^{\Delta} \\ -\gamma_{12}{}^{\Delta} & \rho - \gamma_{22}{}^{\Delta} \end{bmatrix}, \quad \textit{if} \quad \dim \mathcal{S} = 2.$$

It is to be noted that if $\dim \mathcal{S} = 0$, then $\mathcal{S}$ consists of the pairs $(\eta(a), \eta(b))$ satisfying the conditions $1 \cdot \eta(a) + 0 \cdot \eta(b) = 0$, $0 \cdot \eta(a) + 1 \cdot \eta(b) = 0$, and in this case $D(\rho \mid B, \tilde{B})$ is also given as the determinant of the bordered matrix $\Gamma(\rho \mid B, \tilde{B})$ defined as

$$\Gamma(\rho \mid B, \tilde{B}) = \begin{bmatrix} \rho - \gamma_{11}{}^{\Delta} & -\gamma_{12}{}^{\Delta} & 1 & 0 \\ -\gamma_{12}{}^{\Delta} & \rho - \gamma_{22}{}^{\Delta} & 0 & 1 \\ 1 & 0 & 0 & 0 \\ 0 & 1 & 0 & 0 \end{bmatrix}.$$

The determinant $D(\rho \mid B, \tilde{B})$ is the characteristic polynomial of $Q[\eta \mid \tilde{B}] - Q[\eta \mid B]$ on $\mathcal{S}$, and is of degree $d = \dim \mathcal{S}$. All zeros of this polynomial are real, and if a given value ρ_0 is a zero of $D(\rho \mid B, \tilde{B})$ of multiplicity q, then there are exactly q linearly independent solutions of the linear homogeneous algebraic equations in $4 - \text{d}$ unknowns whose coefficients are the elements in the rows of the matrices $\Gamma(\rho \mid B, \tilde{B})$. In case $d = 1, 2$, let $\rho_1 \le \cdots \le \rho_d$ denote these zeros and $w^{\tau} = (w_{\sigma}{}^{\tau})$, $(\sigma = 1, \ldots, 4 - \text{d})$, be a solution of this system of algebraic equations corresponding to $\rho = \rho_{\tau}$, and also orthonormal in the sense that $\sum_{i=1}^{2} w_i{}^{\tau} w_i{}^{\rho} = \delta_{\tau\rho}$, $(\tau, \rho = 1, \ldots, d)$. If ν is an integer such that $0 \le \nu < d$, and $\eta(a)$, $\eta(b)$ is a pair of end-values such that $(\eta(a), \eta(b)) \in \mathcal{S}$ and $w_1{}^{\tau}\eta(a) + w_2{}^{\tau}\eta(b) = 0$ for $1 \le \tau \le \nu$, then $Q[\eta \mid \tilde{B}] - Q[\eta \mid B] \ge \rho_{\nu+1}[\eta^2(a) + \eta^2(b)]$. For a proof of these algebraic properties, the reader is referred to Problem 13 of E.2 in Appendix E.

Theorem 4.3. *If the two problem B and $\tilde{B}$ differ only in the end-forms $Q[\eta \mid B]$ and $Q[\eta \mid \tilde{B}]$, let* **N** *and* **P** *denote, respectively, the number of negative and positive zeros of $D(\rho \mid B, \tilde{B})$, each zero being counted a number of times* *equal to its multiplicity. Then*

$$\tilde{\lambda}_{j+\mathbf{N}} \geq \lambda_j, \quad \textit{and} \quad \lambda_{j+\mathbf{P}} \geq \tilde{\lambda}_j, \quad \textit{for} \quad j = 1, 2, \ldots .$$

If $d = \dim \mathcal{S} = 0$ then the result is trivially true since $\mathbf{N} = 0$, $\mathbf{P} = 0$, and $\lambda_j = \tilde{\lambda}_j$ for $j = 1, 2, \ldots$. Without further comment, we shall suppose that $d > 0$. For a given integer j, let $d_1, \ldots, d_{j+\mathbf{N}}$ be real constants such that the function $\eta(t) = d_1\tilde{u}_1(t) + \cdots + d_{j+\mathbf{N}}\tilde{u}_{j+\mathbf{N}}(t)$ satisfies the $j + \mathbf{N} - 1$ conditions

$$K[\eta, u_i] = 0, \quad \textit{for} \quad 1 \leq i < j,$$
$$w_1^\tau\eta(a) + w_2^\tau\eta(b) = 0, \quad \textit{for} \quad 1 \leq \tau \leq \mathbf{N},$$

and normed so that $1 = d_1^2 + \cdots + d_{j+\mathbf{N}}^2 = K[\eta]$. Then η belongs to the class $D_{Nj}[B \mid K]$ of Theorem 3.1, and $Q^\Delta[\eta] \geq \rho_{\mathbf{N}+1}[\eta^2(a) + \eta^2(b)] \geq 0$ by the remark preceding the statement of the theorem. Hence conclusion (a) of Theorem 3.3, together with the minimizing property of λ_j, implies that

$$\tilde{\lambda}_{j+\mathbf{N}} \geq J[\eta \mid \tilde{B}] = Q^\Delta[\eta] + J[\eta \mid B] \geq J[\eta \mid B] \geq \lambda_j.$$

The remainder of the conclusion of the theorem is a ready consequence of the fact that the zeros of $D(\rho \mid \tilde{B}, B)$ are the negatives of the zeros of $D(\rho \mid B, \tilde{B})$.

If I is any bounded interval on the real line, we *shall denote by* $V(I \mid B)$ *the number of proper values of the problem B on I.* Corresponding to a value L we *shall denote by* $V_L(B)$, $\{W_L(B)\}$, *the number of proper values of B which are less,* $\{$*not greater*$\}$, than L. As an immediate consequence of the above theorem we have the following result.

Corollary. *Under the hypotheses of Theorem* 4.3, *for every real number L the following inequalities hold:*

$$V_L(B) - \mathbf{P} \leq V_L(\tilde{B}) \leq V_L(B) + \mathbf{N},$$
$$W_L(B) - \mathbf{P} \leq W_L(\tilde{B}) \leq W_L(B) + \mathbf{N};$$

moreover, $|V(I \mid B) - V(I \mid \tilde{B})| \leq \mathbf{N} + \mathbf{P}$ *for every bounded sub-interval I of the real line.*

Now suppose that each of the problems B and $\tilde{B}$ satisfies hypothesis $(\hat{\mathbf{H}})$, and they differ only in the spaces $\mathcal{S}[B]$ and $\mathcal{S}[\tilde{B}]$. Problem $\tilde{B}$ is said to be a *sub-problem* of B if $\mathcal{S}[\tilde{B}] \subset \mathcal{S}[B]$; if $d = \dim \mathcal{S}[B]$, $\tilde{d} = \dim \mathcal{S}[\tilde{B}]$, then $d - \tilde{d} \geq 0$ and $\tilde{B}$ is said to be a *sub-problem of B of dimension $d - \tilde{d}$.* If $d > \tilde{d}$ then there exists $d - \tilde{d}$ independent linear forms $\chi_1^{\,\nu}\eta(a) + \chi_2^{\,\nu}\eta(b)$, $(\nu = 1, \ldots, d - \tilde{d})$, in the end-values $\eta(a)$, $\eta(b)$ such that

$$\mathcal{S}[\tilde{B}] = \{(\eta(a), \eta(b)) \mid (\eta(a), \eta(b)) \in \mathcal{S}[B];\ \chi_1^{\,\nu}\eta(a) + \chi_2^{\,\nu}\eta(b) = 0, \quad \nu = 1, \ldots, d - d\}.$$

In particular, for a boundary value problem B of the form (3.1) satisfying hypothesis ($\hat{\mathbf{H}}$), the problem $\tilde{B}$ involving the same differential equation and the *null end-conditions* $\eta(a) = 0 = \eta(b)$ is a sub-problem of B of dimension equal to dim $\mathcal{S}[B]$.

In view of the above remarks, the following results are ready consequences of the minimum properties of the proper values of B and $\tilde{B}$, together with conclusion (a) of Theorem 3.3.

Theorem 4.4. *If each of the problems B and $\tilde{B}$ satisfies hypothesis* ($\hat{\mathbf{H}}$), *and $\tilde{B}$ is a sub-problem of B of dimension $\delta = d - \tilde{d}$, then $\lambda_{j+\delta} \geq \tilde{\lambda}_j \geq \lambda_j$, $(j = 1, 2, \ldots)$.*

Corollary. *Under the hypotheses of Theorem* 4.4, *for every real number L the following inequalities hold:*

$$V_L(B) - \delta \leq V_L(\tilde{B}) \leq V_L(B), \qquad W_L(B) - \delta \leq W_L(\tilde{B}) \leq W_L(B);$$

moreover, $|V(I \mid B) - V(I \mid \tilde{B})| \leq \delta$ for every bounded sub-interval I of the real line.

Now consider two problems B and $\tilde{B}$, each of which satisfies hypothesis ($\hat{\mathbf{H}}$), and which have in common the integrand function $\omega(t, \eta, \zeta \mid B) = \omega(t, \eta, \zeta \mid \tilde{B})$, with respective $Q[\eta \mid B]$, $\mathcal{S}[B]$ and $Q[\eta \mid \tilde{B}]$, $\mathcal{S}[\tilde{B}]$. By $B \times \tilde{B}$ we shall denote the problem with the same integrand function, the quadratic end-form $Q[\eta \mid \tilde{B}]$, and for which $\mathcal{S}[B \times \tilde{B}] = \mathcal{S}[B] \cap \mathcal{S}[\tilde{B}]$. Similarly, $\tilde{B} \times B$ denotes the problem with the same integrand function, the quadratic end-form $Q[\eta \mid B]$, and for which $\mathcal{S}[\tilde{B} \times B] = \mathcal{S}[\tilde{B}] \cap \mathcal{S}[B] = \mathcal{S}[B \times \tilde{B}]$. Let $\{\lambda'_j\}$ and $\{\lambda''_j\}$ denote the sequences of proper values of $B \times \tilde{B}$ and $\tilde{B} \times B$ in nondecreasing order, each repeated a number of times equal to its index. Let d' denote the dimension of $B \times \tilde{B}$ as a sub-problem of $\tilde{B}$, and d'' denote the dimension of $\tilde{B} \times B$ as a subproblem of B. Moreover, let $\mathbf{N}_0$ and $\mathbf{P}_0$ denote, respectively, the number of negative and positive zeros of

$$D(\rho \mid \tilde{B} \times B,\, B \times \tilde{B})\,.$$

In view of Theorem 4.3, applied to problems $\tilde{B} \times B$ and $B \times \tilde{B}$, it follows that

$$\text{(4.1)} \qquad \lambda'_{j+\mathbf{N}_0} \geq \lambda''_j, \qquad \lambda''_{j+\mathbf{P}_0} \geq \lambda'_j, \qquad (j = 1, 2, \ldots).$$

Application of Theorem 4.4 to problems $\tilde{B} \times B$ and B, and to problems $B \times \tilde{B}$ and $\tilde{B}$, then yields the following inequalities:

$$\text{(4.2)} \qquad \lambda_{j+d''} \geq \lambda''_j \geq \lambda_j, \qquad \tilde{\lambda}_{j+d'} \geq \lambda'_j \geq \tilde{\lambda}_j, \qquad (j = 1, 2, \ldots).$$

Therefore, combining the inequalities (4.1), (4.2), one obtains the following result.

Theorem 4.5. *Suppose that each of the problems B and $\tilde{B}$ satisfies hypothesis* ($\hat{\mathbf{H}}$), *and these problems have in common the integrand function* $\omega(t, \eta, \zeta) = \omega(t, \eta, \zeta \mid B) = \omega(t, \eta, \zeta \mid \tilde{B})$. *If d' denotes the dimension of $B \times \tilde{B}$ as a sub-problem of $\tilde{B}$, and d'' denotes the dimension of $\tilde{B} \times B$ as a sub-problem of B, while $\mathbf{N}_0$ and $\mathbf{P}_0$ denote, respectively, the number of negative and positive zeros of $D(\rho \mid \tilde{B} \times B, B \times \tilde{B})$, then*

$$\lambda_{j+d''+\mathbf{P}_0} \geq \tilde{\lambda}_j, \qquad \tilde{\lambda}_{j+d'+\mathbf{N}_0} \geq \lambda_j, \qquad (j = 1, 2, \ldots). \tag{4.3}$$

Now for the problems $\tilde{B} \times B$ and $B \times \tilde{B}$ the Corollary to Theorem 4.3 implies the inequalities

$$V_L(\tilde{B} \times B) - \mathbf{P}_0 \leq V_L(B \times \tilde{B}) \leq V_L(\tilde{B} \times B) + \mathbf{N}_0, \tag{4.4}$$

while the Corollary to Theorem 4.4 implies the inequalities

$$\begin{aligned} V_L(B) - d'' &\leq V_L(\tilde{B} \times B) \leq V_L(B), \\ V_L(\tilde{B}) - d' &\leq V_L(B \times \tilde{B}) \leq V_L(\tilde{B}). \end{aligned} \tag{4.5}$$

Consequently, as a Corollary to Theorem 4.5 one has the following result.

Corollary. *Under the hypotheses of Theorem* 4.5, *for every real number L the following inequalities hold*:

$$\begin{aligned} V_L(B) - \mathbf{P}_0 - d'' &\leq V_L(\tilde{B}) \leq V_L(B) + \mathbf{N}_0 + d', \\ W_L(B) - \mathbf{P}_0 - d'' &\leq W_L(\tilde{B}) \leq W_L(B) + \mathbf{N}_0 + d'; \end{aligned} \tag{4.6}$$

moreover, $|V(I \mid B) - V(I \mid \tilde{B})| \leq \mathbf{N}_0 + \mathbf{P}_0 + d' + d''$ *for every bounded sub-interval I of the real line, and* $\mathbf{N}_0 + \mathbf{P}_0 + d' + d'' \leq 2$.

Relations (4.6) are ready consequences of (4.4), (4.5), and in turn the inequality $|V(I \mid B) - V(I \mid \tilde{B})| \leq \mathbf{N}_0 + \mathbf{P}_0 + d' + d''$ is a direct result of (4.6). Now if dim $\mathcal{S}[B] = d$ then there exist $2 - d$ linear forms $\psi^\tau(\eta \mid B) \equiv \psi_1{}^\tau(B)\eta(a) + \psi_2{}^\tau(B)\eta(b)$ such that

$$\mathcal{S}[B] = \{(\eta(a), \eta(b)) \mid \psi^\tau(\eta \mid B) = 0, \qquad \tau = 1, \ldots, 2 - d\}.$$

Similarly, if dim $\mathcal{S}[\tilde{B}] = \tilde{d}$ then there exist $2 - \tilde{d}$ linear forms $\psi^\sigma(\eta \mid \tilde{B}) \equiv \psi_1{}^\sigma(\tilde{B})\eta(a) + \psi_2{}^\sigma(\tilde{B})\eta(b)$ such that

$$\mathcal{S}[\tilde{B}] = \{(\eta(a), \eta(b)) \mid \psi^\sigma(\eta \mid \tilde{B}) = 0, \qquad \sigma = 1, \ldots, 2 - \tilde{d}\}.$$

Consequently, the dimension d_0 of $\mathcal{S}[B] \cap \mathcal{S}[\tilde{B}]$ is defined by the property that the $(4 - d - \tilde{d}\ ) \times 2$ matrix

$$\begin{bmatrix} \psi_1{}^\tau(B) & \psi_2{}^\tau(B) \\ \psi_1{}^\sigma(\tilde{B}) & \psi_2{}^\sigma(\tilde{B}) \end{bmatrix}, \qquad (\tau = 1, \ldots, 2 - d; \sigma = 1, \ldots, 2 - \tilde{d}),$$

has rank $2 - d_0$, and, therefore,

$$2 - d_0 \leq 4 - d - \tilde{d}, \quad or \quad d + \tilde{d} - d_0 \leq 2.$$

As $D(\rho \mid \tilde{B} \times B, B \times \tilde{B})$ is of degree d_0, it follows that $\mathbf{N}_0 + \mathbf{P}_0 \leq d_0$; moreover, the equalities $d = d_0 + d''$, $\tilde{d} = d_0 + d'$ imply that $d_0 + d' + d'' \leq 2$, and hence $\mathbf{N}_0 + \mathbf{P}_0 + d' + d'' \leq 2$.

Problems VI.4

1. If $p(t)$ is a continuous function on $[0, 1]$, and $\{\lambda_j\}$ denotes the sequence of proper values of the boundary problem

$$u'' + [\lambda - p(t)]u = 0, \qquad u(0) = 0, \qquad u(1) = 0,$$

arranged in increasing order, prove that $\{\lambda_j/j^2\} \to \pi^2$ as $j \to \infty$.

Hint. Use results of Theorem 4.1 to compare the proper values of the given system with those of the related systems

$$u'' + [\lambda \pm \kappa]u = 0, \qquad u(0) = 0, \qquad u(1) = 0,$$

where κ is a constant such that $|p(t)| \leq \kappa$ for $t \in [0, 1]$.

2. For the boundary problems

$$\begin{aligned} B&: u'' + \lambda u = 0, \qquad u(0) = 0, \qquad u'(1) = 0, \\ \tilde{B}&: u'' + \lambda u = 0, \qquad u(0) = 0, \qquad u(1) + u'(1) = 0, \end{aligned}$$

show that $\lambda_j < \tilde{\lambda}_j < \lambda_{j+1}$, $j = 1, 2, \ldots$.

Hint. Use results of Theorem 4.3, and appropriate additional arguments, to exclude the possibility of equality of certain proper values.

3. Let B denote the boundary problem

$$u'' + \lambda u = 0, \qquad u(0) - u(1) = 0, \qquad u'(0) - u'(1) = 0,$$

and $\tilde{B}$ denote a second boundary problem involving the same differential equation $u'' + \lambda u = 0$, and one of the following sets of boundary conditions:

(i) $u(0) = 0, \quad u(1) = 0;$

(ii) $u'(0) = 0, \quad u'(1) = 0;$

(iii) $u(0) = 0, \quad u'(1) = 0.$

In each case verify the results of Theorem 4.5 for the pair B, $\tilde{B}$.

4. If $\lambda_1 \leq \lambda_2 \leq \cdots$ are the proper values of the boundary problem

$$[(1 + e^t)u']' + [2t^2 + \lambda]u = 0,$$
$$B: \qquad u(0) + 2u'(0) = 0,$$
$$u(1) = 0,$$

show that $\lambda_{j+1} \geq j^2\pi^2$, $(j = 1, 2, \ldots)$.

Hint. Note that $J[\eta \mid B] = J_1[\eta] + J_2[\eta]$, where

$$J_1[\eta] = -\eta^2(0) + \int_0^1 \eta'^2\, dt,$$

$$J_2[\eta] = \int_0^1 (e^t\eta'^2 - 2t^2\eta^2)\, dt \geq \int_0^1 (\eta'^2 - 2\eta^2)\, dt.$$

5. Suppose that B and $\tilde{B}$ are boundary problems of the form (3.1) which satisfy hypothesis ($\hat{\mathbf{H}}$), with $\omega(t, \eta, \zeta \mid \tilde{B}) = \omega(t, \eta, \zeta \mid B)$, $Q[\eta \mid \tilde{B}] = Q[\eta \mid B]$, $\mathcal{S}[\tilde{B}] = \mathcal{S}[B]$, and $J[\eta \mid \tilde{B}] = J[\eta \mid B]$ is positive definite on $\mathcal{S}[\tilde{B}] = \mathcal{S}[B]$. Moreover, let $\tilde{\tilde{B}}$ denote the boundary problem with $\omega(t, \eta, \zeta \mid \tilde{\tilde{B}})$, $Q[\eta \mid \tilde{\tilde{B}}]$, $\mathcal{S}[\tilde{\tilde{B}}]$ equal to the respective common elements of B and $\tilde{B}$ while $k(t \mid \tilde{\tilde{B}}) = k(t \mid B) + k(t \mid \tilde{B})$. If $\{\lambda_i, u_i\}$, $\{\tilde{\lambda}_i, \tilde{u}_i\}$, $\{\tilde{\tilde{\lambda}}_i, \tilde{\tilde{u}}_i\}$ denote sets of proper values and corresponding proper functions for the respective problems B, $\tilde{B}$ and $\tilde{\tilde{B}}$, which individually satisfy the conditions of Theorem 3.1, prove that

$$\lambda_\alpha^{-1} + \tilde{\lambda}_\beta^{-1} \geq \tilde{\tilde{\lambda}}_{\alpha+\beta-1}^{-1}, \qquad \alpha, \beta = 1, 2, \ldots.$$

Generalize this result to the case of a finite number of boundary problems (3.1) which differ only in the coefficient functions $k(t)$.

6. Suppose that B and $\tilde{B}$ are boundary problems of the form (3.1) which satisfy hypothesis ($\hat{\mathbf{H}}$), with $\omega(t, \eta, \zeta \mid \tilde{B}) = \omega(t, \eta, \zeta \mid B)$, $Q[\eta \mid \tilde{B}] = Q[\eta \mid B]$, $\mathcal{S}[\tilde{B}] = \mathcal{S}[B]$, and $k(t \mid \tilde{B}) \geq k(t \mid B) > 0$ on $[a, b]$. Let $\{\lambda_i, u_i\}$, $\{\tilde{\lambda}_i, \tilde{u}_i\}$ denote sets of proper values and corresponding proper functions for the respective problems B, $\tilde{B}$ which individually satisfy the conditions of Theorem 3.1. Prove that if $\lambda_p < 0 < \lambda_q$, then $\lambda_j \leq \tilde{\lambda}_j < 0$ for $j = 1, \ldots, p$ and $0 < \tilde{\lambda}_j \leq \lambda_j$ for $j = q, q + 1, \ldots$; moreover, if $q > p + 1$ and $\lambda_j = 0$ for $p < j < q$ then $\tilde{\lambda}_j = 0$ for $p < j < q$.

7. If a boundary problem B of the form (3.1) satisfies hypothesis ($\hat{\mathbf{H}}$), and $\{\lambda_i, u_i\}$ denotes a set of proper values and corresponding proper functions which satisfies the conditions of Theorem 3.1, prove that there exist positive constants c_0, c_1, c_0', c_1' such that

$$j^2c_0 - c_1 \leq \lambda_j \leq j^2c_0' + c_1', \qquad (j = 1, 2, \ldots).$$

In particular, the infinite series $\sum' |\lambda_j|^{-p}$ converges for $p > \frac{1}{2}$, but diverges for $p = \frac{1}{2}$, where $\sum'$ denotes summation over indices j for which $\lambda_j \neq 0$.

Hint. Note that from (3.7), and a similar inequality, there exist positive constants k_0, k_1, k_0', k_1' such that if $\eta \in D'[a, b]$ then

$$J[\eta \mid B] \geq J_0[\eta] \equiv k_0 \int_a^b \eta'^2\, dt - k_1 \int_a^b \eta^2\, dt,$$

$$J[\eta \mid B] \leq J_0'[\eta] \equiv k_0' \int_a^b \eta'^2\, dt + k_1' \int_a^b \eta^2\, dt.$$

Let k_m and k_M be positive constants such that $k_m \leq k(t) \leq k_M$ for $t \in [a, b]$, and set $K_m[\eta] = k_m \int_a^b \eta^2\, dt$, $K_M[\eta] = k_M \int_a^b \eta^2\, dt$. Let B_m denote the boundary problem of the form (3.1) involving the functionals $J_0'[\eta]$, $K_m[\eta]$, with boundary conditions $\eta(a) = 0 = \eta(b)$, and let B_M denote the corresponding boundary problem involving the functionals $J_0[\eta]$, $K_M[\eta]$, and with boundary conditions $\eta(a) = 0 = \eta(b)$. Obtain the stated inequalities by formulating suitable boundary problems, which with the aid of the results of Theorems 4.1, 4.4, and Problem 6 above, yield a comparison of the proper values of B with those of B_m and B_M.

8. Suppose that on an interval $I = [a_0, \infty)$ the coefficients of the differential equation (3.1-a) satisfy the conditions specified in ($\hat{\mathbf{H}}$), and for $[a, b] \subset I$ consider the four boundary problems $B_{00}[a, b]$, $B_{0*}[a, b]$, $B_{*0}[a, b]$, $B_{**}[a, b]$ involving the differential equation $l[u; \lambda](t) = 0$ and the respective boundary conditions:

$$\Delta_{00}[a, b]: u(a) = 0,\ u(b) = 0; \qquad \Delta_{*0}[a, b]: v(a) = 0,\ u(b) = 0;$$

$$\Delta_{0*}[a, b]: u(a) = 0,\ v(b) = 0; \qquad \Delta_{**}[a, b]: v(a) = 0,\ v(b) = 0.$$

For (p, q) any one of the sets $(0, 0)$, $(0, *)$, $(*, 0)$, $(*, *)$, let $\{\lambda_j{}^{pq}(a, b), u_j{}^{pq}(t; a, b)\}$ denote a set of proper values and corresponding proper functions for $B_{pq}[a, b]$, satisfying the conditions of Theorem 3.1.

(i) Prove that if $[a, b] \subset I$, then $B_{00}[a, b]$ is a subproblem of dimension one of each of the problems $B_{0*}[a, b]$, $B_{*0}[a, b]$, while $B_{0*}[a, b]$ and $B_{*0}[a, b]$ are individually subproblems of dimension one of $B_{**}[a, b]$. Moreover, for $j = 1, 2, \ldots,$

$$\lambda_j^{0*}(a, b) < \lambda_j^{00}(a, b) < \lambda_{j+1}^{0*}(a, b); \qquad \lambda_j^{*0}(a, b) < \lambda_j^{00}(a, b) < \lambda_{j+1}^{*0}(a, b);$$

$$\lambda_j^{**}(a, b) < \lambda_j^{0*}(a, b) < \lambda_{j+1}^{**}(a, b); \qquad \lambda_j^{**}(a, b) < \lambda_j^{*0}(a, b) < \lambda_{j+1}^{**}(a, b).$$

(ii) If b, c, d are points of I satisfying $b < c < d$, then

$$\lambda_{r+s-1}^{00}(b, d) \leq \max\,\{\lambda_r^{0*}(b, c), \lambda_s^{*0}(c, d)\}.$$

(iii) Suppose that for arbitrary $c \in I$ there exists a $d > c$ such that $\lambda_1^{*0}(c, d) \leq 0$. For $a \in I$, prove that a necessary and sufficient condition for

$l[u; 0](t) = 0$ to be nonoscillatory on $[a, \infty)$ is that $\lambda_1^{0*}(a, c) > 0$ for all $c > a$.

Hints. (i) Use the result of Theorem 4.4, and appropriate additional arguments to exclude the possibility of equality of certain proper values.

(ii) Determine constants $e_1, \ldots, e_r, k_1, \ldots, k_s$ such that the function $\eta(t)$ defined by

$$\eta(t) = \sum_{\alpha=1}^{r} e_\alpha u_\alpha^{0*}(t; b, c), \quad \text{for} \quad t \in [b, c],$$

$$\eta(t) = \sum_{\beta=1}^{s} k_\beta u_\beta^{*0}(t; c, d), \quad \text{for} \quad t \in [c, d],$$

belongs to $D_0'[b, d]$, satisfies the orthogonality conditions

$$\int_b^d \eta(t) u_i^{00}(t; b, d)\, dt = 0, \quad \text{for} \quad i < r + s - 1,$$

and has $e_1^2 + \cdots + e_r^2 + k_1^2 + \cdots + k_s^2 = 1$. Use appropriate extremizing properties of the proper values of $B_{0*}[b, c]$, $B_{*0}[c, d]$ and $B_{00}[b, d]$ to conclude that

$$\lambda_{r+s-1}^{00}(b, d) \le J[\eta; b, d] \le \max\{\lambda_r^{0*}(b, c), \lambda_s^{*0}(c, d)\}.$$

(iii) Use certain results of (i) and (ii).

5. EXPANSION THEOREMS

The results of the preceding sections will be used now to establish expansion theorems for the functionals $K[\eta]$ and $J[\eta \mid B]$, as well as certain pointwise convergence theorems. It will be supposed that (3.1) is a self-adjoint system for which hypothesis $(\hat{\mathfrak{H}})$ holds, and $\{\lambda_j, u_j\}$ is a sequence of proper values and corresponding proper functions as determined in Theorem 3.1.

***Lemma* 5.1.** *If h is a real-valued piecewise continuous function on $[a, b]$, and $c_j[h] = K[h, u_j]$, $(j = 1, 2, \ldots)$, and d_j are arbitrary real constants, then*

$$K\left[h - \sum_{j=1}^{m} d_j u_j\right] \ge K\left[h - \sum_{j=1}^{m} c_j[h] u_j\right] = K[h] - \sum_{j=1}^{m} c_j^2[h], \tag{5.1}$$

with the equality sign holding if and only if $d_j = c_j[h]$, $(j = 1, \ldots, m)$; in particular

$$\sum_{j=1}^{m} c_j^2[h] \le K[h]. \tag{5.2}$$

Relative to the set of functions $\{u_j\}$ which is orthonormal with respect to the functional K in the sense that $K[u_i, u_j] = \delta_{ij}$, the equality of the last two expressions in (5.1) is known as *Bessel's equality*, and (5.2) is called

Bessel's inequality. For a more general formulation, the reader is referred to Appendix A.

In view of the orthonormal relations $K[u_i, u_j] = \delta_{ij}$ of (3.10-a), relation (5.1) is an immediate consequence of the identity

$$\begin{aligned} K\left[h - \sum_{j=1}^{m} d_j u_j\right] &= K[h] - 2\sum_{j=1}^{m} d_j\, K[h, u_j] + \sum_{i,j=1}^{m} d_i\, d_j\, K[u_i, u_j], \\ &= K[h] - 2\sum_{j=1}^{m} d_j c_j[h] + \sum_{j=1}^{m} d_j^2, \\ &= K\left[h - \sum_{j=1}^{m} c_j[h] u_j\right] + \sum_{j=1}^{m} (d_j - c_j[h])^2. \end{aligned}$$

The following properties of the functional K are immediate consequences of its definition and the non-negativeness of $k(t)$.

***Lemma* 5.2.** *If h_1 and h_2 are piecewise continuous functions on $[a, b]$, then*

$$4K[h_1, h_2] = K[h_1 + h_2] - K[h_1 - h_2], \tag{5.3}$$

$$K[h_1 + h_2] \le 2K[h_1] + 2K[h_2]. \tag{5.4}$$

Theorem 5.1. *If h is a piecewise continuous function on $[a, b]$, then*

$$K[h] = \sum_{j=1}^{\infty} c_j^2[h]. \tag{5.5}$$

If h_1 and h_2 are piecewise continuous functions on $[a, b]$, then

$$K[h_1, h_2] = \sum_{j=1}^{\infty} c_j[h_1]\, c_j[h_2]. \tag{5.6}$$

If $\eta \in D[B]$, and $\eta_m(t) = \eta(t) - \sum_{j=1}^{m} c_j[\eta] u_j(t)$, *then* $K[\eta_m, u_i] = 0$ for $i = 1, \ldots, m$, and $K[\eta_m] = K[\eta] - \sum_{j=1}^{m} c_j^2[\eta]$. If λ_0 is a real number such that $\lambda_0 < \lambda_1$, then the minimizing property of λ_1 implies that $J[\eta; \lambda_0 \mid B]$ is positive definite on $D[B]$. In view of the minimizing property of λ_{m+1} as established in Theorem 3.1, it then follows that

$$(\lambda_{m+1} - \lambda_0)\, K[\eta_m] \le J[\eta_m; \lambda_0 \mid B]. \tag{5.7}$$

As the orthonormal character of the u_i implies that $J[u_i, u_j; \lambda_0 \mid B] = (\lambda_j - \lambda_0)\delta_{ij}$, $(i, j = 1, 2, \ldots)$, we also have

$$J[\eta_m; \lambda_0 \mid B] = J[\eta; \lambda_0 \mid B] - \sum_{j=1}^{m} (\lambda_j - \lambda_0)\, c_j^2[\eta] \le J[\eta; \lambda_0 \mid B]. \tag{5.8}$$

Consequently, since the Corollary to Theorem 3.3 implies that $\lambda_{m+1} - \lambda_0 \to \infty$ as $m \to \infty$, it follows that $K[\eta] - \sum_{j=1}^{m} c_j^2[\eta] = K[\eta_m] \to 0$ as $m \to \infty$, so that (5.5) holds if $h \in D[B]$.

Now if h is an arbitrary real-valued piecewise continuous function on $[a, b]$, and $\varepsilon > 0$, there exists a function η_ε such that $\eta_\varepsilon \in D[B]$ and $K[h - \eta_\varepsilon] < \varepsilon/4$. Indeed, by an elementary argument one may establish the existence of such a function η_ε which is zero throughout neighborhoods of the end-values $t = a$, $t = b$, and each point of discontinuity of h. In turn, the results of the preceding paragraph imply that there exists an m_ε such that if $m > m_\varepsilon$ then $K[\eta_\varepsilon - \eta_{\varepsilon m}] < \varepsilon/4$. In view of Lemma 5.1, and relation (5.4), it follows that if $m > m_\varepsilon$ then

$$\begin{aligned} K[h_m] &\leq K[h - \eta_{\varepsilon m}], \\ &\leq 2K[h - \eta_\epsilon] + 2K[\eta_\varepsilon - \eta_{\varepsilon m}], \\ &< 2\left(\frac{\varepsilon}{4}\right) + 2\left(\frac{\varepsilon}{4}\right) = \varepsilon. \end{aligned}$$

Consequently, $K[h] - \sum_{j=1}^m c_j^2[h] = K[h_m] \to 0$ as $m \to \infty$, so that (5.5) holds for arbitrary piecewise continuous functions h.

Finally, if h_1, h_2 are piecewise continuous functions the relation (5.6) is a consequence of (5.5) for $h = h_1 \pm h_2$, together with the identity (5.3).

***Corollary* 1.** *If h is a piecewise continuous function on $[a, b]$ such that $c_j[h] = 0$ for $j = 1, 2, \ldots$, then $h(t) = 0$ at each point of continuity of h.*

If h is a continuous function on $[a, b]$ such that the series

$$\sum_{j=1}^{\infty} c_j[h]\, u_j(t)$$

converges uniformly on $[a, b]$, then the sum of this infinite series is a continuous function $\phi(t)$ such that $c_j[h - \phi] = 0$, $(j = 1, 2, \ldots)$, and hence we have the following result.

***Corollary* 2.** *If h is a continuous function on $[a, b]$ such that the infinite series $\sum_{j=1}^{\infty} c_j[h]\, u_j(t)$ converges uniformly on this interval, then the sum of this series is equal to $h(t)$ for $t \in [a, b]$.*

***Lemma* 5.3.** *If $\lambda_0 < \lambda_1$, then for $\eta \in D[B]$ the infinite series*

$$\sum_{j=1}^{\infty} (\lambda_j - \lambda_0)\, c_j^2[\eta]$$

converges, and

$$\sum_{j=1}^{\infty} (\lambda_j - \lambda_0)\, c_j^2[\eta] \leq J[\eta; \lambda_0 \mid B] \quad \text{for} \quad \eta \in D[B]. \tag{5.9}$$

If $\eta \in D[B]$, and $\eta_m(t) = \eta(t) - \sum_{j=1}^m c_j[\eta]\, u_j(t)$, then $\eta_m \in D[B]$, and as

$$0 \leq J[\eta_m; \lambda_0 \mid B] = J[\eta; \lambda_0 \mid B] - \sum_{j=1}^{m} (\lambda_j - \lambda_0)\, c_j^2[\eta],$$

it follows that the infinite series $\sum_{j=1}^{\infty}(\lambda_j - \lambda_0)\, c_j^2[\eta]$ converges and satisfies the inequality (5.9). As $\sum_{j=1}^{\infty} c_j^2[\eta] = K[\eta]$ by Theorem 5.1, it is to be noted that the result of the above lemma is equivalent to the convergence of the infinite series $\sum_{j=1}^{\infty} \lambda_j\, c_j^2[\eta]$, and the inequality

$$\text{(5.9')} \qquad \sum_{j=1}^{\infty} \lambda_j\, c_j^2[\eta] \leq J[\eta \mid B], \quad \textit{for} \quad \eta \in D[B].$$

It is actually true that the equality sign holds in each of the relations (5.9) and (5.9'). In order to prove this more precise result, however, the following two lemmas will be established.

***Lemma* 5.4.** *If $\lambda_0 < \lambda_1$, then there exists a $\mu > 0$ such that*

$$\text{(5.10)} \quad J[\eta; \lambda_0 \mid B] \geq \mu\left[\eta^2(a) + \eta^2(b) + \int_a^b (\eta'^2 + \eta^2)\, dt\right], \quad \textit{for} \quad \eta \in D[B].$$

If the results of the third paragraph in Section 3 are applied to the functional $J[\eta; \lambda_0 \mid B] - [\eta^2(a) + \eta^2(b)]$, it follows that there exist constants $k_0 > 0$, $k_1 > 0$ such that

$$J[\eta; \lambda_0 \mid B] - [\eta^2(a) + \eta^2(b)] \geq k_0 \int_a^b \eta'^2\, dt - k_1 \int_a^b \eta^2\, dt, \quad \textit{for} \quad \eta \in D'[a, b],$$

and, hence for $k_0' = \min\{1, k_0\}$,

$$J[\eta; \lambda_0 \mid B] \geq k_0'\left[\eta^2(a) + \eta^2(b) + \int_a^b \eta'^2\, dt\right] - k_1 \int_a^b \eta^2\, dt, \quad \textit{for} \quad \eta \in D'[a, b].$$

Now in view of the minimum property of λ_1 we have $J[\eta; \lambda_0 \mid B] \geq k_1' \int_a^b \eta^2\, dt$ for $\eta \in D[B]$, where $k_1' = (\lambda_1 - \lambda_0)\kappa_0$, and κ_0 is a positive constant such that $k(t) \geq \kappa_0$ for $t \in [a, b]$. Consequently, for $0 < \theta < 1$, we have

$$J[\eta; \lambda_0 \mid B] \geq \theta k_0'\left[\eta^2(a) + \eta^2(b) + \int_a^b \eta'^2\, dt\right] + [(1 - \theta)k_1' - \theta k_1] \int_a^b \eta^2\, dt,$$

for $\eta \in D[B]$, and the choice $\theta = k_1'/(k_0' + k_1' + k_1)$ provides the inequality (5.10) with $\mu = k_0' k_1'/(k_0' + k_1' + k_1)$.

***Lemma* 5.5.** *Suppose that $h_j \in D'[a, b]$ for $j = 1, 2, \ldots$, while $\{h_j(t)\} \to 0$ for $t \in [a, b]$ and $\{\int_a^b (h_j' - h_k')^2\, dt\} \to 0$ as $j, k \to \infty$. Then $\{h_j(t)\} \to 0$ uniformly for $t \in [a, b]$, and $\{\int_a^b h_j'^2\, dt\} \to 0$ as $j \to \infty$.*

Let $h_{jk}(t) = h_j(t) - h_k(t)$, $(j, k = 1, 2, \ldots)$. Then the equation $h_{jk}(t) = h_{jk}(a) + \int_a^t h_{jk}'(s)\, ds$ and elementary inequalities yield the relation

$$h_{jk}^2(t) \leq 2h_{jk}^2(a) + 2\left(\int_a^t h_{jk}'(s)\, ds\right)^2 \leq 2h_{jk}^2(a) + 2(b - a)\int_a^t h_{jk}'^2(s)\, ds.$$

As the hypotheses of the lemma imply that $\{h_{jk}(a)\} \to 0$ and $\{\int_a^t h_{jk}'^2\, ds\} \to 0$ as $j, k \to \infty$, uniformly for $t \in [a, b]$, the convergence of $\{h_j(t)\}$ is uniform on

$[a, b]$. Since $\{\int_a^b h_{jk}'^2\,dt\} \to 0$ as $j, k \to \infty$, for a given $\varepsilon > 0$ there exists an $m = m_\varepsilon$ such that

$$\int_a^b h_{jk}'^2\,dt < \varepsilon \quad \textit{for} \quad j \geq m, \qquad k \geq m, \tag{5.11}$$

and there exists a $\delta = \delta_\varepsilon$ such that $0 < \delta < (b - a)/2$ and

$$\int_a^{a+\delta} h_m'^2\,dt + \int_{b-\delta}^b h_m'^2\,dt < \varepsilon.$$

Now $h_j'^2 = (h_m' + h_{jm}')^2 \leq 2h_m'^2 + 2h_{jm}'^2$, and hence

$$\int_a^b h_j'^2\,dt < \int_{a+\delta}^{b-\delta} h_j'^2\,dt + 4\varepsilon, \quad \textit{for} \quad j \geq m. \tag{5.12}$$

As $h_m'(t)$ is piecewise continuous on $[a, b]$, it follows that

$$\lim_{s\to 0} \int_{a+\delta}^{b-\delta} [h_m'(t) - h_m'(t + s)]^2\,dt = 0. \tag{5.13}$$

Moreover, since

$$\begin{aligned}\int_{a+\delta}^{b-\delta} \left[h_m'(t) - \frac{1}{2d}\int_{-d}^d h_m'(t + s)\,ds\right]^2 dt \\ &= \int_{a+\delta}^{b-\delta} \left[\frac{1}{2d}\int_{-d}^d \{h_m'(t) - h_m'(t + s)\}\,ds\right]^2 dt, \\ &\leq \frac{1}{2d}\int_{-d}^d \left(\int_{a+\delta}^{b-\delta} [h_m'(t) - h_m'(t + s)]^2\,dt\right) ds,\end{aligned}$$

there exists a $d = d_\varepsilon$ such that $0 < d < \delta$, and

$$\int_{a+\delta}^{b-\delta} \left[h_m'(t) - \frac{1}{2d}\int_{-d}^d h_m'(t + s)\,ds\right]^2 dt < \varepsilon. \tag{5.14}$$

If $u_j = u_j(t; d)$ is defined as

$$u_j(t; d) = \frac{1}{2d}\int_{-d}^d h_j(t + s)\,ds, \quad \textit{for} \quad j = 1, 2, \ldots, \qquad t \in [a + \delta, b - \delta],$$

then u_j is of class $\mathfrak{C}^1$ on $[a + \delta, b - \delta]$, and

$$u_j'(t; d) = \frac{1}{2d}[h_j(t + d) - h_j(t - d)] = \frac{1}{2d}\int_{-d}^d h_j'(t + s)\,ds.$$

If the function $u_j(t; d) - u_k(t; d)$ is denoted by $u_{jk}(t; d)$, then

$$\begin{aligned}[u_{jk}'(t; d)]^2 &= \left[\frac{1}{2d}\int_{-d}^d h_{jk}'(t + s)\,ds\right]^2 \\ &\leq \frac{1}{2d}\int_{-d}^d h_{jk}'^2(t + s)\,ds, \quad \textit{for} \quad t \in [a + \delta, b - \delta],\end{aligned} \tag{5.15}$$

by the Schwarz inequality, and hence

$$[u'_{jk}(t; d)]^2 \le \frac{1}{2d} \int_a^b h'^2_{jk}\, dt.$$

Since $\{\int_a^b h'^2_{jk}\, dt\} \to 0$ as $h, k \to \infty$, the sequence $\{u'_j(t; d)\}$ converges uniformly on $[a + \delta, b - \delta]$ to a function $\phi(t)$, which is necessarily continuous on this interval. Moreover,

$$u_j^2(t; d) = \left[\frac{1}{2d} \int_{-d}^{d} h_j(t + s)\, ds\right]^2 \le \frac{1}{2d} \int_{-d}^{d} h_j^2(t + s) \le \frac{1}{2d} \int_a^b h_j^2(t)\, dt,$$

and, as $\{h_j(t)\} \to 0$ uniformly on $[a, b]$, it follows that $\{u_j(t; d)\} \to 0$ uniformly on $[a + \delta, b - \delta]$. Since

$$u_j(t; d) = u_j(a + \delta; d) + \int_{a+\delta}^{t} u'_j(s; d)\, ds, \quad for \quad t \in [a + \delta, b - \delta],$$

it then follows that $\int_{a+\delta}^{t} \phi(s)\, ds = 0$ for $t \in [a + \delta, b - \delta]$, and consequently $\phi(t) \equiv 0$ on this interval. In particular, there exists an $n = n'_\varepsilon$, with $n_\varepsilon \ge m_\varepsilon$ and such that

$$\int_{a+\delta}^{b-\delta} [u'_j(t; d)]^2\, dt < \varepsilon, \quad for \quad j \ge n. \tag{5.16}$$

Inequality (5.15), with $k = m$, then implies that

$$\int_{a+\delta}^{b-\delta} u'^2_{jm}(t; d)\, dt \le \frac{1}{2d} \int_{-d}^{d} \left(\int_{a+\delta}^{b-\delta} h'^2_{jm}(t + s)\, dt\right) ds \le \int_a^b h'^2_{jm}(t)\, dt. \tag{5.17}$$

As $h'^2_j \le 2u'^2_j + 2(h'_j - u'_j)^2$, and $h'_j - u'_j = (h'_{jm} - u'_{jm}) + (h'_m - u'_m)$, it follows that if $j \ge n_\varepsilon$ then

$$\int_{a+\delta}^{b-\delta} h'^2_j\, dt \le 2\int_{a+\delta}^{b-\delta} u'^2_j\, dt + 6\int_{a+\delta}^{b-\delta} [h'^2_{jm} + u'^2_{jm} + (h'_m - u'_m)^2]\, dt.$$

In view of inequalities (5.11), (5.14), (5.16), (5.17) it then follows that

$$\int_{a+\delta}^{b-\delta} h'^2_j\, dt < 20\varepsilon, \quad if \quad j \ge n_\varepsilon.$$

Consequently, in view of inequality (5.12) we have that

$$\int_a^b h'^2_j\, dt < 24\varepsilon, \quad if \quad j \ge n_\varepsilon,$$

thus completing the proof that $\{\int_a^b h'^2_j\, dt\} \to 0$ as $j \to \infty$.

Theorem 5.2. *If $\eta \in D[B]$, then the infinite series $\sum_{j=1}^{\infty} c_j[\eta]\, u_j(t)$ converges to $\eta(t)$ uniformly for $t \in [a, b]$; moreover,*

$$\int_a^b \left[\eta'(t) - \sum_{j=1}^{m} c_j[\eta]\, u_j'(t)\right]^2 dt \to 0 \quad as \quad m \to \infty, \tag{5.18}$$

$$J[\eta \mid B] = \sum_{j=1}^{\infty} \lambda_j\, c_j^2[\eta]. \tag{5.19}$$

For $\eta \in D[B]$, let $\eta_j(t) = \eta(t) - \sum_{i=1}^{j} c_i[\eta]\, u_i(t)$, and $\eta_{jk}(t) = \eta_j(t) - \eta_k(t)$, $j, k = 1, 2, \ldots$. Then $\eta_{jk} \in D[B]$, and for $t \in [a, b]$ we have

$$\begin{aligned} \eta_{jk}^2(t) &= \left[\eta_{jk}(a) + \int_a^t \eta_{jk}'(s)\, ds\right]^2, \\ &\leq 2\eta_{jk}^2(a) + 2(b - a)\int_a^b \eta_{jk}'^2(s)\, ds, \end{aligned}$$

where the last inequality is a consequence of an application of the Schwarz inequality. With the aid of Lemma 5.4 it then follows that for $t \in [a, b]$ and $j < k$,

$$\eta_{jk}^2(t) + \int_a^b \eta_{jk}'^2(t)\, dt \leq \frac{\mu_0}{\mu} J[\eta_{jk}; \lambda_0 \mid B] = \frac{\mu_0}{\mu} \sum_{i=j+1}^{k} (\lambda_i - \lambda_0)\, c_i^2[\eta], \tag{5.20}$$

where $\mu_0 = \max\{2, 1 + 2(b - a)\}$. As the series $\sum_{i=1}^{\infty}(\lambda_i - \lambda_0)\, c_i^2[\eta]$ converges by Lemma 5.3, inequality (5.20) implies that $\int_a^b \eta_{jk}'^2\, dt \to 0$ as $j, k \to \infty$, and also that the sequence $\{\eta_j(t)\}$ converges uniformly on $[a, b]$. Consequently, the series $\sum_{i=1}^{\infty} c_i[\eta]\, u_i(t)$ converges uniformly on $[a, b]$, and by Corollary 2 to Theorem 5.1 the sum of this series is equal to $\eta(t)$. In turn, the sequence $\{\eta_m(t)\}$ converges to zero for $t \in [a, b]$, and by Lemma 5.5 we have that $\int_a^b \eta_m'^2\, dt \to 0$, which is conclusion (5.18).

Finally, for a given $\lambda_0 < \lambda_1$, elementary inequalities imply the existence of a constant μ_1 such that

$$0 \leq J[\eta; \lambda_0 \mid B] \leq \mu_1\left[\eta^2(a) + \eta^2(b) + \int_a^b (\eta'^2 + \eta^2)\, dt\right], \quad for \quad \eta \in D[B].$$

In view of the uniform convergence of $\{\eta_m(t)\}$ to zero on $[a, b]$, and relation (5.18), it then follows that $J[\eta_m; \lambda_0 \mid B] \to 0$ as $m \to \infty$. Moreover, since

$$J[\eta_m; \lambda_0 \mid B] = J[\eta; \lambda_0 \mid B] - \sum_{j=1}^{m}(\lambda_j - \lambda_0)\, c_j^2[\eta],$$

the series $\sum_{j=1}^{\infty}(\lambda_j - \lambda_0)\, c_j^2[\eta]$ converges, and

$$J[\eta; \lambda_0 \mid B] = \sum_{j=1}^{\infty}(\lambda_j - \lambda_0)\, c_j^2[\eta],$$

a relation which is equivalent to (5.19), since $\sum_{j=1}^{\infty} c_j^2[\eta] = K[\eta]$ by Theorem 5.1.

Corresponding to (5.19), one has the bilinear relation

$$(5.21)\quad J[\eta_1, \eta_2 \mid B] = \sum_{j=1}^{\infty} \lambda_j\, c_j[\eta_1] c_j[\eta_2], \quad \textit{for} \quad \eta_\alpha \in D[B], \qquad (\alpha = 1, 2),$$

which is an immediate consequence of (5.19) and the relations

$$4J[\eta_1, \eta_2 \mid B] = J[\eta_1 + \eta_2 \mid B] - J[\eta_1 - \eta_2 \mid B],$$
$$c_j[\eta_1 \pm \eta_2] = c_j[\eta_1] \pm c_j[\eta_2].$$

Moreover, although throughout the above discussion attention has been restricted to real-valued functions, in view of the bilinear character of (5.6) and (5.21), these relations remain valid for complex-valued functions. Specifically, if

$$\Gamma(B) = \{\eta(t) \mid \eta(t) = \eta_R(t) + i\eta_I(t), \qquad \eta_R \in D[B], \qquad \eta_I \in D[B]\},$$

then (5.6) and (5.21) hold for $\eta_\alpha \in \Gamma(B)$, ($\alpha = 1, 2$). In particular, if $\eta \in \Gamma(B)$ then the series $\sum_{j=1}^{\infty} |c_j[\eta]|^2$, $\sum_{j=1}^{\infty} \lambda_j\, |c_j[\eta]|^2$ converge, and

$$(5.22)\quad \begin{aligned} &\text{(a)}\ K[\eta, \bar{\eta}] = \int_a^b k(t)\, |\eta(t)|^2\, dt = \sum_{j=1}^{\infty} |c_j[\eta]|^2, \\ &\text{(b)}\ J[\eta, \bar{\eta} \mid B] = \sum_{j=1}^{\infty} \lambda_j\, |c_j[\eta]|^2. \end{aligned}$$

Theorem 5.3. *If λ is not a proper value of* (3.1), *then the infinite series $\sum_{j=1}^{\infty} (\lambda_j - \lambda)^{-1}\, u_j(t)\, u_j(s)$ converges absolutely and uniformly for $(t, s) \in [a, b] \times [a, b]$, and*

$$(5.23)\qquad \sum_{j=1}^{\infty} (\lambda_j - \lambda)^{-1}\, u_j(t) u_j(s) = -g(t, s; \lambda),$$

where $g(t, s; \lambda)$ is the Green's function for the incompatible differential system

$$(5.24)\qquad l[u; \lambda] = 0, \qquad s_\alpha[u, v] = 0, \qquad (\alpha = 1, 2).$$

Since $u = u_j$, $v = v_j = ru_j' + qu_j$ is a solution of the differential system

$$l[u; \lambda] = -(\lambda_j - \lambda)\, k(t) u_j, \qquad s_\alpha[u, v] = 0, \qquad (\alpha = 1, 2),$$

it follows that

$$(5.25)\quad \begin{aligned} &\text{(a)}\quad u_j(t) = -(\lambda_j - \lambda) \int_a^b g(t, s; \lambda) k(s) u_j(s)\, ds, \\ &\text{(b)}\quad v_j(t) = -(\lambda_j - \lambda) \int_a^b g_1(t, s; \lambda) k(s) u_j(s)\, ds, \end{aligned}$$

where $g_1(t, s; \lambda) = r(t)g_t(t, s; \lambda) + q(t)g(t, s; \lambda)$. Now, since all proper values of (3.1) are real, if λ is not a proper value then $\bar{\lambda}$ is also not a proper value, and $\overline{g(t, s; \lambda)} = g(s, t; \bar{\lambda})$ in view of the self-adjointness of the system (3.1). In particular, for λ not a proper value, and t a fixed value on $[a, b]$, the function $g(t, \cdot; \lambda)$ belongs to class $\Gamma[B]$ as defined above, and (5.25-a) states that

$$c_j[g(t, \cdot; \lambda)] = -(\lambda_j - \lambda)^{-1} u_j(t).$$

Consequently, the first conclusion of Theorem 5.2 applied to $\eta = g(t, \cdot; \lambda)$ provides the result that for each $t \in [a, b]$ the infinite series

$$\sum_{j=1}^{\infty} (\lambda_j - \lambda)^{-1} u_j(t)\, u_j(s)$$

converges uniformly in s on $[a, b]$, and has sum equal to $-g(t, s; \lambda)$.

For λ_0 a real value such that $\lambda_0 < \lambda_1$, the above result for $\lambda = \lambda_0$ and $s = t$ implies that the series of non-negative real terms $\sum_{j=1}^{\infty}(\lambda_j - \lambda_0)^{-1} u_j^2(t)$ converges and has sum equal to $-g(t, t; \lambda_0)$. Since $g(t, s; \lambda_0)$ is continuous in (t, s) on $[a, b] \times [a, b]$, the function $g(t, t; \lambda_0)$ is continuous on $[a, b]$, and from the theorem of Dini on monotone convergence (see Theorem 2.1 of Appendix F) it follows that the series of non-negative real terms

$$\sum_{j=1}^{\infty} (\lambda_j - \lambda_0)^{-1} u_j^2(t)$$

converges uniformly on $[a, b]$ to $-g(t, t; \lambda_0)$.

Now for a given λ that is not a proper value of (3.1) there exists a constant κ such that

$$|\lambda_j - \lambda|^{-1} < \kappa(\lambda_j - \lambda_0)^{-1}, \qquad (j = 1, 2, \ldots).$$

Therefore, since

$$|(\lambda_j - \lambda)^{-1} u_j(t)u_j(s)| \leq 2\kappa(\lambda_j - \lambda_0)^{-1}[u_j^2(t) + u_j^2(s)],$$
$$(j = 1, 2, \ldots ; (t, s) \in [a, b] \times [a, b]),$$

it follows by the comparison test that the infinite series in (5.23) converges absolutely and uniformly for $(t, s) \in [a, b] \times [a, b]$.

In particular, since $\sum_{j=1}^{\infty}(\lambda_j - \lambda_0)^{-1} u_j^2(t)$ converges uniformly to $-g(t, t; \lambda_0)$, and $K[u_j] = 1$ for $j = 1, 2, \ldots$, it follows that

$$\text{(5.26)} \qquad \sum_{j=1}^{\infty}(\lambda_j - \lambda_0)^{-1} = -\int_a^b k(t)g(t, t; \lambda_0)\, dt.$$

Theorem 5.4. *If $h(t)$ is a piecewise continuous function on $[a, b]$, and $u(t)$ is a solution of*

$$\text{(5.27)} \qquad \begin{aligned} l[u](t) + k(t)h(t) &= 0, \qquad t \in [a, b], \\ s_\alpha[u, v] &= 0, \qquad (\alpha = 1, 2), \end{aligned}$$

then the infinite series

$$\sum_{j=1}^{\infty} c_j[u]\, u_j(t), \qquad \sum_{j=1}^{\infty} c_j[u]\, v_j(t), \tag{5.28}$$

converge absolutely and uniformly for $t \in [a, b]$ and have sums equal to $u(t)$ and $v(t) = r(t)u'(t) + q(t)u(t)$, respectively.

If λ_0 is a real value satisfying $\lambda_0 < \lambda_1$, and $h_1(t) = h(t) - \lambda_0 u(t)$, then (5.27) is equivalent to the system

$$\begin{aligned} l[u; \lambda_0](t) + k(t)h_1(t) &= 0, \qquad t \in [a, b], \\ s_\alpha[u, v] &= 0, \qquad (\alpha = 1, 2). \end{aligned} \tag{5.27'}$$

Consequently, if $g(t, s; \lambda_0)$ is the Green's matrix of the incompatible system $l[u; \lambda_0] = 0$, $s_\alpha[u, v] = 0$, $(\alpha = 1, 2)$, then

$$u(t) = -\int_a^b g(t, s; \lambda_0)k(s)h_1(s)\, ds = -K[g(t, \cdot\,; \lambda_0), h_1],$$

$$v(t) = -\int_a^b g_1(t, s; \lambda_0)k(s)h_1(s)\, ds = -K[g_1(t, \cdot\,; \lambda_0), h_1],$$

where $g_1(t, s; \lambda_0) = r(t)g_t(t, s; \lambda_0) + q(t)g(t, s; \lambda_0)$. Now

$$\begin{aligned} c_j[g(t, \cdot\,; \lambda_0)] &= -(\lambda_j - \lambda_0)^{-1}\, u_j(t), \\ c_j[g_1(t, \cdot\,; \lambda_0)] &= -(\lambda_j - \lambda_0)^{-1}\, v_j(t), \end{aligned} \tag{5.29}$$

by relations (5.25). Moreover,

$$c_j[h_1] = -\int_a^b u_j l[u; \lambda_0]\, dt = -\int_a^b u l[u_j; \lambda_0]\, dt = (\lambda_j - \lambda_0)\, c_j[u],$$

and by Theorem 5.1 we have the relations

$$u(t) = -\sum_{j=1}^{\infty} c_j[h_1]\, c_j[g(t, \cdot\,; \lambda_0)] = \sum_{j=1}^{\infty} c_j[u]\, u_j(t),$$

$$v(t) = -\sum_{j=1}^{\infty} c_j[h_1]\, c_j[g_1(t, \cdot\,; \lambda_0)] = \sum_{j=1}^{\infty} c_j[u]\, v_j(t).$$

Moreover, in view of (5.29) and Theorem 3.2 there exist positive constants κ, κ_1 such that

$$\sum_{j=1}^{\infty} c_j^2[g(t, \cdot\,; \lambda_0)] \leq \kappa^2, \qquad \sum_{j=1}^{\infty} c_j^2[g_1(t, \cdot\,; \lambda_0)] \leq \kappa_1{}^2, \quad \textit{for} \quad t \in [a, b].$$

Consequently, with the aid of the Lagrange-Cauchy inequality we have

$$\begin{aligned} \sum_{j=m+1}^{k} |c_j[u]\, u_j(t)| &= \sum_{j=m+1}^{k} |c_j[h_1]\, c_j[g(t, \cdot\,; \lambda_0)]| \\ &\leq \kappa \left[\sum_{j=m+1}^{k} |c_j[h_1]|^2 \right]^{1/2}, \end{aligned}$$

and as $\sum_{j=1}^{\infty} |c_j[h_1]|^2$ converges it follows that the series $\sum_{j=1}^{\infty} c_j[u]\, u_j(t)$ converges absolutely and uniformly for $t \in [a, b]$, and has sum equal to $u(t)$.

In a similar fashion, it follows that the series $\sum_{j=1}^{\infty} c_j[u]\, v_j(t)$ converges absolutely and uniformly for $t \in [a, b]$, and has sum equal to $v(t)$.

Problems VI.5

1. Apply the result of Theorem 5.1 to the boundary problem

$$u'' + \lambda u = 0, \qquad u(0) = 0, \qquad u(\pi) = 0,$$

for $h(t) \equiv 1$ and $h(t) = t$, to show that

$$\sum_{k=1}^{\infty} (2k - 1)^{-2} = \frac{\pi^2}{8}, \qquad \sum_{j=1}^{\infty} j^{-2} = \frac{\pi^2}{6}.$$

2. For the boundary problem of Problem 1, use Theorem 5.3 to show that

$$\sum_{j=1}^{\infty} j^{-2} \sin^2 jt = \frac{t(\pi - t)}{\pi}, \quad \textit{for} \quad t \in [0, \pi].$$

3. Use (5.26) to determine the sum of the series $\sum_{j=1}^{\infty} \lambda_j^{-1}$, where $\lambda_1 \le \lambda_2 \le \cdots$ are the proper values of the boundary problem

$$u'' + \lambda u = 0, \qquad u(0) = 0, \qquad ku(1) + u'(1) = 0,$$

where k is real and $k \neq -1$.

Ans. $[k + 3]/[6(k + 1)]$.

4. Suppose that a self-adjoint problem (3.1) satisfies hypothesis **($\hat{\mathbf{H}}$)**, $\{\lambda_j, u_j(t)\}$ is a sequence of proper values and corresponding proper functions as determined in Theorem 3.1, and λ is not a proper value of this system. Show that the infinite series $\sum_{j=1}^{\infty} (\lambda_j - \lambda)^{-2}\, u_j(t)\, u_j(s)$ converges absolutely and uniformly for $(t, s) \in [a, b] \times [a, b]$, and

$$\sum_{j=1}^{\infty} (\lambda_j - \lambda)^{-2}\, u_j(t) u_j(s) = \int_a^b g(t, r; \lambda) k(r) g(r, s; \lambda)\, dr,$$

$$\sum_{j=1}^{\infty} (\lambda_j - \lambda)^{-2} = \int_a^b \int_a^b k(t) g(t, r; \lambda) k(r) g(r, t; \lambda)\, dr\, dt.$$

Hint. Note that for an arbitrary piecewise continuous function $h(t)$ on $[a, b]$ there are unique functions u, $\hat{u}$ which satisfy with corresponding canonical variables v, $\hat{v}$ the differential system

$$l[u:\lambda](t) + k(t)\hat{u}(t) = 0, \qquad s_\alpha[u, v] = 0, \qquad (\alpha = 1, 2),$$

$$l[\hat{u}:\lambda](t) + k(t)h(t) = 0, \qquad s_\alpha[\hat{u}, \hat{v}] = 0, \qquad (\alpha = 1, 2),$$

and apply certain results of Theorems 5.3, 5.4.

5. Apply the result of Problem 4 to the differential system of Problem 1 to compute the sums of the infinite series

$$\sum_{j=1}^{\infty} j^{-4} \sin jt \sin js, \quad \sum_{j=1}^{\infty} j^{-4}.$$

6. Under the hypothesis of Problem 4, determine the sums of the infinite series

$$\sum_{j=1}^{\infty} (\lambda_j - \lambda)^{-m} u_j(t) u_j(s), \quad \sum_{j=1}^{\infty} (\lambda_j - \lambda)^{-m}, \qquad m = 3, 4, \ldots .$$

7. Use the theory of Lebesgue integration to obtain a brief proof of the result of Lemma 5.3.

Hint. Under the hypothesis of Lemma 5.3, use appropriate results in Lebesgue integration theory to conclude that there exists a function g such that g^2 is integrable in $[a, b]$, and $\int_a^b (h_j' - g)^2 \, dt \to 0$ as $j \to \infty$. From the relation $h_j(t) = h_j(a) + \int_a^t h_j'(s) \, ds$ for $t \in [a, b]$, deduce that $\int_a^t g(s) \, ds = 0$ for $t \in [a, b]$; conclude that $g = 0$ almost everywhere on $[a, b]$, and $\int_a^b h_j'^2 \, dt \to 0$ as $j \to \infty$.

8. Consider a family of boundary problems $B\{s\}$, $s \in S$, in which the real quadratic integrand form

$$2\omega(t, \eta, \zeta) = 2\omega(t, \eta, \zeta : s) = r(t, s)\zeta^2 + 2q(t, s)\zeta\eta + p(t, s)\eta^2$$

and the real quadratic end-form

$$Q[\eta : s] = \gamma_{11}(s)\, \eta^2(a) + 2\gamma_{12}(s)\, \eta(a)\, \eta(b) + \gamma_{22}(s)\, \eta^2(b)$$

depend upon the parameter s, while the end-space $\mathcal{S}$ and the coefficient function $k(t)$ are independent of s. It will be supposed that the following conditions hold:

(i) $\gamma_{\alpha\beta}(s)$, $(\alpha, \beta = 1, 2)$, are continuous in s on S;
(ii) $p(t, s)$, $q(t, s)$ and $r(t, s)$ are continuous in (t, s) on $[a, b] \times S$;
(iii) for each $s \in S$ the boundary problem

$$\begin{aligned} l_s[u; \lambda](t) \equiv {} & [r(t, s)u'(t) + q(t, s)u(t)]' \\ & - [q(t, s)u'(t) + p(t, s)u(t)] + \lambda k(t)u(t) = 0, \\ \hat{u} \in \mathcal{S}, \quad & (Q_1[u : s] - v(a : s), \quad Q_2[u : s] + v(b : s)) \in \mathcal{S}^{\perp} \end{aligned} \tag{5.30}$$

satisfies hypothesis ($\hat{\mathbf{H}}$). If $\{\lambda_j(s), u_j(t : s)\}$ is a set of proper values and corresponding proper functions of (5.30), satisfying the conditions of Theorem 3.1, prove that each $\lambda_j(s)$ is a continuous function of s on S.

Hint. Note that $D[B\{s\}]$ is independent of s, and for $s_0 \in S$, $\lambda_0 < \lambda_1(s_0)$, use the result of Lemma 5.4 to show that for a given $\varepsilon > 0$ there exists a $\delta > 0$ such that if s is a point of S in a δ-neighborhood of s_0 then

$$(1 + \varepsilon)^{-1} J[\eta; \lambda_0 \mid B\{s_0\}] \leq J[\eta; \lambda_0 \mid B\{s\}] \leq (1 + \varepsilon) J[\eta; \lambda_0 \mid B\{s_0\}],$$

for $\eta \in D[B\{s\}] = D[B\{s_0\}]$. Deduce the continuity of each $\lambda_j(s)$ at $s = s_0$ from these inequalities and the result of Theorem 4.1 for certain allied boundary problems. {*Comment.* It is to be noted that the domain S of the parameter s has not been specified as being a subset of the real line. The above argument is valid for s a vector parameter $s = (s_1, \ldots, s_q)$ with domain a subset of $\mathbf{R}_q$, and, indeed, for much more general cases.}

6. NOTES AND REMARKS

The boundary problems of this chapter involve a scalar linear second order differential equation, and are considered under hypotheses that correspond to those of Chapter V. In particular, the treatment contains as an important special case the principal results of the Sturm-Liouville theory, dating from the classical works of Sturm [1] and Liouville [1]. Pertinent references are Bôcher [4, 6], Ince [1, Chs. X, XI], Kamke [1, Ch. VIII-§27], Sansone [1-I, Ch. IV], Birkhoff and Rota [1, Chs. X, XI], Atkinson [1, Ch. 8], Hartman [7, Ch. XI], and Hille [4, Ch. 8]. The derivation in Section 1 of the canonical form for the boundary conditions is a special case of the more general problem to be considered in Section 8 of Chapter VII, and for this particular topic the references of Jackson [1], Latshaw [1], Bliss [3], Morse [2, Ch. IV], Kamke [4], and Reid [9] are pertinent.

The central feature of the treatment of this chapter is the proof of the existence of proper values by a minimizing principle, and the consistent use of the extremizing properties of proper values to establish expansion and representation theorems. Historically, this method of approach goes back to Hilbert [1], and some of the earlier discussions of these properties for self-adjoint boundary value problems relied upon integral equation theory. For the case of a real second-order differential equation with self-adjoint boundary conditions, the first such treatment independent of integral equation theory appears to be that of Mason [1]. For a survey of various methods that have been used for such boundary problems, and their generalizations, reference is again made to Reid [5].

For treatments of oscillation phenomena and proper value problems for certain nonlinear differential equations of the second order, the reader is referred to Whyburn [2], Nehari [3], Moroney [1], and Hooker [1].

It is to be emphasized that the entire discussion of Chapters V, VI is by "real variable methods," and requires an introduction to mathematical analysis of only the level of the usual upper undergraduate course in "Advanced Calculus", with no knowledge of integration theory beyond that of the Riemann integral. The only exceptions to this statement are the comments following the Corollary to Theorem 3.2, noting that the result of this Corollary also follows from the discussion of Section IV.2, and Problem VI.5:7, showing how the result of Lemma 5.3 may be established readily with the use of Lebesgue integration theory.

RELATED REFERENCES AND COMMENTS FOR SPECIFIC PROBLEMS

VI.4:5,6 The results of these problems are special instances of more general results of Morse [1, 2, Ch. IV].

VI.4:8 This problem is the two-dimensional case of a more general result of Reid [22]; as a special instance, the result includes that of Theorem I of Nehari [2].

VII

Self-Adjoint Differential Systems

1. INTRODUCTION

This chapter will be devoted to the study of self-adjoint differential systems and associated boundary problems, with particular emphasis on the use of variational principles for the extension to such systems of the oscillation and comparison theorems of the Sturmian theory for a scalar differential equation of the second order, as discussed in Chapters V and VI. Historically, the use of variational principles for the extension of the Sturmian theory to such differential systems dates from the basic work of Marston Morse, ([1],[2]). In this connection, however, it is to be emphasized that no formal knowledge of the calculus of variations is required for the understanding of the present discussion. All needed details are presented herein, and, indeed, because of the quadratic nature of the involved functionals, the basic variational principles are present as functional concepts and identities. The reader conversant with variational theory, (e.g., see Bliss [5]; in particular, Part II on the so-called Bolza problem), will recognize such fundamental aspects in the notion of conjoined solutions introduced in Section 2 below, and in the general identity presented in Lemma 4.2.

For generality, and also for applications in the "control formulation of certain variational problems," the considered system is written in terms of so-called "canonical variables." Moreover, the assumption of continuity of coefficients is weakened, so that a solution is now in the Carathéodory sense as introduced in Chapter II. It is to be commented that this weakening of the assumption of continuity of coefficients increases the details of proof in only a very few instances, and in these cases it is indicated how the proofs are simplified if the involved coefficient functions are continuous.

At this point, it is perhaps worthwhile for the reader to remind himself of the earlier preliminary definitions, notably in Section II.2. In particular, for a given compact interval $[a, b]$ on the real line symbols $\mathfrak{C}_{nr}[a, b]$, $\mathfrak{C}_{nr}{}^k[a, b]$, $\mathfrak{L}_{nr}[a, b]$, $\mathfrak{L}_{nr}{}^p[a, b]$, $\mathfrak{L}_{nr}{}^\infty[a, b]$, $\mathfrak{A}_{nr}[a, b]$ are used to denote the classes of $n \times r$ matrix functions $M(t) = [M_{\alpha\beta}(t)]$, $(\alpha = 1, \ldots, n; \beta = 1, \ldots, r)$, which on $[a, b]$ are respectively continuous, continuous and possessing continuous derivatives of the first k orders, (Lebesgue) integrable, (Lebesgue) measurable and $|M_{\alpha\beta}(t)|^p$ integrable, measurable and essentially bounded, and a.c. (absolutely continuous). Also for brevity the symbols $\mathfrak{C}_n[a, b]$, $\mathfrak{C}_n{}^k[a, b]$, $\mathfrak{L}_n[a, b]$, $\mathfrak{L}_n{}^p[a, b]$, $\mathfrak{L}_n{}^\infty[a, b]$, $\mathfrak{A}_n[a, b]$ are written for the respective classes designated by indices $n, r = 1$, and $\mathfrak{C}[a, b]$, $\mathfrak{C}^k[a, b]$, $\mathfrak{L}[a, b]$, $\mathfrak{L}^p[a, b]$, $\mathfrak{L}^\infty[a, b]$, $\mathfrak{A}[a, b]$ for the respective classes indicated by $n = 1$, $r = 1$. Moreover, if $M(t)$ and $N(t)$ are matrix functions which are equal a.e. (almost everywhere), we write simply $M(t) = N(t)$.

2. PRELIMINARY RESULTS

Throughout the present chapter it will be supposed that $A(t)$, $B(t)$ and $C(t)$ are $n \times n$ matrix functions satisfying the following hypothesis.

($\mathfrak{H}$) *On a given interval I on the real line the $n \times n$ matrix functions $B(t)$, $C(t)$ are hermitian, and $A(t)$, $B(t)$, $C(t)$ each belongs to $\mathfrak{L}_{nn}{}^\infty[a, b]$ for arbitrary compact subintervals $[a, b]$ of I.*

For brevity, we write $L[u]$ and $L^\star[v]$ for the respective adjoint expressions

$$L[u] = u' - A(t)u, \qquad L^\star[v] = -v' - A^*(t)v. \tag{2.1}$$

Corresponding to the scalar system (V.1.2), attention will now be centered on the vector system

$$\begin{aligned} L_1[u, v](t) &\equiv L^\star[v](t) + C(t)u(t) = 0, \\ L_2[u, v](t) &\equiv L[u](t) - B(t)v(t) = 0. \end{aligned} \qquad t \in I, \tag{2.2}$$

A solution of (2.2) is a pair of n-dimensional vector functions u, v, which individually are of class $\mathfrak{A}_n$ on arbitrary compact subintervals of I, and equations (2.2) hold a.e. on I. In view of hypothesis ($\mathfrak{H}$), it follows readily that the individual vector functions $u(t)$, $v(t)$ belonging to a solution of (2.2) are Lipschitzian on arbitrary compact subintervals of I. In particular, if $y = (y_\alpha)$, $(\alpha = 1, \ldots, 2n)$, with $y_i = u_i$, $y_{n+i} = v_i$, $(i = 1, \ldots, n)$, then (2.2) may be written as the vector equation

$$\mathfrak{L}[y](t) = \mathfrak{J}y'(t) + \mathcal{A}(t)y(t) = 0, \qquad t \in I, \tag{2.2'}$$

where $\mathfrak{J}$ and $\mathcal{A}(t)$ are the $2n \times 2n$ matrices

$$\mathfrak{J} = \begin{bmatrix} 0 & -E_n \\ E_n & 0 \end{bmatrix}, \qquad \mathcal{A}(t) = \begin{bmatrix} C(t) & -A^*(t) \\ -A(t) & -B(t) \end{bmatrix}. \tag{2.3}$$

As $\mathcal{A}(t)$ is hermitian, and $\mathfrak{J}$ is skew-hermitian, the vector operator $\mathfrak{L}[y](t)$ is clearly identical with its adjoint $\mathfrak{L}^\star[y](t) = -\mathfrak{J}^*y'(t) + \mathcal{A}^*(t)y(t)$.

If $M \equiv [M_{\alpha j}]$, $N \equiv [N_{\alpha j}]$, $(\alpha = 1, \ldots, n;\ j = 1, \ldots, r)$, are $n \times r$ matrices, for typographical simplicity the symbol $[M; N]$ will denote the $2n \times r$ matrix $[S_{\sigma j}]$ with $S_{\alpha j} = M_{\alpha j}$, $S_{n+\alpha, j} = N_{\alpha j}$, $(\alpha = 1, \ldots, n;\ j = 1, \ldots, r)$. Thus in referring to a solution of (2.2) or (2.2′) in terms of $u(t)$ and $v(t)$ we write $y(t) = (u(t); v(t))$.

Before presenting some properties of solutions of (2.2), attention will be directed to certain special instances of such vector differential systems. For t on a given interval I, let $2\omega(t, \eta, \zeta)$ denote the form

$$2\omega(t, \eta, \zeta) = \zeta^*[R(t)\zeta + Q(t)\eta] + \eta^*[Q^*(t)\zeta + P(t)\eta] \tag{2.4}$$

in (η, ζ), where $\eta = (\eta_i)$, $\zeta = (\zeta_i)$, $(i = 1, \ldots, n)$. It will be assumed that the coefficient matrices of (2.4) satisfy the following hypothesis on the given interval I.

($\mathfrak{H}_\omega$) *The $n \times n$ matrix functions $R(t)$, $P(t)$ are hermitian, $R(t)$ is non-singular, and each of the matrix functions $P(t)$, $Q(t)$, $R(t)$, $R^{-1}(t)$ belongs to $\mathfrak{L}_{nn}^\infty[a, b]$ for arbitrary compact sub-intervals $[a, b]$ of I.*

If $[a, b] \subset I$ the symbol $J[\eta; a, b]$ will denote the functional

$$J[\eta; a, b] = \int_a^b 2\omega(t, \eta(t), \eta'(t))\, dt. \tag{2.5}$$

For (2.5) the corresponding "vector Euler differential equation" is

$$[R(t)u' + Q(t)u]' - [Q^*(t)u' + P(t)u] = 0, \tag{2.6}$$

which in terms of the "canonical variables"

$$u(t), \qquad v(t) = R(t)u'(t) + Q(t)u(t), \tag{2.7}$$

may be written as a system (2.2) with the matrix functions $A(t)$, $B(t)$, $C(t)$ defined as

$$A = -R^{-1}Q, \qquad B = R^{-1}, \qquad C = P - Q^*R^{-1}Q. \tag{2.8}$$

In particular, the matrix functions $A(t)$, $B(t)$, $C(t)$ satisfy hypothesis ($\mathfrak{H}$) whenever the coefficients of (2.4) satisfy the condition ($\mathfrak{H}_\omega$).

A still more general example of a system (2.2) is presented by a differential system that is of the form of the accessory differential equations for a variational problem of Bolza type, (e.g., see Bliss [5; §81]). In addition to the hermitian form (2.4) in (η, ζ), consider a vector linear form

$$\Phi(t, \eta, \zeta) = \phi(t)\zeta + \theta(t)\eta, \tag{2.9}$$

where $\phi(t)$, $\theta(t)$ are $m \times n$, $(m < n)$, matrix functions. For this problem we shall assume that the following condition holds on the given interval I.

($\mathfrak{H}_\Omega$) (a) *The $n \times n$ matrix functions $R(t)$, $P(t)$ are hermitian.*
(b) *The $(n + m) \times (n + m)$ matrix function*

$$\begin{bmatrix} R(t) & \phi^*(t) \\ \phi(t) & 0 \end{bmatrix} \tag{2.10}$$

is nonsingular for $t \in I$. {From elementary properties of linear algebraic equations, (e.g., see Appendix B), it follows that the inverse of (2.10) is of the form

$$\begin{bmatrix} T(t) & \tau^*(t) \\ \tau(t) & \chi(t) \end{bmatrix}, \tag{2.10'}$$

where $T(t)$ and $\chi(t)$ are hermitian matrix functions of orders n and m, respectively, and $\tau(t)$ is an $m \times n$ matrix function.}

(c) *For arbitrary compact subintervals $[a, b]$ of I each of the matrix functions $P(t)$, $Q(t)$, $R(t)$, $T(t)$, belongs to $\mathfrak{L}_{nn}{}^\infty[a, b]$, each of the matrix functions $\phi(t)$, $\theta(t)$, $\tau(t)$ belongs to $\mathfrak{L}_{mn}{}^\infty[a, b]$, and $\chi(t)$ belongs to $\mathfrak{L}_{mm}{}^\infty[a, b]$.*

For the variational problem involving the functional (2.5) subject to the auxiliary m-dimensional vector differential equation

$$\Phi(t,\eta(t),\eta'(t)) = 0, \qquad t \in [a, b], \tag{2.11}$$

the corresponding "Euler-Lagrange system" may be written in vector form as

$$\begin{aligned} &[R(t)u' + Q(t)u + \phi^*(t)\mu]' - [Q^*(t)u' + P(t)u + \theta^*(t)\mu] = 0, \\ &\Phi(t, u, u') = 0, \end{aligned} \tag{2.12}$$

where $u = u(t)$ is an n-dimensional vector function and $\mu(t)$ is an m-dimensional "multiplier" vector function. In terms of the "canonical variables"

$$u(t), \qquad v(t) = R(t)u'(t) + Q(t)u(t) + \phi^*(t)\mu(t), \tag{2.13}$$

the system (2.12) reduces to a system (2.2), with now

$$\begin{aligned} A &= -(TQ + \tau^*\theta), \qquad B = T, \\ C &= P - Q^*TQ - Q^*\tau^*\theta - \theta^*\tau Q - \theta^*\chi\theta. \end{aligned} \tag{2.14}$$

Moreover, it follows readily that hypothesis ($\mathfrak{H}_\Omega$) implies that the matrix functions A, B, C of (2.14) satisfy hypothesis ($\mathfrak{H}$). The generality of the system (2.2) with coefficient matrix functions of the form (2.14) is illustrated by some of the problems in the set at the end of this section.

Returning to the differential system (2.2) or (2.2′), from Theorem III.2.1 it follows that if y_1 and y_2 are two solutions of (2.2′) then $y_2^*\mathfrak{J}y_1$ is constant on I. Translated in terms of solutions of (2.2), this property is the following.

Lemma **2.1.** *If* $y_\alpha = (u_\alpha; v_\alpha)$, $(\alpha = 1, 2)$, *are solutions of* (2.2), *then*

$$\{u_1; v_1 \mid u_2; v_2\}(t) = v_2^*(t)u_1(t) - u_2^*(t)v_1(t) \tag{2.15}$$

is constant on I.

Corresponding to the vector system (2.2), we have the general matrix differential system

$$\begin{aligned} L_1[U, V](t) &\equiv L^\star[V](t) + C(t)U(t) = 0, \\ L_2[U, V](t) &\equiv L[U](t) - B(t)V(t) = 0, \end{aligned} \qquad t \in I, \tag{2.2$_M$}$$

where $U(t)$, $V(t)$ are $n \times r$ matrix functions, and $L[U](t) = U'(t) - A(t)U(t)$, $L^\star[V](t) = -V'(t) - A^*(t)V(t)$. Therefore, if $(U_\alpha; V_\alpha)$, $(\alpha = 1, 2)$, are two solutions of (2.2_M), Lemma 2.1 implies that the matrix function

$$\{U_1; V_1 \mid U_2; V_2\}(t) = V_2^*(t)U_1(t) - U_2^*(t)V_1(t)$$

is constant on I.

If $y_\alpha = (u_\alpha; v_\alpha)$, $(\alpha = 1, 2)$, are solutions of (2.2) such that the constant function $y_2^*\mathfrak{J}y_1 = \{u_1; v_1 \mid u_2; v_2\}$ is zero, these solutions are said to be *(mutually) conjoined*. As the matrix $\mathfrak{J}$ is skew-hermitian the matrix $\mathfrak{K} = i\mathfrak{J}$ is hermitian, and hence for any $2n$-dimensional vector ρ the form $\mathfrak{K}[\rho] = \rho^*\mathfrak{K}\rho$ is real-valued. Since $\mathfrak{J}$ is real-valued, in case the coefficient matrices $A(t)$, $B(t)$, $C(t)$ are real-valued any real solution $y = (u; v)$ of (2.2) is *self-conjoined*; that is, $y^*\mathfrak{J}y = \{u; v \mid u; v\}$ is the zero function. In general, if ρ is a $2n$-dimensional vector such that $\mathfrak{K}[\rho] > 0$, then $\rho_1 = Q\rho$, where Q is the $2n$-dimensional square matrix $Q = \text{diag}\{-E_n, E_n\}$, is such that $\mathfrak{K}[\rho_1] = -\mathfrak{K}[\rho] < 0$. In particular, ρ and ρ_1 are linearly independent vectors of $\mathbf{C}_{2n}$, and if $\rho(\theta) = (1 - \theta)\rho + \theta\rho_1$ then there exists a θ_0, $0 < \theta_0 < 1$, such that $\mathfrak{K}[\rho(\theta_0)] = 0$. As $\rho(\theta_0)$ is a nonzero vector, it follows that the solution $y = (u; v)$ of (2.2) such that $y(t_0) = \rho(\theta_0)$ for some $t_0 \in I$ is a nontrivial solution that is self-conjoined.

If $Y(t) = (U(t); V(t))$ is a $2n \times r$ matrix whose column vectors are r linearly independent solutions of (2.2) which are mutually conjoined, these solutions form a basis for a *conjoined family of solutions of dimension* r, consisting of the set of all solutions of (2.2) which are linear combinations of these column vectors.

Theorem 2.1. *The maximal dimension of a conjoined family of solutions of* (2.2) *is n; moreover, a given conjoined family of solutions of dimension r < n is contained in a conjoined family of dimension n.*

If $Y(t) = (U(t); V(t))$ is a $2n \times r$ matrix function whose column vectors form a basis for a conjoined family of solutions for (2.2) of dimension r, then $Y(t)$ is of rank r for each $t \in I$ and $Y^*(t)\mathfrak{J}Y(t) \equiv 0$. In particular, the column vectors of the $2n \times r$ matrix $\mathfrak{J}Y(t)$ are orthogonal to the column vectors of the $2n \times r$ matrix $Y(t)$ that is of rank r. Consequently, $r \leq 2n - r$ and $r \leq n$, so that the first part of the lemma is proved. In order to establish the second part, it suffices to show that a given conjoined family of solutions of dimension $r < n$ is contained in a conjoined family of dimension $r + 1$. As before, let $Y(t) = (U(t); V(t))$ be a $2n \times r$ matrix function whose column vectors form a basis for a conjoined family of solutions of (2.2) of dimension r. Now if $r < n$, and $t_1 \in I$, it follows from elementary results on linear algebraic equations, (e.g., Appendix B), that there exists a $2n \times (2n - 2r)$ matrix M such that the $2n \times (2n - r)$ matrix $[Y(t_1) \quad M]$ is of rank $2n - r$ and $Y^*(t_1)\mathfrak{J}M = 0$. For σ and τ arbitrary vectors of dimensions r and $2n - 2r$, respectively, let $\rho = Y(t_1)\sigma + M\tau$. Since $n < 2n - r$, it follows that if F_1 denotes the $n \times 2n$ matrix $F_1 = [iE_n \quad E_n]$ then there exists a nonzero vector ρ such that $F_1\rho = 0$. Then there exists a nonzero n-dimensional vector ξ such that $\xi = (\rho_\alpha)$, $-i\xi = (\rho_{n+\alpha})$, $(\alpha = 1, \ldots, n)$, and by direct substitution it follows that $\mathcal{K}[\rho] = i\rho^*\mathfrak{J}\rho = -2\,|\xi|^2 < 0$. Correspondingly, if $F_2 = [-iE_n \quad E_n]$ there exists a nonzero vector ρ^0 such that $F_2\rho^0 = 0$, and there is a nonzero n-dimensional vector ξ^0 such that $\xi^0 = (\rho_\alpha{}^0)$, $i\xi^0 = (\rho^0_{n+\alpha})$, $(\alpha = 1, \ldots, n)$, and $\mathcal{K}[\rho^0] = i\rho^{0*}\mathfrak{J}\rho^0 = 2\,|\xi^0|^2 > 0$. Now if $\rho = Y(t_1)\sigma + M\tau$ and $\rho^0 = Y(t_1)\sigma^0 + M\tau^0$ it follows that $0 > \mathcal{K}[\rho] = \mathcal{K}[M\tau]$ and $0 < \mathcal{K}[\rho^0] = \mathcal{K}[M\tau^0]$. Consequently, the $(2n - 2r)$-dimensional vectors τ and τ^0 are linearly independent, and there exists a θ_1, $0 < \theta_1 < 1$, such that if $\tau_1 = (1 - \theta_1)\tau + \theta_1\tau^0$ then τ_1 is a non-zero vector for which $\mathcal{K}[M\tau_1] = 0$. If $y(t) = (u(t); v(t))$ is the solution of (2.2) satisfying the initial condition $y(t_1) = M\tau_1$, it then follows that the $r + 1$ solutions of (2.2) consisting of $y(t)$ and the column vectors of $Y(t)$ form a basis for a conjoined family of solutions of dimension $r + 1$.

If $Y(t) = (U(t); V(t))$ is a solution of (2.2_M) on I whose column vectors form a basis for an n-dimensional conjoined family of solutions, then for brevity we shall say that $Y(t)$ is a *conjoined basis for* (2.2). In particular, if $c \in I$ we shall denote by $Y(t; c) = (U(t; c); V(t; c))$ the solution of (2.2_M) satisfying the initial conditions

$$U(c; c) = 0, \qquad V(c; c) = E. \tag{2.16}$$

As $Y^*(c; c)\mathfrak{J}Y(c; c) = 0$, it follows that $Y(t; c)$ is a conjoined basis for (2.2). Correspondingly, if $Y_0(t; c) = (U_0(t; c); V_0(t; c))$ is the solution of (2.2_M) satisfying the initial conditions

$$U_0(c; c) = E, \qquad V_0(c; c) = 0, \tag{2.17}$$

then $Y_0(t; c)$ is also a conjoined basis for (2.2).

The following result, which is an analogue of that of Theorem V.2.3, is of basic importance for the theory of systems (2.2).

Theorem 2.2. *Suppose that $Y_1(t) = (U_1(t); V_1(t))$ is a conjoined basis for (2.2), and $U_1(t)$ is nonsingular on a subinterval I_0 of I. If $c \in I_0$ then $Y(t) = (U(t); V(t))$ is a solution of (2.2_M) on I_0 if and only if on this interval*

$$U(t) = U_1(t)H(t), \qquad V(t) = V_1(t)H(t) + U_1^{*-1}(t)K_1, \tag{2.18}$$

where

$$H(t) = K_0 + \left[\int_c^t U_1^{-1}(s)B(s)U_1^{*-1}(s)\,ds\right]K_1, \tag{2.19}$$

and K_0, K_1 are constant matrices.

If we set $U(t) = U_1(t)H(t)$, $V(t) = V_1(t)H(t) + W(t)$, then substitution in the system (2.2_M) yields the equations

$$U_1H' = BW, \qquad V_1H' + W' = -A^*W. \tag{2.20}$$

Since $0 \equiv \{U_1; V_1 \mid U_1; V_1\} = V_1^*U_1 - U_1^*V_1$ it then follows that $(U_1^*W)' = 0$, and hence there exists a constant matrix K_1 such that $W(t) = U_1^{*-1}(t)K_1$, thus providing the second relation of (2.18). In turn, the equation $H' = U_1^{-1}BW$ yields the equation (2.19), where $K_0 = H(c)$. Conversely, if $Y(t) = (U(t); V(t))$ is defined by (2.18), with $H(t)$ given by (2.19), it follows upon substitution that this matrix function is a solution of (2.2_M). It is to be noted that the constant matrix K_1 is equal to $-Y_1^*\mathfrak{J}Y = -\{U; V \mid U_1; V_1\}$.

Problems VII.2

1. For $n = 2$ and $m = 1$, let

$$2\omega(t, \eta, \zeta) = r(t)\zeta_2^2 + q(t)\eta_2^2 + p(t)\eta_1^2, \qquad \Phi(t, \eta, \zeta) = \zeta_1 - \eta_2,$$

where $p(t)$, $q(t)$, $r(t)$ are real-valued functions with $r(t) \neq 0$ for $t \in I$, while $p(t)$, $q(t)$, $r(t)$, $1/r(t)$ are functions in $\mathfrak{L}^\infty[a, b]$ for arbitrary compact subintervals $[a, b]$ of I. Show that the system (2.2) with coefficient matrices given by (2.14) is equivalent to the *quasi-differential equation*

$$[(r(t)u'')' - q(t)u']' + p(t)u = 0.$$

In particular, if $q(t) \in \mathfrak{C}^1(I)$ and $r(t) \in \mathfrak{C}^2(I)$, then this equation is equivalent to a differential equation $\sum_{j=0}^4 p_j(t)\,u^{[j]} = 0$, with continuous coefficients $p_4 = r$, $p_3 = 2r'$, $p_2 = r'' - q$, $p_1 = -q'$, $p_0 = p$.

2. Suppose $r_j(t)$, $(j = 0, 1, \ldots, n)$, are real-valued functions with $r_n(t) \neq 0$ for $t \in I$, while $r_j(t)$, $(j = 0, 1, \ldots, n)$, and $1/r_n(t)$ are functions in $\mathfrak{L}^\infty[a, b]$ for arbitrary compact subintervals $[a, b]$ of I. If

$$2\omega(t, \eta, \zeta) = r_n(t)\zeta_n^2 + r_{n-1}(t)\eta_n^2 + \cdots + r_0(t)\eta_1^2,$$

and $m = n - 1$ with $\Phi(t, \eta, \zeta) = (\zeta_\beta - \eta_{\beta+1})$, $(\beta = 1, \ldots, n-1)$, show that $(u; v)$ is a solution of the system (2.2), with coefficient matrices given by the corresponding relations (2.14), if and only if there exists a function $w(t) \in \mathfrak{C}^{n-1}(I)$ with $w^{[n-1]}(t)$ a.c. on arbitrary compact subintervals $[a, b]$ of I, and

$$(2.21)\qquad \begin{aligned} u_j(t) &= w^{[j-1]}(t), \qquad (j = 1, \ldots, n),\\ v_n(t) &= r_n(t)\, w^{[n]}(t),\\ v_1'(t) &= r_0(t)\, w(t),\\ v_{\beta+1}'(t) &= r_\beta(t)\, w^{[\beta]}(t) - v_\beta(t), \qquad (\beta = 1, \ldots, n-1). \end{aligned}$$

In particular, if $r_j(t) \in \mathfrak{C}^j[a, b]$ for $j = 0, 1, \ldots, n$, then (2.21) is satisfied by $u = (u_i(t))$, $v = (v_i(t))$, $w(t)$ if and only if $w(t)$ is a solution of the $2n$-th order scalar differential equation

$$(2.22)\qquad l_{2n}[w](t) \equiv \sum_{j=0}^{n} (-1)^j (r_j(t) w^{[j]}(t))^{[j]} = 0;$$

in this case $u_j(t) = w^{[j-1]}(t)$, and

$$(2.23)\qquad v_j(t) = \sum_{\alpha=j}^{n} (-1)^{\alpha-j} (r_\alpha(t) w^{[\alpha]}(t))^{[\alpha-j]}, \qquad (j = 1, \ldots, n).$$

3.^{s} If $T(t)$ is a fundamental matrix solution of $L[T](t) = 0$, show that under the substitution

$$(2.24)\qquad u(t) = T(t)u^0(t), \qquad v(t) = T^{*-1}(t)v^0(t),$$

we have $L[u](t) = T(t)L^0[u^0](t)$, $L^\star[v](t) = T^{*-1}(t)L^{0\star}[v^0](t)$, where $L^0[u^0] = u^{0\prime}$, $L^{0\star}[v^0] = -v^{0\prime}$. Moreover, $L_1[u, v](t) = T^{*-1}(t)\, L_1{}^0[u^0, v^0](t)$, $L_2[u, v](t) = T(t)\, L_2{}^0[u^0, v^0](t)$, where

$$\begin{aligned} L_1{}^0[u^0, v^0](t) &= L^{0\star}[v^0](t) + C^0(t)\, u^0(t),\\ L_2{}^0[u^0, v^0](t) &= L^0[u^0](t) - B^0(t)\, v^0(t), \end{aligned}$$

and $B^0 = T^{-1}BT^{*-1}$, $C^0 = T^*CT$. In particular, $(u; v)$ is a solution of (2.2) if and only if $(u^0; v^0)$ is a solution of

$$(2.25)\qquad L_1{}^0[u^0, v^0](t) = 0, \qquad L_2{}^0[u^0, v^0](t) = 0,$$

and the function (2.15) is invariant under the transformation (2.24); that is, if $u_\alpha(t) = T(t)u_\alpha{}^0(t)$, $v_\alpha(t) = T^{*-1}(t)v_\alpha{}^0(t)$, $(\alpha = 1, 2)$, then $\{u_1; v_1 \mid u_2; v_2\} = \{u_1{}^0; v_1{}^0 \mid u_2{}^0; v_2{}^0\}$.

4.^{ss} Suppose that $Y_1(t) = (U_1(t); V_1(t))$ is a solution of (2.2_M) on I, with $U_1(t)$ non-singular on a subinterval I_0 of I, and $K = -\{U_1; V_1 \mid U_1; V_1\}$. Prove the following generalization of the result of Theorem 2.2. The matrix function $Y(t) = (U(t); V(t))$ is a solution of (2.2_M) on I_0 if and only if on this interval

$$(2.26) \quad U(t) = U_1(t)H(t), \qquad V(t) = V_1(t)H(t) + U_1^{*-1}(t)[K_1 - KH(t)],$$

where K_1 is a constant matrix and $H(t)$ is a solution of the matrix differential equation

$$(2.27) \qquad H'(t) = U_1^{-1}(t)B(t)U_1^{*-1}(t)[K_1 - KH(t)], \qquad t \in I_0.$$

Moreover, if $c \in I_0$ and $T = T(t, c \mid U_1)$ is the solution of the matrix differential system

$$(2.28) \qquad T' = -U_1^{-1}(t)B(t)U_1^{*-1}(t)KT, \qquad T(c) = E,$$

then $H(t)$ is of the form

$$(2.29) \qquad H(t) = T(t, c \mid U_1)[K_0 + S(t, c \mid U_1)K_1]$$

where

$$(2.30) \quad S(t, c \mid U_1) = \int_c^t T^{-1}(s, c \mid U_1)\, U_1^{-1}(s)B(s)U_1^{*-1}(s)\, ds, \qquad t \in I_0,$$

and $K_1 = -\{U; V \mid U_1; V_1\}$.

Hint. Show that equations (2.20) now imply $[KH(t) + U_1^*(t)W(t)]' = 0$.

5.[ss] Suppose that $Y_1(t) = (U_1(t); V_1(t))$ is a solution of (2.2_M) on an interval I with $U_1(t)$ nonsingular on a subinterval I_0 of I, and $K = -\{U_1; V_1 \mid U_1; V_1\}$. If $Y(t) = (U(t); V(t))$ is any solution of (2.2_M), show that for $(t, c) \in I_0 \times I_0$ one has the relation

$$(2.31) \quad \{U; V \mid U_1; V_1\} + KU_1^{-1}(t)U(t) = T^{*-1}(t, c \mid U_1)[\{U; V \mid U_1; V_1\} + KU_1^{-1}(c)U(c)],$$

where $T(t, c \mid U_1)$ is the solution of the differential system (2.28).

Hint. Note that $K = -K^*$, and $F(t) = T^{*-1}(t, c \mid U_1)$ is the solution of the differential system

$$F' = -KU_1^{-1}(t)B(t)U_1^{*-1}(t)F, \qquad F(c) = E.$$

For $H(t) = U_1^{-1}(t)U(t)$ and $K_1 = -\{U; V \mid U_1; V_1\}$, show that (2.27) implies that $F_1(t) = K_1 - KH(t)$ is the solution of the system

$$F_1' = -KU_1^{-1}(t)B(t)U_1^{*-1}(t)F_1, \qquad F_1(c) = K_1 - KH(c),$$

and conclude that $F_1(t) = F(t)F_1(c)$.

6.[s] Suppose that the $n \times n$ matrix functions $A(t)$, $B(t)$, $C(t)$ satisfy hypothesis ($\mathfrak{H}$) on an interval $I = (a_0, b_0)$, and for

$$(2.32)\quad Q(t; \Phi, \Psi) = \Psi(t)B(t)\Psi^*(t) + \Psi(t)A(t)\Phi^*(t) + \Phi(t)\,A^*(t)\Psi^*(t) - \Phi(t)C(t)\Phi^*(t),$$

consider the differential system

$$(2.33)\quad \begin{aligned} \Lambda_1{}^0[\Phi, \Psi](t) &\equiv -\Psi'(t) - Q(t; \Phi, \Psi)\Phi(t) = 0, \\ \Lambda_2{}^0[\Phi, \Psi](t) &\equiv \Phi'(t) - Q(t; \Phi, \Psi)\Psi(t) = 0, \end{aligned} \qquad t \in I,$$

$$(2.34)\quad \Lambda^0[\Phi, \Psi, R](t) \equiv R'(t) - M(t; \Phi, \Psi)R(t) = 0,$$

where

$$(2.35)\quad M(t; \Phi, \Psi) = \Phi(t)A(t)\Phi^*(t) + \Psi(t)C(t)\Phi^*(t) + \Phi(t)B(t)\Psi^*(t) - \Psi(t)A^*(t)\Psi^*(t).$$

For a given $\tau \in I$, show that if $Y(t) = (U(t); V(t))$ is a conjoined basis for the differential system (2.2) satisfying $Y(\tau) = (U_0; V_0)$ then

$$(2.36)\quad U_0^*U_0 + V_0^*V_0 > 0, \qquad V_0^*U_0 - U_0^*V_0 = 0.$$

Moreover, if Φ_0, Ψ_0, R_0 are matrices satisfying

$$(2.37)\quad R_0^*R_0 = U_0^*U_0 + V_0^*V_0, \qquad U_0 = \Phi_0^*R_0, \qquad V_0 = \Psi_0^*R_0,$$

then

$$(2.38)\quad \Phi_0\Phi_0^* + \Psi_0\Psi_0^* = E, \qquad \Psi_0\Phi_0^* - \Phi_0\Psi_0^* = 0,$$

and the solution $(\Phi; \Psi; R)$ of (2.33), (2.34) satisfying the initial conditions

$$(2.39)\quad \Phi(\tau) = \Phi_0, \qquad \Psi(\tau) = \Psi_0, \qquad R(\tau) = R_0,$$

is such that

$$(2.40)\quad U(t) = \Phi^*(t)R(t), \qquad V(t) = \Psi^*(t)R(t), \qquad \textit{for} \quad t \in I.$$

Conversely, if $(\Phi; \Psi; R)$ is a solution of (2.33), (2.34) with initial values $(\Phi_0; \Psi_0; R_0)$ at $t = \tau$, where R_0 is nonsingular and (Φ_0, Ψ_0) satisfy conditions (2.38), then $(U(t); V(t))$ defined by (2.40) is a conjoined basis for (2.2) with

$$(2.41)\quad R^*(t)R(t) = U^*(t)U(t) + V^*(t)V(t), \quad \textit{for} \quad t \in I.$$

(Because of its analogy with the polar coordinate transformation for a real self-adjoint linear homogeneous differential equation of the second order, the system (2.33), (2.34) is referred to as the "polar coordinate transform of (2.2_M).")

Hints. Note that conditions (2.36), (2.37) imply that R_0 is nonsingular and (2.38) holds. With the aid of Problem II.3:6 show that the solution of (2.33) satisfying the initial condition $(\Phi(\tau);\Psi(\tau)) = (\Phi_0;\Psi_0)$ has maximal interval of existence equal to I, and that throughout this interval one has the identities

$$\text{(2.42)}\quad \begin{aligned} &\Phi(t)\Phi^*(t) + \Psi(t)\Psi^*(t) \equiv E, \qquad \Psi(t)\Phi^*(t) - \Phi(t)\Psi^*(t) \equiv 0,\\ &\Phi^*(t)\Phi(t) + \Psi^*(t)\Psi(t) \equiv E, \qquad \Phi^*(t)\Psi(t) - \Psi^*(t)\Phi(t) \equiv 0. \end{aligned}$$

Moreover, if $(U(t); V(t))$ and $(\Phi(t);\Psi(t); R(t))$ are related by equations (2.40), then (2.41) holds. For $U(t)$, $V(t)$, $\Phi(t)$, $\Psi(t)$, $R(t)$ $n \times n$ matrix functions which are a.c. on arbitrary compact subintervals of I, and which satisfy equations (2.40), verify the identities

$$\text{(2.43)}\quad \begin{aligned} L_1{}^0[U, V] &= (\Lambda_1{}^0[\Phi,\Psi])^*R + G_1[\Phi,\Psi]R - \Psi^*\Lambda^0[\Phi,\Psi, R],\\ L_2{}^0[U, V] &= (\Lambda_2{}^0[\Phi,\Psi])^*R + G_2[\Phi,\Psi]R + \Phi^*\Lambda^0[\Phi,\Psi, R], \end{aligned}$$

where

$$\text{(2.44)}\quad \begin{aligned} G_1[\Phi,\Psi] &= [E - \Phi^*\Phi - \Psi^*\Psi][C\Phi^* - A^*\Psi^*]\\ &\qquad + [\Phi^*\Psi - \Psi^*\Phi][A\Phi^* + B\Psi^*],\\ G_2[\Phi,\Psi] &= [\Phi^*\Psi - \Psi^*\Phi][C\Phi^* - A^*\Psi^*]\\ &\qquad - [E - \Phi^*\Phi - \Psi^*\Psi][A\Phi^* + B\Psi^*]. \end{aligned}$$

7. For a given system (2.2_M) whose coefficient matrix functions satisfy hypothesis $(\mathfrak{H})$ on an interval $I = (a_0, b_0)$, first apply the substitution of Problem 3 above and then the results of Problem 6, to obtain a related but different "polar coordinate transform of (2.2_M)".

8.[s] If $Q(t)$ is an hermitian matrix function of class $\mathfrak{L}_{nn}{}^\infty[a, b]$ on arbitrary compact subintervals $[a, b]$ of an interval I, and for a given $t \in I$ the solution of the differential system

$$\text{(2.45)}\quad \begin{aligned} -\Psi' - Q(t)\Phi &= 0, \qquad \Psi(\tau) = E,\\ \Phi' - Q(t)\Psi &= 0, \qquad \Phi(\tau) = 0, \end{aligned}$$

is denoted by $\Phi = S(t; \tau)$, $\Psi = C(t; \tau)$, prove that the following identities hold for $(t, \tau, \sigma) \in I \times I \times I$:

$$\text{(2.46)}\quad \begin{aligned} S^*S + C^*C &\equiv E, \qquad S^*C - C^*S \equiv 0,\\ SS^* + CC^* &\equiv E, \qquad SC^* - CS^* \equiv 0, \end{aligned}$$

$$\text{(2.47)}\quad C(t; \sigma) \equiv C(t; \tau)C^*(\sigma; \tau) + S(t; \tau)S^*(\sigma; \tau),$$

$$\text{(2.48)}\quad S(t; \sigma) \equiv S(t; \tau)C^*(\sigma; \tau) - C(t; \tau)S^*(\sigma; \tau).$$

Hints. Show that if $\Phi = \Phi_0(t)$, $\Psi = \Psi_0(t)$ is a solution of (2.45) then $\Phi = \Psi_0(t)$, $\Psi = -\Phi_0(t)$ is also a solution of this system.

3. NORMALITY AND ABNORMALITY

For a nondegenerate subinterval I_0 of I, let $\Lambda(I_0)$ denote the vector space of n-dimensional vector functions $v(t)$ which are solutions of $L^\star[v](t) = 0$ and $Bv = 0$ on I_0. It is to be noted that in accordance with usage throughout this chapter, this latter statement means that $B(t)v(t) = 0$ for t a.e. on I_0. Clearly $v \in \Lambda(I_0)$ if and only if $(u(t); v(t)) = (0; v(t))$ is a solution of (2.2) on I_0. If $\Lambda(I_0)$ is zero-dimensional the system (2.2) is said to be *normal on* I_0, or to have *abnormality of order zero* on I_0, whereas if $\Lambda(I_0)$ has dimension $d = d(I_0) > 0$ the system is said to be *abnormal*, with *order of abnormality* d *on* I_0. If $I_0 = [a, b]$, for brevity we write $d[a, b]$ instead of the more precise $d([a, b])$, with similar contractions in case I_0 is of the form $[a, b)$, $(a, b]$, or (a, b).

If I_0 is a nondegenerate subinterval of I, then clearly $0 \leq d(I_0) \leq n$. Moreover, if $I_0 \subset I_0^1 \subset I$ then $d(I_0) \geq d(I_0^1)$. In particular, if $I_0 = [a, b]$ then by continuity we have $d[a, b] = d[a, b) = d(a, b] = d(a, b)$. If (2.2) is normal on every nondegenerate subinterval of I, then this system is said to be *identically normal* on I.

It may be verified readily that whenever hypothesis $(\mathfrak{H}_\omega)$ holds a system (2.2) with coefficient matrices (2.8) is identically normal. Another example of an identically normal system is that of Problem VII.2:2.

Two distinct points r and s on I are said to be (*mutually*) *conjugate* with respect to (2.2) if there exists a solution $y(t) = (u(t); v(t))$ of this system with $u(t) \not\equiv 0$ on the subinterval with endpoints r and s, while $u(r) = 0 = u(s)$. The system is called *disconjugate* on a subinterval I_0 of I provided no two distinct points of this interval are conjugate. If there exists a subinterval (c, ∞) of I on which (2.2) is disconjugate, then the system is said to be *disconjugate for large* t.

If $[a, b] \subset I$ we shall denote by $\Omega_0[a, b]$ the vector space of solutions $y(t) = (u(t); v(t))$ of (2.2) which satisfy the end-conditions

$$u(a) = 0 = u(b). \tag{3.1}$$

If $k[a, b]$ is the dimension of $\Omega_0[a, b]$, then $k[a, b]$ is the index of compatibility of the two-point boundary problem (2.2), (3.1). Clearly $k[a, b] \geq d[a, b]$, and $k[a, b] > d[a, b]$ if and only if a and b are conjugate, in which case the integer $k[a, b] - d[a, b]$ is called the *order of* b, $\{a\}$, *as a conjugate point to* a, $\{b\}$. It follows readily that the problem (2.2), (3.1) is self-adjoint, and the following result is a direct consequence of Theorem III.6.2.

***Lemma* 3.1.** *If* $[a, b] \subset I$, *and* u^a, u^b *are given n-dimensional vectors, then there exists a solution* $y_0(t) = (u_0(t); v_0(t))$ *of* (2.2) *satisfying* $u_0(a) = u^a$,

$u_0(b) = u^b$, *if and only if*

(3.2)
$$v^*(a)u^a - v^*(b)u^b = 0, \quad \textit{for arbitrary} \quad y(t) = (u(t); v(t)) \in \Omega_0[a, b].$$

For I_0 a nondegenerate subinterval of I, let $\mathfrak{D}(I_0)$ denote the set of vector functions η which are solutions on I_0 of $L_2[\eta, \zeta] = 0$, where ζ is an n-dimensional vector function that belongs to $\mathfrak{L}_n^2$ on arbitrary compact subintervals of I_0. If $I_0 = [a, b]$, for brevity we write $\mathfrak{D}[a, b]$ instead of the more precise $\mathfrak{D}([a, b])$. The fact that ζ is thus associated with η is denoted by the symbol $\eta \in \mathfrak{D}(I_0){:}\,\zeta$. Now if $v \in \Lambda(I_0)$ and $\eta \in \mathfrak{D}(I_0){:}\,\zeta$ one has the identity $0 = v^*L_2[\eta, \zeta] - (L^*[v])^*\eta = (v^*\eta)'$. This relation, together with the above lemma, implies the following result.

Lemma 3.2. *If I_0 is a nondegenerate subinterval of I and $\eta \in \mathfrak{D}(I_0){:}\,\zeta$, then for $v \in \Lambda(I_0)$ the function $v^*(t)\eta(t)$ is constant on I_0. Moreover, if $[a, b] \subset I_0$ and a and b are not mutually conjugate, then for given n-dimensional vectors u^a, u^b there exists a solution $y(t) = (u(t); v(t))$ of (2.2) satisfying $u(a) = u^a$, $u(b) = 0$, $\{u(a) = 0, u(b) = u^b\}$, if and only if $v^*(a)\, u^a = 0$, $\{v^*(b)u^b = 0\}$, for arbitrary $v \in \Lambda[a, b]$.*

In case c is a point of I such that (2.2) is normal on every subinterval of I that has c as an end-point, and $Y(t; c) = (U(t; c); V(t; c))$ is the solution of (2.2_M) satisfying the initial conditions (2.16), a value $t_1 \in I$, $t_1 \neq c$, is conjugate to c relative to (2.2) if and only if $U(t_1; c)$ is singular, and the order of t_1 as a conjugate point to c is equal to $n - r(t_1)$, where $r(t_1)$ is the rank of $U(t_1; c)$.

If $I_0 \subset I$ and $d(I_0) = d > 0$, then for $a \in I_0$ we shall denote by $\Delta = \Delta(a)$ an $n \times d$ matrix such that the column vectors of the solution matrix $V(t)$ of $L^\star[V] = 0$, $V(a) = \Delta(a)$ form a basis for $\Lambda(I_0)$. For brevity, this association is indicated by the symbol $\Delta(a) \sim \Lambda(I_0)$. If the column vectors of Δ are required to be mutually orthogonal, (i.e., $\Delta^*\Delta = E_d$), as may be done without loss of generality, we write $\Delta(a) \approx \Lambda(I_0)$.

In view of the properties of the involved matrices, the following result is a ready consequence of the definition of conjugate point.

Lemma 3.3. *Suppose that $[a, b] \subset I$ and c is a point of $[a, b)$ such that $d[a, x] = d[a, b] = d$ for $x \in (c, b]$, $\Delta(a) \sim \Lambda[a, b]$, while N is an $n \times (n - d)$ matrix such that $[\Delta(a) \quad N]$ is nonsingular. Moreover, let*

$$Y_\alpha(t) = (U_\alpha(t);\ V_\alpha(t)),\ (\alpha = 0, 1, 2, 3),$$

be the solutions of (2.2_M) satisfying the respective initial conditions

$$Y_0(a) = (0; \Delta(a)), \qquad Y_1(a) = (0; N),$$
$$Y_2(a) = (\Delta(a); 0), \qquad Y_3(a) = (N; 0).$$

Then a value $t_1 \in (c, b]$ is conjugate to $t = a$ relative to (2.2) if and only if one of the following conditions is satisfied:

1°. *$U_1(t_1)$ has rank less than $n - d$;*
2°. *the $n \times n$ matrix $[U_2(t_1) \quad U_1(t_1)]$ is singular;*
3°. *the $2n \times (2n - d)$ matrix*

$$\begin{bmatrix} U_1(a) & U_2(a) & U_3(a) \\ U_1(t_1) & U_2(t_1) & U_3(t_1) \end{bmatrix}$$

has rank less than $2n - d$.

In particular, if $N^*\Delta(a) = 0$ then $([U_2(t) \quad U_1(t)]; [V_2(t) \quad V_1(t)])$ is a conjoined basis for (2.2).

Attention will now be directed to some preliminary results concerning the behavior of solutions of a system (2.2) on a neighborhood of an end-point of a noncompact interval of existence.

Theorem 3.1. *Suppose that $Y_1(t) = (U_1(t); V_1(t))$ is a solution of (2.2_M) on an interval I, with $U_1(t)$ nonsingular on a subinterval I_0 of I, and let $S(t, c \mid U_1)$ denote the matrix function defined in Problem* VII.2:4.

(a) *If c is a point of I_0 such that (2.2) is normal on every subinterval of I_0 that has c as an end-point, and $t_1 \in I_0$, $t_1 \neq c$, then $S(t_1, c \mid U_1)$ is singular if and only if t_1 is conjugate to c, relative to (2.2).*

(b) *If I is an open interval (a_0, b_0), $(-\infty \le a_0 < b_0 \le \infty)$, on which (2.2) is identically normal, while (2.2) is disconjugate on a subinterval $I_0 = (c_0, b_0)$ of I and $Y_1(t) = (U_1(t); V_1(t))$ is a solution of (2.2_M) with $U_1(t)$ nonsingular on I_0, then for $c \in I_0$ the matrix $S(t, c \mid U_1)$ is nonsingular for $t \in I_0$, $t \neq c$. Moreover, if there exists a $c \in I_0$ such that $S^{-1}(t, c \mid U_1) \to 0$ as $t \to b_0$, then $S^{-1}(t, b \mid U_1) \to 0$ as $t \to b_0$ for arbitrary $b \in I_0$.*

If $Y(t; c) = (U(t; c); V(t; c))$ is the solution of (2.2_M) satisfying the initial conditions (2.16), then from Problem VII.2:4 it follows that

$$U(t; c) = T(t, c \mid U_1)S(t, c \mid U_1)K_1, \tag{3.3}$$

where $K_1 = -\{U; V \mid U_1; V_1\}$ is the nonsingular matrix $U_1^*(c)$. As $T(t, c \mid U_1)$ is nonsingular for all t it then follows that if $t_1 \in I_0$ and $t_1 \neq c$ then $S(t_1, c \mid U_1)$ is singular if and only if $U(t_1; c)$ is singular, and in view of the assumption that (2.2) is normal on every subinterval of I_0 that has c as an end-point, this latter condition is equivalent to t_1 being conjugate to c.

In order to establish conclusion (b), it is to be noted first that the fundamental matrix solution $T(t, c \mid U_1)$ of (2.28) satisfies for t, b and c on I the identity $T(t, c \mid U_1) = T(t, b \mid U_1)T(b, c \mid U_1)$. With the aid of this relation, it may be verified readily that for such t, b, c one also has the equation

$$S(t, c \mid U_1) = T(c, b \mid U_1)[S(t, b \mid U_1) - S(c, b \mid U_1)]. \tag{3.4}$$

Now for a general nonsingular $n \times n$ matrix M let $\nu[M]$ and $\mu[M]$ denote, respectively, the maximum and minimum of $|M\xi|$ on the unit ball $\{\xi \mid \xi \in \mathbf{C}_n, |\xi| \leq 1\}$. The relations

$$\begin{aligned}\nu[M^{-1}]\,|M\xi| \geq |M^{-1}(M\xi)| &= |\xi| \\ &= |M(M^{-1}\xi)| \geq \mu[M]\,|M^{-1}\xi|, \quad \textit{for} \quad \xi \in \mathbf{C}_n,\end{aligned}$$

then imply that $\nu[M^{-1}]\,\mu[M] = 1$. Since the condition that $S^{-1}(t, c \mid U_1) \to 0$ as $t \to b_0$ is equivalent to $\nu[S^{-1}(t, c \mid U_1)] \to 0$ as $t \to b_0$, this condition holds if and only if $\mu[S(t, c \mid U_1)] \to \infty$ as $t \to b_0$. Since $T(c, b \mid U_1)$ is nonsingular, it follows from (3.4) that for b and c points on I_0 it is true that $\mu[S(t, c \mid U_1)] \to \infty$ as $t \to b_0$ if and only if $\mu[S(t, b \mid U_1)] \to \infty$ as $t \to b_0$.

In view of conclusion (b) of Theorem 3.1, if an equation (2.2) is identically normal on an open interval $I = (a_0, b_0)$, and disconjugate on a subinterval $I_0 = (c_0, b_0)$ of I, then a solution $Y_1(t) = (U_1(t); V_1(t))$ of (2.2_M) will be called a *principal solution of* (2.2_M) *at* b_0 if $U_1(t)$ is nonsingular for t on some subinterval $I\{Y_1\} = (c_1, b_0)$ of I and $S^{-1}(t, c \mid U_1) \to 0$ for at least one $c \in I\{Y_1\}$, and consequently for all such values c. The concept of a principal solution of (2.2_M) at a_0 is defined in a corresponding manner. If $Y_1(t)$ is a conjoined basis for (2.2) with $U_1(t)$ nonsingular on some subinterval $I\{Y_1\} = (c_1, b_0)$, this definition of principal solution reduces to that given by Hartman [5] for a system (2.2) with coefficients (2.8) as determined by an equation (2.6) satisfying condition $(\mathfrak{H}_\omega)$. In the generality presented above, the definition is due to Reid [16]. In a later section it will be shown that if in addition to the hypotheses of (b) of Theorem 3.1 the condition $B(t) \geq 0$ holds for t a.e. on (a_0, b_0), then in case (2.2) is disconjugate on a subinterval (c_0, b_0) there does exist a principal solution, which is uniquely determined up to multiplication on the right by a nonsingular constant matrix. In general, however, one has the following result, which shows that if (2.2) is disconjugate on a subinterval (c_0, b_0) then a solution $Y_1(t)$ which is principal in the sense defined above posesses a property corresponding to that used as a definitive property by Leighton and Morse [1] for a real scalar second order differential equation.

Theorem 3.2. *Suppose that* (2.2) *is identically normal on an open interval* $I = (a_0, b_0)$, $(-\infty \leq a_0 < b_0 \leq \infty)$. *If* (2.2) *is disconjugate on a subinterval* $I_0 = (c_0, b_0)$ *of* I, *then a solution* $Y_1(t) = (U_1(t); V_1(t))$ *of* (2.2_M) *is a principal solution of* (2.2_M) *at* b_0 *if* $U_1(t)$ *is nonsingular on some subinterval* $I\{Y_1\} = (c_1, b_0)$ *of* I, *and there exists a solution* $Y_2(t) = (U_2(t); V_2(t))$ *of* (2.2_M) *with* $U_2(t)$ *nonsingular on some subinterval* $I\{Y_2\} = (c_2, b_0)$, *and such that for some* $c \in (c_1, b_0)$,

$$U_2^{-1}(t)U_1(t)T(t, c \mid U_1) \to 0 \quad \textit{as} \quad t \to b_0; \tag{3.5}$$

moreover, $\{U_2; V_2 \mid U_1; V_1\}$ *is nonsingular for any such* $Y_2(t)$. *Conversely, if* (2.2) *is disconjugate on a subinterval* (c_0, b_0), *and* $Y_1(t) = (U_1(t); V_1(t))$ *is a principal solution with* $U_1(t)$ *nonsingular on* (c_1, b_0), *then any solution* $Y_2(t) = (U_2(t); V_2(t))$ *of* (2.2_M) *with* $\{U_2; V_2 \mid U_1; V_1\}$ *nonsingular is such that* $U_2(t)$ *is nonsingular on some subinterval* (c_2, b_0) *and* (3.5) *holds for arbitrary* $c \in (c_1, b_0)$.

Suppose that (2.2) is disconjugate on a subinterval (c_0, b_0), and that there is a solution $Y_1(t) = (U_1(t); V_1(t))$ of (2.2_M) with $U_1(t)$ nonsingular on a subinterval (c_1, b_0) of I. If $Y_2(t) = (U_2(t); V_2(t))$ is also a solution of (2.2_M), then from Problem VII.2:4 it follows that for $c \in (c_1, b_0)$,

$$(3.6_\circ) \qquad U_2(t) = U_1(t)T(t, c \mid U_1)[U_1^{-1}(c)U_2(c) + S(t, c \mid U_1)K_1],$$

with $K_1 = -\{U_2; V_2 \mid U_1; V_1\}$, and hence for $t \in (c_1, b_0)$ we have

$$(3.6) \qquad \{U_1(t)T(t, c \mid U_1)\}^{-1}U_2(t) = U_1^{-1}(c)U_2(c) - S(t, c \mid U_1)\{U_2; V_2 \mid U_1; V_1\}.$$

Now if $U_2(t)$ is nonsingular and satisfies (3.5) for some $c \in (c_1, b_0)$, then (3.5) also holds for arbitrary $c \in (c_1, b_0)$ in view of the identity $T(t, c \mid U_1) = T(t, b \mid U_1)T(b, c \mid U_1)$; moreover, (3.5) is equivalent to the condition

$$(3.7) \qquad \mu[\{U_1(t)T(t, c \mid U_1)\}^{-1}U_2(t)] \to \infty \quad \textit{as} \quad t \to b_0.$$

Relation (3.6) then implies that the matrix $\{U_2; V_2 \mid U_1; V_1\}$ is nonsingular, and $\mu[S(t, c \mid U_1)] \to \infty$ as $t \to b_0$, so that $Y_1(t)$ is a principal solution.

On the other hand, if (2.2) is disconjugate on a subinterval (c_0, b_0) of I, and $Y_1(t) = (U_1(t); V_1(t))$ is a principal solution of (2.2_M) at b_0, then $U_1(t)$ is nonsingular throughout a subinterval (c_1, b_0) of (c_0, b_0) and there exists a $c \in (c_1, b_0)$ such that $\mu[S(t, c \mid U_1)] \to \infty$ as $t \to b_0$. For such a value c, and $Y_2(t) = (U_2(t); V_2(t))$ a solution of (2.2_M) with $\{U_2; V_2 \mid U_1; V_1\}(c)$ nonsingular, we have that $\mu[U_1^{-1}(c)U_2(c) - S(t, c \mid U_1)\{U_2; V_2 \mid U_1; V_1\}] \to \infty$ as $t \to b_0$. From $(3.6_\circ)$ it then follows that there exists a $c_2 \in (a_0, b_0)$ such that $U_2(t)$ is nonsingular on (c_2, b_0) and condition (3.7) holds, so that (3.5) holds for this value c.

Now Problem II.3:8 implies that there is a solution $Y_1(t) = (U_1(t); V_1(t))$ of (2.2_M) on an interval $I = (a_0, b_0)$, with $U_1(t)$ nonsingular on a subinterval $I\{Y_1\} = (c_1, b_0)$ of I, if and only if on (c_1, b_0) there is a solution $W = W_1(t)$ of the Riccati matrix differential equation

$$(3.8) \qquad \mathfrak{R}[W] \equiv W' + WA(t) + A^*(t)W + WB(t)W - C(t) = 0,$$

with $W_1(t) = V_1(t)U_1^{-1}(t)$ for $t \in (c_1, b_0)$; moreover, $U = U_1(t)$ is a fundamental matrix solution of the corresponding first order homogeneous matrix differential equation

$$(3.9) \qquad U' = [A(t) + B(t)W_1(t)]U.$$

For a solution $W = W_1(t)$ of (3.8) on (c_1, b_0), and $c \in (c_1, b_0)$, let $H = H(t, c \mid W_1)$, $G = G(t, c \mid W_1)$ be the solutions of the respective differential systems

$$\text{(3.10)} \qquad \begin{aligned} &\text{(a)}\ H' + H[A(t) + B(t)W_1(t)] = 0, \qquad H(c) = E, \\ &\text{(b)}\ G' + [A^*(t) + W_1(t)B(t)]G = 0, \qquad G(c) = E. \end{aligned}$$

If $Y_1(t) = (U_1(t); V_1(t))$ is a corresponding solution of (2.2_M) such that $W_1(t) = V_1(t)U_1^{-1}(t)$, it follows readily that $H(t, c \mid W_1) = U_1(c)U_1^{-1}(t)$. Now $W_1^*(t) - W_1(t) = -U_1^{*-1}(t)\,KU_1^{-1}(t)$, where $K = -\{U_1; V_1 \mid U_1; V_1\}$, and equation (3.10-b) may be written as

$$G' + [A^*(t) + W_1^*(t)B(t)]G = [W_1^*(t) - W_1(t)]\,B(t)G.$$

As (3.10-a) is equivalent to the system

$$H^{*\prime} + [A^*(t) + W_1^*(t)B(t)]H^* = 0, \qquad H^*(c) = E,$$

the substitution $G(t, c \mid W_1) = H^*(t, c \mid W_1)M(t, c \mid W_1)$ leads to the condition that $F = F(t, c \mid W_1) = U_1^*(c)M(t, c \mid W_1)U_1^{*-1}(c)$ is a solution of the differential system

$$F' = -KU_1^{-1}(t)B(t)U_1^{*-1}(t)F, \qquad F(c) = E,$$

so that from a result of the Hint to Problem VII.2:5 we have that

$$F(t, c \mid W_1) = T^{*-1}(t, c \mid U_1),$$

where $T(t, c \mid U_1)$ is the solution of the differential systems (2.28). Consequently,

$$G(t, c \mid W_1) = H^*(t, c \mid W_1)U_1^{*-1}(c)T^{*-1}(t, c \mid U_1)U_1^*(c),$$

and the associated matrix function

$$\text{(3.11)} \qquad Z(t, c \mid W_1) = \int_c^t H(r, c \mid W_1)B(r)G(r, c \mid W_1)\, dr$$

of Problem II.3:9 has the value

$$\text{(3.12)} \qquad Z(t, c \mid W_1) = U_1(c)S^*(t, c \mid U_1)U_1^*(c),$$

where $S(t, c \mid U_1)$ is the matrix function (2.30), as defined in Problem VII.2:4.

Now if $W_1(t)$ and $W(t)$ are both solutions of (3.8) on a subinterval (c_0, b_0) of (a_0, b_0), then for $c \in (c_0, b_0)$ a result of Problem II.3:11 implies that $\Gamma = W(c) - W_1(c)$ is such that $E + \Gamma Z(t, c \mid W_1)$ is nonsingular for $t \in (c_0, b_0)$ and

$$Z(t, c \mid W) = Z(t, c \mid W_1)[E + \Gamma Z(t, c \mid W_1)]^{-1}.$$

Therefore, if $Z(t, c \mid W_1)$ is nonsingular then $Z(t, c \mid W)$ is also nonsingular, and

$$Z^{-1}(t, c \mid W) = Z^{-1}(t, c \mid W_1) + W(c) - W_1(c). \tag{3.13}$$

Consequently, if $Z^{-1}(t, c \mid W_1) \to 0$ as $t \to b_0$, then $Z^{-1}(t, c \mid W) \to W(c) - W_1(c)$ as $t \to b_0$; in particular, if also $Z^{-1}(t, c \mid W) \to 0$ as $t \to b_0$ then $W(c) = W_1(c)$ and $W(t) \equiv W_1(t)$ on (c_0, b_0).

If $W_1(t)$ is a solution of (3.8) on a subinterval (c_0, b_0) of (a_0, b_0), then for any value $c \in (c_0, b_0)$ the condition that $Z^{-1}(t, c \mid W_1) \to 0$ as $t \to b_0$ is equivalent to the condition that $S^{-1}(t, c \mid U_1) \to 0$ as $t \to b_0$, where $(U_1(t); V_1(t))$ is a corresponding solution of (2.2_M) with $U_1(t)$ nonsingular and $W_1(t) = V_1(t)U_1^{-1}(t)$. Since by Theorem 3.1 the validity of this latter limit relation for one value of c on (c_0, b_0) implies the validity of the relation for arbitrary c on this subinterval, the same is true for the limit relation involving $Z^{-1}(t, c \mid W_1)$. If $W_1(t)$ is a solution of (3.8) on a subinterval (c_0, b_0), then $W_1(t)$ is called a *distinguished solution of* (3.8) *at* b_0 if $Z^{-1}(t, c \mid W_1) \to 0$ as $t \to b_0$ for at least one $c \in (c_0, b_0)$, and consequently for all such values c. The concept of a distinguished solution of (3.8) at a_0 is defined in a corresponding manner. These remarks, together with the above relation (3.13), imply the following basic result.

Theorem 3.3. *Suppose that* (2.2) *is identically normal on an open interval* $I = (a_0, b_0)$, $(-\infty \leq a_0 < b_0 \leq \infty)$, *and disconjugate on a subinterval* $I_0 = (c_0, b_0)$ *of* I. *Then there exists a principal solution* $Y_1(t) = (U_1(t); V_1(t))$ *of* (2.2_M) *at* b_0 *if and only if the Riccati matrix differential equation* (3.8) *has a distinguished solution* $W = W_1(t)$ *at* b_0, *and* $W_1(t) = V_1(t)\, U_1^{-1}(t)$ *for* t *on some subinterval* (c_0, b_0) *of* I. *In case such a principal solution exists the equation* (3.8) *has a unique distinguished solution at* b_0, *and the most general principal solution* $Y(t)$ *of* (2.2_M) *at* b_0 *is of the form* $Y(t) = Y_1(t)C$, *where* C *is a nonsingular* $n \times n$ *matrix.*

Since hypothesis ($\mathfrak{H}$) requires $B(t)$ and $C(t)$ to be hermitian matrix functions on I, if $W = W_1(t)$ is a solution of the Riccati matrix differential equation (3.8) on a subinterval I_0 of I then $W = W_1^*(t)$ is also a solution of (3.8) on the same subinterval. Moreover, from the definitions of $H(t, c \mid W_1)$ and $G(t, c \mid W_1)$ as solutions of the differential systems of (3.10), it follows that $H^*(t, c \mid W_1) = G(t, c \mid W_1^*)$, $G^*(t, c \mid W_1) = H(t, c \mid W_1^*)$, and, therefore, $Z^*(t, c \mid W_1) = Z(t, c \mid W_1^*)$. In particular, if $W = W_1(t)$ is a distinguished solution of (3.8) at b_0 or a_0 then $W = W_1^*(t)$ is also a distinguished solution of (3.8) at the same end-point. In view of the above established results on the uniqueness of a distinguished solution, and the equivalence of the existence of a distinguished solution of (3.8) to the existence of a principal solution of (2.2_M), we have the following result on principal solutions at b_0. A corresponding result holds for principal solutions at a_0.

Corollary. *If the hypotheses of Theorem 3.3 are satisfied, and $Y_1(t) = (U_1(t); V_1(t))$ is a principal solution of (2.2_M) at b_0, then $Y_1(t)$ is a conjoined basis for (2.2), with $U_1(t)$ nonsingular on some subinterval $I\{Y_1\} = (c_1, b_0)$ of I.*

A more specific characterization of a principal solution is provided by the following theorem.

Theorem 3.4. *Suppose that (2.2) is identically normal on $I = (a_0, b_0)$, $(-\infty \leq a_0 < b_0 \leq \infty)$, and disconjugate on a subinterval $I_0 = (c_0, b_0)$ of I. If $W(t)$ is a solution of (3.8) on I_0, then for $c \in I_0$ the matrix function $Z(t, c \mid W)$ is non-singular for $t \in I_0$, $t \neq c$. Moreover,*

$$(3.14) \qquad Z^{-1}(s, c \mid W) = W(c) - W_s(c), \quad \textit{for} \quad s \in I_0, \qquad s \neq c,$$

where $W_s(t) = V_s(t)U_s^{-1}(t)$ and $Y_s(t) = (U_s(t); V_s(t))$ is the solution of (2.2_M) determined by the initial conditions

$$(3.15) \qquad U_s(s) = 0, \qquad V_s(s) = E.$$

In particular, there exists a principal solution $Y_{b_0}(t) = (U_{b_0}(t); V_{b_0}(t))$ of (2.2_M) at b_0 if and only if for the solution $Y_s(t)$ of (2.2_M) determined by (3.15) the corresponding $W_s(t) = V_s(t)U_s^{-1}(t)$ is such that $W_{b_0}(t) = \lim_{s \to b_0} W_s(t)$ exists for $t \in (c_0, b_0)$ and $W = W_{b_0}(t)$ is a solution of (3.8) on (c_0, b_0), in which case $W_{b_0}(t)$ is necessarily the distinguished solution of (3.8) at b_0. Correspondingly, if (2.2) is disconjugate on a subinterval (a_0, c_0) of I, then (2.2_M) has a principal solution $Y_{a_0}(t) = (U_{a_0}(t); V_{a_0}(t))$ at a_0 if and only if the above defined solution $Y_s(t)$ of (2.2_M) is such that $W_{a_0}(t) = \lim_{s \to a_0} W_s(t)$ exists for $t \in (a_0, c_0)$ and $W = W_{a_0}(t)$ is a solution of (3.8) on (a_0, c_0), in which case $W_{a_0}(t)$ is necessarily the distinguished solution of (3.8) at a_0.

In view of the assumption that (2.2) is identically normal on I and disconjugate on I_0, the fact that $Z(t, c \mid W)$ is nonsingular for $t \in I_0$, $t \neq c$, is a direct consequence of conclusion (a) of Theorem 3.1 and the equation of the form (3.12) connecting $Z(t, c \mid W)$ and the associated $S(t, c \mid U)$. Also, on each of the subintervals $I_0^-(s) = \{t \mid t \in I_0,\ t < s\}$ and $I_0^+(s) = \{t \mid t \in I_0,$ $t > s\}$ the matrix function $U_s(t)$ is nonsingular, so that $W(t)$ and $W_s(t) = V_s(t)U_s^{-1}(t)$ are solutions of (3.8) on each of these subintervals. Moreover, $H(t, c \mid W_s) = U_s(c)U_s^{-1}(t)$ for $(t, c) \in I_0^-(s) \times I_0^-(s)$ and $(t, c) \in I_0^+(s) \times I_0^+(s)$, so that by Problem II.3:11 we have the relation

$$U_s(c)U_s^{-1}(t) = [E + Z(t, c \mid W)\{W_s(c) - W(c)\}]^{-1}H(t, c \mid W).$$

Consequently, for such values of (t, c) we have

$$[E + Z(t, c \mid W)\{W_s(c) - W(c)\}]U_s(c) = H(t, c \mid W)U_s(t),$$

and by continuity this latter relation also holds for $t = s$ and $c \neq s$. As $U_s(s) = 0$ and $U_s(c)$ is nonsingular for $c \in I_0$, $c \neq s$, it follows that if c and s are distinct points of I_0 then $E + Z(s, c \mid W)\{W_s(c) - W(c)\} = 0$, so that (3.14) holds. The conclusion of the theorem concerning the existence of a principal solution at b_0 is then a direct consequence of (3.14) and the above established equivalence of the existence of a distinguished solution of (3.8) at b_0 and the existence of a principal solution of (2.2_M) at this end-point. The conclusion concerning the existence of a principal solution at a_0 is established by a similar argument.

Problems VII.3

1. For $n = 2$, $A(t) \equiv 0$, $B(t) \equiv \text{diag}\{1, 2\}$, $C(t) = -B(t)$, show that the solution $Y(t) = (U(t; c); V(t; c))$ of (2.2_M) determined by the initial conditions (2.16) is given by

$$U(t; c) = \text{diag}\{\sin(t - c), \sin 2(t - c)\},$$

$$V(t; c) = \text{diag}\{\cos(t - c), \cos 2(t - c)\}.$$

Conclude that $t = c + (2k + 1)\pi/2$, $(k = 0, \pm 1, \pm 2, \ldots)$, is a conjugate point to $t = c$ of order 1, and that $t = c + m\pi$, $(m = \pm 1, \pm 2, \ldots)$ is a conjugate point to $t = c$ of order 2.

Find also the solution $Y_0(t; c) = (U_0(t; c); V_0(t; c))$ of (2.2_M) determined by the initial conditions (2.17). Show that:

(i) $U^2(t; c) + U_0^2(t; c) \equiv E$.

(ii) $U(t; c)$ and $U_0(t; c)$ are both singular for $t = c + (2k + 1)\pi/2$, $(k = 0, \pm 1, \pm 2, \ldots)$.

2. For $n = 4$, $A(t) \equiv 0$, $C(t) \equiv 0$, and

$$B(t) = \begin{bmatrix} 0 & 0 & 1 & 0 \\ 0 & 0 & t & 0 \\ 1 & t & 0 & 0 \\ 0 & 0 & 0 & 0 \end{bmatrix},$$

show that $d[a, b] = 1$ for each compact interval $[a, b]$, and that a basis for $\Lambda[a, b]$ is given by the constant vector function $v(t) \equiv e^{(4)} \equiv (\delta_{\alpha 4})$, $(\alpha = 1, 2, 3, 4)$. Moreover, arbitrary distinct values $t = a$ and $t = b$ are conjugate relative to the corresponding system (2.2), with a corresponding determining solution of (2.2) given by $(u_\alpha(t); v_\alpha(t))$, $(\alpha = 1, 2, 3, 4)$, with $u_\beta(t) \equiv 0$ for $\beta = 1, 2, 4$, $u_3(t) = -(b - a)(t - a) + (t - a)^2$, $v_1(t) \equiv -(a + b)$, $v_2(t) \equiv 2$, $v_3(t) \equiv 0$, $v_4(t)$ an arbitrary constant.

3. For $n = 4$, $A(t) \equiv 0$, $C(t) \equiv 0$, and

$$B(t) = \begin{bmatrix} g(t) & 0 & 1 & 0 \\ 0 & g(t) & t & 0 \\ 1 & t & g(t) & 0 \\ 0 & 0 & 0 & g(t) \end{bmatrix},$$

where $g(0) = 0$, $g(t) = t^4[\sin(1/t) + |\sin(1/t)|]$ for $t \neq 0$, show that the corresponding system (2.2) is normal on every interval for which $t = 0$ is an end-point, but has order of abnormality equal to 1 on every subinterval of the form $([(2k+2)\pi]^{-1}, [(2k+1)\pi]^{-1})$ or $(-[(2k+1)\pi]^{-1}, -[(2k+2)\pi]^{-1})$, $(k = 1, 2, \ldots)$.

4. AN ASSOCIATED FUNCTIONAL

In this section we shall be concerned with a given compact subinterval $[a, b]$ of I, and vector functions η belonging to the class $\mathfrak{D}[a, b]$; that is, $\eta \in \mathfrak{A}_n[a, b]$ and there exists a $\zeta \in \mathfrak{L}_n^2[a, b]$ such that $L_2[\eta, \zeta] = 0$ on $[a, b]$. The subclass of $\mathfrak{D}[a, b]$ on which $\eta(a) = 0 = \eta(b)$ will be denoted by $\mathfrak{D}_0[a, b]$. As introduced in the preceding section, the fact that η belongs to $\mathfrak{D}[a, b]$ or $\mathfrak{D}_0[a, b]$ with an associated ζ is indicated by $\eta \in \mathfrak{D}[a, b]:\zeta$ or $\eta \in \mathfrak{D}_0[a, b]:\zeta$.

If $(\eta_\alpha, \zeta_\alpha) \in \mathfrak{L}_n^2[a, b] \times \mathfrak{L}_n^2[a, b]$ for $\alpha = 1, 2$, we shall denote by $J[\eta_1:\zeta_1, \eta_2:\zeta_2; a, b]$ the functional

$$J[\eta_1:\zeta_1, \eta_2:\zeta_2; a, b] = \int_a^b \{\zeta_2^*(t)B(t)\zeta_1(t) + \eta_2^*(t)C(t)\eta_1(t)\}\, dt. \tag{4.1}$$

As $B(t)$ and $C(t)$ are hermitian matrix functions, (4.1) defines an hermitian form on $\mathfrak{L}_n^2[a, b] \times \mathfrak{L}_n^2[a, b]$. That is, if $(\eta_\beta, \zeta_\beta) \in \mathfrak{L}_n^2[a, b] \times \mathfrak{L}_n^2[a, b]$ for $\beta = 1, 2, 3$, then

(4.2)

(a) $J[\eta_1:\zeta_1, \eta_2:\zeta_2; a, b] = \overline{J[\eta_2:\zeta_2, \eta_1:\zeta_1; a, b]}$;

(b) $J[c\eta_1:c\zeta_1, \eta_2:\zeta_2; a, b] = cJ[\eta_1:\zeta_1, \eta_2:\zeta_2; a, b]$, *for c a complex constant;*

(c) $J[\eta_1 + \eta_2:\zeta_1 + \zeta_2, \eta_3:\zeta_3; a, b]$
$= J[\eta_1:\zeta_1, \eta_3:\zeta_3; a, b] + J[\eta_2:\zeta_2, \eta_3:\zeta_3; a, b\ ]$

Now if $\eta_\alpha \in \mathfrak{D}[a, b]:\zeta_\alpha$, $(\alpha = 1, 2)$, the vector functions ζ_α are in general not determined uniquely. The value of the corresponding functional $J[\eta_1:\zeta_1, \eta_2:\zeta_2; a, b]$ is independent of the particular ζ_α's, however, and consequently the symbol for this integral is reduced to $J[\eta_1, \eta_2; a, b]$. That is, if

$\eta_\alpha \in \mathfrak{D}[a, b]:\zeta_\alpha$, $(\alpha = 1, 2)$, then we write

$$J[\eta_1, \eta_2; a, b] = \int_a^b \{\zeta_2^*(t)B(t)\zeta_1(t) + \eta_2^*(t)C(t)\eta_1(t)\}\, dt. \tag{4.3}$$

The following results, which correspond to conclusion (ii) of Lemma V.3.1, are ready consequences of the definitions of the involved functions.

Lemma 4.1. (a) *If* $\eta_\alpha \in \mathfrak{D}[a, b]:\zeta_\alpha$, $(\alpha = 1, 2)$, *and* $\zeta_1 \in \mathfrak{A}_n[a, b]$, *then*

$$J[\eta_1, \eta_2; a, b] = \eta_2^*\zeta_1\Big|_a^b + \int_a^b \eta_2^* L_1[\eta_1, \zeta_1]\, dt, \tag{4.4'}$$

$$J[\eta_1; a, b] = \eta_1^*\zeta_1\Big|_a^b + \int_a^b \eta_1^* L_1[\eta_1, \zeta_1]\, dt. \tag{4.4''}$$

(b) *If* $\eta_\alpha \in \mathfrak{D}[a, b]:\zeta_\alpha$ *and* $\zeta_\alpha \in \mathfrak{A}_n[a, b]$ *for* $\alpha = 1, 2$, *then*

$$\int_a^b \eta_2^* L_1[\eta_1, \zeta_1]\, dt - \int_a^b (L_1[\eta_2, \zeta_2])^*\eta_1\, dt = \{\zeta_2^*\eta_1 - \eta_2^*\zeta_1\}\Big|_a^b, \tag{4.5}$$

$$= \{\eta_1; \zeta_1 \mid \eta_2; \zeta_2\}\Big|_a^b.$$

In particular, if $a \leq t_1 < t_2 \leq b$, and t_1, t_2 are mutually conjugate with respect to (2.2), then there exists a solution $(u; v)$ of (2.2) with $u(t) \not\equiv 0$ on $[t_1, t_2]$, and $u(t_1) = 0 = u(t_2)$. If $(\eta(t), \zeta(t)) = (u(t), v(t))$ for $t \in [t_1, t_2]$, and $(\eta(t), \zeta(t)) = (0, 0)$ for $t \in [a, t_1) \cup (t_2, b]$, then $\eta \in \mathfrak{D}_0[a, b]:\zeta$, and with the aid of relation (4.4″) for $(\eta_1, \zeta_1) = (u, v)$ on $[t_1, t_2]$ it follows that

$$J[\eta; a, b] = J[u; t_1, t_2] = u^*v\Big|_{t_1}^{t_2} = 0.$$

Consequently, we have the following result.

Corollary. *If* $[a, b] \subset I$, *and there is a pair of points on* $[a, b]$ *which are conjugate with respect to* (2.2), *then there exists an* $\eta \in \mathfrak{D}_0[a, b]$ *such that* $\eta(t) \not\equiv 0$ *on* $[a, b]$ *and* $J[\eta; a, b] = 0$.

Theorem 4.1. *If* $[a, b] \subset I$ *and* $u \in \mathfrak{A}_n[a, b]$, *then there exists a* v *such that* $(u; v)$ *is a solution of* (2.2) *on* $[a, b]$ *if and only if there exists a* $v_1 \in \mathfrak{L}_n^2[a, b]$ *such that* $u \in \mathfrak{D}[a, b]:v_1$ *and*

$$J[u:v_1, \eta:\zeta; a, b] = 0, \quad \textit{for} \quad \eta \in \mathfrak{D}_0[a, b]:\zeta. \tag{4.6}$$

If $(u; v)$ is a solution of (2.2) on $[a, b]$, and $\eta \in \mathfrak{D}_0[a, b]:\zeta$, then $u \in \mathfrak{D}[a, b]:v$ and (4.6) is a consequence of (4.4′) for $(\eta_1, \zeta_1) = (u, v)$, $(\eta_2, \zeta_2) = (\eta, \zeta)$.

On the other hand, suppose that $u \in \mathfrak{D}[a, b]:v_1$, and (4.6) holds; that is,

$$\int_a^b \{\zeta^*(t)B(t)v_1(t) + \eta^*(t)C(t)u(t)\}\, dt = 0, \quad \textit{for} \quad \eta \in \mathfrak{D}_0[a, b]:\zeta. \tag{4.7}$$

If $v_0(t)$ is any solution of the vector differential equation $L^\star[v_0] + Cu = 0$, then (4.7) may be written as

$$\int_a^b \{(v_1, L[\eta]) - (L^\star[v_0], \eta)\}\, dt = 0,$$

and since $(v_0, L[\eta]) - (L^\star[v_0], \eta) = (\eta^* v_0)'$ it follows that

$$\int_a^b (v_1 - v_0, L[\eta])\, dt = 0, \quad \textit{for} \quad \eta \in \mathfrak{D}_0[a, b]:\zeta. \tag{4.7'}$$

Now if $U(t)$ is a fundamental matrix solution of $L[U] = 0$, then by the method of variation of parameters we have that $\eta \in \mathfrak{D}_0[a, b]:\zeta$ if and only if

$$\eta(t) = -U(t)\int_t^b U^{-1}(s)B(s)\zeta(s)\, ds, \qquad t \in [a, b], \tag{4.8}$$

and

$$\int_a^b U^{-1}(s)B(s)\zeta(s)\, ds = 0. \tag{4.8'}$$

From (4.7′) and (4.8′) it then follows that

$$\int_a^b \zeta^*(s)B(s)[v_1(s) - v_0(s)]\, ds = 0,$$

if $\zeta \in \mathfrak{L}_n^2[a, b]$ and

$$\int_a^b \zeta^*(s)B(s)U^{*-1}(s)\, ds = 0.$$

Therefore, (see Problem 7 of F.1 in Appendix F), there exists a constant n-dimensional vector λ such that $B(t)[v_1(t) - v_0(t)] = B(t)\, U^{*-1}(t)\lambda$. As $L^\star[U^{*-1}] = 0$ it then follows that $v(t) = v_0(t) + U^{*-1}(t)\lambda$ is also a solution of $L^\star[v] + Cu = 0$ on $[a, b]$. Moreover $B(t)v_1(t) = B(t)v(t)$, and, therefore, $u \in \mathfrak{D}[a, b]:v$, so that $(u; v)$ is a solution of (2.2).

Corresponding to Corollaries 1 and 2 to Theorem V.3.1, we now have the following results.

***Corollary* 1.** *If $[a, b] \subset I$, $J[\eta; a, b]$ is non-negative definite on $\mathfrak{D}_0[a, b]$, and u is an element of $\mathfrak{D}_0[a, b]$ satisfying $J[u; a, b] = 0$, then there exists a $v \in \mathfrak{A}_n[a, b]$ such that $(u; v)$ is a solution of (2.2) on $[a, b]$. In particular, if $u(t) \not\equiv 0$ on $[a, b]$ then b is conjugate to a.*

***Corollary* 2.** *Suppose that* $[a, b] \subset I$, *and* $J[\eta; a, b]$ *is non-negative definite on* $\mathfrak{D}_0[a, b]$. *If* $(u; v)$ *is a solution of* (2.2), *and* $u_0 \in \mathfrak{D}[a, b]$ *with* $u_0(a) = u(a)$, $u_0(b) = u(b)$, *then* $J[u_0; a, b] \geq J[u; a, b]$; *moreover, if* $J[\eta; a, b]$ *is positive definite on* $\mathfrak{D}_0[a, b]$ *then* $J[u_0; a, b] \geq J[u; a, b]$ *with the equality sign holding only if* $u_0(t) = u(t)$ *for* $t \in [a, b]$.

Preliminary to the study of necessary and sufficient conditions for a system (2.2) to be disconjugate on a subinterval of I, the following result is stated without proof, as it may be established by direct verification.

***Lemma* 4.2.** *Suppose that* $[a, b] \subset I$, *and* $U(t)$, $V(t)$ *are* $n \times r$ *a.c. matrix functions on* $[a, b]$. *If* $\eta_\alpha \in \mathfrak{A}_n[a, b]$, $\zeta_\alpha \in \mathfrak{L}_n^2[a, b]$ *for* $\alpha = 1, 2$, *and there exist* a.c. *r-dimensional vector functions* $h_\alpha(t)$ *such that* $\eta_\alpha(t) = U(t)\, h_\alpha(t)$ *on* $[a, b]$, *then on* $[a, b]$ *we have the identity*

$$\begin{aligned} \zeta_2^* B\zeta_1 + \eta_2^* C\eta_1 = {} & \{\zeta_2 - Vh_2\}^* B\{\zeta_1 - Vh_1\} \\ & - h_2^*\, V^* L_2[\eta_1, \zeta_1] - (L_2[\eta_2, \zeta_2])^* Vh_1 \\ & + h_2^*(V^* L_2[U, V] + U^*\, L_1[U, V])h_1 \\ & - h_2^*\{U^* V - V^* U\}h_1' + [h_2^* U^* Vh_1]'. \end{aligned} \tag{4.9}$$

The following result is an immediate consequence of the identity (4.9).

***Corollary*.** *If* $[a, b] \subset I$, *and the column vectors of* $Y(t) = (U(t); V(t))$ *form a basis for an r-dimensional conjoined family of solutions of* (2.2), *while* $\eta \in \mathfrak{D}[a, b]\colon \zeta$ *and there exists an r-dimensional* a.c. *vector function h such that* $\eta(t) = U(t)\, h(t)$ *for* $t \in [a, b]$, *then*

$$J[\eta; a, b] = \eta^* Vh \Big|_a^b + \int_a^b [\zeta - Vh]^* B[\zeta - Vh]\, dt. \tag{4.10}$$

Theorem 4.2. *If* $[a, b] \subset I$, *and* $J[\eta; a, b]$ *is non-negative definite on* $\mathfrak{D}_0[a, b]$, *then* $B(t)$ *is non-negative definite for* t a.e. *on* $[a, b]$.

If it is not true that $B(t)$ is non-negative definite for t a.e. on $[a, b]$, then there exists a constant vector ζ_0 with $|\zeta_0| = 1$, and positive constants k_1, k_2 such that $I_0 = \{t \mid t \in [a, b], \nu[B(t)] < k_1, \zeta_0^* B(t)\zeta_0 < -k_2\}$ is of positive measure. Let $\phi_0(t)$ be the characteristic function of I_0; that is, $\phi_0(t) = 1$ for $t \in I_0$, and $\phi_0(t) = 0$ for $t \notin I_0$. Moreover, let t_0 be a point of the open interval (a, b) such that the function $\int_a^t \phi_0(s)\, ds$, $t \in [a, b]$, has derivative equal to 1 at $t = t_0$, and consequently if $t_0 \in (c,d) \subset [a, b]$ then the set of points $I_0 \cap (c, d)$ has positive measure. In particular, if $B(t)$ is continuous on $[a, b]$, then for t_0 any point of (a, b) at which $B(t_0)$ fails to be non-negative definite there exists a vector ζ_0 and constants k_1, k_2 satisfying the conditions specified above.

For $U(t)$ a fundamental matrix solution of $L[U] = 0$, let k_3 be a constant such that $\nu[U(t)\, U^{-1}(s)] \leq k_3$ for $(t, s) \in [a, b] \times [a, b]$, and let (c, d) be an open subinterval of $[a, b]$ containing t_0 and such that

$$(d - c)^{-1} > \frac{k_1^2 k_3^2}{k_2} \int_a^b \nu[C(t)]\, dt. \tag{4.11}$$

Now let $g(t)$ be a continuous scalar function which is not identically zero on $[c, d]$, and such that the solution $u(t)$ of the differential system $L_2[u, \zeta_0\phi_0 g] = 0$, $u(c) = 0$, satisfies $u(d) = 0$. Indeed, $g(t)$ may be chosen as a polynomial $g(t) = c_0 + c_1 t + \cdots + c_n t^n$ with $|c_0|^2 + \cdots + |c_n|^2 = 1$. Then $u(t) \not\equiv 0$ on $[c, d]$,

$$u(t) = \int_c^t U(t) U^{-1}(s) B(s) \zeta_0 \phi_0(s) g(s)\, ds, \qquad t \in [c, d],$$

and in view of the definitive properties of k_1 and k_3 we have

$$|u(t)| \leq k_1 k_3 \int_c^d |\phi_0(s) g(s)|\, ds, \quad \text{for} \quad t \in [c, d].$$

If $\eta(t) = u(t)$, $\zeta(t) = \zeta_0\, \phi_0(t)\, g(t)$ for $t \in [c, d]$, and $\eta(t) \equiv 0$, $\zeta(t) \equiv 0$ on $[a, c) \cup (d, b]$, then $\eta \in \mathfrak{D}_0[a, b] : \zeta$ and

$$J[\eta; a, b] \leq -k_2 \int_c^d |\phi_0(t) g(t)|^2\, dt$$
$$+ k_1^2 k_3^2 \left(\int_c^d |\phi_0(t) g(t)|\, dt \right)^2 \left(\int_a^b \nu[C(t)]\, dt \right).$$

As the Schwarz inequality implies that

$$\left(\int_c^d |\phi_0(t) g(t)|\, dt \right)^2 \leq (d - c) \int_c^d |\phi_0(t) g(t)|^2\, dt,$$

while $\int_c^d |\phi_0(t)\, g(t)|^2\, dt > 0$ in view of the above described choice of t_0 and (c, d), it follows that

$$J[\eta; a, b] \leq -\{k_2 - k_1^2 k_3^2 (d - c) \int_a^b \nu[C(t)]\, dt\} \int_c^d |\phi_0(t) g(t)|^2\, dt < 0,$$

contrary to the hypothesis that $J[\eta; a, b]$ is non-negative definite on $\mathfrak{D}_0[a, b]$.

Theorem 4.3. *Suppose that $[a, b] \subset I$, and $J[\eta; a, b]$ is positive definite on $\mathfrak{D}_0[a, b]$. If $d[a, b] = d$, $\Delta(a) \approx \Lambda[a, b]$, and N is an $n \times (n - d)$ matrix such that $N^*\Delta(a) = 0$ and $[\Delta(a) \quad N]$ is nonsingular, then there exists a unique solution $Y_b(t) = (U_b(t); V_b(t))$ of (2.2_M) such that*

$$U_b(a) = N, \qquad U_b(b) = 0, \qquad V_b^*(a)\Delta(a) = 0. \tag{4.12}$$

The column vectors of $Y_b(t)$ form a basis for a conjoined family of solutions of (2.2) of dimension $n - d$, and if $Y_4(t) = (U_4(t); V_4(t))$ is a second solution of (2.2_M) whose column vectors form a basis for a conjoined family of solutions of (2.2) of dimension $n - d$, and satisfying

$$(4.13) \qquad U_4(a) = N, \qquad V_4^*(a)\Delta(a) = 0, \qquad U_4^*(a)V_4(a) > U_b^*(a)V_b(a),$$

then $U_4(t)$ is of rank $n - d$ on $[a, b]$. Moreover, if $Y_2(t) = (U_2(t); V_2(t))$ is the solution of (2.2_M) satisfying the initial conditions $U_2(a) = \Delta(a)$, $V_2(a) = 0$, then

$$Y(t) = ([U_2(t) \quad U_4(t)]; [V_2(t) \quad V_4(t)]) = (U(t); V(t))$$

is a conjoined basis for (2.2) with $U(t)$ nonsingular on $[a, b]$.

In view of the Corollary to Theorem 4.1, the condition that $J[\eta; a, b]$ is positive definite on $\mathfrak{D}_0[a, b]$ implies that a and b are not mutually conjugate. In turn, since $N^*\Delta(a) = 0$, it follows from Lemma 3.2 that there is a solution $(U_{b0}(t); V_{b0}(t))$ of (2.2_M) satisfying $U_{b0}(a) = N$, $U_{b0}(b) = 0$. Moreover, if $Y_0(t) = (0; V_0(t))$ is the solution of (2.2_M) with $V_0(a) = \Delta(a)$, then the most general solution $(U_b(t); V_b(t))$ of (2.2_M) satisfying $U_b(a) = N$, $U_b(b) = 0$ is of the form

$$(4.14) \qquad U_b(t) = U_{b0}(t), \qquad V_b(t) = V_{b0}(t) + V_0(t)\Gamma,$$

where Γ is an arbitrary $d \times d$ matrix, and as $\Delta^*(a)\,\Delta(a) = E_d$ it follows that the unique solution satisfying (4.12) is given by (4.14) with

$$\Gamma = -\Delta^*(a)V_{b0}(a);$$

also, the matrix $U_b^*(a)V_b(a)$ is hermitian, so that the column vectors of $Y_b(t)$ form a basis for a conjoined family of solutions of (2.2) of dimension $n - d$.

Now if $Y_4(t) = (U_4(t); V_4(t))$ is a solution of (2.2_M) satisfying (4.13) then the matrix $U_4(b)$ is of rank $n - d$. Indeed, if $U_4(b)\xi = 0$ then $(u(t); v(t)) = ([U_4(t) - U_b(t)]\xi; [V_4(t) - V_b(t)]\xi)$ is a solution of (2.2) with $u(a) = 0 = u(b)$. As b and a are not mutually conjugate it then follows that $u(t) \equiv 0$ on $[a, b]$ and $v(t) \in \Lambda[a, b]$. In turn, (4.12) and (4.13) imply that $\Delta^*(a)\,v(a) = 0$, and since $\Delta(a) \approx \Lambda[a, b]$ it follows that $v(t) \equiv 0$ on $[a, b]$. Then $0 = \xi^*N^*v(a) = \xi^*[U_4^*(a)V_4(a) - U_b^*(a)V_b(a)]\xi$, and $\xi = 0$ in view of the last condition of (4.13).

Now if $c \in (a, b)$, and there exists a nonzero vector ξ such that $U_4(c)\xi = 0$, let $(\eta(t), \zeta(t)) = ([U_4(t) - U_b(t)]\xi, [V_4(t) - V_b(t)]\xi)$ for $t \in [a, c]$, and $(\eta(t), \zeta(t)) = (-U_b(t)\xi, -V_b(t)\xi)$ for $t \in (c, b]$. Then $\eta \in \mathfrak{D}_0[a, b]:\zeta$, and application of (4.4″) to the individual subintervals $[a, c]$ and $[c, b]$ yields the relation

$$(4.15) \quad J[\eta; a, b] = \xi^*[U_4^*(c) - U_b^*(c)][V_4(c) - V_b(c)]\xi - \xi^*U_b^*(c)V_b(c)\xi.$$

Since $U_4(c)\xi = 0$, the right-hand member of (4.15) is equal to

$$-\xi^* U_b^*(c)V_4(c)\xi = -\xi^*[U_b^*(c)V_4(c) - V_b^*(c)U_4(c)]\xi.$$

Moreover, in view of Lemma 2.1 we have that

$$\begin{aligned} U_b^*(c)V_4(c) - V_b^*(c)U_4(c) &= U_b^*(a)V_4(a) - V_b^*(a)U_4(a) \\ &= U_4^*(a)V_4(a) - U_b^*(a)V_b(a), \end{aligned}$$

where the last relation follows from $U_b(a) = U_4(a) = N$, and the fact that the matrix $V_b^*(a)U_b(a)$ is hermitian. Thus under the assumption that $U_4(c)\xi = 0$ we have, in view of (4.13), that

$$J[\eta; a, b] = -\xi^*[U_4^*(a)V_4(a) - U_b^*(a)V_b(a)]\xi < 0,$$

whereas the positive definiteness of $J[\eta; a, b]$ on $\mathfrak{D}_0[a, b]$ implies that $J[\eta; a, b] \geq 0$. Consequently the assumption that there is a value on $[a, b]$ at which $U_4(t)$ has rank less than $n - d$ has led to a contradiction.

Now if $(U_2(t); V_2(t))$ and $(U(t); V(t))$ are defined as in the statement of the theorem the matrix function $U(t)$ is nonsingular on $[a, b]$. Indeed, if $c \in [a, b]$ and $U(c)$ is singular, then there exist vectors ρ and σ of respective dimensions d and $n - d$, with ρ and σ not both zero and $(u_0(t); v_0(t)) = (U_2(t)\rho + U_4(t)\sigma; V_2(t)\rho + V_4(t)\sigma)$ a solution of (2.2) with $u_0(c) = 0$. Then $v^*(t)u_0(t) \equiv 0$ on $[a, b]$ for arbitrary $v(t) \in \Lambda[a, b]$ by Lemma 3.2, and hence $0 = \Delta^*(a)u_0(a) = \Delta^*(a)\Delta(a)\rho = \rho$. Therefore $u(t) = U_4(t)\sigma$, and $U_4(c)$ is of rank less than $n - d$, contrary to the preceding result. Consequently, $U(t)$ is nonsingular on $[a, b]$. Finally, since $U_4^*(a)V_4(a) = N^*V_4(a)$ is hermitian by hypothesis, one may verify directly that $U^*(a)V(a) - V^*(a)U(a) = 0$, and hence $(U(t); V(t))$ is a conjoined basis for (2.2) with $U(t)$ nonsingular on $[a, b]$.

Theorem 4.4. *If* $[a, b] \subset I$ *then* $J[\eta; a, b]$ *is positive definite on* $\mathfrak{D}_0[a, b]$ *if and only if* $B(t) \geq 0$ *for* t a.e. *on* $[a, b]$, *and there exists a conjoined basis* $Y(t) = (U(t); V(t))$ *for* (2.2) *with* $U(t)$ *nonsingular on* $[a, b]$.

If $J[\eta; a, b]$ is positive definite on $\mathfrak{D}_0[a, b]$, then the results of Theorems 4.2, 4.3 imply that $B(t) \geq 0$ for t a.e. on $[a, b]$, and the existence of a conjoined basis $Y(t) = (U(t); V(t))$ for (2.2) with $U(t)$ nonsingular on $[a, b]$. Conversely, when such a basis $Y(t) = (U(t); V(t))$ exists, for $\eta \in \mathfrak{D}_0[a, b]:\zeta$ and $h(t) = U^{-1}(t)\eta(t)$ it follows from the Corollary to Lemma 4.2 that

$$J[\eta; a, b] = \int_a^b [\zeta - Vh]^* B(t)[\zeta - Vh]\, dt.$$

The condition that $B(t) \geq 0$ for t a.e. on $[a, b]$ then implies that $J[\eta; a, b] \geq 0$ for $\eta \in \mathfrak{D}_0[a, b]:\zeta$, and $J[\eta; a, b] = 0$ if and only if $B(t)[\zeta(t) - V(t)h(t)] = 0$

for t a.e. on $[a, b]$. As $L_2[\eta, \zeta] = 0$ and $L_2[U, V] = 0$, it then follows that $Uh' = B[\zeta - Vh] = 0$ a.e., so that $h'(t) = 0$ a.e. and $h(t) \equiv h(a) = 0$ for $t \in [a, b]$. Consequently, $J[\eta; a, b] = 0$ if and only if $\eta(t) = U(t)h(t) = 0$ for $t \in [a, b]$, so that $J[\eta; a, b]$ is positive definite on $\mathfrak{D}_0[a, b]$.

Theorem 4.5. *If $[a, b] \subset I$, then $J[\eta; a, b]$ is positive definite on $\mathfrak{D}_0[a, b]$ if and only if $B(t) \geq 0$ for t a.e. on $[a, b]$, and there is no value on $(a, b]$ which is conjugate to $t = a$.*

From the Corollary to Lemma 4.1 and Theorem 4.2 it follows that the positive definiteness of $J[\eta; a, b]$ on $\mathfrak{D}_0[a, b]$ implies the disconjugacy of (2.2) on $[a, b]$, and the condition $B(t) \geq 0$ for t a.e. on $[a, b]$.

Conversely, suppose that $B(t) \geq 0$ for t a.e. on $[a, b]$ and that there is no value on $(a, b]$ which is conjugate to $t = a$. Let c be the supremum of values $b_1 \in (a, b]$ such that $J[\eta; a, b_1]$ is positive definite on $\mathfrak{D}_0[a, b_1]$. Such values b_1 exist, since if $Y_0(t) = (U_0(t); V_0(t))$ is the conjoined basis for (2.2) determined by the initial conditions $U_0(a) = E$, $V_0(a) = 0$, and $b_1 \in (a, b]$ is such that $U_0(t)$ is nonsingular on $[a, b_1]$, then by Theorem 4.4 the functional $J[\eta; a, b_1]$ is positive definite on $\mathfrak{D}_0[a, b_1]$. It will be shown firstly that $J[\eta; a, c] \geq 0$ for arbitrary $\eta \in \mathfrak{D}_0[a, c]$. Indeed, suppose $\eta_1 \in \mathfrak{D}_0[a, c]$ and $J[\eta_1; a, c] < 0$. Let $Y_1(t) = (U_1(t); V_1(t))$ be the conjoined basis for (2.2) determined by the initial conditions $U_1(c) = E$, $V_1(c) = 0$; also, for $d_1 = \lim_{t \to c} d[t, c]$ let ε_1 be a positive constant such that $0 < \varepsilon_1 < c - a$, and so small that $d[c - \varepsilon_1, c] = d_1$ and $U_1(t)$ is non-singular on $[c - \varepsilon_1, c]$. Now let $\Delta(c)$ be such that $\Delta(c) \sim \Lambda[c - \varepsilon_1, c]$, and ε_0 so small that $0 < \varepsilon_0 < \varepsilon_1$ and $d[c - \varepsilon_1, c - \varepsilon_0] = d[c - \varepsilon_1, c] = d_1$. In particular, the results of the Corollary to Lemma 4.1 and Theorem 4.4 imply that (2.2) is disconjugate on $[c - \varepsilon_1, c]$. If $\eta_1 \in \mathfrak{D}_0[a, c]\colon \zeta_1$, and $0 \leq \varepsilon \leq \varepsilon_0$, it then follows with the aid of Lemma 3.2 that there is a unique solution $(u_\varepsilon(t); v_\varepsilon(t))$ of (2.2) such that

$$u_\varepsilon(c - \varepsilon_1) = \eta(c - \varepsilon_1), \qquad u_\varepsilon(c - \varepsilon) = 0, \qquad \Delta^*(c)v_\varepsilon(c) = 0.$$

Moreover, in view of criterion 3° of Lemma 3.3, the functions $(u_\varepsilon(t); v_\varepsilon(t))$ tend uniformly to $(u_0(t); v_0(t))$ on $[c - \varepsilon_1, c]$ as $\varepsilon \to 0$. Now for $0 \leq \varepsilon \leq \varepsilon_0$ define

$$\begin{aligned}(\eta_\varepsilon(t), \zeta_\varepsilon(t)) &= (\eta_1(t)\ \zeta_1(t)), \quad \textit{for} \quad t \in [a, c - \varepsilon_1),\\ &= (u_\varepsilon(t), v_\varepsilon(t)), \quad \textit{for} \quad t \in [c - \varepsilon_1, c - \varepsilon],\\ &= (0, 0), \quad \textit{for} \quad t \in (c - \varepsilon, c].\end{aligned}$$

Then $\eta_\varepsilon \in \mathfrak{D}_0[a, c]\colon \zeta_\varepsilon$, $\eta_\varepsilon \in \mathfrak{D}_0[a, c - \varepsilon]\colon \zeta_\varepsilon$, and $J[\eta_\varepsilon; a, c] = J[\eta_\varepsilon; a, c - \varepsilon]$. Moreover, the definitive property of c assures us that $J[\eta_\varepsilon; a, c] \geq 0$ for $0 < \varepsilon \leq \varepsilon_0$, and the limit behavior of $(u_\varepsilon(t); v_\varepsilon(t))$ implies that $J[\eta_\varepsilon; a, c] \to J[\eta_0; a, c]$ as $\varepsilon \to 0$. Consequently, we have the inequality

$$0 \leq J[\eta_0; a, c] = J[\eta_1; a, c - \varepsilon_1] + J[u_0; c - \varepsilon_1, c].$$

Since $U_1(t)$ is nonsingular on $[c - \varepsilon_1, c]$, from the results of Theorem 4.4 and Corollary 2 to Theorem 4.1 it follows that $J[u_0; c - \varepsilon_1, c] \leq J[\eta_1, c - \varepsilon_1, c]$, and hence we have the contradictory result $J[\eta_1; a, c] \geq 0$. Thus it has been established that $J[\eta; a, c]$ is non-negative for arbitrary $\eta \in \mathfrak{D}_0[a, c]$. Indeed, $J[\eta; a, c]$ is positive definite on $\mathfrak{D}_0[a, c]$, since if $\eta(t)$ were an element of $\mathfrak{D}_0[a, c]$ that did not vanish identically on $[a, c]$, and for which $J[\eta; a, c] = 0$, Theorem 4.1 would imply the existence of a solution $(u(t); v(t))$ of (2.2) with $u(t) \equiv \eta(t)$ on $[a, c]$. It would then follow that $t = c$ is conjugate to $t = a$, contrary to the hypothesis that there is no point on $(a, b]$ conjugate to a.

As $J[\eta; a, c]$ is positive definite on $\mathfrak{D}_0[a, c]$, it then follows from Theorem 4.4 that there exists a conjoined basis $(U(t); V(t))$ for (2.2) with $U(t)$ nonsingular on $[a, c]$. If $c \in (a, b)$, then by continuity there would be a $c_1 \in (c, b]$ such that $U(t)$ is nonsingular on $[a, c_1]$. Theorem 4.4 would then imply that $J[\eta; a, b_1]$ is positive definite on $\mathfrak{D}_0[a, b_1]$ for $b_1 \in (a, c_1]$, contrary to the definition of c. Hence $c = b$, and $J_2[\eta; a, b]$ is positive definite on $\mathfrak{D}_0[a, b]$, thus completing the proof of the theorem.

If the roles of $t = a$ and $t = b$ are interchanged in the above proof, one has the following result.

Corollary. *If* $[a, b] \subset I$, *then* $J[\eta; a, b]$ *is positive definite on* $\mathfrak{D}_0[a, b]$ *if and only if* $B(t) \geq 0$ *for* t a.e. *on* $[a, b]$, *and there is no value on* $[a, b)$ *which is conjugate to* $t = b$.

For each integer q, $0 \leq q \leq n$, let I_q denote the subset of $[a, b]$ on which $B(t)$ is of rank $n - q$. The set I_q is measurable, as it is the sum of subsets on which individual minors of $B(t)$ of order $n - q$ are nonsingular, while all minors of higher order are singular. In view of this dichotomy of I_q it follows readily that on I_q there exists an $n \times q$ measurable matrix $\pi(t; q)$ such that $B(t)\pi(t; q) = 0$, $\pi^*(t; q)\pi(t; q) = E_q$, and the $(n + q) \times (n + q)$ matrix

$$(4.16) \qquad M(t; q) = \begin{bmatrix} B(t) & \pi(t; q) \\ \pi^*(t; q) & 0 \end{bmatrix}, \qquad t \in I_q,$$

is nonsingular. For $t \in I_q$ let $R(t; q)$ denote the $n \times n$ matrix such that

$$(4.17) \qquad M^{-1}(t; q) = \begin{bmatrix} R(t; q) & \pi(t; q) \\ \pi^*(t; q) & 0 \end{bmatrix}, \qquad t \in I_q.$$

The matrix $R(t; q)$ is hermitian, measurable on I_q, and is the E. H. Moore generalized inverse of $B(t)$, (see Appendix B, Section 3); in particular, $R(t; q)B(t)R(t; q) = R(t; q)$ for $t \in I_q$. On $[a, b]$ define $\pi(t)$, $M(t)$, and $R(t)$ as $\pi(t; q)$, $M(t; q)$, and $R(t; q)$ for $t \in I_q$; the matrix $R(t)$ is $n \times n$ for all $t \in [a, b]$, but $M(t)$ is $(n + q) \times (n + q)$ and $\pi(t)$ is $n \times q$ for $t \in I_q$. Now if

$\eta \in \mathfrak{D}[a, b]:\zeta$, then $\check{\zeta}(t) = \zeta(t) - \pi(t)\pi^*(t)\zeta(t)$ is such that also $\eta \in \mathfrak{D}[a, b]:\check{\zeta}$. Moreover, $\check{\zeta}(t) = R(t)L[\eta](t)$ for $t \in [a, b]$ and

$$\zeta^*(t)B(t)\zeta(t) = \{L[\eta](t)\}^*R(t)L[\eta](t), \quad \textit{for} \quad t \in [a, b]. \tag{4.18}$$

The conditions of hypothesis ($\mathfrak{H}$) are not strong enough to imply that $R(t; q)$ is integrable on I_q, and in general the matrix function $R(t)$ is not integrable on the interval $[a, b]$. Since the elements of $M(t; q)$ belong to $\mathfrak{L}^\infty$ on I_q, however, it follows readily that $R(t) \in \mathfrak{L}_{nn}{}^\infty[a, b]$ whenever the following additional hypothesis is satisfied.

($\mathfrak{H}_1[a, b]$) *There exists a positive constant* $k = k[a, b]$ *such that* $|\det M(t)| \geq k$, *for* t a.e. *on* $[a, b]$.

In connection with this hypothesis it is to be noted that $|\det M(t)|$ is the absolute value of the product of the nonzero roots of the characteristic equation $\det[\lambda E - B(t)] = 0$, (e.g., see Problem 12 of E.2 in Appendix E). As $B(t) \in \mathfrak{L}_{nn}{}^\infty[a, b]$ for arbitrary compact subintervals $[a, b]$ of I, hypothesis ($\mathfrak{H}_1[a, b]$) is therefore equivalent to the condition that for a given $[a, b] \subset I$ there exists a corresponding positive constant k_0 such that $B^2(t) \pm k_0 B(t) \geq 0$ for t a.e. on $[a, b]$. If $B(t) \geq 0$ this condition reduces to the restriction that there is a corresponding positive constant k_0 such that $B^2(t) - k_0 B(t) \geq 0$ for t a.e. on $[a, b]$.

Problems VII.4

1.[ss] Suppose that $[a, b] \subset I$, $B(t) \geq 0$ for t a.e. on $[a, b]$, and (2.2) is normal on every subinterval $[a, b_1]$, $b_1 \in (a, b]$, and on every subinterval $[a_1, b]$, $a_1 \in [a, b)$. By a method similar to that employed in Theorem V.3.3 for the second order system, show that $J[\eta; a, b]$ is positive definite on $\mathfrak{D}_0[a, b]$ if and only if one of the following conditions hold:

(i) (2.2) is disconjugate on $[a, b]$;

(ii) if $Y_a(t) = (U_a(t); V_a(t))$ is a conjoined basis for (2.2) with $U_a(a) = 0$, $V_a(a)$ non-singular, then $U_a(t)$ is non-singular for $t \in (a, b]$;

(iii) if $Y_b(t) = (U_b(t); V_b(t))$ is a conjoined basis for (2.2) with $U_b(b) = 0$, $V_b(b)$ non-singular, then $U_b(t)$ is non-singular for $t \in [a, b)$;

(iv) there exists a conjoined basis $Y(t) = (U(t); V(t))$ for (2.2) with $U(t)$ non-singular on $[a, b]$.

Hints. Note that the positive definiteness of $J[\eta; a, b]$ on $\mathfrak{D}_0[a, b]$ implies condition (i) by the Corollary to Lemma 4.1. Moreover, as (i) implies that there is no point on $(a, b]$ conjugate to $t = a$, under the added assumption that (2.2) is normal on every subinterval $[a, b_1]$, $b_1 \in (a, b]$, condition (ii) is a consequence of (i). As in the proof of Theorem 4.4, show that whenever (ii) holds then $J[\eta; a_1, b]$ is positive definite on $\mathfrak{D}_0[a_1, b]$ for $a_1 \in (a, b)$, and under the added assumption that (2.2) is normal on every subinterval $[a_1, b]$, $a_1 \in [a, b)$, conclude that $U_b(t)$ is nonsingular for $t \in (a, b)$. From the fact that the matrix function $\{U_a; V_a \mid U_b; V_b\}$ is constant on $[a, b]$ conclude that the nonsingularity of $U_b(a)$ is a consequence of the nonsingularity of $U_a(b)$, thus completing the proof that (ii) implies (iii). To show that (iii) implies (iv), first note that by an argument similar to that just indicated the conclusion (iii) implies (ii). Moreover, if $Y_a(t)$ and $Y_b(t)$ are as in (ii) and (iii), respectively, then the constant matrix function $\{U_a; V_a \mid U_b; V_b\}$ is nonsingular, and upon suitable modification of the initial values $V_a(a)$ or $V_b(b)$ one may obtain $\{U_a; V_a \mid U_b; V_b\}(t) \equiv -E$. With such a choice, by direct computation show that $Y(t) = (U(t); V(t)) = (U_a(t) + U_b(t); V_a(t) + V_b(t))$ is a conjoined basis for (2.2), and $U(a) = U_b(a)$, $U(b) = U_a(b)$ are individually nonsingular. Proceed to show that $U(t)$ is nonsingular on (a, b) by the following argument. If $c \in (a, b)$ and ξ is a vector such that $U(c)\xi = 0$, show that if $(\eta(t), \zeta(t)) = (U_a(t)\xi, V_a(t)\xi)$ for $t \in [a, c]$, $(\eta(t), \zeta(t)) = (-U_b(t)\xi, -V_b(t)\xi)$ for $t \in (c, b]$, then $\eta \in \mathfrak{D}_0[a, b]:\zeta$, and application of (4.4″) to the individual intervals $[a, c]$, $[c, b]$ yields the relation

$$J[\eta; a, b] = \xi^* U_a^*(c) V_a(c)\xi - \xi^* U_b^*(c) V_b(c)\xi.$$

Utilize appropriately the relations $U_a^* V_a = V_a^* U_a$ and $U_a\xi = -U_b\xi$ for $t = c$ to conclude that

$$J[\eta; a, b] = \xi^*[U_a^*(c) V_b(c) - V_a^*(c) U_b(c)]\xi = -|\xi|^2 \leq 0.$$

On the other hand, if $h(t) = \xi$ for $t \in [a, c]$, and $h(t) = -U_a^{-1}(t) U_b(t)\xi$ for $t \in (c, b]$, show that Lemma 4.2 implies that

$$J[\eta; a, b] = \int_a^b [\zeta - Vh]^* B(t)[\zeta - Vh]\, dt,$$

and hence $J[\eta; a, b] \geq 0$, in view of the hypothesis that $B(t) \geq 0$ for t a.e. on $[a, b]$. Finally, as in Theorem 4.4, conclude that the positive definiteness of $J[\eta; a, b]$ on $\mathfrak{D}_0[a, b]$ is a consequence of (iv) and the condition that $B(t) \geq 0$ for t a.e. on $[a, b]$.

2.[ss] If $[a, b] \subset I$, $B(t) \geq 0$ for t a.e. on $[a, b]$, and there exists a conjoined basis $Y(t) = (U(t); V(t))$ for (2.2) with $U(t)$ nonsingular on the open interval (a, b), show that $J[\eta; a, b]$ is non-negative definite on $\mathfrak{D}_0[a, b]$; moreover, if $\eta_1 \in \mathfrak{D}_0[a, b]:\zeta_1$ with $\eta_1(t) \not\equiv 0$ on $[a, b]$ and $J[\eta_1; a, b] = 0$, then there exists a solution $(u(t); v(t))$ of (2.2) with $u(t) = \eta_1(t)$ for $t \in [a, b]$, so that $t = b$ and $t = a$ are mutually conjugate points. If, in addition, (2.2) is normal on every subinterval $[a, b_1]$, $b_1 \in (a, b]$, and on every subinterval $[a_1, b]$, $a_1 \in [a, b)$, then for an η_1 satisfying the above conditions one has that $|\eta_1(t)| > 0$ for $t \in (a, b)$.

Hint. To prove that the stated conditions imply that $J[\eta; a, b]$ is non-negative definite on $\mathfrak{D}_0[a, b]$, suppose that there exists an $\eta_0 \in \mathfrak{D}_0[a, b]$ such that $J[\eta_0; a, b] < 0$, and obtain a contradiction by an argument similar to that used in the proof of Theorem 4.5, now modifying $\eta_0(t)$ in neighborhoods of both $t = a$ and $t = b$. With the aid of Corollary 1 to Theorem 4.1, conclude that if $\eta_1 \in \mathfrak{D}_0[a, b]:\zeta_1$ with $\eta_1(t) \not\equiv 0$ on $[a, b]$, and $J[\eta_1; a, b] = 0$, then there exists a solution $(u(t); v(t))$ of (2.2) with $u(t) \equiv \eta_1(t)$, so that $t = b$ and $t = a$ are conjugate points. If $c \in (a, b)$ and $\eta_1(c) = 0$, consider the associated vector functions $\eta_a(t) = \eta_1(t)$ for $t \in [a, c]$, $\eta_a(t) = 0$ for $t \in (c, b]$, and $\eta_b(t) = \eta_1(t)$ for $t \in [c, b]$, $\eta_b(t) = 0$ for $t \in [a, c)$. Conclude that $J[\eta_a; a, b] = 0 = J[\eta_b; a, b]$, and that not both $\eta_a(t)$ and $\eta_b(t)$ are identically zero on $[a, b]$. Use Corollary 1 to Theorem 4.1 to conclude that there exists a solution $(u_1(t); v_1(t))$ of (2.2) with $u_1(t) \not\equiv 0$ on $[a, b]$, while $u_1(a) = 0 = u_1(b)$ and either $u(t) \equiv 0$ on $[a, c]$, or $u(t) \equiv 0$ on $[c, b]$, and show that the existence of such a solution is impossible when the added normality conditions hold.

3. Suppose that $B(t) \geq 0$ for t a.e. on a compact subinterval $[a, b]$ of I, and that there is no point on $(a, b]$ which is conjugate to $t = a$. Let $t_1, \ldots, t_l$, where $a < t_l < t_{l-1} < \cdots < t_1 < b$, be the points of discontinuity of the function $d[a, t]$, and for brevity set $t_{l+1} = a$, $t_0 = b$, and $d_\gamma = d[a, t_\gamma]$, $(\gamma = 0, 1, \ldots, l)$. Moreover, let $Y(t) = (U(t); V(t))$ be a conjoined basis for (2.2) satisfying

$$U(a) = 0, \qquad V^*(a)V(a) = E_n, \qquad U_{i\beta}(t) \equiv 0 \quad \text{for} \quad t \in [a, t_\gamma],$$
$$(\beta = 1, \ldots, d_\gamma; \gamma = 0, 1, \ldots, l),$$

and $Y_2(t) = (U_2(t); V_2(t))$ the conjoined basis for (2.2) satisfying

$$U_2(a) = V(a), \qquad V_2(a) = 0.$$

If $Y(t; \rho) = (U(t; \rho); V(t; \rho))$ is defined for real values ρ as

$$U_{i\beta}(t; \rho) = U_{2i\beta}(t) + \rho U_{i\beta}(t), \qquad V_{i\beta}(t; \rho) = V_{2i\beta}(t) + \rho V_{i\beta}(t),$$
$$(i = 1, \ldots, n; \beta = 1, \ldots, d_l),$$

$$U_{ij}(t; \rho) = U_{ij}(t), \qquad V_{ij}(t; \rho) = V_{ij}(t),$$
$$(i = 1, \ldots, n; j = d_l + 1, \ldots, n),$$

prove that $Y(t; \rho)$ is a conjoined basis for (2.2), and there exists a ρ_0 such that if $\rho > \rho_0$ then $U(t; \rho)$ is nonsingular on $(a, b]$.

Hint. For $\gamma = 0, 1, \ldots, l$, let $(U(t \mid \gamma); V(t \mid \gamma))$ be the solution of (2.2_M) specified by

$$U_{i\beta}(t \mid \gamma) = U_{2i\beta}(t), \qquad V_{i\beta}(t \mid \gamma) = V_{2i\beta}(t), \qquad (i = 1, \ldots, n; \beta = 1, \ldots, d_\gamma),$$

$$U_{ij}(t \mid \gamma) = U_{ij}(t), \qquad V_{ij}(t \mid \gamma) = V_{ij}(t), \qquad (i = 1, \ldots, n; j = d_\gamma + 1, \ldots, n),$$

and note that from Lemma 3.3 it follows that a value $t = \tau$ on $(t_{\gamma+1}, t_\gamma]$, $(\gamma = 0, 1, \ldots, l)$, is conjugate to $t = a$ if and only if one of the following conditions holds:

(i) the $n \times (n - d_\gamma)$ matrix $[U_{ik}(\tau)]$, $(i = 1, \ldots, n;\ k = d_\gamma + 1, \ldots, n)$, has rank less than $n - d_\gamma$;

(ii) the matrix $U(\tau \mid \gamma)$ is singular;

(iii) the $2n \times 2n$ matrix

$$\begin{bmatrix} U(a) & U_2(a) \\ U(\tau) & U_2(\tau) \end{bmatrix}$$

has rank less than $2n - d_\gamma$.

Proceed to prove the stated result by induction, noting first that the nonexistence on $(a, b]$ of a point conjugate to $t = a$ implies that $U(t \mid \gamma)$ is nonsingular on $(t_{\gamma+1}, t_\gamma]$, $(\gamma = 0, 1, \ldots, l)$. In particular, $U(t; \rho) = U(t \mid l)$ on $(a, t_l]$, and hence on this interval $U(t; \rho)$ is nonsingular for arbitrary ρ. Suppose that for $\gamma = \sigma$, $(1 \leq \sigma \leq l)$, there exists a value ρ_σ such that $U(t; \rho_\sigma)$ is nonsingular on $(a, t_\sigma]$, and let ε be such that $t_\sigma + \varepsilon \in (t_\sigma, t_{\sigma-1})$ and $U(t; \rho_\sigma)$ is nonsingular on $(a, t_\sigma + \varepsilon]$. If $s \in (a, t_\sigma + \varepsilon]$ and ξ is a vector such that $U(s; \rho)\xi = 0$, let $(u(t); v(t)) = (U(t; \rho)\xi; V(t; \rho)\xi)$, and note that $u(t) = U(t; \rho_\sigma)h(t)$, with $h(t)$ a.c. on $[a, s]$. With the aid of the Corollary to Lemma 4.2 show that $J[u; a, s] + u^*(a)V(a; \rho_\sigma)\xi \geq 0$. On the other hand, note that direct integration yields $J[u; a, s] = -u^*(a)v(a)$, and from the initial values of $U(t; \rho)$, $V(t; \rho)$ at $t = a$ deduce that

$$J[u; a, s] + u^*(a)V(a; \rho_\sigma)\xi = -(\rho - \rho_\sigma)(|\xi_1|^2 + \cdots + |\xi_{d_l}|^2).$$

Conclude that if $\rho > \rho_\sigma$ then $U(t; \rho)$ is nonsingular on $(a, t_\sigma + \varepsilon]$. Note that for $t \in [t_\sigma + \varepsilon, t_{\sigma-1}]$ the determinant of $U(t; \rho)$ is a polynomial in ρ of degree $d_l - d_{\sigma-1}$, with leading coefficient equal to $\det U(t \mid \sigma - 1)$, which is different from zero; conclude that for ρ sufficiently large we have $\det U(t; \rho) \neq 0$ for $t \in [t_\sigma + \varepsilon, t_{\sigma-1}]$, and consequently $U(t; \rho)$ is nonsingular on $(a, t_{\sigma-1}]$.

4.[ss] If $B(t) \geq 0$ for t a.e. on a compact subinterval $[a, b]$ of I, prove that there exist corresponding constants $l_0 > 0$, $l_1 \geq 0$, such that

$$(4.19) \quad J[\eta; a, b] \geq l_0 \int_a^b |\eta'(t)|^2\, dt - l_1 \int_a^b |\eta(t)|^2\, dt, \quad \textit{for} \quad \eta \in \mathfrak{D}[a, b].$$

Hint. As $B(t) \in \mathfrak{L}_{nn}{}^{\infty}[a, b]$, there exists a constant k such that $\nu[B(t)] \leq 1/k$ for t a.e. on $[a, b]$. For k such a constant, show that the condition $B(t) \geq 0$ a.e. on $[a, b]$ implies that $B(t) - kB^2(t) \geq 0$ a.e. on $[a, b]$, and conclude that if $\eta \in \mathfrak{D}[a, b]:\zeta$ then $\zeta^* B\zeta \geq k\,|\eta' - A\eta|^2$ a.e. on $[a, b]$. Use elementary inequalities and the fact that $C(t) \in \mathfrak{L}_{nn}{}^{\infty}[a, b]$ to obtain the existence of constants $l_0 > 0$, $l_1 \geq 0$ satisfying (4.19).

5. If $B(t)$ is continuous and of constant rank on I, show that the matrix $\pi(t)$ defined at the end of Section 4 may be chosen as continuous on this interval; moreover, in this case $R(t)$ is continuous on I and hypothesis $(\mathfrak{H}_1[a, b])$ holds.

6.[ss] If $B(t) \geq 0$ for t a.e. on a compact subinterval $[a, b]$ of I, and there exists a positive constant k_0 such that $B^2(t) - k_0 B(t) \geq 0$ for t a.e. on $[a, b]$, show that there exists a positive constant l such that

$$J[\eta; a, b] \leq l \int_a^b \{|\eta'(t)|^2 + |\eta(t)|^2\}\, dt, \quad \textit{for} \quad \eta \in \mathfrak{D}[a, b]. \tag{4.20}$$

7. Suppose that hypothesis $(\mathfrak{H})$ holds, and $A(t) \equiv 0$, $B(t) \geq 0$ for t a.e. on I. If $[a, b]$ is a compact subinterval of I, prove that the following conditions are equivalent:

(i) the differential system

$$-v' + C(t)u = 0, \qquad u' - B(t)v = 0,$$

has order of abnormality q on $[a, b]$;

(ii) the linear manifold of constant n-dimensional vectors π satisfying $B(t)\pi = 0$ a.e. on $[a, b]$ has dimension q;

(iii) the rank of the hermitian matrix $\int_a^b B(t)\, dt$ is $n - q$.

8.[s] Suppose that $R(t)$ is a nonsingular hermitian matrix function such that $R(t)$ and $R^{-1}(t)$ belong to $\mathfrak{L}_{nn}{}^{\infty}[a, b]$, with $R(t) > 0$ for t a.e. on $[a, b]$. Show that if $\eta \in \mathfrak{A}_n[a, b]$ with $\eta' \in \mathfrak{L}_n{}^2[a, b]$, then

$$\int_a^b \eta^{*\prime}(t)R(t)\eta'(t)\, dt \geq 4\eta^*(c)\left[\int_a^b R^{-1}(s)\, ds\right]^{-1} \eta(c), \quad \textit{for} \quad c \in (a, b]; \tag{4.21}$$

moreover, the inequality holds in (4.21) if $\eta(t) \not\equiv 0$ and for each $t_0 \in (a, b)$ with $|\eta(t_0)| \neq 0$ there is a neighborhood $(t_0)_\delta = (t_0 - \delta, t_0 + \delta)$ on which there is defined a continuous vector function ζ such that $\zeta(t) = R(t)\eta'(t)$ for t a.e. on $(t_0)_\delta$.

Hint. Note that hypothesis $(\mathfrak{H}_\omega)$ holds for $R(t)$ satisfying the conditions of the problem, with $Q(t) \equiv 0$, $P(t) \equiv 0$, and the corresponding functional (2.5) is $J_0[\eta; a, b] = \int_a^b \eta^{*\prime}(t)R(t)\eta'(t)\, dt$. Moreover, for the associated system (2.2) conjoined bases are given by $Y_a(t) = (U_a(t); V_a(t)) = (U(t); E)$, $Y_b(t) = (U_b(t); V_b(t)) = (U(b) - U(t), -E)$, where $U(t) = \int_a^t R^{-1}(s)\, ds$, and each of the matrix functions $U_a(t)$, $U_b(t)$ is nonsingular for $t \in (a, b)$. For $c \in (a, b)$, let $u_a(t) = U_a(t)U_a^{-1}(c)\eta(c)$, $u_b(t) = U_b(t)U_b^{-1}(c)\eta(c)$, and with the aid of Corollary 2 to Theorem 4.1 conclude that

$$J_0[\eta; a, c] \geq J_0[u_a; a, c] = \eta^*(c)U^{-1}(c)\eta(c), \tag{4.22$'$}$$

$$J_0[\eta; c, b] \geq J_0[u_b; c, b] = \eta^*(c)[U(b) - U(c)]^{-1}\eta(c), \tag{4.22$''$}$$

and the equality sign in (4.22′) or (4.22″) holds only if $\eta(t) \equiv u_a(t)$ or $\eta(t) \equiv u_b(t)$ on the respective interval. Consequently, relations (4.22′), (4.22″) imply the inequality

$$J_0[\eta; a, b] \geq \eta^*(c)\{U^{-1}(c) + [U(b) - U(c)]^{-1}\}\eta(c),$$

and (4.21) follows from the fact that if A, B are hermitian matrices such that $A > 0$, $B - A > 0$, then the hermitian matrix $A^{-1} + (B - A)^{-1} - 4B^{-1}$ is non-negative definite. Show that the stated result for hermitian matrices is equivalent to this result for the special case $B = E$, and if A and $E - A$ are positive definite hermitian matrices the desired result follows from the identity $A^{-1} + (E - A)^{-1} - 4E = (E - 2A)^2 A^{-1}(E - A)^{-1}$, and the fact that $(E - 2A)^2$, A^{-1} and $(E - A)^{-1}$ are individually non-negative definite hermitian matrices that are mutually commutative under multiplication. It is to be remarked that for the proofs of the stated matrix relations the results of Problems 2, 8, 11 of E.2 in Appendix E are pertinent.

9.[s] Suppose that $\eta_j \in \mathfrak{D}[a, b]: \zeta_j, j = 1, 2, \ldots$, and there exists a constant k such that $\int_a^b |\zeta_j(t)|^2 \, dt \leq k$, $(j = 1, 2, \ldots)$; moreover, the sequence $\{\eta_j(a)\}$ is bounded. Prove that there exists a subsequence $\{\eta_m(t)\}$ of $\{\eta_j(t)\}$ and a vector function $\eta_0(t) \in \mathfrak{D}[a, b]$ such that $\{\eta_m(t)\} \to \eta_0(t)$ uniformly on $[a, b]$. Moreover, if $B(t) \geq 0$ for t a.e. on $[a, b]$, then

$$\liminf_{m \to \infty} J[\eta_m; a, b] \geq J[\eta_0; a, b]. \tag{4.23}$$

Hint. From results on the Lebesgue integral, conclude that there exists a subsequence $\{\eta_m(t)\}$ of $\{\eta_j(t)\}$ satisfying with the associated vector functions $\{\zeta_m(t)\}$ the following conditions: $\{\eta_m(a)\}$ converges to a vector ξ; the sequence $\{\zeta_m(t)\}$ converges weakly in $\mathfrak{L}_n^2[a, b]$ to a vector function $\zeta_0(t) \in \mathfrak{L}_n^2[a, b]$; that is, if $w(t) \in \mathfrak{L}_n^2[a, b]$ then $\lim_{m \to \infty} \int_a^b w^*(t)\, \zeta_m(t) \, dt = \int_a^b w^*(t)\, \zeta_0(t) \, dt$. For $U(t)$ a fundamental matrix of $L[u](t) = 0$, and $\hat{\eta}_m(t) = U^{-1}(t)\eta_m(t)$, use the relations

$$\hat{\eta}_m(t) = \hat{\eta}_m(t_1) + \int_{t_1}^{t} U^{-1}(s)B(s)\zeta_m(s) \, ds, \qquad (t, t_1) \in [a, b] \times [a, b],$$

to show that the vector functions $\hat{\eta}_m(t)$ satisfy an inequality of the form $|\hat{\eta}_m(t_2) - \hat{\eta}_m(t_1)|^2 \leq k_1 \, |t_2 - t_1| \int_a^b |\zeta_m(s)|^2 \, ds$, and hence are uniformly equi-continuous on $[a, b]$. Conclude that if

$$\hat{\eta}_0(t) = U^{-1}(a)\xi + \int_a^t U^{-1}(s)B(s)\zeta_0(s)\,ds$$

then $\{\hat{\eta}_m(t)\}$ converges uniformly to $\hat{\eta}_0(t)$ on $[a, b]$, and hence $\{\eta_m(t)\} = \{U(t)\hat{\eta}_m(t)\}$ converges uniformly to $\eta_0(t) = U(t)\hat{\eta}_0(t)$ on $[a, b]$; moreover, $\eta_0 \in \mathfrak{D}[a, b]:\zeta_0$. Proceed to show that if $B(t) \geq 0$ for t a.e. on $[a, b]$ then

$$J[\eta_m; a, b] \geq J[\eta_0; a, b] + J[\eta_m - \eta_0, \eta_0; a, b] + J[\eta_0, \eta_m - \eta_0; a, b] + \int_a^b [\eta_m - \eta_0]^* C(t)[\eta_m - \eta_0]\,dt,$$

and deduce relation (4.23).

10.[ss] Suppose that on a compact subinterval $[a, b]$ of I the system (2.2) is normal and disconjugate; moreover, $B(t) \geq 0$ for t a.e. on $[a, b]$. If $Y_{ab}(t) = (U_{ab}(t); V_{ab}(t))$ is the solution of (2.2_M) determined by the initial conditions $U_{ab}(a) = E$, $U_{ab}(b) = 0$, and $Y(t) = (U(t); V(t))$ is a solution of (2.2_M) with $U(a) = E$, $V(a) > V_{ab}(a)$, prove that $Y(t)$ is a conjoined basis for (2.2) with $U(t)$ non-singular on $[a, b]$. Similarly, if $Y_{ba}(t) = (U_{ba}(t); V_{ba}(t))$ is the solution of (2.2_M) determined by the initial conditions $U_{ba}(b) = E$, $U_{ba}(a) = 0$, and $Y(t) = (U(t); V(t))$ is a solution of (2.2_M) with $U(b) = E$, $V(b) < V_{ba}(b)$, prove that $Y(t)$ is a conjoined basis for (2.2) with $U(t)$ nonsingular on $[a, b]$.

Hint. With the aid of Theorem 4.5, show that the first conclusion is a direct consequence of the result to which Theorem 4.3 reduces when (2.2) is normal on $[a, b]$.

5. DISCONJUGACY CRITERIA

The results of the previous section may be coalesced to present the following conditions for disconjugacy.

Theorem 5.1. *If $[a, b] \subset I$, and $B(t) \geq 0$ for t a.e. on $[a, b]$, then each of the following conditions is necessary and sufficient for (2.2) to be disconjugate on $[a, b]$.*

(i) *$J[\eta; a, b]$ is positive definite on $\mathfrak{D}_0[a, b]$.*
(ii) *There is no point on $(a, b]$ conjugate to $t = a$.*
(iii) *There is no point on $[a, b)$ conjugate to $t = b$.*
(iv) *There exists a conjoined basis $Y(t) = (U(t); V(t))$ for (2.2) with $U(t)$ nonsingular on $[a, b]$.*

(v) *There exists an* a.c. $n \times n$ *hermitian matrix function* $W(t)$, $t \in [a, b]$, *which is a solution of the Riccati matrix differential equation*

(5.1)
$$\mathfrak{R}[W] \equiv W' + WA(t) + A^*(t)W + WB(t)W - C(t) = 0, \qquad t \in [a, b].$$

Under the hypothesis that $[a, b] \subset I$, and $B(t) \geq 0$ for t a.e. on $[a, b]$, we have the following results: (a) Theorem 4.5 implies that (i) is equivalent to each of the conditions (ii) and (iii); (b) Theorem 4.4 implies that (i) is equivalent to (iv); (c) disconjugacy on $[a, b]$ implies (ii), and hence (a) implies (i), which in turn implies disconjugacy on $[a, b]$ by the Corollary to Theorem 4.1. Consequently, under the hypothesis that $[a, b] \subset I$, and $B(t) \geq 0$ for t a.e. on $[a, b]$, we have that each of the conditions (i)–(iv) is necessary and sufficient for (2.2) to be disconjugate on $[a, b]$. Finally, irrespective of the condition that $B(t) \geq 0$ for t a.e. on $[a, b]$, (v) is equivalent to (iv). Indeed, from Problem II.3:8 it follows that there is a solution $(U(t); V(t))$ of (2.2_M) with $U(t)$ nonsingular on $[a, b]$ if and only if there is on this interval a solution $W(t)$ of (5.1) such that $W(t) = V(t)U^{-1}(t)$ for one value, and consequently for all values, of t on $[a, b]$. Moreover, $W(t)$ is hermitian for all $t \in [a, b]$ if and only if it is hermitian for one value τ, which is equivalent to the constant matrix function $V^*U - U^*V$ being equal to 0, so that $(U(t); V(t))$ is a conjoined basis for (2.2).

Now suppose that for $\alpha = 1, 2$ the $n \times n$ matrix functions $A_\alpha(t)$, $B_\alpha(t)$, $C_\alpha(t)$ satisfy hypothesis $(\mathfrak{H})$ on a given interval I on the real line. If $I_0 \subset I$ the corresponding classes $\mathfrak{D}(I_0)$, $\mathfrak{D}_0(I_0)$ will be denoted by $\mathfrak{D}_\alpha(I_0)$, $\mathfrak{D}_{\alpha 0}(I_0)$. In particular, if

$$(5.2) \qquad A_1(t) \equiv A_2(t), \qquad B_1(t) \equiv B_2(t), \quad \text{for} \quad t \in I,$$

then clearly $\mathfrak{D}_1(I_0) = \mathfrak{D}_2(I_0)$ and $\mathfrak{D}_{10}(I_0) = \mathfrak{D}_{20}(I_0)$ for arbitrary $I_0 \subset I$. However, these classes may be identical without relations (5.2) holding. Indeed, for a system (2.2) equivalent under the substitution (2.7) to a second order matrix differential equation (2.6) satisfying the condition $(\mathfrak{H}_\omega)$, the class $\mathfrak{D}(I_0)$ consists of vector functions $\eta(t)$ which for arbitrary compact subintervals $[a, b]$ of I_0 are such that $\eta \in \mathfrak{A}_n[a, b]$ and $\eta' \in \mathfrak{L}_n{}^2[a, b]$. Similarly, for a system (2.2) that is equivalent under the substitution (2.13) to a system (2.12) which satisfies hypothesis $(\mathfrak{H}_\Omega)$, the class $\mathfrak{D}(I_0)$ consists of vector functions $\eta(t)$ which for arbitrary compact subintervals $[a, b]$ of I_0 are such that $\eta \in \mathfrak{A}_n[a, b]$, $\eta' \in \mathfrak{L}_n{}^2[a, b]$, and $\Phi(t, \eta(t), \eta'(t)) = 0$ for $t \in I_0$. A similar statement holds for a general system (2.2) which satisfies hypotheses $(\mathfrak{H})$, $(\mathfrak{H}_1[a, b])$, with now $\Phi(t, \eta(t), \eta'(t)) \equiv \pi^*(t)L[\eta](t) = 0$, where $\pi(t)$ is defined as in Section 4.

For $\alpha = 1, 2$ we have the respective differential systems

$$(5.3_\alpha) \quad \begin{aligned} L_1^\alpha[u, v](t) &\equiv -v'(t) - A_\alpha^*(t)v(t) + C_\alpha(t)u(t) = 0, \\ L_2^\alpha[u, v](t) &\equiv u'(t) - A_\alpha(t)u(t) - B_\alpha(t)v(t) = 0, \end{aligned} \qquad t \in I,$$

and corresponding functionals

$$(5.4_\alpha) \qquad J_\alpha[\eta; a, b] = \int_a^b [\zeta^* B_\alpha(t)\zeta + \eta^* C_\alpha(t)\eta]\, dt.$$

In particular, if

$$(5.5) \qquad \mathfrak{D}_1(I) = \mathfrak{D}_2(I),$$

then for $[a, b] \subset I$ the difference functional

$$(5.6) \qquad J_{1,2}[\eta; a, b] = J_1[\eta; a, b] - J_2[\eta; a, b]$$

is well-defined for $\eta \in \mathfrak{D}[a, b]$, where for brevity we write $\mathfrak{D}[a, b]$ for the common value of $\mathfrak{D}_1[a, b]$ and $\mathfrak{D}_2[a, b]$.

Theorem 5.2. *Suppose that for $\alpha = 1, 2$ the $n \times n$ matrix functions $A_\alpha(t)$, $B_\alpha(t)$, $C_\alpha(t)$ satisfy $(\mathfrak{H})$, and $B_2(t) \geq 0$ for t a.e. on I; also, $\mathfrak{D}_1[a, b] = \mathfrak{D}_2[a, b] = \mathfrak{D}[a, b]$ for arbitrary subintervals $[a, b]$ of I, and $J_{1,2}[\eta; a, b]$ is non-negative definite on $\mathfrak{D}_0[a, b] = \mathfrak{D}_{10}[a, b] = \mathfrak{D}_{20}[a, b]$. If $[a_0, b_0] \subset I$ and (5.3_2) is disconjugate on $[a_0, b_0]$, then (5.3_1) is also disconjugate on $[a_0, b_0]$. Moreover, if $J_{1,2}[\eta; a, b]$ is positive definite on $\mathfrak{D}_0[a, b]$ for arbitrary $[a, b] \subset I$ then the solutions of (5.3_2) oscillate more rapidly than the solutions of (5.3_1) in the following sense: if t_1 and t_2 are mutually conjugate with respect to the system (5.3_1), then any conjoined basis $(U(t); V(t))$ for (5.3_2) is such that $U(t)$ is singular for at least one value on the open interval (t_1, t_2).*

As in the proof of Theorem 4.2, the condition that $J_{1,2}[\eta; a, b]$ is non-negative definite on $\mathfrak{D}_0[a, b]$ implies that $B_1(t) - B_2(t) \geq 0$ for t a.e. on $[a, b]$, and thus the hypotheses of the theorem imply that $B_1(t) \geq B_2(t) \geq 0$ for t a.e. on I. Now if (5.3_2) is disconjugate on $[a_0, b_0]$ then $J_2[\eta; a_0, b_0]$ is positive definite on $\mathfrak{D}_0[a_0, b_0]$ by Theorem 5.1. The added condition that $J_{1,2}[\eta; a_0, b_0]$ is non-negative definite on $\mathfrak{D}_0[a_0, b_0]$ then implies that $J_1[\eta; a_0, b_0]$ is positive definite on $\mathfrak{D}_0[a_0, b_0]$, and in turn Theorem 5.1 implies that (5.3_1) is disconjugate on $[a_0, b_0]$.

Now suppose that t_1 and t_2 are mutually conjugate points with respect to system (5.3_1). If $t_1 < t_2$, and $(u(t); v(t))$ is a solution of (5.3_1) with $u(t) \not\equiv 0$ on $[t_1, t_2]$, $u(t_1) = 0 = u(t_2)$, then $u \in \mathfrak{D}_0[t_1, t_2]$ and $J_1[u; t_1, t_2] = 0$. Consequently, if $J_{1,2}[\eta; a, b]$ is positive definite on $\mathfrak{D}_0[a, b]$ for arbitrary $[a, b] \subset I$, we have that $J_2[u; t_1, t_2] < 0$, and the final conclusion of the theorem is a direct consequence of the result of Problem VII.4:2.

Theorem 5.3. *If* $[a, b] \subset I$, *and* $B(t) \geq 0$ *for* t a.e. *on* $[a, b]$, *then* (2.2) *is disconjugate on* $[a, b]$ *if and only if one of the following conditions holds:*

(i) *there exists on* $[a, b]$ *a nonsingular* $n \times n$ *matrix function* $U(t)$ *such that* $U \in \mathfrak{D}[a, b]: V$ *with a Lipschitzian matrix function* $V(t)$, *while* $V^*(t)U(t) - U^*(t)V(t) \equiv 0$, *and* $U^*(t)L_1[U, V](t) \geq 0$ *for* t a.e. *on* $[a, b]$;

(ii) *there exists an* $n \times n$ *hermitian matrix function* $W(t)$ *which is Lipschitzian and satisfies* $\mathfrak{R}[W](t) \leq 0$ *for* t a.e. *on* $[a, b]$.

If (2.2) is disconjugate on $[a, b]$ then by Theorem 5.1 there exists a conjoined basis $(U(t); V(t))$ for (2.2) with $U(t)$ nonsingular on $[a, b]$, so that conclusion (i) holds for such a matrix function $U(t)$, and conclusion (ii) holds for $W(t)$ equal to the corresponding matrix function $V(t)U^{-1}(t)$. On the other hand, if $U(t)$ is a matrix function satisfying condition (i), and $M(t) = U^*(t)L_1[U, V](t)$, then $M(t) \in \mathfrak{L}_{nn}^{\infty}[a, b]$ and $(U(t); V(t))$ is a conjoined basis for a system (5.3_2) with

$$A_2(t) \equiv A(t), \qquad B_2(t) \equiv B(t),$$
$$C_2(t) = C(t) - U^{*-1}(t)M(t)U^{-1}(t) \leq C(t).$$

As $U(t)$ is nonsingular on $[a, b]$, Theorem 5.1 then implies that this system (5.3_2) is disconjugate on $[a, b]$. Since $C(t) \geq C_2(t)$, if (2.2) is identified with system (5.3_1) then $J_{1,2}[\eta; a, b] = \int_a^b \eta^*(t)[C(t) - C_2(t)]\eta(t)\, dt \geq 0$ for $\eta \in \mathfrak{D}_0[a, b]$, and Theorem 5.2 then implies that (2.2) is disconjugate on $[a, b]$.

In turn, if $W(t)$ satisfies condition (ii), and $\Psi(t) = \mathfrak{R}[W](t)$, then $\Psi(t) \in \mathfrak{L}_{nn}^{\infty}[a, b]$ and the solution $U(t)$ of the system

$$U' = [A(t) + B(t)W(t)]U, \qquad U(a) = E,$$

and $V(t) = W(t)U(t)$ are $n \times n$ Lipschitzian matrix functions on $[a, b]$ with $U(t)$ nonsingular. Moreover,

$$V^*(t)U(t) - U^*(t)V(t) = U^*(t)[W^*(t) - W(t)]U(t) \equiv 0,$$

and $U^*(t)L_1[U, V](t) = -U^*(t)\Psi(t)\, U(t) \geq 0$, so that (2.2) is disconjugate on $[a, b]$ by the preceding result.

For the following theorems of this section concerning the existence of principal solutions, as well as many of the results in subsequent sections, it will be supposed that for certain intervals I the following hypothesis is satisfied.

$\mathfrak{H}_N(I)$ *On the interval* I *on the real line the* $n \times n$ *matrix functions* $A(t)$, $B(t)$, $C(t)$ *satisfy hypothesis* $(\mathfrak{H})$, $B(t) \geq 0$ *for* t a.e. *on this interval, and the system is identically normal on* I.

For results concerning principal solutions for differential systems that are not required to be self-adjoint, and without the assumption of identical normality, the reader is referred to Reid [24].

Theorem 5.4. *Suppose that* (2.2) *satisfies* $\mathfrak{H}_N(a_0, b_0)$, *where* $-\infty \leq a_0 < b_0 \leq \infty$, *and there is a subinterval* $I_0 = (c_0, b_0)$ *on which* (2.2) *is disconjugate. For* c *and* s *distinct points of* I_0 *let* $Y_{cs}(t) = (U_{cs}(t); V_{cs}(t))$ *be the solution of* (2.2_M) *satisfying the conditions* $U_{cs}(c) = E$, $U_{cs}(s) = 0$. *Then* $Y_{cb_0}(t) = (U_{cb_0}(t); V_{cb_0}(t)) = \lim_{s\to b_0} Y_{cs}(t)$ *exists and is a principal solution of* (2.2_M) *at* b_0, *with* $U_{cb_0}(t)$ *nonsingular on* I_0; *in particular,* $Y_{cb_0}(t)$ *is a conjoined basis for* (2.2) *on* I_0. *and* $Y_{bb_0}(t) = Y_{cb_0}(t)U_{bb_0}(c)$ *for* $(b, c, t) \in I_0 \times I_0 \times I_0$.

As (2.2) is identically normal and disconjugate on I_0, from Lemma 3.1 it follows that $Y_{cs}(t)$ is uniquely determined by the boundary conditions $U_{cs}(c) = E$, $U_{cs}(s) = 0$, at distinct points c and s of I_0. Moreover, $Y_{cs}(t)$ is a conjoined basis for (2.2), so that $U^*_{cs}(t)V_{cs}(t)$ is hermitian for $t \in I$; in particular, $V_{cs}(c)$ is hermitian. Now for a given $c \in I_0$, let a and b be points on I_0 satisfying $a < c < b$, and for an arbitrary n-dimensional vector ξ let $\eta(t)$ be the vector function on $[a, b]$ defined as

$$\eta(t) = U_{ca}(t)\xi, \quad \textit{for} \quad t \in [a, c]; \qquad \eta(t) = U_{cb}(t)\xi, \quad \textit{for} \quad t \in [c, b].$$

Then $\eta \in \mathfrak{D}_0[a, b]$, and since on $[a, b]$ the system (2.2) is disconjugate and $B(t) \geq 0$ for t a.e. on this interval, Theorem 5.1 implies that

$$\begin{aligned} 0 \leq J[\eta; a, b] &= \xi^* U^*_{ca}(c)V_{ca}(c)\xi - \xi^* U^*_{cb}(c)V_{cb}(c)\xi, \\ &= \xi^*[V_{ca}(c) - V_{cb}(c)]\xi. \end{aligned}$$

As this relation holds for arbitrary $\xi \in \mathbf{C}_n$, we have

(5.7) $\quad V_{cb}(c) < V_{ca}(c), \quad \textit{for} \quad (a, b, c) \in I_0 \times I_0 \times I_0, \qquad a < c < b.$

Now for $c < b < s < b_0$, and ξ an arbitrary nonzero vector in $\mathbf{C}_n$, let $(u(t); v(t)) = (U_{cs}(t)\xi; V_{cs}(t)\xi)$ and $\eta_0(t) = U_{cb}(t)\xi$ for $t \in [c, b]$, $\eta_0(t) = 0$ for $t \in [b, s]$. Then $y(t) = (u(t); v(t))$ is a solution of (2.2), and $\eta_0 \in \mathfrak{D}[c, s]$ with $\eta_0(c) = u(c)$, $\eta_0(s) = u(s)$; also, the identical normality of (2.2), and the fact that $\eta_0(t) = 0$ for $t \in [b, s]$, imply that $\eta_0(t) \not\equiv u(t)$ on $[c, s]$. The results of Theorem 5.1 and Corollary 2 to Theorem 4.1 then imply that

$$-\xi^* V_{cs}(c)\xi = J[u; c, s] < J[\eta_0; c, s] = J[\eta_0; c, b] = -\xi^* V_{cb}(c)\xi.$$

Consequently, we have that

(5.8) $\quad V_{cb}(c) < V_{cs}(c), \quad \textit{for} \quad (c, b, s) \in I_0 \times I_0 \times I_0, \qquad c < b < s.$

By a similar argument it follows that

(5.9) $\quad V_{cr}(c) < V_{ca}(c), \quad \textit{for} \quad (r, a, c) \in I_0 \times I_0 \times I_0, \qquad r < a < c.$

From (5.7), (5.8) it follows that for fixed $c \in I_0$ the one-parameter family of hermitian matrices $V_{cs}(c)$, $s \in (c, b_0)$, is monotone and bounded, and by Problem 7 of E.2 in Appendix E we have that there is an hermitian matrix V_{cb_0} such that $\{V_{cs}(c)\} \to V_{cb_0}$ as $s \to b_0$. Moreover, in view of relations (5.7), (5.8), (5.9) it follows that

$$(5.10) \quad V_{cs}(c) < V_{cb_0} < V_{cr}(c), \quad \text{for} \quad (r, c, s) \in I_0 \times I_0 \times I_0, \qquad r < c < s.$$

If $Y_{cb_0}(t) = (U_{cb_0}(t); V_{cb_0}(t))$ is the solution of (2.2_M) determined by the initial condition $Y_{cb_0} = (E; V_{cb_0})$, then clearly $\{Y_{cs}(t)\} \to Y_{cb_0}(t)$ as $s \to b_0$, for $t \in I$, and indeed the convergence is uniform on any compact subinterval of I. Moreover, in view of the result of Problem VII.4:10, inequality (5.10) implies that $U_{cb_0}(t)$ is nonsingular on each subinterval $[r, s]$ of I_0 with $r < c < s$, and hence $U_{cb_0}(t)$ is nonsingular on I_0. Now if $Y_s(t) = (U_s(t); V_s(t))$ is the solution of (2.2_M) determined by (3.15), then for $(s, c) \in I_0 \times I_0$, and $s \neq c$, we have $Y_s(t) = Y_{cs}(t)\, V_{cs}^{-1}(s)$, and $W_s(t) = V_s(t)U_s^{-1}(t)$ is identical with $W_{cs}(t) = V_{cs}(t)U_{cs}^{-1}(t)$ for $(s, c, t) \in I_0 \times I_0 \times I_0$, $t \neq s$. As $U_{cb_0}(t)$ is nonsingular on I_0, the matrix function $W_{cb_0}(t) = V_{cb_0}(t)U_{cb_0}^{-1}(t)$ is a solution of the Riccati matrix differential equation (3.8) on this interval, which on an arbitrary compact subinterval of I_0 is the uniform limit of $W_{cs}(t) = W_s(t)$ as $s \to b_0$. From Theorem 3.4 it then follows that $W_{cb_0}(t)$ is the distinguished solution of (3.8) at b_0, and $Y_{cb_0}(t)$ is a principal solution of (2.2_M) at this end-point. The fact that $Y_{cb_0}(t)$ is a conjoined basis for (2.2) follows from the Corollary to Theorem 3.3; this fact is also a direct consequence of the hermitian character of $V_{cb_0} = U_{cb_0}^*(c)\, V_{cb_0}(c)$. The final conclusion of the theorem is an immediate consequence of the fact that the determining initial values imply that if b, c, and s belong to I_0 then $Y_{bs}(t)$, $Y_{cs}(t)U_{bs}(c)$ are individually solutions of (2.2_M) with the corresponding $U_{bs}(t)$, $U_{cs}(t)U_{bs}(c)$ equal for $t = c$ and $t = s$, so that $Y_{bs}(t) = Y_{cs}(t)\, U_{bs}(c)$ for arbitrary $t \in I$ and hence $Y_{bb_0}(t) = Y_{cb_0}(t)\, U_{bb_0}(c)$ for $(b, c, t) \in I_0 \times I_0 \times I_0$.

In general, suppose that on the interval $I = (a_0, b_0)$ the system (2.2) is identically normal, and is not disconjugate. Then there exists a $b \in I$ such that there are points of I which precede b and are conjugate to b, and consequently there is a largest such conjugate point $t = \gamma(b)$ preceding b. Now in view of the assumption of identical normality, if $Y(t; b) = (U(t; b); V(t; b))$ is the solution of (2.2_M) determined by the initial conditions $U(b; b) = 0$, $V(b; b) = E$, a value $t \in I$, $t \neq b$, is conjugate to b if and only if $U(t; b)$ is singular. From Problem VII.4:2 it then follows that if $\gamma(b)$ exists for a value $b \in (a_0, b_0)$ then $\gamma(s)$ exists and is a strictly increasing function of s on $[b, b_0)$. In accordance with the terminology introduced by Leighton and Morse [1] for a scalar second order linear differential equation, the *first conjugate point* $\gamma(b_0)$ *of* $t = b_0$ *on* (a_0, b_0) is defined to be the limit of $\gamma(s)$ as $s \to b_0$. Clearly such a system (2.2) is disconjugate on a subinterval of I of the form (c_0, b_0) if and only if either this system is disconjugate on (a_0, b_0), or $\gamma(b_0)$ exists and $\gamma(b_0) < b_0$. In this latter case the interval I_0 of Theorem 5.4 may be chosen as $(\gamma(b_0), b_0)$, and, in particular, if $c \in (\gamma(b_0), b_0)$ the conjoined basis $Y_{cb_0}(t)$ has $U_{cb_0}(t)$ nonsingular on $(\gamma(b_0), b_0)$. On the other hand, the definition of $\gamma(b_0)$ implies that (2.2) fails to be disconjugate on each subinterval (c_0, b_0) with $a_0 < c_0 < \gamma(b_0)$, and, therefore, if $c \in (\gamma(b_0), b_0)$ then $U_{cb_0}(t)$ is singular for $t = \gamma(b_0)$.

Theorem 5.5. *If* (2.2) *satisfies hypothesis* $\mathfrak{H}_N(a_0, b_0)$, *where* $-\infty \leq a_0 < b_0 \leq \infty$, *is disconjugate on a subinterval* $I_0 = (c_0, b_0)$, *and* $Y(t) = (U(t); V(t))$ *is a solution of* (2.2_M) *with* $U(t)$ *nonsingular on a subinterval* $I\{Y\} = (a, b_0)$, *then for* $r \in I_0 \cap I\{Y\}$ *the matrix*

$$M(r, U) = \lim_{t \to \infty} S^{-1}(t, r \mid U) \tag{5.11}$$

exists and is finite.

For a value r common to the intervals I_0 and $I\{Y\}$ it follows from Problem VII.2:4 that we have the relations

$$\begin{aligned} U_{rb_0}(t) &= U(t)T(t, r \mid U)[U^{-1}(r) - S(t, r \mid U)\{U_{rb_0}; V_{rb_0} \mid U; V\}], \\ U_{rs}(t) &= U(t)T(t, r \mid U)[E - S(t, r \mid U)\, S^{-1}(s, r \mid U)]U^{-1}(r), \end{aligned} \tag{5.12}$$

and since $Y_{rs}(t) \to Y_{rb_0}(t)$ as $s \to b_0$ it follows that the matrix $M(r; U)$ defined by (5.11) exists and has the finite value

$$M(r; U) = \{U_{rb_0}; V_{rb_0} \mid U; V\}\, U(r). \tag{5.13}$$

Now from the Corollary to Theorem 3.3 it follows that if $Y_1(t) = (U_1(t); V_1(t))$ is a principal solution of (2.2_M) at b_0 then $Y_1(t)$ is a conjoined basis for (2.2), and therefore $T(t, c \mid U_1) \equiv E$. Moreover, the first conclusion of Theorem 3.2 for $Y_2(t) = Y_1(t)$ implies that if (2.2_M) has a solution $Y_1(t)$ with $U_1(t)$ nonsingular on a subinterval (c, b_0), and $T(t, c \mid U_1) \to 0$ as $t \to b_0$, then $Y_1(t)$ is a principal solution of (2.2_M) at b_0. Therefore the following corollary is a direct consequence of the results of Theorems 3.2, 5.4, and formula (5.13).

Corollary. *If* (2.2) *satisfies the hypotheses of Theorem* 5.5, *then:*

(i) *if* $Y_1(t) = (U_1(t); V_1(t))$ *is a solution of* (2.2_M) *with* $U_1(t)$ *nonsingular on a subinterval* $I_0 = (c_0, b_0)$ *of* I, *and* $r \in I_0$, *then it is not true that* $T(t, r \mid U_1) \to 0$ *as* $t \to b_0$;

(ii) *if* $Y_1(t) = (U_1(t); V_1(t))$ *is a principal solution of* (2.2_M) *at* b_0, *then for a solution* $Y_2(t) = (U_2(t); V_2(t))$ *the matrix* $\{U_1; V_1 \mid U_2; V_2\}$ *is nonsingular if and only if* $U_2(t)$ *is nonsingular for* t *on some subinterval* $I\{Y_2\} = (c_0, b_0)$ *and*

$U_2^{-1}(t)U_1(t) \to 0$ as $t \to b_0$; moreover, if $\{U_1; V_1 \mid U_2; V_2\}$ is nonsingular then for r on some subinterval (c, b_0) of I we have that $\lim_{t\to b_0} S(t, r \mid U_2)$ *exists and is nonsingular.*

Problems VII.5

1.[ss] If $[a, b] \subset I$, and $B(t) \geq 0$ for t a.e. on $[a, b]$, prove that there exists a $\delta > 0$ such that if $t_\alpha \in [a, b]$, $\alpha = 1, 2$, and $|t_1 - t_2| < \delta$, then t_1 and t_2 are not mutually conjugate.

Hint. For $c \in [a, b]$, consider the conjoined basis determined by the initial conditions (2.17), and use appropriate results of Theorem 5.1.

2. Suppose that hypothesis $(\mathfrak{H})$ holds, with $A(t) \equiv 0$, $B(t) \geq 0$ for t a.e. on I. Prove that the following condition is sufficient for the differential system

$$(5.14) \qquad \begin{aligned} -v'(t) + C(t)u(t) &= 0, \\ u'(t) - B(t)v(t) &= 0, \end{aligned}$$

to be disconjugate on I:

(a) There exist hermitian $n \times n$ matrix functions $M(t)$ and $C_1(t)$ which are respectively Lipschitzian and of class $\mathfrak{L}_{nn}^\infty$ on arbitrary compact subintervals of I such that $M'(t) + C(t) = C_1(t)$, $C_1(t) \geq M(t)B(t)M(t)$ for t a.e. on I. In particular, this condition holds in each of the following cases:

(a') $I = [0, 1]$, $B(t) \equiv E$, $C(t) \leq -4\left[\int_0^t C(s)\,ds\right]^2$ for t a.e. on I;

(a'') $I = (0, \infty)$, $B(t) \equiv E$, $\int_1^\infty C(s)\,ds = \lim_{t\to\infty} \int_1^t C(s)\,ds$

exists and is finite, and the matrix $M_1(t) = -\int_t^\infty C(s)\,ds$ satisfies either (α): $C(t) \leq -4\,M_1^2(t)$ for t a.e. on I, or (β): $-3E \leq 4t\,M_1(t) \leq E$ for t a.e. on I.

Hints. Show that $W(t) = -M(t)$ satisfies condition (ii) of Theorem 5.3. Verify that (a') implies condition (a) with $C_1(t) = -C(t)$, $M(t) = -2\int_0^t C(s)\,ds$; similarly, (α) of (a'') implies (a) with $C_1(t) = -C(t)$, $M(t) = 2\int_t^\infty C(s)\,ds$, while (β) of (a'') implies condition (a) with $C_1(t) = (4t^2)^{-1}E$, $M(t) = -M_1(t) - (4t)^{-1}E$.

3. Suppose that hypothesis $(\mathfrak{H})$ holds, with $A(t) \equiv 0$, $B(t)$ nonsingular, $B^{-1}(t) \in \mathfrak{L}_{nn}^\infty[a, b]$ and $B(t) > 0$ for t a.e. on a compact interval $[a, b]$; moreover, there exists a non-negative real-valued function $\theta(t)$ of class $\mathfrak{L}^\infty[a, b]$ such that $\theta(t)E + C(t) \geq 0$ for t a.e. on $[a, b]$. Show that the system

(5.14) is disconjugate on $[a, b]$ if the constant hermitian matrix

$$D = 4\left[\int_a^b B(t)\,dt\right]^{-1} - \left(\int_a^b \theta(t)\,dt\right)E$$

is non-negative definite.

Hint. Suppose that t_1 and t_2 are conjugate points with respect to (5.14), with $a \leq t_1 < t_2 \leq b$, and that $(u; v)$ is a solution of (5.14) with $u(t_1) = 0 = u(t_2)$ and $u(t) \not\equiv 0$ for $t \in (t_1, t_2)$. For $\eta(t) = u(t)$ for $t \in [t_1, t_2]$, $\eta(t) = 0$ for $t \in [a, t_1) \cup (t_2, b]$, show that

$$0 = I[\eta; a, b] = \int_a^b \{\eta^{*\prime}(t)B^{-1}(t)\eta'(t) + \eta^*(t)C(t)\eta(t)\}\,dt,$$

and with the aid of Problem VII.4:8 obtain a contradiction.

4.[s] Suppose that hypothesis ($\mathfrak{H}$) holds on $I = [a, \infty)$, with $A(t) \equiv 0$, $B(t) \geq 0$ and $C(t) \leq 0$ for t a.e. on I; moreover,

(α) $\int_a^t B(s)\,ds > 0$ for $t \in (a, \infty)$, and $\lambda\,[\int_a^t B(s)\,ds] \to \infty$ as $t \to \infty$, where in general the symbol $\lambda[M]$ is used to denote the smallest proper value of an hermitian matrix M;

(β) if $t_1 \in I$, then there exists a $t_2 > t_1$ such that $\int_{t_1}^{t_2} C(s)\,ds < 0$. Prove that the following conditions are equivalent, where $(U(t; a); V(t; a))$ is the solution of (2.2_M) satisfying the initial conditions

$$U(t; a) = 0, \qquad V(t; a) = E: \tag{5.15}$$

(i) the system (5.14) is disconjugate on I;
(ii) $U(t; a)$ is nonsingular on (a, ∞);
(iii) $V(t; a)$ is nonsingular on I.

Hints. Conclude from Problems 1 and 7 above that (i) is equivalent to (ii). Proceed to prove that (ii) and (iii) are equivalent by the following argument. If $V(t; a)$ is nonsingular on I, show that $W_0(t) = U(t; a)\,V^{-1}(t; a)$ is such that

$$W_0(a) = 0, \quad W_0'(t) = B(t) - W_0(t)C(t)W_0(t), \quad \text{for} \quad t \in I. \tag{5.16}$$

Conclude that

$$W_0(t) \geq \int_a^t B(s)\,ds > 0, \quad \text{for} \quad t \in (a, \infty),$$

so that $U(t; a)$ is nonsingular on (a, ∞) and (ii) holds. Conversely, if (ii) holds, then the fact that $W(t) = V(t; a)U^{-1}(t, a)$ satisfies $W'(t) = C(t) - W(t)B(t)W(t)$ on (a, ∞) implies that

$$W(t_1) - W(t_2) = -\int_{t_1}^{t_2} C(s)\,ds + \int_{t_1}^{t_2} W(s)B(s)W(s)\,ds \geq 0, \quad \text{for} \quad a < t_1 < t_2 < \infty. \tag{5.17}$$

Conclude that the hypotheses of the problem imply that all proper values of $W(t)$ are nonincreasing, and that no proper value can remain constant on an interval of the form (b, ∞), so that there exists a $c \in (a, \infty)$ such that for $t \in [c, \infty)$ all proper values are different from zero, and hence $W(t)$ is nonsingular on $[c, \infty)$. As $W_0(t) = W^{-1}(t)$ satisfies the differential equation of (5.16) on $[c, \infty)$, conclude that $W_0(t_2) \geq W_0(t_1)$ for $c \leq t_1 < t_2 < \infty$, and $\lambda[W_0(t)] \to \infty$ as $t \to \infty$; in particular, there exists a value $t = b$ such that if $t \geq b$ then $W_0(t) > 0$, and hence also $W(t) > 0$. From the monotone nonincreasing nature of $W(t)$, conclude that $W(t) > 0$ for $t \in (a, \infty)$, and therefore $V(t; a)$ is nonsingular for $t \in [a, \infty)$.

5. If the coefficient matrix functions in (2.2) are constant $n \times n$ matrices A, B, C with B and C hermitian, prove that (2.2) is identically normal on $(-\infty, \infty)$ if and only if the $n \times n^2$ matrix

$$[B \quad AB \quad \cdots \quad A^{n-1}B]$$

has rank n.

Hint. Note that $u(t) \equiv 0$, $v(t)$ is a solution of (2.2) if and only if $v(t) = [\exp\{-A^*t\}]\xi$, where ξ is a constant vector, and $\xi^*[\exp\{-At\}]B \equiv 0$ for $t \in (-\infty, \infty)$; with the aid of the Cayley-Hamilton theorem, (Corollary 2 to Theorem 3.1 in Appendix E), show that there exists a nonzero vector ξ satisfying this condition if and only if the matrix of the problem has rank less than n.

6.[s] Suppose that hypothesis ($\mathfrak{H}$) holds on $[a, \infty)$, with $A(t) \equiv 0$, $B(t) \geq 0$ and $C(t) \leq 0$ for t a.e. on I. Moreover,

(α') $\lambda[\int_a^t B(s)\,ds] \to \infty$ as $t \to \infty$, where, as in Problem 4, the symbol $\lambda[M]$ is used to denote the smallest proper value of an hermitian matrix M.

Prove that:

(i) If there exists a subinterval $I_0 = [a_0, \infty)$ of I on which (5.14) is disconjugate then the improper matrix integral $\int_a^\infty C(s)\,ds = \lim_{t\to\infty}\int_a^t C(s)\,ds$ is convergent.

(ii) If in addition to the above conditions, the coefficient matrix function $C(t)$ satisfies condition (β) of Problem 4, then (5.14) is disconjugate for large t if and only if $\int_a^\infty C(s)\,ds$ converges and on some subinterval (c_0, ∞) of I there exists a continuous hermitian matrix function $W(t)$ such that

$$\int_t^\infty W(s)B(s)W(s)\,ds$$

converges and

$$W(t) + \int_t^\infty C(s)\,ds - \int_t^\infty W(s)B(s)W(s)\,ds = 0, \qquad t \in (c_0, \infty). \tag{5.18}$$

Hints. (i) Prove by contrapositive argument, noting that if the improper matrix integral $\int_a^\infty C(s)\,ds$ does not converge then for $c_1 \in I$, and k a given positive constant, there exists a $c_2 = c_2\{c_1, k\}$ such that $\nu[\int_{c_1}^{c_2} C(s)\,ds] > k$. Moreover, condition ($\alpha'$) implies that if $c \in I$ then there exists a τ_c such that $\int_c^t B(s)\,ds > 0$ if $t \geq \tau_c$, and $\{\int_c^t B(s)\,ds\}^{-1} \to 0$ as $t \to \infty$. For $c \in I$, let c_1, c_2, d be such that $c < c_1 < c_2 < d$, and

$$\nu\left[\left\{\int_c^{c_1} B(s)\,ds\right\}^{-1}\right] < 1, \qquad \nu\left[\int_{c_1}^{c_2} C(s)\,ds\right] > 2, \qquad \nu\left[\left\{\int_{c_2}^{d} B(s)\,ds\right\}^{-1}\right] < 1,$$

and denote by ξ a unit vector in $\mathbf{C}_n$ such that $\xi^*[\int_{c_1}^{c_2} C(s)\,ds]\xi < -2$. For brevity, set $D_1 = \{\int_c^{c_1} B(s)\,ds\}^{-1}$, $D_2 = \{\int_{c_2}^{d} B(s)\,ds\}^{-1}$, and

$$\begin{aligned}(\eta(t), \zeta(t)) &= \left(\left\{\int_c^t B(s)\,ds\right\}D_1\xi,\ D_1\xi\right), \quad \textit{for} \quad t \in [c, c_1),\\ &= (\xi, 0), \quad \text{for} \quad t \in [c_1, c_2],\\ &= \left(\xi - \left\{\int_{c_2}^t B(s)\,ds\right\}D_2\xi,\ -D_2\xi\right), \quad \textit{for} \quad t \in (c_2, d].\end{aligned}$$

Show that $\eta \in \mathfrak{D}_0[c, d]:\zeta$, and

$$J[\eta; c, d] \leq \xi^* D_1 \xi + \xi^*\left[\int_{c_1}^{c_2} C(s)\,ds\right]\xi + \xi^* D_2 \xi < 0.$$

From Problem VII.4:2 conclude that an arbitrary conjoined basis $Y(t) = (U(t); V(t))$ for (2.2) is such that $U(t)$ is singular for at least one value t on (c, d); in particular, (5.14) fails to be disconjugate on $[c, d]$.

(ii) If $\int_a^\infty C(s)\,ds$ converges, and there is a continuous hermitian matrix function $W(t)$ for which $\int_t^\infty W(s)B(s)W(s)\,ds$ converges for $t \in (c_0, \infty)$ and (5.14) holds, then $W(t)$ is an hermitian solution of the corresponding Riccati matrix differential equation

$$W' + WB(t)W - C(t) = 0, \qquad t \in (c_0, \infty),$$

so that (5.14) is disconjugate on (c_0, ∞) by Theorem 5.1. Conversely, if (5.14) is disconjugate on a subinterval $[c, \infty)$ then $\int_a^\infty C(s)\,ds$ converges by conclusion (i). Moreover, if $Y(t; c) = (U(t; c); V(t; c))$ is the solution of (5.14) satisfying $Y(c; c) = (0; E)$, and τ_c is such that $\int_c^t B(s)\,ds > 0$ for $t \geq \tau_c$, then on each subinterval $[c, d]$ with $d \geq \tau_c$ the system (5.14) is normal and $U(d; c)$ is nonsingular. Show that $W(t) = V(t; c)U^{-1}(t; c)$ is an hermitian matrix function which satisfies an equation of the form (5.17) for $\tau_c \leq t_1 < t_2 < \infty$, and proceed as in Problem 4 to show that there is a subinterval $[c_1, \infty)$ of $[c, \infty)$ on which $W(t)$ is nonsingular. Note that on $[c_1, \infty)$ the hermitian matrix function $W_0(t) = W^{-1}(t)$ satisfies the differential equation

$W_0' = B(t) - W_0C(t)W_0 \geq B(t)$, and consequently

$$W_0(t) \geq W_0(c_1) + \int_{c_1}^t B(s)\,ds, \qquad t \in (c_1, \infty).$$

With the aid of condition (α') show that there is a subinterval $[c_0, \infty)$ of $[c_1, \infty)$ on which $W_0(t)$ is positive definite and $\lambda[W_0(t)] \to \infty$ as $t \to \infty$. Conclude that $W(t)$ is also positive definite on $[c_0, \infty)$ and $W(t) \to 0$ as $t \to \infty$. From (5.17) for $c_0 \leq t_1 < t_2$ conclude that $\int_t^\infty W(s)B(s)W(s)\,ds$ is convergent and (5.18) holds for $t \in [c_0, \infty)$.

7. Suppose that (2.2) satisfies hypothesis $\mathfrak{H}_N(a_0, b_0)$, where $-\infty \leq a_0 < b_0 \leq \infty$, while there exists a subinterval $I_0 = (c_0, b_0)$ on which this system is disconjugate and $Y_{cb_0}(t) = (U_{cb_0}(t); V_{cb_0}(t))$, $c \in I_0$, is the solution of (2.2_M) as determined in Theorem 5.4. If $Y(t) = (U(t); V(t))$ is a solution of (2.2_M) with $U(t)$ nonsingular on I_0, and $S(b_0, b \mid U) = \lim_{t \to b_0} S(t, b \mid U)$ exists and is finite for some $b \in I_0$, show that for arbitrary $c \in I_0$ we have:

(a) $S(b_0, c \mid U)$ exists, and

$$S(b_0, c \mid U) = T(c, b \mid U)[S(b_0, b \mid U) - S(c, b \mid U)]; \tag{5.19}$$

(b) $\{U_{cb_0}; V_{cb_0} \mid U; V\}$ is nonsingular;
(c) $U^{-1}(t)U_{cb_0}(t) \to 0$ as $t \to b_0$;
(d) $\{U_{cb_0}; V_{cb_0} \mid U; V\} - \{U; V \mid U; V\}\, U^{-1}(c)$ is nonsingular, and $T(b_0, c \mid U) = \lim_{t \to b_0} T(t, c \mid U)$ exists and is equal to the nonsingular matrix $\{U; V \mid U_{cb_0}; V_{cb_0}\}^{-1}[\{U; V \mid U_{cb_0}; V_{cb_0}\} - U^{*-1}(c)\{U; V \mid U; V\}]$;
(e) $U_{cb_0}(t) = U(t)\, S(b_0, t \mid U)\{U_{cb_0}; V_{cb_0} \mid U; V\}$.

Hints. (a) Use relation (3.4).
(b) Use the value (5.13) for $M(c; U) = \lim_{t \to \infty} S^{-1}(t, c \mid U)$ to conclude that whenever $S(b_0, c \mid U)$ exists and is finite we have

$$E = S(b_0, c \mid U)\{U_{cb_0}; V_{cb_0} \mid U; V\}U(c), \tag{5.20}$$

and hence $\{U_{cb_0}; V_{cb_0} \mid U; V\}$ is nonsingular.
(c) Use (ii) of the Corollary to Theorem 5.5, to show that conclusion (b) implies (c).
(d) Use the result of Problem VII.2:5 to obtain the relation

$$\begin{aligned} &\{U_{cb_0}; V_{cb_0} \mid U; V\} - \{U; V \mid U; V\}\, U^{-1}(t)U_{cb_0}(t) \\ &\quad = T^{*-1}(t, c \mid U)[\{U_{cb_0}; V_{cb_0} \mid U; V\} - \{U; V \mid U; V\}U^{-1}(c)] \end{aligned} \tag{5.21}$$

for $(t, c) \in I_0 \times I_0$. If $\xi \in \mathbf{C}_n$ and

$$[\{U_{cb_0}; V_{cb_0} \mid U; V\} - \{U; V \mid U; V\}\, U^{-1}(c)]\xi = 0,$$

use the above conclusions (b) and (c) to prove that $\xi = 0$. Deduce from (5.21) that $T(b_0, c \mid U) = \lim_{t\to b_0} T(t, c \mid U)$ does exist and has the value given in (d).

(e) Note that (5.20) is equivalent to the condition

$$E = U(t)S(b_0, t \mid U)\{U_{tb_0}; V_{tb_0} \mid U; V\}, \quad \textit{for} \quad t \in I_0, \tag{5.22}$$

and with the aid of the relation $Y_{cb_0}(s) = Y_{tb_0}(s)U_{cb_0}(t)$ established in Theorem 5.4 conclude that $\{U_{tb_0}; V_{tb_0} \mid U; V\}U_{cb_0}(t) = \{U_{cb_0}; V_{cb_0} \mid U; V\}$, and hence (5.22) implies conclusion (e).

8.[s] For a system (2.2) which satisfies hypothesis $\mathfrak{H}_N(a_0, b_0)$, where $-\infty \le a_0 < b_0 \le \infty$, establish the following results.

(i) If (2.2) is disconjugate on a subinterval $I_0 = (c_0, b_0)$ of $I = (a_0, b_0)$, then:

(α) if $W_0(t)$ is an hermitian solution of (3.8) on a subinterval $[c, b_0)$ of I_0, and $W(t)$ is a solution of (3.8) satisfying $W(c) = W_0(c) + \Gamma$, then $W(t)$ exists on $[c, b_0)$ if either $\mathfrak{Im}\,\Gamma$ is definite, or if there exist real constants $\lambda_0 > 0$, λ_1 such that $\lambda_0 \mathfrak{Re}\,\Gamma + \lambda_1 \mathfrak{Im}\,\Gamma \ge 0$; in particular, if Γ is an hermitian matrix satisfying $\Gamma \ge 0$ then $W(t) - W_0(t) \ge 0$ for $t \in [c, b_0)$;

(β) if $W_{b_0}(t)$ is the distinguished solution of (3.8) at b_0, and $W(t)$ is a solution of this equation which for a value c satisfies $W(c) = W_{b_0}(t) + \Gamma$, where Γ is an hermitian matrix that fails to be non-negative definite, then $W(t)$ does not exist throughout the interval $[c, b_0)$.

(ii) If (2.2) is disconjugate on $I = (a_0, b_0)$, then the distinguished solutions $W_{b_0}(t)$ and $W_{a_0}(t)$ are individually hermitian solutions of (3.8) on I such that:

(α) if $Y_s(t) = (U_s(t); V_s(t))$ is the solution of (2.2_M) determined by the initial conditions (3.15), and $W_s(t) = V_s(t)U_s^{-1}(t)$, then $W_s(t) \to W_{b_0}(t)$ as $s \to b_0$ and $W_s(t) \to W_{a_0}(t)$ as $s \to a_0$;

(β) if $W(t)$ is an hermitian solution of (3.8) which exists throughout I, then $W(t) - W_{b_0}(t) \ge 0$ and $W_{a_0}(t) - W(t) \ge 0$ for $t \in I$, while if $W(t)$ is an hermitian solution of (3.8) for which at some value $c \in I$ the matrix $W(c) - W_{b_0}(c)$, $\{W_{a_0}(c) - W(c)\}$, fails to be non-negative definite, then $W(t)$ does not exist throughout the interval $[c, b_0)$, $\{(a_0, c]\}$.

Hints. (i-α). Note that for an hermitian solution $W_0(t)$ of (3.8) the solutions of the corresponding differential systems (3.10) satisfy $G(t, c \mid W_0) = H^*(t, c \mid W_0)$, and in view of hypothesis $\mathfrak{H}_N(a_0, b_0)$ and the disconjugacy of (2.2) on I_0 deduce that $Z(t, c \mid W_0) > 0$ for $t \in I_0^+(c) = \{t \mid t \in I_0, t > c\}$. If $W(t)$ is a solution of (3.8) satisfying $W(c) = W_0(c) + \Gamma$, conclude from Problem II.3:9 that $W(t)$ exists on $[c, b_0)$ if and only if $E + Z(t, c \mid W_0)\Gamma$ is nonsingular on $[c, b_0)$, and this latter condition holds if and only if

$Z^{-1}(t, c \mid W_0) + \Gamma$ is nonsingular on (c, b_0). If $s \in (c, b_0)$ and

$$[Z^{-1}(s, c \mid W_0) + \Gamma]\xi = 0,$$

show that $\xi^*[Z^{-1}(s, c \mid W_0) + \Re e\, \Gamma]\xi = 0$, $\xi^*[\Im m\, \Gamma]\xi = 0$, and proceed to establish that each of the conditions stated in (i-α) implies that $\xi = 0$, so that $E + Z(s, c \mid W_0)\Gamma$ is nonsingular and $W(t)$ exists on $[c, b_0)$. In particular, if $W(c) = W_0(c) + \Gamma$ and $\Gamma \geq 0$, use the representation (II.3.14) to prove that $W(t) - W_0(t) \geq 0$ for $t \in [c_0, b_0)$, noting that if $\Gamma \geq 0$, $Z > 0$ and $E + Z\Gamma$ is nonsingular, then

$$\Gamma[E + Z\Gamma]^{-1} = [E + Z\Gamma]^{*-1}[\Gamma + \Gamma Z \Gamma][E + Z\Gamma]^{-1} \geq 0.$$

(i-β) If $c \in I_0$ show that $Z^{-1}(t, c \mid W_{b_0}) > 0$ for $t \in (c, b_0)$, and that for an hermitian Γ the hermitian matrix function $Z^{-1}(t, c \mid W_{b_0}) + \Gamma$ tends to Γ as $t \to b_0$, while this matrix function is positive definite for $t > c$, and sufficiently close to c. Show that if Γ fails to be non-negative definite then there exists a value $s \in (c, b_0)$ such that $Z^{-1}(s, c \mid W_{b_0}) + \Gamma$ is singular, and conclude that if $W(t)$ is the solution of (3.8) satisfying $W(c) = W_{b_0}(c) + \Gamma$ then $W(t)$ does not exist throughout the interval $[c, s]$.

(ii-α) Recall results of Theorems 3.4 and 5.4 for conclusions involving the end-point b_0, and establish similar conclusions involving the end-point a_0.

(ii-β) To the distinguished solution of (3.8) at b_0 apply appropriate results from conclusion (i) above. Establish the results involving the distinguished solution at a_0 by similar argument.

9. For the system (2.2) with $n = 2$, $A(t) \equiv 0$, $B(t) \equiv E_2$, $C(t) = \text{diag}\{0, -1/(4t^2)\}$ on $I = (0, \infty)$, show that the system satisfies hypothesis $\mathfrak{H}_N(I)$ and is disconjugate on I. For $c \in I$ determine: $Y_{cs}(t)$ as defined in Theorem 5.4; $Y_{c\infty}(t) = \lim_{s\to\infty} Y_{cs}(t)$; $Y_{c0}(t) = \lim_{s\to 0} Y_{cs}(t)$. Show that there exists a solution $Y(t) = (U(t); V(t))$ of the corresponding system (2.2_M) with

$$U(t) = \begin{bmatrix} 1 & t \\ 0 & t^{1/2} \end{bmatrix},$$

and compute the corresponding matrix functions $T(t, 1 \mid U)$, $S(t, 1 \mid U)$, and $M(1; U) = \lim_{t\to\infty} S^{-1}(t, 1 \mid U)$.

10. Suppose that $r(t)$ and $p(t)$ are real-valued continuous functions on $I = (a_0, b_0)$, $-\infty \leq a_0 < b_0 \leq \infty$, with $r(t) > 0$ on I, and the second-order linear differential equation

$$(r(t)u')' - p(t)u = 0, \tag{5.23}$$

is disconjugate on I. Show that a nontrivial solution $u = u_0(t)$ is a principal solution of this equation at b_0, {i.e., $(u(t); v(t)) = (u_0(t); r(t)u_0'(t))$ is a principal solution of the corresponding first order system

$$-v'(t) + p(t)\,u(t) = 0, \;\; u'(t) - \{1/r(t)\}\,v(t) = 0$$

at $b_0\}$, if and only if one of the following conditions holds:

(i) if $u(t)$ is a solution of (5.23) which is not a multiple of $u_0(t)$, then $u_0(t)/u(t) \to 0$ at $t \to b_0$;

(ii) if c is such that $u_0(t) \neq 0$ for $t \in [c, b_0)$, then the improper integral $\int_c^{b_0} |u_0(s)|^{-2}\,ds = \lim_{t\to b_0} \int_c^t |u_0(s)|^{-2}\,ds$ diverges to $+\infty$.

Formulate, and establish, corresponding results for principal solutions of (5.23) at the end-point a_0.

11. Show that each of the following differential equations is disconjugate on the indicated interval I, and determine a principal solution of the equation at each of the end-points of I.

(i) $u'' - u = 0, \quad I = (-\infty, \infty)$;
(ii) $u'' + \{1/[4t^2]\}u = 0, \quad I = (0, \infty)$;
(iii) $([e^t + 2]^{-1}u')' - 2[e^t + 2]^{-2}u = 0, \quad I = (-\infty, \infty)$.

Hints. (iii) Show that the general solution of the differential equation is $u(t) = c_1(1 + e^{-t}) + c_2e^t$.

12. Suppose that the matrix coefficient functions in (2.2) are constant $n \times n$ matrices A, B, C with B and C hermitian, $B \geq 0$, while the system is identically normal on $(-\infty, \infty)$. Prove that the system is disconjugate on $(-\infty, \infty)$ if and only if there exists an hermitian constant matrix W satisfying the algebraic matrix equation

$$WA + A^*W + WBW - C = 0. \tag{5.24}$$

If such a system is disconjugate on $(-\infty, \infty)$ then the distinguished solutions $W_\infty(t)$, $W_{-\infty}(t)$ of (2.2_M) at ∞ and $-\infty$, respectively, are constant matrices W_∞, $W_{-\infty}$. Moreover, if $(U_0(t); V_0(t))$ is the solution of the corresponding system (2.2_M) with $U_0(0) = 0$, $V_0(0) = E$, then $W_0(t) = V_0(t)U_0^{-1}(t)$ converges to W_∞ and $W_{-\infty}$ as $t \to -\infty$ and $t \to \infty$, respectively. Also, if W is any hermitian solution of (5.24), then $W_\infty \leq W \leq W_{-\infty}$.

Hint. Use the results of (ii) of Problem 8 above, showing that the $W_s(t)$ of that problem is equal to $W_0(t - s)$.

13.[s] If (2.2) satisfies hypothesis $\mathfrak{H}_N(a_0, b_0)$, where $-\infty \leq a_0 < b_0 \leq \infty$, and there exists a subinterval $I_0 = (c_0, b_0)$ on which this system is disconjugate, prove that there exists a conjoined basis $Y_2(t) = (U_2(t); V_2(t))$ for (2.2) such that $U_2(t)$ is nonsingular for t on a subinterval $[c, b_0)$ of I_0 and the hermitian matrix integral

$$\int_c^b U_2^{-1}(t)B(t)U_2^{*-1}(t)\,dt \tag{5.25}$$

is convergent. Also, for any such conjoined basis $Y_2(t) = (U_2(t); V_2(t))$ the scalar integral

$$\int_c^{b_0} \frac{\nu[B(t)]\,dt}{\{\nu[U_2(t)]\}^2} \tag{5.26}$$

is convergent. In particular, if such a system (2.2) is disconjugate on a subinterval $I_0 = (c_0, b_0)$, and all solutions of (2.2) remain bounded as $t \to b_0$, then

$$\int_c^{b_0} \nu[B(t)]\,dt < \infty, \quad \textit{for} \quad c \in (a_0, b_0). \tag{5.27}$$

Also, in view of the hermitian non-negative character of $B(t)$, condition (5.27) is equivalent to each of the following conditions:

$$\int_c^{b_0} \operatorname{Tr} B(t)\,dt < \infty, \quad \textit{for} \quad c \in (a_0, b_0); \tag{5.27$'$}$$

(5.27″) *the matrix integral* $\int_c^{b_0} B(t)\,dt$ *is convergent, for* $c \in (a_0, b_0)$.

Hints. If $Y_1(t) = (U_1(t); V_1(t))$ is a principal solution of (2.2_M) at b_0, and $Y_2(t) = (U_2(t); V_2(t))$ is the solution of (2.2_M) such that $U_2(\tau) = V_1(\tau)$, $V_2(\tau) = -U_1(\tau)$ for some $\tau \in (a_0, b_0)$, show that $Y_2(t)$ is a conjoined basis for (2.2) with $\{U_1; V_1 \mid U_2; V_2\}$ nonsingular, and deduce from conclusion (ii) of the Corollary to Theorem 5.5 that the matrix integral (5.25) is convergent. Infer the convergence of the scalar integral (5.26) from the inequalities

$$\nu[B] \leq \nu[U_2]\,\nu[U_2^{-1}BU_2^{*-1}]\nu[U_2^*] = \{\nu[U_2]\}^2\,\nu[U_2^{-1}BU_2^{*-1}].$$

Note that if $\nu[U_2(t)] \leq M$ for $t \in [c, b_0)$, then $\nu[B(t)]/\{\nu[U_2(t)]\}^2 \geq M^{-2}\,\nu[B(t)]$ on this subinterval; also, $\nu[B(t)]$ is the greatest proper value of $B(t)$, and hence $\nu[B(t)]E \geq B(t)$ and $\nu[B(t)] \leq \operatorname{Tr} B(t) \leq n\,\nu[B(t)]$.

14.[s] Suppose that on an interval $I = (a_0, b_0)$, $-\infty \leq a_0 < b_0 \leq \infty$, the $n \times n$ matrix function $Q(t)$ is hermitian, belongs to $\mathfrak{L}_{nn}^{\infty}[a, b]$ for arbitrary compact subintervals $[a, b]$ of I, and $Q(t) > 0$ for t a.e. on I. Prove that there exists a subinterval $I_0 = (c_0, b_0)$ of I such that the differential system

$$\begin{aligned} -v' - Q(t)u &= 0, \\ u' - Q(t)v &= 0, \end{aligned} \tag{5.28}$$

is disconjugate on I_0 if and only if

(5.29) $$\int_c^{b_0} \nu[Q(t)]\,dt < \infty, \quad for \quad c \in (a_0, b_0);$$

moreover, (5.29) is equivalent to each of the following conditions:

(5.29′) $$\int_c^{b_0} \operatorname{Tr} Q(t)\,dt < \infty, \quad for \quad c \in (a_0, b_0);$$

(5.29″) *the matrix integral* $\int_c^{b_0} Q(t)\,dt$ *is convergent, for* $c \in (a_0, b_0)$.

Hints. If (5.28) is disconjugate on a subinterval $I_0 = (c_0, b_0)$ of I, use the results of Problem VII.2:8 and the above Problem 13 to establish the convergence of the scalar integral (5.29). Show that (5.29) is equivalent to each of the conditions (5.29′), (5.29″) by an argument similar to that suggested for the corresponding results in Problem 13. If $r(t) = \nu[Q(t)]$ is such that condition (5.29) holds, choose $\hat{r}(t)$ such that $\hat{r}(t) - r(t)$ is continuous, $\hat{r}(t) > r(t)$ and $\int_c^{b_0} \hat{r}(t)\,dt < \infty$; for example one may choose $\hat{r}(t) = r(t) + 1/(1 + t^2)$. Note that for $[a, b] \subset I$ the functional

(5.30) $$J[\eta; a, b] = \int_a^b \{\zeta^*(t)Q(t)\zeta(t) - \eta^*(t)Q(t)\eta(t)\}\,dt$$

corresponding to (5.28) is such that

(5.31) $$J[\eta; a, b] \geq \int_a^b \{[\hat{r}(t)]^{-1}\,|\eta'(t)|^2 - \hat{r}(t)\,|\eta(t)|^2\}\,dt$$

if η belongs to the corresponding class $\mathfrak{D}[a, b]$ with $\zeta(t) = Q^{-1}(t)\eta'(t)$. From the fact that the scalar differential equation $\{[\hat{r}(t)]^{-1}\,u'(t)\}' + \hat{r}(t)\,u(t) = 0$ has the solution $u(t) = \sin(\int_\tau^t \hat{r}(s)\,ds)$, conclude that the differential system corresponding to the functional of the right-hand member of (5.31) is disconjugate on subintervals (c_0, b_0) with c_0 sufficiently close to b_0 whenever condition (5.29) holds. With the aid of Theorem 5.1, deduce that the system (5.28) is also disconjugate on such a subinterval.

15.[s] Suppose that (2.2) satisfies hypothesis $\mathfrak{H}_N(a_0, b_0)$, where $-\infty \leq a_0 < b_0 \leq \infty$, and that $Y(t) = (U(t); V(t))$ is a conjoined basis for (2.2) for which there exists a subinterval $I_1 = (c_1, b_0)$ such that the hermitian matrix function

$$P(t; Y) = V^*(t)B(t)V(t) + V^*(t)A(t)U(t) + U^*(t)A^*(t)V(t) - U^*(t)C(t)U(t),$$

is positive definite for t a.e. on I_1. If $S(t)$ is a matrix function such that $S^*(t)S(t) = U^*(t)U(t) + V^*(t)V(t)$, prove that there exists a subinterval

$I_0 = (c_0, b_0)$ of I_1 on which (2.2) is disconjugate if and only if the matrix function $Q(t)$ defined as $Q(t) = S^{*-1}(t)P(t; Y)S^{-1}(t)$ satisfies condition (5.29), {equivalently, condition (5.29′) or (5.29″)}.

Hint. To (2.2) apply the polar coordinate transformation, as presented in Problem VII.2:6, and for the resulting system of the form (5.28) use the result of Problem 14. It is to be noted that in view of the results of Problem 7 of D.2 in Appendix D, and Problems 5, 6 of F.7 in Appendix F, the most general matrix function $S(t)$ satisfying the specified condition is $S(t) = H(t)R_0(t)$, where $R_0(t)$ is the positive definite hermitian square root of $U^*(t)U(t) + V^*(t)V(t)$, and $H(t)$ is unitary for $t \in (a_0, b_0)$; moreover, $R_0(t)$ is a.c. on arbitrary compact subintervals of (a_0, b_0).

6. COMPARISON THEOREMS

Corresponding to the terminology of Chapter V for a scalar equation, if $[a, b]$ is a given subinterval of I we shall denote by $\mathfrak{D}_{*0}[a, b]$ the set of vector functions η belonging to $\mathfrak{D}[a, b]$ and satisfying $\eta(b) = 0$. Similarly, $\mathfrak{D}_{0*}[a, b]$ will denote the set of vector functions η belonging to $\mathfrak{D}[a, b]$ and satisfying $\eta(a) = 0$, so that $\mathfrak{D}_0[a, b] = \mathfrak{D}_{*0}[a, b] \cap \mathfrak{D}_{0*}[a, b]$. We shall now consider functionals of the form

$$\begin{aligned} \hat{J}[\eta_1:\zeta_1, \eta_2:\zeta_2; a, b] &= \eta_2^*(a)\Gamma\eta_1(a) \\ &\quad + \int_a^b \{\zeta_2^*(t)B(t)\zeta_1(t) + \eta_2^*(t)C(t)\eta_1(t)\}\, dt, \\ &= \eta_2^*(a)\,\Gamma\eta_1(a) + J[\eta_1:\zeta_1, \eta_2:\zeta_2; a, b], \end{aligned} \tag{6.1}$$

where the coefficient matrix functions $A(t)$, $B(t)$, $C(t)$ satisfy hypothesis ($\mathfrak{H}$) and Γ is an $n \times n$ hermitian matrix. Clearly the functional $\hat{J}$ is an hermitian form on $\mathfrak{L}_n^2[a, b] \times \mathfrak{L}_n^2[a, b]$; that is, it possesses properties analogous to (4.2). As in the case of the functional (4.1), if $\eta_\alpha \in \mathfrak{D}[a, b]:\zeta_\alpha$, $(\alpha = 1, 2)$, then the value of (6.1) is independent of the particular ζ_α's, so that the notation is abbreviated to $\hat{J}[\eta_1, \eta_2; a, b]$, and $\hat{J}[\eta; a, b] = \hat{J}[\eta, \eta; a, b]$.

The following results correspond to those of Theorem V.6.1 and its Corollary, and will be stated without proof, since they may be established in an analogous manner, now using the results of Theorem 4.1 and its Corollary 1.

Theorem 6.1. *If* $[a, b] \subset I$ *and* $u \in \mathfrak{A}_n[a, b]$, *then there exists a* v *such that* $(u; v)$ *is a solution of* (2.2) *on* $[a, b]$ *which satisfies the initial condition*

$$\Gamma u(a) - v(a) = 0, \tag{6.2}$$

if and only if there exists a $v_1 \in \mathfrak{L}_n^2[a, b]$ *such that* $u \in \mathfrak{D}[a, b]:v_1$ *and* $\hat{J}[u:v_1, \eta:\zeta; a, b] = 0$ *for* $\eta \in \mathfrak{D}_{*0}[a, b]:\zeta$.

Corollary. *If* $\hat{J}[\eta; a, b]$ *is non-negative definite on* $\mathfrak{D}_{*0}[a, b]$, *and* u *is an element of* $\mathfrak{D}_{*0}[a, b]$ *satisfying* $\hat{J}[u; a, b] = 0$, *then there exists a* $v \in \mathfrak{A}_n[a, b]$ *such that* $(u; v)$ *is a solution of* (2.2) *on* $[a, b]$ *which satisfies the boundary conditions*

$$(6.3) \qquad \Gamma u(a) - v(a) = 0, \qquad u(b) = 0.$$

As the matrix Γ is hermitian, the solution $Y(t) = (U(t); V(t))$ of (2.2_M) determined by the initial conditions

$$(6.4) \qquad U(a) = E, \qquad V(a) = \Gamma,$$

is a conjoined basis for (2.2).

Corresponding to Theorem 4.4 we now have the following result.

Theorem 6.2. *The functional* $\hat{J}[\eta; a, b]$ *is positive definite on* $\mathfrak{D}_{*0}[a, b]$ *if and only if* $B(t) \geq 0$ *for* t a.e. *on* $[a, b]$, *and the conjoined basis* $Y(t) = (U(t); V(t))$ *for* (2.2) *determined by the initial conditions* (6.4) *is such that* $U(t)$ *is nonsingular on* $[a, b]$.

The conclusion that $B(t) \geq 0$ for t a.e. on $[a, b]$ whenever $\hat{J}[\eta; a, b]$ is positive definite on $\mathfrak{D}_{*0}[a, b]$ follows from Theorem 4.2, and the fact that $\mathfrak{D}_0[a, b] \subset \mathfrak{D}_{*0}[a, b]$. If the conjoined basis $(U(t); V(t))$ for (2.2) determined by (6.4) is such that there exists a $c \in (a, b]$ for which $U(c)$ is singular, let ξ be a nonzero vector satisfying $U(c)\xi = 0$. Then $u(t) = U(t)\xi$, $v(t) = V(t)\xi$ is a solution of (2.2) satisfying (6.2), and $u(t) \not\equiv 0$ on $[a, c]$. If $(\eta(t), \zeta(t)) = (u(t), v(t))$ for $t \in [a, c]$, $(\eta(t), \zeta(t)) = (0, 0)$ for $t \in (c, b]$, then $\eta \in \mathfrak{D}_{*0}[a, b]{:}\zeta$, and since $\hat{J}[\eta; a, b] = \hat{J}[u; a, c]$ it follows with the aid of relation (4.4″) that $\hat{J}[\eta; a, b] = 0$, contrary to the assumption that $\hat{J}[\eta; a, b]$ is positive definite on $\mathfrak{D}_{*0}[a, b]$.

Conversely, if $B(t) \geq 0$ for t a.e. on $[a, b]$ and $U(t)$ is nonsingular on $[a, b]$, for $\eta \in \mathfrak{D}_{*0}[a, b]{:}\zeta$ let $h(t) = U^{-1}(t)\eta(t)$. With the aid of the Corollary to Lemma 4.2 it then follows that

$$\hat{J}[\eta; a, b] = \int_a^b [\zeta - Vh]^* B(t)[\zeta - Vh]\, dt,$$

and proceeding as in the proof of Theorem 4.4 it follows that $\hat{J}[\eta; a, b]$ is positive definite on $\mathfrak{D}_{*0}[a, b]$.

Relative to the functional (6.1), or relative to the differential system (2.2) with initial condition (6.2), a value $\tau \in I$ is a *right-hand* {*left-hand*} *focal point to* $t = a$ if $\tau > a$, $\{\tau < a\}$, and there is a solution $(u(t); v(t))$ of (2.2) which satisfies (6.2), has $u(\tau) = 0$, and $u(t) \not\equiv 0$ for t on the interval with end-points $t = a$ and $t = \tau$.

Analogous to the first result of Theorem 5.2 we now have the following theorem, which will be stated without proof as it may be established by an argument similar to that occurring in the proof of Theorem 5.2, and using the result of Theorem 6.2.

Theorem 6.3. *Suppose that for* $\alpha = 1, 2$ *the* $n \times n$ *matrix functions* $A_\alpha(t), B_\alpha(t), C_\alpha(t)$ *satisfy hypothesis* ($\mathfrak{H}$) *and* $B_2(t) \geq 0$ *for* t a.e. *on* I. *Moreover, for arbitrary compact subintervals* $[a, b]$ *of* I *we have* $\mathfrak{D}_1[a, b] = \mathfrak{D}_2[a, b] = \mathfrak{D}[a, b]$, *and* Γ_α, $(\alpha = 1, 2)$, *are hermitian matrices such that*

$$\begin{aligned} \hat{J}_{1,2}[\eta; a, b] &= \hat{J}_1[\eta; a, b] - \hat{J}_2[\eta; a, b], \\ &= \eta^*(a)[\Gamma_1 - \Gamma_2]\eta(a) + J_{1,2}[\eta; a, b], \end{aligned}$$

is non-negative definite on $\mathfrak{D}_{*0}[a, b]$. *If relative to* $\hat{J}_2[\eta; a, b]$ *there is no right-hand focal point to* $t = a$ *on* $(a, b]$, *then relative to* $\hat{J}_1[\eta; a, b]$ *there is also no right-hand focal point to* $t = a$ *on* $(a, b]$.

7. MORSE FUNDAMENTAL HERMITIAN FORMS FOR DIFFERENTIAL SYSTEMS

It is to be noted that the theorems of the preceding section do not extend to differential systems (2.2) all the results of corresponding theorems for second order systems in Chapter V. In particular, in the above Theorem 6.2 we did not show that the non-negative definiteness of $\hat{J}[\eta; a, b]$ on $\mathfrak{D}_{*0}[a, b]$ implies that the conjoined basis $(U(t); V(t))$ determined by the initial conditions (6.4) is such that $U(t)$ is nonsingular for $t \in (a, b)$. Indeed, for differential systems (2.2) this result is in general not true without some additional normality restrictions. The non-negativeness of $\hat{J}[\eta; a, b]$ on $\mathfrak{D}_{*0}[a, b]$ does not exclude the possible existence of a solution $(u(t); v(t))$ of (2.2) which satisfies the boundary conditions (6.3), has $u(t) \not\equiv 0$ on $[a, b]$, while there exists a $c \in (a, b)$ such that $u(t) \equiv 0$ on $(c, b]$. This possibility is illustrated by the simple example of $n = 1$, $a = 0$, $b = \pi$, $\Gamma = 0$, $A(t) \equiv 0$, $C(t) \equiv -1$, while $B(t) = 1$ for $t \in [0, \pi/2]$, $B(t) = 0$ for $t \in (\pi/2, \pi]$. This system has a solution $(u(t); v(t)) = (\cos t; -\sin t)$ for $t \in (0, \pi/2]$, $(u(t); v(t)) = (0; -1)$ for $t \in (\pi/2, \pi]$.

Also, in the above Theorem 6.3 there is no generalization of the part of Theorem V.6.3 dealing with the comparative positions of a sequence of right-hand focal points of $t = a$ relative to $\hat{J}_1[\eta; a, b]$, and a sequence of such focal points relative to $\hat{J}_2[\eta; a, b]$. In this instance, a basic property is that for differential systems (2.2) we may have t_1' and t_2' as right-hand focal points to $t = a$ relative to $\hat{J}_1[\eta; a, b]$ with $a < t_1' < t_2'$, and yet have $\hat{J}_1[\eta; t_1', t_2']$ remain positive definite on the class $\mathfrak{D}_0[t_1', t_2']$. For example, if $n = 2$, $\Gamma = 0$, $A(t) \equiv 0$, $B(t) \equiv E_2$, $C(t) \equiv \operatorname{diag}\{-1, -9/4\}$, then $t = \pi/2$ and $t = \pi/3$ are both focal points to $t = 0$ relative to the functional

$$\int_0^b \{|\eta_1'|^2 + |\eta_2'|^2 - |\eta_1|^2 - \tfrac{9}{4}|\eta_2|^2\}\, dt,$$

with the conjoined basis (6.4) given by

$$U(t) = \operatorname{diag}\{\cos t, \cos \tfrac{3}{2}t\}, \qquad V(t) = \operatorname{diag}\{-\sin t, -\tfrac{3}{2}\sin \tfrac{3}{2}t\}.$$

Moreover,

$$U_0(t) = \operatorname{diag}\{\sin t, \cos [\tfrac{3}{4}(2t - \pi)]\},$$
$$V_0(t) = \operatorname{diag}\{\cos t, -\tfrac{3}{2}\sin [\tfrac{3}{4}(2t - \pi)]\}$$

is a conjoined basis with $U_0(t)$ nonsingular on the interval $(\pi/6, 5\pi/6)$, which contains both $t = \pi/2$ and $t = \pi/3$.

This second phenomenon is one that is inherent in the extension of the study of differential systems from the second order case of Chapter V to the more general systems of this chapter. The difficulty mentioned in the first paragraph of this section, however, is removed if attention is restricted to systems that satisfy the additional hypothesis $\mathfrak{H}_N(I)$, as formulated in Section 5 above. That is, on I the $n \times n$ matrix functions $A(t)$, $B(t)$, $C(t)$ satisfy hypothesis $(\mathfrak{H})$, $B(t) \geq 0$ for t a.e. on I, and on this interval the system (2.2) is identically normal.

If hypothesis $\mathfrak{H}_N(I)$ holds and $Y(t) = (U(t); V(t))$ is a conjoined basis for (2.2), a value $t = c$ is called a *focal point of the family of order* k, if $U(c)$ has rank $n - k$. The following lemma presents a basic property of the set of focal points belonging to a given conjoined basis for such an equation (2.2).

***Lemma* 7.1.** *Suppose that hypothesis $\mathfrak{H}_N(I)$ holds on a given interval I, and that* $[a, b]$ *is a compact subinterval of I on which* (2.2) *is disconjugate. If* $Y(t) = (U(t); V(t))$ *is a conjoined basis for* (2.2), *then on* $(a, b]$ *and* $[a, b)$ *there are at most n focal points of* $Y(t)$, *each focal point being counted a number of times equal to its order. In particular, the focal points of a conjoined basis for* (2.2) *are isolated.*

If $t = c$ is a focal point of a conjoined basis $Y(t) = (U(t); V(t))$, and ξ is a nonzero vector such that $U(c)\xi = 0$, then $(u(t); v(t)) = (U(t)\xi; V(t)\xi)$ is a solution of (2.2) such that $u(c) = 0$, and $u(t)$ is not identically zero throughout any nondegenerate subinterval of I. Also, if $t = c$ is a focal point of $Y(t)$ of order k and $\xi = \xi^{(j)}$, $(j = 1, \ldots, k)$, are k linearly independent vectors satisfying $U(c)\xi = 0$, then $(u^{(j)}(t); v^{(j)}(t)) = (U(t)\xi^{(j)}; V(t)\xi^{(j)})$, $(j = 1, \ldots, k)$, are linearly independent solutions of (2.2) such that on any nondegenerate subinterval of I the n-dimensional vector functions $u^{(1)}(t), \ldots, u^{(k)}(t)$ are linearly independent. Now suppose that (2.2) is disconjugate on $[a, b]$, and that $(a, b]$ contains $n + q$, $(q \geq 1)$, focal points $t_1 \leq t_2 \leq \cdots \leq t_{n+q}$ of a conjoined basis $Y(t) = (U(t); V(t))$, where each focal point is counted a number of times equal to its order. Then there exist $n + q$ solutions $(u^{(j)}(t); v^{(j)}(t)) = (U(t)\xi^{(j)}; V(t)\xi^{(j)})$, $(j = 1, \ldots, n + q)$, such that $u^{(j)}(t_j) = 0$, and if $(\eta^{(j)}(t), \zeta^{(j)}(t)) = (u^{(j)}(t), v^{(j)}(t))$ for $t \in [a, t_j]$, $(\eta^{(j)}(t), \zeta^{(j)}(t)) = (0, 0)$ for $t \in (t_j, b]$, then $\eta^{(1)}(t), \ldots, \eta^{(n+q)}(t)$ are linearly independent on $[a, b]$. Moreover, $\eta^{(j)} \in \mathfrak{D}_{*0}[a, b]:\zeta^{(j)}$, and one may verify directly that

$$J[\eta^{(j)}, \eta^{(i)}; a, b] = -\eta^{(i)*}(a)V(a)\xi^{(j)}, \qquad (i, j = 1, \ldots, n + q).$$

Now since $q > 0$ there exist constants $d_1, \ldots, d_{n+q}$ not all zero, and such that $\eta(t) = \eta^{(1)}(t)\, d_1 + \cdots + \eta^{(n+q)}(t)\, d_{n+q}$ satisfies the condition $\eta(a) = 0$, so that

$$\begin{aligned} J[\eta; a, b] &= \sum_{i,j=1}^{n+q} \bar{d}_i\, d_j J[\eta^{(j)}, \eta^{(i)}; a, b] \\ &= -\eta^*(a)V(a)[\xi^{(1)}\, d_1 + \cdots + \xi^{(n+q)}\, d_{n+q}] = 0. \end{aligned}$$

On the other hand, the vector function $\eta(t)$ is a nonidentically vanishing element of $\mathfrak{D}_0[a, b]$, and in view of Theorem 5.1 we have the contradictory result $J[\eta; a, b] > 0$.

By a similar argument it follows that on $[a, b)$ there are at most n focal points of $Y(t)$, with each focal point being counted a number of times equal to its order.

The final statement of the lemma on the isolated nature of the focal points of a conjoined basis for (2.2) is an immediate consequence of the preceding results, since if $t = c$ were a non-isolated focal point of such a family there would exist a positive δ so small that (2.2) would be disconjugate on each of the subintervals $[c - \delta, c]$, $[c, c + \delta]$ of I, while one of these intervals would contain infinitely many focal points of the family.

Throughout the following discussion of the Morse hermitian forms for differential systems associated with (2.2) *it will be assumed, without further comment, that hypothesis $\mathfrak{H}_N(I)$ holds for the involved interval I.*

Corresponding to the terminology and notation of Section V.8, a partition

$$\text{(7.1)} \qquad \Pi: a = t_0 < t_1 < \cdots < t_m < t_{m+1} = b,$$

of a compact subinterval $[a, b]$ of I will be called a *fundamental partition of* $[a, b]$, {*relative to* (2.2)}, provided this system is disconjugate on each of the subintervals $[t_{j-1}, t_j]$, $(j = 1, \ldots, m + 1)$. The result of Problem VII.5:1 assures the existence of fundamental partitions of a given subinterval $[a, b]$. Again it is to be emphasized that if Π is a fundamental partition of $[a, b]$, then any refinement of Π is also a fundamental partition of this interval.

For a fixed positive integer m let T_m denote the set of $(m + 2)$-tuples $T = \{t_0, t_1, \ldots, t_{m+1}\}$ belonging to fundamental partitions (7.1) of compact subintervals $[a, b]$ of I, and signify by $\mathcal{S}(\Pi)$ the totality of sequences $\xi = (\xi_0, \xi_1, \ldots, \xi_m, \xi_{m+1})$ of n-dimensional vectors $\xi_\sigma = (\xi_{i\sigma})$, $(i = 1, \ldots, n;\ \sigma = 0, 1, \ldots, m + 1)$. Since (2.2) is disconjugate on each subinterval $[t_{j-1}, t_j]$ of a fundamental partition, and this system is required to be identically normal, it follows that there exists a unique solution

$$(u^{(j)}(t); v^{(j)}(t)) = (u^{(j)}(t; T, \xi); v^{(j)}(t; T, \xi))$$

of (2.2) which satisfies the end-conditions

$$(7.2) \qquad u^{(j)}(t_{j-1}) = \xi_{j-1}, \qquad u^{(j)}(t_j) = \xi_j, \qquad (j = 1, \ldots, m + 1).$$

Moreover, these solutions are expressible in terms of a given set of $2n$ linearly independent solutions of (2.2) in a manner that is a direct extension of formulas (V.8.2), (V.8.3), and, in particular, the functions $u^{(j)}(t; T, \xi)$, $v^{(j)}(t; T, \xi)$ are linear functions in ξ with coefficients that are continuous functions of $t, t_0, \ldots, t_{m+1}$ on $I \times T_m$.

For $\xi \in \mathcal{S}(\Pi)$ the vector function on $[a, b]$ defined by

$$(7.3) \qquad u_\xi(t) = u^{(j)}(t), \qquad t \in [t_{j-1}, t_j], \qquad (j = 1, \ldots, m + 1),$$

is such that $u_\xi \in \mathfrak{D}[a, b]:v_\xi$, where $v_\xi(t)$ is a vector function of $\mathfrak{L}_n^2[a, b]$ such that

$$(7.4) \qquad v_\xi(t) = v^{(j)}(t), \qquad t \in (t_{j-1}, t_j), \qquad (j = 1, \ldots, m + 1).$$

The subclass of elements $\xi = (\xi_0, \xi_1, \ldots, \xi_{m+1})$ of $\mathcal{S}(\Pi)$ for which $\xi_{m+1} = 0$ will be denoted by $\mathcal{S}_{*0}(\Pi)$. Similarly, we introduce the notations $\mathcal{S}_{0*}(\Pi) = \{\xi \mid \xi \in \mathcal{S}(\Pi), \xi_0 = 0\}$, and $\mathcal{S}_0(\Pi) = \mathcal{S}_{*0}(\Pi) \cap \mathcal{S}_{0*}(\Pi)$, so that u_ξ belongs to $\mathfrak{D}_0[a, b]$, $\mathfrak{D}_{*0}[a, b]$, or $\mathfrak{D}_{0*}[a, b]$, according as ξ is a member of the respective class $\mathcal{S}_0(\Pi)$, $\mathcal{S}_{*0}(\Pi)$, or $\mathcal{S}_{0*}(\Pi)$.

Now if $\xi \in \mathcal{S}_0(\Pi)$ and $\zeta \in \mathcal{S}_0(\Pi)$ it follows from the conditions $u_\xi \in \mathfrak{D}_0[a, b]:v_\xi$, $u_\zeta \in \mathfrak{D}_0[a, b]:v_\zeta$, and $J[u_\xi, u_\zeta; a, b] = \overline{J[u_\zeta, u_\xi; a, b]}$, that the functional

$$(7.5) \qquad Q^0[\xi, \zeta \mid \Pi] = J[u_\xi, u_\zeta; a, b], \quad \textit{for} \quad \xi \in \mathcal{S}_0(\Pi), \qquad \zeta \in \mathcal{S}_0(\Pi),$$

is of the form

$$(7.6) \qquad Q^0[\xi, \zeta \mid \Pi] = \sum_{\alpha,\beta=1}^{m} \zeta_\alpha^* Q_{\alpha\beta}^0[\Pi]\xi_\beta,$$

where the $n \times n$ matrices $Q_{\alpha\beta}^0[\Pi]$ are such that

$$(7.6') \qquad Q_{\alpha\beta}^0[\Pi] = (Q_{\beta\alpha}^0[\Pi])^*, \qquad (\alpha, \beta = 1, \ldots, m).$$

That is, if $N = nm$ and $\check{\xi} = (\check{\xi}_\rho)$, $\check{\zeta} = (\check{\zeta}_\rho)$, $(\rho = 1, \ldots, N)$, where $\check{\xi}_{(p-1)n+i} = \xi_{ip}$, $\check{\zeta}_{(p-1)n+i} = \zeta_{ip}$, $(i = 1, \ldots, n;\ p = 1, \ldots, m)$, then

$Q^0[\xi, \zeta \mid \Pi]$ is an hermitian form in $\check{\xi}$, $\check{\zeta}$. Moreover, corresponding to (V.8.8) we now have the relation

$$(7.7)\quad Q^0[\xi, \zeta \mid \Pi] = \sum_{\alpha=1}^{m} \zeta_\alpha^*[v_\xi(t_\alpha^-) - v_\xi(t_\alpha^+)], \quad for \quad \xi \in \mathcal{S}_0(\Pi), \qquad \zeta \in \mathcal{S}_0(\Pi).$$

As with other hermitian functionals, we write $Q^0[\xi \mid \Pi]$ for $Q^0[\xi, \xi \mid \Pi]$.

The following analogue of Lemma V.8.2 may then be established by the same method of proof as used for the two-dimensional real system of Chapter V, and will be stated here without proof.

***Lemma* 7.2.** *If Π is a fundamental partition* (7.1) *of* $[a, b]$, *then the hermitian form $Q^0[\xi \mid \Pi]$ is singular if and only if $t = b$ is conjugate to $t = a$. Moreover, if $Q^0[\xi \mid \Pi]$ is singular then its nullity is equal to the order of $t = b$ as a conjugate point to $t = a$.*

Proceeding as in Chapter V, one establishes the following results, which are the direct generalizations of Theorem V.8.1 and Lemma V.8.3.

P_1. *If Π^1 and Π^2 are both fundamental partitions of a compact subinterval $[a, b]$ of I, then the two hermitian forms $Q^0[\xi \mid \Pi^1]$ and $Q^0[\xi \mid \Pi^2]$ have the same index, and the same nullity.*

P_2. *If $[a, b_1] \subset [a, b_2] \subset I$, and Π^1, Π^2 are fundamental partitions of $[a, b_1]$ and $[a, b_2]$, respectively, then the indices i_α and nullities n_α of the corresponding hermitian forms $Q^0[\xi \mid \Pi^\alpha]$,* $(\alpha = 1, 2)$, *are such that $i_1 \leq i_2$ and $i_1 + n_1 \leq i_2 + n_2$.*

Moreover, by the same type of argument as employed in the case considered in Chapter V, we have the following basic result.

Theorem 7.1. *If Π is a fundamental partition of a compact subinterval $[a, b]$ of I, then the index of $Q^0[\xi \mid \Pi]$ is equal to the number of points on the open interval (a, b) which are conjugate to $t = a$, where each point conjugate to $t = a$ is counted a number of times equal to its order as a conjugate point.*

Now an argument similar to that employed for the proof of the above theorem, but with the roles of $t = a$ and $t = b$ interchanged, yields the result that the index of $Q^0[\xi \mid \Pi]$ is also equal to the number of points on the open interval (a, b) which are conjugate to $t = b$. Consequently, for the general system (2.2) satisfying hypothesis $\mathfrak{H}_N(I)$ we have the following result which is far from trivial, although its counterpart for the system considered in Chapter V is a ready consequence of the result of Theorem V.2.4 on the separation of zeros of two linearly independent solutions.

Theorem 7.2. *If $[a, b] \subset I$, then the number of points on the interval (a, b), $\{(a, b]\}$, conjugate to $t = a$ is equal to the number of points on the interval (a, b), $\{[a, b)\}$, which are conjugate to $t = b$, where in each case a point is counted a number of times equal to its order as a conjugate point.*

If $Q^0[\xi \mid \Pi]$ is negative definite, {non-positive definite}, on a k-dimensional subspace spanned by vectors $\check{\xi}^\beta = (\check{\xi}_\rho^{\ \beta})$, $(\rho = 1, \ldots, N = mn)$, where $\check{\xi}^\beta_{(p-1)n+i} = \xi^\beta_{ip}$, $(i = 1, \ldots, n;\ p = 1, \ldots, m)$, then $J[\eta; a, b]$ is negative definite, {non-positive definite}, on a k-dimensional subspace of $\mathfrak{D}_0[a, b]$ spanned by $u_{\xi^1}(t), \ldots, u_{\xi^k}(t)$. Conversely, if $J[\eta; a, b]$ is negative definite, {non-positive definite}, on a k-dimensional subspace of $\mathfrak{D}_0[a, b]$ spanned by $\eta_1(t), \ldots, \eta_k(t)$, then for Π a fundamental partition (7.1) of $[a, b]$, and $\xi^\beta_\sigma = \eta_\beta(t_\sigma)$, $(\beta = 1, \ldots, k;\ \sigma = 0, 1, \ldots, m + 1)$, we have for arbitrary constants $d_1, \ldots, d_k$ that

$$\eta(t) = \sum_{\beta=1}^{k} d_\beta\, \eta_\beta(t), \qquad \xi_\sigma = \sum_{\beta=1}^{k} d_\beta \xi^\beta_\sigma,$$

are such that $u_\xi = \sum_{\beta=1}^{k} d_\beta\, u_{\xi\beta}(t)$, and in view of the results of Corollary 2 to Theorem 4.1 and Theorem 4.5 it follows that the disconjugacy of (2.2) on individual subintervals $[t_{j-1}, t_j]$ implies $J[\eta; t_{j-1}, t_j] \geq J[u_\xi; t_{j-1}, t_j]$, $(j = 1, \ldots, m + 1)$. Therefore $Q^0[\xi \mid \Pi]$ is correspondingly negative definite, {non-positive definite}, on the k-dimensional subspace spanned by the $\check{\xi}^\beta$, $(\beta = 1, \ldots, k)$, and we have the following result.

Theorem 7.3. *If $[a, b] \subset I$, and Π is a fundamental partition of $[a, b]$, then the index, {index plus nullity}, of $Q^0[\xi \mid \Pi]$ is equal to the largest non-negative integer k such that there exists a k-dimensional manifold in $\mathfrak{D}_0[a, b]$ on which the functional $J[\eta; a, b]$ is negative definite, {non-positive definite}.*

For a given $c \in I$, the set of points of I which are right-hand conjugate points to $t = c$ will be ordered as a sequence $\{t_p^+(c)\}$ with $t_p^+(c) \leq t_{p+1}^+(c)$, and each repeated a number of times equal to its order as a conjugate point. In case there are no points of I which are right-hand conjugate points to $t = c$ this sequence is vacuous. Otherwise, it may be finite or infinite, and in the latter case the result of Lemma 7.1 implies that only a finite number of elements of the sequence belong to any given compact subinterval of I. Correspondingly, the set of points of I which are left-hand conjugate points to $t = c$ are ordered as a sequence $\{t_p^-(c)\}$, with $t_{p+1}^-(c) \leq t_p^-(c)$ and similar convention as to repetitions.

Theorem 7.4. *If $t_p^+(c)$, $\{t_p^-(c)\}$, exists for $c = c_0$, then there exists a $\delta > 0$ such that $t_p^+(c)$, $\{t_p^-(c)\}$, exists for $c \in (c_0 - \delta, c_0 + \delta)$; moreover, $t_p^+(c)$, $\{t_p^-(c)\}$, is continuous at $c = c_0$.*

Explicit proof will be limited to the consideration of $t_p^+(c)$, as the treatment of $t_p^-(c)$ is entirely analogous in detail. For a fundamental partition

$$\Pi^0 : c_0 = t_0^0 < t_1^0 < \cdots < t_m^0 < t_{m+1}^0 = t_p^+(c_0)$$

of the interval $[c_0, t_p^+(c_0)]$, let $\delta_0 > 0$ be so small that

$$\delta_0 < t_1^0 - t_0^0, \qquad \delta_0 < t_{m+1}^0 - t_m^0, \qquad [c_0 - \delta_0, t_p^+(c_0) + \delta_0] \subset I,$$

the system (2.2) is disconjugate on each of the subintervals $[c_0 - \delta_0, t_1^0]$, $[t_m^0, t_p^+(c_0) + \delta_0]$, and the interval $[t_p^+(c_0) - \delta_0, t_p^+(c_0) + \delta_0]$ contains no points conjugate to $t = c_0$ distinct from $t_p^+(c_0)$. Now for $a \in (c_0 - \delta_0, c_0 + \delta_0)$ and $b \in (t_p^+(c_0) - \delta_0, t_p^+(c_0) + \delta_0)$, let $\Pi\{a, b\}: a = t_0 < t_1 < \cdots < t_m < t_{m+1} = b$ be the fundamental partition of $[a, b]$ specified by $t_j = t_j^0$, $(j = 1, \ldots, m)$. If $b_2 \in (t_p^+(c_0), t_p^+(c_0) + \delta_0)$, then $Q^0[\xi \mid \Pi\{c_0, b_2\}]$ has its index equal to the sum of the index and nullity of $Q^0[\xi \mid \Pi^0]$, which is not less than p. As the coefficients of $Q^0[\xi \mid \Pi\{a, b\}]$ are continuous functions of a and b it follows that if $b_2 \in (t_p^+(c_0), t_p^+(c_0) + \delta_0)$ then there exists a δ_1 satisfying $0 < \delta_1 < \delta_0$, and such that if $|a - c_0| < \delta_1$ then the index of $Q^0[\xi \mid \Pi\{a, b_2\}]$ is at least p, so that $t_p^+(a)$ exists and $t_p^+(a) < b_2$. On the other hand, if $b_1 \in (t_p^+(c_0) - \delta_0, t_p^+(c_0))$ then the index of $Q^0[\xi \mid \Pi\{a, b_1\}]$ does not exceed $p - 1$, and by a similar continuity argument it follows that there exists a δ_2 satisfying $0 < \delta_2 < \delta_0$, and such that if $|a - c_0| < \delta_2$ then $t_p^+(a) > b_1$ whenever $t_p^+(a)$ exists. If δ denotes the smaller of δ_1 and δ_2, we then have for $|a - c_0| < \delta$ that $t_p^+(a)$ exists and $b_1 < t_p^+(a) < b_2$, and hence the conclusion of the theorem is a consequence of the arbitrariness of b_1 and b_2.

Finally, we shall establish the following monotoneity property of conjugate points.

Theorem 7.5. *If $a_\alpha \in I$, $(\alpha = 1, 2)$, and $a_1 < a_2$, then whenever $t_p^+(a_2)$ $\{t_p^-(a_1)\}$, exists the corresponding p-*th *conjugate point $t_p^+(a_1)$, $\{t_p^-(a_1)\}$, exists, and $t_p^+(a_1) < t_p^+(a_2)$, $\{t_p^-(a_2) > t_p^-(a_1)\}$.*

Again, specific details of proof will be limited to the case of right-hand conjugate points, since the case of left-hand conjugate points may be considered by a similar argument.

The stated result will be established by induction. If $t_1^+(a_2)$ exists and $(u_1(t); v_1(t))$ is a solution of (2.2) determining $t = t_1^+(a_2)$ as a conjugate point to $t = a_2$, let $(\eta(t), \zeta(t)) = (u_1(t), v_1(t))$ for $t \in [a_2, t_1^+(a_2)]$, $(\eta(t), \zeta(t)) = (0, 0)$ for $t \in [a_1, a_2)$. Then $\eta \in \mathfrak{D}_0[a_1, t_1^+(a_2)]: \zeta$, and $J[\eta; a_1, t_1^+(a_2)] = J[u_1; a_2, t_1^+(a_2)] = 0$, so that $J[\eta; a_1, t_1^+(a_2)]$ is non-positive definite on a one-dimensional manifold in $\mathfrak{D}_0[a_1, t_1^+(a_2)]$. In view of Theorem 7.3, it then follows that $t_1^+(a_1)$ exists and $t_1^+(a_1) \leq t_1^+(a_2)$. Now if $t_1^+(a_1)$ were equal to $t_1^+(a_2)$ then $t = a_2$ would be a point of the interval $(a_1, t_1^+(a_2))$ conjugate to $t = t_1^+(a_2)$, while on $(a_1, t_1^+(a_1)) = (a_1, t_1^+(a_2))$ there would be no point conjugate to $t = a_1$, in contradiction to the result of Theorem 7.2. Hence $t_1^+(a_1) < t_1^+(a_2)$, and the conclusion of the theorem is established for $p = 1$.

Now suppose that the conclusion of the theorem is not true for all positive integers p, and denote by q the smallest positive integer such that $t_q^+(a_2)$

exists, and it is not true that $t_q^+(a_1)$ exists and satisfies the inequality $t_q^+(a_1) < t_q^+(a_2)$. Let $(u_j(t); v_j(t))$, $(j = 1, \ldots, q)$, be solutions of (2.2) determining $t_j^+(a_2)$ as a conjugate point to $t = a_2$, and such that if a conjugate point has order greater than one it is repeated a number of times equal to its order, and the corresponding determining solutions are chosen to be linearly independent. If $(\eta_j(t), \zeta_j(t)) = (u_j(t), v_j(t))$ for $t \in [a_2, t_j^+(a_2)]$, and $(\eta_j(t), \zeta_j(t)) = (0, 0)$ for $t \in [a_1, a_2) \cup (t_j^+(a_2), t_q^+(a_2)]$, then $\eta_j \in \mathfrak{D}_0[a_1, t_q^+(a_2)] : \zeta_j$ for $j = 1, \ldots, q$, and the vector functions $\eta_1(t), \ldots, \eta_q(t)$ are linearly independent elements of $\mathfrak{D}_0[a_1, t_q^+(a_2)]$. With the aid of the identities of Lemma 4.1 it follows that $J[\eta_i, \eta_j; a_1, t_q^+(a_2)] = 0$ for $i, j = 1, \ldots, q$, and hence $J[\eta; a_1, t_q^+(a_2)]$ is non-positive definite on the q-dimensional manifold in $\mathfrak{D}_0[a_1, t_q^+(a_2)]$ spanned by these elements. In view of Theorem 7.3 it then follows that $t_q^+(a_1)$ exists and $t_q^+(a_1) \le t_q^+(a_2)$. As the definitive property of the integer q requires that it is not true that $t_q^+(a_1)$ exists and $t_q^+(a_1) < t_q^+(a_2)$, it then follows that $t_q^+(a_1) = t_q^+(a_2)$, while $t_j^+(a_1) < t_j^+(a_2)$ for $j < q$. In particular, $t_{q-1}^+(a_1) < t_{q-1}^+(a_2) \le t_q^+(a_2) = t_q^+(a_1)$, so that the number of points on $(a_1, t_q^+(a_2))$ which are conjugate to $t = a_1$ is equal to $q - 1$. On the other hand, from Theorem 7.2 it follows that there are q points on $[a_2, t_q^+(a_2))$ which are conjugate points to $t = t_q^+(a_2)$, and as $[a_2, t_q^+(a_2)) \subset (a_1, t_q^+(a_2))$ we have the contradictory result that the number of points on $(a_1, t_q^+(a_2))$ conjugate to $t = t_q^+(a_2)$ is not equal to the number of points on this interval conjugate to $t = a_1$. Thus the assumption that the conclusion of the theorem is not true for all positive integers p has led to a contradiction.

Returning to the consideration of comparison theorems of the type introduced in Theorem 5.2, the combined results of Theorems 7.1, 7.2, and 7.3, now imply the following theorem.

Theorem 7.6. *Suppose that for* $\alpha = 1, 2$ *the system* (5.3_α) *satisfies hypothesis* $\mathfrak{H}_N(I)$; *also, for arbitrary compact subintervals* $[a, b]$ *of* I *we have* $\mathfrak{D}_1[a, b] = \mathfrak{D}_2[a, b] = \mathfrak{D}[a, b]$, *and the functional* $J_{1,2}[\eta; a, b]$ *of* (5.6) *is non-negative definite on* $\mathfrak{D}_0[a, b]$. *If* $t_{p\alpha}^+(c)$ *and* $t_{p\alpha}^-(c)$, $(p = 1, 2, \ldots)$, *denote the sequences of right- and left-hand conjugate points to* $t = c$ *relative to the respective system* (5.3_α), *then whenever the conjugate point* $t_{p1}^+(c)$, $\{t_{p1}^-(c)\}$, *exists the conjugate point* $t_{p2}^+(c)$, $\{t_{p2}^-(c)\}$, *also exists and*

$$t_{p2}^+(c) \le t_{p1}^+(c), \qquad \{t_{p2}^-(c) \ge t_{p1}^-(c)\}; \tag{7.8}$$

moreover, if $J_{1,2}[\eta; a, b]$ *is positive definite on* $\mathfrak{D}_0[a, b]$ *for arbitrary* $[a, b] \subset I$, *then strict inequalities hold in* (7.8).

For a system (2.2) satisfying hypothesis $\mathfrak{H}_N(I)$, we return to the consideration of the functional

$$\hat{J}[\eta; a, b] = \eta^*(a)\Gamma\eta(a) + \int_a^b \{\zeta^*(t)B(t)\zeta(t) + \eta^*(t)C(t)\eta(t)\}\, dt,$$

defined on compact subintervals $[a, b]$ of I by (6.1), where Γ is an $n \times n$ hermitian matrix. If Π is a fundamental partition (7.1) of $[a, b]$ and ξ, ζ are $(m + 2)$-tuples of n-dimensional vectors belonging to $\mathcal{S}_{*0}[\Pi]$, then the corresponding vector functions $(u_\xi(t); v_\xi(t))$, $(u_\zeta(t); v_\zeta(t))$ defined by (7.3), (7.4) are such that $u_\xi \in \mathfrak{D}_{*0}[a, b]:v_\xi$ and $u_\zeta \in \mathfrak{D}_{*0}[a, b]:v_\zeta$. The functional

$$Q^{*0}[\xi, \zeta \mid \Pi] = \hat{J}[u_\xi, u_\zeta; a, b], \tag{7.9}$$

now may be written as

$$Q^{*0}[\xi, \zeta \mid \Pi] = \sum_{\alpha,\beta=0}^{m} \zeta_\alpha^* Q_{\alpha\beta}^{*0}[\Pi]\xi_\beta, \tag{7.10}$$

where the $n \times n$ matrices $Q_{\alpha\beta}^{*0}$ are such that

$$Q_{\alpha\beta}^{*0}[\Pi] = (Q_{\beta\alpha}^{*0}[\Pi])^*, \qquad (\alpha, \beta = 0, 1, \ldots, m).$$

Corresponding to (7.7) we now have the representation

$$Q^{*0}[\xi, \zeta \mid \Pi) = \zeta_0^*[\Gamma\xi_0 - v_\xi(a)] + \sum_{j=1}^{m} \zeta_j^*[v_\xi(t_j^-) - v_\xi(t_j^+)], \tag{7.11}$$
$$\text{for} \quad \xi \in \mathcal{S}_{*0}(\Pi), \qquad \zeta \in \mathcal{S}_{*0}(\Pi).$$

Thus $Q^{*0}[\xi, \zeta \mid \Pi]$ is an hermitian form in the $(m + 1)n$-dimensional vectors $\check{\xi} = (\check{\xi}_\tau)$, $\check{\zeta} = (\check{\zeta}_\tau)$, $(\tau = 1, \ldots, (m + 1)n)$, where $\check{\xi}_{pn+i} = \xi_{ip}$, $\check{\zeta}_{pn+i} = \zeta_{ip}$, $(i = 1, \ldots, n; p = 0, 1, \ldots, m)$.

Corresponding to Lemma 7.2 and Theorems 7.1, 7.3 we now have the following results which will be stated without proof, as the details are analogous to those presented in establishing the earlier results.

***Lemma* 7.3.** *If Π is a fundamental partition of $[a, b]$, then $Q^{*0}[\xi \mid \Pi]$ is singular if and only if $t = b$ is a right-hand focal point to $t = a$ relative to the functional $\hat{J}[\eta; a, b]$. Moreover, if $Q^{*0}[\xi \mid \Pi]$ is singular then its nullity is equal to the order of $t = b$ as a focal point.*

Theorem 7.7. *If Π is a fundamental partition of a compact subinterval $[a, b]$ of I, then the index of $Q^{*0}[\xi \mid \Pi]$ is equal to the number of points on the open interval (a, b) which are right-hand focal points to $t = a$ relative to the functional $\hat{J}[\eta; a, b]$, where each focal point is counted a number of times equal to its order.*

Theorem 7.8. *If $[a, b] \subset I$, and Π is a fundamental partition of $[a, b]$, then the index, {index plus nullity}, of $Q^{*0}[\xi \mid \Pi]$ is equal to the largest non-negative integer k such that there exists a k-dimensional manifold in $\mathfrak{D}_{*0}[a, b]$ on which the functional $\hat{J}[\eta; a, b]$ is negative definite, {non-positive definite}.*

For a general conjoined basis $Y_0(t) = (U_0(t); V_0(t))$ of (2.2), the designation of a value c at which $U_0(c)$ is singular as a focal point of the basis is consistent with the characterization of a focal point with respect to the functional $\hat{J}$. Indeed, if $t = a$ is a value at which $U_0(a)$ is nonsingular, then the conjoined nature of $Y_0(t)$ implies that $W_0(a) = V_0(a)U_0^{-1}(a)$ is hermitian, and $(U(t); V(t)) = (U_0(t)U_0^{-1}(a); V_0(t)U_0^{-1}(a))$ is the conjoined basis determined by the initial condition (6.4) for the functional $\hat{J}[\eta; a, b]$ of (6.1) with $\Gamma = W_0(a)$; moreover, a value $c > a$ is a right-hand focal point of order k relative to this functional if and only if $U(c)$ is singular and has rank $n - k$.

For a given $a \in I$, the set of points on I which are right-hand focal points to $t = a$ relative to the functional $\hat{J}[\eta; a, b]$ will be ordered as a sequence $\tau_p^+(\Gamma)$, $(p = 1, 2, \ldots)$, numbered so that $\tau_p^+(\Gamma) \leq \tau_{p+1}^+(\Gamma)$, and each repeated a number of times equal to its order as a focal point. For focal points we have the following basic separation theorem.

Theorem 7.9. *Suppose that* (2.2) *satisfies hypothesis* $\mathfrak{H}_N(I)$, *and for* $\alpha = 1, 2$ *and* $[a, b] \subset I$ *let*

$$(7.12)\quad \hat{J}_\alpha[\eta; a, b] = \eta^*(a)\Gamma_\alpha\eta(a) + \int_a^b \{\zeta^*(t)B(t)\zeta(t) + \eta^*(t)C(t)\eta(t)\}\, dt,$$

where Γ_1 *and* Γ_2 *are* $n \times n$ *hermitian matrices. Moreover, let* **P** *and* **N** *denote the number of positive and negative proper values of the hermitian matrix* $\Gamma_1 - \Gamma_2$, *where each proper value is repeated a number of times equal to its multiplicity. If for a positive integer* p *the focal point* $\tau_{p+\mathbf{P}}^+(\Gamma_2)$ *exists, then* $\tau_p^+(\Gamma_1)$ *exists and* $\tau_p^+(\Gamma_1) \leq \tau_{p+\mathbf{P}}^+(\Gamma_2)$; *if* $\tau_{p+\mathbf{N}}^+(\Gamma_1)$ *exists then* $\tau_p^+(\Gamma_2)$ *exists and* $\tau_p^+(\Gamma_2) \leq \tau_{p+\mathbf{N}}^+(\Gamma_1)$.

Indeed, if $\tau_{p+\mathbf{P}}^+(\Gamma_2)$ exists then by Theorem 7.8 there exists a set of $p + \mathbf{P}$ linearly independent elements $\eta_j(t)$, $(j = 1, \ldots, p + \mathbf{P})$, of $\mathfrak{D}_{*0}[a, \tau_{p+\mathbf{P}}^+(\Gamma_2)]$ such that $\hat{J}_2[\eta; a, \tau_{p+\mathbf{P}}^+(\Gamma_2)]$ is non-positive definite on the $(p + \mathbf{P})$-dimensional subspace of $\mathfrak{D}_{*0}[a, \tau_{p+\mathbf{P}}^+(\Gamma_2)]$ spanned by $\eta_1(t), \ldots, \eta_{p+\mathbf{P}}(t)$. Now let $\xi_1, \ldots, \xi_{\mathbf{P}}$ be a set of linearly independent n-dimensional vectors which are proper vectors of the hermitian matrix $\Gamma_1 - \Gamma_2$ corresponding to the **P** positive proper values of this matrix. The set of vector functions $\eta(t)$ of the form

$$(7.13)\quad \eta(t) = \sum_{\gamma=1}^{p+\mathbf{P}} d_\gamma \eta_\gamma(t), \quad \textit{satisfying} \quad \xi_\beta^*\eta(a) = 0, \qquad (\beta = 1, \ldots, \mathbf{P}),$$

is then a linear manifold $\mathfrak{M}$ in $\mathfrak{D}_{*0}[a, \tau_{p+\mathbf{P}}^+(\Gamma_2)]$ of dimension at least p. Since the orthogonality conditions of (7.13) imply that

$$\eta^*(a)[\Gamma_1 - \Gamma_2]\eta(a) \leq 0 \text{ for } \eta \in \mathfrak{M},$$

it follows from the relation

$$\hat{J}_1[\eta; a, b] = \eta^*(a)[\Gamma_1 - \Gamma_2]\eta(a) + \hat{J}_2[\eta; a, b]$$

that $\hat{J}_1[\eta; a, \tau^+_{p+\mathbf{P}}(\Gamma_2)]$ is nonpositive for $\eta \in \mathfrak{M}$. From Theorem 7.8 it then follows that the functional $\hat{J}_1[\eta; a, \tau^+_{p+\mathbf{P}}(\Gamma_2)]$ has index plus nullity equal to at least p, and hence $\tau^+_p(\Gamma_1)$ exists and $\tau^+_p(\Gamma_1) \leq \tau^+_{p+\mathbf{P}}(\Gamma_2)$.

The final conclusion of the theorem is a consequence of the first conclusion, when the roles of $\hat{J}_1$ and $\hat{J}_2$ are interchanged.

***Corollary* 1.** *For a given subinterval I_0 of I the number of focal points on I_0 of any conjoined basis for* (2.2) *differs from that of any other conjoined basis for this system by at most n.*

Let $Y_\alpha(t) = (U_\alpha(t); V_\alpha(t))$, $(\alpha = 1, 2)$, be two conjoined bases for (2.2), and I_0 a compact subinterval $[a_0, b_0]$ of I. In view of the isolated nature of focal points, there exists a value $a < a_0$ such that each $U_\alpha(a)$ is nonsingular and the focal points of $Y_\alpha(t)$ on $[a, b_0]$ all occur on the interval $[a_0, b_0]$. As pointed out above, $(U_\alpha(t)U_\alpha^{-1}(a); V_\alpha(t)U_\alpha^{-1}(a))$, $(\alpha = 1, 2)$, is a conjoined basis determined by the initial conditions (6.4) for the corresponding functional $\hat{J}_\alpha$ with $\Gamma_\alpha = V_\alpha(a)U_\alpha^{-1}(a)$. Since each of the integers $\mathbf{P}$, $\mathbf{N}$ of the above theorem for the matrix $\Gamma_1 - \Gamma_2$ does not exceed n, the conclusion of the above theorem implies the result of the corollary in case I_0 is a compact subinterval $[a_0, b_0]$. In turn, if I_0 is of the form $(a_0, b_0]$, $[a_0, b_0)$, or (a_0, b_0), the conclusion of the corollary is a consequence of this first result, and the fact that in each case there is a compact subinterval $[a, b]$ of I_0 such that the number of focal points of each family on I_0 is equal to the number of focal points of that family on $[a, b]$.

For the differential system considered in Chapter V, any non-identically vanishing real solution defines a one-dimensional conjoined family, and the result of the above corollary is that of the Sturm comparison theorem.

Application of the above corollary to the particular conjoined basis which determines the points conjugate to a given point $t = a$, yields the following result.

***Corollary* 2.** *If $[a, b] \subset I$, and relative to* (2.2) *there are p conjugate points to $t = a$ on the interval $(a, b]$ or (a, b), then any conjoined basis for* (2.2) *has at most $p + n$ focal points on this interval.*

Problems VII.7

1. Suppose that $Y_\alpha(t) = (U_\alpha(t); V_\alpha(t))$, $(\alpha = 1, 2)$, are conjoined bases for a system (2.2) that satisfies hypothesis $\mathfrak{H}_N(I)$, and $t = a$ is a point such that $U_\alpha(a)$ is nonsingular for $\alpha = 1, 2$. Moreover, let the set of right-hand focal points of $Y_\alpha(t)$ on $\{t \mid t \in I,\ t > a\}$ be ordered as a sequence $\{\tau^+_{p,\alpha}\}$, $(p = 1, 2, \ldots; \alpha = 1, 2)$, numbered so that $\tau^+_{p,\alpha} \leq \tau^+_{p+1,\alpha}$, and each repeated a number of times equal to its order as a focal point. Show that if

$$V_1(a)U_1^{-1}(a) \geq V_2(a)U_2^{-1}(a)$$

then whenever $\tau_{p,1}^+$ exists the focal point $\tau_{p,2}^+$ also exists and $\tau_{p,2}^+ \leq \tau_{p,1}^+$. Formulate and prove the corresponding result for left-hand focal points.

2. If (2.2) satisfies hypothesis $\mathfrak{H}_N(I)$, and two conjoined bases for this system have in common h linearly independent solutions, prove that for a given subinterval I_0 of I the number of focal points of one basis on I_0 differs from that of the other basis by at most $n - h$.

Hint. If $Y_\alpha(t) = (U_\alpha(t); V_\alpha(t))$, $(\alpha = 1, 2)$, designates these conjoined families of solutions, and $t = a$ is a point at which each $U_\alpha(a)$ is nonsingular, show that for $\Gamma_\alpha = V_\alpha(a)U_\alpha^{-1}(a)$ the matrix $\Gamma_1 - \Gamma_2$ has nullity equal to h when these conjoined families have in common h linearly independent solutions, and conclude, in the notation of Theorem 7.9, that $\mathbf{P} + \mathbf{N} \leq n - h$.

3.[s] Suppose that $Q(t)$ is an $n \times n$ hermitian matrix function of class $\mathfrak{L}_{nn}^\infty[a, b]$ on arbitrary compact subintervals $[a, b]$ of a given interval I on the real line, and that $Q(t) > 0$ for t a.e. on I. For $\tau \in I$, let $\Phi = S(t; \tau)$, $\Psi = C(t; \tau)$ denote the solution of the matrix differential system

$$-\Psi'(t) - Q(t)\,\Phi(t) = 0, \qquad \Psi(\tau) = E,$$
$$\Phi'(t) - Q(t)\Psi(t) = 0, \qquad \Phi(\tau) = 0,$$

as in Problem VII.2:8. Moreover, for $\alpha = 1, 2$ let M_α, N_α be $n \times n$ matrices such that $M_\alpha^* N_\alpha - N_\alpha^* M_\alpha = 0$, and the $n \times 2n$ matrix $[M_\alpha^* \;\; N_\alpha^*]$ is of rank n. Show that the matrix functions $\Phi_\alpha(t) = S(t; \tau)M_\alpha + C(t; \tau)N_\alpha$, $\Psi_\alpha(t) = C(t; \tau)M_\alpha - S(t; \tau)N_\alpha$ are such that each $(\Phi_\alpha(t); \Psi_\alpha(t))$, $(\alpha = 1, 2)$, is a conjoined basis for the vector differential system $-\psi' - Q(t)\phi = 0$, $\phi' - Q(t)\psi = 0$, and conclude that for an arbitrary subinterval I_0 of I the number of focal points of $(\Phi_1(t); \Psi_1(t))$ on I_0 differs from the number of focal points of $(\Phi_2(t); \Psi_2(t))$ on I_0 by at most n. In particular, for $M_1 = E$, $N_1 = 0$, $M_2 = 0$, $N_2 = E$, it follows that on a given subinterval I_0 of I the number of zeros of $\det S(t; \tau)$ differs from the number of zeros of $\det C(t; \tau)$ by at most n, where a zero t_0 of $\det S(t; \tau)$ or $\det C(t; \tau)$ is counted k times if $S(t_0; \tau)$ or $C(t_0; \tau)$, respectively, is of rank $n - k$.

Hints. Use the hint provided for Problem VII.2:8, and the result of Corollary 1 to Theorem 7.9.

4. Suppose that (2.2) satisfies hypothesis $\mathfrak{H}_N(a_0, b_0)$, where $-\infty \leq a_0 < b_0 \leq \infty$, and that $Y(t) = (U(t); V(t))$ is a conjoined basis for (2.2) for which there is a subinterval $I_1 = (c_1, b_0)$ of $I = (a_0, b_0)$ such that the hermitian matrix function

$$P(t; Y) = V^*(t)B(t)V(t) + V^*(t)A(t)U(t) \\ + U^*(t)A^*(t)V(t) - U^*(t)C(t)U(t)$$

is positive definite for t a.e. on I_1. For $\alpha = 1, 2$ let M_α, N_α be $n \times n$ matrices such that $M_\alpha^* N_\alpha - N_\alpha^* M_\alpha = 0$, and the $n \times 2n$ matrix $[M_\alpha^* \quad N_\alpha^*]$ is of rank n. If $F_\alpha(t) = M_\alpha^* U(t) + N_\alpha^* V(t)$, $(\alpha = 1, 2)$, prove that for an arbitrary subinterval I_0 of I_1 the number of zeros of $\det F_1(t)$ on I_0 differs from the number of zeros of $\det F_2(t)$ on I_0 by at most n, where a zero t_0 of $\det F_\alpha(t)$ is counted k times if $F_\alpha(t_0)$ is of rank $n - k$.

Hints. To (2.2) apply the polar coordinate transformation, as presented in Problem VII.2:6. For the resulting system use the result of the above Problem 3, and interpret the results in terms of the given conjoined basis $(U(t); V(t))$ for (2.2).

8. SELF-ADJOINT DIFFERENTIAL SYSTEMS

Corresponding to the two dimensional systems treated in Chapter VI, we now consider two-point self-adjoint boundary problems involving the differential equations (2.2). With the notation $L[u](t) = u'(t) - A(t)u(t)$, $L^\star[v](t) = -v'(t) - A^*(t)v(t)$ as used in the earlier sections of this chapter, such a boundary problem is of the form

$$(\mathcal{B}) \qquad \begin{aligned} &\text{(a)} \begin{array}{l} L_1[u, v] \equiv L^\star[v](t) + C(t)u(t) = 0, \\ L_2[u, v] \equiv L[u](t) - B(t)v(t) = 0, \end{array} \qquad t \in [a, b], \\ &\text{(b)}\ M_1 u(a) + M_2 v(a) + M_3 u(b) + M_4 v(b) = 0, \end{aligned}$$

where the coefficient matrices M_β, $(\beta = 1, 2, 3, 4)$, are of dimension $2n \times n$ and such that the $2n \times 4n$ matrix

$$\mathcal{M} = [M_1 \quad M_2 \quad M_3 \quad M_4]$$

has rank $2n$. *It will be assumed that the* $n \times n$ *coefficient matrix functions* $A(t)$, $B(t)$, $C(t)$ *satisfy hypothesis* $(\mathfrak{H})$ *on a given compact interval* $[a, b]$.

In terms of the $2n$ dimensional vector function $y = (y_\alpha)$, $(\alpha = 1, \ldots, 2n)$, with $y_i = u_i$, $y_{n+i} = v_i$, $(i = 1, \ldots, n)$, system $(\mathcal{B})$ may be written as

$$(\mathcal{B}') \qquad \begin{aligned} &\text{(a)}\ \mathfrak{L}[y](t) \equiv \mathfrak{J}y'(t) + \mathcal{A}(t)y(t) = 0, \qquad t \in [a, b], \\ &\text{(b)}\quad s[y] \equiv \mathcal{M}\hat{y} = 0, \end{aligned}$$

where the hermitian matrix function $\mathcal{A}(t)$ and the constant skew hermitian matrix $\mathfrak{J}$ are specified as in (2.3), and $\hat{y}$ denotes the $4n$-dimensional vector $(\hat{y}_\sigma)$ with $\hat{y}_\alpha = y_\alpha(a)$, $\hat{y}_{2n+\alpha} = y_\alpha(b)$, $\alpha = 1, \ldots, 2n$. As noted in Section 2, the differential expression $\mathfrak{L}[y]$ is formally self-adjoint, and if $y(t)$ and $z(t)$ are $2n$-dimensional a.c. vector functions on $[a, b]$, then

$$(8.1)\qquad \int_a^b (\mathfrak{L}[y], z)\,dt - \int_a^b (y, \mathfrak{L}[z])\,dt = z^*(t)\,\mathfrak{J}y(t)\Big|_{t=a}^{t=b}, \\ = \hat{z}^*[\text{diag}\{-\mathfrak{J}, \mathfrak{J}\}]\hat{y}.$$

Consequently, the above defined problem $(\mathcal{B})$ is self-adjoint if and only if

$$(8.2)\qquad z^*[\text{diag}\{-\mathfrak{J}, \mathfrak{J}\}]y = 0, \quad \textit{whenever} \quad \mathcal{M}\hat{y} = 0, \qquad \mathcal{M}\hat{z} = 0.$$

Moreover, in a manner similar to that used in the derivation of relation (VI.1.3″), it follows that (8.2) is equivalent to the matrix equation

$$(8.3)\qquad \mathcal{M}[\text{diag}\{-\mathfrak{J}, \mathfrak{J}\}]\mathcal{M}^* = 0.$$

For brevity, let $S_{\sigma\tau}$ denote the $2n \times 2n$ matrix

$$S_{\sigma\tau} = [M_\sigma \quad M_\tau], \qquad (\sigma, \tau = 1, 2, 3, 4),$$

and introduce the abbreviations

$$(8.4)\qquad D = \text{diag}\{-E_n, E_n\}, \qquad N = S_{24}D.$$

By direct verification it follows that the matrix equation (8.3) may be written as

$$(8.3')\qquad S_{13}N^* - NS_{13}^* = 0;$$

that is, the $2n \times 2n$ matrix $S_{13}N^* = S_{13}DS_{24}^*$ is hermitian. Moreover, since $S_{24} = ND$, the boundary conditions $(\mathcal{B}\text{-b})$ may be written as

$$(8.5)\qquad S_{13}\hat{u} + ND\hat{v} = 0.$$

Theorem 8.1. *A necessary and sufficient condition for a differential system* $(\mathcal{B})$ *to be self-adjoint is that there exists a* $2n \times 2n$ *hermitian matrix* $Q = Q[\mathcal{B}]$ *and a linear subspace* $\mathcal{S} = \mathcal{S}[\mathcal{B}]$ *of* $\mathbf{C}_{2n}$ *such that* $\hat{u}$, $\hat{v}$ *satisfies the boundary conditions* $(\mathcal{B}\text{-b})$ *if and only if*

$$(8.6)\qquad \hat{u} \in \mathcal{S}, \qquad T[u, v] \equiv Q\hat{u} + D\hat{v} \in \mathcal{S}^\perp,$$

where $\mathcal{S}^\perp = \mathcal{S}^\perp[\mathcal{B}]$ *denotes the orthogonal complement of* $\mathcal{S} = \mathcal{S}[\mathcal{B}]$ *in* $\mathbf{C}_{2n}$.

The following proof of Theorem 8.1 is basically the same as that given for Theorem VI.1.1, made concise through matrix algebra. If the matrix N has rank zero, then the boundary conditions reduce to $S_{13}\hat{u} = 0$ with S_{13} nonsingular, so that (8.5) is equivalent to (8.6) with $\mathcal{S}$ the zero-dimensional subspace of $\mathbf{C}_{2n}$. If N is of rank $2n$, then (8.5) is equivalent to $N^{-1}S_{13}\hat{u} + D\hat{v} = 0$, with $N^{-1}S_{13}$ hermitian in view of (8.3′), and consequently (8.5) is equivalent to (8.6) with $\mathcal{S} = \mathbf{C}_{2n}$, $Q = N^{-1}S_{13}$.

Now if N has rank equal to d, $1 \leq d \leq 2n - 1$, let ψ be a $2n \times (2n - d)$ matrix of rank $2n - d$ and such that $\psi^* N = 0$, and let χ be a $2n \times d$ matrix such that the $2n \times 2n$ matrix $[\psi \quad \chi]$ is nonsingular. The boundary conditions (8.5) are then equivalent to the equations

$$(8.5') \qquad \begin{aligned} \psi^* S_{13} \hat{u} &= 0,. \\ \chi^* S_{13} \hat{u} + \chi^* N D \hat{v} &= 0, \end{aligned}$$

and since the matrix $\mathcal{M}$ is of rank $2n$ it follows that the $(2n - d) \times 2n$ matrix $\psi^* S_{13}$ is of rank $2n - d$, and the $d \times 2n$ matrix $\chi^* N$ is of rank d. Moreover, equation (8.3′) implies that $(\psi^* S_{13})(\chi^* N)^* = \psi^* S_{13} N^* \chi = (\psi^* N) S_{13}^* \chi = 0$, and as the matrices $\chi^* N$ and N are each of rank d it follows that $\psi^* S_{13} N^* = 0$, and any $k \times 2n$ matrix R satisfying $RN^* = 0$ is of the form $R = \lambda \psi^* S_{13}$ for some $k \times (2n - d)$ matrix λ.

Let H be the E. H. Moore generalized inverse of the hermitian matrix $S_{13} N^*$, (see Appendix B; Section 3). Then H is an hermitian $2n \times 2n$ matrix such that $0 = (S_{13} N^*) - (S_{13} N^*) H (S_{13} N^*) = (S_{13} - N S_{13}^* H S_{13}) N^*$, and by the above comment there is a $2n \times (2n - d)$ matrix Λ such that

$$(8.7) \qquad S_{13} - N S_{13}^* H S_{13} = \Lambda \psi^* S_{13}.$$

Now the boundary conditions (8.5′) are equivalent to the conditions

$$\begin{aligned} \psi^* S_{13} \hat{u} &= 0, \\ (\chi^* S_{13} - \chi^* \Lambda \psi^* S_{13}) \hat{u} + \chi^* N D \hat{v} &= 0. \end{aligned}$$

In turn, relation (8.7) implies that these boundary conditions may be written as

$$\begin{aligned} \psi^* S_{13} \hat{u} &= 0, \\ \chi^* N [S_{13}^* H S_{13} \hat{u} + D \hat{v}] &= 0, \end{aligned}$$

which is a set of conditions of the form (8.6) with $\mathcal{S}$ the d-dimensional subspace of $\mathbf{C}_{2n}$ defined as $\{\hat{u} \mid \psi^* S_{13} \hat{u} = 0\}$, and $Q = S_{13}^* H S_{13}$.

If the $2n \times 2n$ matrix $Q = Q[\mathcal{B}]$ has been determined as in Theorem 8.1, we write $Q[\eta_1, \eta_2] = Q[\eta_1, \eta_2 \mid \mathcal{B}]$ for the hermitian form $\hat{\eta}_2^* Q \hat{\eta}_1$, and for $\eta_\alpha \in \mathfrak{D}[a, b] : \zeta_\alpha$, $(\alpha = 1, 2)$, the functional $J[\eta_1, \eta_2 \mid \mathcal{B}]$ is defined as

$$(8.8) \quad J[\eta_1, \eta_2 \mid \mathcal{B}] = Q[\eta_1, \eta_2 \mid \mathcal{B}] + \int_a^b \{\zeta_2^*(t) B(t) \zeta_1(t) + \eta_2^*(t) C(t) \eta_1(t)\}\, dt.$$

Corresponding to earlier abbreviations, $J[\eta_1 \mid \mathcal{B}]$ is used to denote $J[\eta_1, \eta_1 \mid \mathcal{B}]$.

The following preliminary result, which will be used frequently in the treatment of boundary problems $(\mathcal{B})$, is a ready consequence of relation (4.4′).

Lemma 8.1. *If* $\eta_\alpha \in \mathfrak{D}[a,b]:\zeta_\alpha$, $(\alpha=1,2)$, *then*

$$(8.9)\quad \begin{aligned} &\text{(a)} \quad J[\eta_1, \eta_2 \mid \mathcal{B}] = \hat{\eta}_2^* T[\eta_1, \zeta_1] + \int_a^b \eta_2^* L_1[\eta_1, \zeta_1]\, dt, \\ &\text{(b)} \quad J[\eta_1 \mid \mathcal{B}] = \hat{\eta}_1^* T[\eta_1, \zeta_1] + \int_a^b \eta_1^* L_1[\eta_1, \zeta_1]\, dt. \end{aligned}$$

The symbol $\mathfrak{D}[\mathcal{B}]$ is also employed for the set $\{\eta \mid \eta \in \mathfrak{D}[a, b],\ \hat{\eta} \in \mathcal{S}[\mathcal{B}]\}$; moreover, we write $\eta \in \mathfrak{D}[\mathcal{B}]:\zeta$ if $\eta \in \mathfrak{D}[\mathcal{B}]$ and $L[\eta] = B\zeta$. Finally, for $\eta_\alpha(t) \in \mathfrak{C}_n[a, b]$, $(\alpha = 1, 2)$, we introduce the notations

$$(8.10)\quad \begin{aligned} N[\eta_1, \eta_2] &= \hat{\eta}_2^* \hat{\eta}_1 + \int_a^b \eta_2^*(t)\, \eta_1(t)\, dt, \\ N[\eta_1] = N[\eta_1, \eta_1] &= |\hat{\eta}_1|^2 + \int_a^b |\eta_1(t)|^2\, dt. \end{aligned}$$

Theorem 8.2. *If* $(\mathcal{B})$ *is self-adjoint and has only the identically vanishing solution* $(u(t); v(t)) \equiv (0; 0)$, *then there exist* $n \times n$ *matrix functions* $G(t, s)$, $G_0(t, s)$ *for* $(t, s) \in \square = [a, b] \times [a, b]$ *such that:*

(i) $G(t, s)$ *is continuous in* (t, s) *on* $\square$, *is* a.c. *in each argument on* $[a, b]$ *for fixed values of the other argument, and* $G(t, s) \equiv [G(s, t)]^*$ *on* $\square$.

(ii) $G_0(t, s)$ *is continuous in* (t, s) *on each of the triangular domains* $\Delta_1 = \{(t, s) \mid (t, s) \in \square,\ s < t\}$ *and* $\Delta_2 = \{(t, s) \mid (t, s) \in \square,\ t < s\}$, *is bounded on* $\square$, *and the restriction of* G_0 *to* Δ_α, $(\alpha = 1, 2)$, *has a finite limit at each* (t, t) *with* $t \in [a, b]$.

(iii) *If* $s \in (a, b)$, *and* ξ *is an arbitrary n-dimensional vector, then* $(u(t); v(t)) = (G(t, s)\xi;\ G_0(t, s)\xi)$ *is a solution of the differential equations* ($\mathcal{B}$-a) *on each of the subintervals* $[a, s)$ *and* $(s, b]$, *and also satisfies the boundary conditions* ($\mathcal{B}$-b); *in particular*, $u \in \mathfrak{D}[\mathcal{B}]:v$.

(iv) *If* $f \in \mathfrak{L}_n^2[a, b]$, *then the unique solution of the differential system*

$$(8.11)\quad \begin{aligned} &\text{(a)} \quad L_1[u, v](t) = f(t), \qquad L_2[u, v](t) = 0, \qquad t \in [a, b], \\ &\text{(b)} \quad \hat{u} \in \mathcal{S}[\mathcal{B}], \qquad T[u, v] \in \mathcal{S}^\perp[\mathcal{B}], \end{aligned}$$

is given by

$$(8.12)\quad \begin{aligned} u(t) &= \int_a^b G(t, s) f(s)\, ds, \\ v(t) &= \int_a^b G_0(t, s) f(s)\, ds. \end{aligned} \qquad t \in [a, b],$$

(v) *If* $\eta \in \mathfrak{D}[\mathcal{B}]:\zeta$, *and* $(u(t); v(t))$ *is the unique solution of the differential system*

$$(8.13)\quad \begin{aligned} &\text{(a)} \quad L_1[u, v](t) = \eta(t), \qquad L_2[u, v](t) = 0, \qquad t \in [a, b], \\ &\text{(b)} \quad \hat{u} \in \mathcal{S}[\mathcal{B}], \qquad T[u, v] - \hat{\eta} \in \mathcal{S}^\perp[\mathcal{B}], \end{aligned}$$

then there exists a $k > 0$ *such that* $N[u] \le k^2 N[\eta]$; *moreover*

$$J[u, \eta \mid \mathcal{B}] = N[\eta], \qquad J[u \mid \mathcal{B}] = N[\eta, u]. \tag{8.14}$$

Conclusions (i)–(iv) are direct consequences of the results of Section III.7 on the properties of the Green's matrix for an incompatible boundary problem. If $Y(t) = (U(t); V(t))$ is a fundamental matrix of solutions of the differential equations ($\mathcal{B}$-a), then there exists a $2n \times 2n$ matrix R such that the solution of (8.13) is given by

$$\begin{aligned} u(t) &= U(t)R\hat{\eta} + \int_a^b G(t, s)\eta(s)\,ds, \\ v(t) &= V(t)R\hat{\eta} + \int_a^b G_0(t, s)\eta(s)\,ds. \end{aligned} \qquad t \in [a, b], \tag{8.15}$$

By elementary inequalities it follows that there exists a constant k_1 such that

$$|u(t)|^2 \le k_1 N[\eta], \quad \textit{for} \quad t \in [a, b], \tag{8.16}$$

which implies in turn that $N[u] \le k_1[2 + (b - a)]N[\eta]$. Moreover, if $(u(t); v(t))$ is the solution of (8.13), and $\eta_1 \in \mathfrak{D}[\mathcal{B}]$, with the aid of relation (8.9-a) it follows that

$$J[u, \eta_1 \mid \mathcal{B}] = \hat{\eta}_1^* \hat{\eta} + \int_a^b \eta_1^* \eta\,dt = N[\eta, \eta_1],$$

and the relations (8.14) follow from this identity for $\eta_1 = \eta$ and $\eta_1 = u$.

9. NORMALITY AND ABNORMALITY OF BOUNDARY PROBLEMS ($\mathcal{B}$)

As in Section 3, let $\Lambda[a, b]$ denote the vector space of n-dimensional vector functions $v(t)$ which on $[a, b]$ are solutions of the differential equation $L^\star[v] = 0$, and satisfy $B(t)v(t) = 0$ for $t \in [a, b]$; that is, $v \in \Lambda[a, b]$ if and only if $u(t) \equiv 0$, $v(t)$ is a solution of the differential system ($\mathcal{B}$-a). Now for a given problem ($\mathcal{B}$) involving a subspace $\mathcal{S}$ of $\mathbf{C}_{2n}$, the subspace of $\Lambda[a, b]$ on which the $2n$ dimensional vector $D\hat{v}$ belongs to $\mathcal{S}^\perp$ will be denoted by $\Lambda\{\mathcal{S}\}$. Clearly, $v \in \Lambda\{\mathcal{S}\}$ if and only if $u(t) \equiv 0$, $v(t)$ is a solution of ($\mathcal{B}$). If $\Lambda\{\mathcal{S}\}$ is zero dimensional the *boundary problem* ($\mathcal{B}$) *is said to be normal*, or to have *order of abnormality zero*, whereas if $\Lambda\{\mathcal{S}\}$ has dimension $\delta > 0$ the problem ($\mathcal{B}$) is said to be abnormal, with order of abnormality equal to δ. If w_ν, $(\nu = 1, \ldots, 2n - d)$, is a basis for $\mathcal{S}^\perp$, and $v_\beta(t)$, $(\beta = 1, \ldots, d_a)$, is a basis for $\Lambda[a, b]$, then ($\mathcal{B}$) is normal if and only if the $2n \times 2n - d + d_a$ matrix

$$[w_\nu \quad D\hat{v}_\beta] \tag{9.1}$$

is of rank $2n - d + d_a$. If $(\mathcal{B})$ has order of abnormality equal to δ then (9.1) has rank $2n - d + d_a - \delta$, and upon deleting a suitable set $w_\nu^* \hat{\eta} = 0$, $(\nu = \nu_1, \ldots, \nu_\delta)$, of the conditions defining $\mathcal{S}$ the remaining conditions $w_\nu^* \hat{\eta} = 0$, $(\nu \neq \nu_j, j = 1, \ldots, \delta)$, defines a linear subspace $\mathcal{S}_\mu$ that is of dimension $d + \delta$, is such that $\mathcal{S} \subset \mathcal{S}_\mu$, and the problem

$$(\mathcal{B}_\mu) \qquad \begin{array}{ccc} L_1[u, v](t) = 0, & L_2[u, v](t) = 0, & t \in [a, b], \\ \hat{u} \in \mathcal{S}_\mu, & T[u, v] \in \mathcal{S}_\mu^\perp, & \end{array}$$

is normal. Moreover, since Lemma 3.2 implies that $\hat{v}_\beta^* D\hat{\eta} = 0$ for arbitrary $\eta \in \mathfrak{D}[a, b]$, an n-dimensional vector function $\eta(t)$ belongs to $\mathfrak{D}[\mathcal{B}]$ if and only if $\eta(t)$ belongs to $\mathfrak{D}[\mathcal{B}_\mu]$. Also $(\mathcal{B}_\mu)$ is a normal problem equivalent to $(\mathcal{B})$ in the following sense. If $(u(t); v(t))$ is a nonidentically vanishing solution of $(\mathcal{B}_\mu)$ then $u(t) \not\equiv 0$ on $[a, b]$, and $(u(t); v(t))$ is a solution of $(\mathcal{B})$. Moreover, if $(u(t); v(t))$ is a solution of $(\mathcal{B})$ then there exist unique constants c_β, $(\beta = 1, \ldots, d_a)$, such that $(u(t); v(t) + \sum_\beta c_\beta v_\beta(t))$ is a solution of $(\mathcal{B}_\mu)$.

For brevity, let $\mathcal{S}_a$ denote the $(2n - d_a)$-dimensional subspace of $\mathbf{C}_{2n}$ defined as

$$(9.2) \qquad \mathcal{S}_a = \{\hat{\eta} \mid \hat{v}^* D\hat{\eta} = 0, \quad \textit{for} \quad v \in \Lambda[a, b]\}.$$

In particular, if $(\mathcal{B}^0)$ denotes the problem $(\mathcal{B})$ with $\mathcal{S}$ the zero-dimensional subspace of $\mathbf{C}_{2n}$, then a corresponding normal problem $(\mathcal{B}_\mu{}^0)$ determined by the above process is $(\mathcal{B})$ with $\mathcal{S} = \mathcal{S}_a{}^\perp$; that is, the system involving the differential equations ($\mathcal{B}$-a) and the boundary conditions

$$(\mathcal{B}_\mu{}^0\text{-b}) \qquad \hat{u} \in \mathcal{S}_a{}^\perp, \qquad T[u, v] \in \mathcal{S}_a.$$

Now for a normal problem $(\mathcal{B})$ the condition that the matrix (9.1) has rank $2n - d + d_a$ is equivalent to the condition that

$$\dim\, [\mathcal{S}[\mathcal{B}] \cap \mathcal{S}_a]^\perp = \dim \mathcal{S}^\perp[\mathcal{B}] + \dim \mathcal{S}_a{}^\perp = (2n - d) + d_a,$$

so that $\dim[\mathcal{S}[\mathcal{B}] \cap \mathcal{S}_a] = d - d_a$. Consider with $(\mathcal{B})$ a second problem $(\mathcal{B}_\#)$ which involves the same differential equations ($\mathcal{B}_\#$-a) as ($\mathcal{B}$-a), and the boundary conditions

$$(\mathcal{B}_\#\text{-b}) \qquad \hat{u} \in \mathcal{S}_\#, \qquad T[u, v] \in \mathcal{S}_\#{}^\perp,$$

where $\mathcal{S}_\#$ is a second subspace of $\mathbf{C}_{2n}$. If $\dim \mathcal{S} = d$, $\dim \mathcal{S}_\# = d_\#$, and each of the systems $(\mathcal{B})$, $(\mathcal{B}_\#)$ is normal, then $\dim[\mathcal{S} \cap \mathcal{S}_a] = d - d_a$ and $\dim[\mathcal{S}_\# \cap \mathcal{S}_a] = d_\# - d_a$. If $\mathcal{S}_\# \cap \mathcal{S}_a \subset \mathcal{S} \cap \mathcal{S}_a$, then $d \geq d_\#$ and $(\mathcal{B}_\#)$ is called a *subproblem* of $(\mathcal{B})$ of *dimension* $d - d_\#$. If $d > d_\#$ then there exist $d - d_\#$ linear forms $\theta_\tau[\hat{\eta}] = \theta_\tau^* \hat{\eta}, = 0$, $(\tau = 1, \ldots, d - d_\#)$, such that

$$(9.3) \qquad \mathcal{S}_\# \cap \mathcal{S}_a = \{\eta \mid \eta \in \mathcal{S} \cap \mathcal{S}_a, \theta_\tau[\hat{\eta}] = 0, \tau = 1, \ldots, d - d_\#\}.$$

In particular, if $(\mathfrak{B})$ is normal then the problem $(\mathfrak{B}_\mu{}^0)$, consisting of the differential equations $(\mathfrak{B}\text{-a})$ and the above boundary conditions $(\mathfrak{B}_\mu{}^0\text{-b})$, is a subproblem of $(\mathfrak{B})$ of dimension $d - d_a$.

10. PRELIMINARY EXISTENCE THEOREMS

The major portion of the remainder of this chapter will be devoted to various aspects of the theory of differential systems

$$\text{(10.1)}\qquad \begin{aligned} L_1[u, v](t) = \lambda K(t)u(t), \qquad L_2[u, v](t) = 0, \qquad t \in [a, b], \\ \hat{u} \in \mathcal{S}, \qquad T[u, v] \equiv Q\hat{u} + D\hat{v} \in \mathcal{S}^\perp, \end{aligned}$$

involving the characteristic parameter λ, and under the following hypothesis.

$(\mathfrak{H}_\mathfrak{B})$ *On a given interval* $[a, b]$ *of the real line the* $n \times n$ *matrix functions* $A(t)$, $B(t)$, $C(t)$ *satisfy hypothesis* $(\mathfrak{H})$, $\mathcal{S}$ *is a subspace of* $\mathbf{C}_{2n}$ *such that* (10.1) *is normal,* Q *is an hermitian* $2n \times 2n$ *matrix,* $D = \operatorname{diag}\{-E_n, E_n\}$, *and* $K(t)$ *is an hermitian matrix function of class* $\mathfrak{L}_{nn}{}^\infty[a, b]$ *such that* $K(t) \not\equiv 0$ *for* t *on a subset of* $[a, b]$ *of positive measure.*

For brevity, we also introduce the notation

$$\text{(10.2)}\qquad K[\eta_1, \eta_2] = \int_a^b \eta_2^*(t)K(t)\eta_1(t)\,dt, \qquad K[\eta_1] = K[\eta_1, \eta_1];$$

clearly $K[\eta_1, \eta_2]$ is an hermitian form on $\mathfrak{L}_n{}^2[a, b] \times \mathfrak{L}_n{}^2[a, b]$.

It is to be noted that (10.1) has in common with $(\mathfrak{B})$ of Section 8 the equation $L_2[u, v](t) = 0$ and the subspace $\mathcal{S}$ of $\mathbf{C}_{2n}$, so that for (10.1) the classes of vector functions $\mathfrak{D}[a, b]$, $\mathfrak{D}[\mathfrak{B}]$ and $\mathfrak{D}_0[\mathfrak{B}] = \mathfrak{D}_0[a, b]$ have the same meaning as for the system $(\mathfrak{B})$. Moreover, the condition that (10.1) is normal is the same as the condition that $(\mathfrak{B})$ is normal; that is, if $(u(t) \equiv 0, v(t))$, $t \in [a, b]$, is a solution of (10.1) for a value of λ, (and hence for all values of λ), then $v(t) \equiv 0$ on $[a, b]$.

As a preliminary result, the following theorem is established.

Theorem 10.1. *If hypothesis* $(\mathfrak{H}_\mathfrak{B})$ *holds, and* $K[u] > 0$ *whenever* $(u(t); v(t))$ *is a proper solution of* (10.1), *then all proper values of this system are real, and solutions* $(u(t); v(t))$, $(u_0(t); v_0(t))$ *corresponding to distinct proper values* λ, λ_0 *are K-orthogonal in the sense that* $K[u_0, u] = 0$. *Moreover, if* λ *is a proper value of index* m_λ, *then the linear vector space of solutions of* (10.1) *for this value* λ *has a basis* $(u_i(t); v_i(t))$, *which is K-orthonormal in the sense that* $K[u_i, u_j] = \delta_{ij}$, $(i, j = 1, \ldots, m_\lambda)$.

If $(u(t); v(t))$ and $(u_0(t); v_0(t))$ are solutions of (10.1) corresponding to respective proper values λ and λ_0, from relation (8.9-a) it follows that $J[u, u_0 \mid \mathfrak{B}] = \lambda K[u, u_0]$, $J[u_0, u \mid \mathfrak{B}] = \lambda_0 K[u_0, u]$, and consequently

$$\text{(10.3)}\qquad (\lambda - \bar{\lambda}_0)\, K[u, u_0] = 0.$$

If λ is nonreal, then this relation for $\lambda_0 = \lambda$ and $(u_0(t); v_0(t)) = (u(t); v(t))$ implies that $K[u] = 0$, so that the assumption that $K[u]$ is positive for an arbitrary proper solution $(u(t); v(t))$ of (10.1) implies that $u(t) \equiv 0$ on $[a, b]$. The condition that (10.1) is normal then requires $v(t) = 0$ for $t \in [a, b]$, and therefore a nonreal value λ is not a proper value of such a system. If λ and λ_0 are distinct proper values of (10.1), then the above established relation (10.3) implies that $K[u, u_0] = 0$. Finally, if λ is a proper value of index $m = m_\lambda$ and $(u_i^0(t); v_i^0(t))$, $(i = 1, \ldots, m)$, are linearly independent solutions of (10.1) for this value of λ, then $[K[u_i^0, u_j^0]]$, $(i, j = 1, \ldots, m)$, is a positive definite hermitian matrix. Consequently, (see Appendix E; especially Problem 3 of E.1, and Theorem E.2.2), there exists a unitary matrix $\Gamma = [\Gamma_{ij}]$ such that $\Gamma^*[K[u_i^0, u_j^0]]\Gamma$ is a diagonal matrix $[\mu_i \delta_{ij}]$ with $\mu_i > 0$, and the stated result holds for $u_j(t) = (1/\sqrt{\mu_j}) \sum_{i=1}^m \Gamma_{ij} u_i^0(t)$, $(j = 1, \ldots, m)$.

In particular, if $\mathcal{S}_a$ is defined by (9.2) and the $n \times n$ matrix functions $A(t), B(t), C(t)$ satisfy hypothesis $(\mathfrak{H})$, then for an arbitrary $2n \times 2n$ hermitian matrix Q the boundary problem

$$
(10.4) \qquad \begin{aligned} L_1[u, v](t) &= \lambda u(t), \qquad L_2[u, v](t) = 0, \qquad t \in [a, b], \\ \hat{u} &\in \mathcal{S}_a^{\perp}, \qquad T[u, v] \in \mathcal{S}_a, \end{aligned}
$$

is a special case of (10.1) with $K(t) \equiv E_n$. It will be left to the reader to verify that (10.4) is normal, so that this system satisfies the above hypothesis $(\mathfrak{H}_{\mathscr{B}})$. In particular, (10.4) has only real proper values, and hence by a general result of Section IV.2 on the distribution of proper values of such a system the set of all proper values is at most denumerably infinite. Let λ_0 be a real number which is not a proper value of (10.4). Then for $\lambda = \lambda_0$ the system (10.4) has only the identically vanishing solution, and for arbitrary $w \in \mathbf{C}_{2n}$ the nonhomogeneous system

$$
(10.5) \qquad \begin{aligned} L_1[u, v](t) &= \lambda_0 u(t), \qquad L_2[u, v](t) = 0, \qquad t \in [a, b], \\ \hat{u} - w &\in \mathcal{S}_a^{\perp}, \qquad T[u, v] \in \mathcal{S}_a, \end{aligned}
$$

has a unique solution $(u(t); v(t))$. Since $u \in \mathfrak{D}[a, b]{:}v$ we have that $\hat{u} \in \mathcal{S}_a$. Consequently, if $w \in \mathcal{S}_a$ then $\hat{u} - w \in \mathcal{S}_a \cap \mathcal{S}_a^{\perp}$, and therefore $\hat{u} = w$. In particular, the following corollary is a consequence of this result and the initial conclusion of Lemma 3.2.

Corollary. *If hypothesis* $(\mathfrak{H})$ *holds for the coefficient matrix functions of a system* $(\mathscr{B})$ *as defined in Section* 8, *then a vector* $w \in \mathbf{C}_{2n}$ *is such that there exists a vector function* $\eta(t) \in \mathfrak{D}[a, b]$ *with* $\hat{\eta} = w$ *if and only if* $w \in \mathcal{S}_a$.

Corresponding to the results of Theorems 4.1 and 6.1, we now have the following result for general differential systems of the form $(\mathscr{B})$.

Theorem 10.2. *If hypothesis* ($\mathfrak{H}$) *holds for the coefficient matrix functions of* ($\mathfrak{B}$) *and* $u \in \mathfrak{A}_n[a, b]$, *then there exists a* v *such that* $(u; v)$ *is a solution of* ($\mathfrak{B}$) *if and only if there exists a* $v_1 \in \mathfrak{L}_n^2[a, b]$ *such that* $u \in \mathfrak{D}[\mathfrak{B}]:v_1$ *and*

$$(10.6) \qquad J[u:v_1, \eta:\zeta; \mathfrak{B}] = 0, \quad \text{for} \quad \eta \in \mathfrak{D}[\mathfrak{B}]:\zeta.$$

If $(u; v)$ is a solution of ($\mathfrak{B}$) then $u \in \mathfrak{D}[\mathfrak{B}]:v$, and with the aid of Lemma 8.1 it follows readily that $J[u:v, \eta:\zeta; \mathfrak{B}] = J[u, \eta; \mathfrak{B}] = 0$ for $\eta \in \mathfrak{D}[\mathfrak{B}]:\zeta$. On the other hand, if $u \in \mathfrak{D}[\mathfrak{B}]:v_1$ and relation (10.6) holds, then since $\mathfrak{D}_0[\mathfrak{B}] \subset \mathfrak{D}[\mathfrak{B}]$ it follows from Theorem 4.1 that there exists a $v_0(t)$ such that $(u(t); v_0(t))$ is a solution of the differential system ($\mathfrak{B}$-a) on $[a, b]$. If $\eta \in \mathfrak{D}[\mathfrak{B}]:\zeta$, Lemma 8.1 then yields the condition

$$\hat{\eta}^* T[u, v_0] = 0 \quad \text{for} \quad \eta \in \mathfrak{D}[\mathfrak{B}]:\zeta.$$

As in Section 9, let w_ν, $(\nu = 1, \ldots, 2n - d)$, be a basis for $\mathcal{S}^\perp$, and $v_\beta(t)$, $(\beta = 1, \ldots, d_a)$, a basis for $\Lambda[a, b]$. Moreover, let w be a $2n$-dimensional vector such that

$$(10.7) \qquad w_\nu^* w = 0, \quad (\nu = 1, \ldots, 2n - d); \quad \hat{v}_\beta^* D w = 0, \quad (\beta = 1, \ldots, d_a).$$

As the equations $\hat{v}_\beta^* D w = 0$ imply that $w \in \mathcal{S}_a$, it follows from the above Corollary to Theorem 10.1 that there exists a vector function $\eta(t)$ in $\mathfrak{D}[a, b]$ such that $\hat{\eta} = w$. The equations $w_\nu^* \hat{\eta} = 0$ then imply that $\hat{\eta} \in \mathcal{S}$ and hence $\eta \in \mathfrak{D}[\mathfrak{B}]$. Therefore, we have established that $w^* T[u, v_0] = 0$ if $w \in \mathbf{C}_{2n}$ and satisfies the conditions (10.7). Consequently there exist constants d_ν, $(\nu = 1, \ldots, 2n - d)$, and c_β, $(\beta = 1, \ldots, d_a)$, such that

$$(10.8) \qquad T[u, v_0] = \sum_\nu d_\nu w_\nu + \sum_\beta c_\beta D \hat{v}_\beta.$$

Now if $v(t) = v_0(t) - \sum_\beta c_\beta v_\beta(t)$, then $(u(t); v(t))$ is a solution of the differential equations ($\mathfrak{B}$-a), and (10.8) is equivalent to the condition that $T[u, v] \in \mathcal{S}^\perp$. As the initial condition that $u \in \mathfrak{D}[\mathfrak{B}]:v_1$ implies that $\hat{u} \in \mathcal{S}$, it then follows that the boundary conditions ($\mathfrak{B}$-b) hold for $(u(t); v(t))$, thus completing the proof of the theorem.

For $N[\eta]$ defined by (8.10), we shall denote by $\mathfrak{D}_N[\mathfrak{B}]$ the class of vector functions $\{\eta \mid \eta \in \mathfrak{D}[\mathfrak{B}], N[\eta] = 1\}$. Prefatory to the detailed consideration of boundary problems of the form (10.1), we shall establish the following two theorems on the existence of extrema for the functional $J[\eta \mid \mathfrak{B}]$.

Theorem 10.3. *Suppose that* ($\mathfrak{B}$) *is normal, and the functional* $J[\eta \mid \mathfrak{B}]$ *is non-negative on* $\mathfrak{D}[\mathfrak{B}]$. *Then the infimum of* $J[\eta \mid \mathfrak{B}]$ *on* $\mathfrak{D}_N[\mathfrak{B}]$ *is zero if and only if there exists a non-identically vanishing solution* $(u(t); v(t))$ *of* ($\mathfrak{B}$); *moreover, if* $u \in \mathfrak{D}[\mathfrak{B}]$ *and* $J[u \mid \mathfrak{B}] = 0$, *then there exists a* $v(t)$ *such that* $(u(t); v(t))$ *is a solution of* ($\mathfrak{B}$).

If $(u(t); v(t))$ is a nonidentically vanishing solution of $(\mathcal{B})$ then $u(t) \not\equiv 0$ on $[a, b]$ in view of the normality of this system, and one may suppose that $N[u] = 1$ so that $u \in \mathfrak{D}_N[\mathcal{B}]:v$. Then $J[u \mid \mathcal{B}] = 0$, so that the infimum of $J[\eta \mid \mathcal{B}]$ on $\mathfrak{D}_N[\mathcal{B}]$ is zero. Now if $(\mathcal{B})$ has only the identically vanishing solution, and $\eta \in \mathfrak{D}[\mathcal{B}]:\zeta$, let $(u(t); v(t))$ be the unique solution of the corresponding system (8.13). For k as in conclusion (v) of Theorem 8.2, and with the aid of the relations (8.14), it then follows that

$$\begin{aligned} 0 \leq J[\eta - (1/k)u \mid \mathcal{B}] &= J[\eta \mid \mathcal{B}] - \frac{1}{k} J[\eta, u \mid \mathcal{B}] \\ &\quad - \frac{1}{k} J[u, \eta \mid \mathcal{B}] + \frac{1}{k^2} J[u \mid \mathcal{B}], \\ &= J[\eta \mid \mathcal{B}] - \frac{2}{k} N[\eta] + \frac{1}{k^2} N[\eta, u]. \end{aligned}$$

As $N[\eta_1, \eta_2]$ is a non-negative hermitian form on $\mathfrak{C}_n[a, b] \times \mathfrak{C}_n[a, b]$, the Cauchy-Bunyakovsky-Schwarz inequality $|N[\eta, u]|^2 \leq N[\eta]\, N[u]$, (see Appendix A, Section 5), implies that $|N[\eta, u]|^2 \leq k^2N^2[\eta]$ and $N[\eta, u] \leq kN[\eta]$. Therefore $J[\eta \mid \mathcal{B}] \geq (1/k)N[\eta]$, and the infimum of $J[\eta \mid \mathcal{B}]$ on $\mathfrak{D}_N[\mathcal{B}]$ is not less than $1/k$.

Now the hypothesis that $J[\eta \mid \mathcal{B}]$ is non-negative on $\mathfrak{D}[\mathcal{B}]$ implies that $J[\eta_1, \eta_2 \mid \mathcal{B}]$ is a non-negative hermitian form on $\mathfrak{D}[\mathcal{B}] \times \mathfrak{D}[\mathcal{B}]$, and hence we have the Cauchy-Bunyakovsky-Schwarz inequality $|J[\eta_1, \eta_2 \mid \mathcal{B}]|^2 \leq J[\eta_1 \mid \mathcal{B}]\, J[\eta_2 \mid \mathcal{B}]$ for $\eta_\alpha \in \mathfrak{D}[\mathcal{B}]$, $\alpha = 1, 2$. Consequently, if $u \in \mathfrak{D}[\mathcal{B}]$ and $J[u \mid \mathcal{B}] = 0$, then $J[u, \eta \mid \mathcal{B}] = 0$ for arbitrary $\eta \in \mathfrak{D}[\mathcal{B}]$, and from Theorem 10.2 it follows that there exists a $v(t)$ such that $(u(t); v(t))$ is a solution of $(\mathcal{B})$.

Let $\mathcal{F}$ denote a finite set $\mathcal{F} = \{f_1(t), \ldots, f_k(t)\}$ of n-dimensional vector functions $f_\sigma(t)$, $(\sigma = 1, \ldots, k)$, which are of class $\mathfrak{L}_n^2[a, b]$, and for which there exists an associated set of vector functions $\eta_\sigma(t) \in \mathfrak{D}[\mathcal{B}]$ such that the $k \times k$ matrix

$$\left[\int_a^b \eta_\sigma^*(t) f_\tau(t)\, dt\right], \qquad (\sigma, \tau = 1, \ldots, k), \tag{10.9}$$

is nonsingular.

Associated with such a set $\mathcal{F}$ and the above discussed problem $(\mathcal{B})$ is the differential system

$$(\mathcal{B}/\mathcal{F}) \quad \begin{aligned} &\text{(a)} \quad L_1[u, v](t) - \sum_{\tau=0}^{k} e_\tau f_\tau(t) = 0, \qquad L_2[u, v](t) = 0, \qquad t \in [a, b], \\ &\text{(b)} \quad \hat{u} \in \mathcal{S}, \qquad T[u, v] \in \mathcal{S}^\perp, \\ &\text{(c)} \quad \int_a^b f_\sigma^*(t) u(t)\, dt = 0, \qquad (\sigma = 1, \ldots, k), \end{aligned}$$

where $e_1, \ldots, e_k$ are constants. The classes of functions η in $\mathfrak{D}[\mathfrak{B}]$ and $\mathfrak{D}_N[\mathfrak{B}]$ which satisfy ($\mathfrak{B}/\mathfrak{F}$-c) will be denoted by $\mathfrak{D}[\mathfrak{B}/\mathfrak{F}]$ and $\mathfrak{D}_N[\mathfrak{B}/\mathfrak{F}]$, respectively.

Theorem 10.4. *Suppose that the self-adjoint system* ($\mathfrak{B}$) *is normal, and that* $J[\eta \mid \mathfrak{B}]$ *is non-negative definite on* $\mathfrak{D}[\mathfrak{B}/\mathfrak{F}]$, *where* $\mathfrak{F}$ *is a family of vector functions possessing the properties described above. Then the infimum of* $J[\eta \mid \mathfrak{B}]$ *on* $\mathfrak{D}_N[\mathfrak{B}/\mathfrak{F}]$ *is zero if and only if there exist constants* $e_1, \ldots, e_k$ *such that the system* ($\mathfrak{B}/\mathfrak{F}$) *has a nonidentically vanishing solution* $(u(t); v(t))$; *moreover, if* $u \in \mathfrak{D}[\mathfrak{B}/\mathfrak{F}]$ *and* $J[u \mid \mathfrak{B}] = 0$, *then there exists a* $v(t)$ *such that* $(u(t); v(t))$ *is a solution of* ($\mathfrak{B}/\mathfrak{F}$) *for suitable constants* $e_1, \ldots, e_k$. *In particular, if the vector functions* f_σ *of* $\mathfrak{F}$ *are such that there exist constants* γ_σ *and vector functions* $(u_\sigma(t); v_\sigma(t))$ *satisfying the differential systems*

$$(10.10)\quad \begin{aligned} &L_1[u_\sigma, v_\sigma](t) - \gamma_\sigma f_\sigma(t) = 0, \qquad L_2[u_\sigma, v_\sigma](t) = 0, \qquad t \in [a, b], \\ &\hat{u}_\sigma \in \mathcal{S}, \qquad T[u_\sigma, v_\sigma] \in \mathcal{S}^\perp, \end{aligned}$$

and the $k \times k$ *matrix* $[\int_a^b u_\tau^*(t) f_\sigma(t)\, dt]$, $(\sigma, \tau = 1, \ldots, k)$, *is nonsingular, then whenever* $(u(t); v(t))$ *is a solution of* ($\mathfrak{B}/\mathfrak{F}$) *with constants* $e_1, \ldots, e_k$ *one must have* $e_\sigma = 0$, $(\sigma = 1, \ldots, k)$.

The condition that ($\mathfrak{B}$) is normal implies that if $u(t) \equiv 0$, $v(t)$, $t \in [a, b]$, is a solution of ($\mathfrak{B}/\mathfrak{F}$) for parameter values $e = (e_1, \ldots, e_k)$, then also $v(t) \equiv 0$ and $e_\sigma = 0$, $(\sigma = 1, \ldots, k)$. Indeed, if $u(t) \equiv 0$ the equation $L_2[u, v](t) = 0$ implies that $B(t)v(t) = 0$ for $t \in [a, b]$, and hence $J[u, \eta \mid \mathfrak{B}] = 0$ for arbitrary $\eta \in \mathfrak{D}[\mathfrak{B}]$. On the other hand, as $\hat{u} \in \mathcal{S}$ and $T[u, v] \in \mathcal{S}^\perp$, for $\eta_\sigma(t)$, $(\sigma = 1, \ldots, k)$, the elements of $\mathfrak{D}[\mathfrak{B}]$ for which the matrix (10.9) is nonsingular, we have

$$J[u, \eta_\sigma \mid \mathfrak{B}] = \sum_{\tau=1}^{k} \left(\int_a^b \eta_\sigma^*(t) f_\tau(t)\, dt \right) e_\tau = 0, \qquad (\sigma = 1, \ldots, k),$$

and hence $e_\tau = 0$, $(\tau = 1, \ldots, k)$. Consequently, if $(u(t); v(t))$ is a solution of ($\mathfrak{B}/\mathfrak{F}$) with parameter values $e_1, \ldots, e_k$, and $(u(t); v(t); e) \not\equiv (0; 0; 0)$, then $N[u] > 0$, and we may assume that $u \in \mathfrak{D}_N[\mathfrak{B}/\mathfrak{F}]$. Then

$$\begin{aligned} J[u \mid \mathfrak{B}] &= \hat{u}^* T[u, v] + \int_a^b u^*(t) L_1[u, v](t)\, dt, \\ &= \sum_{\tau=1}^{k} \left(\int_a^b u^*(t) f_\tau(t)\, dt \right) e_\tau = 0, \end{aligned}$$

and the infimum of $J[\eta \mid \mathfrak{B}]$ on $\mathfrak{D}_N[\mathfrak{B}/\mathfrak{F}]$ is equal to zero.

Now if ($\mathfrak{B}/\mathfrak{F}$) has only the solution $(u(t); v(t); e) \equiv (0; 0; 0)$, and $\eta \in \mathfrak{D}[\mathfrak{B}/\mathfrak{F}]$, then there exists a unique solution $(u(t); v(t);\ e)$ of the differential system

$$
(10.11)\quad \begin{aligned} &L_1[u, v](t) - \sum_{\tau=1}^{k} e_\tau f_\tau(t) = \eta(t), \qquad L_2[u, v](t) = 0, \qquad t \in [a, b], \\ &\hat{u} \in \mathcal{S}, \qquad T[u, v] \in \mathcal{S}^\perp, \qquad \int_a^b f_\sigma^*(t) u(t)dt = 0, \quad (\sigma = 1, \ldots, k), \end{aligned}
$$

and there is a constant $k_1 > 0$ such that $N[u] \le k_1^2 N[\eta]$; moreover, $J[u, \eta \mid \mathcal{B}] = N[\eta]$, $J[u \mid \mathcal{B}] = N[\eta, u]$. This result may be established by a method paralleling the proof of the corresponding result in Theorem VI.2.2, or by the transformation presented in Problem 2 below. Corresponding to the analogous step in the proof of Theorem 10.3,

$$
0 \le J[\eta - (1/k_1)u \mid \mathcal{B}] \le J[\eta \mid \mathcal{B}] - (1/k_1)\, N[\eta],
$$

and consequently the infimum of $J[\eta \mid \mathcal{B}]$ on $\mathfrak{D}_N[\mathcal{B}/\mathcal{F}]$ is not less than $1/k_1$. Moreover, if $u \in \mathfrak{D}[\mathcal{B}/\mathcal{F}]$ and $J[u \mid \mathcal{B}] = 0$, then the Cauchy-Bunyakovsky-Schwarz inequality implies that $J[u, \eta \mid \mathcal{B}] = 0$ for arbitrary $\eta \in \mathfrak{D}[\mathcal{B}/\mathcal{F}]$. As presented in Problem 3 below, from the result of Theorem 10.2 for the associated differential system formulated in the preceding Problem 2 it follows that there exists a $v(t)$ and constants $e = (e_1, \ldots, e_k)$ such that $(u(t); v(t); e)$ is a solution of $(\mathcal{B}/\mathcal{F})$.

Finally, if there exist constants γ_σ and vector functions satisfying the differential systems (10.10), the fact that $e_\sigma = 0$, $(\sigma = 1, \ldots, k)$, for any solution $(u(t); v(t); e)$ of $(\mathcal{B}/\mathcal{F})$ follows from the fact that $J[u_\sigma, u \mid \mathcal{B}]$ and $J[u, u_\sigma \mid \mathcal{B}]$ are complex conjugates, while the evaluation of these quantities with the aid of Lemma 8.1 yields the relations

$$
\begin{aligned} J[u_\sigma, u \mid \mathcal{B}] &= \hat{u}^* T[u_\sigma, v_\sigma] + \gamma_\sigma \int_a^b u^*(t) f_\sigma(t)\, dt = 0, \\ J[u, u_\sigma \mid \mathcal{B}] &= \hat{u}_\sigma^* T[u, v] + \sum_{\tau=1}^{k} \left[\int_a^b u_\sigma^*(t) f_\tau(t)\, dt \right] e_\tau, \\ &= \sum_{\tau=1}^{k} \left[\int_a^b u_\sigma^*(t) f_\tau(t)\, dt \right] e_\tau. \end{aligned}
$$

Problems VII.10

1. Verify that under the conditions stated in the text the system (10.4) is normal, and this system satisfies hypothesis $(\mathfrak{H}_{\mathcal{B}})$.

2. If the hypotheses of Theorem 10.4 hold, and the system $(\mathcal{B}/\mathcal{F})$ has only the solution $(u(t); v(t); e) \equiv (0; 0; 0)$, prove that for $\eta \in \mathfrak{D}[\mathcal{B}/\mathcal{F}]$ the differential system (10.11) possesses a unique solution $(u(t); v(t); e)$, and $J[u, \eta \mid \mathcal{B}] = N[\eta]$, $J[u \mid \mathcal{B}] = N[\eta, u]$; moreover, there exists a constant $k_1 > 0$ such that $N[u] \le k_1^2 N[\eta]$.

Hint. Consider the differential system

$$\begin{aligned} &-\mathbf{v}'(t) - \mathbf{A}^*(t)\mathbf{v}(t) + \mathbf{C}(t)\mathbf{u}(t) = 0, \\ (10.12)\qquad &\mathbf{u}'(t) - \mathbf{A}(t)\mathbf{u}(t) - \mathbf{B}(t)\mathbf{v}(t) = 0, \qquad t \in [a, b], \\ \hat{\mathbf{u}} \in \mathbf{S}, \qquad &\mathbf{Q}\hat{\mathbf{u}} + [\operatorname{diag}\{-E_{n+k}, E_{n+k}\}]\hat{\mathbf{v}} \in \mathbf{S}^{\perp} \end{aligned}$$

in the $(n + k)$-dimensional vector functions $\mathbf{u}(t) \equiv (\mathbf{u}_\alpha(t))$, $\mathbf{v}(t) \equiv (\mathbf{v}_\alpha(t))$, $(\alpha = 1, \ldots, n + k)$, where $\mathbf{A}(t)$, $\mathbf{B}(t)$, $\mathbf{C}(t)$ are the $(n + k) \times (n + k)$ matrix functions defined by

$$\mathbf{A}(t) = \operatorname{diag}\{A(t), 0\}, \qquad \mathbf{B}(t) = \operatorname{diag}\{B(t), 0\},$$

$$\mathbf{C}(t) = \begin{bmatrix} C_{ij}(t) & f_{i\tau}(t) \\ \bar{f}_{j\sigma}(t) & 0_{\sigma\tau} \end{bmatrix},$$

while $\mathbf{S}$ is the subspace of $\mathbf{C}_{2n+2k}$ and $\mathbf{Q}$ is the $2(n + k) \times 2(n + k)$ hermitian matrix defined by the following properties in terms of $u(t) = (\mathbf{u}_i(t))$, $(i = 1, \ldots, n)$, and $\rho(t) = (\mathbf{u}_{n+\sigma}(t))$, $(\sigma = 1, \ldots, k)$: $\mathbf{S} = \{\mathbf{u} \mid \hat{u} \in \mathcal{S}\}$, and if

$$Q = \begin{bmatrix} Q_1 & Q_2 \\ Q_2^* & Q_3 \end{bmatrix},$$

where Q_β, $(\beta = 1, 2, 3)$, are $n \times n$ matrices, then

$$\mathbf{Q} = \begin{bmatrix} \mathbf{Q}_1 & \mathbf{Q}_2 \\ \mathbf{Q}_2^* & \mathbf{Q}_3 \end{bmatrix},$$

where $\mathbf{Q}_\beta$ are the $(n + k) \times (n + k)$ matrices $\mathbf{Q}_\beta = \operatorname{diag}\{Q_\beta, 0\}$, $(\beta = 1, 2, 3)$. Show that $(u(t); v(t); e)$ is a solution of $(\mathcal{B}/\mathcal{F})$ if and only if $(\mathbf{u}(t); \mathbf{v}(t))$ with $\mathbf{u}_i(t) = u_i(t)$, $\mathbf{v}_i(t) = v_i(t)$, $(i = 1, \ldots, n)$, $\rho(t) = (\mathbf{u}_{n+\sigma}(t)) = (-e_\sigma)$, $\mathbf{v}_{n+\sigma}(t) = \int_a^t f_\sigma^*(s)\, u(s)ds$, $(\sigma = 1, \ldots, k)$, is a solution of (10.12). Note that if $\mathbf{D}$ is the class of vector functions for (10.12) corresponding to the class $\mathfrak{D}[\mathcal{B}]$ for $\mathcal{B}$, then $\boldsymbol{\eta} \in \mathbf{D}$ if and only if $(\boldsymbol{\eta}_i) \equiv \eta \in \mathfrak{D}[\mathcal{B}]$ and $\boldsymbol{\eta}_{n+\sigma}'(t) = 0$, $(\sigma = 1, \ldots, k)$. If $\eta \in \mathfrak{D}[\mathcal{B}/\mathcal{F}]$, apply (v) of Theorem 8.2 to a system related to (10.12) as (8.13) is related to $(\mathcal{B})$, with $(\boldsymbol{\eta}_i) = \eta$ and $\boldsymbol{\eta}_{n+\sigma}(t) \equiv 0$.

3. If the hypotheses of Theorem 10.4 hold, and $u(t)$ is a vector function in $\mathfrak{D}[\mathcal{B}/\mathcal{F}]$ such that $J[u, \eta \mid \mathcal{B}] = 0$ for arbitrary $\eta \in \mathfrak{D}[\mathcal{B}/\mathcal{F}]$, prove that there exists a vector function $v(t)$ and constant $e = (e_1, \ldots, e_k)$ such that $(u(t); v(t); e)$ is a solution to $(\mathcal{B}/\mathcal{F})$.

Hint. Let $\mathbf{J}[\boldsymbol{\eta}_1, \boldsymbol{\eta}_2]$ be the functional associated with (10.12) in the manner that $J[\eta_1, \eta_2 \mid \mathcal{B}]$ is associated with $(\mathcal{B})$, and let $\mathbf{D}$ be as defined in the Hint to Problem 2. Show that if $u \in \mathfrak{D}[\mathcal{B}/\mathcal{F}]$ then $\mathbf{u}$ with $\mathbf{u}_i(t) = u_i(t)$, $(i = 1, \ldots, n)$, $\mathbf{u}_{n+\sigma}(t) \equiv 0$, $(\sigma = 1, \ldots, k)$, is such that $\mathbf{u} \in \mathbf{D}$, and if $\boldsymbol{\eta} \in \mathbf{D}$ and $\eta_i(t) = \boldsymbol{\eta}_i(t)$, $(i = 1, \ldots, n)$, then $\eta \in \mathfrak{D}[\mathcal{B}/\mathcal{F}]$ and $\mathbf{J}[\mathbf{u}, \boldsymbol{\eta}] = J[u, \eta \mid \mathcal{B}]$.

11. EXISTENCE OF PROPER VALUES

We shall now establish results on the existence and distribution of proper values for a boundary problem

$$(11.1)\qquad \begin{aligned} &\text{(a)}\quad L_1[u, v](t) = \lambda K(t)u(t), \quad L_2[u, v](t) = 0, \qquad t \in [a, b],\\ &\text{(b)}\quad \hat{u} \in \mathcal{S}, \qquad T[u, v] \equiv Q\hat{u} + D\hat{v} \in \mathcal{S}^{\perp}, \end{aligned}$$

as formulated at the beginning of Section 10. *It will be assumed throughout our discussion that the hypothesis* ($\mathfrak{H}_{\mathcal{B}}$) *of Section* 10 *holds.* That is, on a given interval $[a, b]$ on the real line the $n \times n$ matrix functions $A(t)$, $B(t)$, $C(t)$ satisfy hypothesis ($\mathfrak{H}$) of Section 2, $\mathcal{S}$ is a subspace of $\mathbf{C}_{2n}$ such that (11.1) is normal, Q is an hermitian $2n \times 2n$ matrix, $D = \operatorname{diag}\{-E_n, E_n\}$, and $K(t)$ is an hermitian matrix function of class $\mathfrak{L}_{nn}{}^{\infty}[a, b]$ such that $K(t) \not\equiv 0$ for t on a subset of $[a, b]$ of positive measure. As before, $\mathfrak{D}[\mathcal{B}]$ will denote the vector space $\{\eta \mid \eta \in \mathfrak{D}[a, b], \hat{\eta} \in \mathcal{S}\}$, and the symbols $K[\eta_1, \eta_2]$, $K[\eta_1]$ are defined by (10.2).

For the treatment of the present section, the further hypotheses are introduced:

($\mathfrak{H}_K$)
- (a) *$K(t)$ is non-negative definite for $t \in [a, b]$;*
- (b) *there exists a real number λ_0 such that $J[\eta; \lambda_0 \mid \mathcal{B}] = J[\eta \mid \mathcal{B}] - \lambda_0 K[\eta]$ is positive definite on $\mathfrak{D}[\mathcal{B}]$;*
- (c) *$K[\eta]$ is positive definite on subspaces of $\mathfrak{D}[\mathcal{B}]$ of arbitrarily high dimension.*

In particular, it is to be noted that hypotheses ($\mathfrak{H}$) and ($\mathfrak{H}_K$-b) imply the following conditions:

($\mathfrak{H}_2$)
- (a) $B(t) \geq 0$ *for t* a.e. *on* $[a, b]$;
- (b) *if* $\eta \in \mathfrak{D}[\mathcal{B}]$, $\eta(t) \not\equiv 0$ on $[a, b]$, *and* $K[\eta] = 0$, *then* $J[\eta \mid \mathcal{B}] > 0$;
- (c) *$J[\eta \mid \mathcal{B}]$ is bounded below on the set $\mathfrak{D}_N[\mathcal{B}; K]$, defined as*

$$(11.2)\qquad \mathfrak{D}_N[\mathcal{B}; K] = \{\eta \mid \eta \in \mathfrak{D}[\mathcal{B}], K[\eta] = 1\}.$$

As $\mathfrak{D}_0[a, b] \subset \mathfrak{D}[\mathcal{B}]$, condition ($\mathfrak{H}_2$-a) is a consequence of the result of Theorem 4.2 for the functional $J[\eta; \lambda_0 \mid \mathcal{B}]$, and conditions ($\mathfrak{H}_2$-b, c) are obvious consequences of ($\mathfrak{H}_K$-b). Property ($\mathfrak{H}_2$-c) is used directly in the proof of the following Theorem 11.1. It is to be remarked that Problem 1 of the set of problems at the end of this section presents the result that hypotheses ($\mathfrak{H}_{\mathcal{B}}$), ($\mathfrak{H}_K$-a), ($\mathfrak{H}_2$-a, b), together with ($\mathfrak{H}_1[a, b]$) of Section 4, imply ($\mathfrak{H}_2$-c).

Now if $(u(t); v(t))$ is a proper solution of (11.1) for a proper value λ, then $u(t) \not\equiv 0$ in view of the assumption that this system is normal. Moreover, with the aid of Lemma 8.1 it follows that $J[u \mid \mathcal{B}] = \lambda K[u]$, and if $K[u]$ were zero then $u(t)$ would be a nonidentically vanishing element of $\mathfrak{D}[\mathfrak{B}]$ such that $J[u; \lambda_0 \mid \mathfrak{B}] = 0$, contrary to ($\mathfrak{H}_K$-b). Therefore $K[u] > 0$, and from Theorem 10.1 it follows that all proper values of (11.1) are real, and the further conclusions of Theorem 10.1 hold for the proper solutions of this system.

The basic result on the existence of proper values is that of the following theorem.

Theorem 11.1. *If hypotheses* ($\mathfrak{H}_{\mathfrak{B}}$) *and* ($\mathfrak{H}_K$-a, b, c) *hold, then for* (11.1) *the proper values may be ordered as a sequence* $\lambda_1 \leq \lambda_2 \leq \cdots$ *with corresponding proper solutions* $(u(t); v(t)) = (u_j(t); v_j(t))$ *such that*

(a) $K[u_i, u_j] = \delta_{ij}$, $(i, j = 1, 2, \ldots)$;
(b) $\lambda_1 = J[u_1 \mid \mathfrak{B}]$ *is the minimum of* $J[\eta \mid \mathfrak{B}]$ *on the class* $\mathfrak{D}_N[\mathfrak{B}; K]$;
(c) *for* $j = 2, 3, \ldots$ *the class*

$$(11.2_j) \quad \mathfrak{D}_{Nj}[\mathfrak{B}; K] = \{\eta \mid \eta \in \mathfrak{D}_N[\mathfrak{B}; K], \quad K[\eta, u_i] = 0, \quad (i = 1, \ldots, j-1)\},$$

is nonempty, and $\lambda_j = J[u_j \mid \mathfrak{B}]$ *is the minimum of* $J[\eta \mid \mathfrak{B}]$ *on* $\mathfrak{D}_{Nj}[\mathfrak{B}; K]$.

Moreover, $\{\lambda_j\} \to \infty$ *as* $j \to \infty$.

In view of hypotheses ($\mathfrak{H}_K$-b, c) the class $\mathfrak{D}_N[\mathfrak{B}; K]$ is nonempty, the infimum λ_1 of $J[\eta \mid \mathfrak{B}]$ on this class is finite, and $J[\eta; \lambda_1 \mid \mathfrak{B}] = J[\eta \mid \mathfrak{B}] - \lambda_1 K[\eta]$ is non-negative on $\mathfrak{D}[\mathfrak{B}]$. In view of the normality hypothesis, it then follows from Theorem 10.3 applied to $J[\eta; \lambda_1 \mid \mathfrak{B}]$ that $\lambda = \lambda_1$ is a proper value of (11.1). Moreover, as $\lambda K[u] = J[u \mid \mathfrak{B}]$ and $K[u] > 0$ for any proper solution $(u(t); v(t))$ corresponding to a proper value λ, it follows that λ_1 is the smallest proper value of this system. If proper values $\lambda_1 \leq \lambda_2 \leq \cdots \leq \lambda_{m-1}$ exist, and there are corresponding proper solutions $(u_j(t); v_j(t))$ of (11.1) satisfying the conclusions (a), (b), (c) of the theorem for $j = 1, \ldots, m-1$, then in view of ($\mathfrak{H}_K$-c) the class $\mathfrak{D}_{Nm}[\mathfrak{B}; K]$ defined by (11.2_m) is nonempty. The infimum λ_m of $J[y \mid \mathfrak{B}]$ on $\mathfrak{D}_{Nm}[\mathfrak{B}; K]$ is such that $\lambda_m \geq \lambda_{m-1}$, and $J[\eta; \lambda_m \mid \mathfrak{B}] = J[\eta \mid \mathfrak{B}] - \lambda_m K[\eta]$ is non-negative on the linear manifold $\{\eta \mid \eta \in \mathfrak{D}[\mathfrak{B}], \quad K[\eta, u_\sigma] = 0, \quad \sigma = 1, \ldots, m-1\}$. Moreover, since $K[u_\tau, u_\sigma] = \delta_{\sigma\tau}$, $(\sigma, \tau = 1, \ldots, m-1)$, application of Theorem 10.4 to $J[\eta; \lambda_m \mid \mathfrak{B}]$ with $k = m-1$, and $f_\sigma(t) = K(t)\, u_\sigma(t)$, $\eta_\sigma(t) = u_\sigma(t)$, $\gamma_\sigma = \lambda_\sigma$ for $\sigma = 1, \ldots, m-1$, yields the result that $\lambda = \lambda_m$ is a proper value of (11.1), and there exists a corresponding proper solution $(u_m(t); v_m(t))$ such that the conclusions of the theorem hold for $j = 1, 2, \ldots, m$. Consequently, conclusions (a), (b), (c) of the theorem are established by mathematical induction.

Now from the discussion of Section IV.2, it follows that the proper values of (11.1) form the set of zeros of an entire function of λ. As this system possesses only real proper values, it then follows that the set of proper values has no finite accumulation point, and therefore $\{\lambda_j\} \to \infty$ as $j \to \infty$. Moreover, since $K[u] > 0$ and $\lambda K[u] = J[u \mid \mathfrak{B}]$ for $(u(t); v(t))$ a proper solution of (11.1) corresponding to a proper value λ, it follows from Theorem 10.1 that each proper value λ of (11.1) appears in the sequence $\{\lambda_j\}$, and occurs a number of times equal to its index.

It is to be remarked that if $(\mathfrak{H}_{\mathfrak{B}})$, $(\mathfrak{H}_K$-a, b) are satisfied, but $(\mathfrak{H}_K$-c) does not hold, then the method of proof of Theorem 11.1 leads to the existence of the finite set of proper values of (11.1), and the stated extremizing properties of these proper values and corresponding proper solutions.

Corresponding to Theorem VI.3.3, one now has the following result.

Theorem 11.2. *Suppose that hypotheses* $(\mathfrak{H}_{\mathfrak{B}})$ *and* $(\mathfrak{H}_K$-a, b, c) *hold, and* $\{\lambda_j; u_j(t); v_j(t)\}$, $(j = 1, 2, \ldots)$, *is a sequence of proper values and corresponding proper vector functions as determined in Theorem* 11.1.

(a) *If* r *is a positive integer, and* $d_1, \ldots, d_r$ *are constants such that* $|d_1|^2 + \cdots + |d_r|^2 = 1$, *then for* $\eta(t) = \sum_{j=1}^r d_j\, u_j(t)$, $\zeta(t) = \sum_{j=1}^r d_j\, v_j(t)$, *we have* $\eta \in \mathfrak{D}[\mathfrak{B}]:\zeta$ *and*

$$J[\eta \mid \mathfrak{B}] = \sum_{j=1}^{r} \lambda_j\, |d_j|^2 \le \lambda_r.$$

(b) (*Courant-Hilbert max-min property*) *If* $\mathcal{F} = \{f_1(t), \ldots, f_k(t)\}$ *is a set of n-dimensional vector functions* $f_j(t)$ *of class* $\mathfrak{L}_n{}^2[a, b]$, *and* $\lambda[\mathcal{F}]$ *denotes the minimum of* $J[\eta \mid \mathfrak{B}]$ *on*

$$\mathfrak{D}_N[\mathfrak{B}; K/\mathcal{F}] = \{\eta \mid \eta \in \mathfrak{D}_N[\mathfrak{B}; K],\quad \int_a^b f_j^*(t)\eta(t)\,dt = 0,\quad (j = 1, \ldots, k)\},$$

then λ_{k+1} *is the maximum of* $\lambda[\mathcal{F}]$, *and this maximum is attained for* $f_j(t) = K(t)u_j(t)$, $(j = 1, \ldots, k)$.

***Lemma* 11.1.** *If hypotheses* $(\mathfrak{H}_{\mathfrak{B}})$ *and* $(\mathfrak{H}_K$-a, b, c) *hold and* $\lambda_0 < \lambda_1$, *where* $\lambda = \lambda_1$ *is the smallest proper value of* (11.1), *then there exists an* $l > 0$ *such that if* $\eta \in \mathfrak{D}[\mathfrak{B}]$ *then*

$$(11.3)\qquad J[\eta; \lambda_0 \mid \mathfrak{B}] \ge l\left[|\eta(a)|^2 + |\eta(b)|^2 + \int_a^b \{|\eta'(t)|^2 + |\eta(t)|^2\}\,dt\right].$$

In view of the minimizing property of λ_1, if $\lambda_0 < \lambda_1$ then $J[\eta; \lambda_0 \mid \mathfrak{B}] \ge 0$ for $\eta \in \mathfrak{D}[\mathfrak{B}]$. Since by hypothesis the system (11.1) is normal, and $\lambda = \lambda_0$ is not a proper value, it then follows from the result of Theorem 10.3 applied to $J[\eta; \lambda_0 \mid \mathfrak{B}]$ that the infimum of this functional on $\mathfrak{D}_N[\mathfrak{B}]$ is positive. That is, there exists an $l_2 > 0$ such that

$$(11.4)\qquad J[\eta; \lambda_0 \mid \mathfrak{B}] \ge l_2\left[|\eta(a)|^2 + |\eta(b)|^2 + \int_a^b |\eta(t)|^2\,dt\right].$$

Now, as pointed out above, the hypotheses of Theorem 11.1 imply that $B(t) \geq 0$ for t a.e. on $[a, b]$. Consequently, by Problem VII.4:4 there exist real constants $l_0 > 0$, $l_1 \geq 0$ such that

$$(11.5)\quad J[\eta; \lambda_0 \mid \mathcal{B}] \geq l_0 \int_a^b |\eta'(t)|^2\, dt - l_1 \int_a^b |\eta(t)|^2\, dt, \quad \textit{for} \quad \eta \in \mathfrak{D}[\mathcal{B}].$$

Therefore, upon evaluating $J[\eta; \lambda_0 \mid \mathcal{B}]$ as $\theta J[\eta; \lambda_0 \mid \mathcal{B}] + (1 - \theta) J[\eta; \lambda_0 \mid \mathcal{B}]$, $0 < \theta < 1$, and applying inequalities (11.4) and (11.5) to these respective terms, it follows that (11.3) holds for $\theta = (l_0 + l_1)/(l_0 + l_1 + l_2)$ and $l = (1 - \theta) l_0$.

Now if $\eta \in \mathfrak{D}[\mathcal{B}]$ then $\eta(t) = \eta(a) + \int_a^t \eta'(s)\, ds = \eta(b) - \int_t^b \eta'(s)\, ds$, and by elementary inequalities it follows that

$$(11.6)\quad \tfrac{1}{2}|\eta(t)|^2 \leq |\eta(a)|^2 + |\eta(b)|^2 + (b - a) \int_a^b |\eta'(s)|^2\, ds.$$

Consequently, we have the following additional relation.

Corollary. *Under the hypotheses of Lemma* 11.1, *there exists an* $l' > 0$ *such that if* $\eta \in \mathfrak{D}[\mathcal{B}]$, *and* $t \in [a, b]$, *then*

$$(11.7)\quad J[\eta; \lambda_0 \mid \mathcal{B}] \geq l' \left[|\eta(t)|^2 + \int_a^b \{|\eta'(s)|^2 + |\eta(s)|^2\}\, ds \right].$$

Now suppose that the hypotheses of Theorem 11.1 are satisfied, and for $\eta \in \mathfrak{D}[\mathcal{B}]{:}\zeta$ let

$$\eta_m(t) = \eta(t) - \sum_{j=1}^m c_j[\eta]\, u_j(t), \qquad \zeta_m(t) = \zeta(t) - \sum_{j=1}^m c_j[\eta] v_j(t),$$

where $c_j[\eta]$ is the j-th Fourier coefficient $c_j[\eta] = K[\eta;\, u_j]$ with respect to the K-orthonormal system $\{u_j(t)\}$. Then $\eta_m \in \mathfrak{D}[\mathcal{B}]{:}\zeta_m$ and $0 \leq K[\eta_m] = K[\eta] - \sum_{j=1}^m |c_j[\eta]|^2$. Moreover, if $\lambda_0 < \lambda_1$ then

$$0 \leq J[\eta_m; \lambda_0 \mid \mathcal{B}] = J[\eta; \lambda_0 \mid \mathcal{B}] - \sum_{j=1}^m (\lambda_j - \lambda_0)\, |c_j[\eta]|^2 \leq J[\eta; \lambda_0 \mid \mathcal{B}].$$

Now $K[\eta_m, u_i] = 0$, $(i = 1, \ldots, m)$, and in view of the extremizing property of λ_{m+1} we have $J[\eta_m \mid \mathcal{B}] \geq \lambda_{m+1} K[\eta_m]$, and therefore

$$(\lambda_{m+1} - \lambda_0) K[\eta_m] \leq J[\eta_m; \lambda_0 \mid \mathcal{B}] \leq J[\eta; \lambda_0 \mid \mathcal{B}], \quad \textit{for} \quad m = 1, 2, \ldots.$$

As $\{\lambda_j\} \to \infty$ as $j \to \infty$, it then follows that $K[\eta_m] = K[\eta] - \sum_{j=1}^m |c_j[\eta]|^2$ tends to zero as $m \to \infty$, and hence we have the following result.

Theorem 11.3. *Under the hypotheses of Theorem* 11.1, *if* $\lambda_0 < \lambda_1$ *and* $\eta \in \mathfrak{D}[\mathcal{B}]$, *then the infinite series* $\sum_{j=1}^{\infty} |c_j[\eta]|^2$, $\sum_{j=1}^{\infty} (\lambda_j - \lambda_0) |c_j[\eta]|^2$ *converge, and*

$$\sum_{j=1}^{\infty} |c_j[\eta]|^2 = K[\eta], \tag{11.8}$$

$$\sum_{j=1}^{\infty} (\lambda_j - \lambda_0) |c_j[\eta]|^2 \leq J[\eta; \lambda_0 \mid \mathcal{B}]. \tag{11.9}$$

In view of (11.8), inequality (11.9) is obviously equivalent to

$$\sum_{j=1}^{\infty} \lambda_j |c_j[\eta]|^2 \leq J[\eta \mid \mathcal{B}]. \tag{11.9'}$$

Moreover, equation (11.8) clearly implies the corresponding result

$$\sum_{j=1}^{\infty} c_j[\eta_1] \overline{c_j[\eta_2]} = K[\eta_1, \eta_2], \quad \textit{for} \quad \eta_\alpha \in \mathfrak{D}[\mathcal{B}], \qquad (\alpha = 1, 2). \tag{11.8'}$$

Corresponding to the result (iv) of Theorem 8.2, if λ is not a proper value of (11.1) then there exist $n \times n$ matrix functions $G(t, s; \lambda)$, $G_0(t, s; \lambda)$ which satisfy conditions analogous to (i), (ii), (iii) of Theorem 8.2, and if $f \in \mathfrak{L}_n^2[a, b]$ then the unique solution of the differential system

$$\begin{gathered} L_1[u, v](t) - \lambda K(t)u(t) = f(t), \qquad L_2[u, v](t) = 0, \qquad t \in [a, b], \\ \hat{u} \in \mathcal{S}, \qquad T[u, v] \in \mathcal{S}^{\perp}, \end{gathered} \tag{11.10}$$

is given by

$$\begin{aligned} &\text{(a)} \qquad u(t) = \int_a^b G(t, s; \lambda) f(s)\, ds, \\ &\text{(b)} \qquad v(t) = \int_a^b G_0(t, s; \lambda) f(s)\, ds. \end{aligned} \qquad t \in [a, b], \tag{11.11}$$

Moreover, in view of the self-adjoint nature of (11.1), it follows from Lemma IV.2.4 that $G(t, s; \lambda) = [G(s, t; \bar{\lambda})]^*$. Now if $\lambda = \lambda_0$ is not a proper value of (11.1), then $(u(t); v(t)) = (u_j(t); v_j(t))$ is the solution of (11.10) for $\lambda = \bar{\lambda}_0$ and $f(t) = (\lambda_j - \bar{\lambda}_0) K(t)u_j(t)$, and therefore the corresponding equation (11.11-a) yields the relation

$$(\lambda_j - \lambda_0)^{-1} u_j^*(t) = \int_a^b u_j^*(s)K(s)G(s, t; \lambda_0)\, ds, \quad \textit{for} \quad t \in [a, b]. \tag{11.12}$$

Since for arbitrary n-dimensional vectors ξ, and $t \in [a, b]$, we have $G(\cdot, t; \lambda_0)\xi \in \mathfrak{D}[\mathcal{B}]\colon G_0(\cdot, t; \lambda_0)\xi$, equation (11.12) gives the j-th Fourier coefficients of the column vectors of $G(\cdot, t; \lambda_0)$, so that the following result is a consequence of the above relation (11.8′).

Theorem 11.4. *Under the hypotheses of Theorem* 11.1, *if* λ *is not a proper value of* (11.1) *then*

$$(11.13)\quad \sum_{j=1}^{\infty} |\lambda_j - \lambda|^{-2} u_j(\tau)u_j^*(t) = \int_a^b [G(s, \tau; \lambda)]^* K(s) G(s, t; \lambda)\, ds$$

for $(\tau, t) \in [a, b] \times [a, b]$; *in particular*,

$$(11.14)\quad \sum_{j=1}^{\infty} |\lambda_j - \lambda|^{-2} |u_j(t)|^2 = \operatorname{Tr}\left\{\int_a^b [G(s, t; \lambda)]^* K(s) G(s, t; \lambda)\, ds\right\}.$$

Preliminary to the proof of Theorem 11.5, the following two lemmas will be established.

***Lemma* 11.2.** *Suppose that* $w_j(t)$, $(j = 1, 2, \ldots)$, *are n-dimensional vector functions with* $w_j \in \mathfrak{A}_n[a, b]$, $w_j' \in \mathfrak{L}_n^2[a, b]$, *while* $\{w_j(t)\} \to 0$ *for* $t \in [a, b]$ *and* $\{\int_a^b |w_j'(t) - w_k'(t)|^2\, dt\} \to 0$ *as* $j, k \to \infty$. *Then* $\{w_j(t)\} \to 0$ *uniformly for* $t \in [a, b]$, *and* $\{\int_a^b |w_j'(t)|^2\, dt\} \to 0$ *as* $j \to \infty$.

If $w_{jk}(t) = w_j(t) - w_k(t)$ then the equation $w_{jk}(t) = w_{jk}(a) + \int_a^t w_{jk}'(s)\, ds$ and elementary inequalities yield the relation $|w_{jk}(t)|^2 \le 2\,|w_{jk}(a)|^2 + 2\,|\int_a^t w_{jk}'(s)\, ds|^2 \le 2\,|w_{jk}(a)|^2 + 2(b - a) \int_a^t |w_{jk}'(s)|^2\, ds$. As the hypotheses of the lemma imply that $\{w_{jk}(a)\} \to 0$ and $\{\int_a^t |w_{jk}'(s)|^2\, ds\} \to 0$ as $j, k \to \infty$, uniformly for $t \in [a, b]$, it follows that $\{w_{jk}(t)\} \to 0$ as $j, k \to \infty$, uniformly for $t \in [a, b]$, and hence the convergence of the sequence $\{w_j(t)\}$ to 0 is uniform on this interval. Now, in view of the completeness of $\mathfrak{L}_n^2[a, b]$, the fact that $\{\int_a^b |w_{jk}'(s)|^2\, ds\} \to 0$ as $j, k \to \infty$ implies the existence of a vector function $v(t) \in \mathfrak{L}_n^2[a, b]$ such that

$$\left\{\int_a^b |w_j'(s) - v(s)|^2\, ds\right\} \to 0.$$

Since

$$\left|w_j(t) - w_j(a) - \int_a^t v(s)\, ds\right|^2 = \left|\int_a^t [w_j'(s) - v(s)]\, ds\right|^2$$
$$\le (t - a) \int_a^t |w_j'(s) - v(s)|^2\, ds \le (b - a) \int_a^b |w_j'(s) - v(s)|^2\, ds,$$

it follows that

$$\left|\int_a^t v(s)\, ds\right|^2 = \lim_{j \to \infty} \left|w_j(t) - w_j(a) - \int_a^t v(s)\, ds\right|^2 = 0,$$

and therefore

$$\int_a^t v(s)\, ds = 0 \quad \text{for} \quad t \in [a, b].$$

Consequently, $v(t) = 0$ for t a.e. on $[a, b]$, and

$$\left\{\int_a^b |w_j'(s)|^2\,ds\right\} = \left\{\int_a^b |w_j'(s) - v(s)|^2\,ds\right\} \to 0 \quad \text{as} \quad j \to \infty.$$

***Lemma* 11.3.** *Under the hypotheses of Theorem* 11.1, *if* $\eta \in \mathfrak{D}[\mathfrak{B}]$ *then the sequence* $\eta_m(t) = \eta(t) - \sum_{j=1}^m c_j[\eta]\, u_j(t)$, $m = 1, 2, \ldots$, *converges uniformly on* $[a, b]$ *to a continuous vector function* $\eta_0(t)$ *which is such that* $K[\eta_0] = 0$; *moreover,*

$$\left\{\int_a^b |\eta_j'(t) - \eta_k'(t)|^2\,dt\right\} \to 0, \quad as \quad j, k \to \infty. \tag{11.15}$$

For $\eta \in \mathfrak{D}[\mathfrak{B}]:\zeta$, let $\eta_m(t) = \eta(t) - \sum_{j=1}^m c_j[\eta]\, u_j(t)$, $\zeta_m(t) = \zeta(t) - \sum_{j=1}^m c_j[\eta]\, v_j(t)$, and $\eta_{jk}(t) = \eta_j(t) - \eta_k(t)$, $\zeta_{jk}(t) = \zeta_j(t) - \zeta_k(t)$, for $j < k$. Then $\eta_{jk} \in \mathfrak{D}[\mathfrak{B}]:\zeta_{jk}$, and

$$J[\eta_{jk}; \lambda_0 \mid \mathfrak{B}] = \sum_{i=j+1}^{k} (\lambda_i - \lambda_0)\, |c_i[\eta]|^2.$$

If $\lambda_0 < \lambda_1$, then from the convergence of the series of (11.9) it follows that $J[\eta_{jk}; \lambda_0 \mid \mathfrak{B}] \to 0$ as $j, k \to \infty$. Therefore, since the Corollary to Lemma 11.1 implies the inequality

$$|\eta_{jk}(t)|^2 + \int_a^b |\eta_{jk}'(t)|^2\,dt \le \left(\frac{1}{l'}\right) J[\eta_{jk}; \lambda_0 \mid \mathfrak{B}], \qquad t \in [a, b],$$

it follows that $\{\int_a^b |\eta_j'(t) - \eta_k'(t)|^2\,dt\} = \{\int_a^b |\eta_{jk}'(t)|^2\,dt\} \to 0$ as $j, k \to \infty$, and the sequence $\{\eta_j(t)\}$ converges uniformly on $[a, b]$ to a limit vector function $\eta_0(t)$, which is necessarily continuous. Moreover, $K[\eta_0] = \lim_{m\to\infty} K[\eta_m] = K[\eta] - \sum_{j=1}^\infty |c_j[\eta]|^2 = 0$, by conclusion (11.8) of Theorem 11.3.

We shall now introduce the following hypothesis, which is stronger than condition $(\mathfrak{H}_K\text{-a})$.

$(\mathfrak{H}_K\text{-a}')$ *The hermitian matrix function* $K(t)$ *is such that* $K[\eta]$ *is a positive definite functional on* $\mathfrak{C}_n[a, b]$; *that is, if* $\eta(t) \in \mathfrak{C}_n[a, b]$ *and* $\eta(t) \not\equiv 0$ *on* $[a, b]$, *then* $K[\eta] > 0$.

Corresponding to Theorems VI.5.2 and VI.5.3 we now have the following results.

Theorem 11.5. *Suppose that the boundary problem* (11.1) *satisfies hypotheses* $(\mathfrak{H}_\mathfrak{B})$, $(\mathfrak{H}_K\text{-a}', \text{b, c})$, $(\mathfrak{H}_1[a, b])$ *of Section* 4, *and* $\{\lambda_j; u_j(t); v_j(t)\}$, $(j = 1, 2, \ldots)$, *is a sequence of proper values and corresponding proper vector functions as determined in Theorem* 11.1.

(a) *If* $\eta \in \mathfrak{D}[\mathfrak{B}]$, *then*

(i) $\sum_{j=1}^\infty c_j[\eta]\, u_j(t)$ *converges to* $\eta(t)$, *uniformly on* $[a, b]$;
(ii) $\{\int_a^b |\eta'(t) - \sum_{j=1}^m c_j[\eta]\, u_j'(t)|^2\,dt\} \to 0$, *as* $m \to \infty$;
(iii) $J[\eta \mid \mathfrak{B}] = \sum_{j=1}^\infty \lambda_j\, |c_j[\eta]|^2$.

(b) *If λ is not a proper value of* (11.1), *then the corresponding Green's matrix $G(t, s; \lambda)$ has the expansion*

$$(11.16)\quad G(t, s; \lambda) = \sum_{j=1}^{\infty}(\lambda_j - \lambda)^{-1}\, u_j(t)u_j^*(s), \quad \textit{for} \quad (t, s) \in [a, b] \times [a, b],$$

and the series in (11.16) *converges uniformly on* $[a, b] \times [a, b]$.

If we set $\eta_m(t) = \eta(t) - \sum_{j=1}^{m} c_j[\eta]\, u_j(t)$, then conclusion (a-i) is an immediate consequence of Lemma 11.3, as the strengthened hypothesis ($\mathfrak{H}_K$-a′) implies that the limit vector function $\eta_0(t)$ of that lemma is zero. Also, conclusion (a-ii) is a consequence of the result of (a-i), together with the condition (11.15) and the result of Lemma 11.2.

If $\lambda_0 < \lambda_1$, then in view of hypothesis ($\mathfrak{H}_1[a, b]$) and the result of Problem VII.4:6 we have that there exists a constant μ_1 such that

$$0 \le J[\eta; \lambda_0 \mid \mathcal{B}] \le \mu_1\left[|\eta(a)|^2 + |\eta(b)|^2 + \int_a^b \left\{|\eta'(s)|^2 + |\eta(s)|^2\right\} ds\right],$$

$$\textit{for} \quad \eta \in \mathfrak{D}[\mathcal{B}].$$

Consequently, in view of the uniform convergence of $\{\eta_m(t)\}$ to zero on $[a, b]$, and conclusion (a-ii), we have that $J[\eta_m; \lambda_0 \mid \mathcal{B}] \to 0$ as $m \to \infty$. As $J[\eta_m; \lambda_0 \mid \mathcal{B}] = J[\eta; \lambda_0 \mid \mathcal{B}] - \sum_{j=1}^{m} (\lambda_j - \lambda_0)\, |c_j[\eta]|^2$, it then follows that the infinite series $\sum_{j=1}^{\infty} (\lambda_j - \lambda_0)\, |c_j[\eta]|^2$ converges and has sum $J[\eta; \lambda_0 \mid \mathcal{B}]$. That is, inequality (11.9) is now an equality, so that in view of (11.8) we also have (11.9′) with the equality sign holding, which is conclusion (a-iii).

Now in view of the properties of the Green's matrix presented previous to the statement of Theorem 11.4, and relation (11.12) specifically, equation (11.16) is a consequence of conclusion (i) for the column vectors of $G(\cdot, s; \lambda)$. Indeed, from (i) it follows that for fixed $s \in [a, b]$ the convergence of the series in (11.16) is uniform with respect to t on $[a, b]$. Now for $\lambda_0 < \lambda_1$ the continuous scalar function $\mathrm{Tr}[G(t, t; \lambda_0)]$ is the sum of the series

$$(11.17)\qquad \sum_{j=1}^{\infty}(\lambda_j - \lambda_0)^{-1}\, |u_j(t)|^2,$$

whose terms are real-valued, non-negative and continuous on $[a, b]$. By Dini's Theorem, (see Theorem 2.1 of Appendix F), it then follows that the convergence of the series (11.17) is uniform on $[a, b]$. If λ is not a proper value of (11.1), then there exists a constant k such that $|\lambda_j - \lambda|^{-1} \le k\, |\lambda_j - \lambda_0|^{-1}$ for $j = 1, 2, \ldots$. Moreover, the absolute value of each element of the matrix $u_j(t)u_j^*(s)$ does not exceed $|u_j(t)|\, |u_j(s)|$, which in turn is dominated by $(|u_j(t)|^2 + |u_j(s)|^2)/2$. Consequently, the absolute value of each element of the matrix $(\lambda_j - \lambda)^{-1} u_j(t)u_j^*(s)$ does not exceed

$$\frac{k}{2} |\lambda_j - \lambda_0|^{-1}(|u_j(t)|^2 + |u_j(s)|^2),$$

and the uniform convergence of the series of (11.16) on $[a, b] \times [a, b]$ is a consequence of the uniform convergence of the series (11.17) on $[a, b]$.

Problems VII.11

1. Show that if the boundary problem (11.1) satisfies hypotheses $(\mathfrak{H}_\mathcal{B})$, $(\mathfrak{H}_K\text{-a})$, $(\mathfrak{H}_2\text{-a, b})$, together with $(\mathfrak{H}_1[a, b])$ of Section 4, then condition $(\mathfrak{H}_2\text{-c})$ holds.

Hint. Establish the stated result by indirect argument, noting that if it is not true then there exists a sequence $\eta_j \in \mathfrak{D}[\mathcal{B}]:\zeta_j$, $(j = 1, 2, \ldots)$, satisfying the conditions

$$(11.18)\quad N[\eta_j] = 1, \qquad J[\eta_j \mid \mathcal{B}] + jK[\eta_j] \leq 0, \qquad (j = 1, 2, \ldots),$$

where the involved norm functional $N[\eta]$ is defined as in equation (8.1). With the aid of the result of Problem VII.4:4 conclude that there exist real constants $l_0' > 0$, $l_1' \geq 0$ such that

$$J[\eta \mid \mathcal{B}] \geq l_0' \int_a^b |\eta'(t)|^2\, dt - l_1' N[\eta], \quad \textit{for} \quad \eta \in \mathfrak{D}[a, b],$$

and hence

$$\int_a^b |\eta_j'(t)|^2\, dt \leq \frac{l_1'}{l_0'}, \qquad (j = 1, 2, \ldots).$$

Conclude that $\{\eta_j'(t)\}$ is a bounded sequence in $\mathfrak{L}_n^2[a, b]$, and $\{\eta_j(t)\}$ is a uniformly bounded sequence on $[a, b]$, so that the functions of the sequence $\{\eta_j(t)\}$ are uniformly bounded and uniformly equi-continuous on $[a, b]$. From the discussion at the end of Section 4 deduce that one may also suppose that $\pi^*(t)\zeta_j(t) = 0$ on $[a, b]$, so that $\{\zeta_j(t)\} = \{R(t)L[\eta_j](t)\}$ is also a bounded sequence in $\mathfrak{L}_n^2[a, b]$. With the aid of Problem VII.4:9 conclude that there exists a subsequence $\{\eta_m(t), \zeta_m(t)\}$ of $\{\eta_j(t), \zeta_j(t)\}$ such that $\{\zeta_m(t)\}$ converges weakly in $\mathfrak{L}_n^2[a, b]$ to $\zeta_0(t)$, and $\{\eta_m(t)\}$ converges uniformly on $[a, b]$ to $\eta_0(t)$, while $\eta_0 \in \mathfrak{D}[\mathcal{B}]:\zeta_0$ and $J[\eta_0 \mid \mathcal{B}] \leq \liminf_{m\to\infty} J[\eta_m \mid \mathcal{B}] \leq 0$. If $k > 0$ is such that $J[\eta_m \mid \mathcal{B}] \geq -k$ for $m = 1, 2, \ldots$, conclude with the aid of (11.18) that $0 \leq K[\eta_m] \leq k/m$, so that $K[\eta_0] = \lim_{m\to\infty} K[\eta_m] = 0$. As $N[\eta_0] = \lim_{m\to\infty} N[\eta_m] = 1$, deduce that $\eta_0(t) \not\equiv 0$ and $K[\eta_0] = 0$, $J[\eta_0 \mid \mathcal{B}] \leq 0$, contrary to $(\mathfrak{H}_2\text{-b})$.

2. If the hypotheses of Theorem 11.1 hold, and $\eta_j(t) \in \mathfrak{D}[\mathcal{B}]$, $(j = 1, 2, \ldots)$, with $K[\eta_i, \eta_j] = \delta_{ij}$, $(i, j = 1, 2, \ldots)$, prove that the sequence

$\{J[\eta_j \mid \mathcal{B}]\}$ is unbounded. This result provides an alternative proof of the result that the sequence $\{\lambda_j\}$ of Theorem 11.1 diverges to ∞.

Hint. With the aid of Lemma 11.1, show that the assumption that the sequence $\{J[\eta_j \mid \mathcal{B}]\}$ is bounded implies that the vector functions $\{\eta_j(t)\}$ are uniformly bounded and equi-continuous on $[a, b]$, and hence that there exists a subsequence $\{\eta_m(t)\}$ of $\{\eta_j(t)\}$ which converges uniformly on $[a, b]$ to a continuous vector function $\eta_0(t)$. Proceed to establish the contradictory results

$$K[\eta_0] = \lim_{m\to\infty} K[\eta_m] = 1, \qquad K[\eta_0] = \lim_{m\to\infty} K[\eta_m, \eta_{m+1}] = 0.$$

3^s. Suppose that the $n \times n$ matrix functions $A(t)$, $B(t)$, $C(t)$ satisfy hypothesis $(\mathfrak{H})$ on $[a, b]$, $B(t) \geq 0$ for t a.e. on this interval, $K(t)$ is an hermitian matrix function of class $\mathfrak{L}_{nn}^{\infty}[a, b]$ satisfying condition ($\mathfrak{H}_K$-a′), and the differential system (11.1-a) is identically normal on $[a, b]$. Prove that the number of conjugate points to $t = a$ on (a, b), $\{(a, b]\}$, relative to (11.1-a) for a given real value $\lambda = l$ is equal to the number of proper values of the boundary problem

$$\begin{aligned} &L_1[u, v](t) = \lambda K(t)u(t), \qquad L_2[u, v](t) = 0, \qquad t \in [a, b], \\ &u(a) = 0, \qquad u(b) = 0, \end{aligned} \tag{11.19}$$

which are less than, {not greater than}, l, where each proper value is counted a number of times equal to its index.

Hint. Use results of Theorems 7.3 and 11.2.

12. COMPARISON THEOREMS

A boundary problem $\mathcal{B}$ of the form (11.1) depends upon the differential operator $L[u \mid \mathcal{B}](t) = u'(t) - A(t \mid \mathcal{B})\, u(t)$, $L_2[u, v \mid \mathcal{B}](t) = L[u \mid \mathcal{B}](t) - B(t \mid \mathcal{B})\, u(t)$, the function space $\mathfrak{D}[a, b]$ consisting of those $\eta \in \mathfrak{A}_n[a, b]$ for which there is a corresponding $\zeta \in \mathfrak{L}_n^2[a, b]$ such that $L_2[\eta, \zeta \mid \mathcal{B}](t) = 0$, the end-manifold $\mathcal{S}[\mathcal{B}]$ in $\mathbf{C}_{2n}$, the domain $\mathfrak{D}[\mathcal{B}] = \{\eta \mid \eta \in \mathfrak{D}[a, b], \hat{\eta} \in \mathcal{S}[\mathcal{B}]\}$, the end-form $Q[\eta \mid \mathcal{B}] = \hat{\eta}^* Q[\mathcal{B}]\eta$, and the hermitian functionals

$$J[\eta \mid \mathcal{B}] = Q[\eta \mid \mathcal{B}] + \int_a^b 2\omega(t, \eta(t), \eta'(t) \mid \mathcal{B})\, dt,$$

$$K[\eta \mid \mathcal{B}] = \int_a^b \eta^*(t) K(t \mid \mathcal{B}) \eta(t)\, dt,$$

where

$$2\omega(t, \eta, \zeta \mid \mathcal{B}) = \zeta^* B(t \mid \mathcal{B})\zeta + \eta^* C(t \mid \mathcal{B})\eta.$$

It will be supposed throughout this section that the problems considered satisfy the hypotheses $(\mathfrak{H}_{\mathcal{B}})$ *and* ($\mathfrak{H}_K$-a, b, c) *occurring in Theorem* 11.1; *in particular, these conditions imply that each problem considered is normal.* The results to be established in this section are generalizations of the results of Section VI.4, and in most cases the proofs are direct matrix generalizations of those presented for the two-dimensional problems.

If I is a finite interval on the real line, the symbol $V(I \mid \mathfrak{B})$ will denote the number of proper values of $\mathfrak{B}$ on I; for a real number L the symbol $V_L(\mathfrak{B})$, $\{W_L(\mathfrak{B})\}$, will denote the number of proper values of this problem which are less, {not greater}, than L, where in each case a proper value is counted a number of times equal to its index. Most of the following results involve two problems $\mathfrak{B}$ and $\tilde{\mathfrak{B}}$ and, corresponding to the notation of Section VI.4, the respective sequences of proper values and proper functions satisfying the conditions of Theorem 11.1 will be designated by $\{\lambda_j; u_j(t); v_j(t)\}$ and $\{\tilde{\lambda}_j; \tilde{u}_j(t); \tilde{v}_j(t)\}$.

As an immediate consequence of the extremizing properties of the proper values of $\mathfrak{B}$ and $\tilde{\mathfrak{B}}$, as given specifically in Theorem 11.2, one has the following comparison theorem.

Theorem 12.1. *Suppose that $\mathfrak{B}$ and $\tilde{\mathfrak{B}}$ have in common $\mathcal{S}[\mathfrak{B}] = \mathcal{S}[\tilde{\mathfrak{B}}]$, $\mathfrak{D}[\mathfrak{B}] = \mathfrak{D}[\tilde{\mathfrak{B}}]$, and $K(t \mid \mathfrak{B}) \equiv K(t \mid \tilde{\mathfrak{B}}) \equiv K(t)$. If $\Delta J[\eta \mid \mathfrak{B}, \tilde{\mathfrak{B}}] = J[\eta \mid \tilde{\mathfrak{B}}] - J[\eta \mid \mathfrak{B}]$ is non-negative definite on $\mathfrak{D}[\mathfrak{B}] = \mathfrak{D}[\tilde{\mathfrak{B}}]$, then $\tilde{\lambda}_j \geq \lambda_j$, $(j = 1, 2, \dots)$; moreover, if $\Delta J[\eta \mid \mathfrak{B}, \tilde{\mathfrak{B}}]$ is positive definite on $\mathfrak{D}[\mathfrak{B}] = \mathfrak{D}[\tilde{\mathfrak{B}}]$, then $\tilde{\lambda}_j > \lambda_j$, $(j = 1, 2, \dots)$.*

Clearly $\mathfrak{D}[\mathfrak{B}] = \mathfrak{D}[\tilde{\mathfrak{B}}]$ if $\mathcal{S}[\mathfrak{B}] = \mathcal{S}[\tilde{\mathfrak{B}}]$, and the matrix functions $A(t)$, $B(t)$ for the two problems are identical. This latter condition is not necessary, however, in view of the comments in the third paragraph of Section 5. Indeed, if the matrices $\pi(t) = \pi(t \mid \mathfrak{B})$ and $\pi(t) = \pi(t \mid \tilde{\mathfrak{B}})$ for the respective problems are defined as in Section 4 above, then a necessary and sufficient condition for $\mathfrak{D}[\mathfrak{B}] = \mathfrak{D}[\tilde{\mathfrak{B}}]$ is that $\mathcal{S}[\mathfrak{B}] = \mathcal{S}[\tilde{\mathfrak{B}}]$ and for arbitrary $\eta(t) \in \mathfrak{A}_n[a, b]$ we have $\pi^*(t \mid \mathfrak{B})L[\eta \mid \tilde{\mathfrak{B}}](t) = 0$ for $t \in [a, b]$ if and only if $\pi^*(t \mid \tilde{\mathfrak{B}})L[\eta \mid \tilde{\mathfrak{B}}](t) = 0$ for $t \in [a, b]$. In particular, for boundary problems associated with the accessory differential system for a variational problem of so-called Bolza type, (see, for example, Bliss [5, §91]), the differential equation restraints appear naturally in the latter form.

Now consider two problems $\mathfrak{B}$ and $\tilde{\mathfrak{B}}$ which differ only in the end-forms $Q[\eta \mid \mathfrak{B}]$ and $Q[\eta \mid \tilde{\mathfrak{B}}]$. That is, $A(t \mid \mathfrak{B}) \equiv A(t \mid \tilde{\mathfrak{B}})$, $B(t \mid \mathfrak{B}) \equiv B(t \mid \tilde{\mathfrak{B}})$, $C(t \mid \mathfrak{B}) \equiv C(t \mid \tilde{\mathfrak{B}})$, $K(t \mid \mathfrak{B}) \equiv K(t \mid \tilde{\mathfrak{B}})$ and $\mathcal{S}[\mathfrak{B}] = \mathcal{S}[\tilde{\mathfrak{B}}]$; in particular, $\mathfrak{D}[\mathfrak{B}] = \mathfrak{D}[\tilde{\mathfrak{B}}]$. As for the corresponding problem in Section VI.4, let $Q^{\Delta}[\eta] = Q[\eta \mid \tilde{\mathfrak{B}}] - Q[\eta \mid \mathfrak{B}] = \hat{\eta}^* Q^{\Delta} \hat{\eta}$. Moreover, let M be a $2n \times (2n - d)$ matrix whose column vectors form a basis for $\mathcal{S}^{\perp}[\mathfrak{B}] = \mathcal{S}^{\perp}[\tilde{\mathfrak{B}}]$, and $V_0(t)$ an $n \times d_a$ matrix function whose column vectors $v_\beta(t)$ provide a basis for $\Lambda[a, b]$. From the discussion of Section 9, the condition of normality for the problems $\mathfrak{B}$ and $\tilde{\mathfrak{B}}$ insures that the $2n \times (2n - d + d_a)$ matrix $[M \quad D\hat{V}_0]$ has rank $2n - d + d_a$, where $D = \operatorname{diag}\{-E_n, E_n\}$.

Now consider the polynomial $D(\rho \mid \mathcal{B}, \tilde{\mathcal{B}})$ of degree $d - d_a$ defined as follows: if $d = d_a$ then $D(\rho \mid \mathcal{B}, \tilde{\mathcal{B}}) \equiv 1$; if $1 \leq d - d_a \leq 2n$, then $D(\rho \mid \mathcal{B}, \tilde{\mathcal{B}}) = \det \Gamma(\rho \mid \mathcal{B}, \tilde{\mathcal{B}})$, where $\Gamma(\rho \mid \mathcal{B}, \tilde{\mathcal{B}})$ is the matrix

$$\Gamma(\rho \mid \mathcal{B}, \tilde{\mathcal{B}}) = \begin{bmatrix} \rho E_{2n} - Q^{\Delta} & M & D\hat{V}_0 \\ M^* & 0 & 0 \\ \hat{V}_0^* D & 0 & 0 \end{bmatrix}, \quad \textit{whenever } 1 \leq d - d_a \leq 2n - 1;$$

$$\Gamma(\rho \mid \mathcal{B}, \tilde{\mathcal{B}}) = \rho E_{2n} - Q^{\Delta}, \quad \textit{whenever} \quad d - d_a = 2n.$$

In view of the extremizing properties of the zeros of $D(\rho \mid \mathcal{B}, \tilde{\mathcal{B}})$, (see Problem 13 of E.2 in Appendix E), we have the following result.

Theorem 12.2. *Suppose that $\mathcal{B}$ and $\tilde{\mathcal{B}}$ differ only in the end-forms $Q[\eta \mid \mathcal{B}]$, $Q[\eta \mid \tilde{\mathcal{B}}]$, and let* **N** *and* **P** *denote, respectively, the number of negative and positive zeros of the polynomial $D(\rho \mid \mathcal{B}, \tilde{\mathcal{B}})$ defined above, where each zero is counted a number of times equal to its multiplicity. Then $\tilde{\lambda}_{j+\mathbf{N}} \geq \lambda_j$, $\lambda_{j+\mathbf{P}} \geq \tilde{\lambda}_j$, $(j = 1, 2, \ldots)$, $V_L(\mathcal{B}) - \mathbf{P} \leq V_L(\tilde{\mathcal{B}}) \leq V_L(\mathcal{B}) + \mathbf{N}$ and $W_L(\mathcal{B}) - \mathbf{P} \leq W_L(\tilde{\mathcal{B}}) \leq W_L(\mathcal{B}) + \mathbf{N}$, for arbitrary real L; moreover,*

$$|V(I \mid \mathcal{B}) - V(I \mid \tilde{\mathcal{B}})| \leq \mathbf{N} + \mathbf{P}$$

for every bounded subinterval I on the real line.

Now suppose that $\mathcal{B}$ and $\tilde{\mathcal{B}}$ differ only in the end-manifolds $\mathcal{S}[\mathcal{B}]$ and $\mathcal{S}[\tilde{\mathcal{B}}]$, which are of respective dimensions d and $\tilde{d}$. Since each of the problems is normal, from the discussion of Section 9 it follows that $\dim(\mathcal{S}[\mathcal{B}] \cap \mathcal{S}_a) = d - d_a$, and $\dim(\mathcal{S}[\tilde{\mathcal{B}}] \cap \mathcal{S}_a) = \tilde{d} - d_a$. Moreover, $\tilde{\mathcal{B}}$ is a subproblem of $\mathcal{B}$ whenever $\mathcal{S}[\tilde{\mathcal{B}}] \cap \mathcal{S}_a \subset \mathcal{S}[\mathcal{B}] \cap \mathcal{S}_a$, in which case $d \geq \tilde{d}$ and $\tilde{\mathcal{B}}$ is said to be a sub-problem of $\mathcal{B}$ of dimension $d - \tilde{d}$. Also, if $d > \tilde{d}$ then there exist $d - \tilde{d}$ linear forms $\theta_\tau[\hat{\eta}] = \theta_\tau^* \hat{\eta}$, $(\tau = 1, \ldots, d - \tilde{d})$, such that

$$\mathcal{S}[\tilde{\mathcal{B}}] \cap \mathcal{S}_a = \{\hat{\eta} \mid \hat{\eta} \in \mathcal{S}[\mathcal{B}] \cap \mathcal{S}_a, \quad \theta_\tau[\hat{\eta}] = 0, \quad \tau = 1, \ldots, d - \tilde{d}\}.$$

In view of these remarks, the following results are ready consequences of the minimum properties of the proper values of $\mathcal{B}$ and $\tilde{\mathcal{B}}$ as presented in Theorems 11.1 and 11.2.

Theorem 12.3. *If the problems $\mathcal{B}$ and $\tilde{\mathcal{B}}$ differ only in the end-manifolds $\mathcal{S}[\mathcal{B}]$, $\mathcal{S}[\tilde{\mathcal{B}}]$, and $\tilde{\mathcal{B}}$ is a subproblem of $\mathcal{B}$ of dimension $\delta = d - \tilde{d}$, then $\lambda_{j+\delta} \geq \tilde{\lambda}_j \geq \lambda_j$, $(j = 1, 2, \ldots)$, and $V_L(\mathcal{B}) - \delta \leq V_L(\tilde{\mathcal{B}}) \leq V_L(\mathcal{B})$, $W_L(\mathcal{B}) - \delta \leq W_L(\tilde{\mathcal{B}}) \leq W_L(\mathcal{B})$ for arbitrary real L; moreover,*

$$|V(I \mid \mathcal{B}) - V(I \mid \tilde{\mathcal{B}})| \leq \delta$$

for every bounded subinterval I on the real line.

Problems VII.12

1. For problems $\mathfrak{B}$ and $\tilde{\mathfrak{B}}$ formulate, and establish, results corresponding to those of Theorem VI.4.5 and its Corollary.

2.[s] Suppose that on an interval $I = [a_0, \infty)$ the $n \times n$ coefficient matrix functions $A(t)$, $B(t)$, $C(t)$ satisfy hypothesis $(\mathfrak{H})$ and $B(t) \geq 0$ for t a.e. on I, the hermitian matrix function $K(t)$ is such that $K(t) \in \mathfrak{L}_{nn}{}^{\infty}[a, b]$ and $K[\eta]$ is positive definite on $\mathfrak{C}_n[a, b]$ for arbitrary compact subintervals $[a, b]$ of I, while the differential system (11.1-a) is identically normal on I. Corresponding to Problem VI.4:8, for $[a, b] \subset I$ consider the four boundary problems $\mathfrak{B}_{00}[a, b]$, $\mathfrak{B}_{0*}[a, b]$, $\mathfrak{B}_{*0}[a, b]$, $\mathfrak{B}_{**}[a, b]$, involving the differential equations (11.1-a), and the respective boundary conditions:

$$\Delta_{00}[a, b]: u(a) = 0, \quad u(b) = 0; \qquad \Delta_{*0}[a, b]: v(a) = 0, \quad u(b) = 0;$$
$$\Delta_{0*}[a, b]: u(a) = 0, \quad v(b) = 0; \qquad \Delta_{**}[a, b]: v(a) = 0, \quad v(b) = 0.$$

Also, for (p, q) any one of the sets $(0, 0)$, $(*, 0)$, $(0, *)$, $(*, *)$, let

$$\{\lambda_j{}^{pq}[a, b]; u_j{}^{pq}(t; a, b); v_j{}^{pq}(t; a, b)\}$$

denote a set of proper values and corresponding proper solutions for $\mathfrak{B}_{pq}[a, b]$ satisfying the conditions of Theorem 11.1.

(i) Prove that if $[a, b] \subset I$, then $\mathfrak{B}_{00}[a, b]$ is a subproblem of dimension n of each of the problems $\mathfrak{B}_{0*}[a, b]$, $\mathfrak{B}_{*0}[a, b]$, while $\mathfrak{B}_{0*}[a, b]$ and $\mathfrak{B}_{*0}[a, b]$ are individually subproblems of dimension n of $\mathfrak{B}_{**}[a, b]$. Moreover, for $j = 1, 2, \ldots,$

$$\lambda_j^{0*}[a, b] \leq \lambda_j^{00}[a, b] \leq \lambda_{j+n}^{0*}[a, b]; \qquad \lambda_j^{*0}[a, b] \leq \lambda_j^{00}[a, b] \leq \lambda_{j+n}^{*0}[a, b],$$

$$\lambda_j^{**}[a, b] \leq \lambda_j^{0*}[a, b] \leq \lambda_{j+n}^{**}[a, b]; \quad \lambda_j^{**}[a, b] \leq \lambda_j^{*0}[a, b] \leq \lambda_{j+n}^{**}[a, b].$$

(ii) If b, c, d are points of I satisfying $b < c < d$, and $r + s > n$, then

$$\lambda_{r+s-n}^{00}[b, d] \leq \max\{\lambda_r^{0*}[b, c], \lambda_s^{*0}[c, d]\}.$$

(iii) Suppose that for arbitrary $c \in I$ there exists a $d > c$ such that $\lambda_n^{*0}[c, d] \leq 0$. For $a \in I$, prove that a necessary and sufficient condition for the differential system (11.1-a) to be disconjugate on $[a, \infty)$ is that $\lambda_1^{0*}[a, c] > 0$ for all $c > a$.

3.[s] Suppose that the hypotheses of Problem 2 hold, and also the additional conditions:

(i) $C(t) \leq 0$ for t a.e. on $I = [a_0, \infty)$.

(ii) For $T(t)$ a fundamental matrix solution of $L[T](t) \equiv T'(t) - A(t)T(t) = 0$ on I the following conditions hold:

(α) if $b \in I$, then there exists a $c > b$ such that $\int_b^c T^*(s)C(s)T(s)\,ds < 0$;

(β) if $c \in I$, then $\lambda[\int_c^t T^{-1}(s)B(s)T^{*-1}(s)\,ds] \to \infty$ as $t \to \infty$, where in general $\lambda[M]$ denotes the smallest proper value of an hermitian matrix M.

Establish the following results, where the notation employed is that introduced in Problem 2:

(a) if $b \in I$, then there exists a $d > b$ such that $\lambda_n^{*0}[b, d] < 0$;

(b) for $a \in I$, a necessary and sufficient condition for (2.2) to be disconjugate on $[a, \infty)$ is that $\lambda_1^{0*}[a, c] > 0$ for all $c > a$.

Hints. With the aid of the result of Problem VII.2:3, reduce the general case to the special one in which $A(t) \equiv 0$, and note that in this instance the $T(t)$ of condition (ii) may be chosen as the identity matrix E_n. By an argument similar to that presented in the hint for (i) in Problem VII.5:6, show that if $b \in I$ then there exists a $d > b$ such that $J[\eta; b, d]$ is negative definite on a manifold in $\mathfrak{D}_{*0}[b, d]$ of dimension n, and conclude that $\lambda_n^{*0}[b, d] < 0$ for such values d. Use the result of (iii) of Problem 2.

4.[s] Suppose that in addition to the hypotheses of Problem 2 the conditions (i) and (ii-β) of Problem 3 hold, while (ii-α) of this latter problem is replaced by the following condition:

(ii-α') There exists a non-null constant vector π such that

$$\pi^*[\textstyle\int_b^c T^*(s)C(s)T(s)\,ds]\pi \to -\infty$$

as $c \to \infty$ for at least one, (and consequently all), $b \in I$.

With the aid of conclusion (ii) of Problem VII.5:6, show that the system (2.2) is not disconjugate for large t, that is, there is no value $a \in I$ such that (2.2) is disconjugate on $[a, \infty)$.

5. Consider the differential system (2.2) determined as in Problem VII.2:2, which, as there established, is such that if $r_j(t) \in \mathfrak{C}^j[a, b]$, $(j = 0, 1, \ldots, n)$, then $(u; v)$ is a solution of (2.2) if and only if $w(t) = u_1(t)$ is a solution of the scalar differential equation

$$l_{2n}[w](t) \equiv \sum_{j=0}^{n} (-1)^j (r_j(t)\, w^{[j]}(t))^{[j]} = 0.$$

In addition to the hypotheses of Problem VII.2:2, suppose that the following conditions hold on an interval $I = [a_0, \infty)$:

(i) $r_n(t) > 0$ for t a.e. on I, and $\int_{a_0}^{\infty} [r_n(s)]^{-1}\, ds = \infty$;

(ii) $r_j(t) \le 0$ for t a.e. on I, and $j = 0, 1, \ldots, n-1$.

Show that:

(a) If there exists a subinterval $I_1 = [a_1, \infty)$ of I such that the system (2.2) is disconjugate on I_1, then each of the integrals

$$\int_a^{\infty} r_j(s) s^{2n-2j-2}\, ds, \qquad (j = 0, 1, \ldots, n-1),$$

is convergent.

(b) If in addition to the above conditions there is no subinterval $I_2 = [a_2, \infty)$ of I on which $r_0(t) = 0$ for t a.e. on I_2, then for $a \in I$ the system (2.2) is disconjugate on $[a, \infty)$ if and only if for arbitrary $c > a$ the functional

$$\int_a^c \sum_{\alpha=0}^{n} r_\alpha(s)\, |\eta^{[\alpha]}(s)|^2\, ds$$

is positive definite on the class of scalar functions $\eta(t) \in \mathfrak{C}^{n-1}[a, c]$, with $\eta^{[n-1]}(t)$ a.c. and $\eta^{[n]} \in \mathfrak{L}^2[a, c]$, while $\eta^{[\alpha-1]}(a) = 0$, $(\alpha = 1, \ldots, n)$.

Hints. To the involved system (2.2) apply the results of the above Problems 3 and 4, noting that if $c \in I$ then the $T(t)$ occurring in (ii) of Problem 3, and satisfying the initial condition $T(c) = E_n$, is given by

$$T_{\alpha\beta}(t) \equiv 0, \text{ for } 1 \le \beta < \alpha \le n; \quad T_{\alpha\beta}(t) = \frac{(t-c)^{\beta-\alpha}}{(\beta-\alpha)!}, \textit{ for } 1 \le \alpha \le \beta \le n,$$

and hence the elements of $S(t) = T^{*-1}(t)$ are

$$S_{\alpha\beta}(t) \equiv 0, \text{ for } 1 \le \alpha < \beta \le n; \quad S_{\alpha\beta}(t) = \frac{(-1)^{\alpha-\beta}(t-c)^{\alpha-\beta}}{(\alpha-\beta)!}, \quad \textit{for } 1 \le \beta \le \alpha \le n.$$

Moreover, if for $\pi \in \mathbf{C}_n$ the vector function $\eta(t \mid \pi) = (\eta_\alpha(t \mid \pi))$ is defined by

$$\eta_\alpha(t \mid \pi) = \left[\sum_{\beta=1}^{n} \frac{\pi_\beta (t-c)^{\beta-1}}{(\beta-1)\,!}\right]^{[\alpha-1]}, \qquad (\alpha = 1, \ldots, n),$$

then $\pi^* C(t)\pi = \sum_{\alpha=1}^{n} r_{\alpha-1}(t)\, |\pi_\alpha|^2$, and

$$\pi^*\left[\int_a^c T^*(s)C(s)T(s)\, ds\right]\pi = \int_a^c \left[\sum_{\alpha=1}^{n} r_{\alpha-1}(s)\mid \eta_\alpha(s \mid \pi)|^2\right] ds,$$

$$\pi^*\left[\int_c^t T^{-1}(s)B(s)T^{*-1}(s)\, ds\right]\pi = \int_c^t [r_n(s)]^{-1}\, |\eta_1(s \mid \pi)|^2\, ds.$$

Note that for a suitable choice of $\pi = (\pi_\alpha)$ the component $\eta_1(t \mid \pi)$ is equal to a given polynomial of degree less than n, and for the choice of π such that $\eta_1(t \mid \pi) = t^{n-1}$ deduce conclusion (a) from the result of Problem 4. Use Problem 8 of F.1 in Appendix F to show that hypotheses (i) and (ii), together with the added restriction that there is no subinterval $I_2 = [a_2, \infty)$ on which $r_0(t) = 0$ for t a.e. on I_2, imply the conditions of Problem 3, and deduce (b) from conclusion (b) of Problem 3 and the extremizing property of the smallest proper value $\lambda_1^{0*}[a, c]$ of $\mathfrak{B}_{0*}[a, c]$.

6. For two problems $\mathfrak{B}$ and $\tilde{\mathfrak{B}}$ which have in common $\mathcal{S}[\mathfrak{B}] = \mathcal{S}[\tilde{\mathfrak{B}}]$, $\mathfrak{D}[\mathfrak{B}] = \mathfrak{D}[\tilde{\mathfrak{B}}]$, and $K(t \mid \mathfrak{B}) \equiv K(t \mid \tilde{\mathfrak{B}}) \equiv K(t)$, formulate a result analogous to that of Problem VI.5:8, and use Lemma 11.1 to establish this result.

7. Suppose that the $n \times n$ matrix functions $A(t)$, $B(t)$, $C(t)$ satisfy hypothesis $(\mathfrak{H})$ on $[a, b]$, $B(t) \geq 0$ for t a.e. on this interval, $K(t)$ is an hermitian matrix of class $\mathfrak{L}_{nn}{}^{\infty}[a, b]$ satisfying condition ($\mathfrak{H}_K$-a′), and the differential system (11.1-a) is identically normal on $[a, b]$. If the associated "null end-point problem" (11.19) is a sub-problem of the boundary problem (11.1) of dimension δ, use results of Problem VII.11:3 and the above Theorem 12.3, to show that for a given real-valued L the number of conjugate points to $t = a$ on (a, b) relative to the differential equation (11.1-a) for $\lambda = L$ is at least $V_L(\mathcal{B}) - \delta$ and at most $V_L(\mathcal{B})$.

8. Consider a boundary problem

(12.1)
$$L[y](t) \equiv A_1(t)[A_2(t)y(t)]' + A_0(t)y(t) = \mu B_0(t)y(t), \qquad t \in [a, b],$$
$$M\hat{u}_y = 0,$$

of the form (IV.5.1), which satisfies the following conditions:

(i) The $n \times n$ matrix functions $B_0(t)$, $A_j(t)$, $(j = 0, 1, 2)$, are continuous on $[a, b]$, with $B_0(t) \not\equiv 0$, $A_1(t)$ and $A_2(t)$ non-singular, and $B_0(t) \geq 0$ for $t \in [a, b]$, while M is an $n \times 2n$ matrix of rank n.

(ii) The system (12.1) is self-adjoint, so that $\mathfrak{D}(L) = \mathfrak{D}(L^\star)$ and $L[y](t) = L^\star[y](t)$ for $y \in \mathfrak{D}(L) = \mathfrak{D}(L^\star)$.

(iii) If $[a_0, b_0] \subset [a, b]$, and $L[y](t) = 0$, $B_0(t)y(t) = 0$ for $t \in [a_0, b_0]$, then $y(t) \equiv 0$ on $[a, b]$.

(iv) $\mu = 0$ is not a proper value of (12.1).

Show that (12.1) possesses infinitely many proper values. Moreover, if $Y = Y(t, \mu)$ is the solution of the matrix differential system

$$L[Y](t) = \mu B_0(t)Y(t), \qquad Y(a) = E, \qquad t \in [a, b], \tag{12.2}$$

and for arbitrary $L > 0$ the symbol V_L denotes the number of proper values of (12.1) satisfying $|\mu| \leq L$, then for arbitrary real $\mu \neq 0$ the determinant of the matrix function $F(t, \mu) = Y(t, \mu) - Y(t, -\mu)$ has on (a, b) at least $V_{|\mu|-n}$ zeros and at most $V_{|\mu|}$ zeros, where each proper value of (12.1) is counted a number of times equal to its index and a zero $t = t_1$ of the determinant of $F(t, \mu)$ is counted $n - r$ times if $F(t_1, \mu)$ has rank r.

Hint. Note that $(y; z)$ is a solution of the differential system

(12.3)
$$\text{(a)}\ L[z](t) = \lambda B_0(t)y(t), \qquad L[y](t) = B_0(t)z(t),$$
$$\text{(b)}\ M\hat{u}_y = 0, \qquad M\hat{u}_z = 0$$

if and only if $u = u_y = A_2 y$, $v = v_z = A_1^* z$ is a solution of the system

(12.4)
$$\text{(a)}\ -v'(t) - A^*(t)v(t) = \lambda K(t)u(t), \qquad u'(t) - A(t)u(t) - B(t)v(t) = 0,$$
$$\text{(b)}\ M\hat{u} = 0, \qquad P^*[\operatorname{diag}\{-E, E\}]\hat{v} = 0,$$

where $A = -A_1^{-1}A_0A_2^{-1}$, $B = A_1^{-1}B_0A_1^{*-1}$, $K = A_2^{*-1}B_0A_2^{-1}$ and P is a $2n \times n$ matrix of rank n which satisfies $MP = 0$. Show that (12.4) is a boundary problem of the form (11.1) which satisfies hypotheses ($\mathfrak{H}_{\mathfrak{B}}$) and ($\mathfrak{H}_K$-a, b, c). In particular, show that ($\mathfrak{H}_K$-b) holds with $\lambda_0 = 0$, so that all proper values of the boundary problem (12.4) are positive. In order to establish ($\mathfrak{H}_K$-c), consider a vector π such that $B(t)\pi \not\equiv 0$ on $[a, b]$. If $[a_1, b_1]$ is a subinterval of $[a, b]$ throughout which $B(t)\pi \neq 0$, show that there exists a set of constants $c_0, c_1, \ldots, c_n$ not all zero, and such that if $p(t)$ denotes the polynomial $c_0 + c_1t + \cdots + c_nt^n$ then there is a solution $u(t)$ of the system $u'(t) - A(t)u(t) = B(t)\pi p(t)$, $u(a_1) = 0 = u(b_1)$. Show that if $(y_1(t); z_1(t))$ is a solution of (12.3) for a value $\lambda > 0$ then $y(t) = y_1(t) + \lambda^{-1/2} z_1(t)$ is a solution of (12.1) for $\mu = \lambda^{1/2}$, and $y(t) = y_1(t) - \lambda^{-1/2} z_1(t)$ is a solution of (12.1) for $\mu = -\lambda^{1/2}$. Moreover, λ is a proper value of (12.3) if and only if either $\mu = \lambda^{1/2}$ or $\mu = -\lambda^{1/2}$ is a proper value of (12.1), and the index of λ as a proper value of (12.3) is equal to the sum of the indices of $-\lambda^{1/2}$ and $\lambda^{1/2}$ as proper values of (12.1).

For $Y(t, \mu)$ the solution of (12.2) and $\mu \neq 0$, set

$$Y_0(t, \mu) = (2\mu)^{-1}[Y(t, \mu) - Y(t, -\mu)],$$
$$Z_0(t, \mu) = 2^{-1}[Y(t, \mu) + Y(t, -\mu)],$$

and show that $Y = Y_0(t, \mu)$, $Z = Z_0(t, \mu)$ is the solution of the matrix differential system

(12.5)

$$\text{(a)}\ L[Z_0](t) = \mu^2 B_0(t) Y_0(t), \qquad L[Y_0](t) = B_0(t)Z_0(t), \qquad t \in [a, b],$$
$$\text{(b)}\ Y_0(a) = 0, \qquad Z_0(a) = E,$$

and $U_0(t) = A_2(t)Y_0(t)$, $V_0(t) = A_1^*(t)Z_0(t)$ is the solution of the differential system

(12.5′)

$$\text{(a)}\ -V_0'(t) - A^*(t)V_0(t) = \mu^2 K(t)U_0(t),$$
$$U_0'(t) - A(t)U_0(t) = B(t)V_0(t),$$
$$\text{(b)}\ U_0(a) = 0, \qquad V_0(a) = A_1^*(a).$$

Note that in view of condition (iii) above a value $t = t_1$ on $(a, b]$ is conjugate to $t = a$ relative to (12.3-a) for $\lambda = \mu^2$ if and only if $Y_0(t_1, \mu)$ is singular, and that the order of t_1 as a conjugate point to $t = a$ is equal to $n - r$ if $Y_0(t_1, \mu)$ has rank r. Use the results of Problem VII.11:3 and Theorem 12.3 to establish the stated conclusion.

13. NOTES AND REMARKS

As indicated in the initial section, the treatment of this chapter emphasizes the use of variational principles for the extension to general self-adjoint differential systems of the oscillation and comparison theorems of the classical Sturmian theory. Significant references for the overall coverage of this chapter are Morse [1,2, Ch. IV], Hu [1], Birkhoff and Hestenes [1], Hestenes [1], and Reid [5,12,30]. Throughout this chapter the assumption of continuity of the coefficient matrix functions is weakened, so that a solution is to be understood in the Carathéodory sense. In spite of this generality of problem setting, however, in reference to previous literature no distinction will be made in regard to the hypotheses underlying that given discussion, or in regard to the class of differentially admissible vector functions employed by the cited author, since the corresponding results for the more general setting may be obtained by obvious modification of argument, or are readily derivable from the stated results by well-known approximation theorems for the Lebesgue integral.

The notion of conjugate, or conjoined, solutions of a real self-adjoint differential system (2.2) dates from von Escherisch [1]. For the concept of a principal solution of a disconjugate system, and the related concept of a distinguished solution for the associated Riccati matrix differential equation, as appearing in Theorems 3.1, 3.2, 3.3, 3.4, 5.4 and 5.5, the reader is referred to Hartman [5], Sandor [1], Reid [16,23,24], and Hartman [7, Ch. XI-Sec. 10]. For the case of a real scalar linear differential equation of the second order, related papers are Leighton and Morse [1], Leighton [1], Hartman and Wintner [2,5] and Hartman [3].

For the reader familiar with variational theory, the identity of Lemma 4.2, and equation (4.10) of its Corollary, are seen to embody the so-called *Legendre* or *Clebsch transformation* of the functional $J[\eta; a, b]$, and the result of Theorem 4.2 is the *necessary condition of Legendre*, or *Clebsch*, for this functional. Also, the result of Theorem 4.5 is one of the usual forms of the *Jacobi condition* for the functional $J[\eta; a, b]$. In this connection, the reader is referred to Carathéodory [2, Ch. XV], and Bliss [5, Ch. II-Sec. 23, Ch. VIII-Sec. 81, and Ch. IX, Secs. 89–91].

The results of Theorems 4.4 and 4.5 are the most basic tools for the study of oscillation phenomena for self-adjoint vector differential equations, and under varying degrees of generality are to be found in Morse [1,2, Ch. IV], Birkhoff and Hestenes [1], Bliss [5], and Reid [12,20,24]. For discussion of the Riccati matrix differential equation, and its relation to oscillation phenomena, pertinent references are Radon [1,2], Levin [1] and Reid [10,23,25].

Section 7 presents the essential properties of the Morse fundamental Hermitian forms, and the basic relation between the negative index of these forms and the number of focal points of a given conjoined family of solutions of the associated differential equation. For an extension of this procedure to generalized differential systems, the reader is referred to Reid [20]. In particular, such generalized differential systems include as special instances corresponding systems of second order difference equations, (see Reid [20, Sec. 6]); a direct treatment of certain central oscillation and comparison theorems for such systems of difference equations is to be found in Harris [1].

For the development of the treatment of boundary problems as presented in Sections 8–12, the basic papers again are Morse [1,2, Ch. IV], Bliss [3,4], Reid [3,17,30]. In particular, results of Jackson [1], Latshaw [1], and Kamke [4] are historical antecedents of the derivation of a canonical form for the boundary conditions in Section 8. For a more detailed discussion of normality and abnormality, the reader is referred to Bliss [5, Sec. 77] and Reid [24,30]. An earlier discussion of comparison theorems in the format of Section 12, and indeed for an integro-differential system, is to be found in Reid [6].

Significant papers for various aspects of the subject area covered by this chapter are Hartman [1], Hartman and Wintner [6], Sternberg [1,2], Kaufman and Sternberg [1], Atkinson [1, Ch. 10], Etgen [1,2,3], Tomastik [1], and Ahlbrandt [1]. For treatments of related self-adjoint boundary problems which involve the characteristic parameter in the boundary conditions, see Reid [3,5], Bobonis [1], and Zimmerberg [1,2,3]. For papers dealing with oscillation and comparison phenomena for self-adjoint scalar differential equations of the fourth order, attention is directed to Leighton and Nehari [1], Barrett [4,5,6], Hinton [1], Howard [1], Sternberg and Sternberg [1]. For relatively recent papers dealing with self-adjoint scalar differential equations of general order, see Schubert [1], Kaufman and Sternberg [2], Weinberger [1], Sloss [1], Hunt [1], and Sherman [1,2].

It is to be emphasized that when the general results of this chapter are applied to special systems which are equivalent to self-adjoint scalar differential, or quasi-differential, equations of even order, then for scalar equations the concepts of conjugate point and oscillation are the restrictive ones ensuing from the corresponding concepts for the associated differential system. In particular, these concepts lack the generality corresponding to the definitions introduced by Leighton and Nehari [1] for real fourth order differential equations of the form $[r(t)u'']'' - p(t)u = 0$. The reader is referred to Barrett [7] for a presentation of results on oscillation phenomena, in terms of the more general definitions for higher order differential equations, especially those of the third order, together with an extensive bibliography for this area.

A very important topic that has been omitted from the discussion of this chapter is the Weinstein method for the determination of lower bounds for the proper values of a boundary problem, which consists of finding a base problem, constructing intermediate problems, and solving the intermediate problems in terms of the base problem. For a comprehensive treatment of the works of Weinstein, Aronszajn, and others in this field, the reader is referred to Gould [1].

RELATED REFERENCES AND COMMENTS FOR SPECIFIC PROBLEMS

VII.2:2 Reid [9].

VII.2:3 Reid [17,29].

VII.2:4,5 Reid [16].

VII.2:6,7,8 Barrett [3]; Reid [18]; Etgen [1,2,3].

VII.4:1 Under varying degrees of normality assumptions, this result is present in Bliss [5, Secs. 89, 90], Morse [1,2], Reid [13,23,27].

VII.4:3 Reid [4].

VII.4:4,6 Inequalities of the form (4.19), (4.20) are present in most estimates of hermitian integral functionals. Under the stated generality of these problems, the inequalities are presented in Reid [30].

VII.4:8 This generalization of the Liapunov inequality (V.3.16) appears in Reid [12, Th. 2.3].

VII.4:9 The first conclusion of this problem is a special case of Reid [28, Th. 3.2].

VII.4:10 Morse [2, Ch. IV-Th. 8.5]; Reid [16, Th. 4.3].

VII.5:2,3 Reid [12, Sec. 4].

VII.5:4 Ahlbrandt [1, Th. 3.3].

VII.5:5 Reid [25].

VII.5:6 This problem extends results of Hille [1, p. 243], and Sternberg [1, p. 316]. Conclusion (i) follows from Reid [22, Th. 3.3]; conclusion (ii) has been established by Ahlbrandt [1, Th. 4.1].

VII.5:7 Reid [16, Th. 6.2].

VII.5:8 Reid [23, Ths. 5.3 and 6.1].

VII.5:10 For a real linear homogeneous differential equation of the second order, the results of this problem establish the equivalence of the concept of a principal solution as defined in the text with that introduced by Leighton and Morse [1], and by Hartman and Wintner [5].

VII.5:12 Reid [23, Th. 7.2].

VII.5:13,14,15 The results of these problems are essentially those of the respective Ths. 5.2, 5.3, 5.4 of Reid [18].

VII.7:1,2 The results of these problems are direct generalizations of results of Morse [2, Ch. IV-Th. 8.5].

VII.11:1 Reid [30, Lemma 5.1].

VII.12:3 Reid [22, Corollary to Th. 3.2]; this problem generalizes a result of Howard [1].

VII.12.4 Reid [22, Th. 3.3].

VII.12.5 Reid [22, Th. 4.2 for (a), Th. 4.1 for (b)]. For $n = 1$ the result of conclusion (a) was established by Leighton [3]; for $n = 2$ the result of conclusion (a) is a special case of a theorem of Sternberg and Sternberg [1].

VII.12:8 Reid [7, Th. 6.3]; in more general setting, Reid [30, Th. 7.3].

VIII

Stability and Asymptotic Behavior of Differential Equations

1. INTRODUCTION

Suppose that

$$(1.1) \qquad y = \phi(t), \qquad t \in [a, \infty),$$

is an n-dimensional vector function of class $\mathfrak{C}_n{}^1[a, \infty)$ that is a solution of the vector differential equation

$$(1.2) \qquad y' = f(t, y),$$

where the vector function $f(t, y) \equiv (f_\alpha(t, y)) \equiv (f_\alpha(t, y_1, \ldots, y_n))$, $(\alpha = 1, \ldots, n)$, is continuous in an open region $\mathfrak{R}$ of (t, y)-space containing the graph of (1.1). The solution (1.1) of (1.2) is said to be *stable*, (more precisely, *stable on* $[a, \infty)$ or *stable as* $t \to \infty$), if for each $\tau \in [a, \infty)$ and $\varepsilon > 0$ there is a positive $\delta = \delta(\varepsilon, \tau)$ such that if $|\eta - \phi(\tau)| < \delta$, and $y = y(t)$ is any solution of (1.2) satisfying $y(\tau) = \eta$, then the interval of definition of $y(t)$ may be extended to $[\tau, \infty)$, and for any such extension $|y(t) - \phi(t)| < \varepsilon$ on $[\tau, \infty)$. In particular, if (1.1) is a stable solution, and $\tau \in [a, \infty)$, then for each $\tau_1 > \tau$ the unique solution $y = y(t)$ of (1.2) on $[\tau, \tau_1]$ satisfying $y(\tau) = \phi(\tau)$ is $y(t) \equiv \phi(t)$, $t \in [\tau, \tau_1]$.

If the solutions of (1.2) are locally unique in $\mathfrak{R}$, then by Theorem I.5.7 the totality of solutions of (1.2) in $\mathfrak{R}$ may be represented as

$$y = \phi(t; \tau, \eta), \qquad a(\tau, \eta) < t < b(\tau, \eta),$$

where $\phi(\tau; \tau, \eta) = \eta$, and the vector function $\phi(t; \tau, \eta)$ is continuous in (t, τ, η) on the set $(\tau, \eta) \in \mathfrak{R}$, $a(\tau, \eta) < t < b(\tau, \eta)$. In particular, if τ_1, τ_2 are two given values on $[a, \infty)$ then for each $\varepsilon > 0$ there is a $\delta' = \delta'(\varepsilon, \tau_1, \tau_2)$ such that if $|\eta - \phi(\tau_1)| < \delta'$ then the closed interval I with endpoints τ_1 and

τ_2 belongs to $a(\tau_1, \eta) < t < b(\tau_1, \eta)$ and $|\phi(t; \tau_1, \eta) - \phi(t)| < \varepsilon$ on I. Consequently, if the solutions of (1.2) are locally unique in $\mathfrak{R}$ then the definitive condition for stability holds for all τ on $[a, \infty)$ whenever it holds for any one value $\tau = \tau_0$.

A solution (1.1) of (1.2) is said to be *uniformly stable* on $[a, \infty)$ if in the definition of stability the quantity $\delta(\varepsilon, \tau)$ may be chosen independent of τ; that is, if for each $\varepsilon > 0$ there is a $\delta(\varepsilon) > 0$ such that if $\tau \in [a, \infty)$ and $|\eta - \phi(\tau)| < \delta(\varepsilon)$, then for $y = y(t)$ any solution of (1.2) satisfying $y(\tau) = \eta$ the interval of definition of $y(t)$ may be extended to $[\tau, \infty)$, and for any such extension $|y(t) - \phi(t)| < \varepsilon$ on $[\tau, \infty)$. If the solutions of (1.2) are locally unique in $\mathfrak{R}$ the comments of the preceding paragraph imply that the condition of stability of (1.1) is equivalent to the statement that if $\tau \in [a, \infty)$ and $\varepsilon > 0$ then there is a $\delta'(\varepsilon, \tau) > 0$ such that if $|\eta - \phi(\tau)| < \delta'(\varepsilon, \tau)$ then $a(\tau, \eta) < a$, $b(\tau, \eta) = \infty$, and $|\phi(t; \tau, \eta) - \phi(t)| < \varepsilon$ on $[a, \infty)$. A corresponding modification of phraseology in the definition of uniform stability produces the condition that for each $\varepsilon > 0$ there is a $\delta'(\varepsilon) > 0$ such that if $\tau \in [a, \infty)$ and $|\eta - \phi(\tau)| < \delta'(\varepsilon)$ then for $y = y(t)$ any solution of (1.2) satisfying $y(\tau) = \eta$ the interval of definition of $y(t)$ may be extended to $[a, \infty)$ and $|y(t) - \phi(t)| < \varepsilon$ on $[a, \infty)$. This new condition is stronger than the condition of uniform stability, and a solution (1.1) which satisfies this condition is said to be *strictly stable*. For example, as the solution of the scalar differential equation

$$y' = -y, \tag{1.3}$$

passing through the point (τ, η) is $y = \phi(t; \tau, \eta) \equiv \eta \exp\{\tau - t\}$, it follows that $|\phi(t; \tau, \eta)| \leq |\eta|$ on $[\tau, \infty)$ and the solution $\phi(t) \equiv 0$ of (1.3) satisfies the condition of uniform stability on $[0, \infty)$ with $\delta(\varepsilon) = \varepsilon$; however, it is clear that this solution is not strictly stable since $\phi(0; \tau, \eta) = \eta e^{\tau}$. That uniform stability is stronger than stability is illustrated by the scalar differential equation

$$y' = (t \sin t - \cos t - 2k)y, \tag{1.4}$$

with $2k > 1$. The solution of (1.4) passing through (τ, η) is $y = \phi(t; \tau, \eta) \equiv \eta \exp\{\tau \cos \tau - t \cos t + 2k(\tau - t)\}$, and for the solution $\phi(t) \equiv 0$ the condition of stability holds with $\delta(\varepsilon, \tau) = \varepsilon \exp\{-\tau \cos \tau - 2k\tau\}$; on the other hand, the relation

$$\phi((2m + 1)\pi; (2m + \tfrac{1}{2})\pi, \eta) = \eta \exp\{(2m + 1 - k)\pi\}, \qquad (m = 1, 2, \ldots),$$

shows that it is impossible to satisfy the condition of stability with a $\delta(\varepsilon, \tau)$ that is independent of τ.

A solution (1.1) of (1.2) is said to be *asymptotically stable* if it is stable and moreover the $\delta(\varepsilon, \tau)$ appearing in the definition of stability may be so chosen that $|y(t) - \phi(t)| \to 0$ as $t \to \infty$ for every solution $y = y(t)$ of (1.2) satisfying $|y(\tau) - \phi(\tau)| < \delta(\varepsilon, \tau)$. For example, $\phi(t) \equiv 0$ is an asymptotically stable solution of each of the equations (1.3) and (1.4). On the other hand, $\phi(t) \equiv 0$ is a strictly stable solution of

$$y' = iy, \tag{1.5}$$

but this solution is not asymptotically stable.

Finally, a solution (1.1) of (1.2) is said to be *uniformly asymptotically stable* if it is uniformly stable and there exists a $\delta_0 > 0$ such that for $\varepsilon > 0$ there is a $T(\varepsilon) > 0$ with the property that if $\tau \in [a, \infty)$, and $y(t)$ is a solution of (1.2) satisfying $|y(\tau) - \phi(\tau)| < \delta_0$, then $|y(t) - \phi(t)| \leq \varepsilon$ for $t \in [\tau + T(\varepsilon), \infty)$. Clearly the solutions of (1.3) are uniformly asymptotically stable on $[0, \infty)$.

Problems VIII.1

1. If $f(t, y)$ is independent of t, in which case the differential equation (1.2) is said to be *autonomous*, and $y = \phi(t)$, $t \in [a, \infty)$, is a solution of this equation which is constant, show that this solution is stable if and only if it is uniformly stable, and asymptotically stable if and only if it is uniformly asymptotically stable.

Hint. Note that if $y = y_0(t)$, $t \in [a, \infty)$, is a solution of an autonomous differential equation (1.2), and $c > 0$, then $y = y_0(t - c)$, $t \in [a + c, \infty)$, is also a solution of this equation.

2. LINEAR DIFFERENTIAL EQUATIONS

Of particular interest and importance is the case of a linear vector differential equation

$$y' = A(t)y + g(t), \tag{2.1}$$

where the elements of the $n \times n$ matrix $A(t) \equiv [A_{\alpha\beta}(t)]$, $(\alpha, \beta = 1, \ldots, n)$, and the n-dimensional vector function $g(t) \equiv (g_\alpha(t))$, $(\alpha = 1, \ldots, n)$, are continuous on $[a, \infty)$. As established in Section I.9, for arbitrary (τ, η) with $\tau \in [a, \infty)$ there is a unique solution $y = \phi(t; \tau, \eta)$ passing through the point (τ, η), and this solution is defined on $[a, \infty)$. Moreover, if $Y(t)$ is a fundamental matrix for the homogeneous linear vector differential equation

$$y' = A(t)y, \tag{2.2}$$

and $y = \phi(t)$ is any particular solution of (2.1), then

$$\phi(t; \tau, \eta) = \phi(t) + Y(t)Y^{-1}(\tau)[\eta - \phi(\tau)]. \tag{2.3}$$

From (2.3) it is clear that for a linear differential equation (2.1) all solutions possess any one of the five types of stability defined in Section 1 whenever a single solution has this type of stability, and the type of stability is that possessed by $\phi(t) \equiv 0$ as a solution of (2.2). Consequently, *in the discussion of linear differential equations we shall for brevity say that the equation* (2.2) *possesses the indicated stability property.*

Theorem 2.1. *If $Y(t)$ is a fundamental matrix for* (2.2), *and* $Y(t, s) = Y(t)Y^{-1}(s)$ *for* $(t, s) \in [a, \infty) \times [a, \infty)$, *then on* $[a, \infty)$ *this equation is:*

(a) *stable if and only if there exists a* $\kappa_0 > 0$ *such that*

$$\nu[Y(t)] \leq \kappa_0, \quad \textit{for} \quad t \in [a, \infty); \tag{2.4}$$

(b) *uniformly stable if and only if there exists a* $\kappa_1 > 0$ *such that*

$$\nu[Y(t, \tau)] \leq \kappa_1 \quad \textit{for} \quad \tau \in [a, t], \qquad t \in [a, \infty); \tag{2.5}$$

(c) *strictly stable if and only if there exists a* $\kappa_2 > 0$ *such that*

$$\nu[Y(t, \tau)] \leq \kappa_2 \quad \textit{for} \quad \tau \in [a, \infty), \qquad t \in [a, \infty); \tag{2.6}$$

(d) *asymptotically stable if and only if*

$$\nu[Y(t)] \to 0 \quad \textit{as} \quad t \to \infty; \tag{2.7}$$

(e) *uniformly asymptotically stable if and only if there exist constants* $\kappa_3 > 0$, $\mu > 0$ *such that*

$$\nu[Y(t, \tau)] \leq \kappa_3 \exp\{-\mu(t - \tau)\}, \quad \textit{for} \quad \tau \in [a, t], \qquad t \in [a, \infty). \tag{2.8}$$

Conclusions (a)–(d) are immediate from the definitions, since every solution of (2.2) is of the form $y(t) = Y(t)\xi$, where ξ is an n-dimensional vector; in particular, $y(t) = Y(t, \tau)\, y(\tau)$ for $(t, \tau) \in [a, \infty) \times [a, \infty)$. If condition (2.8) holds then (2.2) is uniformly stable, since (2.5) holds with $\kappa_1 = \kappa_3$; also, for $T(\varepsilon) = -(1/\mu)\ln(\varepsilon/\kappa_3)$ the inequality (2.8) implies that $\nu[Y(t, \tau)y(\tau)] \leq \varepsilon\, |y(\tau)|$ if $t \in [\tau + T(\varepsilon), \infty)$, so that the condition of uniform asymptotic stability holds with $\delta_0 = 1$, $T(\varepsilon) = -(1/\mu)\ln(\varepsilon/\kappa_3)$. Conversely, if (2.2) is uniformly asymptotically stable, then for δ_0 as in the definition of uniform asymptotic stability and $0 < \varepsilon_1 < \delta_0$, there exists a $T_1 = T(\varepsilon_1) > 0$ such that if $\tau \in [a, \infty)$ and $|\xi_1| < \delta_0$, then $|Y(t, \tau)\xi_1| \leq \varepsilon_1$ for $t \in [\tau + T_1, \infty)$. Consequently, if $|\xi| \leq 1$ then $|Y(\tau + T_1, \tau)\xi| \leq (\varepsilon_1/\delta_0) < 1$, so that

$$\nu[Y(\tau + T_1, \tau)] \leq \frac{\varepsilon_1}{\delta_0}, \quad \textit{for} \quad \tau \in [a, \infty).$$

Moreover, since uniform asymptotic stability implies uniform stability, from conclusion (b) it follows that there exists a constant κ_1 such that

$$\nu[Y(t, \tau)] \leq \kappa_1 \quad \textit{for} \quad t \in [\tau, \tau + T_1], \qquad \tau \in [a, \infty).$$

Now for $t \in [\tau, \infty)$, $\tau \in [a, \infty)$, let m denote the non-negative integer such that $t \in [\tau + mT_1, \tau + (m + 1)T_1)$. Then

$$Y(t, \tau) = Y(t, \tau + mT_1)Y(\tau + mT_1, \tau + (m - 1)T_1) \cdots Y(\tau + T_1, \tau),$$

and hence $\nu[Y(t, \tau)] \leq \kappa_1(\varepsilon_1/\delta_0)^m = (\kappa_1\delta_0/\varepsilon_1)(\varepsilon_1/\delta_0)^{m+1}$. Consequently, for $\mu = -(1/T_1)\ln(\varepsilon_1/\delta_0)$ and $\kappa_3 = \kappa_1\delta_0/\varepsilon_1$ the relation (2.8) holds.

The condition that a fundamental matrix $Y(t)$ of (2.2) be such that $Y(t)Y^{-1}(\tau)$ is bounded for $(t, \tau) \in [a, \infty) \times [a, \infty)$, is clearly equivalent to the condition that $Y(t)$ and $Y^{-1}(t)$ are both bounded on $[a, \infty)$. Moreover, as shown in Section III.2, if $Y(t)$ is a fundamental matrix for (2.2) then $Y^{*-1}(t)$ is a fundamental matrix for the adjoint homogeneous differential equation

$$z' = -A^*(t)z, \tag{2.9}$$

and consequently the following result is immediate.

Corollary. *The equation* (2.2) *is strictly stable on* $[a, \infty)$ *if and only if both this equation and its adjoint* (2.9) *are stable on this interval.*

As shown in Theorem I.9.2, if $Y(t)$ is a fundamental matrix for (2.2) then

$$\det Y(t) = [\det Y(\tau)] \exp\left\{\int_\tau^t \operatorname{Tr} A(s)\, ds\right\} \tag{2.10}$$

and consequently

$$|\det Y(t)| = |\det Y(\tau)| \exp\left\{\int_\tau^t \Re[\operatorname{Tr} A(s)]\, ds\right\}. \tag{2.10'}$$

Now if all solutions of (2.2) are bounded on $[a, \infty)$, then $|\det Y(t)|$ is bounded on this interval, and consequently $\int_\tau^t \Re[\operatorname{Tr} A(s)]\, ds$ has a finite upper bound on $[a, \infty)$. As $\Re[\operatorname{Tr} A^*(s)] = \Re[\operatorname{Tr} A(s)]$, the corresponding condition for the adjoint equation is that if all solutions of (2.9) are bounded on $[a, \infty)$ then $\int_\tau^t \Re[\operatorname{Tr} A(s)]\, ds$ has a finite lower bound on this interval. Moreover, if all solutions of (2.2) are bounded and $|\det T(t)|$ has a positive lower bound on $[a, \infty)$, then $Y^{-1}(t) \equiv [\text{cofactor } Y_{\beta\alpha}(t)]/\det Y(t)]$ is bounded on $[a, \infty)$, so that all solutions of (2.9) are bounded on this interval; clearly an analogous statement, with the roles of (2.2) and (2.9) interchanged, also holds. These results may be formulated as the following theorem.

Theorem 2.2. *A necessary condition for the equation* (2.2) {*equation* (2.9)} *to be stable on* $[a, \infty)$ *is that* $\int_\tau^t \Re[\operatorname{Tr} A(s)]\, ds$ *has a finite upper bound* {*lower bound*} *on this interval; moreover, if this integral has a finite lower bound* {*upper bound*} *and equation* (2.2) {*equation* (2.9)} *is stable then equation* (2.9) {*equation* (2.2)} *is stable also.*

Now if $y(t)$ is a solution of (2.2) then $(|y(t)|^2)' = y^*(t)[A(t) + A^*(t)]\,y(t) = 2y^*(t)[\Re e\, A(t)]\,y(t)$; similarly, if $z(t)$ is a solution of (2.9) then $(|z(t)|^2)' = -z^*(t)[A(t) + A^*(t)]\,z(t) = -2z^*(t)[\Re e\, A(t)]\,z(t)$. Thus we have the following result as an immediate consequence of the extremizing properties of the proper values of an hermitian matrix. In this connection the reader is referred to Appendix E. In particular, the continuity of the proper values $\lambda_g(t)$ and $\lambda_l(t)$ follows from Problem 6 of E.2.

Theorem 2.3. *If for each value of t on $[a, \infty)$ the greatest and the least proper values of the hermitian matrix $\Re e\, A(t)$ are denoted by $\lambda_g(t)$ and $\lambda_l(t)$, respectively, then for any solution $y(t)$ of* (2.2),

$$(2.11)\qquad |y(\tau)| \exp\left\{\int_\tau^t \lambda_l(s)\,ds\right\} \le |y(t)| \le |y(\tau)| \exp\left\{\int_\tau^t \lambda_g(s)\,ds\right\},$$

$$a \le \tau \le t < \infty.$$

Correspondingly, if $z(t)$ is a solution of (2.9), *then*

$$(2.12)\qquad |z(\tau)| \exp\left\{-\int_\tau^t \lambda_g(s)\,ds\right\} \le |z(t)| \le |z(\tau)| \exp\left\{-\int_\tau^t \lambda_l(s)\,ds\right\},$$

$$a \le \tau \le t < \infty.$$

In particular, (2.2) *is stable on $[a, \infty)$ if $\int_a^t \lambda_g(s)\,ds$ has a finite upper bound on this interval; correspondingly,* (2.9) *is stable if $\int_a^t \lambda_l(s)\,ds$ has a finite lower bound on $[a, \infty)$.*

If $T(t)$ is a nonsingular matrix of class $\mathfrak{C}^1[a, \infty)$, and

$$(2.13)\qquad y(t) = T(t)u(t), \qquad t \in [a, \infty),$$

then $y(t)$ is a solution of (2.2) if and only if $u(t)$ is a solution of

$$(2.14)\qquad u' = B(t)u,$$

where

$$(2.15)\qquad B = T^{-1}(AT - T'), \qquad t \in [a, \infty).$$

Moreover, for such a matrix $T(t)$ the vector functions $z(t)$, $v(t)$ satisfying

$$(2.16)\qquad z(t) = T^{*-1}(t)v(t), \qquad t \in [a, \infty),$$

are such that $z(t)$ is a solution of (2.9) if and only if $v(t)$ is a solution of

$$(2.17)\qquad v' = -B^*(t)v,$$

with $B(t)$ given by (2.15).

If $A(t)$ and $B(t)$ are continuous $n \times n$ matrix functions on $[a, \infty)$, then $B(t)$ is said to be *kinematically similar*, (or *t-similar*), to $A(t)$, and we write $B(t) \overset{\text{ks}}{=} A(t)$, if there exists a nonsingular matrix function $T(t)$ of class $\mathfrak{C}^1$ on a subinterval $[c, \infty)$ of $[a, \infty)$ which satisfies (2.15), and for which both $T(t)$ and $T^{-1}(t)$ are bounded on $[c, \infty)$. If T is a constant nonsingular matrix, then (2.15) reduces to $B(t) = T^{-1}A(t)T$, so that for all $t \in [c, \infty)$ the matrices $A(t)$ and $B(t)$ are similar, under the same similarity transformation. Correspondingly, the differential system (2.14) is kinematically similar, (or t-similar), to (2.2) if $B(t) \stackrel{ks}{=} A(t)$. It may be established directly that kinematic similarity is indeed an equivalence, that is:

(i) $A(t) \stackrel{ks}{=} A(t)$;
(ii) $A(t) \stackrel{ks}{=} B(t)$ if $B(t) \stackrel{ks}{=} A(t)$;
(iii) $C(t) \stackrel{ks}{=} A(t)$ if $C(t) \stackrel{ks}{=} B(t)$ and $B(t) \stackrel{ks}{=} A(t)$.

From the above remarks it is seen that if (2.14) is kinematically similar to (2.2) then (2.14) possesses one of the above defined types of stability if and only if (2.2) possesses the same type of stability.

If $A(t)$ is a continuous $n \times n$ matrix function on $[a, \infty)$ for which there exists a constant matrix C such that $C \stackrel{ks}{=} A(t)$, then $A(t)$, and the corresponding vector differential equation (2.2), are said to be *reducible*. In particular, if $A(t)$ is kinematically similar to the zero matrix then this matrix, and the corresponding equation (2.2) are termed *reducible to zero*. If $Y(t)$ is a fundamental matrix of $Y'(t) = A(t)Y(t)$, then clearly (2.2) is reducible to zero if and only if both $Y(t)$ and its inverse $Y^{-1}(t)$ are bounded on $[a, \infty)$. Consequently, in view of Theorem 2.1 and its Corollary, it follows that (2.2) is reducible to zero if and only if (2.2) is strictly stable.

Of particular interest is the case of $A(t)$ a constant matrix A. In this instance, by Theorem 1.1 of Appendix E, there is a nonsingular constant matrix T such that $B = T^{-1}AT$ is superdiagonal, and (2.14) is of the form

$$
\begin{aligned}
u_1' &= \lambda_1 u_1 + B_{12}u_2 + \cdots + B_{1n}u_n, \\
u_2' &= \qquad\quad\ \lambda_2 u_2 + \cdots + B_{2n}u_n, \\
&\qquad\qquad\quad \cdots \\
u_n' &= \qquad\qquad\qquad\qquad\quad \lambda_n u_n,
\end{aligned} \tag{2.18}
$$

where $\lambda_1, \ldots, \lambda_n$ is any prescribed arrangement of the proper values of A with each proper value occurring a number of times equal to its multiplicity.

If one employs the result of Theorem 3.3 of Appendix E it follows that if the distinct proper values of A are $\lambda_1, \ldots, \lambda_r$, and $\lambda = \lambda_j$, $(j = 1, \ldots, r)$, is of multiplicity m_j, then T may be chosen so that

$$B = \operatorname{diag}\{B^{(1)}, \ldots, B^{(r)}\}, \tag{2.19}$$

where $B^{(j)}$, $(j = 1, \ldots, r)$, is a superdiagonal matrix with diagonal elements equal to λ_j. Indeed, the cited theorem shows that each $B^{(j)}$ may be chosen as an elementary Jordan matrix. Now if B has the form (2.19) it follows that the fundamental matrix solution $U(t)$ of (2.18) satisfying $U(0) = E$ is given by

$$U(t) = \operatorname{diag}\{U^{(1)}(t), \ldots, U^{(r)}(t)\}, \tag{2.20}$$

where $U^{(j)}(t)$, $(j = 1, \ldots, r)$, is the superdiagonal matrix

(2.21)

$$U^{(j)}(t) = e^{\lambda_j t}\left[E + t(B^{(j)} - \lambda_j E) + \cdots + \frac{t^{m_j-1}}{(m_j-1)!}(B^{(j)} - \lambda_j E)^{m_j-1}\right].$$

Moreover, $B^{(j)} - \lambda_j E$ is the zero matrix if and only if the proper value $\lambda = \lambda_j$ of A has its index equal to its multiplicity.

Now if $A(t)$ is a constant matrix A and $Y(t)$ is a fundamental matrix of (2.2) satisfying $Y(0) = E$, then $Y(t)Y^{-1}(\tau) = Y(t - \tau)$, and the condition of stability is equivalent to the condition of uniform stability. The following theorem follows from the form (2.20), (2.21) of a fundamental matrix for the corresponding equation (2.14) with B a constant matrix of the form (2.19), together with the fact that $\lambda = \lambda_0$ is a proper value of A with index k and multiplicity m if and only if $\lambda = -\bar{\lambda}_0$ is a proper value of $-A^*$ with index k and multiplicity m.

Theorem 2.4. *If $A(t)$ is a constant matrix A then the equation* (2.2) *is stable,* {*uniformly stable also*}, *if and only if all proper values λ of A have $\Re\lambda \leq 0$, and all proper values λ with $\Re\lambda = 0$ have index equal to multiplicity. Moreover, the equation* (2.2) *is strictly stable if and only if all proper values λ of A have $\Re\lambda = 0$ and index equal to multiplicity. Finally, the equation* (2.2) *is asymptotically stable if and only if $\Re\lambda < 0$ for all proper values λ of A.*

For general linear homogeneous differential equations (2.2) there exists a unitary matrix $T(t)$ of class $\mathfrak{C}^1[a, \infty)$ and such that the $B(t)$ of (2.14) is superdiagonal. Prefatory to the proof of this result the following two lemmas will be established.

***Lemma* 2.1.** *Suppose that $\mathfrak{R}$ is a class of $n \times n$ matrices such that:* (i) *$\mathfrak{R}$ is closed under multiplication;* (ii) *if $M \in \mathfrak{R}$, and is nonsingular, then $-M^{-1} \in \mathfrak{R}$;* (iii) *if the elements of $M(t)$ are continuously differentiable on $[a, \infty)$, and $M(t) \in \mathfrak{R}$ for t on $[a, \infty)$, then $M'(t) \in \mathfrak{R}$ for $t \in [a, \infty)$. If $Y(t)$ is a fundamental matrix for* (2.2), *and $T(t)$ is a nonsingular matrix of class $\mathfrak{C}^1[a, \infty)$ and such that $Y^{-1}(t)T(t) \in \mathfrak{R}$ for $t \in [a, \infty)$, then $B(t) = T^{-1}(t)[A(t)T(t) - T'(t)] \in \mathfrak{R}$ for $t \in [a, \infty)$.*

To prove this lemma, one need note only that if $V(t) = Y^{-1}(t)T(t) \in \mathfrak{R}$ for $t \in [a, \infty)$ then $T(t) = Y(t)V(t)$ and $B = V^{-1}Y^{-1}(-YV') = (-V^{-1})V' \in \mathfrak{R}$ for $t \in [a, \infty)$.

***Lemma* 2.2.** *If $T(t)$ is a nonsingular matrix of class $\mathfrak{C}^1[a, \infty)$ then the hermitian matrix $W(t) = T^*(t)T(t)$ satisfies with*

$$B(t) = T^{-1}(t)[A(t)T(t) - T'(t)]$$

the matrix differential equation

$$WB + B^*W + W' = T^*(A + A^*)T. \tag{2.22}$$

Indeed the relation $B = T^{-1}[AT - T']$ implies that $WB = T^*AT - T^*T'$, from which it follows that $B^*W = T^*A^*T - T^{*\prime}T$, and (2.22) is immediate.

If M is a nonsingular $n \times n$ matrix, and U is the unitary matrix whose sequence of column vectors $u^{(1)}, \ldots, u^{(n)}$ is the set of vectors obtained by applying the Gram-Schmidt orthonormalization process, (see Appendix D), to the sequence of column vectors of M, then for brevity we shall write $U = gs[M]$. Clearly $gs[M] = MV$, where V is a superdiagonal matrix with real diagonal elements; moreover, if $M(t)$ is a nonsingular matrix of class $\mathfrak{C}^1[a, \infty)$ then $V(t)$ defined by $gs[M(t)] = M(t)V(t)$ is also of class $\mathfrak{C}^1[a, \infty)$. With these remarks, we shall proceed to prove the following result, which was established initially by Perron.

Theorem 2.5. *If $Y(t)$ is a fundamental matrix for* (2.2), *and $T(t) = gs[Y(t)]$, then the corresponding matrix $B(t)$ of* (2.14) *is such that*: (i) *$B(t)$ is superdiagonal, and has real-valued diagonal elements*: (ii) *if $A(t)$ is bounded on $[a, \infty)$ then $B(t)$ is bounded on this interval.*

The matrix $T(t) = gs[Y(t)]$ is of the form $T(t) = Y(t)V(t)$ where the elements of $V(t)$ are of class $\mathfrak{C}^1[a, \infty)$, and for each t on this interval $V(t)$ belongs to the class $\mathfrak{R}$ of superdiagonal matrices with real-valued diagonal elements. As such a class $\mathfrak{R}$ satisfies the conditions (i), (ii), (iii) of Lemma 2.1 it follows that $B(t)$ belongs to $\mathfrak{R}$; that is, $B(t)$ satisfies conclusion (i) of Theorem 2.5. As $T(t) = gs[Y(t)]$ is unitary the matrix $W = T^*T$ is the identity matrix and hence

$$B + B^* \equiv T^*(A + A^*)T, \qquad t \in [a, \infty). \tag{2.23}$$

As $B(t)$ is superdiagonal the matrix equation (2.23) is equivalent to

$$B_{\alpha\alpha} = \tfrac{1}{2}u^{(\alpha)*}(A + A^*)u^{(\alpha)}, \; B_{\alpha\beta} = u^{(\alpha)*}(A + A^*)u^{(\beta)} \quad \textit{for} \quad \alpha < \beta, \; t \in [a, \infty), \tag{2.24}$$

where $u^{(\alpha)}(t)$, $(\alpha = 1, \ldots, n)$, denote the column vectors of the unitary matrix $T(t)$. As $|u^{(\alpha)}(t)| \equiv 1$, $(\alpha = 1, \ldots, n)$, conclusion (ii) of Theorem 2.5 is a ready consequence of (2.24); indeed, for the boundedness of $B(t)$ on $[a, \infty)$ it suffices for $A(t) + A^*(t)$ to be bounded on this interval.

If $\phi(t)$ is a complex-valued function on $[a, \infty)$, the *type number* of $\phi(t)$ is the number $\mathbf{n}\{\phi\} = \mathbf{n}\{\phi(t)\}$ of the extended real number system defined as

$$\mathbf{n}\{\phi\} = \limsup_{t \to \infty} t^{-1} \ln |\phi(t)|,$$

where it is to be understood that if t_0 is any value such that $\phi(t_0) = 0$ then in the above formula $\ln |\phi(t_0)|$ is taken to be equal to $-\infty$. In particular, a necessary and sufficient condition for $\mathbf{n}$ to be the type number of $\phi(t)$ is that if k is a real number greater than $\mathbf{n}$ then $\phi(t)\exp\{-kt\} \to 0$ as $t \to \infty$, while for each real number h less than $\mathbf{n}$ there exists a sequence $\{t_n\}$ on $[a, \infty)$ such that $\{t_n\} \to \infty$ and $|\phi(t_n)| \exp\{-ht_n\} \to +\infty$. It may be verified readily that if $\phi_\alpha(t)$, $(\alpha = 1, 2)$, are functions on $[a, \infty)$ such that $\mathbf{n}\{\phi_1\} > \mathbf{n}\{\phi_2\}$ then $\mathbf{n}\{\phi_1 + \phi_2\} = \mathbf{n}\{\phi_1\}$; also, if $\mathbf{n}\{\phi_1\} = \mathbf{n}\{\phi_2\}$ then $\mathbf{n}\{\phi_1 + \phi_2\} \leq \mathbf{n}\{\phi_1\}$. Moreover, if $\phi(t)$, $t \in [a, \infty)$, is a function such that $\phi(t) \neq 0$ for $t \in [a, \infty]$, then $\mathbf{n}\{\phi\} + \mathbf{n}\{\phi^{-1}\} \geq 0$, and $\mathbf{n}\{\phi\} + \mathbf{n}\{\phi^{-1}\} = 0$ if and only if $\lim_{t\to\infty} t^{-1} \ln |\phi(t)|$ exists and is finite.

If $V_n[a, \infty)$ denotes the class of n-dimensional vector functions $v(t)$ defined on $[a, \infty)$, then for $v \in V_n[a, \infty)$ the type number $\mathbf{n}\{v\}$ is defined as equal to the type number of the norm function $|v(t)|$ of $v(t)$. For example if $v(t) = \eta\, p(t)\exp\{\lambda t\}$, where η is a nonzero vector of $\mathbf{C}_n$, $p(t)$ is a non-identically vanishing polynomial in t, and λ is a complex number, then $\mathbf{n}\{v\} = \Re e\, \lambda$. It may be established readily that if $v(t) \equiv (v_\alpha(t))$, $(\alpha = 1, \ldots, n)$, is a member of $V_n[a, \infty)$, then the type number of the vector function $v(t)$ is the greatest of the type numbers $\mathbf{n}\{v_\alpha\}$ of the scalar components $v_\alpha(t)$, $(\alpha = 1, \ldots, n)$.

The results of the following lemma are ready consequences of the definition of type number.

***Lemma* 2.3.** (a) *If $v(t) \in V_n[a, \infty)$, and k is a nonzero constant, then $kv(t) \in V_n[a, \infty)$ and $\mathbf{n}\{kv\} = \mathbf{n}\{v\}$;* (b) *if $v_\alpha(t) \in V_n[a, \infty)$, $(\alpha = 1, 2)$, with $\mathbf{n}\{v_2\} > \mathbf{n}\{v_1\}$, and $v(t) = k_1v_1(t) + k_2v_2(t)$, where k_1, k_2 are constants with $k_2 \neq 0$, then $v(t) \in V_n[a, \infty)$ and $\mathbf{n}\{v\} = \mathbf{n}[v_2\}$;* (c) *if $v_\alpha(t) \in V_n[a, \infty)$, $(\alpha = 1, 2)$, and $\mathbf{n}\{v_1\} = \mathbf{n}\{v_2\}$, then $\mathbf{n}\{v_1 + v_2\} \leq \mathbf{n}\{v_1\} = \mathbf{n}\{v_2\}$;* (d) *if $v_\alpha(t) \in V_n[a, \infty)$, $(\alpha = 1, \ldots, q)$, with $\mathbf{n}\{v_1\} < \mathbf{n}\{v_2\} < \cdots < \mathbf{n}\{v_q\}$, and $v(t) = c_1v_1(t) + \cdots + c_qv_q(t)$, where $c_1, \ldots, c_q$ are constants not all zero, then $v(t) \in V_n[a, \infty)$ and $\mathbf{n}\{v\} = \mathbf{n}\{v_p\}$, where p is the largest of the integers $1, \ldots, q$ for which $c_p \neq 0$.*

For a linear homogeneous vector differential equation (2.2) the aggregate of type numbers of nonzero solutions of this equation is for brevity called the *set of type numbers of* (2.2). The following result is an immediate consequence of the results of Theorem 2.3 and Lemma 2.3.

Theorem 2.6. *For a differential equation* (2.2) *the set of type numbers is at most n in number. Moreover, if $y(t)$ is a nonzero solution of this equation, and $\lambda_g(t)$, $\lambda_l(t)$ are defined as in Theorem* 2.3, *then for $\tau \in [a, \infty)$,*

$$\mathbf{n}\left\{\exp\left[\int_\tau^t \lambda_l(s)\,ds\right]\right\} \leq \mathbf{n}\{y\} \leq \mathbf{n}\left\{\exp\left[\int_\tau^t \lambda_g(s)\,ds\right]\right\}.$$

In particular, if $A(t)$ is bounded on $[a, \infty)$ then the inequalities $|\lambda_g(t)| \leq \nu[A(t)]$, $|\lambda_l(t)| \leq \nu[A(t)]$, imply that for each nonzero solution $y(t)$ of (2.2) the type number $\mathbf{n}\{y\}$ is finite and

$$|\mathbf{n}\{y\}| \leq \limsup_{t\to\infty} \nu[A(t)].$$

Now if $y(t)$ and $u(t)$ are n-dimensional vector functions which are nonzero for $t \in [a, \infty)$, and $y(t) = T(t)u(t)$, where $T(t)$ is bounded on $[a, \infty)$, then from the inequality $|y(t)| \leq \nu[T(t)]\,|u(t)|$ and the boundedness of $\nu[T(t)]$ it follows that $\mathbf{n}\{y\} \leq \mathbf{n}\{u\}$. If $T(t)$ is nonsingular, and $T^{-1}(t)$ is also bounded on $[a, \infty)$, then $u(t) = T^{-1}(t)y(t)$ and correspondingly $\mathbf{n}\{u\} \leq \mathbf{n}\{y\}$. Consequently, for linear homogeneous differential equations of the form (2.2) we have the following result.

Theorem 2.7. *If* (2.14) *is kinematically similar to* (2.2) *under the transformation* (2.13), *then for a nonzero solution $y(t)$ of* (2.2) *the corresponding solution $u(t) = T^{-1}(t)y(t)$ of* (2.14) *is such that $\mathbf{n}\{u\} = \mathbf{n}\{y\}$. In particular, the type numbers for* (2.14) *are the same as the type numbers for* (2.2).

Now consider an equation in which $A(t)$ is continuous and bounded on $[a, \infty)$, so that all of the type numbers of (2.2) are finite. A given fundamental set of solutions of this equation is said to be *normal*, according to Liapunov [1], if any linear combination with nonzero coefficients of all or part of the solutions of this set has type number that is equal to the largest type number of the individual solutions appearing in this linear combination. That normality is indeed a property of individual fundamental sets of solutions is illustrated by the equation (2.2) with $n = 2$, $A(t) = \text{diag}\{2, 1\}$, and the fundamental sets of solutions

$$Y_1(t) = \begin{bmatrix} e^{2t} & 0 \\ 0 & e^t \end{bmatrix}, \qquad Y_2(t) = \begin{bmatrix} e^{2t} & e^{2t} \\ 0 & e^t \end{bmatrix},$$

since the first is normal, while the second is not.

Corresponding to a fundamental set of solutions $y^{(1)}(t), \ldots, y^{(n)}(t)$ of (2.2), let

$$S\{y^{(1)}, \ldots, y^{(n)}\} = \sum_{j=1}^{n} \mathbf{n}\{y^{(j)}\}.$$

Under the assumption that $A(t)$ is bounded in $[a, \infty)$, it follows that $S\{y^{(1)}, \ldots, y^{(n)}\}$ is bounded on the class of fundamental sets of solutions of (2.2). Indeed, in view of Theorem 2.6, $S\{y^{(1)}, \ldots, y^{(n)}\}$ can assume only a finite number of distinct values. Moreover, in view of the result of Problem 15 of the following problem set VIII.2, it follows that $S\{y^{(1)}, \ldots, y^{(n)}\}$ is not less than the left-hand member of (2.28). Consequently,

$$S\{y^{(1)}, \ldots, y^{(n)}\} \geq \liminf_{t\to\infty} t^{-1} \int_a^t \Re e\{\operatorname{Tr} A(s)\}\, ds$$
$$= -\mathbf{n}\left\{\exp\left[-\int_a^t \Re e\{\operatorname{Tr} A(s)\}\, ds\right]\right\}$$

for an arbitrary fundamental set of solutions of (2.2).

For a given differential equation (2.2), let $S_{\min}(A)$ denote the minimum of $S\{y^{(1)}, \ldots, y^{(n)}\}$ for fundamental sets of solutions of this equation. Then

$$\sigma(A) = S_{\min}(A) - \liminf_{t\to\infty} t^{-1} \int_a^t \Re e\{\operatorname{Tr} A(s)\}\, ds$$

is a non-negative constant, called the *constant of irregularity* of (2.2). If $\sigma(A) = 0$, then the differential equation (2.2) is said to be *regular*, in the sense of Liapunov. In particular, if (2.2) is regular then

$$\lim_{t\to\infty} t^{-1} \int_a^t \Re e\{\operatorname{Tr} A(s)\}\, ds$$

exists, in view of the inequality (2.28).

For a given real number r let $\mathfrak{M}_r$ denote the set of solutions $y(t)$ of (2.2) with $\mathbf{n}\{y\} \leq r$. As a consequence of the results of Lemma 2.3 and Theorem 2.6 we have that $\mathfrak{M}_r$ is a linear vector space whose dimension d_r satisfies $0 \leq d_r \leq n$, and if $r < s$ then $\mathfrak{M}_r$ is a subspace of $\mathfrak{M}_s$. Moreover, in view of the remark following Theorem 2.6, if $A(t)$ is bounded on $[a, \infty)$ and $p = \limsup_{t\to\infty} \nu[A(t)]$, then $d_r = 0$ if $r < -p$ and $d_r = n$ if $r \geq p$. The conclusions of the following lemma are direct consequences of the definitions of the involved concepts and the results of Lemma 2.3 and Theorem 2.6.

***Lemma* 2.4.** *If $A(t)$ is bounded on $[a, \infty)$, and $\tau_1 < \tau_2 < \cdots < \tau_q$ denotes the set of distinct type numbers of* (2.2) *indexed by order, then $1 \leq q \leq n$ and if $\delta_j = d_{\tau_j} = \dim \mathfrak{M}_{\tau_j}$ then:*

(a) *there exists a fundamental set of solutions $y_0^{(1)}(t), \ldots, y_0^{(n)}(t)$ of* (2.2) *such that*

$$\mathbf{n}\{y_0^{(j)}\} = \tau_1 \quad \textit{for} \quad 1 \leq j \leq \delta_1,$$
$$\mathbf{n}\{y_0^{(j)}\} = \tau_k \quad \textit{for} \quad \delta_{k-1} < j \leq \delta_k, \qquad k = 2, \ldots, q,$$

and $S\{y_0^{(1)}, \ldots, y_0^{(n)}\} = S_{\min}(A)$;

(b) *a fundamental set of solutions* $y^{(1)}(t), \ldots, y^{(n)}(t)$ *of* (2.2) *is normal if and only if* $S\{y^{(1)}, \ldots, y^{(n)}\} = S_{\min}(A)$, *and in this case if the solutions of this set are indexed so that* $\mathbf{n}\{y^{(i)}\} \leq \mathbf{n}\{y^{(i+1)}\}$, $(i = 1, \ldots, n-1)$, *then* $\mathbf{n}\{y^{(j)}\} = \mathbf{n}\{y_0^{(j)}\}$, $(j = 1, \ldots, n)$.

If in the notation of this lemma we set $\chi_1 = \delta_1$, $\chi_k = \delta_k - \delta_{k-1}$, $k = 2, \ldots, q$, then χ_i is called the *multiplicity* of the type number τ_i; in particular, we have $\chi_1 + \chi_2 + \cdots + \chi_q = n$ and $\chi_1\tau_1 + \cdots + \chi_q\tau_q = S_{\min}(A)$.

If $T(t)$ is a nonsingular matrix function of class $\mathfrak{C}^1$ on $[a, \infty)$, and such that $T(t)$, $T^{-1}(t)$ and $T'(t)$ are all bounded on this interval, then (2.13) is called a *Liapunov transformation*, and (2.14) is said to be *Liapunov similar* to (2.2). Now if $A(t)$ is bounded on $[a, \infty)$, and (2.14) is Liapunov similar to (2.2), then $B(t)$ is also bounded on $[a, \infty)$. Moreover, if $Y(t)$ is a fundamental matrix for (2.2) then $U(t) = T^{-1}(t)Y(t)$ is a fundamental matrix for (2.14), and from the formula (2.10′) for the determinants of these fundamental matrices it follows that there exists a real constant c such that

$$\int_a^t \Re\{\mathrm{Tr}\, A(s)\}\, ds = c + \int_a^t \Re\{\mathrm{Tr}\, B(s)\}\, ds, \qquad t \in [a, \infty).$$

Consequently,

$$\liminf_{t\to\infty} t^{-1}\int_a^t \Re\{\mathrm{Tr}\, A(s)\}\, ds = \liminf_{t\to\infty} t^{-1}\int_a^t \Re\{\mathrm{Tr}\, B(s)\}\, ds,$$

and we have the following result.

Theorem 2.8. *If* $A(t)$ *is bounded on* $[a, \infty)$ *and* (2.14) *is Liapunov similar to* (2.2), *then* $\sigma(B) = \sigma(A)$; *in particular, if* (2.2) *is regular then* (2.14) *is regular.*

For a constant matrix A let B be a superdiagonal matrix (2.19) that is similar to A. If $U(t)$ is the particular fundamental matrix (2.20) for the corresponding equation (2.14), then in view of (2.21) it follows readily that the column vectors $u^{(1)}(t), \ldots, u^{(n)}(t)$ of $U(t)$ are such that

$$S\{u^{(1)}, \ldots, u^{(n)}\} = \sum_{j=1}^{r} m_j \,\Re\, \lambda_j = \Re\{\mathrm{Tr}\, B\} = \lim_{t\to\infty} t^{-1}\int_a^t \Re\{\mathrm{Tr}\, B(s)\}\, ds.$$

Consequently (2.14) is regular, and since the transformation (2.13) with $T(t)$ a nonsingular constant matrix is a Liapunov transformation we have that (2.2) is also regular. In general, if $A(t)$ is an $n \times n$ continuous matrix function on $[a, \infty)$ which is bounded, and the equation (2.2) is reducible, then for C a constant matrix such that $C \overset{\text{ks}}{\equiv} A(t)$ under (2.13), the corresponding matrix function $T(t)$ has $T'(t)$ also bounded on $[a, \infty)$, so that (2.13) is a Liapunov transformation. Consequently we have the following corollary to the above theorem.

Corollary. *If $A(t)$ is an $n \times n$ continuous matrix function which is bounded on $[a, \infty)$, and (2.2) is reducible, then (2.2) is regular.*

Problems VIII.2

1.[s] Show that for $1 < 2k < 3/2$, $h \neq 0$, the differential system

$$\begin{aligned} y_1' &= -ky_1, \\ y_2' &= he^{-kt}y_1 + (2t \sin 2t - \cos 2t - 2k)y_2, \end{aligned}$$

is not stable on $[0, \infty)$, whereas if $1 < 2k$ and $h = 0$ the system is asymptotically stable.

Hint. Show that the system has the general solution

$$y_1 = c_1 e^{-kt},\ y_2 = e^{-t(\cos 2t + 2k)}\left(c_2 + hc_1 \int_0^t e^{s \cos 2s}\, ds\right).$$

For $t = (2n + 1)\pi/2$ note that $\cos 2s \geq \frac{1}{2}$ on $t - 2\pi/3 \leq s \leq t - \pi/3$, and proceed to show that $\int_0^t e^{s \cos 2s}\, ds > e^{-\pi/3} e^{t/2}$ for such values of t.

2. Show that for $1 < 2k < 1 + e^{-\pi}$, $h \neq 0$, the differential system

$$\begin{aligned} y_1' &= -ky_1, \\ y_2' &= he^{-kt}y_1 + (\sin \ln t - \cos \ln t - 2k)y_2, \end{aligned}$$

is not stable on $[1, \infty)$, whereas if $1 < 2k$ and $h = 0$ the system is asymptotically stable.

Hint. Show that the system has the general solution

$$y_1 = c_1 e^{-kt},\ y_2 = e^{-t(\cos \ln t + 2k)}\left(c_2 + hc_1 \int_1^t e^{s \cos \ln s}\, ds\right).$$

If ε is such that $0 < \varepsilon < \frac{1}{2}$ and $r(\varepsilon) = \cos \varepsilon\pi\, e^{-(1+\varepsilon)\pi}$ satisfies $1 + r(\varepsilon) > 2k$, for $t = e^{(2n+1)\pi}$ note that $\cos \ln s \geq \cos \varepsilon\pi$ on $[t_1, t_2]$, where $t_1 = te^{-(1+\varepsilon)\pi}$, $t_2 = te^{-(1-\varepsilon)\pi}$, and conclude that

$$\int_1^t e^{s \cos \ln s}\, ds > \int_{t_1}^{t_2} e^{s \cos \varepsilon\pi}\, ds > (t_2 - t_1)e^{tr(\varepsilon)}.$$

3. For the differential equation $y' = 1 - y^2$ show that: (i) the solution $\phi(t) \equiv 1$ is an asymptotically stable solution; (ii) the solution $\phi(t) \equiv -1$ is not stable.

4. For the differential equation $y' = \frac{1}{2} \cos t\, e^{\sin t}(1 + y^2)$ show that the solution $\phi(t) = \tan(\frac{1}{2}e^{\sin t} - \frac{1}{2})$ is strictly stable, but not asymptotically stable.

5. If $r(t)$, $p(t)$ are continuous real-valued scalar functions on $[a, \infty)$, and $r(t) > 0$ on this interval, show that any solution $u(t)$ of

$$[r(t)u'(t)]' + p(t)u(t) = 0, \qquad t \in [a, \infty) \tag{2.25}$$

is such that $|u(t)|$ and $|u'(t)|$ are bounded on $[a, \infty)$ whenever the integral $\int_a^\infty |[1/r(t)] - p(t)|\, dt$ is convergent.

Hint. Consider a first order system equivalent to (2.25), and use Theorem 2.3.

6. If the $n \times n$ matrix function $A(t)$ is continuous on $[a, \infty)$, and there exists a unitary matrix $T(t)$ of class $\mathfrak{C}^1[a, \infty)$ such that the matrix function $B(t)$ is related to $A(t)$ by (2.15), show that $\Re e\{\text{Tr } B(t)\} \equiv \Re e\{\text{Tr } A(t)\}$.

Hint. Use (2.10′).

7.^{ss} If $A(t)$ is such that $\int_a^\infty \nu[A(s)]\, ds < \infty$ show that:

(i) (2.2) is strictly stable;
(ii) if $y(t)$ is a solution of (2.2), then $y(\infty) = \lim_{t\to\infty} y(t)$ exists;
(iii) if $y(t)$ is a solution of (2.2) and $y(t) \not\equiv 0$, then $y(\infty) \neq 0$;
(iv) there exists a fundamental matrix $Y(t)$ of (2.2) such that $Y(\infty) = E$;
(v) for an arbitrary constant vector η there is a unique solution $y(t)$ of (2.2) such that $y(\infty) = \eta$.

Hints. (i) Use Theorem 2.3, noting that $\lambda_g(t) \leq \nu[A(t)]$, $\lambda_l(t) \geq -\nu[A(t)]$; (ii) if $y(t)$ is a solution of (2.2), and κ is such that $|y(t)| \leq \kappa$ on $[a, \infty)$, show that $|y(t_2) - y(t_1)| \leq \kappa \left|\int_{t_1}^{t_2} \nu[A(s)]\, ds\right|$;
(iii) if $Y(t)$ is a fundamental matrix of (2.2) for which $y(t)$ is a column vector, note that $Y(t)$ and $Y^{-1}(t)$ are bounded on $[a, \infty)$;
(iv) use the result of (iii).

8. The system (2.2) is strictly stable if $A_1(t) = \int_a^t A(s)\, ds$ is such that

$$A(t)A_1(t) = A_1(t)A(t), \qquad t \in [a, \infty), \tag{2.26}$$

and there is a constant $\kappa_1 > 0$ such that $\nu[A_1(t)] \leq \kappa_1$ on $[a, \infty)$.

Hint. Show that (2.26) implies that the fundamental matrix $Y(t)$ of (2.2) satisfying $Y(a) = E$ is given by $Y(t) = \exp\{A_1(t)\}$.

9. The system (2.2) is strictly stable if $A_1(t) = \int_a^t A(s)\, ds$ is such that there exist constants $0 \leq \kappa_1 < 1$, $\kappa_2 \geq 0$ satisfying the inequalities

$$\nu[A_1(t)] \leq \kappa_2, \int_a^t \nu\left[\left(\int_s^t A(r)\, dr\right) A(s)\right] ds \leq \kappa_1, \qquad t \in [a, \infty). \tag{2.27}$$

Hint. Show that for $Y(t)$ the fundamental matrix of (2.2) satisfying $Y(a) = E$ we have

$$Y(t) = E + A_1(t) + \int_a^t [A_1(t) - A_1(s)]A(s)Y(s)\,ds.$$

For a given interval $[a, t]$, $(a < t)$, let t_0 be a value on this interval such that $\nu[Y(t_0)] = \max_{a \le s \le t} \nu[Y(s)]$; from the above relation conclude that $\nu[Y(t_0)] \le (1 + \kappa_2)/(1 - \kappa_1)$, and consequently that (2.2) is stable. Show that

$$\left| \int_a^t [\operatorname{Tr} A(s)]\,ds \right| \le n\nu[A_1(t)], \qquad t \in [a, \infty),$$

and with the aid of Theorem 2.2 conclude that the conditions of the problem imply that (2.4) is stable also.

10. Prove that the first order linear homogeneous system equivalent to $y'' + (2/t)y' + y = 0$ under the substitution $y_1 = y$, $y_2 = y'$ is uniformly stable and asymptotically stable on $[1, \infty)$, but is not strictly stable on this interval.

Hint. Note that $y = (1/t)\sin t$ and $y = (1/t)\cos t$ are solutions of the given second order differential equation.

11. For Λ and Γ constant $n \times n$ matrices such that $\Lambda = \operatorname{diag}\{\lambda_1, \ldots, \lambda_n\}$ and $\Lambda - \Gamma$ is skew hermitian, let $S(t)$ be a solution of the matrix differential equation $S'(t) + S(t)[\Lambda - \Gamma] = 0$ with $S(0)$ unitary, and let C be a nonsingular constant $n \times n$ matrix. If $A(t) = C[S'(t) + S(t)\Lambda]\,S^*(t)C^{-1}$, prove that: (i) $\Delta_{A(t)}(\lambda)$ is equal to the characteristic polynomial $\Delta_\Gamma(\lambda)$ for $t \in (-\infty, \infty)$; (ii) $Y(t) = CS(t)\exp\{t\Lambda\}$ is a fundamental matrix solution of $Y'(t) = A(t)Y(t)$, and $V(t) = CS(t)$ is such that $V(t)$ and $V^{-1}(t)$ are bounded on $(-\infty, \infty)$; (iii) in particular, if $n = 2$, $\lambda_1 > 0 > \lambda_2$ with $\lambda_1 + \lambda_2 < 0$, and ω a positive constant such that $\lambda_1\lambda_2 + \omega^2 > 0$, and $2\omega < \lambda_1 - \lambda_2$ then for $\Gamma = \Lambda + \Omega$ with $\Omega_{11} = \Omega_{22} = 0$, $\Omega_{12} = -\Omega_{21} = \omega$, all zeros of $\Delta_{A(t)}(\lambda) \equiv \Delta_\Gamma(\lambda)$ are negative; however, the column vectors $y^{(1)}(t)$ and $y^{(2)}(t)$ of $Y(t)$ are such that $|y^{(1)}(t)| \to \infty$ and $|y^{(2)}(t)| \to 0$ as $t \to \infty$.

Hint. Show that $S(t)$ is unitary for $t \in (-\infty, \infty)$.

12. If the $n \times n$ matrix function $A(t)$ is nonsingular and of class $\mathfrak{C}^1$ on $[a, \infty)$, with both $A(t)$ and $A^{-1}(t)$ bounded on this interval, prove that (2.2) is strictly stable in case one of the following conditions holds:

(i) $\int_a^\infty \nu[A^2(s) + A'(s)]\,ds < \infty$,

(ii) $\int_a^\infty \nu[A^2(s) - A'(s)]\,ds < \infty$.

Hints. (i) Note that under the transformation (2.13) with $T(t) = A^{-1}(t)$ the equation (2.2) is kinematically similar to (2.14) with

$$B(t) = [A^2(t) + A'(t)]\, A^{-1}(t).$$

(ii) Apply the conclusion of (i) to the adjoint equation (2.9), and use the result of the Corollary to Theorem 2.1.

13. Give the details of the proofs of the results of Lemmas 2.3 and 2.4.

14.[s] For $n = 2$, and

$$A(t) = \begin{bmatrix} \cos \ln t & -\sin \ln t \\ -\sin \ln t & \cos \ln t \end{bmatrix}, \qquad t \in (0, \infty),$$

show that two independent solutions of (2.2) are given by

$$y^{(1)}(t) = \begin{pmatrix} \exp\{t \cos \ln t\} \\ \exp\{t \cos \ln t\} \end{pmatrix}, \qquad y^{(2)}(t) = \begin{pmatrix} \exp\{t \sin \ln t\} \\ -\exp\{t \sin \ln t\} \end{pmatrix},$$

and that $\mathbf{n}\{y^{(j)}\} = 1$, for $j = 1, 2$.

15.[ss] If $y^{(1)}(t), \ldots, y^{(n)}(t)$ are linearly independent solutions of (2.2), and $\tau \in [a, \infty)$, prove that

$$\text{(2.28)} \qquad \mathbf{n}\left\{\exp\left[\int_a^t \Re\{\operatorname{Tr} A(s)\}\, ds\right]\right\} = \limsup_{t\to\infty} t^{-1} \int_a^t \Re\{\operatorname{Tr} A(s)\}\, ds \leq \sum_{j=1}^{n} \mathbf{n}\{y^{(j)}\}.$$

Hint. If $Y(t)$ is the fundamental matrix for (2.2) with column vectors $y^{(j)}(t)$, $(j = 1, \ldots, n)$, use the relation (I.9.6), and note that the Hadamard determinant inequality, (see Problem 4 of D.4 in Appendix D), implies that $|\det Y(t)| \leq \prod_{j=1}^{n} |y^{(j)}(t)|$.

16. Prove that the differential system of Problem 14 is not regular.

Hint. Show that

$$\int_1^t \Re\{\operatorname{Tr} A(s)\}\, ds = t[\cos \ln t + \sin \ln t] - 1,$$

$$= \sqrt{2}\, t \cos\left(\ln t - \frac{\pi}{4}\right) - 1,$$

and conclude that $\lim_{t\to\infty} t^{-1} \int_1^t \Re\{\operatorname{Tr} A(s)\}\, ds$ does not exist.

17. Suppose that the continuous matrix function $A(t)$ is bounded on $[a, \infty)$, and that the column vectors $y^{(1)}(t), \ldots, y^{(n)}(t)$ of $Y(t)$ and the column vectors $z^{(1)}(t), \ldots, z^{(n)}(t)$ of $Z(t)$ are fundamental sets of solutions of (2.2) and the adjoint system (2.9), respectively.

(i) If the following order relations

$$(2.29)\qquad \begin{aligned} &\text{(a)}\ \mathbf{n}\{y^{(1)}\} \le \mathbf{n}\{y^{(2)}\} \le \cdots \le \mathbf{n}\{y^{(n)}\},\\ &\text{(b)}\ \mathbf{n}\{z^{(1)}\} \ge \mathbf{n}\{z^{(2)} \ge \cdots \ge \mathbf{n}\{z^{(n)}\},\end{aligned}$$

hold, prove that:

(a) $\mathbf{n}\{y^{(j)}\} + \mathbf{n}\{z^{(j)}\} \ge 0, \qquad (j = 1, \ldots, n);$

(b) if $Z_0(t) = [z_0^{(1)}(t) \cdots z_0^{(n)}(t)]$ is the fundamental set of solutions of (2.9) defined by $Z_0^*(t)Y(t) \equiv E_n$, then for $j = 1, \ldots, n$,

$$\mathbf{n}\{z_0^{(j)}\} + \mathbf{n}\{y^{(j)}\} \le S\{y^{(1)}, \ldots, y^{(n)}\} - \liminf_{t\to\infty} t^{-1}\int_a^t \Re\{\mathrm{Tr}\, A(s)\}\, ds;$$

in particular, if $y^{(1)}(t), \ldots, y^{(n)}(t)$ is a normal fundamental set of solutions of (2.2), then

$$(2.30)\qquad \begin{aligned} &\mathbf{n}\{z_0^{(j)}\} + \mathbf{n}\{y^{(j)}\} \le \sigma(A), \quad \text{and}\\ &S\{z_0^{(1)}, \ldots, z_0^{(n)}\} + S_{\min}(A) \le n\sigma(A);\end{aligned}$$

(c) if $\{y^{(1)}(t), \ldots, y^{(n)}(t)\}$ and $\{z^{(1)}(t), \ldots, z^{(n)}(t)\}$ are normal fundamental sets of solutions of (2.2) and (2.9), respectively, satisfying (2.29), and (2.2) is regular, then

$$(2.31)\qquad \mathbf{n}\{y^{(j)}\} + \mathbf{n}\{z^{(j)}\} = 0, \qquad j = 1, \ldots, n.$$

(ii) If the relations (2.29-a) and (2.31) hold, then (2.2) and (2.9) are both regular, $\{y^{(1)}(t), \ldots, y^{(n)}(t)\}$, $\{z^{(1)}(t), \ldots, z^{(n)}(t)\}$ are individually normal fundamental sets of solutions of these respective equations, and (2.29-b) also holds.

(iii) If (2.2) is regular then (2.9) is also regular.

(iv) If $y^{(1)}(t), \ldots, y^{(n)}(t)$ is a normal fundamental set of solutions of (2.2), then this equation is regular if and only if

$$(2.32)\qquad \sum_{j=1}^{n} \mathbf{n}\{y^{(j)}\} = \mathbf{n}\left\{\exp\left[\int_a^t \Re\{\mathrm{Tr}\, A(s)\}\, ds\right]\right\},$$

and also

$$(2.33)\qquad \mathbf{n}\left\{\exp\left[\int_a^t \Re\{\mathrm{Tr}\, A(s)\}\, ds\right]\right\} + \mathbf{n}\left\{\exp\left[-\int_a^t \Re\{\mathrm{Tr}\, A(s)\}\, ds\right]\right\} = 0.$$

Hints. (i-a) To prove conclusion (a) note that since $Z^*(t)Y(t) = [(y^{(\beta)}, z^{(\alpha)})]$, $(\alpha, \beta = 1, \ldots, n)$, is a nonsingular constant matrix we have in the case of given fundamental sets $\{y^{(1)}(t), \ldots, y^{(n)}(t)\}$, $\{z^{(1)}(t), \ldots, z^{(n)}(t)\}$ satisfying (2.29-a, b) that for each $j = 1, \ldots, n$ there exist integers $\beta \le j$, $\alpha \ge j$ such that $C_{\alpha\beta} = (y^{(\beta)}, z^{(\alpha)}) \ne 0$, and with the aid of the relation $|C_{\alpha\beta}| =$

$|(y^{(\beta)}, z^{(\alpha)})| \leq |y^{(\beta)}|\, |z^{(\alpha)}|$ show that

$$0 \leq \mathbf{n}\{y^{(\beta)}\} + \mathbf{n}\{z^{(\alpha)}\} \leq \mathbf{n}\{y^{(j)}\} + \mathbf{n}\{z^{(j)}\}.$$

(i-b) From the representation of the elements of $Y^{-1}(t)$ as cofactors of $Y(t)$ divided by det $Y(t)$, note that the cofactors of $Y(t)$ appearing in the j-th row of $Y^{-1}(t)$ are determinants of minors of $Y(t)$ that do not contain elements from the j-th column vector of $Y(t)$. With the aid of the Hadamard determinant inequality, (see Problem 4 of D.4 in Appendix D), and formula (2.10′) for det $Y(t)$, conclude that for $j = 1, \ldots, n$ there exists a positive constant κ_j such that

$$|z_0^{(j)}(t)| \leq \kappa_j \left[\prod_{\substack{\alpha=1 \\ \alpha \neq j}}^{n} |y^{(\alpha)}(t)| \right] \exp\left\{ -\int_a^t \Re\{\mathrm{Tr}\, A(s)\}\, ds \right\},$$

and hence

$$\mathbf{n}\{z_0^{(j)}\} \leq S\{y^{(1)}, \ldots, y^{(n)}\} - \mathbf{n}\{y^{(j)}\} + \limsup_{t\to\infty} \left[-t^{-1} \int_a^t \Re\{\mathrm{Tr}\, A(s)\}\, ds \right].$$

Note also that the relation $(y^{(j)}, z_0^{(j)}) = 1$ implies $0 \leq \mathbf{n}\{z_0^{(j)}\} + \mathbf{n}\{y^{(j)}\}$.

(i-c) Use results of Lemma 4.3, (i-a) and (i-b) to conclude that

$$\begin{aligned} 0 \leq \sum_{\alpha=1}^{n} [\mathbf{n}\{z^{(j)}\} + \mathbf{n}\{y^{(j)}\}] &\leq \sum_{\alpha=1}^{n} [\mathbf{n}\{z_0^{(j)}\} + \mathbf{n}\{y^{(j)}\}] \\ &\leq S\{z_0^{(1)}, \ldots, z_0^{(n)}\} + S_{\min}(A) \leq n\sigma(A) = 0. \end{aligned}$$

(ii) Note that (2.31) implies

$$\begin{aligned} 0 &= S\{y^{(1)}, \ldots, y^{(n)}\} + S\{z^{(1)}, \ldots, z^{(n)}\} \\ &= \left[S\{y^{(1)}, \ldots, y^{(n)}\} - \limsup_{t\to\infty} t^{-1} \int_a^t \Re\{\mathrm{Tr}\, A(s)\}\, ds \right] \\ &\quad + \left[S\{z^{(1)}, \ldots, z^{(n)}\} - \limsup_{t\to\infty} t^{-1} \int_a^t \Re\{\mathrm{Tr}[-A^*(s)]\}\, ds \right] \\ &\quad + \left[\limsup_{t\to\infty} t^{-1} \int_a^t \Re\{\mathrm{Tr}\, A(s)\}\, ds - \liminf_{t\to\infty} t^{-1} \int_a^t \Re\{\mathrm{Tr}\, A(s)\}\, ds \right], \end{aligned}$$

where each of the bracketed terms is non-negative, so that these terms are individually zero. Conclude that $\lim_{t\to\infty} t^{-1} \int_a^t \Re\{\mathrm{Tr}\, A(s)\}\, ds$ exists, that $S\{y^{(1)}, \ldots, y^{(n)}\} = S_{\min}(A)$, $S\{z^{(1)}, \ldots, z^{(n)}\} = S_{\min}(-A^*)$, and $\sigma(A) = 0 = \sigma(-A^*)$. Use induction to establish that (2.29-b) also holds, noting as an initial step that if $\hat{z}^{(1)}(t), \ldots, \hat{z}^{(n)}(t)$ is a re-ordering of the normal fundamental set $z^{(1)}(t), \ldots, z^{(n)}(t)$ such that $\mathbf{n}\{\hat{z}^{(1)}\} \geq \mathbf{n}\{\hat{z}^{(2)}\} \geq \cdots \geq \mathbf{n}\{\hat{z}^{(n)}\}$, then in view of conclusion (i-a) and (2.31) we have

$$0 \leq \mathbf{n}\{y^{(n)}\} + \mathbf{n}\{\hat{z}^{(n)}\} \leq \mathbf{n}\{y^{(n)}\} + \mathbf{n}\{z^{(n)}\} = 0,$$

and hence $\mathbf{n}\{z^{(n)}\} = \mathbf{n}\{\hat{z}^{(n)}\}$, the minimum type number of a solution of (2.9).

(iii) Use results of (i-c) and (ii).

(iv) Use appropriately some of the above results, together with the comment that (2.33) holds if and only if $\lim_{t\to\infty} t^{-1} \int_a^t \Re\{\mathrm{Tr}\, A(s)\}\, ds$ exists and is finite.

18. Discuss the nature of real solutions of the two-dimensional system

$$y' = Ay, \quad \text{where} \quad A = \begin{bmatrix} a & b \\ c & d \end{bmatrix},$$

and a, b, c, d are real constants. In particular, sketch roughly in the y-plane the projections of the solution curves $y = y(t)$, $t \in (-\infty, \infty)$.

Hint. Note that the characteristic polynomial for A is $\Delta_A(\lambda) = \lambda^2 - p\lambda + q$ where $p = a + d$, $q = ad - bc$, and if λ_1, λ_2 denote the roots of $\Delta_A(\lambda) = 0$ then $p = \lambda_1 + \lambda_2$, $q = \lambda_1\lambda_2$. Moreover, by Theorem 3.3 of Appendix E there is a nonsingular 2×2 matrix T such that $B = T^{-1}AT$ has one of the following canonical forms

$$B = \begin{bmatrix} \lambda_1 & 0 \\ 0 & \lambda_2 \end{bmatrix}, \quad \text{or} \quad B = \begin{bmatrix} \lambda_1 & 1 \\ 0 & \lambda_1 \end{bmatrix},$$

and the second case occurs only when λ_1 is a double root of $\Delta_A(\lambda) = 0$ whose index is not equal to its multiplicity, and hence for which there exists a vector η satisfying $(\lambda_1 E - A)^2\eta = 0$ while $(\lambda_1 E - A)\eta \neq 0$. Moreover, the matrix T may be chosen as real except for the case of nonreal complex conjugate roots. In general, the matrix T has column vectors τ_1, τ_2, where in the first case τ_α is a proper vector for A corresponding to the proper value λ_α, $(\alpha = 1, 2)$, while in the second case τ_1 is a proper vector corresponding to the proper value λ_1 for which there is an associated vector τ_2 such that $(A - \lambda_1 E)\tau_2 = \tau_1$. When the first case holds with λ_1, λ_2 real, consider the corresponding system (2.14), written in component form as

$$u' = \lambda_1 u, \qquad v' = \lambda_2 v,$$

with general solution

$$u = k_1 e^{\lambda_1 t}, \qquad v = k_2 e^{\lambda_2 t}.$$

In the second case the system (2.14) may be written as

$$u' = \lambda_1 u + v, \qquad v' = \lambda_1 v,$$

with general solution

$$u = k_1 e^{\lambda_1 t} + k_2(te^{\lambda_1 t}), \qquad v = k_2 e^{\lambda_1 t}.$$

When the first case holds with nonreal complex conjugate roots $\lambda_1 = \mu + i\nu$, $\lambda_2 = \mu - i\nu$, $\nu \neq 0$, and $\tau_1 = \rho + i\sigma$, with ρ, σ real-valued vectors, is a proper vector corresponding to the proper value λ_1, then τ_2 may be chosen as $\rho - i\sigma$. In this case the real matrix T with column vectors ρ, σ is nonsingular, with $AT = TB$ and $T^{-1}AT = B$, where B is the real skew matrix

$$B = \begin{bmatrix} \mu & \nu \\ -\nu & \mu \end{bmatrix}.$$

Under the corresponding real constant transformation (2.13) the vector equation becomes (2.14), which in component form is

$$u' = \mu u + \nu v, \qquad v' = -\nu u + \mu v,$$

with general solution $u = k_1 e^{\mu t} \sin(\nu t + k_2)$, $v = k_1 e^{\mu t} \cos(\nu t + k_2)$. Consider the following separate cases, where p, q are defined as above and $\delta = p^2 - 4q = (a - d)^2 + 4bc$. In each case the name in parenthesis is applied to the origin in the real y or (u, v)-plane to describe the behavior in the neighborhood of this point of the projections onto this (phase) plane of the graphs of solutions.

1. $q < 0$; (saddle point);
2. $q > 0$, $\delta > 0$; (node);
3. $\delta < 0$, $p \neq 0$; (focus, or spiral point);
4. $q > 0$, $p = 0$; (center);
5. $\delta = 0$, $p \neq 0$, (degenerate node);
6. $q = 0$, $p \neq 0$;
7. $q = 0$, $p = 0$.

3. PRELIMINARY RESULTS FOR NONHOMOGENEOUS LINEAR SYSTEMS

In this section we shall be concerned with linear vector differential equations

$$y' = A(t)y + g(t), \tag{3.1}$$

where the $n \times n$ matrix function $A(t) \equiv [A_{\alpha\beta}(t)]$, $(\alpha, \beta = 1, \ldots, n)$, and the n-dimensional vector function $g(t) \equiv (g_\alpha(t))$, $(\alpha = 1, \ldots, n)$, are continuous on $[a, \infty)$.

Theorem 3.1. *If* (2.2) *is uniformly stable on* $[a, \infty)$ *and* $|g(t)|$ *is integrable on this interval, then each solution of* (3.1) *is bounded on* $[a, \infty)$.

The general solution $y(t)$ of (3.1) is given by

$$y(t) = Y(t)\xi + \int_a^t Y(t, s)g(s)\, ds, \qquad t \in [a, \infty), \tag{3.2}$$

where $Y(t)$ is an arbitrary fundamental matrix of (2.2), $Y(t, s) = Y(t)Y^{-1}(s)$ is the fundamental matrix for (2.2) which is equal to E when $t = s$, and ξ is an arbitrary constant n-dimensional vector. In particular, $y(a) = Y(a)\xi$, and

$$(3.2') \qquad y(t) = Y(t, a)y(a) + \int_a^t Y(t, s)g(s)\, ds, \qquad t \in [a, \infty).$$

If (2.2) is uniformly stable on $[a, \infty)$, then there is a constant $\kappa > 0$ such that $\nu[Y(t, s)] \leq \kappa$ on $s \in [a, t]$, $t \in [a, \infty)$, and hence by (3.2′),

$$(3.3) \qquad \begin{aligned} |y(t)| &\leq \kappa\left(|y(a)| + \int_a^t |g(s)|\, ds\right) \\ &\leq \kappa\left(|y(a)| + \int_a^\infty |g(s)|\, ds\right), \qquad t \in [a, \infty). \end{aligned}$$

Theorem 3.2. *If* (2.2) *is uniformly stable on* $[a, \infty)$, *and also asymptotically stable, then whenever* $|g(t)|$ *is integrable on this interval each solution* $y(t)$ *of* (3.1) *is such that* $y(t) \to 0$ *as* $t \to \infty$.

For a given $\varepsilon > 0$ let $t_1 = t_{1\varepsilon} > a$ be such that $\kappa \int_{t_1}^\infty |g(s)|\, ds < \varepsilon/2$, where κ is a positive constant such that $\nu[Y(t, s)] < \kappa$ for $s \in [a, t]$, $t \in [a, \infty)$; in view of the asymptotic stability of (2.2) there is a $t_2 = t_{2\varepsilon} \geq t_{1\varepsilon}$ such that

$$\nu[Y(t)]\, \nu[Y^{-1}(t_1)]\kappa \int_a^{t_1} |g(s)|\, ds < \frac{\varepsilon}{2} \quad \textit{for} \quad t \geq t_2.$$

From the relation

$$\int_a^t Y(t, s)g(s)\, ds = \int_{t_1}^t Y(t, s)g(s)\, ds + Y(t, t_1) \int_a^{t_1} Y(t_1, s)g(s)\, ds$$

it then follows that $\nu[\int_a^t Y(t, s)g(s)\, ds] < \varepsilon$ whenever $t \geq t_2$, and the conclusion of the theorem is an immediate consequence of (3.2).

It is to be noted that in Theorems 3.1 and 3.2 the uniformity of the stability of (2.2) is of basic importance. For example, in view of Problem 1 of VIII.2 the nonhomogeneous equation

$$y' = (2t \sin 2t - \cos 2t - 2k)y + e^{-2kt}, \qquad 1 < 2k < 3/2,$$

has all of its solutions unbounded on $[0, \infty)$, while the corresponding homogeneous equation is asymptotically stable on this interval.

Theorem 3.3. *If the adjoint equation* (2.9) *is uniformly stable on* $[a, \infty)$, *then for each* $g(t)$ *with* $|g(t)|$ *integrable on* $[a, \infty)$ *there is a solution* $y(t)$ *of* (3.1) *such that* $y(t) \to 0$ *as* $t \to \infty$.

In view of Theorem 2.1 the equation (2.9) is uniformly stable if and only if there is a constant κ such that $\nu[Y(t, s)] \leq \kappa$ on $a \leq t \leq s < \infty$, where $Y(t)$ is a fundamental matrix for (2.2). Consequently $|Y(t, s)g(s)| \leq \kappa|g(s)|$ for $a \leq t \leq s < \infty$, and whenever $|g(t)|$ is integrable on $[a, \infty)$ the improper integral

$$y(t) = -\int_t^\infty Y(t, s)g(s)\,ds, \qquad t \in [a, \infty), \tag{3.4}$$

exists, and $y(t) \to 0$ as $t \to \infty$. Therefore $\int_t^\infty Y^{-1}(s)g(s)\,ds$ exists on $[a, \infty)$ and (3.4) is of the form (3.2) with $\xi = -\int_a^\infty Y^{-1}(s)g(s)\,ds$.

Problems VIII.3

1. If $g(t)$ is a continuous scalar function on $[a, \infty)$ such that $|g(t)|$ is integrable on this interval, use Theorems 3.1 and 3.3 to prove the following results for the differential equation $y'' + y = g(t)$, $t \in [a, \infty)$:

(i) all solutions are bounded;
(ii) there exists a unique solution for which $y \to 0$ as $t \to \infty$.

2. If $g(t)$ is a continuous scalar function on $[1, \infty)$ such that $|g(t)|$ is integrable on this interval, show that:

(i) each solution of $y'' + (2/t)y' + y = g(t)$, $t \in [1, \infty)$, satisfies $y \to 0$, $y' \to 0$ as $t \to \infty$;
(ii) there is a unique solution of $y'' - (2/t)y' + y = g(t)$, $t \in [1, \infty)$, such that $y \to 0$ as $t \to \infty$.

Hints. (i) Use the result of Problem VIII.2:11, and Theorem 3.2 above; (ii) use the result of Problem VIII.2:10, Theorem 3.3 above, and the fact that the homogeneous equation $y'' - (2/t)y' + y = 0$ has solutions $y = \sin t - t \cos t$ and $y = \cos t + t \sin t$.

3. Prove the following converse of Theorem 3.1: If $A(t)$ is continuous on $[a, \infty)$, and for arbitrary continuous vector functions $g(t)$ with $|g(t)|$ integrable on this interval all solutions of (3.1) are bounded on $[a, \infty)$, then (2.2) is uniformly stable on $[a, \infty)$.

4. If $A(t)$ is continuous on $[a, \infty)$, prove that each of the following conditions implies the other:

(i) the function $\int_a^t \nu[Y(t, s)]\,ds$, $t \in [a, \infty)$, is bounded;
(ii) for arbitrary continuous vector functions $g(t)$ which are bounded on $[a, \infty)$, all solutions $y(t)$ of (3.1) are bounded on this interval.

4. STABILITY OF RELATED HOMOGENEOUS DIFFERENTIAL SYSTEMS

We shall now investigate relations between stability properties of solutions of the homogeneous equation (2.2), and a related homogeneous equation

$$y' = [A(t) + C(t)]y, \qquad t \in [a, \infty), \tag{4.1}$$

where $A(t)$ and $C(t)$ are $n \times n$ matrix functions which are continuous on $[a, \infty)$.

One might surmise that (4.1) is stable whenever (2.2) is stable, and the equation

$$y' = C(t)y, \qquad t \in [a, \infty), \tag{4.2}$$

is also stable. That such a result is false, however, is shown by the example with

$$A(t) = \begin{bmatrix} 0 & 2 \\ -1 & 0 \end{bmatrix}, \qquad C(t) = \begin{bmatrix} 0 & -1 \\ 2 & 0 \end{bmatrix};$$

for this example both (2.2) and (4.2) are strictly stable on $[0, \infty)$, while (4.1) has the solution $y_1 = y_2 = e^t$. Indeed, the equations (2.2) and (4.2) may be individually strictly stable on $[a, \infty)$, in addition $C(t) \to 0$ as $t \to \infty$, and also the improper integral $\int_a^\infty C(s)\, ds = \lim_{t\to\infty} \int_a^t C(s)\, ds$ exist, without the equation (4.1) being stable. Such is illustrated by equation (4.1) with

$$A(t) = \begin{bmatrix} 0 & 1 \\ -1 & 0 \end{bmatrix}, \qquad C(t) = \begin{bmatrix} 0 & 0 \\ \dfrac{-8 \sin 2t}{2t + \sin 2t} & 0 \end{bmatrix}, \tag{4.3}$$

in which case (4.1) has the unbounded solution $y_1 = (t + \frac{1}{2} \sin 2t)\cos t$, $y_2 = -(t + \frac{1}{2} \sin 2t)\sin t + 2 \cos^3 t$.

Theorem 4.1. *If (2.2) is uniformly stable on $[a, \infty)$, and $\nu[C(t)]$ is integrable on $[a, \infty)$, then (4.1) is uniformly stable on $[a, \infty)$.*

If $Y(t)$ is a fundamental matrix of (2.2), then in view of (3.2′) a vector function $y(t)$ is a solution of (4.1) if and only if $y(t)$ satisfies the integral equation

$$y(t) = Y(t, \tau)y(\tau) + \int_\tau^t Y(t, s)C(s)y(s)\, ds, \quad (\tau, t) \in [a, \infty) \times [a, \infty). \tag{4.4}$$

If (2.2) is uniformly stable on $[a, \infty)$ there is a constant $\kappa > 0$ such that $\nu[Y(t, s)] \le \kappa$ on $a \le s \le t < \infty$, and the inequality

$$|y(t)| \le \kappa\, |y(\tau)| + \int_\tau^t \kappa\nu[C(s)]\, |y(s)|\, ds, \quad \textit{for} \quad \tau \in [a, t], \quad t \in [a, \infty), \tag{4.5}$$

is a ready consequence of the equation (4.4). From Theorem I.2.1 it then follows that $|y(t)|$ satisfies the further inequalities

$$|y(t)| \le \kappa\, |y(\tau)| \exp\left\{\kappa \int_\tau^t \nu[C(s)]\, ds\right\}, \tag{4.6′}$$

$$\le \kappa\, |y(\tau)| \exp\left\{\kappa \int_a^\infty \nu[C(s)]\, ds\right\}, \quad \textit{for} \quad \tau \in [a, t], \quad t \in [a, \infty), \tag{4.6″}$$

and consequently (4.1) is uniformly stable on $[a, \infty)$.

If (2.2) is strictly stable on $[s, \infty)$, then there exists a constant κ such that $\nu[Y(t, s)] \leq \kappa$ for $(t, s) \in [a, \infty) \times [a, \infty)$, and corresponding to (4.5), (4.6′) and (4.6″) we have the following respective inequalities for $(\tau, t) \in [a, \infty) \times [a, \infty)$:

$$|y(t)| \leq \kappa\,|y(\tau)| + \left|\int_\tau^t \kappa\nu[C(s)]\,|y(s)|\,ds\right|, \tag{4.7}$$

$$|y(t)| \leq \kappa\,|y(\tau)| \exp\left\{\kappa\left|\int_\tau^t \nu[C(s)]\,ds\right|\right\}, \tag{4.8′}$$

$$|y(t)| \leq \kappa\,|y(\tau)| \exp\left\{\kappa\int_a^\infty \nu[C(s)]\,ds\right\}. \tag{4.8″}$$

As such inequalities hold for an arbitrary solution $y(t)$ of (4.1), we have the result of the following theorem.

Theorem 4.2. *If* (2.2) *is strictly stable on* $[a, \infty)$, *and* $\nu[C(t)]$ *is integrable on* $[a, \infty)$, *then* (4.1) *is strictly stable on* $[a, \infty)$.

Theorem 4.3. *If* (2.2) *is uniformly stable on* $[a, \infty)$, *and also asymptotically stable, then whenever* $\nu[C(t)]$ *is integrable on* $[a, \infty)$ *the equation* (4.1) *is asymptotically stable.*

By Theorem 4.1, uniform stability of (2.2) and integrability of $\nu[C(t)]$ imply that (4.1) is uniformly stable. Consequently, if $y(t)$ is a solution of (4.1) the condition that $\nu[C(t)]$ is integrable on $[a, \infty)$ implies that $g(t) = C(t)y(t)$ is such that $|g(t)|$ is integrable on this interval, and the conclusion of the theorem is an immediate consequence of Theorem 3.2.

Theorem 4.4. *If* (2.2) *is uniformly asymptotically stable on* $[a, \infty)$, *then there exists a constant* $\kappa > 0$ *such that if* $\nu[C(t)] \leq \kappa$ *on* $[a, \infty)$ *then all solutions* $y(t)$ *of* (4.1) *are such that* $y(t) \to 0$ *as* $t \to \infty$.

Let κ_3 and μ be positive constants such that the fundamental matrix $Y(t, s)$ for (2.2) satisfies $\nu[Y(t, s)] \leq \kappa_3 \exp\{-\mu(t - s)\}$ for $s \in [a, t]$, $t \in [a, \infty)$. For a solution $y(t)$ of (4.1), let $y_0(t)$ denote the corresponding solution of (2.2) such that $y_0(a) = y(a)$. Then $y_0(t) = Y(t, a)y(a)$, and $|y_0(t)| \leq \kappa_4 \exp\{-\mu t\}$, with $\kappa_4 = |y(a)|\,\kappa_3 \exp\{\mu a\}$. Since the variation of parameters formula yields the equation

$$y(t) = y_0(t) + \int_a^t Y(t, s)C(s)y(s)\,ds, \qquad t \in [a, \infty),$$

it follows that if $\nu[C(t)] \leq \kappa$ for $t \in [a, \infty)$ then

$$|y(t)| \leq \kappa_4 \exp\{-\mu t\} + \int_a^t \kappa\kappa_3 \exp\{-\mu(t - s)\}\,|y(s)|\,ds, \qquad t \in [a, \infty).$$

The function $\rho(t) = |y(t)| \exp\{\mu t\}$ then satisfies the inequality

$$0 \leq \rho(t) \leq \kappa_4 + \int_a^t \kappa\kappa_3\, \rho(s)\, ds, \qquad t \in [a, \infty),$$

and from Theorem I.2.1 it follows that

$$0 \leq \rho(t) \leq \kappa_4 \exp\{\kappa\kappa_3 (t - a)\}, \qquad t \in [a, \infty).$$

Consequently, whenever $\kappa < \mu/\kappa_3$ the constant $h = \mu - \kappa\kappa_3$ is positive and $|y(t)| \leq \kappa_4 \exp\{-\mu a\} \exp\{-h(t - a)\} = |y(a)|\kappa_3 \exp\{-h\,(t - a)\}$, so that $y(t) \to 0$ as $t \to \infty$.

***Lemma* 4.1.** *Suppose that A is a constant $n \times n$ matrix with distinct proper values $\lambda_1, \cdots, \lambda_n$, and $D(t)$ is an $n \times n$ continuous matrix function on $[a, \infty)$ such that $D(t) \to 0$ as $t \to \infty$. Then there exists an $a_1 \geq a$ such that on $[a_1, \infty)$ the proper values of $A + D(t)$ are distinct, and may be labelled $\lambda_1(t), \cdots, \lambda_n(t)$ with $\lambda_\gamma(t)$ continuous and such that $\lambda_\gamma(t) \to \lambda_\gamma$ as $t \to \infty$; moreover, the corresponding proper vectors $v = v^{(\gamma)}(t)$ satisfying $[\lambda_\gamma(t)E - A - D(t)]v = 0$ may be chosen to be continuous and to approach as $t \to \infty$ nonzero limit vectors $\eta^{(\gamma)}$ satisfying $(\lambda_\gamma E - A)\eta^{(\gamma)} = 0$. If $D(t)$ is of class $\mathfrak{C}^1$ on $[a, \infty)$ then the above determined $\lambda_\gamma(t)$, $v^{(\gamma)}(t)$ are of class $\mathfrak{C}^1$ on $[a_1, \infty)$; furthermore, if $\int_a^\infty \nu[D'(t)]\, dt < \infty$ then $\int_{a_1}^\infty |\lambda_\gamma'(t)|\, dt < \infty$ and $\int_{a_1}^\infty |v^{(\gamma)\prime}(t)|\, dt < \infty$.*

The result of this lemma is in the general domain of implicit function theorems, and may be established by various arguments. In particular, the fact that $\det[\lambda E - A - D(t)]$ is a polynomial in λ permits one to use complex variable theory, (e.g., see Hille [2, Sec. 9.2]), to prove the basic conclusion that on a suitable subinterval $[a_0, \infty)$ the proper values of $A + D(t)$ are distinct, and may be labelled as $\lambda_1(t), \ldots, \lambda_n(t)$ with $\lambda_\gamma(t)$ continuous on $[a_0, \infty)$ and $\lambda_\gamma(t) \to \lambda_\gamma$ as $t \to \infty$.

If $M[\lambda; t] = \lambda E - A - D(t)$, with the understanding that $M[\lambda; \infty] = \lambda E - A$, let $C_{\alpha\beta}[\lambda; t]$ denote the cofactor of the element $M_{\alpha\beta}[\lambda; t]$, $(\alpha, \beta = 1, \ldots, n)$, and $\phi[\lambda; t] = \det M[\lambda; t]$. The condition that $\lambda_1 \ldots, \lambda_n$ are distinct zeros of the polynomial $\phi[\lambda; \infty]$ implies that $M[\lambda_\gamma; \infty]$, $(\gamma = 1, \ldots, n)$, is of rank $n - 1$, and hence there is a corresponding index α_γ for which $\sum_{\beta=1}^n |C_{\alpha_\gamma\beta}[\lambda_\gamma; \infty]| > 0$. By continuity there is a value $a_1 \geq a_0$ such that $\sum_{\beta=1}^n |C_{\alpha_\gamma\beta}[\lambda_\gamma(t); t]| > 0$ for $\gamma = 1, \ldots, n$ and $t \in [a_1, \infty)$. As $M[\lambda_\gamma(t); t]$ is singular the vectors $v^{(\gamma)}(t) = (v_\beta^{(\gamma)}(t))$, with $v_\beta^{(\gamma)}(t) = C_{\alpha_\gamma\beta}[\lambda_\gamma(t); t]$, are continuous and satisfy $M[\lambda_\gamma(t); t]v^{(\gamma)}(t) = 0$ on $[a_1, \infty)$. Moreover, as $t \to \infty$ the vector $v^{(\gamma)}(t)$ tends to the nonzero vector $v^{(\gamma)}(\infty) = \eta^{(\gamma)} = (\eta_\beta^{(\gamma)})$, with $\eta_\beta^{(\gamma)} = C_{\alpha_\gamma\beta}[\lambda_\gamma; \infty]$. Since the proper values $\lambda_1(t), \ldots, \lambda_n(t)$ are distinct on $[a_0, \infty)$, for each $\gamma = 1, \ldots, n$ the partial derivative $\phi_\lambda[\lambda_\gamma(t); t]$ is different from zero on this interval, and by the usual type of argument it follows that if $D(t)$ is of class $\mathfrak{C}^1$ on $[a, \infty)$ then the functions $\lambda_\gamma(t)$ are of class $\mathfrak{C}^1$ and satisfy $\lambda_\gamma'(t) = -\phi_t[\lambda_\gamma(t); t]/\phi_\lambda[\lambda_\gamma(t); t]$ on this interval. Moreover, since $\phi_\lambda[\lambda_\gamma(t); t] \to \phi_\lambda[\lambda_\gamma; \infty] \neq 0$ as $t \to \infty$, and $\phi_t[\lambda; t]$ is linear and homogeneous in the derivative elements $D'_{\alpha\beta}(t)$, the condition $\int_a^\infty \nu[D'(t)]\, dt < \infty$ implies that

$$\int_{a_0}^\infty |\lambda_\gamma'(t)|\, dt < \infty, \qquad (\gamma = 1, \ldots, n).$$

Finally, as the components of the vectors $v^{(\gamma)}(t)$ determined above are cofactors of elements of the matrix $M[\lambda_\gamma(t); t]$, it follows that the components of the derivative vectors $v^{(\gamma)\prime}(t)$ are linear and homogeneous in the elements $D'_{\alpha\beta}(t)$, $\lambda_\gamma'(t)$, and consequently $\int_{a_1}^\infty |v^{(\gamma)\prime}(t)|\, dt < \infty$. It is to be remarked that by a simple argument one may extend the definition of the vectors $v^{(\gamma)}(t)$ to the interval $[a_0, a_1]$ in such a manner that the conclusions of the lemma hold also on the interval $[a_0, \infty)$. For the subsequent application of the above lemma to stability problems, however, such extension would provide no additional results as one is concerned only with the behavior of solutions in a neighborhood of $t = \infty$.

Theorem 4.5. *Suppose that the $n \times n$ matrix A and the $n \times n$ matrix functions $C(t)$, $D(t)$ satisfy the following conditions:*

(i) *the matrix A has distinct proper values;*
(ii) *$C(t)$ is continuous on $[a, \infty)$ with $\int_a^\infty \nu[C(t)]\, dt < \infty$;*
(iii) *$D(t)$ is of class $\mathfrak{C}^1$ on $[a, \infty)$, $D(t) \to 0$ as $t \to \infty$, and*

$$\int_a^\infty \nu[D'(t)]\, dt < \infty.$$

Then on a suitable subinterval $[a_1, \infty)$ there exists a nonsingular matrix $V(t)$ of class $\mathfrak{C}^1$, such that $V(t)$ tends to a nonsingular matrix V as $t \to \infty$, and under the transformation $y(t) = V(t)u(t)$ the differential equation

$$y' = [A + D(t) + C(t)]y, \qquad t \in [a_1, \infty), \tag{4.9}$$

is equivalent to

$$u' = [\Lambda(t) + F(t)]u, \qquad t \in [a_1, \infty), \tag{4.10}$$

where $\Lambda(t) = [\lambda_\alpha(t)\delta_{\alpha\beta}]$, $(\alpha, \beta = 1, \ldots, n)$, with $\lambda_1(t), \ldots, \lambda_n(t)$ the proper values of $A + D(t)$, and $\int_{a_1}^\infty \nu[F(t)]\, dt < \infty$.

Let $[a_1, \infty)$ be a subinterval on which the conclusions of Lemma 4.1 hold, and denote by $V(t)$ the matrix with column vectors $v^{(1)}(t), \ldots, v^{(n)}(t)$ as determined in Lemma 4.1. Then $V^{-1}(t)[A + D(t)]V(t) = \Lambda(t)$, and under the transformation $y(t) = V(t)u(t)$ the equation (4.9) is equivalent to (4.10) with $F(t) = V^{-1}(t)[C(t)V(t) - V'(t)]$. Since $V(t)$ approaches a nonsingular matrix V at $t \to \infty$, and $\int_{a_1}^\infty \nu[V'(t)]\, dt < \infty$ by Lemma 4.1, in view of hypothesis (ii) it follows that $\int_{a_1}^\infty \nu[F(t)]\, dt < \infty$.

Theorem 4.6. *The equation* (4.9) *is uniformly stable if the hypotheses of Theorem* 4.5 *hold, and in addition all proper values of* $A + D(t)$ *have non-positive real parts on* $[a_1, \infty)$.

As (4.9) is equivalent to (4.10) under the transformation $y(t) = V(t)u(t)$, where $V(t)$ approaches a nonsingular matrix V as $t \to \infty$, the equation (4.9) is uniformly stable on $[a_1, \infty)$ if and only if (4.10) is uniformly stable on this interval. Moreover, since $\int_a^\infty \nu[F(t)]\,dt < \infty$, it follows from Theorem 4.1 that (4.10) is uniformly stable on $[a_1, \infty)$ whenever

$$(4.11) \qquad u' = \Lambda(t)u, \qquad t \in [a_1, \infty),$$

is uniformly stable. Now if $u(t) \equiv (u_\alpha(t))$, $(\alpha = 1, \ldots, n)$, is a solution of (4.11), then

$$u_\alpha(t) = u_\alpha(\tau)\exp\{\textstyle\int_\tau^t \lambda_\alpha(s)\,ds\}, \quad \textit{and} \quad |u_\alpha(t)| = |u_\alpha(\tau)|\exp\{\textstyle\int_\tau^t \Re\lambda_\alpha(s)\,ds\}.$$

In particular, if $\Re\lambda_\alpha(t) \leq 0$ on $[a_1, \infty)$ then $|u_\alpha(t)| \leq |u_\alpha(\tau)|$ for $\tau \in [a_1, t]$, $t \in [a_1, \infty)$, so that (4.11) is uniformly stable on $[a_1, \infty)$.

Corollary. *If the hypotheses of Theorem* 4.6 *hold, and* $\int_{a_1}^t \Re\lambda_\alpha(s)\,ds \to -\infty$ *as* $t \to \infty$ *for* $\alpha = 1, \ldots, n$, *then* (4.9) *is also asymptotically stable.*

Problems VIII.4

1. Suppose that $a_0, a_1, \ldots, a_{n-1}$ are constants such that the roots of the polynomial equation

$$(4.12) \qquad \lambda^n + a_{n-1}\lambda^{n-1} + \cdots + a_1\lambda + a_0 = 0,$$

all have $\Re\lambda \leq 0$, and all roots λ of this equation for which $\Re\lambda = 0$ are simple roots. Show that each solution $u(t)$ of the differential equation

(4.13)

$$u^{[n]} + [a_{n-1} + g_{n-1}(t)]u^{[n-1]} + \cdots + [a_0 + g_0(t)]u = 0, \qquad t \in [a, \infty),$$

is such that $|u(t)| + \cdots + |u^{[n-1]}(t)|$ is bounded on $[a, \infty)$ provided that on this interval the functions $g_j(t)$ are continuous and

$$(4.14) \qquad \int_a^\infty |g_j(t)|\,dt < \infty, \qquad (j = 0, 1, \ldots, n-1);$$

moreover, if all roots λ of (4.12) have $\Re\lambda < 0$ then $|u(t)| + \cdots + |u^{[n-1]}(t)| \to 0$ as $t \to \infty$. Show also that if $\Re\lambda < 0$ for all roots λ of (4.12) then there is a constant $\kappa > 0$ such that if the functions $g_j(t)$ are continuous and satisfy $|g_j(t)| \leq \kappa$ on $[a, \infty)$ then each solution $u(t)$ of (4.13) is such that $|u(t)| + \cdots + |u^{[n-1]}(t)| \to 0$ as $t \to \infty$.

2. Prove that all solutions of the scalar equation

$$u'' + [1 + g(t) + h(t)]u = 0,$$

are such that $|u(t)| + |u'(t)|$ is bounded on $[a, \infty)$ provided that on this interval $g(t)$ is continuous, $h(t)$ is a real-valued function of class $\mathfrak{C}^1$ and $h(t) \to 0$ as $t \to \infty$, while

$$\int_a^\infty |g(t)|\, dt < \infty, \qquad \int_a^\infty |h'(t)|\, dt < \infty.$$

3. Show that (4.1) is stable provided (2.2) is stable and the equation

$$u' = Y^{-1}(t)C(t)Y(t)u, \qquad t \in [a, \infty), \tag{4.15}$$

is also stable, where $Y(t)$ denotes a fundamental matrix of (2.2). Moreover, if (2.2) is strictly stable then (4.1) is uniformly stable whenever (4.15) is uniformly stable, and (4.1) is strictly stable whenever (4.15) is strictly stable.

Hint. Show that if $U(t)$ is a fundamental matrix for (4.15) then $Y_1(t) = Y(t)U(t)$ is a fundamental matrix for (4.1) and

$$Y_1(t)Y_1^{-1}(s) = Y(t)U(t)U^{-1}(s)Y^{-1}(s).$$

4. If $A(t)$ and $B(t)$ are continuous $n \times n$ matrix functions on $[a, \infty)$, and there exists a nonsingular matrix function $T(t)$ of class $\mathfrak{C}^1$ on a subinterval $[c, \infty)$ of $[a, \infty)$ such that

$$B = T^{-1}[AT - T' + F], \quad \text{for} \quad t \in [c, \infty),$$

where $F(t)$ is continuous with $\int_c^\infty \nu[F(t)]\, dt < +\infty$, and both $T(t)$ and $T^{-1}(t)$ are bounded on $[c, \infty)$, then $B(t)$ is said to be *t_∞-similar* to $A(t)$, and equation (2.14) is said to be t_∞-similar to (2.2). If the involved transformation matrix $T(t)$ is to be emphasized, the phrase "with respect to $T(t)$" is used. Prove that:

(i) If $B(t)$ is t_∞-similar to $A(t)$ with respect to $T(t)$, then $A(t)$ is t_∞-similar to $B(t)$ with respect to $T^{-1}(t)$.

(ii) If $B(t)$ is t_∞-similar to $A(t)$ with respect to $T_1(t)$, and $C(t)$ is t_∞-similar to $B(t)$ with respect to $T_2(t)$, then $C(t)$ is t_∞-similar to $A(t)$ with respect to $T_1(t)T_2(t)$.

(iii) If (2.2) is uniformly stable on $[a, \infty)$, then every equation (2.14) that is t_∞-similar to (2.2) is also uniformly stable.

Hints. (iii) Show that if $y(t)$ is a solution of (2.2) and $y(t) = T(t)u(t)$, where (2.14) is t_∞-similar to (2.2) with respect to $T(t)$, then $u(t)$ is a solution of an equation $u' = [B(t) + C(t)]u$ on an interval $[c, \infty)$, with

$$\int_c^\infty \nu[C(t)]\, dt < \infty;$$

apply the result of Theorem 4.1.

5. FURTHER RESULTS ON THE ASYMPTOTIC BEHAVIOR OF SOLUTIONS OF HOMOGENEOUS DIFFERENTIAL EQUATIONS

The following lemma on integral equations will be used in the subsequent theorem on differential equations.

Lemma 5.1. *Suppose that on $t \in [a, s]$, $s \in [a, \infty)$, the matrix function $K(t, s) \equiv [K_{\alpha\beta}(t, s)]$, $(\alpha, \beta = 1, \ldots, n)$, is continuous in (t, s) and there exists a non-negative continuous function $\theta(s)$ such that*

$$\int_a^\infty \theta(s)\, ds < \infty, \qquad \nu[K(t, s)] \leq \theta(s), \quad \text{for} \quad t \in [a, s], \quad s \in [a, \infty).$$

If $u(t)$ is a continuous n-dimensional vector function for which $\theta(t)\,|u(t)|$ is integrable on $[a, \infty)$, then there exists a unique solution $y(t)$ of the integral equation

$$y(t) = u(t) + \int_t^\infty K(t, s)y(s)\, ds, \qquad t \in [a, \infty), \tag{5.1}$$

for which $\int_a^\infty \theta(t)\,|y(t)|\, dt < \infty$, and this solution is such that $y(t) - u(t) \to 0$ as $t \to \infty$.

Let $\phi(t) = \theta(t) \max\{1, |u(t)|\}$; in view of the hypotheses of the lemma the function $\phi(t)$ is integrable on $[a, \infty)$. As $|K(t, s)u(s)| \leq \theta(s)\,|u(s)| \leq \phi(s)$ if $s \in [t, \infty)$, for each $t \in [a, \infty)$ the vector function $K(t, s)u(s)$ is integrable on $[t, \infty)$. Now let $y^{(0)}(t) = u(t)$, and consider the recursion relations

$$y^{(j+1)}(t) = u(t) + \int_t^\infty K(t, s)y^{(j)}(s)\, ds, \qquad (j = 0, 1, \ldots). \tag{5.2}$$

By induction it follows that for $t \in [a, \infty)$, and $j = 1, 2, \ldots,$

$$\begin{aligned}
&\text{(a)} \quad |y^{(j)}(t) - y^{(j-1)}(t)| \leq \frac{1}{j!}\left(\int_t^\infty \phi(s)\, ds\right)^j, \\
&\text{(b)} \quad |y^{(j)}(t) - u(t)| \leq \sum_{k=1}^{j} \frac{1}{k!}\left(\int_t^\infty \phi(s)\, ds\right)^k, \\
&\qquad\qquad \leq \exp\left\{\int_t^\infty \phi(s)\, ds\right\} - 1, \\
&\text{(c)} \quad |y^{(j)}(t)| \leq |u(t)| + \exp\left\{\int_a^\infty \phi(s)\, ds\right\}, \\
&\qquad\qquad \leq \left[1 + \exp\left\{\int_a^\infty \phi(s)\, ds\right\}\right] \max\{1, |u(t)|\}.
\end{aligned} \tag{5.3}$$

In particular, the condition that $y^{(j)}(t)$ is continuous and satisfies inequality (5.3-c) implies that $\nu[K(t, s)y^{(j)}(s)] \leq [1 + \exp\{\int_a^\infty \phi(r)\, dr\}]\, \phi(s)$ for $s \in$

$[t, \infty)$, and consequently the $y^{(j+1)}(t)$ of (5.2) is a continuous vector function on $[a, \infty)$. Inequality (5.3-a) implies that the sequence $\{y^{(j)}(t)\}$ converges uniformly on the interval $[a, \infty)$ to a limit vector function $y(t)$, which is continuous on this interval; moreover,

$$\tag{5.4} \begin{aligned} |y(t) - y^{(j)}(t)| &\leq \sum_{k=1}^{\infty} \frac{1}{(j+k)!} \left(\int_t^\infty \phi(s)\, ds \right)^{j+k}, \\ &\leq \frac{\kappa_1}{(j+1)!} \left(\int_t^\infty \phi(s)\, ds \right)^{j+1} \quad \textit{for} \quad t \in [a, \infty), \end{aligned}$$

where $\kappa_1 = \exp\{\int_a^\infty \phi(s)\, ds\}$. In view of (5.3-c) it follows that $\theta(t)\, |y(t)| \leq (1 + \kappa_1)\, \phi(t)$, so that $\theta(t)\, |y(t)|$ is integrable on $[a, \infty)$; moreover, (5.4) implies that for $j = 0, 1, \ldots$ we have the inequality

$$\left| \int_t^\infty K(t, s) y^{(j)}(s)\, ds - \int_t^\infty K(t, s) y(s)\, ds \right| \leq \frac{\kappa_1}{(j+2)!} \left(\int_t^\infty \phi(s)\, ds \right)^{j+2}, \qquad t \in [a, \infty),$$

and consequently $y(t)$ is a solution of (5.1) for which $\theta(t)\, |y(t)|$ is integrable on $[a, \infty)$. From (5.3-b) it then follows that $|y(t) - u(t)| \leq \exp\{\int_t^\infty \phi(s)\, ds\} - 1$, and hence $y(t) - u(t) \to 0$ as $t \to \infty$.

Now suppose that $y = z(t)$ is also a solution of (5.1) such that $\theta(t)\, |z(t)|$ is integrable on $[a, \infty)$. Then

$$|y(t) - z(t)| \leq \int_t^\infty \theta(s)\, |y(s) - z(s)|\, ds, \qquad t \in [a, \infty),$$

and the non-negative function $\rho(t) = \int_t^\infty \theta(s)\, |y(s) - z(s)|\, ds$ is such that $\rho(t) \to 0$ as $t \to \infty$, and $[\rho(t) \exp\{\int_a^t \theta(s)\, ds\}]' \geq 0$, so that $\rho(t) \equiv 0$ and $z(t) \equiv y(t)$ on $[a, \infty)$.

As an immediate application of the preceding lemma one has the following result.

Theorem 5.1. *If $C(t)$ is a continuous $n \times n$ matrix function on $[a, \infty)$ such that for a fundamental matrix $Y(t)$ of* (2.2) *we have $\nu[Y(t) Y^{-1}(s) C(s)] \leq \theta(s)$ for $t \in [a, s]$, $s \in [a, \infty)$, with $\theta(t)$ integrable on $[a, \infty)$, and $y = u(t)$ is a solution of* (2.2) *for which $\theta(t)\, |u(t)|$ is also integrable on this interval, then there exists a solution $y(t)$ of* (4.1) *for which $y(t) - u(t) \to 0$ as $t \to \infty$.*

Indeed, for $u(t)$ a solution of (2.2) satisfying the hypotheses of the theorem, there is by Lemma 5.1 a unique solution of the integral equation

$$\tag{5.5} y(t) = u(t) - \int_t^\infty Y(t) Y^{-1}(s) C(s)\, y(s)\, ds, \qquad t \in [a, \infty),$$

which is such that $\int_a^\infty \theta(t)\,|y(t)|\,dt < \infty$ and $y(t) - u(t) \to 0$ as $t \to \infty$. Consequently, the proof of the theorem is completed by the remark that any solution of (5.5) is a solution of the differential equation (4.1).

It is to be remarked that the hypotheses of the theorem do not imply that there is a unique solution $y(t)$ of (4.1) such that $y(t) - u(t) \to 0$ as $t \to \infty$. For example, if (2.2) is uniformly and asymptotically stable, and $C(t)$ is such that $\nu[C(t)]$ is integrable on $[a, \infty)$, then by Theorem 4.3 the equation (4.1) is also asymptotically stable, so that all solutions of (2.2) and (4.1) tend to the zero vector as $t \to \infty$. Moreover, for $Y(t)$ a fundamental matrix of an arbitrary system (2.2) one may determine continuous matrices $C(t)$ such that $\nu[C(t)]$ is integrable on $[a, \infty)$ and $\nu[Y(t)Y^{-1}(s)C(s)] \le \theta(s)$ for $t \in [a, s]$, $s \in [a, \infty)$, with $\theta(t)$ integrable on $[a, \infty)$.

Theorem 5.2. *If the differential equation* (2.9) *adjoint to* (2.2) *is uniformly stable on* $[a, \infty)$, *and* $\nu[C(t)]$ *is integrable on this interval, then the bounded solutions of* (2.2) *and* (4.1) *are in asymptotic one-to-one correspondence; that is, if* $y = u(t)$ *is a bounded solution of* (2.2), *then there is a unique bounded solution* $y(t)$ *of* (4.1) *such that* $y(t) - u(t) \to 0$ *as* $t \to \infty$, *and if* $y(t)$ *is a bounded solution of* (4.1) *then there is a unique bounded solution* $y = u(t)$ *of* (2.2) *such that* $y(t) - u(t) \to 0$ *as* $t \to \infty$.

If $Y(t)$ is a fundamental matrix for (2.2), then the condition that the equation (2.9) adjoint to (2.2) is uniformly stable on $[a, \infty)$ implies that there is a constant κ such that $\nu[Y(t)Y^{-1}(s)] \le \kappa$ for $t \in [a, s]$, $s \in [a, \infty)$, and in view of the hypothesis that $\nu[C(t)]$ is integrable on $[a, \infty)$ it follows that the hypothesis of Theorem 5.1 holds with $\theta(t) = \kappa\nu[C(t)]$. Moreover, if $y = u(t)$ is a bounded solution of (2.2) then $\theta(t)\,|u(t)|$ is integrable on $[a, \infty)$, and by Theorem 5.1 there is a solution $y(t)$ of (4.1) for which $y(t) - u(t) \to 0$ as $t \to \infty$. Now suppose that $y = z(t)$ is any solution of (4.1) such that $z(t) - u(t) \to 0$ as $t \to \infty$. Since $z(t)$ is bounded on $[a, \infty)$, the integral $\int_t^\infty Y(t)Y^{-1}(s)C(s)z(s)\,ds$ exists for $t \in [a, \infty)$, and $y = z(t)$ is a solution of (5.5) with $u(t)$ of that equation replaced by the solution $u^{(0)}(t)$ of (2.2) given by

$$u^{(0)}(t) = Y(t)\left\{Y^{-1}(a)z(a) + \int_a^\infty Y^{-1}(s)C(s)z(s)\,ds\right\};$$

moreover, $z(t) - u^{(0)}(t) \to 0$ as $t \to \infty$. Hence $u(t)$ and $u^{(0)}(t)$ are solutions of (2.2) such that $u(t) - u^{(0)}(t) \to 0$, and we shall proceed to show that the uniform stability of the adjoint of (2.2) implies that $u(t) \equiv u^{(0)}(t)$. Indeed, if the constant vector ξ is such that $Y(t)\xi = u(t) - u^{(0)}(t)$, then $|Y(a)\xi| = |Y(a)Y^{-1}(s)\,[u(s) - u^{(0)}(s)]| \le \kappa\,|u(s) - u^{(0)}(s)|$, and since $u(s) - u^{(0)}(s) \to 0$ as $s \to \infty$ it follows that $\xi = 0$ and $u(t) \equiv u^{(0)}(t)$. That is, $y(t)$ and $z(t)$ are both solutions of (5.5) with the given bounded solution $u(t)$ of (2.2), and in view of the uniqueness property of solutions of (5.5) we have that $z(t) \equiv y(t)$. Thus it has been proved that for a bounded solution $y = u(t)$ of (2.2) there is a unique solution $y(t)$ of (4.1) such that $y(t) - u(t) \to 0$ as $t \to \infty$. Now by Theorem 4.1 the equation adjoint to (4.1) is uniformly stable whenever (2.9) is uniformly stable, and the final conclusion of the theorem is a consequence of the result established above, with the roles of (2.2) and (4.1) interchanged.

Theorem 5.3. *Suppose that there is a nonsingular matrix $T(t)$ of class $\mathfrak{C}^1$ on $[a, \infty)$ for which $B(t) \equiv T^{-1}(t)[A(t)T(t) - T'(t)] = \text{diag}\{A_1(t), A_2(t)\}$, where $A_1(t)$ is a $\rho \times \rho$ matrix, $A_2(t)$ is a $\sigma \times \sigma$ matrix, and such that the ρ-dimensional system*

$$(5.6) \qquad u' = A_1(t)u, \qquad t \in [a, \infty),$$

is uniformly and asymptotically stable, while the system adjoint to the σ-dimensional system

$$(5.7) \qquad v' = A_2(t)v, \qquad t \in [a, \infty),$$

is uniformly stable. If $\theta(t) = \nu[T^{-1}(t)C(t)T(t)]$ is integrable on $[a, \infty)$, and $y = y^{(0)}(t)$ is a solution of (2.2) for which $\phi(t) = \theta(t)\max\{1, |T^{-1}(t)y^{(0)}(t)|\}$ is integrable on $[a, \infty)$, then there is a solution $y(t)$ of (4.1) such that

$$T^{-1}(t)[y(t) - y^{(0)}(t)] \to 0 \quad as \quad t \to \infty.$$

Under the transformation $y(t) = T(t)w(t)$ the equation (4.1) reduces to

$$(5.8) \qquad w' = [B(t) + D(t)]w, \qquad t \in [a, \infty),$$

where $B = T^{-1}(AT - T')$ and $D = T^{-1}CT$. If $w^{(0)}(t) = T^{-1}(t)y^{(0)}(t)$, then $\phi(t) = \theta(t)\max\{1, |w^{(0)}(t)|\}$, and the conclusion of the theorem is equivalent to the existence of a solution $w(t)$ of (5.8) such that $w(t) - w^{(0)}(t) \to 0$ as $t \to \infty$. Let $U(t)$ be a fundamental matrix for (5.6), and $V(t)$ a fundamental matrix for (5.7). Then the uniform stability of (5.6) implies that there is a constant κ_1 such that $\nu[U(t)U^{-1}(s)] \le \kappa_1$ for $s \in [a, t]$, $t \in [a, \infty)$, while the uniform stability of the equation adjoint to (5.7) implies that there is a constant κ_2 such that $\nu[V(t)V^{-1}(s)] \le \kappa_2$ for $t \in [a, s]$, $s \in [a, \infty)$. Consequently if

$$K_1(t, s) = \text{diag}\{U(t)U^{-1}(s), 0^{(\sigma, \sigma)}\},$$

and

$$K_2(t, s) = \text{diag}\{0^{(\rho, \rho)}, V(t)\, V^{-1}(s)\},$$

then $\nu[K_1(t, s)] \le \kappa_1$ for $s \in [a, t]$, $t \in [a, \infty)$, and $\nu[K_2(t, s)] \le \kappa_2$ for $t \in [a, s]$, $s \in [a, \infty)$. In particular, as

$$|K_2(t, s)D(s)w^{(0)}(s)| \le \kappa_2\, \theta(s)\, |w^{(0)}(s)| \le \kappa_2\, \phi(s),$$

for each $t \in [a, \infty)$ the vector function $K_2(t, s)D(s)w^{(0)}(s)$ is integrable on $[t, \infty)$.

Now let τ be a value on $[a, \infty)$ such that $c = \max\{\kappa_1, \kappa_2\} \int_\tau^\infty \phi(s)\,ds < 1$, and consider the recursion relation

$$w^{(j+1)}(t) = w^{(0)}(t) + \int_\tau^t K_1(t, s)D(s)w^{(j)}(s)\,ds$$

(5.9)

$$- \int_t^\infty K_2(t, s)D(s)w^{(j)}(s)\,ds, \qquad (j = 0, 1, \ldots).$$

In view of the above comment on the integrability of $K_2(t, s)D(s)w^{(0)}(s)$, it follows that for $j = 0$ the right-hand member of (5.9) exists for $t \in [\tau, \infty)$, and defines a continuous vector function $w^{(1)}(t)$. Moreover, for $t \in [\tau, \infty)$,

$$|w^{(1)}(t) - w^{(0)}(t)| \leq \int_\tau^t \kappa_1\, \phi(s)\,ds + \int_t^\infty \kappa_2\, \phi(s)\,ds \leq c,$$

$$|w^{(1)}(t)| \leq |w^{(0)}(t)| + c \leq (1 + c)\max\{1, |w^{(0)}(t)|\}.$$

By induction it follows that for $t \in [\tau, \infty)$, and $j = 1, 2, \ldots,$

(a) $|w^{(j)}(t) - w^{(j-1)}(t)| \leq c^j,$

(5.10) (b) $|w^{(j)}(t) - w^{(0)}(t)| \leq \sum_{k=1}^{j} c^k \leq \dfrac{c}{1 - c},$

(c) $|w^{(j)}(t)| \leq |w^{(0)}(t)| + \dfrac{c}{1 - c} \leq \left[1 + \dfrac{c}{1 - c}\right] \max\{1, |w^{(0)}(t)|\}.$

In particular, the condition that $w^{(j)}(t)$ is continuous and satisfies inequality (5.10-c) on $[\tau, \infty)$ implies that the right-hand member of (5.9) exists for $t \in [\tau, \infty)$, and defines a continuous vector function $w^{(j+1)}(t)$. From (5.10-a) it follows that the sequence $\{w^{(j)}(t)\}$ converges uniformly on $[\tau, \infty)$ to a limit vector function $w(t)$ that is continuous on this interval; moreover

(5.11) $$|w(t) - w^{(j)}(t)| \leq \textstyle\sum_{k=1}^\infty c^{j+k} = \dfrac{c^{j+1}}{1 - c}, \qquad t \in [\tau, \infty).$$

In view of inequality (5.10-c) it then follows that

$$\theta(t)\,|w(t)| \leq \left[1 + \frac{c}{1 - c}\right] \phi(t),$$

so that $\theta(t)\,|w(t)|$ is integrable on $[\tau, \infty)$, and for each $t \in [\tau, \infty)$ the vector function $K_2(t, s)D(s)w(s)$ is integrable on $[t, \infty)$. Inequality (5.11) implies that

$$\left|\int_\tau^t K_1(t, s)D(s)[w(s) - w^{(j)}(s)]\,ds - \int_t^\infty K_2(t, s)D(s)[w(s) - w^{(j)}(s)]\,ds\right|$$

$$\leq \frac{c^{j+1}}{1 - c}\left[\int_\tau^t \kappa_1\, \phi(s)\,ds + \int_t^\infty \kappa_2\, \phi(s)\,ds\right] \leq \frac{c^{j+2}}{1 - c},$$

and consequently $w(t)$ is a solution of the integral equation

$$(5.12)\qquad \begin{aligned} w(t) = w^{(0)}(t) &+ \int_\tau^t K_1(t, s)D(s)w(s)\,ds \\ &- \int_t^\infty K_2(t, s)D(s)w(s)\,ds, \qquad t \in [\tau, \infty). \end{aligned}$$

If $u^{(0)}(t) \equiv (w_\gamma^{(0)}(t))$, $u(t) \equiv (w_\gamma(t))$, $(\gamma = 1, \ldots, \rho)$, and

$$v^{(0)}(t) = (w_{\rho+\mu}^{(0)}(t)), \qquad v(t) = (w_{\rho+\mu}(t)), \qquad (\mu = 1, \ldots, \sigma),$$

the vector equation (5.12) is equivalent to the vector system

$$(5.13)\qquad \begin{aligned} &\text{(a)}\quad u(t) = u^{(0)}(t) + \int_\tau^t U(t)U^{-1}(s)D_1(s)w(s)\,ds, \\ &\text{(b)}\quad v(t) = v^{(0)}(t) - \int_t^\infty V(t)V^{-1}(s)D_2(s)w(s)\,ds, \end{aligned}$$

where $D_1(t) = [D_{\gamma\beta}(t)]$, $(\gamma = 1, \ldots, \rho; \beta = 1, \ldots, n)$, and $D_2(t) = [D_{\rho+\mu,\beta}(t)]$, $(\mu = 1, \ldots, \sigma; \beta = 1, \ldots, n)$. Now the vector functions $u(t)$, $v(t)$ of (5.13) satisfy the differential equations

$$(5.14)\qquad \begin{aligned} u' &= A_1 u + D_1(t)w(t), \\ v' &= A_2 v + D_2(t)w(t), \end{aligned} \qquad t \in [\tau, \infty),$$

and consequently $w(t)$ is a solution of (5.8). Moreover,

$$|V(t)V^{-1}(s)D_2(s)w(s)| \le \kappa_2\theta(s)\,|w(s)| \le \kappa_2\left[1 + \frac{c}{1-c}\right]\phi(s) \quad \textit{for}$$

$$t \in [\tau, s],\ s \in [\tau, \infty),$$

and from (5.13-b) it follows that $v(t) - v^{(0)}(t) \to 0$ as $t \to \infty$. Now the condition that (5.9) is asymptotically stable implies that $u^{(0)}(t) \to 0$ as $t \to \infty$. Moreover, in view of the inequality

$$|D_1(t)w(t)| \le \theta(t)\,|w(t)| \le \left[1 + \frac{c}{1-c}\right]\phi(t)$$

and Theorem 3.2, it follows that also $u(t) \to 0$, and hence $u(t) - u^{(0)}(t) \to 0$ as $t \to \infty$, thus completing the proof of the conclusion that $w(t) - w^{(0)}(t) \to 0$ as $t \to \infty$.

Theorem 5.4. *Suppose that $A(t)$ is a constant matrix A, and that $C(t)$ is such that $\nu[C(t)]$ is integrable on $[a, \infty)$. If λ_0 is a proper value of A such that each proper value λ of A with $\Re\lambda = \Re\lambda_0$ has index equal to its multiplicity, and η is a vector satisfying $(\lambda_0 E - A)\eta = 0$, then there exists a solution $y(t)$ of the differential equation* (4.1) *such that $e^{-\lambda_0 t}y(t) \to \eta$ as $t \to \infty$.*

Let $\lambda_1, \ldots, \lambda_r$ denote the distinct proper values of A, and denote by m_j the multiplicity of λ_j, $(j = 1, \ldots, r)$. It will be supposed that these proper values are so ordered that $\Re\,\lambda_j \leq \Re\,\lambda_k$ for $j < k$. Let S be a constant matrix such that $S^{-1}AS = \text{diag}\{A^{(1)}, \ldots, A^{(r)}\}$, where $A^{(j)}$ is an $m_j \times m_j$ superdiagonal matrix with diagonal elements equal to λ_j. In view of the hypotheses of the theorem, the matrix $A^{(j)}$ is diagonal if $\Re\,\lambda_j = \Re\,\lambda_0$. If $T(t) = e^{\lambda_0 t}S$, then under the transformation $y(t) = T(t)w(t)$ the equation (4.1) reduces to

$$(5.15) \qquad w' = [B + D(t)]w, \qquad t \in [a, \infty),$$

where $B = \text{diag}\{A^{(1)} - \lambda_0 E^{(m_1)}, \ldots, A^{(r)} - \lambda_0 E^{(m_r)}\}$, and $D(t) = S^{-1}C(t)S$; moreover, in view of the integrability of $\nu[C(t)]$, it follows that $\nu[D(t)]$ is also integrable on $[a, \infty)$. If ξ is a constant vector such that $\eta = S\xi$, then $B\xi = 0$, $w = w^{(0)}(t) \equiv \xi$ is a solution of the differential equation

$$(5.16) \qquad w' = Bw, \qquad t \in [a, \infty),$$

and the conclusion of the theorem is equivalent to the existence of a solution $w(t)$ of (5.15) such that $w(t) \to \xi$ as $t \to \infty$.

Let $p + 1$ denote the smallest positive integer such that $\Re\,\lambda_{p+1} = \Re\,\lambda_0$. If $p = 0$, that is, if $\Re\,\lambda_1 = \Re\,\lambda_0$, then all proper values μ of the constant matrix B have $\Re\,\mu \geq 0$, and if $\Re\,\mu = 0$ then the index of μ is equal to its multiplicity. In this case it follows from Theorem 2.4 that the equation adjoint to (5.16) is uniformly stable, and by Theorem 5.2 there is a unique solution $w(t)$ of (5.15) such that $w(t) \to \xi$ as $t \to \infty$.

If $p > 0$, let $\rho = m_1 + \cdots + m_p$, and $\sigma = n - \rho$. If $B^{(1)} = [B_{\alpha\beta}]$, $(\alpha, \beta = 1, \ldots, \rho)$, and $B^{(2)} = [B_{\rho+\alpha,\rho+\beta}]$, $(\alpha, \beta = 1, \ldots, \sigma)$, then $B = \text{diag}\{B^{(1)}, B^{(2)}\}$ and all proper values μ of $B^{(1)}$ have $\Re\,\mu < 0$, while all proper values μ of $B^{(2)}$ have $\Re\,\mu \geq 0$; moreover, if μ is a proper value of $B^{(2)}$ with $\Re\,\mu = 0$, then the index of μ is equal to its multiplicity. Consequently the ρ-dimensional equation

$$u' = B^{(1)}u, \qquad t \in [a, \infty),$$

is uniformly and asymptotically stable, while the adjoint of the σ-dimensional equation

$$v' = B^{(2)}v, \qquad t \in [a, \infty),$$

is uniformly stable. From Theorem 5.3 it then follows that there is a solution $w(t)$ of (5.15) such that $w(t) \to \xi$ as $t \to \infty$, thus establishing the conclusion of the theorem in this case.

Of particular importance is the case of the following corollary.

Corollary. *Suppose that $A(t)$ is a constant matrix A such that each proper value of A has index equal to its multiplicity, and that $C(t)$ is such that $\nu[C(t)]$*

is integrable on $[a, \infty)$. *If* $\lambda_1, \ldots, \lambda_n$ *denote the proper values of* A, *with each repeated a number of times equal to its multiplicity, and* $\eta^{(1)}, \ldots, \eta^{(n)}$ *is a corresponding set of linearly independent vectors satisfying* $(\lambda_\beta E - A)\eta^{(\beta)} = 0$, $(\beta = 1, \ldots, n)$, *then there exist n linearly independent solutions* $y^{(\beta)}(t)$ *of* (4.1) *such that* $e^{-\lambda_\beta t} y^{(\beta)}(t) \to \eta^{(\beta)}$ *as* $t \to \infty$.

The existence of the solutions $y^{(\beta)}(t)$ follows from the above theorem. In order to establish the linear independence of these solutions let

$$Y(t) = [y_\alpha^{(\beta)}(t)], \Lambda = [\delta_{\alpha\beta} e^{-\lambda_\beta t}], W(t) = [e^{-\lambda_\beta t} y_\alpha^{(\beta)}(t)], N = [\eta_\alpha^{(\beta)}], \quad (\alpha, \beta = 1, \ldots, n).$$

Since N is nonsingular, and $W(t) \to N$ as $t \to \infty$, the matrix $W(t)$ is nonsingular for t sufficiently large. The equation $W(t) = Y(t)\Lambda(t)$ then implies that $Y(t)$ is also nonsingular for t sufficiently large, and hence $Y(t)$ is a fundamental matrix for (4.1).

Theorem 5.5. *Suppose that* $\lambda_1(t), \ldots, \lambda_n(t)$ *are continuous functions on* $[a_1, \infty)$, *and there is an integer* k, $(1 \le k \le n)$, *such that for each* j $(j = 1, \ldots, n)$, *either* $\lambda_j(t)$ *satisfies the conditions*

$$(5.17) \quad \begin{aligned} &\text{(a)} \quad \int_{a_1}^{t} \Re[\lambda_j(s) - \lambda_k(s)]\, ds \to -\infty \quad as \quad t \to \infty, \\ &\text{(b)} \quad \int_{t_1}^{t_2} \Re[\lambda_j(s) - \lambda_k(s)]\, ds < c_j, \quad for \quad a_1 \le t_1 \le t_2 < \infty, \end{aligned}$$

or $\lambda_j(t)$ *satisfies the condition*

$$(5.18) \quad \int_{t_1}^{t_2} \Re[\lambda_j(s) - \lambda_k(s)]\, ds > d_j, \quad for \quad a_1 \le t_1 \le t_2 < \infty,$$

where c_j *in* (5.17-b) *and* d_j *in* (5.18) *are suitable constants. If* $\Lambda(t)$ *is the diagonal matrix* $[\delta_{\alpha\beta}\lambda_\beta(t)]$, $(\alpha, \beta = 1, \ldots, n)$, *and* $F(t)$ *is a continuous* $n \times n$ *matrix for which* $\nu[F(t)]$ *is integrable on* $[a_1, \infty)$, *then there exists a solution* $y(t)$ *of the differential equation*

$$(5.19) \quad y' = [\Lambda(t) + F(t)]y, \qquad t \in [a_1, \infty),$$

such that $y(t) \exp\{-\int_{a_1}^{t} \lambda_k(s)\, ds\} \to e^{(k)}$ *as* $t \to \infty$.

Clearly $y(t)$ is a solution of (5.19) if and only if $z(t) = y(t)\exp\{-\int_{a_1}^{t} \lambda_k(s)\, ds\}$ is a solution of

$$(5.20) \quad z' = [B(t) + F(t)]z, \qquad t \in [a_1, \infty),$$

where $B(t)$ is the diagonal matrix with $B_{\alpha\alpha}(t) = \lambda_\alpha(t) - \lambda_k(t)$, $(\alpha = 1, \ldots, n)$;

moreover, the equation

$$z' = B(t)z, \qquad t \in [a_1, \infty), \tag{5.21}$$

has the particular solution $z = z^{(0)}(t) \equiv e^{(k)}$, and the conclusion of the theorem is equivalent to the existence of a solution $z(t)$ of (5.20) such that $z(t) \to e^{(k)}$ as $t \to \infty$.

Now if condition (5.18) holds for all integers $j = 1, \ldots, n$ then the differential equation adjoint to (5.21) is uniformly stable, and from Theorem 5.2 it follows that there is a unique solution $z(t)$ of (5.20) such that $z(t) \to e^{(k)}$ as $t \to \infty$. On the other hand, if (5.18) does not hold for all integers $j = 1, \ldots, n$, let ρ denote the number of values of j for which $\lambda_j(t)$ satisfies the conditions of (5.17); as (5.18) holds for $\lambda_j(t)$ with $j = k$, we have $1 \leq \rho < n$. Then there exists a constant matrix T such that $T^{-1} B(t)T = \operatorname{diag}\{B^{(1)}(t), B^{(2)}(t)\}$, where $B^{(1)}(t) = [B^{(1)}_{\alpha\beta}(t)]$, $(\alpha, \beta = 1, \ldots, \rho)$, is a diagonal matrix with diagonal elements those values of $\lambda_j(t) - \lambda_k(t)$ for which conditions (5.17) hold, while $B^{(2)}(t) = [B^{(2)}_{\rho+\alpha,\rho+\beta}(t)]$, $(\alpha, \beta = 1, \ldots, \sigma = n - \rho)$, is a diagonal matrix with diagonal elements those values of $\lambda_j(t) - \lambda_k(t)$ for which condition (5.18) holds. The ρ-dimensional differential equation $u' = B^{(1)}(t)u$, $t \in [a_1, \infty)$, is asymptotically and uniformly stable in view of the respective conditions (5.17-a) and (5.17-b); correspondingly, the equation adjoint to $v' = B^{(2)}(t)v$, $t \in [a_1, \infty)$, is uniformly stable in view of the condition (5.18). Consequently, by Theorem 5.3 there is a solution $z(t)$ of (5.20) such that $T^{-1}[z(t) - e^{(k)}] \to 0$, and hence $z(t) \to e^{(k)}$ as $t \to \infty$.

Corollary. *Suppose that the $n \times n$ matrix functions A, $C(t)$ and $D(t)$ satisfy the following conditions:* (i) *the constant matrix A has distinct proper values $\lambda_1, \ldots, \lambda_n$;* (ii) *$D(t)$ is of class $\mathfrak{C}^1$ on $[a, \infty)$, $D(t) \to 0$ as $t \to \infty$, and $\nu[D'(t)]$ is integrable on $[a, \infty)$;* (iii) *if $[a_1, \infty)$ is a subinterval of $[a, \infty)$ on which the conclusions of Lemma* 4.1 *hold for the proper values $\lambda_1(t), \ldots, \lambda_n(t)$ of $A + D(t)$, then for each pair of integers j, $k = 1, \ldots, n$ either* (5.17) *or* (5.18) *holds;* (iv) *$C(t)$ is continuous and $\nu[C(t)]$ is integrable on $[a, \infty)$. Then there exist n linearly independent solutions $y^{(1)}(t), \ldots, y^{(n)}(t)$ of* (4.9) *such that $y^{(\beta)}(t)\exp\{-\int_{a_1}^t \lambda_\beta(s)\,ds\} \to \eta^{(\beta)}$ as $t \to \infty$, where $\eta^{(\beta)}$ is a proper vector for A satisfying $(\lambda_\beta E - A)\eta^{(\beta)} = 0$, $(\beta = 1, \ldots, n)$.*

Under the hypotheses of the above corollary it is a ready consequence of Theorems 4.5 and 5.5 that there are n solutions $y^{(\beta)}(t)$, $(\beta = 1, \ldots, n)$, of (4.9) such that $y^{(\beta)}(t)\exp\{-\int_{a_1}^t \lambda_\beta(s)\,ds\} \to \eta^{(\beta)}$, where $\eta^{(\beta)}$ is a proper vector of A satisfying $(\lambda_\beta E - A)\eta^{(\beta)} = 0$. As $\eta^{(1)}, \ldots, \eta^{(n)}$ are linearly independent vectors, the conclusion that $y^{(1)}(t), \ldots, y^{(n)}(t)$ are linearly independent solutions of (4.9) may be established by the same type of argument as that used for the proof of the corresponding result in the Corollary to Theorem 5.4.

Problems VIII.5

1. If $p(t)$ is a continuous function such that $t\,|p(t)|$ is integrable on $[a, \infty)$, then there exists a solution $y = y_1(t)$ of the differential equation

$$y'' + p(t)y = 0, \qquad t \in [a, \infty), \tag{5.22}$$

such that $y_1(t) \to 1$, $y_1'(t) \to 0$ as $t \to \infty$. If $t^2\,|p(t)|$ is integrable on $[a, \infty)$, then there exists a solution $y = y_2(t)$ of (5.22) such that $y_2(t) - t \to 0$, $y_2'(t) - 1 \to 0$ as $t \to \infty$.

Hint. Consider the first order system $y_1' = y_2$, $y_2' = -p(t)y_1$, and show that if $Y(t)$ is a fundamental matrix for the system $y_1' = y_2$, $y_2' = 0$ then there is a constant c such that $\nu[Y(t)Y^{-1}(s)] \le c(1 + s)$ for $t \in [a, s]$, $s \in [a, \infty)$. Use Theorem 5.1.

2. For the second order differential equation

$$(r(t)y')' + p(t)y = 0, \qquad t \in [a, \infty),$$

with $r(t)$, $p(t)$ continuous and $r(t) > 0$ on $[a, \infty)$, obtain results analogous to those of the preceding problem for the equation (5.22).

3. For the n-th order differential equation

$$y^{(n)} + p(t)y = 0,$$

formulate and establish generalizations of the results of Problem 1.

4. If $g(t)$ is a continuous function such that $|g(t)|$ is integrable on $[a, \infty)$, show that the differential equation $y'' + g(t)y = 0$ has unbounded solutions.

5. If $g(t)$ is a continuous function such that $|g(t)|$ is integrable on $[a, \infty)$, show that there exist solutions $y = y_j(t)$, $(j = 1, 2)$, of the differential equation

$$y'' + (k^2 + g(t))y = 0, \qquad (k > 0), \qquad t \in [a, \infty),$$

such that $y_1(t) - \sin kt \to 0$, $y_1'(t) - k\cos kt \to 0$, and $y_2(t) - \cos kt \to 0$, $y_2'(t) + k\sin kt \to 0$ as $t \to \infty$.

Hint. Use Theorem 5.2.

6. Use the result of the preceding problem to obtain asymptotic forms for the solutions of Bessel's equation

$$y'' + \left(\frac{1}{t}\right)y' + \left(1 - \frac{n^2}{t^2}\right)y = 0.$$

Hint. Show that under the transformation $y = t^{-1/2}u$ this equation reduces to $u'' + \{1 + (\frac{1}{4} - n^2)/t^2\}u = 0$.

7. If $a_0, \ldots, a_{n-1}$ are constants such that the polynomial $\lambda^n + a_{n-1}\lambda^{n-1} + \cdots + a_1\lambda + a_0$ has distinct zeros $\lambda_1, \ldots, \lambda_n$, while for $j = 0, 1, \ldots, n-1$ the functions $g_j(t)$ are continuous and such that $|g_j(t)|$ are integrable on $[a, \infty)$, show that for $\beta = 1, \ldots, n$ there is a solution $u = u_\beta(t)$ of the differential equation

$$u^{[n]} + \{a_{n-1} + g_{n-1}(t)\}u^{[n-1]} + \cdots + \{a_0 + g_0(t)\}u = 0, \qquad t \in [a, \infty),$$

such that $e^{-\lambda_\beta t}u_\beta^{[\alpha]} \to \lambda_\beta^\alpha$, $(\alpha = 0, 1, \ldots, n-1)$.

Hint. Use the Corollary of Theorem 5.4.

6. DIFFERENTIAL EQUATIONS WITH PERIODIC COEFFICIENTS

Attention will now be directed to equations of the form (2.2) in which the continuous matrix function $A(t)$ is defined on $(-\infty, \infty)$ and has period $\omega > 0$; that is, $A(t + \omega) \equiv A(t)$ for $t \in (-\infty, \infty)$. It is to be emphasized that it is not required that ω be the smallest value such that $A(t)$ has period ω, in case such exists. In particular, if $A(t)$ is a constant matrix, then ω may be an arbitrary positive number. Now if $Y(t)$ is a fundamental matrix for the differential equation

$$y' = A(t)y, \qquad t \in (-\infty, \infty), \tag{6.1}$$

with $A(t + \omega) \equiv A(t)$, then

$$[Y(t + \omega)]' = A(t + \omega)Y(t + \omega) = A(t)Y(t + \omega).$$

Consequently, $Y(t + \omega)$ is also a fundamental matrix for (6.1), and there exists a nonsingular constant matrix H such that

$$Y(t + \omega) \equiv Y(t)H, \qquad t \in (-\infty, \infty). \tag{6.2}$$

If $Y(t)$, $Y_1(t)$ are two fundamental matrices for (6.1), and H, H_1 are corresponding nonsingular constant matrices satisfying $Y(t + \omega) \equiv Y(t)H$, $Y_1(t + \omega) \equiv Y_1(t)H_1$, then since there is a constant matrix C such that $Y_1(t) = Y(t)C$ it follows that

$$Y_1(t + \omega) = Y(t + \omega)C = Y(t)HC = Y_1(t)[C^{-1}HC],$$

so that $H_1 = C^{-1}HC$. Conversely, if $Y(t)$ is a fundamental matrix for (6.1) satisfying (6.2), and $H_1 = C^{-1}HC$, then $Y_1(t) = Y(t)C$ is a fundamental matrix for (6.1) satisfying $Y_1(t + \omega) \equiv Y_1(t)H_1$. Thus the totality of constant matrices H satisfying (6.2) with some fundamental matrix $Y(t)$ for (6.1) consists of any one such matrix H and all matrices $C^{-1}HC$ similar to H.

If H is a constant matrix satisfying (6.2) with a fundamental matrix $Y(t)$ for (6.1), then a proper value μ for H of index k and multiplicity m is termed a (*characteristic*) *multiplier* for the equation (6.1), or for the periodic matrix $A(t)$, of index k and multiplicity m. In view of the above comments, the multipliers thus defined are uniquely determined by $A(t)$, and are independent of the particular constant matrix H occurring in (6.2); in this connection the reader is referred to the initial paragraphs in Section 1 of Appendix E. For μ a multiplier, and η a corresponding nonzero vector satisfying $(\mu E - H)\eta = 0$, the vector function $y(t) = Y(t)\eta$ is a nontrivial solution of (6.1) satisfying

$$y(t + \omega) \equiv \mu y(t), \qquad t \in (-\infty, \infty). \tag{6.3}$$

In particular, (6.1) possesses a nontrivial solution of period ω if and only if $\mu = 1$ is a multiplier for this equation. In general, by Theorem 3.3 of Appendix E there is a nonsingular matrix C such that $H_1 = C^{-1}HC$ is of the form $\operatorname{diag}\{J(M_1; \tau_1), \ldots, J(M_\rho; \tau_\rho)\}$, where $M_1, \ldots, M_\rho$ is the sequence of multipliers for (6.1) with each repeated a number of times equal to its index, and for μ a multiplier of index k and multiplicity m the set $\beta_1, \ldots, \beta_k$ of values β for which $M_\beta = \mu$ is such that $\tau_{\beta_1} + \cdots + \tau_{\beta_k} = m$. Consequently, if $g_0 = 0$, $g_\gamma = \tau_1 + \cdots + \tau_\gamma$, $(\gamma = 1, \ldots, \rho)$, then the column vectors $y^{(1)}, \ldots, y^{(n)}$ of $Y_1(t) = Y(t)C$ satisfy the following relations for $\gamma = 0, \ldots, \rho - 1$:

$$\begin{aligned} y^{(g_\gamma+1)}(t + \omega) &= M_{\gamma+1}\, y^{(g_\gamma+1)}(t), \\ y^{(j)}(t + \omega) &= M_{\gamma+1}\, y^{(j)}(t) + y^{(j-1)}(t), \qquad g_\gamma + 1 < j \le g_{\gamma+1}. \end{aligned} \tag{6.4}$$

Conditions on the asymptotic nature of solutions of (6.1) as $t \to \infty$, or as $t \to -\infty$, may be phrased in terms of the multipliers for this equation. It is to be remarked, however, that one need consider explicitly only the criteria for behavior as $t \to \infty$, since for $y(t)$ a solution of (6.1) the behavior of $y(t)$ as $t \to -\infty$ is specified by the behavior of $y(-t)$ as $t \to \infty$, while $y(t)$ satisfies (6.1) if and only if $u(t) = y(-t)$ satisfies the equation

$$u' = [-A(-t)]u, \qquad t \in (-\infty, \infty). \tag{6.5}$$

In particular, (6.1) is said to possess a given type of stability on an interval of the form $(-\infty, a_1]$ whenever (6.5) possesses this type of stability on an interval of the form $[a, \infty)$.

Now $A(t)$ has period ω if and only if $-A(-t)$ has period ω, and if $Y(t)$ is a fundamental matrix for (6.1) satisfying (6.2) with the constant matrix H, then $U(t) = Y(-t)$ is a fundamental matrix for (6.5) and $U(t + \omega) \equiv Y(-t - \omega) \equiv Y(-t)H^{-1} \equiv U(t)H^{-1}$. The following result is then an immediate consequence of the fact that $\mu = \mu_0$ is a proper value for H of index k and multiplicity m if and only if $\mu = 1/\mu_0$ is a proper value for H^{-1} of index k and multiplicity m.

Lemma 6.1. *A value* $\mu = \mu_0$ *is a multiplier for* (6.1) *of index* k *and multiplicity* m *if and only if* $\mu = 1/\mu_0$ *is a multiplier for* (6.5) *of index* k *and multiplicity* m.

It is to be remarked also that $A(t)$ has period ω if and only if $-A^*(t)$ has period ω, and that $Y(t)$ is a fundamental matrix for (6.1) if and only if $Z(t) = Y^{*-1}(t)$ is a fundamental matrix for the differential equation

$$z' = -A^*(t)z, \qquad t \in (-\infty, \infty), \tag{6.6}$$

adjoint to (6.1). Moreover, if H is the constant matrix satisfying (6.2) with $Y(t)$, then $Z(t + \omega) \equiv Y^{*-1}(t + \omega) \equiv Y^{*-1}(t)H^{*-1} \equiv Z(t)H^{*-1}$, and the following result is an immediate consequence of the fact that $\mu = \mu_0$ is a proper value for H of index k and multiplicity m if and only if $\mu = 1/\bar{\mu}_0$ is a proper value for H^{*-1} of index k and multiplicity m.

Lemma 6.2. *A value* $\mu = \mu_0$ *is a multiplier for* (6.1) *of index* k *and multiplicity* m *if and only if* $\mu = 1/\bar{\mu}_0$ *is a multiplier for* (6.6) *of index* k *and multiplicity* m.

The basic connection between differential equations (6.1) with periodic $A(t)$ and linear differential equations with constant coefficients is given in the following theorem, which is due to G. Floquet.

Theorem 6.1. *Any fundamental matrix* $Y(t)$ *for* (6.1) *may be written as*

$$Y(t) = P(t)\exp\{tS\}, \tag{6.7}$$

where $P(t)$ *is a nonsingular matrix of period* ω, *and* S *is a constant matrix; moreover, if* $P(t)$ *and* S *satisfy* (6.7) *with a fundamental matrix* $Y(t)$ *of* (6.1), *then* $P'(t) + P(t)S - A(t)P(t) = 0$ *for* $t \in (-\infty, \infty)$, *and under the transformation*

$$y(t) = P(t)w(t), \qquad t \in (-\infty, \infty), \tag{6.8}$$

the differential equation (6.1) *reduces to*

$$w' = Sw, \qquad t \in (-\infty, \infty). \tag{6.9}$$

Let $Y(t)$ be a fundamental matrix for 6.1, and H the corresponding matrix satisfying (6.2). Since H is nonsingular, it follows from Problem 6 in E.3 of Appendix E that there is a constant matrix S such that $H = \exp\{\omega S\}$. Now if $P(t)$ is defined as $P(t) = Y(t)\exp\{-tS\}$, then

$$P(t + \omega) = Y(t + \omega)\exp\{-(t + \omega)S\} =$$
$$Y(t)\exp\{\omega S\}\exp\{-(t + \omega)S\} = Y(t)\exp\{-tS\} = P(t),$$

so that $P(t)$ has period ω and $Y(t)$ is given by (6.7). From the relations $Y'(t) = A(t)Y(t)$ and $[\exp\{-tS\}]' = -[\exp\{-tS\}]S$ it follows that

$$P'(t) + P(t)S - A(t)P(t) = 0,$$

and $S = P^{-1}(t)[A(t)P(t) - P'(t)]$ for $t \in (-\infty, \infty)$, and consequently (6.1) reduces to (6.9) under the transformation (6.8).

Suppose that $Y(t)$ is a fundamental matrix for (6.1) satisfying (6.2) with the nonsingular constant matrix H, and S is such that $H = \exp\{\omega S\}$. By Theorem 3.3 of Appendix E it follows that S is of the form $C^{-1}S_1C$, with S_1 of the canonical form

$$S_1 = \operatorname{diag}\{J(\sigma_1; \tau_1), \ldots, J(\sigma_\rho; \tau_\rho)\}, \tag{6.10}$$

where $\sigma_1, \ldots, \sigma_\rho$ is the sequence of proper values for S with each repeated a number of times equal to its index, and for σ a proper value for S of index k and multiplicity m the set of $\beta_1, \ldots, \beta_k$ of indices β for which $\sigma_\beta = \sigma$ is such that $\tau_{\beta_1} + \cdots + \tau_{\beta_k} = m$. From the relations

$$H = \exp\{\omega S\} = C^{-1}(\exp\{\omega S_1\})C,$$

and

$$\exp\{\omega S_1\} = \operatorname{diag}\{e^{\omega\sigma_1} E(\omega; \tau_1), \ldots, e^{\omega\sigma_\rho} E(\omega; \tau_\rho)\},$$

with $E(\omega; \tau) = \exp\{\omega J(0; \tau)\}$, it then follows that if $\Sigma_1, \ldots, \Sigma_r$ are the distinct proper values of S with Σ_α of index K_α and multiplicity M_α, $(\alpha = 1, \ldots, r)$, and $\alpha_1, \ldots, \alpha_g$ are those values of α for which $e^{\omega\Sigma_\alpha}$ has a common value μ, then μ is a proper value of $H = \exp\{\omega S\}$ of index $K_{\alpha_1} + \cdots + K_{\alpha_g}$ and multiplicity $M_{\alpha_1} + \cdots + M_{\alpha_g}$; in particular, μ is a proper value for H with index equal to multiplicity if and only if $K_{\alpha_j} = M_{\alpha_j}$, $(j = 1, \ldots, g)$. It is to be noted also that if $H = \exp\{\omega S\}$ with $S = C^{-1}S_1C$, and S_1 given by (6.10), then for $\sigma_j^0 = \sigma_j + 2n_j\pi i/\omega$, $(j = 1, \ldots, \rho)$, with n_j an arbitrary integer, and $S_1^0 = \operatorname{diag}\{J(\sigma_1^0; \tau_1), \ldots, J(\sigma_\rho^0; \tau_\rho)\}$, we have that $S^0 = C^{-1}S_1^0C$ is such that $H = \exp\{\omega S^0\}$ also.

If S is a constant matrix such that a fundamental matrix for (6.1) has the representation (6.7) with $P(t)$ a nonsingular matrix that has period ω, then a proper value σ for S of index k and multiplicity m is termed a (*characteristic*) *exponent* for (6.1), or for the periodic matrix $A(t)$, of index k and multiplicity m. In view of the remarks of the preceding paragraph it is clear that the characteristic exponents for $A(t)$ are not determined uniquely, but are specified to within integral multiples of $2\pi i/\omega$. In particular, however, a multiplier μ for $A(t)$ satisfies $|\mu| > 1$, $|\mu| < 1$, or $|\mu| = 1$, if and only if an arbitrary corresponding exponent σ for $A(t)$ satisfies the respective relation $\Re\sigma > 0$, $\Re\sigma < 0$, or $\Re\sigma = 0$. Moreover, a multiplier μ has index equal to multiplicity if and only if each characteristic exponent σ satisfying $e^{\omega\sigma} = \mu$ has index equal to multiplicity.

If (6.7) is a representation for a fundamental matrix $Y(t)$ of (6.1), then $Y(t)Y^{-1}(\tau) = P(t)[\exp\{(t-\tau)S\}]P^{-1}(\tau)$. Now the periodicity of the non-singular matrix $P(t)$ implies the existence of constants κ_1, κ_2 such that

$$(6.11) \qquad \nu[P(t)] \leq \kappa_1, \qquad \nu[P^{-1}(t)] \leq \kappa_2, \qquad t \in (-\infty, \infty),$$

and consequently on a given interval of the form $[a, \infty)$ or $(-\infty, a_1]$ the differential equation (6.1) possesses a specified type of stability if and only if the corresponding differential equation (6.9) has this type of stability on the interval of consideration. In view of Theorem 2.4, we have the following result.

Theorem 6.2. *The differential equation* (6.1) *is stable*, (*uniformly stable also*), *on an interval of the form* $[a, \infty)$, $\{(-\infty, a_1]\}$, *if and only if all multipliers* μ *for* $A(t)$ *are such that* $|\mu| \leq 1$, $\{|\mu| \geq 1\}$, *and all multipliers* μ *with* $|\mu| = 1$ *have index equal to multiplicity; moreover*, (6.1) *is strictly stable on* $[a, \infty)$ *or* $(-\infty, a_1]$ *if and only if all multipliers* μ *have* $|\mu| = 1$ *and index equal to multiplicity. Finally*, (6.1) *is asymptotically stable on* $[a, \infty)$, $\{(-\infty, a_1]\}$, *if and only if all multipliers* μ *for this equation satisfy* $|\mu| < 1$, $\{|\mu| > 1\}$.

The following Corollary is an immediate consequence of the above theorem and Lemmas 6.1 and 6.2.

***Corollary* 1.** *The differential equation* (6.6) *adjoint to* (6.1) *possesses a specified type of stability on an interval of the form* $[a, \infty)$ *if and only if* (6.1) *has this type of stability on an interval of the form* $(-\infty, a_1]$.

For brevity a differential equation (6.1) will be said to be *stable on* $(-\infty, \infty)$ if it is stable on all intervals of the form $[a, \infty)$ or $(-\infty, a_1]$; with this definition the following result is a direct consequence of the above theorem.

***Corollary* 2.** *The differential equation* (6.1) *is stable on* $(-\infty, \infty)$ *if and only if it is strictly stable on some interval of the form* $[a, \infty)$; *moreover, if* (6.1) *is stable on* $(-\infty, \infty)$ *then it is uniformly stable on* $(-\infty, \infty)$ *in the sense that for a given* $\varepsilon > 0$ *there is a* $\delta_\varepsilon > 0$ *such that if* $y(t)$ *is a solution of* (6.1) *for which* $|y(\tau)| < \delta_\varepsilon$ *at some* $\tau \in (-\infty, \infty)$ *then* $|y(t)| < \varepsilon$ *for* $t \in (-\infty, \infty)$.

Now consider a differential equation

$$(6.12) \qquad y' = [A(t) + C(t)]y, \qquad t \in (-\infty, \infty),$$

with $A(t+\omega) \equiv A(t)$, and $A(t)$, $C(t)$ continuous matrix functions on $(-\infty, \infty)$. If (6.7) is a representation for a fundamental matrix $Y(t)$ of (6.1), then under the transformation (6.8) the differential equation (6.12) reduces to

$$(6.13) \qquad w' = [S + D(t)]w, \qquad t \in (-\infty, \infty),$$

with $D(t) = P^{-1}(t)C(t)P(t)$. In particular, if κ_1, κ_2 are constants satisfying (6.11), then $\nu[D(t)] \leq \kappa_1\kappa_2\nu[C(t)]$ and $\nu[C(t)] \leq \kappa_1\kappa_2\nu[D(t)]$, so that $\nu[D(t)]$ is integrable on an interval $[a, \infty)$, $(-\infty, a_1]$, or $(-\infty, \infty)$, if and only if $\nu[C(t)]$ is integrable on this same interval. Consequently the following result follows from Theorem 4.1, together with the above Theorem 6.2 and its Corollary 2.

Theorem 6.3. *If the differential equation* (6.1) *is stable on an interval* $[a, \infty)$, $(-\infty, a_1]$, *or* $(-\infty, \infty)$, *and* $C(t)$ *is a continuous matrix function with* $\nu[C(t)]$ *integrable on this interval, then the equation* (6.12) *is uniformly stable on the interval of consideration.*

Correspondingly, the results of Theorems 4.3 and 6.2 imply the following result.

Theorem 6.4. *If on an interval* $[a, \infty)$ *or* $(-\infty, a_1]$ *the differential equation* (6.1) *is asymptotically stable and* $\nu[C(t)]$ *is integrable, then on this interval the differential equation* (6.12) *is uniformly and asymptotically stable.*

Problems VIII.6

1. Suppose that $A(t) = [A_{\alpha\beta}(t)]$, $(\alpha, \beta = 1, 2)$, is a 2×2 real-valued continuous matrix function which have period ω on $(-\infty, \infty)$, and that $A_{11}(t) + A_{22}(t) \equiv 0$. If $Y(t)$ is the fundamental matrix of

$$y' = A(t)y, \qquad t \in (-\infty, \infty), \tag{6.14}$$

satisfying $Y(0) = E$, and $K(\omega) = \frac{1}{2}[Y_{11}(\omega) + Y_{22}(\omega)]$, show that:

(i) if $-1 < K(\omega) < 1$, then all solutions $y(t)$ of (6.14) are bounded on $(-\infty, \infty)$;

(ii) if $K(\omega) > 1$, or $K(\omega) < -1$, then no solution of (6.14) is bounded on $(-\infty, \infty)$ except $y(t) \equiv 0$;

(iii) if $K(\omega) = 1$, then at least one solution of (6.14) has period ω;

(iv) if $K(\omega) = -1$, then at least one solution of (6.14) has period 2ω.

Hint. Note that $K(\omega)$ is real, $\det Y(t) \equiv 1$, and that μ is a multiplier for (6.14) if and only if $\mu^2 - 2K(\omega)\mu + 1 = 0$.

2. For the linear second order differential equation $(r(t)u')' + p(t)u = 0$, with $r(t)$, $p(t)$ real-valued continuous functions of period ω on $(-\infty, \infty)$ and $r(t) > 0$ on this interval, state in detail the results on the behavior of solutions on $(-\infty, \infty)$ that are obtained by applying the results of Problem 1 to the first order system $y_1' = \{1/r(t)\}y_2$, $y_2' = -p(t)y_1$.

3. If σ and ω are positive real constants such that $\sigma\omega \neq k\pi$, $(k = 1, 2, \ldots)$, show that there exists a $\delta > 0$ such that for $g(t)$, $h(t)$ any real-valued functions of period ω satisfying $|g(t)| \leq \delta$, $|h(t)| \leq \delta$ on $(-\infty, \infty)$ all solutions $y = (y_\alpha(t))$, $(\alpha = 1, 2,)$, of

$$(6.15) \qquad y_1' = [\sigma + g(t)]y_2, \qquad y_2' = -[\sigma + h(t)]y_1,$$

are bounded on $(-\infty, \infty)$.

Hint. For the system $(6.15_\circ)$: $y_1' = \sigma\, y_2$, $y_2' = -\sigma\, y_1$ note that the function $K(\omega)$ of Problem 1 is equal to $\cos \sigma\omega$, and hence for this equation case (i) of that problem holds whenever $\sigma\omega \neq k\pi$, $(k = 1, 2, \ldots)$. With the aid of Corollary 2 to Theorem I.2.1 show that if $y(t)$ and $y^{(0)}(t)$ are solutions of (6.15) and $(6.15_\circ)$, respectively, satisfying $y(0) = y^{(0)}(0)$ then

$$|y(t) - y^{(0)}(t)| \leq \delta\, |y^{(0)}(0)| \exp\{(\sigma + \delta)\omega\}$$

on $[0, \omega]$ whenever $|g(t)| \leq \delta$, $|h(t)| \leq \delta$, and proceed to establish the stated result by a continuity argument.

4. Establish Theorem 6.2 without using the result of Theorem 6.1.

Hint. Note that the relations (6.4) imply

$$y^{(g_\gamma+\beta)}(t + p\omega) = \sum_{j=0}^{\beta-1} {}_pC_j\, M_{\gamma+1}^{p-j}\, y^{(g_\gamma+\beta-j)}(t), \quad \textit{for} \quad 1 \leq \beta \leq \tau_{\gamma+1}$$
$$\textit{and} \quad p = \beta, \beta + 1, \ldots.$$

5. Suppose that all multipliers μ of (6.1) are such that $|\mu| \geq 1$, and all multipliers μ with $|\mu| = 1$ have index equal to multiplicity. If $C(t)$ is continuous and $\nu[C(t)]$ is integrable on $[a, \infty)$, show that the solutions of (6.1) and (6.12) which are bounded on $[a, \infty)$ are in asymptotic one-to-one correspondence.

Hint. Apply Theorem 5.2 to the equation (6.13) and the equation $w' = Sw$.

6. If (6.1) has n independent solutions of the form $e^{\sigma_\beta t}\, p^{(\beta)}(t)$, where $p^{(\beta)}(t)$, $(\beta = 1, \ldots, n)$, are vector functions of period ω, while $C(t)$ is continuous and $\nu[C(t)]$ is integrable on $[a, \infty)$, show that there exist n linearly independent solutions $y^{(\beta)}(t)$ of (6.12) such that $e^{-\sigma_\beta t}\, y^{(\beta)}(t) - p^{(\beta)}(t) \to 0$ as $t \to \infty$, $(\beta = 1, \ldots, n)$.

Hint. Note that the hypotheses of the problem imply that (6.1) has a fundamental matrix of the form $P(t)\exp\{tS\}$, where $P(t)$ is the nonsingular periodic matrix with column vectors $p^{(1)}(t), \ldots, p^{(n)}(t)$, and $S = [\sigma_\alpha \delta_{\alpha\beta}]$. To the corresponding equation (6.13) apply the result of the Corollary to Theorem 5.4.

7. STABILITY OF NONLINEAR DIFFERENTIAL EQUATIONS

We shall consider now some results on the stability of nonlinear vector differential equations

$$(7.1) \qquad y' = f(t, y),$$

where $y \equiv (y_\alpha(t))$ and $f(t, y) \equiv (f_\alpha(t, y))$, $(\alpha = 1, \ldots, n)$. For simplicity of statement of conditions it will be assumed in the following discussion that the vector function $f(t, y)$ is continuous in (t, y) on

$$(7.2) \qquad \mathfrak{D} = \{(t, y) \mid t \in [a, \infty),\, y \in \mathbf{C}_n\},$$

unless specifically stated otherwise.

***Lemma* 7.1.** *Suppose that for each b on (a, ∞) there are corresponding constants $\kappa(b)$, $\kappa_1(b)$ such that*

$$(7.3) \qquad |f(t, y)| \leq \kappa(b)\, |y| + \kappa_1(b) \quad \textit{for} \quad t \in [a, b], \qquad y \in \mathbf{C}_n.$$

If $y = \phi(t)$ is a given solution of equation (7.1) on a nondegenerate subinterval I of $[a, \infty)$, then there exists a solution of (7.1) on $[a, \infty)$ such that $y(t) = \phi(t)$ for $t \in I$; in particular, if $(\tau, \eta) \in \mathfrak{D}$ there is a solution of (7.1) on $[a, \infty)$ passing through the point (τ, η).

If b is a value on (a, ∞) such that $I \subset [a, b]$, then from Theorem I.3.3 it follows that the solution $y = \phi(t)$, $t \in I$, of (7.1) may be extended successively to the intervals $[a, b + m]$, $(m = 0, 1, \ldots)$, and hence there is a solution $y(t)$ of (7.1) on $[a, \infty)$ satisfying $y(t) = \phi(t)$ for $t \in I$. Now if $(\tau, \eta) \in \mathfrak{D}$, then by Theorem I.3.4 there is a solution $y = \phi(t)$ of (7.1) on an interval I containing $t = \tau$, and such that $\phi(\tau) = \eta$; application of the first conclusion of the present lemma to this $\phi(t)$ provides a solution $y(t)$ of (7.1) on $[a, \infty)$ that passes through the initial point (τ, η).

Of particular interest is the case of differential equations (7.1) of the form

$$(7.4) \qquad y' = A(t)y + g(t, y),$$

where $A(t)$ is an $n \times n$ matrix function which is continuous on $[a, \infty)$, while $g(t, y)$ is a vector function that is continuous in (t, y) on the set $\mathfrak{D}$ of (7.2). One would expect that under suitable conditions there would be an intimate relation between the solutions of (7.4) and solutions of the corresponding linear homogeneous equation

$$(7.5) \qquad y' = A(t)y, \qquad t \in [a, \infty);$$

in particular, the equations (3.1) and (4.1) of the present chapter are the special instances of (7.4) with $g(t, y) \equiv g(t)$ and $g(t, y) \equiv C(t)y$, respectively.

***Lemma* 7.2.** *Suppose that there exist non-negative continuous functions* $\theta(t)$, $\theta_1(t)$ *which are integrable on* $[a, \infty)$, *and such that*

$$(7.6)\qquad |g(t, y)| \leq \theta(t)\,|y| + \theta_1(t), \quad \textit{for} \quad (t, y) \in \mathfrak{D}.$$

If the linear differential equation (7.5) *is uniformly stable on* $[a, \infty)$, *then for any solution* $y = \phi(t)$ *of* (7.4) *on a nondegenerate subinterval* I *of* $[a, \infty)$ *the interval of definition may be extended to* $[a, \infty)$, *and the vector function* $y(t)$ *defining any such extension is bounded on* $[a, \infty)$; *moreover, if* (7.5) *is uniformly and asymptotically stable on* $[a, \infty)$, *then a solution* $y(t)$ *of* (7.4) *on* $[a, \infty)$ *is such that* $y(t) \to 0$ *as* $t \to \infty$.

The condition that $g(t, y)$ satisfies (7.6) with continuous functions $\theta(t)$, $\theta_1(t)$ implies that $f(t, y) = A(t)y + g(t, y)$ satisfies (7.3) with $\kappa(b)$ and $\kappa_1(b)$ upper bounds on $[a, b]$ for $\nu[A(t)] + \theta(t)$ and $\theta_1(t)$, respectively. Consequently, by Lemma 7.1 it follows that for any solution $y = \phi(t)$, $t \in I$, the interval of definition may be extended to $[a, \infty)$. For $y(t)$ the vector function defining such an extension we then have

$$(7.7)\qquad y(t) = Y(t)Y^{-1}(\tau)y(\tau) + \int_\tau^t Y(t)Y^{-1}(s)g(s, y(s))\,ds,$$

for $(t, \tau) \in [a, \infty) \times [a, \infty)$, where $Y(t)$ is a fundamental matrix for the homogeneous equation (7.5). Since the condition that (7.5) be uniformly stable on $[a, \infty)$ implies that there is a constant κ such that

$$\nu[Y(t)Y^{-1}(s)] \leq \kappa$$

for $s \in [a, t]$, $t \in [a, \infty)$, it follows from (7.7) that

$$(7.8)\quad |y(t)| \leq \kappa\,|y(\tau)| + \int_\tau^t \kappa[\theta(s)\,|y(s)| + \theta_1(s)]\,ds, \quad \tau \in [a, t], \quad t \in [a, \infty).$$

and application of Theorem I.2.1 then yields the inequality

$$(7.9)\quad |y(t)| \leq \kappa\left[|y(\tau)| + \int_\tau^t \theta_1(s)\,ds\right]\exp\left\{\kappa\int_\tau^t \theta(s)\,ds\right\}, \quad \tau \in [a, t], \quad t \in [a, \infty).$$

In particular, in view of the integrability of $\theta(t)$ and $\theta_1(t)$ on $[a, \infty)$, inequality (7.9) with $\tau = a$ gives the bound

$$(7.10)\quad |y(t)| \leq \kappa\left[|y(a)| + \int_a^\infty \theta_1(s)\,ds\right]\exp\left\{\kappa\int_a^\infty \theta(s)\,ds\right\}, \qquad t \in [a, \infty).$$

Moreover, inequality (7.6) implies for a solution $y(t)$ of (7.4) on $[a, \infty)$ that $g(t) = g(t, y(t))$ is a vector function with $|g(t)|$ integrable on this interval, and hence from Theorem 3.2 it follows that $y(t) \to 0$ as $t \to \infty$ whenever the homogeneous equation (7.5) is uniformly and asymptotically stable on $[a, \infty)$.

Theorem 7.1. *Suppose that there exists a non-negative continuous function* $\theta(t)$ *which is integrable on* $[a, \infty)$ *and such that*

$$|g(t, y)| \leq \theta(t)\,|y|, \quad \textit{for} \quad (t, y) \in \mathfrak{D}. \tag{7.11}$$

If (7.5) *is uniformly stable on* $[a, \infty)$, *then* $y(t) \equiv 0$ *is a solution of* (7.4) *that is uniformly stable on* $[a, \infty)$; *moreover, if* (7.5) *is uniformly and asymptotically stable on* $[a, \infty)$, *then* $y(t) \equiv 0$ *is an asymptotically stable solution of* (7.4).

The condition (7.11) clearly implies that $g(t, 0) \equiv 0$ on $[a, \infty)$, and consequently that $y(t) \equiv 0$ is a solution of (7.4). From Lemma 7.2 it follows that if $(\tau, \eta) \in \mathfrak{D}$ and $y(t)$ is any solution of (7.4) satisfying $y(\tau) = \eta$, then the interval of definition of $y(t)$ may be extended to $[a, \infty)$, and for any such extension $|y(t)|$ is bounded on $[a, \infty)$. Indeed, inequality (7.9) implies that

$$|y(t)| \leq \kappa\, |y(\tau)| \exp\left\{\kappa \int_a^\infty \theta(s)\, ds\right\}, \qquad \tau \in [a, t], \quad t \in [a, \infty),$$

and consequently $|y(t)| < \varepsilon$ on $[\tau, \infty)$ whenever

$$|y(\tau)| < (\varepsilon/\kappa)\exp\left\{-\kappa \int_a^\infty \theta(s)\, ds\right\},$$

thus establishing the conclusion that $y(t) \equiv 0$ is a uniformly stable solution of (7.4) on $[a, \infty)$. The final conclusion of the theorem is an immediate consequence of the above Lemma 7.2, and the fact that $y(t) \equiv 0$ is a solution of (7.4).

Theorem 7.2. *Suppose that* $g(t, y)$ *is such that* $|g(t, 0)|$ *is integrable on* $[a, \infty)$, *and there is a non-negative continuous function* $\theta(t)$ *that is integrable on* $[a, \infty)$ *and for which*

$$|g(t, y) - g(t, z)| \leq \theta(t)\,|y - z| \quad \textit{for} \quad (t, y) \in \mathfrak{D}, \qquad (t, z) \in \mathfrak{D}. \tag{7.12}$$

If (7.5) *is uniformly stable on* $[a, \infty)$, *then each solution* $y(t)$ *of* (7.4) *is bounded and uniformly stable on* $[a, \infty)$; *moreover, if* (7.5) *is uniformly and asymptotically stable on* $[a, \infty)$, *then each solution* $y(t)$ *of* (7.4) *is such that* $y(t) \to 0$ *as* $t \to \infty$.

As (7.12) implies that $g(t, y)$ is locally Lipschitzian on $\mathfrak{D}$, the solutions of (7.4) are locally unique on $\mathfrak{D}$ whenever (7.12) holds. Moreover, since it follows from (7.12) that $|g(t, y)| \leq \theta(t)\,|y| + |g(t, 0)|$, whenever $|g(t, 0)|$ and $\theta(t)$ are integrable on $[a, \infty)$ it is a consequence of Lemma 7.2 that each solution $y(t)$ of (7.4) is bounded on $[a, \infty)$. Now if $y = \phi(t)$ and $y = y(t)$ are solutions of (7.4), then $w(t) = y(t) - \phi(t)$ satisfies the differential equation

$$w' = A(t)w + g(t, y(t)) - g(t, \phi(t)), \qquad t \in [a, \infty).$$

As (7.12) implies that $|g(t, y(t)) - g(t, \phi(t))| \leq \theta(t)\,|w(t)|$ on $[a, \infty)$, by argument similar to that used in the proof of inequality (7.9) it follows that

$$(7.13)\quad |w(t)| \leq \kappa\,|w(\tau)| \exp\left\{\kappa \int_a^\infty \theta(s)\,ds\right\}, \quad \textit{for} \quad \tau \in [a, t], \quad t \in [a, \infty),$$

where κ is a constant such that a fundamental matrix $Y(t)$ of (7.5) satisfies $\nu[Y(t)Y^{-1}(s)] \leq \kappa$ for $s \in [a, t]$, $t \in [a, \infty)$. Inequality (7.13) clearly implies that $|y(t) - \phi(t)| < \varepsilon$ for $t \in [\tau, \infty)$ whenever

$$|y(\tau) - \phi(\tau)| < (\varepsilon/\kappa)\exp\left\{-\kappa \int_a^\infty \theta(s)\,ds\right\},$$

so that $y = \phi(t)$ is a solution of (7.4) which is uniformly stable on $[a, \infty)$. The final conclusion of the theorem is an obvious consequence of the corresponding result of Lemma 7.2.

Theorem 7.3. *Suppose that* $g(t, y)$ *is such that* $g(t, 0) \equiv 0$ *on* $[a, \infty)$, *and there is a non-negative continuous function* $\theta(t)$ *which is integrable on* $[a, \infty)$ *and such that* (7.12) *holds. If the differential equation adjoint to* (7.5) *is uniformly stable, then the solutions* $y(t)$ *of* (7.4) *and* (7.5) *for which* $\theta(t)\,|y(t)|$ *is integrable on* $[a, \infty)$ *are in asymptotic one-to-one correspondence.*

Suppose that $y(t)$ is a solution of (7.4) for which $\theta(t)\,|y(t)|$ is integrable on $[a, \infty)$. If $Y(t)$ is a fundamental matrix for (7.5), then the condition that the equation adjoint to (7.5) be uniformly stable on $[a, \infty)$ implies the existence of a constant κ such that $\nu[Y(t)Y^{-1}(s)] \leq \kappa$ for $t \in [a, s]$, $s \in [a, \infty)$. Now the condition $g(t, 0) \equiv 0$ and (7.12) provide the inequality $|g(t, y)| \leq \theta(t)\,|y|$ on $\mathfrak{D}$, and consequently if $y(t)$ is a solution of (7.4) with $\theta(t)\,|y(t)|$ integrable on $[a, \infty)$ then $Y(t)Y^{-1}(s)g(s, y(s))$ is an integrable vector function of s on each interval $[t, \infty)$. From the relation (7.7) it then follows that

$$(7.14)\qquad y(t) = y^{(0)}(t) - \int_t^\infty Y(t)Y^{-1}(s)g(s, y(s))\,ds, \qquad t \in [a, \infty),$$

where $y^{(0)}(t)$ is the solution of (7.5) given by

$$(7.15)\quad y^{(0)}(t) = Y(t)\xi, \quad \textit{with} \quad \xi = Y^{-1}(\tau)y(\tau) + \int_\tau^\infty Y^{-1}(s)g(s, y(s))\,ds, \qquad \tau \in [a, \infty).$$

Consequently, $|y(t) - y^{(0)}(t)| \leq \kappa \int_t^\infty \theta(s)\,|y(s)|\,ds$, and $|y(t) - y^{(0)}(t)| \to 0$ as $t \to \infty$. Moreover, in view of the inequality

$$\theta(t)\,|y^{(0)}(t)| \leq \theta(t)\,|y^{(0)}(t) - y(t)| + \theta(t)\,|y(t)|$$

and the integrability of $\theta(t)$ and $\theta(t)\,|y(t)|$, it follows that the function $\theta(t)\,|y^{(0)}(t)|$ is integrable on $[a, \infty)$. Finally, as in the proof of Theorem 5.2, the uniform stability of the equation adjoint to (7.5) implies that two solutions $y^{(1)}(t)$ and $y^{(2)}(t)$ of (7.5) are identical whenever $|y^{(1)}(t) - y^{(2)}(t)| \to 0$ as $t \to \infty$, and hence (7.15) is the unique solution $y^{(0)}(t)$ of (7.5) such that $|y(t) - y^{(0)}(t)| \to 0$ as $t \to \infty$. It is to be noted that in the proof of this portion of the theorem one has not used the full implication of (7.12), but merely the inequality (7.11).

Now suppose that $y = y^{(0)}(t)$ is a solution of (7.5) such that $\theta(t)\,|y^{(0)}(t)|$ is integrable on $[a, \infty)$. Under the hypotheses of the theorem the function $\phi(t) = \kappa\theta(t)\max\{1, |y^{(0)}(t)|\}$ is integrable on $[a, \infty)$, and since

$$\nu[Y(t)Y^{-1}(s)g(s, y^{(0)}(s))] \leq \kappa\theta(s)\,|y^{(0)}(s)| \leq \phi(s) \qquad \text{for} \quad t \in [a, s], \qquad s \in [a, \infty),$$

it follows that for $j = 0$ the recursion relation

$$\text{(7.16)} \quad y^{(j+1)}(t) = y^{(0)}(t) - \int_t^\infty Y(t)Y^{-1}(s)g(s, y^{(j)}(s))\,ds, \qquad t \in [a, \infty),$$

defines a vector function $y^{(1)}(t)$ that is continuous on $[a, \infty)$. By induction it follows that for $j = 1, 2, \ldots$ there is a continuous vector function $y^{(j)}(t)$ satisfying (7.16), and the relations

$$\text{(7.17)} \quad \begin{aligned} &\text{(a)} \quad |y^{(j)}(t) - y^{(j-1)}(t)| \leq \frac{1}{j!}\left(\int_t^\infty \phi(s)\,ds\right)^j, \\ &\text{(b)} \quad |y^{(j)}(t) - y^{(0)}(t)| \leq \sum_{k=1}^{j} \frac{1}{k!}\left(\int_t^\infty \phi(s)\,ds\right)^k \leq \exp\left\{\int_t^\infty \phi(s)\,ds\right\} - 1, \\ &\text{(c)} \quad |y^{(j)}(t)| \leq |y^{(0)}(t)| + \exp\left\{\int_a^\infty \phi(s)\,ds\right\} \\ &\qquad \leq \left[1 + \exp\left\{\int_a^\infty \phi(s)\,ds\right\}\right]\max\{1, |y^{(0)}(t)|\}. \end{aligned}$$

In particular, the condition that $y^{(j)}(t)$ is continuous and satisfies (7.17-c) implies that

$$\nu[Y(t)Y^{-1}(s)g(s, y^{(j)}(s))] \leq \left[1 + \exp\left\{\int_a^\infty \phi(r)\,dr\right]\right\}\phi(s), \quad \text{for} \quad s \in [t, \infty),$$

and consequently that $y^{(j+1)}(t)$ is a continuous vector function on $[a, \infty)$. Inequality (7.17-a) implies that the sequence $\{y^{(j)}(t)\}$ converges uniformly on $[a, \infty)$ to a limit vector function $y(t)$, which is consequently continuous on this interval; moreover,

$$\text{(7.18)} \quad \begin{aligned} |y(t) - y^{(j)}(t)| &\leq \sum_{k=1}^{\infty} \frac{1}{(j+k)!}\left(\int_t^\infty \phi(s)\,ds\right)^{j+k}, \\ &\leq \frac{\kappa_1}{(j+1)!}\left(\int_t^\infty \phi(s)\,ds\right)^{j+1}, \qquad t \in [a, \infty), \end{aligned}$$

where $\kappa_1 = \exp\{\int_a^\infty \phi(s)\,ds\}$. In view of (7.17-c) we have that $\theta(t)\,|y(t)| \leq [(1+\kappa_1)/\kappa]\phi(t)$, so that $\theta(t)\,|y(t)|$ is integrable on $[a, \infty)$. Moreover, (7.18) implies that

$$\left|\int_t^\infty Y(t)Y_1^{-1}(s)g(s, y^{(j)}(s))\,ds - \int_t^\infty Y(t)Y^{-1}(s)g(s, y(s))\,ds\right| \leq \frac{\kappa_1}{(j+2)!}\left(\int_t^\infty \phi(s)\,ds\right)^{j+2}, \qquad (j = 0, 1, \ldots).$$

Therefore $y(t)$ is a continuous vector function for which $\theta(t)\,|y(t)|$ is integrable on $[a, \infty)$, and $y(t)$ satisfies the relation (7.14) with the given solution $y^{(0)}(t)$ of (7.5); in particular, $y^{(0)}(t)$ is related to $y(t)$ by (7.15) and $y = y(t)$ is a solution of (7.4). Now suppose that $y = z(t)$ is also a solution of (7.4) which is such that $\theta(t)\,|z(t)|$ is integrable on $[a, \infty)$ and satisfies with the given solution $y^{(0)}(t)$ the condition $|z(t) - y^{(0)}(t)| \to 0$ as $t \to \infty$. By the first part of the above proof it then follows that (7.14) and (7.15) hold with $y(t)$ replaced by $z(t)$, and therefore

$$y(\tau) - z(\tau) = -\int_\tau^\infty Y(\tau)Y^{-1}(s)[g(s, y(s)) - g(s, z(s))]\,ds, \qquad \tau \in [a, \infty).$$

Consequently,

$$|y(\tau) - z(\tau)| \leq \int_\tau^\infty \kappa\theta(s)\,|y(s) - z(s)|\,ds, \qquad \tau \in [a, \infty),$$

and the non-negative function $\rho(t) = \int_t^\infty \kappa\theta(s)\,|y(s) - z(s)|\,ds$ is such that $\rho(t) \to 0$ as $t \to \infty$ and $[\rho(t)\exp\{\int_a^t \kappa\theta(s)\,ds\}]' \geq 0$, so that $\rho(t) \equiv 0$ and $z(t) \equiv y(t)$ on $[a, \infty)$. That is, if $y = y^{(0)}(t)$ is a solution of (7.5) for which $\theta(t)\,|y^{(0)}(t)|$ is integrable on $[a, \infty)$, then there is a unique solution $y(t)$ of (7.4) for which $\theta(t)\,|y(t)|$ is integrable on this interval and $|y(t) - y^{(0)}(t)| \to 0$ as $t \to \infty$, thus completing the proof of the theorem.

***Lemma* 7.3.** *If $A(t)$ is a constant matrix A with $\Re\,\lambda < 0$ for each proper value λ of A, then there exist positive constants κ_1, κ_2 such that if $g(t, y)$ is a continuous vector function satisfying with a constant κ the inequality*

$$(7.19) \qquad |g(t, y)| \leq \kappa\,|y| \quad \textit{for} \quad (t, y) \in \mathfrak{D},$$

then for an arbitrary solution $y(t)$ of the differential equation

$$(7.4') \qquad y' = Ay + g(t, y),$$

on a nondegenerate subinterval the interval of definition may be extended to $[a, \infty)$, and the vector function $y(t)$ defining any such extension satisfies the inequality

(7.20)

$$|y(t)| \leq \kappa_2\,|y(\tau)|\exp\{-(\kappa_1 - \kappa\kappa_2)(t - \tau)\}, \qquad \tau \in [a, t], \qquad t \in [a, \infty).$$

If κ_1 is a positive constant such that $\mathfrak{Re}\,\lambda < -\kappa_1$ for all proper values λ of A, then for $Y(t)$ the fundamental matrix of the differential equation

$$(7.5') \qquad y' = Ay,$$

satisfying $y(0) = E$ there is a constant $\kappa_2 > 0$ such that $\nu[Y(t)] \leq \kappa_2 e^{-\kappa_1 t}$ on $[0, \infty)$. Whenever (7.19) holds it follows from Lemma 7.1 that for any solution of (7.4′) on a nondegenerate subinterval the interval of definition may be extended to $[a, \infty)$, and the vector function $y(t)$ defining any such extension satisfies the equation

$$y(t) = Y(t - \tau)y(\tau) + \int_\tau^t Y(t - s)g(s, y(s))\, ds$$

for $(t, \tau) \in [a, \infty) \times [a, \infty)$. Consequently, for $\tau \in [a, t]$, $t \in [a, \infty)$, we have

$$|y(t)| \leq \kappa_2 e^{-\kappa_1(t-\tau)}|y(\tau)| + \int_\tau^t \kappa\kappa_2 e^{-\kappa_1(t-s)}|y(s)|\, ds,$$

and $\rho(t) = |y(t)|\, e^{\kappa_1(t-\tau)}$ satisfies the inequality

$$0 \leq \rho(t) \leq \kappa_2\, |y(\tau)| + \int_\tau^t \kappa\kappa_2\rho(s)\, ds.$$

Theorem I.2.1 then implies the inequality

$$0 \leq \rho(t) \leq \kappa_2\, |y(\tau)| \exp\{\kappa\kappa_2(t - \tau)\}, \qquad \tau \in [a, t], \qquad t \in [a, \infty),$$

which is equivalent to (7.20).

Theorem 7.4. *Suppose that the vector function $g(t, y)$ is continuous in (t, y) on $[a, \infty) \times \mathbf{C}_n$, and for each $\zeta > 0$ there are values $a_\zeta \geq a$ and $\rho_\zeta > 0$ such that*

$$(7.21) \qquad |g(t, y)| \leq \zeta\, |y| \quad \textit{for} \quad t \in [a_\zeta, \infty), \qquad |y| \leq \rho_\zeta.$$

If $A(t)$ is a constant matrix A, and the corresponding homogeneous equation (7.5′) is asymptotically stable on $[a, \infty)$, then the solution $y(t) \equiv 0$ of (7.4′) is uniformly and asymptotically stable on an interval of the form $[a_1, \infty)$.

If $A(t)$ is a constant matrix and the corresponding homogeneous equation (7.5′) is asymptotically stable, then by Theorem 2.4 all proper values λ of A satisfy $\mathfrak{Re}\,\lambda < 0$. As in the proof of Lemma 7.3, let κ_1, κ_2 be positive constants such that the fundamental matrix $Y(t)$ of (7.5′) with $Y(0) = E$ satisfies $\nu[Y(t)] \leq \kappa_2 e^{-\kappa_1 t}$ on $[0, \infty)$, and for $0 < \zeta < \kappa_1/\kappa_2$ define $g(t, y; \zeta) = g(t, y)$ for $t \in [a_\zeta, \infty)$, $|y| \leq \rho_\zeta$, and $g(t, y; \zeta) = g(t, \rho_\zeta y/|y|)$ for $t \in [a_\zeta, \infty)$, $|y| > \rho_\zeta$. It follows that $g(t, y; \zeta)$ is continuous in (t, y) and satisfies $|g(t, y; \zeta)| \leq \zeta|y|$ for $t \in [a_\zeta, \infty)$, $y \in \mathbf{C}_n$, while $g(t, y; \zeta) \equiv g(t, y)$ on

$t \in [a_\zeta, \infty)$, $|y| \leq \rho_\zeta$. From Lemma 7.3 it follows that for an arbitrary solution of

$$y' = Ay + g(t, y; \zeta),$$

on a nondegenerate subinterval of $[a, \infty)$ the interval of definition may be extended to $[a_\zeta, \infty)$, and the vector function $y(t)$ defining any such extension satisfies the inequality

(7.22)
$$|y(t)| \leq \kappa_2 |y(\tau)| \exp\{-(\kappa_1 - \zeta\kappa_2)(t - \tau)\}, \qquad \tau \in [a_\zeta, t], \qquad t \in [a, \infty).$$

As $0 < \zeta < \kappa_1/\kappa_2$, it follows from (7.22) that $|y(t)| < \varepsilon$ on $[\tau, \infty)$ whenever $|y(\tau)| < \varepsilon/\kappa_2$, and $|y(t)| \to 0$ as $t \to \infty$. In view of the above definition of $g(t, y; \zeta)$, it is seen that if $|y(t)| < \rho_\zeta$ on $[a_\zeta, \infty)$ then $y(t)$ is a solution of (7.4′), and consequently $y(t) \equiv 0$ is a solution of (7.4′) which is uniformly and asymptotically stable on $[a_\zeta, \infty)$ whenever $0 < \zeta < \kappa_1/\kappa_2$.

8. THE DIRECT METHOD OF LIAPUNOV

In this section it will be supposed that the n-dimensional vector function $g(t, u) \equiv (g_\alpha(t, u_1, \ldots, u_n))$, $(\alpha = 1, \ldots, n)$, satisfies the following condition:

(H) $g(t, u)$ is real-valued and continuous on a domain

$$\mathfrak{D}(a, k) = \{(t, u) \mid t \in [a, \infty), u \in \mathbf{R}_n, |u| \leq k\}, \tag{8.1}$$

of $\mathbf{R} \times \mathbf{R}_n$, *with*

$$g(t, 0) \equiv 0, \quad \textit{for} \quad t \in [a, \infty), \tag{8.2}$$

and the solutions of the real vector differential equation

$$u' = g(t, u), \tag{8.3}$$

are locally unique on $\mathfrak{D}(a, k)$.

A scalar function $V(t, u)$ will be termed an *admissible comparison function* for (8.3) if it is real-valued and of class $\mathfrak{C}^1$ in (t, u) on a domain $\mathfrak{D}(a_0, k_0) \subset \mathfrak{D}(a, k)$, and $V(t, 0) \equiv 0$ for $t \in [a_0, \infty)$. An admissible comparison function $W(u)$ which is independent of t is termed *positive*, {*negative*}, *semi-definite* if there exists a $k_0 \leq k$ such that $W(u) \geq 0$, $\{W(u) \leq 0\}$, for $u \in \mathbf{R}_n$ with $0 \leq |u| \leq k_0$; if such a $W(u)$ is zero only for $u = 0$, then it is called *positive*, {*negative*}, *definite*. An admissible comparison function $V(t, u)$ is said to be *positive*, {*negative*}, *definite* on $\mathfrak{D}(a_0, k_0)$ if there exists a positive definite $W(u)$ such that $V(t, u) \geq W(u)$, $\{-V(t, u) \geq W(u)\}$, for $(t, u) \in \mathfrak{D}(a_0, k_0)$.

If $V(t, u)$ is an admissible comparison function for (8.3) on $\mathfrak{D}(a_0, k_0) \subset \mathfrak{D}(a, k)$, then $V^0(t, u)$ is used to denote the scalar function

$$V^0(t, u) = V_t(t, u) + \sum_{\alpha=1}^{n} V_{u_\alpha}(t, u)\, g_\alpha(t, u). \tag{8.4}$$

If it is desired to show the dependence of $V^0(t, u)$ upon the vector function $g(t, u)$, the more descriptive notation $V^0(t, u \mid g)$ is used. The following preliminary result is a ready consequence of the above definitions.

***Lemma* 8.1.** *If $V(t, u)$, $(t, u) \in \mathfrak{D}(a_0, k_0) \subset \mathfrak{D}(a, k)$, is an admissible comparison function for* (8.3), *and $u(t)$, $t \in I \subset [a_0, \infty)$, is a solution of this differential equation with $|u(t)| < k_0$, then for $t \in I$,*

$$\frac{d}{dt} V(t, u(t)) = V^0(t, u(t)). \tag{8.5}$$

***Lemma* 8.2.** *If $W(u)$, $|u| \leq k_0$, is a definite function, and $u(t)$, $t \in [a_0, \infty)$, is a real-valued n-dimensional vector function with $|u(t)| \leq k_0$ on $[a_0, \infty)$ and such that $W(u(t)) \to 0$ as $t \to \infty$, then $u(t) \to 0$ as $t \to \infty$.*

The conclusion of this lemma is immediate, since $W(u) \neq 0$ for $0 < |u| \leq k_0$, and for $0 < \varepsilon < k_0$ the nonzero continuous function $|W(u)|$ has a positive minimum on the compact set $\{u \mid \varepsilon \leq |u| \leq k_0\}$.

The following three theorems are due to Liapunov [1], and comprise the basic results of the so-called "direct method", or "second method" of Liapunov.

Theorem 8.1. *If there exists for* (8.3) *an admissible comparison function $V(t, u)$, $(t, u) \in \mathfrak{D}(a_0, k_0) \subset \mathfrak{D}(a, k)$, which is definite, and whose related function $V^0(t, u)$ is a semi-definite function such that $V(t, u)\, V^0(t, u) \leq 0$ for $(t, u) \in \mathfrak{D}(a_0, k_0)$, then the solution $u(t) \equiv 0$ of* (8.3) *is stable.*

Without loss of generality, it may be assumed that $V(t, u)$ is positive definite on $\mathfrak{D}(a_0, k_0)$. With this understanding, the hypotheses of the theorem imply that there exists a positive definite function $W(u)$ such that $V(t, u) \geq W(u)$ for $(t, u) \in \mathfrak{D}(a_0, k_0)$. For $0 < \varepsilon \leq k_0$, let m_ε denote the minimum of $W(u)$ on the compact set $\{u \mid \varepsilon \leq |u| \leq k_0\}$, and for a given $\tau \in [a_0, \infty)$ let $\delta(\varepsilon, \tau)$ be such that $0 < \delta(\varepsilon, \tau) < k_0$ and $|V(\tau, \eta)| < m_\varepsilon$ for $|\eta| < \delta(\varepsilon, \tau)$. If $u = \phi(t; \tau, \eta)$ is the solution of (8.3) satisfying the initial condition $u(\tau) = \eta$, then for $|\eta| < \delta(\varepsilon, \tau)$ and any compact interval $[\tau, t_2]$ on which $u(t)$ exists and $|u(t)| \leq k_0$, we have $0 \geq V^0(t, u(t)) = [V(t, u(t))]'$, so that $0 \leq V(t, u(t)) \leq V(\tau, \eta) < m_\varepsilon$ on $[\tau, t_2]$, and hence $|u(t)| < \varepsilon$ on this interval. In view of Theorem I.5.7 it then follows that the solution $u = \phi(t; \tau, \eta)$ is extensible to $[\tau, \infty)$, and $|u(t)| \leq \varepsilon$ on this interval, so that the solution $u(t) \equiv 0$ of (8.3) is stable.

A function $V(t, u)$ which satisfies the hypotheses of Theorem 8.1, and which may therefore be used for testing the stability behavior of solutions of a differential equation (8.3), is termed a *Liapunov function for the differential equation* (8.3).

A real-valued function $V(t, u)$ is said to have an *infinitesimal upper bound* on $\mathfrak{D}(a, b)$ if $V(t, u) = o(|u|)$ at $|u| = 0$, uniformly with respect to t on $[a, \infty)$; that is, if for an arbitrary $\varepsilon > 0$ there exists a k_ε satisfying $0 < k_\varepsilon \leq k$, and

$$|V(t, u)| < \varepsilon \quad \textit{for} \quad t \in [a, \infty), \qquad |u| \leq k_\varepsilon.$$

Theorem 8.2. *If there exists for* (8.3) *an admissible comparison function* $V(t, u)$ *which is definite, has an infinitesimal upper bound, and* $V^0(t, u)$ *is a definite function of sign opposite to that of* $V(t, u)$, *then the solution* $u(t) \equiv 0$ *of* (8.3) *is uniformly and asymptotically stable.*

Let $\mathfrak{D}(a_0, k_0)$ be a sub-domain of $\mathfrak{D}(a, k)$ on which there exist positive definite functions $W_1(u)$, $W_2(u)$ such that

(8.6)
$$V(t, u) \geq W_1(u), \qquad -V^0(t, u) \geq W_2(u) \quad \textit{for} \quad (t, u) \in \mathfrak{D}(a_0, k_0).$$

If $u(t) = \phi(t; \tau, \eta)$, $t \in [\tau, t_2]$, is a solution of (8.3) with $|u(\tau)| \leq \delta(\varepsilon, \tau)$, where $\delta(\varepsilon, \tau)$ is defined as in the proof of Theorem 8.1, then $u(t)$ is extensible to $[a_0, \infty)$ and $|u(t)| < \varepsilon$ throughout this interval. Moreover, since $V(t, u)$ has an infinitesimal upper bound, $\delta(\varepsilon, \tau)$ may be chosen independent of τ on $[a_0, \infty)$, and the stability of the solution $u(t) \equiv 0$ of (8.3) is uniform. As

$$[V(t, u(t))]' = V^0(t, u(t)) \leq -W_2(u(t)),$$

the function $V(t, u(t))$ is positive and monotone non-increasing on $[a_0, \infty)$, and hence there exists a value $L \geq 0$ such that $V(t, u(t)) \to L$ as $t \to \infty$. In view of the existence of an infinitesimal upper bound, if $L > 0$ then there would exist a $k_L > 0$ such that $|u(t)| \geq k_L$ for $t \in [\tau, \infty)$, and for $m(k_L)$ the positive minimum of $W_2(u)$ on $\{u \mid k_L \leq |u| \leq k_0\}$ we would have for $t \in [\tau, \infty)$ the inequality

$$\begin{aligned} V(t, u(t)) &\leq V(\tau, u(\tau)) - \int_\tau^t W_2(u(s))\, ds \\ &\leq V(\tau, u(\tau)) - m(k_L)[t - \tau], \end{aligned}$$

which contradicts the non-negative character of $V(t, u(t))$. Consequently $L = 0$, so that $V(t, u(t)) \to 0$ as $t \to \infty$, and hence $W_1(u(t)) \to 0$ and $u(t) \to 0$ by Lemma 8.2. For $V(t, u)$ negative definite a like proof may be used.

Theorem 8.3. *If there exists for* (8.3) *an admissible comparison function* $V(t, u)$, $(t, u) \in \mathfrak{D}(a_0, k_0) \subset \mathfrak{D}(a, k)$, *which has an infinitesimal upper bound,*

$V^0(t, u)$ is definite, and there exists a value $a_1 \geq a$ such that for each $\tau \geq a_1$ and $0 < \kappa < k_0$ there is at least one η with $|\eta| < \kappa$ and $V(\tau, \eta)V^0(\tau, \eta) > 0$, then the solution $u(t) \equiv 0$ of (8.3) is not stable.

Suppose that $V^0(t, u)$ is positive definite, and let $W(u)$ be a positive definite function such that $V^0(t, u) \geq W(u)$ *on* $\mathfrak{D}(a_0, k_0)$. Since $V(t, u)$ has an infinitesimal upper bound, one may suppose that a_0 and k_0 are such that $V(t, u)$ is bounded on $\mathfrak{D}(a_0, k_0)$, and γ is a positive constant such that $|V(t, u)| < \gamma$ on $\mathfrak{D}(a_0, k_0)$. For a_1 a value satisfying the conditions of the theorem and $\tau \geq a_1$, if $u(t)$, $t \in [\tau, t_2]$, is a solution of (8.3) satisfying the inequality $|u(t)| \leq k_0$ then $V(t, u(t))$ is a nondecreasing function on $[\tau, t_2]$. In particular, for $\tau > a_1$ and $0 < \kappa < k_0$ there is an η such that $|\eta| < \kappa$ and $V(\tau, \eta) > 0$, and $V(t, \phi(t; \tau, \eta))$ is a positive nondecreasing function on any interval $[\tau, t_2]$ throughout which $|\phi(t; \tau, \eta)| \leq k_0$. Now let $\mathfrak{M}(\tau, \eta; k_0)$ denote the set of vectors $u \in \mathbf{R}_n$ such that $|u| \leq k_0$, and for which there exists corresponding $t_0 \geq \tau$ such that $V(t_0, u) \geq V(\tau, \eta)$; clearly $\eta \in \mathfrak{M}(\tau, \eta; k_0)$. In view of the existence of an infinitesimal upper bound for $V(t, u)$, there exists a positive κ_0 such that $|u| \geq \kappa_0$ if $u \in \mathfrak{M}(\tau, \eta; k_0)$. Finally, let m_0 denote the minimum of $W(u)$ on the set $\{u \mid \kappa_0 \leq |u| \leq k_0\}$. If the solution $u = \phi(t; \tau, \eta)$ exists and satisfies the inequality $|\phi(t; \tau, \eta)| \leq k_0$ for $t \in [\tau, t_2]$, then

$$V(t, \phi(t; \tau, \eta)) \geq V(\tau, \eta) + (t - \tau)m_0, \quad \textit{for} \quad t \in [\tau, t_2],$$

and, in view of the bound $V(t, \phi(t; \tau, \eta)) \leq \gamma$ for $t \in [\tau, t_2]$, it follows that $(t_2 - \tau) < \gamma/m_0$ and $t_2 < \tau + \gamma/m_0$. That is, there exists a value $t_3 > \tau$ such that $|\phi(t_3; \tau, \eta)| \geq k_0$, and consequently the solution $u(t) \equiv 0$ of (8.3) is not stable. The case of $V^0(t, u)$ negative definite may be treated similarly.

Problems VIII.8

1.[s] Suppose that A is a constant $n \times n$ matrix whose proper values λ all have $\Re\lambda < 0$. Show that for C an arbitrary $n \times n$ matrix the improper matrix integral

$$\int_0^\infty e^{tA^*}Ce^{tA}\,dt = \lim_{b\to\infty}\int_0^b e^{tA^*}Ce^{tA}\,dt \tag{8.7}$$

exists and defines a matrix R which is the unique solution of the algebraic matrix equation

$$RA + A^*R = -C. \tag{8.8}$$

Show that if A and C are real then R is real, and if C is hermitian then R is hermitian; also, if C is definite hermitian then R is correspondingly definite hermitian. In particular, if A is real and C is real, symmetric, and positive definite, then R is also real, symmetric, and positive definite.

Hint. Note that the solution of the matrix differential system $T' = A^*T + TA$, $T(0) = C$, is $T(t) = Y^*(t)CY(t)$, where $Y(t) = \exp\{tA\}$ and

$$T(b) - T(0) = \int_0^b T'(s)\,ds = A^*\left[\int_0^b T(s)\,ds\right] + \left[\int_0^b T(s)\,ds\right]A,$$

Moreover, if all proper values of A have negative real parts, then equation (2.2) is uniformly asymptotically stable by Theorem 2.4, so that $T(b) \to 0$ as $b \to \infty$ and the improper matrix integral (8.7) exists and is finite. From the solvability theory of systems of linear algebraic equations, (see, for example, Section 2 of Appendix B), conclude that the existence of a solution R of (8.8) for arbitrary C implies that for each individual matrix C the solution is unique.

2.[s] For a vector differential equation

$$u' \equiv Au, \tag{8.9}$$

where A is a real constant $n \times n$ matrix whose proper values all have $\Re\,\lambda < 0$, show that there is a quadratic admissible comparison function $V(t, u) = \tilde{u}Ru$, with R a real, symmetric, positive definite matrix such that the corresponding function $V^0(t, u)$ is equal to $-c\,|u|^2$, with c a given positive constant.

Hint. Determine R as in Problem 1, with $C = cE$.

3. Suppose that $g(u) \equiv (g_\alpha(u_1, \ldots, u_n))$, $(\alpha = 1, \ldots, n)$ is continuous and has continuous partial derivative matrix $g_u(u) = [g_{\alpha u_\beta}(u)]$, $(\alpha, \beta = 1, \ldots, n)$, for $|u| \leq k$, and $g(0) = 0$, while all proper values λ of the constant matrix $A = g_u(0)$ are such that $\Re\,\lambda < 0$. If $V(t, u) = \tilde{u}Ru$ is determined for the corresponding linear system (8.9) as in Problem 2 above, show that there exists a value k_1 such that $0 < k_1 \leq k$ and the corresponding function $V^0(t, u \mid g)$ is such that $V^0(t, u \mid g) \leq -(c/2)\,|u|^2$ for $|u| \leq k_1$.

4. Suppose that $r(u) = r(u_1, \ldots, u_n)$ is a real-valued scalar function which is continuous and continuously differentiable for $|u| \leq k$, and $r(0) > 0$. Prove that the solution $u(t) \equiv 0$ of the vector differential equation

$$u' = -ur(u), \tag{8.10}$$

is uniformly and asymptotically stable.

Hint. Show that $V(t, u) = |u|^2$ is a Liapunov function for (8.10) on a suitable domain.

5. Suppose that the real-valued vector function $g(t, u) \equiv (g_\alpha(t, u))$, $(\alpha = 1, \ldots, n)$, is of class $\mathfrak{C}^{(1)}$ in u on an open region

$$\{(t, u) \mid t \in (a_0, \infty), \qquad u \in \mathbf{R}_n, \qquad |u| < k_0\},$$

of (t, u) space, $g(t, 0) \equiv 0$ for $t \in (a_0, \infty)$, and on this region the vector differential equation (8.3) has a first integral $W(t, u)$ that is also a definite admissible comparison function. Show that the solution $u(t) \equiv 0$ of (8.3) is stable.

Hint. Recall Theorem I.10.6.

6. Suppose that $k(s)$ is a real-valued continuously differentiable function of s on $(-\infty, \infty)$, with $k(0) = 0$ and $sk(s) > 0$. Show that for the differential system

$$u_1' = u_2,$$
$$u_2' = -k(u_1),$$

the function $W(t, u) = \frac{1}{2}u_2^2 + \int_0^{u_1} k(s)\, ds$ is a first integral which satisfies the conditions of Problem 5.

7. Suppose that $n = 2$, and

$$g(t, u) = \begin{bmatrix} a & 0 \\ 0 & -b \end{bmatrix} u + r(t, u),$$

where a and b are positive constants, and the two-dimensional vector function $r(t, u) = (r_\alpha(t, u_1, u_2))$, $(\alpha = 1, 2)$, is continuously differentiable on a domain $\{(t, u) \mid t \in [a_1, \infty), |u| \leq k_1\}$; moreover, $r(t, 0) \equiv 0$ for $t \in [a_1, \infty)$, and there exist constants $\kappa > 0$, $p > 2$ such that $|\tilde{u}r(t, u)| \leq \kappa\, |u|^p$ for $t \in [a, \infty)$ and $|u| \leq k_1$. Show that $V(t, u) = u_1^2 - u_2^2$ satisfies the conditions of Theorem 8.3 on a suitable domain $\mathfrak{D}(a_0, k_0) = \{(t, u) \mid t \in [a_0, \infty),\ |u| \leq k_0\}$, and hence the solution $u(t) \equiv 0$ of the corresponding differential equation (8.3) is not stable.

9. NOTES AND REMARKS

This chapter is devoted to a brief introduction to the asymptotic behavior of solutions, and stability problems for ordinary vector differential equations. For an excellent survey of this topic, and a comprehensive bibliography, the reader is referred to Cesari [2]. Other suggested works for collateral reading are Bellman [3], Coddington and Levinson [1, Chs. 3, 13], Antosiewicz [1], Nemetyskiĭ and Stepanov [1], Lefschetz [1], LaSalle and Lefschetz [1], Sansone and Conti [1], Hahn [1, 2], Massera and Schäffer [1], and Coppel [2].

The concepts of stability and asymptotic stability are due to Liapunov [1], and the definition of uniform stability was given by Persidskiĭ [1]. The notions of strict (or strong) stability, and uniformly asymptotic stability, were given by Ascoli [1] and Malkin [1], respectively. Liapunov [1] introduced the concept of kinematic or t-similarity, and also the notion of a reducible differential system; the specific terminology "kinematic similarity" is due to Markus [1].

The inequality estimate (2.3) is usually attributed to Ważewski [1], (see also Wintner [2, Appendix]). The transformation result of Theorem 2.5 was first established by Perron [4]. Subsequent simplification of details of proof were presented by Diliberto [1, 2] and Reid [15]; the proof given in the text is that of Reid [15]. The definition of the type number of a vector function is due to Perron [3], and is the negative of the "characteristic number" of Liapunov [1]. The concept of a normal fundamental set of solutions, and the notion of a regular differential system, are due to Liapunov [1].

Supplementary references for Sections 2, 3, and 4 are Perron [5], Caligo [1], Trjitzinsky [1], Conti [1], and Coppel [1, 2]. The concept of asymptotic equivalence, which forms the central theme of Sections 5 and 7, dates from Levinson [1, 2] and Weyl [1]. Discussions of related material that goes far beyond that of these sections is to be found in Coddington and Levinson [1, Chs. 3, 13], Sansone and Conti [1, Ch. IX] and Coppel [2].

For the Floquet theory the reader is referred to Floquet [1], Sansone [1-I, Ch. VI], Ince [1; Ch. XV], Coddington and Levinson [1, Ch. 3].

As collateral reading for Section 8, attention is directed to Hahn [1, 2], Lefschetz [1], LaSalle and Lefschetz [1], Nemetyskiĭ and Stepanov [1], and Massera and Schäffer [1].

RELATED REFERENCES AND COMMENTS FOR SPECIFIC PROBLEMS

VIII.2:2 Perron [5].
VIII.2:8 Conti [1, Th. 4].
VIII.2:9 Conti [1, Th. 3].
VIII.2:10 Cesari [1].
VIII.2:11 For a special case of conclusion (iii), see an example of Vinograd [1], which is also given in Hahn [2, p. 307].
VIII.2:12 Conti [1, Ths. 6, 6′].
VIII.2:14,16 Liapunov [1, p. 236]; Hahn [2, p. 312].
VIII.2:17 Perron [3]; Diliberto [1].
VIII.3:3,4 Bellman [2]; Coppel [2].
VIII.4:1 Hukuhara [3]; see also Späth [1]. This result is called the Dini-Hukuhara theorem. See Cesari [2, Ch. 2-§3], for a comprehensive survey of related results.
VIII.4:2 Conti [1].
VIII.5:1 Hille [1, Ths. 2,3].
VIII.5:7 Levinson [1,2].
VIII.8:1 See, for example, Hahn [2, p. 119].

Appendices

FOREWORD

The following appendices present results on vector spaces, the algebra of matrices, the theory of quadratic and hermitian forms, and elementary convergence theorems, that are prefatory to the text material on differential equations. In lecturing on the text material over a period of many years, the author has found that although most of his students have had previous acquaintance with much of the material of these appendices, the heterogeneous background of a given class usually made it highly desirable to have available a common source of reference. In particular, with a supplementary treatment accompanying the text, and using the same notations and terminologies as employed therein, at the appropriate time a given auxiliary topic may be injected into the mainstream of the course with minimum distraction. In general, these appendices should serve to focus the attention of the student on these particular topics, and be a source for initiating review and perhaps further study, especially in connection with some of the particular results that are given in the problem sets.

For the student to whom a major portion of the material is new the treatment of the appendices may prove to be too concise, in which case the following references are suggested for supplementary reading:

Halmos, P. R., *Finite Dimensional Vector Spaces*, 2nd edition, Van Nostrand, 1958;

Hoffman, K. and R. Kunze, *Linear Algebra*, Prentice-Hall, 1961;

Nehring, E. D., *Linear Algebra and Matrix Theory*, John Wiley and Sons, 1963.

Appendix A

Vector Spaces

1. INTRODUCTION

In the study of differential equations one is frequently concerned with the concept of a *vector space* $\mathfrak{B}$ over a *scalar field* $\mathfrak{F}$, where $\mathfrak{F}$ is either the set of complex numbers, or the set of real numbers. Such a vector space is a set of elements, called *vectors*, which possesses the following properties.

(P_1) *To each pair* x, y *of vectors in* $\mathfrak{B}$ *there corresponds a unique vector* $x + y$, *the sum of* x *and* y, *such that:*

(i) *addition is commutative*, $(x + y = y + x)$;

(ii) *addition is associative*, $(x + [y + z] = [x + y] + z)$;

(iii) *there exists a unique (zero) vector* θ *such that* $x + \theta = x$ *for all vectors* x;

(iv) *to each vector* x *in* $\mathfrak{B}$ *there corresponds a unique vector* $-x$ *such that* $x + (-x) = \theta$.

(P_2) *To each pair* x, c, *with* x *in* $\mathfrak{B}$ *and* c *in* $\mathfrak{F}$, *there corresponds a unique vector* cx, *the product of* c *and* x, *such that:*

(i) *multiplication by scalars is associative*, $(c[dx] = [cd]x)$;

(ii) $1 \cdot x = x$, *for each vector* x;

(iii) *multiplication by scalars is distributive with respect to vector addition*, $(c[x + y] = cx + cy)$;

(iv) *multiplication by vectors is distributive with respect to scalar addition*, $([c + d]x = cx + dx)$.

The following are examples of vector spaces which are of particular significance for the theory of differential equations.

(1) *The n-dimensional complex number space* $\mathbf{C}_n$, *consisting of the set of all ordered n-tuples of complex numbers* $x = (x_\alpha)$, $(\alpha = 1, \ldots, n)$, *and* $\mathfrak{F}$ *the field of complex numbers, where if* $x = (x_\alpha)$ *and* $y = (y_\alpha)$ *then* $x + y = (x_\alpha + y_\alpha)$, $cx = (cx_\alpha)$, $\theta = (\theta_\alpha)$ *with* $\theta_\alpha = 0$, $(\alpha = 1, \ldots, n)$, *and* $-x = (-x_\alpha)$.

(2) *The n-dimensional real number space* $\mathbf{R}_n$, *consisting of the set of all ordered n-tuples of real numbers* $x = (x_\alpha)$, *and* $\mathfrak{F}$ *the field of real numbers, with the same formal definitions of addition and scalar multiplication as in* $\mathbf{C}_n$, *except that the scalars are now restricted to be real numbers.*

(3) *For* $\mathfrak{F}$ *either the field of real numbers, or the field of complex numbers, the set of all scalar-valued functions* $f : I \rightarrow \mathfrak{F}$, *where I is an interval on the real line, and with vector addition and scalar multiplication defined by*

$$(f + g)(t) = f(t) + g(t), \quad \textit{for} \quad t \in I;$$
$$(cf)(t) = cf(t), \quad \textit{for} \quad t \in I, \quad \textit{and} \quad c \in \mathfrak{F}.$$

(4) *The set of all scalar polynomials* $p(t)$ *with coefficients in a specified field* $\mathfrak{F}$, *where vector addition and scalar multiplication are as in* (3).

(5) *For a given positive integer m, the set of all polynomials* $p(t)$ *of Example* (4), *with degree not exceeding* $m - 1$.

(6) *For* $\mathfrak{F}$ *either the field of real numbers, or the field of complex numbers, the set of all scalar-valued functions y which are continuous, have continuous derivatives of the first n orders, and are solutions of the n-th order linear homogeneous differential equation*

$$a_n(t)y^{[n]} + a_{n-1}(t)y^{[n-1]} + \cdots + a_0(t)y = 0, \qquad t \in (-\infty, \infty),$$

where for $j = 0, 1, \ldots, n$ *the coefficient functions* a_j *are continuous on* $(-\infty, \infty)$ *with functional value* $a_j(t)$ *in* $\mathfrak{F}$.

A non-empty subset $\mathfrak{M}$ of a vector space $\mathfrak{V}$ is called a *subspace* or *linear manifold* if for every pair of vectors x, y in $\mathfrak{M}$ all linear combinations $cx + dy$ are also contained in $\mathfrak{M}$. In particular, if x is in $\mathfrak{M}$ then $1 \cdot x + (-1) \cdot x = 0 \cdot x = \theta$ is also in $\mathfrak{M}$. Clearly the largest subspace of $\mathfrak{V}$ is $\mathfrak{V}$ itself, and the smallest subspace is the set consisting of the single element θ. In general, if $\mathfrak{M}_1$ and $\mathfrak{M}_2$ are subspaces of $\mathfrak{V}$, then the set $\mathfrak{M}_1 \cap \mathfrak{M}_2$ of elements common to $\mathfrak{M}_1$ and $\mathfrak{M}_2$ is a subspace. Two subspaces $\mathfrak{M}_1$ and $\mathfrak{M}_2$ are said to be equal, and we write $\mathfrak{M}_1 = \mathfrak{M}_2$, if each element of $\mathfrak{M}_1$ belongs to $\mathfrak{M}_2$ and each element of $\mathfrak{M}_2$ belongs to $\mathfrak{M}_1$.

2. LINEAR INDEPENDENCE

A finite set of vectors $x^{(1)}, \ldots, x^{(k)}$ of a vector space $\mathfrak{V}$ is said to be *linearly independent* if the only linear combination $c_1x^{(1)} + \cdots + c_kx^{(k)}$ that is the zero vector θ is the zero, or trivial, combination with $c_j = 0$, $(j = 1, \ldots, k)$. If there is a nontrivial combination which is the zero vector then the set $x^{(1)}, \ldots, x^{(k)}$ is said to be *linearly dependent*. The phrase "$x^{(1)}, \ldots, x^{(k)}$ are linearly independent, (or dependent)" is also used instead of "the set $x^{(1)}, \ldots, x^{(k)}$ is linearly independent, (or dependent)." For given vectors $x^{(1)}, \ldots, x^{(k)}$ of a vector space $\mathfrak{V}$ the set of all linear combinations of these vectors is a subspace of $\mathfrak{V}$, which will be denoted by $[x^{(1)}, \ldots, x^{(k)}]$. It follows immediately that if $x^{(1)}, \ldots, x^{(k)}, y^{(1)}, \ldots, y^{(h)}$ are vectors in $\mathfrak{V}$ such that each $y^{(\beta)}$, $(\beta = 1, \ldots, h)$, belongs to $[x^{(1)}, \ldots, x^{(k)}]$, and each $x^{(\alpha)}$, $(\alpha = 1, \ldots, k)$, belongs to $[y^{(1)}, \ldots, y^{(h)}]$, then $[x^{(1)}, \ldots, x^{(k)}] = [y^{(1)}, \ldots, y^{(h)}]$. In general, a set X of vectors in a vector space $\mathfrak{V}$ is said to be linearly independent if every nonempty finite subset of X is linearly independent.

Theorem 2.1. *If $x^{(1)}, \ldots, x^{(k)}$ are vectors of a vector space $\mathfrak{V}$, and $y^{(1)}, \ldots, y^{(h)}$ is a linearly independent set of vectors in $\mathfrak{V}$ such that each $y^{(\beta)}$ belongs to $[x^{(1)}, \ldots, x^{(k)}]$, then $h \leq k$, and among the $x^{(1)}, \ldots, x^{(k)}$ there is a subset of h vectors, which by renumbering may be chosen as $x^{(1)}, \ldots, x^{(h)}$, such that $[x^{(1)}, \ldots, x^{(k)}] = [y^{(1)}, \ldots, y^{(h)}, x^{(h+1)}, \ldots, x^{(k)}]$.*

This result will be proved by mathematical induction. If $h = 1$, then $y^{(1)}$ is a non-zero vector belonging to $[x^{(1)}, \ldots, x^{(k)}]$, so that $y^{(1)} = c_1 x^{(1)} + \cdots + c_k x^{(k)}$, with not all of the scalars $c_1, \ldots, c_k$ equal to zero. By renumbering suitably the vectors $x^{(1)}, \ldots, x^{(k)}$ we obtain $c_1 \neq 0$, so that $x^{(1)}$ is a linear combination of $y^{(1)}, x^{(2)}, \ldots, x^{(k)}$, and $[x^{(1)}, \ldots, x^{(k)}] = [y^{(1)}, x^{(2)}, \ldots, x^{(k)}]$; also, $1 \leq k$. Now suppose that the result of the theorem is true for $h = m$, and that the set of $m + 1$ vectors $y^{(1)}, \ldots, y^{(m+1)}$ is linearly independent, and each $y^{(\beta)}$ belongs to $[x^{(1)}, \ldots, x^{(k)}]$. By the induction hypothesis, $m \leq k$, and upon renumbering suitably the $x^{(1)}, \ldots, x^{(k)}$ we may obtain $[x^{(1)}, \ldots, x^{(k)}] = [y^{(1)}, \ldots, y^{(m)}, x^{(m+1)}, \ldots, x^{(k)}]$. As the set $y^{(1)}, \ldots, y^{(m+1)}$ is linearly independent, $y^{(m+1)}$ does not belong to $[y^{(1)}, \ldots, y^{(m)}]$, so that $m < k$ and $m + 1 \leq k$; moreover, $y^{(m+1)} = d_1 y^{(1)} + \cdots + d_m y^{(m)} + d_{m+1} x^{(m+1)} + \cdots + d_k x^{(k)}$, with $d_{m+1}, \ldots, d_k$ not all zero. If the $x^{(m+1)}, \ldots, x^{(k)}$ are so renumbered that $d_{m+1} \neq 0$, then $x^{(m+1)}$ is a linear combination of $y^{(1)}, \ldots, y^{(m+1)}, x^{(m+2)}, \ldots, x^{(k)}$, and $[y^{(1)}, \ldots, y^{(m+1)}, x^{(m+2)}, \ldots, x^{(k)}] = [y^{(1)}, \ldots, y^{(m)}, x^{(m+1)}, \ldots, x^{(k)}] = [x^{(1)}, \ldots, x^{(k)}]$. Thus by mathematical induction the result of the theorem is established for arbitrary integers h.

The following result is an immediate corollary to the above theorem.

***Corollary* 1.** *If $x^{(1)}, \ldots, x^{(k)}$ and $y^{(1)}, \ldots, y^{(h)}$ are two sets of vectors in a vector space $\mathfrak{V}$, with each set linearly independent, and $[x^{(1)}, \ldots, x^{(k)}] = [y^{(1)}, \ldots, y^{(h)}]$, then $h = k$.*

If $\mathfrak{V}$ is a vector space, and there exists a positive integer d such that $\mathfrak{V}$ contains a linearly independent set of d vectors, while every set of $d + 1$ vectors in $\mathfrak{V}$ is linearly dependent, then $\mathfrak{V}$ is said to be *finite dimensional;* the integer d is called the *dimension* of $\mathfrak{V}$ and we write $d = \dim \mathfrak{V}$. A vector space with just one element, which must then be the zero element, is also called finite dimensional, with dimension zero. If $\mathfrak{V}$ is not finite dimensional, it is called infinite dimensional. If $\mathfrak{V}$ is a vector space of finite dimension $d \geq 1$, then a *basis* in $\mathfrak{V}$ is a linearly independent set X of vectors such that each vector in $\mathfrak{V}$ is a linear combination of vectors belonging to X. In view of the above Corollary 1 to Theorem 2.1, each basis of such a vector space $\mathfrak{V}$ consists of d linearly independent vectors $x^{(1)}, \ldots, x^{(d)}$ such that $[x^{(1)}, \ldots, x^{(d)}] = \mathfrak{V}$. For example, if $\delta_{\alpha\beta}$ denotes the so-called *Kronecker delta* with $\delta_{\alpha\beta} = 1$ for $\alpha = \beta$, and $\delta_{\alpha\beta} = 0$ for $\alpha \neq \beta$, $(\alpha, \beta = 1, \ldots, n)$, and $e^{(\beta)} = (e_\alpha^{(\beta)})$ is the element of $\mathbf{C}_n$ with $e_\alpha^{(\beta)} = \delta_{\alpha\beta}$, then it follows readily that the set $e^{(1)}, \ldots, e^{(n)}$ is linearly independent, and $[e^{(1)}, \ldots, e^{(n)}] = \mathbf{C}_n$. Consequently, $\mathbf{C}_n$ has dimension n according to the above definition; in particular, it is to be noted that there is conformity with the earlier designation of $\mathbf{C}_n$ as n-dimensional complex number space. In a similar fashion it follows that in $\mathbf{R}_n$ the vectors $e^{(\beta)} = (e_\alpha^{(\beta)})$, with $e_\alpha^{(\beta)} = \delta_{\alpha\beta}$, $(\alpha, \beta = 1, \ldots, n)$, form a basis, so that $\mathbf{R}_n$ is also n-dimensional.

The following result is a ready consequence of Theorem 2.1 and the above Corollary 1.

***Corollary* 2.** *If $\mathfrak{V}$ is a vector space of finite dimension n, and $x^{(1)}, \ldots, x^{(d)}$ are linearly independent vectors of $\mathfrak{V}$, then $d \leq n$, and $[x^{(1)}, \ldots, x^{(d)}] = \mathfrak{V}$ if and only if $d = n$; if $d < n$ then there exist $n - d$ vectors $x^{(d+1)}, \ldots, x^{(n)}$ such that the set $x^{(1)}, \ldots, x^{(n)}$ is a basis for $\mathfrak{V}$.*

3. THE CONJUGATE SPACE OF LINEAR FUNCTIONALS

A *linear functional* λ on a vector space $\mathfrak{V}$ is a scalar-valued function defined for all elements of $\mathfrak{V}$ such that $\lambda[cx + dy] = c\lambda[x] + d\lambda[y]$ for arbitrary vectors x, y of $\mathfrak{V}$, and arbitrary scalars c, d. In particular, $\lambda_0[x] = 0$ for all $x \in \mathfrak{V}$ is a linear functional on $\mathfrak{V}$. Now for a given vector space $\mathfrak{V}$ consider the totality of linear functionals on $\mathfrak{V}$. If λ_1 and λ_2 are linear functionals on $\mathfrak{V}$, and c_1, c_2 are scalars, then $\lambda[x] = c_1\lambda_1[x] + c_2\lambda_2[x]$ for $x \in \mathfrak{V}$ defines a linear functional λ, which will be denoted by $c_1\lambda_1 + c_2\lambda_2$. In particular, the linear functional λ_0 defined above is such that $\lambda + \lambda_0 = \lambda$ for arbitrary linear functionals λ on $\mathfrak{V}$; that is, λ_0 is the zero element θ' for the set of linear functionals. With these definitions of addition, scalar multiplication, and zero element, the totality of linear functionals on $\mathfrak{V}$ is itself a vector space, which will be denoted by $\mathfrak{V}'$; the vector space $\mathfrak{V}'$ is called the *conjugate space*, or *dual space*, of $\mathfrak{V}$.

Let $\mathfrak{V}$ be an n-dimensional vector space, with basis $x^{(1)}, \ldots, x^{(n)}$. If $x \in \mathfrak{V}$ and $x = \xi_1 x^{(1)} + \cdots + \xi_n x^{(n)}$, then for any linear functional $\lambda \in \mathfrak{V}'$ we have

$$\lambda[x] = \xi_1\lambda[x^{(1)}] + \cdots + \xi_n\lambda[x^{(n)}]. \tag{3.1}$$

The coefficients $\lambda[x^{(1)}], \ldots, \lambda[x^{(n)}]$ in (3.1) are independent of x, and may be arbitrarily prescribed. Indeed, if $c_1, \ldots, c_n$ are given scalars, then there is a unique $\lambda \in \mathfrak{B}'$ such that $\lambda[x^{(i)}] = c_i$, $(i = 1, \ldots, n)$; this linear function λ is given by

$$\lambda[x] = \sum_{i=1}^{n} \xi_i c_i, \quad \textit{for} \quad x = \sum_{i=1}^{n} \xi_i x^{(i)}. \tag{3.2}$$

In particular, if for $j = 1, \ldots, n$ the scalars c_i are chosen as $c_i = \delta_{ij}$, $(i = 1, \ldots, n)$, then the corresponding functionals λ_j of $\mathfrak{B}'$ are such that

$$\lambda_j[x^{(i)}] = \delta_{ij}, \qquad (i, j = 1, \ldots, n). \tag{3.3}$$

The set of functionals $\lambda_1, \ldots, \lambda_n$ is clearly a linearly independent set in $\mathfrak{B}'$; that is, if $c_1\lambda_1 + \cdots + c_n\lambda_n = \theta'$ then $c_1 = 0 = \cdots = c_n$. Moreover, if λ is any functional of $\mathfrak{B}'$ then $\lambda[x]$ is given by (3.2), with $c_i = \lambda[x^{(i)}]$, $(i = 1, \ldots, n)$, and hence $\lambda = \sum_{j=1}^{n} c_j\lambda_j$. That is, we have established the following basic result.

Theorem 3.1. *If $\mathfrak{B}$ is an n-dimensional vector space, and $x^{(1)}, \ldots, x^{(n)}$ is a basis for $\mathfrak{B}$, then there is a uniquely determined basis $\lambda_1, \ldots, \lambda_n$ for $\mathfrak{B}'$ such that $\lambda_j[x^{(i)}] = \delta_{ij}$, $(i, j = 1, \ldots, n)$. In particular, $\mathfrak{B}'$ is also n-dimensional.*

The basis $\lambda_1, \ldots, \lambda_n$ for $\mathfrak{B}'$ determined by (3.3), is said to be *dual to the basis* $x^{(1)}, \ldots, x^{(n)}$ for $\mathfrak{B}$.

The following results are stated without proofs, since they may be established readily using a basis $x^{(1)}, \ldots, x^{(n)}$ for $\mathfrak{B}$, and the corresponding dual basis $\lambda_1, \ldots, \lambda_n$ for $\mathfrak{B}'$.

***Corollary* 1.** *If the vector space $\mathfrak{B}$ is n-dimensional, and x is an element of $\mathfrak{B}$ such that $\lambda[x] = 0$ for every $\lambda \in \mathfrak{B}'$, then $x = 0$.*

***Corollary* 2.** *Suppose that $\mathfrak{B}$ is a vector space of finite dimension n, and $\mathfrak{M}$ is a proper subspace of $\mathfrak{B}$. If $x^{(0)}$ is an element of $\mathfrak{B}$ not in $\mathfrak{M}$, then there exists a linear functional $\lambda \in \mathfrak{B}'$ such that $\lambda[x^{(0)}] \neq 0$, and $\lambda[x] = 0$ for $x \in \mathfrak{M}$.*

In particular, for a given basis $x^{(1)}, \ldots, x^{(n)}$ of an n-dimensional vector space $\mathfrak{B}$ with scalar field $\mathfrak{F}$, relation (3.2) establishes a one-to-one correspondence between linear functionals λ of $\mathfrak{B}'$ and n-tuples (c_i) of elements in $\mathfrak{F}$; moreover, a set of elements $\lambda_1, \ldots, \lambda_q$ in $\mathfrak{B}'$ is linearly independent if and only if the associated set of n-tuples $(c_i^1), \ldots, (c_i^q)$ is a linearly independent set of elements of the corresponding n-dimensional number space, $\mathbf{C}_n$ or $\mathbf{R}_n$.

If $\mathfrak{S}$ is a nonvacuous subset of a vector space $\mathfrak{B}$, then $\mathfrak{S}^0$ denotes the *annihilator* of $\mathfrak{S}$ in $\mathfrak{B}'$, consisting of all $\lambda \in \mathfrak{B}'$ such that $\lambda[x] = 0$ for all x in $\mathfrak{S}$. The following properties of annihilators are ready consequences of the definition.

(i) *For any nonvacuous subset* $\mathfrak{S} \subset \mathfrak{V}$, $\mathfrak{S}^0$ *is a linear manifold in* $\mathfrak{V}'$.

(ii) *If* $\mathfrak{S} = (\theta)$, *the set consisting of the single zero element in* $\mathfrak{V}$, *then* $\mathfrak{S}^0 = \mathfrak{V}'$.

(iii) *If* $\mathfrak{S} = \mathfrak{V}$, *then* $\mathfrak{S}^0 = (\theta')$, *the set consisting of the single zero element in* $\mathfrak{V}'$.

(iv) *If* $\mathfrak{S}_1 \subset \mathfrak{S}_2 \subset \mathfrak{V}$, *then* $\mathfrak{S}_2^0 \subset \mathfrak{S}_1^0 \subset \mathfrak{V}'$.

Theorem 3.2. *If* $\mathfrak{V}$ *is an n-dimensional vector space, and* $\mathfrak{M}$ *is a d-dimensional subspace of* $\mathfrak{V}$, *then* $\mathfrak{M}^0$ *is an* $(n - d)$*-dimensional subspace of* $\mathfrak{V}'$.

If $n = 0$ then the result is immediate, so we shall suppose that $n \geq 1$. Moreover, if $d = 0$ or $d = n$ then the result follows from the above properties (ii) and (iii), respectively, so we shall suppose that $0 < d < n$. Let $x_1, \ldots, x_n$ be a basis for $\mathfrak{V}$, chosen such that $x_1, \ldots, x_d$ is a basis for $\mathfrak{M}$; such a choice is clearly possible in view of Corollary 2 to Theorem 2.1. Let $\lambda_1, \ldots, \lambda_n$ be a basis for $\mathfrak{V}'$, dual to the basis $x_1, \ldots, x_n$ of $\mathfrak{V}$. If $x \in \mathfrak{M}$ then $x = \xi_1 x_1 + \cdots + \xi_d x_d$, and consequently for $j = d + 1, \ldots, n$ we have $\lambda_j[x] = \xi_1 \lambda_j[x_1] + \cdots + \xi_d \lambda_j[x_d] = 0$, so that $\mathfrak{M}^0$ has dimension at least $n - d$. On the other hand, if $\lambda \in \mathfrak{M}^0$ and $\eta_1, \ldots, \eta_n$ are constants such that $\lambda = \eta_1 \lambda_1 + \cdots + \eta_n \lambda_n$, then the fact that $x_i \in \mathfrak{M}$ for $i = 1, \ldots, d$ implies that $0 = \eta_1 \lambda_1[x_i] + \cdots + \eta_d \lambda_d[x_i] = \eta_i$, so that λ is a linear combination of $\lambda_{d+1}, \ldots, \lambda_n$, and $\mathfrak{M}^0$ has dimension at most $n - d$. Consequently, $\mathfrak{M}^0$ has dimension equal to $n - d$. Moreover, it is to be noted that in the proof there has been designated a basis $\lambda_{d+1}, \ldots, \lambda_n$ for $\mathfrak{M}^0$.

If $\mathcal{S}$ is a nonvacuous subset of $\mathfrak{V}'$, then the symbol ${}^0\mathcal{S}$ is used to denote the *annihilator* of $\mathcal{S}$ in $\mathfrak{V}$, consisting of all $x \in \mathfrak{V}$ such that $\lambda[x] = 0$ for all $\lambda \in \mathcal{S}$. For such annihilators we have the following properties.

(i) *For any nonvacuous subset* $\mathcal{S}$ *of* $\mathfrak{V}'$, ${}^0\mathcal{S}$ *is a linear manifold in* $\mathfrak{V}$.

(ii) *If* $\mathcal{S} = (\theta')$, *the set consisting of the single zero element in* $\mathfrak{V}'$, *then* ${}^0\mathcal{S} = \mathfrak{V}$.

(iii) *If* $\mathfrak{V}$ *is finite dimensional, then* ${}^0(\mathfrak{V}') = (\theta)$.

(iv) *If* $\mathcal{S}_1 \subset \mathcal{S}_2 \subset \mathfrak{V}'$, *then* ${}^0\mathcal{S}_2 \subset {}^0\mathcal{S}_1 \subset \mathfrak{V}$.

Properties (i), (ii), (iv) are ready consequences of the definitions, and Property (iii) follows from Corollary 2 to Theorem 3.1. The result of that Corollary and property (iii) are also valid for infinite dimensional vector spaces, but transfinite induction is required for the proof in the general case.

Theorem 3.3. *If* $\mathfrak{V}$ *is a finite dimensional vector space, and* $\mathfrak{M}$ *is a subspace of* $\mathfrak{V}$, *then* ${}^0(\mathfrak{M}^0) = \mathfrak{M}$.

If $x \in \mathfrak{M}$, and $\lambda \in \mathfrak{M}^0$, then $0 = \lambda[x]$, and consequently $\mathfrak{M} \subset {}^0(\mathfrak{M}^0)$. Now if $\mathfrak{M} \neq {}^0(\mathfrak{M}^0)$, there exists a $x^{(0)}$ in ${}^0(\mathfrak{M}^0)$ which does not belong to $\mathfrak{M}$. From Corollary 2 to Theorem 3.1 it then follows that there exists a $\lambda \in \mathfrak{V}'$ such that $\lambda[x^{(0)}] \neq 0$ and $\lambda[x] = 0$ for all $x \in \mathfrak{M}$. In particular, $\lambda \in \mathfrak{M}^0$ and

$\lambda[x^{(0)}] \neq 0$, contrary to the given condition that $x^{(0)} \in {}^0(\mathfrak{M}^0)$. Therefore, $\mathfrak{M} = {}^0(\mathfrak{M}^0)$.

4. LINEAR TRANSFORMATIONS

Let $\mathfrak{V}$ and $\mathfrak{W}$ be vector spaces over the same scalar field $\mathfrak{F}$. A *linear transformation* $T: \mathfrak{V} \to \mathfrak{W}$ is a (single-valued) mapping of $\mathfrak{V}$ into $\mathfrak{W}$ which associates to each $x \in \mathfrak{V}$ a unique element $Tx \in \mathfrak{W}$ such that for arbitrary x, y in $\mathfrak{V}$, and arbitrary scalars c, d of $\mathfrak{F}$, we have

$$T(cx + dy) = c[Tx] + d[Ty].$$

The set of all elements $x \in \mathfrak{V}$ such that Tx is the zero element in $\mathfrak{W}$ is a linear subspace in $\mathfrak{V}$, called the *null-space* of T, and denoted by $\mathfrak{N}(T)$; $\dim \mathfrak{N}(T)$ is called the *nullity* of T. Also the set of all elements $y \in \mathfrak{W}$ for which there are corresponding elements $x \in \mathfrak{V}$ such that $Tx = y$ is a linear subspace in $\mathfrak{W}$, called the *range* of T, and denoted by $\mathfrak{R}(T)$; $\dim \mathfrak{R}(T)$ is called the *rank* of T. The transformation T is said to be *invertible* if for each $y \in \mathfrak{R}(T)$ there is a unique $x \in \mathfrak{V}$ such that $Tx = y$; if T is invertible then T^{-1} will denote the *inverse* of T, whose domain of definition is the linear subspace $\mathfrak{R}(T)$ in $\mathfrak{W}$, and whose value $T^{-1}y$ is such that $T^{-1}y \in \mathfrak{V}$ and $T(T^{-1}y) = y$ for each $y \in \mathfrak{R}(T)$. From the definition it also follows that $T^{-1}(Ty) = y$.

Theorem 4.1. *If $\mathfrak{V}$ and $\mathfrak{W}$ are vector spaces over the same scalar field $\mathfrak{F}$, with $\mathfrak{V}$ finite dimensional, and T is a linear transformation of $\mathfrak{V}$ into $\mathfrak{W}$, then*

$$\dim \mathfrak{N}(T) + \dim \mathfrak{R}(T) = \dim \mathfrak{V}. \tag{4.1}$$

If $\mathfrak{W}$ is finite dimensional, then

$$\dim [\mathfrak{R}(T)]^0 + \dim \mathfrak{R}(T) = \dim \mathfrak{W}. \tag{4.2}$$

Suppose that $\mathfrak{V}$ has finite dimension n, and that $\{x^{(1)}, \ldots, x^{(n)}\}$ is a basis for $\mathfrak{V}$. If $\dim \mathfrak{N}(T) = 0$, then $\{Tx^{(1)}, \ldots, Tx^{(n)}\}$ are linearly independent elements of $\mathfrak{R}(T) \in \mathfrak{W}$. As each element of $\mathfrak{R}(T)$ is of the form $T(\sum_{j=1}^n \xi_j x^{(j)}) = \sum_{j=1}^n \xi_j [Tx^{(j)}]$ for suitable scalars $\xi_j \in \mathfrak{F}$, it follows that $\dim \mathfrak{R}(T) = n$ and hence relation (4.1) holds. If $\dim \mathfrak{N}(T) = n$, then T is the zero transformation such that Tx is the zero element of $\mathfrak{W}$ for all $x \in \mathfrak{V}$, and $\dim \mathfrak{R}(T) = 0$, so again relation (4.1) is valid. Now if $\dim \mathfrak{N}(T) = d$, where $0 < d < n$, let $\{x^{(1)}, \ldots, x^{(n)}\}$ be a basis for $\mathfrak{V}$ such that $\{x^{(1)}, \ldots, x^{(d)}\}$ is a basis for the linear subspace $\mathfrak{N}(T)$. It then follows readily that the set $\{Tx^{(d+1)}, \ldots, Tx^{(n)}\}$ is linearly independent and the $n - d$ vectors $Tx^{(k)}$, $d < k \leq n$, form a basis for $\mathfrak{R}(T)$, so that also in this case (4.1) holds. Finally, in case $\mathfrak{W}$ is finite dimensional then relation (4.2) is an immediate consequence of Theorem 3.2.

Of particular importance is the case when $\mathfrak{V}$ and $\mathfrak{W}$ are of the same finite dimension, and for which we have the following basic result.

Theorem 4.2. *If $\mathfrak{V}$ and $\mathfrak{W}$ are finite dimensional vector spaces over the same scalar field $\mathfrak{F}$, with* $\dim \mathfrak{V} = \dim \mathfrak{W} = n$, *and T is a linear transformation of $\mathfrak{V}$ into $\mathfrak{W}$, then* $\dim \mathfrak{N}(T) = \dim [\mathfrak{R}(T)]^0$, *and the following five conditions are equivalent:*

(i) *T is invertible;*

(ii) $\dim \mathfrak{N}(T) = 0$;

(iii) $\dim \mathfrak{R}(T) = n$;

(iv) *if $\{x^{(1)}, \ldots, x^{(n)}\}$ is any basis for $\mathfrak{V}$, then $\{Tx^{(1)}, \ldots, Tx^{(n)}\}$ is a basis for $\mathfrak{W}$;*

(v) *there is some basis $\{x^{(1)}, \ldots, x^{(n)}\}$ for $\mathfrak{V}$ such that $\{Tx^{(1)}, \ldots, Tx^{(n)}\}$ is a basis for $\mathfrak{W}$.*

Suppose that $\mathfrak{V}$ and $\mathfrak{W}$ are finite dimensional vector spaces over the same scalar field $\mathfrak{F}$, of corresponding dimensions n and m, while $X = \{x^{(1)}, \ldots, x^{(n)}\}$ and $Y = \{y^{(1)}, \ldots, y^{(m)}\}$ are bases for $\mathfrak{V}$ and $\mathfrak{W}$, respectively. If $T: \mathfrak{V} \to \mathfrak{W}$ is a linear transformation, then there exists a unique array of scalars $A(T; Y, X) = [A_{\alpha\beta}]$, $(\alpha = 1, \ldots, m; \beta = 1, \ldots, n)$ such that if $x = \sum_{\beta=1}^{n} \xi_\beta x^{(\beta)}$ then $y = \sum_{\alpha=1}^{m} \eta_\alpha y^{(\alpha)}$ is the element of $\mathfrak{W}$ such that $y = Tx$ if and only if

$$\eta_\alpha = \sum_{\beta=1}^{n} A_{\alpha\beta}\xi_\beta, \qquad (\alpha = 1, \ldots, m). \tag{4.3}$$

The scalars $A_{\alpha\beta}$ are clearly determined by the definitive relations

$$Tx^{(\beta)} = \sum_{\alpha=1}^{n} A_{\alpha\beta} y^{(\alpha)}, \qquad (\beta = 1, \ldots, n). \tag{4.4}$$

Conversely, if $[A_{\alpha\beta}]$, $(\alpha = 1, \ldots, m; \beta = 1, \ldots, n)$, is a given array of scalars, and for each $x = \sum_{\beta=1}^{n} \xi_\beta x^{(\beta)}$ of $\mathbf{C}_n$ the corresponding element $y = \sum_{\alpha=1}^{m} \eta_\alpha y^{(\alpha)}$ of $\mathbf{C}_m$ is specified by the (η_α) given in terms of the (ξ_β) by (4.3), then $y = Tx$ is a linear transformation of $\mathfrak{V}$ into $\mathfrak{W}$ with $A(T; Y, X)$ equal to the given $[A_{\alpha\beta}]$.

It is to be emphasized that the $A_{\alpha\beta}$ for the transformation $T: \mathfrak{V} \to \mathfrak{W}$ are dependent upon the involved bases X and Y of the spaces $\mathfrak{V}$ and $\mathfrak{W}$. Moreover, for specified bases X, Y of $\mathfrak{V}$, $\mathfrak{W}$ it is clear that $A(T; Y, X)$ depends linearly upon T in the sense that if $T^{(j)}: \mathfrak{V} \to \mathfrak{W}$, $(j = 1, 2)$, are linear transformations of $\mathfrak{V}$ into $\mathfrak{W}$ with corresponding $A(T^{(j)}; Y, X) = [A^{(j)}_{\alpha\beta}]$, and c_1, c_2 are arbitrary scalars, then $Tx = c_1[T^{(1)}x] + c_2[T^{(2)}x]$ defines a linear transformation $T: \mathfrak{V} \to \mathfrak{W}$ with $A(T; Y, X) = [c_1 A^{(1)}_{\alpha\beta} + c_2 A^{(2)}_{\alpha\beta}]$. In particular, $A^0 = [A^0_{\alpha\beta}]$, with $A^0_{\alpha\beta} = 0$, $(\alpha = 1, \ldots, m; \beta = 1, \ldots, n)$, is such that $A^0 = A(T^0; Y, X)$, where $T^0 x$ is the zero element of $\mathfrak{W}$ for arbitrary

$x \in \mathfrak{V}$, and therefore in the class of linear transformations $T: \mathfrak{V} \to \mathfrak{W}$ the transformation T^0 is the zero transformation satisfying $T + T^0 = T$ for all $T: \mathfrak{V} \to \mathfrak{W}$. Consequently, for specified bases X, Y of $\mathfrak{V}$, $\mathfrak{W}$ the aggregate of $A(T; Y, X)$ corresponding to linear transformations $T: \mathfrak{V} \to \mathfrak{W}$ is itself a vector space over the same scalar field $\mathfrak{F}$.

Suppose that $\mathfrak{V}_1$, $\mathfrak{V}_2$, $\mathfrak{V}_3$ are finite dimensional vector spaces over the same scalar field $\mathfrak{F}$, of corresponding dimensions n, m, p, and with respective bases $X = \{x^{(1)}, \ldots, x^{(n)}\}$, $Y = \{y^{(1)}, \ldots, y^{(m)}\}$, $Z = \{z^{(1)}, \ldots, z^{(p)}\}$. If $T_1: \mathfrak{V}_1 \to \mathfrak{V}_2$ is a linear transformation with $A(T_1; Y, X) = [A_{\alpha\beta}]$, $(\alpha = 1, \ldots, m;$ $\beta = 1, \ldots, n)$, and $T_2: \mathfrak{V}_2 \to \mathfrak{V}_3$ is a linear transformation with $A(T_2; Z, Y) = [B_{\gamma\alpha}]$, $(\gamma = 1, \ldots, p;\ \alpha = 1, \ldots, m)$, then the *product transformation* $T: \mathfrak{V}_1 \to \mathfrak{V}_3$ with value $Tx = T_2[T_1 x]$ has $A(T; Z, X) = [C_{\gamma\beta}]$, with $C_{\gamma\beta} = \sum_{\alpha=1}^{m} B_{\gamma\alpha} A_{\alpha\beta}$, $(\gamma = 1, \ldots, p;\ \beta = 1, \ldots, n)$.

5. HERMITIAN FORMS

If $\mathfrak{V}$ is a vector space over a scalar field $\mathfrak{F}$, then a function ϕ on $\mathfrak{V} \times \mathfrak{V}$ to $\mathfrak{F}$ is called a *bilinear form* on $\mathfrak{V}$ if $\phi(x, y_0)$ is a linear functional in x on $\mathfrak{V}$ for fixed $y_0 \in \mathfrak{V}$, and $\phi(x_0, y)$ is a linear functional in y on $\mathfrak{V}$ for fixed $x_0 \in \mathfrak{V}$. Again, with the specific understanding that the field of scalars is either the field of real numbers or the field of complex numbers, a function ϕ on $\mathfrak{V} \times \mathfrak{V}$ to $\mathfrak{F}$ is called a *sesquilinear form* if $\phi(x, y_0)$ is a linear functional in x on $\mathfrak{V}$ for fixed $y_0 \in \mathfrak{V}$, and $\overline{\phi(x_0, y)}$ is a linear functional in y on $\mathfrak{V}$ for fixed $x_0 \in \mathfrak{V}$. In particular, if $\mathfrak{F}$ is the field of real numbers then ϕ is a sesquilinear form if and only if it is a bilinear form.

If ϕ is a sesquilinear form then for arbitrary elements x, y, z of $\mathfrak{V}$, and arbitrary scalars c, d, we have the following properties:

$$(5.1)\qquad \begin{aligned} \phi(x + y, z) &= \phi(x, z) + \phi(y, z), \\ \phi(x, y + z) &= \phi(x, y) + \phi(x, z), \\ \phi(cx, y) &= c\, \phi(x, y), \\ \phi(x, cy) &= \bar{c}\, \phi(x, y). \end{aligned}$$

A sesquilinear form ϕ which is such that

$$(5.2)\qquad \overline{\phi(x, y)} = \phi(y, x), \quad \textit{for} \quad x \in \mathfrak{V}, y \in \mathfrak{V},$$

is called an *hermitian form*. In particular, if ϕ is an hermitian form then $\phi(x, x)$ is real-valued for $x \in \mathfrak{V}$. An hermitian form ϕ is said to be *non-negative definite* if $\phi(x, x) \geq 0$ for all $x \in \mathfrak{V}$. If ϕ is a non-negative definite form which is such that $\phi(x, x) = 0$ only if $x = \theta$ then ϕ is said to be *positive definite*.

Theorem 5.1. *If ϕ is a non-negative definite hermitian form on the vector space $\mathfrak{V}$, then $\mu[x] = \sqrt{\phi(x, x)}$ has the following properties, where x, y, z denote arbitrary elements of $\mathfrak{V}$ and c is any scalar:*

$$\begin{aligned} &\text{(i)} \ \mu[cx] = |c| \, \mu[x], \\ &\text{(ii)} \ \mu[x] \geq 0, \\ (5.2) \quad &\text{(iii)} \ \mu[x] = 0 \textit{ if and only if } \phi(x, y) = 0 \textit{ for all } y \in \mathfrak{V}, \\ &\text{(iv)} \ |\phi(x, y)| \leq \mu[x] \, \mu[y], \ (\textit{Cauchy-Bunyakovsky-Schwarz inequality}), \\ &\text{(v)} \ \mu[x + y] \leq \mu[x] + \mu[y], \ (\textit{triangle inequality}). \end{aligned}$$

Properties (i) and (ii) are obvious from the definition of μ. To prove (iii), it is to be noted that if x is such that $\phi(x, y) = 0$ for all $y \in \mathfrak{V}$ then, in particular, $\phi(x, x) = 0$ and $\mu[x] = \sqrt{\phi(x, x)} = 0$. Now for arbitrary real numbers s one has the relation

$$(5.3) \quad 0 \leq \mu^2[y + s\phi(y, x)x] = \mu^2[y] + 2s \, |\phi(x, y)|^2 + s^2 \, |\phi(x, y)|^2 \, \mu^2[x].$$

Consequently, if x is such that $\mu[x] = 0$ then

$$0 \leq \mu^2[y] + 2s \, |\phi(x, y)|^2$$

for arbitrary real numbers s, and hence also $\phi(x, y) = 0$.

Now in view of condition (iii) the inequality (iv) holds if $\mu[x] = 0$. Moreover, if $\mu[x] > 0$ then relation (5.3) for $s = -1/\mu^2[x]$ yields the inequality $\mu^2[y] - |\phi(x, y)|^2/\mu^2[x] \geq 0$, which is equivalent to (iv). Finally, inequality (v) results from the identity

$$(5.4) \qquad \mu^2[x + y] = \mu^2[x] + 2 \, \Re\mathrm{e} \, \phi(x, y) + \mu^2[y],$$

and the fact that the inequality $\Re\mathrm{e} \, \phi(x, y) \leq |\phi(x, y)| \leq \mu[x] \, \mu[y]$ implied by (iv) yields the relation

$$\mu^2[x + y] \leq \mu^2[x] + 2\mu[x] \, \mu[y] + \mu^2[y] = (\mu[x] + \mu[y])^2.$$

It is to be remarked that *if ϕ is a positive definite hermitian form on $\mathfrak{V}$, then conclusion* (5.2-iii) *may be replaced by*:

$$\text{(iii)}' \ \mu[x] = 0 \textit{ if and only if } x = \theta.$$

Borrowing geometric terminology, if ϕ is an hermitian form on $\mathfrak{V}$ then two elements x and y of $\mathfrak{V}$ are said to be *ϕ-orthogonal* if $\phi(x, y) = 0$, or equivalently, $\phi(y, x) = 0$. Correspondingly, a set $\{y^{(1)}, \ldots, y^{(k)}\}$ is said to be *ϕ-orthonormal* if

$$(5.5) \qquad \phi(y^{(\beta)}, y^{(\gamma)}) = \delta_{\beta\gamma}, \qquad (\beta, \gamma = 1, \ldots, k).$$

Theorem 5.2. *If ϕ is an hermitian form on $\mathfrak{B}$, $\{y^{(1)}, \ldots, y^{(k)}\}$ is a ϕ-orthonormal system, and $y \in \mathfrak{B}$, then the associated element*

$$z = y - \sum_{\beta=1}^{k} \phi(y, y^{(\beta)})y^{(\beta)} \tag{5.6}$$

satisfies the following relations:

$$\text{(i)} \quad \phi(z, y^{(\gamma)}) = 0, \qquad (\gamma = 1, \ldots, k);$$

$$\text{(ii)} \quad \phi\left(y - \sum_{\beta=1}^{k} c_\beta y^{(\beta)}, y - \sum_{\beta=1}^{k} c_\beta y^{(\beta)}\right) = \phi(z, z) + \sum_{\beta=1}^{k} |c_\beta - \phi(y, y^{(\beta)})|^2, \tag{5.7}$$

for arbitrary scalars $c_1, \ldots, c_k$;

$$\text{(iii)} \quad \phi(y, y) = \phi(z, z) + \sum_{\beta=1}^{k} |\phi(y, y^{(\beta)})|^2.$$

Properties (i) and (ii) are immediate consequences of the definitive properties of an hermitian form, and conclusion (iii) is the special case of (ii) in which $c_\beta = 0$, $(\beta = 1, \ldots, k)$.

Corollary. *If ϕ is a non-negative definite hermitian form on $\mathfrak{B}$, and $\{y^{(1)}, \ldots, y^{(k)}\}$ is a ϕ-orthonormal system, then $\mu[x] = \sqrt{\phi(x, x)}$ satisfies the following conditions:*

(i) *for arbitrary scalars $c_1, \ldots, c_k$, we have*

$$\mu\left[y - \sum_{\beta=1}^{k} c_\beta y^{(\beta)}\right] \geq \mu\left[y - \sum_{\beta=1}^{k} \phi(y, y^{(\beta)})y^{(\beta)}\right],$$

and the equality sign holds if and only if $c_\beta = \phi(y, y^{(\beta)})$, $(\beta = 1, \ldots, k)$;

$$\text{(ii)} \quad \mu^2[y] \geq \sum_{\beta=1}^{k} |\phi(y, y^{(k)})|^2, \tag{5.8}$$

and the equality sign holds if and only if $\mu[y - \sum_{\beta=1}^{k} (y, y^{(\beta)})y^{(\beta)}] = 0$.

Relation (5.7-iii) is known as the *Bessel equality* for ϕ-orthonormal systems $\{y^{(1)}, \ldots, y^{(k)}\}$, and (5.8) is called the *Bessel inequality*.

If $\mathfrak{M}$ is a subspace of $\mathfrak{B}$ of finite nonzero dimension, and ϕ is a positive definite hermitian functional on $\mathfrak{M}$, then $\mathfrak{M}$ has a ϕ-orthonormal basis. Indeed, if $x^{(1)}, \ldots, x^{(d)}$ is any basis for $\mathfrak{M}$, then with the aid of the relations (5.7) it follows that

$$\begin{aligned} y^{(1)} &= \frac{x^{(1)}}{\mu[x^{(1)}]}, \\ y^{(\gamma)} &= \frac{x^{(\gamma)} - \sum_{\beta=1}^{\gamma-1} \phi(x^{(\gamma)}, y^{(\beta)})y^{(\beta)}}{\left[\mu^2[x^{(\gamma)}] - \sum_{\beta=1}^{\gamma-1} |\phi(x^{(\gamma)}, y^{(\beta)})|^2\right]^{1/2}}, \qquad (\gamma = 2, \ldots, d), \end{aligned} \tag{5.9}$$

defines a ϕ-orthonormal basis $\{y^{(1)}, \ldots, y^{(d)}\}$ for $\mathfrak{M}$. Relations (5.9) are known as the *Gram-Schmidt process of orthonormalization of* $x^{(1)}, \ldots, x^{(d)}$.

Problems A.5

1. If $A_{\alpha\beta}$, $(\alpha, \beta = 1, \ldots, n)$, are complex numbers, show that $\phi(x, y) = \sum_{\alpha,\beta=1}^{n} A_{\alpha\beta} x_\beta \bar{y}_\alpha$ is a sesquilinear form on $\mathbf{C}_n$, which is hermitian if and only if $A_{\alpha\beta} = \bar{A}_{\beta\alpha}$, $(\alpha, \beta = 1, \ldots, n)$.

2. Let $\mathfrak{V}$ denote the vector space of all complex-valued continuous functions on $[a, b]$, with vector addition and scalar multiplication by complex numbers defined as in Example (3) of Section 1. If $K(t, s)$ is a continuous complex-valued function of (t, s) on $[a, b] \times [a, b]$, show that

$$\phi(x, y) = \int_a^b \int_a^b K(t, s)\, x(s)\, \overline{y(t)}\, ds\, dt$$

is a sesquilinear form on $\mathfrak{V}$, which is hermitian if and only if $K(t, s) = \overline{K(s, t)}$ for $(t, s) \in [a, b] \times [a, b]$.

3. Let $\mathfrak{V}$ be the vector space of all complex-valued continuously differentiable functions on $[0, \pi]$, with vector addition and scalar multiplication by complex numbers defined as in Example (3) of Section 1. Show that

$$\phi(x, y) = \int_0^\pi [\bar{y}'(t)x'(t) + \bar{y}(t)x(t)]\, dt$$

is an hermitian form on $\mathfrak{V}$, and

$$y^{(\beta)}(t) = \sqrt{2/[\pi(\beta^2 + 1)]}\, \sin \beta t, \qquad \beta = 1, \ldots, k,$$

is a corresponding ϕ-orthonormal set.

Appendix B

Matrices, Systems of Linear Equations

1. ALGEBRA OF MATRICES

In this and following sections it will be supposed that the field $\mathfrak{F}$ of scalars is the field of complex numbers, unless specifically stated otherwise. A rectangular array of scalars

$$(1.1)\quad A = \begin{bmatrix} A_{11} & \cdots & A_{1n} \\ \cdot & \cdots & \cdot \\ \cdot & \cdots & \cdot \\ \cdot & \cdots & \cdot \\ A_{m1} & \cdots & A_{mn} \end{bmatrix} = \begin{bmatrix} A_{1\beta} \\ \cdot \\ \cdot \\ \cdot \\ A_{m\beta} \end{bmatrix} = [A_{\alpha 1} \quad \cdots \quad A_{\alpha n}] = [A_{\alpha\beta}],$$

$$(\alpha = 1, \ldots, m; \beta = 1, \ldots, n),$$

having m rows and n columns, is called an $m \times n$ matrix. If A is an $m \times n$ matrix and B is a $p \times q$ matrix, then A and B are said to be equal, and we write $A = B$, in case $m = p$, $n = q$, and $A_{\alpha\beta} = B_{\alpha\beta}$, $(\alpha = 1, \ldots, m; \beta = 1, \ldots, n)$. The $m \times n$ matrix whose elements are all zero will be denoted by $0^{(m,n)}$, $[0_{\alpha\beta}]$, $(\alpha = 1, \ldots, m;\ \beta = 1, \ldots, n)$, or merely by 0 in case its dimensions are clear from the context. The $n \times n$ identity matrix $[\delta_{\alpha\beta}]$, $(\alpha, \beta = 1, \ldots, n)$ will be denoted by E_n, or merely by E when there is no ambiguity. If $A = [A_{\alpha\beta}]$, $(\alpha = 1, \ldots, m;\ \beta = 1, \ldots, n)$ is an $m \times n$ matrix and c is a scalar, then the matrix $[cA_{\alpha\beta}]$ is designated by cA or Ac; in particular, the symbol $-A$ is used to denote the matrix $(-1)A$. If $A = [A_{\alpha\beta}]$ and $B = [B_{\alpha\beta}]$ are two $m \times n$ matrices then the sum $A + B$ is defined as $A + B = [A_{\alpha\beta} + B_{\alpha\beta}]$. In the class of $m \times n$ matrices addition is clearly commutative, $(A + B = B + A)$, and associative, $(A + [B + C] = [A + B] + C)$. Moreover, $C = B + (-1)A = B + (-A)$ is the unique matrix such that $A + C = B$, and we write $C = B - A$.

If $A = [A_{\alpha\beta}]$ is an $m \times n$ matrix and $M = [M_{\beta\gamma}]$ is an $n \times p$ matrix, then the product matrix AM is defined as the $m \times p$ matrix $AM = [\sum_{\beta=1}^{n} A_{\alpha\beta}M_{\beta\gamma}]$, $(\alpha = 1, \ldots, m; \gamma = 1, \ldots, p)$. If A, B are $m \times n$ matrices and M, N are $n \times p$ matrices, then $(A + B)M = (AM) + (BM)$, and $A(M + N) = (AM) + (AN)$; that is, multiplication is distributive with respect to addition. Moreover, if A is an $m \times n$ matrix, M an $n \times p$ matrix, and R a $p \times q$ matrix, then for multiplication of matrices the associative law $A(MR) = (AM)R$ holds. In particular, the closing paragraphs of Section A.4 define the *matrix A of a linear transformation $T: \mathfrak{B} \to \mathfrak{W}$ relative to prescribed bases X and Y of the respective vector spaces $\mathfrak{B}$ and $\mathfrak{W}$.*

If A is an $m \times n$ matrix and B an $n \times m$ matrix, the two products AB and BA are both defined. If $m \neq n$ these two products are not equal, for one is an $m \times m$ matrix and the other is an $n \times n$ matrix. If $m = n$, it is not true in general that the commutative relation $AB = BA$ holds, as is exemplified by

$$A = \begin{bmatrix} 0 & 1 \\ 0 & 0 \end{bmatrix}, \qquad B = \begin{bmatrix} 0 & 0 \\ 1 & 0 \end{bmatrix}.$$

The matrix obtained upon replacing each element $A_{\alpha\beta}$ by its complex conjugate $\bar{A}_{\alpha\beta}$ is denoted by $\bar{A} = [\bar{A}_{\alpha\beta}]$, and called the *conjugate* of A. If $C = A + B$ is defined, then $\bar{C} = \bar{A} + \bar{B}$; if $D = AB$ is defined, then $\bar{D} = \bar{A}\bar{B}$. For A an $m \times n$ matrix the $n \times m$ matrix obtained by interchanging rows and columns of A is denoted by $\tilde{A}$, and called the *transpose* of A. If $C = A + B$ is defined, then $\tilde{C} = \tilde{A} + \tilde{B}$; if $D = AB$ is defined, then $\tilde{D} = \tilde{B}\tilde{A}$. Finally, for a given matrix A the transpose of its conjugate is equal to the conjugate of its transpose, and is denoted by A^*. A square matrix $A = [A_{\alpha\beta}]$, $(\alpha, \beta = 1, \ldots, n)$, is said to be *symmetric* if $A = \tilde{A}$, *skew-symmetric* if $A = -\tilde{A}$, *hermitian* if $A = A^*$, and *skew-hermitian* if $A = -A^*$. If A is an $n \times n$ matrix, it follows readily that $B = \frac{1}{2}(A + A^*)$ and $C = -\frac{1}{2}i(A - A^*)$ are the unique hermitian matrices such that $A = B + iC$; B is called the *real part* of A, and abbreviated $B = \mathfrak{Re}\, A$; similarly, C is called the *pure imaginary* part of A, and abbreviated $C = \mathfrak{Im}\, A$.

In the following we shall identify the vector $x = (x_\alpha)$, $(\alpha = 1, \ldots, n)$, of $\mathbf{C}_n$ with the $n \times 1$ matrix $[A_{\alpha 1}]$ for which $A_{\alpha 1} = x_\alpha$, $(\alpha = 1, \ldots, n)$. In particular, if A is an $m \times n$ matrix of complex elements, and $x \in \mathbf{C}_n$, then $Ax \in \mathbf{C}_m$ and

$$y = Ax, \qquad x \in \mathbf{C}_n, \tag{1.2}$$

defines a linear transformation $T: \mathbf{C}_n \to \mathbf{C}_m$.

For a given $m \times n$ matrix $A = [A_{\alpha\beta}]$, $(\alpha = 1, \ldots, m; \beta = 1, \ldots, n)$ the vectors $y^{(\beta)}$, $(\beta = 1, \ldots, n)$, of $\mathbf{C}_m$ with $y^{(\beta)}_\alpha = A_{\alpha\beta}$, $(\alpha = 1, \ldots, m)$, are called the *column vectors* of A. As the linear transformation T defined by (1.2) is such that $[y^{(1)}, \ldots, y^{(n)}] = \mathfrak{R}(T)$, the number $r(A) = \dim [y^{(1)}, \ldots, y^{(n)}]$ is clearly the rank of this transformation T by the definition of Section A.4. Moreover, $\mathfrak{N}(A) = \{x \mid x \in \mathbf{C}_n, Ax = 0\}$ is the null-space of this transformation T, and if $k(A) = \dim \mathfrak{N}(A)$ then $r(A) + k(A) = n$ by Theorem A.4.1.

The transpose $\tilde{A}$ of the $m \times n$ matrix $A = [A_{\alpha\beta}]$ has m column vectors $z^{(\alpha)}$, $(\alpha = 1, \ldots, m)$, where $z^{(\alpha)} \in \mathbf{C}_n$ and $z^{(\alpha)}_\beta = A_{\alpha\beta}$, $(\beta = 1, \ldots, n)$. Correspondingly, for $r(\tilde{A}) = \dim [z^{(1)}, \ldots, z^{(m)}]$ and $k(\tilde{A})$ the dimension of $\mathfrak{N}(\tilde{A}) = \{y \mid y \in \mathbf{C}_m, \tilde{A}y = 0\}$, we have that $r(\tilde{A}) + k(\tilde{A}) = m$. Now for the particular basis $\{e^{(1)}, \ldots, e^{(m)}\}$ of $\mathbf{C}_m$, where $e^{(\alpha)} = (e^{(\alpha)}_\gamma)$ with $e^{(\alpha)}_\gamma = \delta_{\alpha\gamma}$, $(\alpha, \gamma = 1, \ldots, m)$, an element $y = (y_\alpha)$ of $\mathbf{C}_m$ has the unique representation $y = \sum_{\alpha=1}^m y_\alpha e^{(\alpha)}$. Correspondingly, a linear functional λ of $\mathbf{C}'_m$ has a unique representation

$$\lambda[y] = \sum_{\alpha=1}^{m} \eta_\alpha y_\alpha, \quad for \quad y = (y_\alpha) \in \mathbf{C}_m.$$

In particular, $\lambda \in [\mathfrak{R}(T)]^0$ if and only if

$$0 = \sum_{\alpha=1}^{m} \eta_\alpha y^{(\beta)}_\alpha = \sum_{\alpha=1}^{m} \eta_\alpha A_{\alpha\beta} = \sum_{\alpha=1}^{m} \eta_\alpha z^{(\alpha)}_\beta, \qquad (\beta = 1, \ldots, n),$$

and consequently $\dim [\mathfrak{R}(T)]^0 = \dim \mathfrak{N}(\tilde{A}) = k(\tilde{A})$. As conclusion (4.2) of Theorem A.4.1 implies that $m = \dim \mathbf{C}_m = \dim [\mathfrak{R}(T)]^0 + \dim \mathfrak{R}(T)$, it then follows that $m = k(\tilde{A}) + r(A) = m - r(\tilde{A}) + r(A)$, and therefore $r(A) = r(\tilde{A})$. The common value r of $r(A)$ and $r(\tilde{A})$ is known as the *rank* of the matrix A, and denoted by $rk\ A$; clearly $rk\ A \leq \min \{m, n\}$.

A square $n \times n$ matrix A is called *nonsingular* or *singular* according as its rank is equal to n or less than n. In particular, if A is nonsingular then for each n-dimensional vector $e^{(\beta)} = (\delta_{\alpha\beta})$, $(\alpha, \beta = 1, \ldots, n)$, there is a unique $x^{(\beta)} \in \mathbf{C}_n$ such that $Ax^{(\beta)} = e^{(\beta)}$, and if B is the $n \times n$ matrix with ordered column vectors $x^{(1)}, \ldots, x^{(n)}$, then $AB = E_n$. Moreover, $A(BA - E_n) = (AB)A - A = A - A = 0$, so that each column vector of $BA - E_n$ belongs to the zero dimensional subspace $\mathfrak{N}(A)$, and hence also $BA = E_n$. Conversely, if there exists a matrix B such that $AB = E_n$, then $A(Bx) = x$ for arbitrary $x \in \mathbf{C}_n$, so that the column vectors of A form a basis for $\mathbf{C}_n$ and A is nonsingular. The matrix B is called the *inverse* of the nonsingular matrix A, and denoted by A^{-1}. In particular, if A is a nonsingular $n \times n$ matrix and $T: \mathbf{C}_n \to \mathbf{C}_n$ is the linear transformation of $\mathbf{C}_n$ into $\mathbf{C}_n$ defined by (1.2), then the transformation $T^{-1}: \mathbf{C}_n \to \mathbf{C}_n$ defined by $x = A^{-1}y$ is the inverse of the transformation T.

Problems B.1

1. If A is an $m \times n$ matrix, show that $rk\ A = rk\ \bar{A} = rk\ A^*$; moreover, $A = 0$ if and only if $A^*A = 0$.

2. Show that if A and B are matrices such that the product matrix AB is defined, then $rk\ (AB) \leq \min\{rk\ A, rk\ B\}$.

3. Show that an $m \times n$ matrix A has rank r if and only if there exists an $m \times r$ matrix P of rank r and an $n \times r$ matrix Q of rank r such that $A = P\tilde{Q}$.

4. Show that the rank of a nonzero matrix $A = [A_{\alpha\beta}]$, $(\alpha = 1, \ldots, m;$ $\beta = 1, \ldots, n)$, is equal to the largest integer k such that there is a $k \times k$ minor

$$\begin{bmatrix} A_{\alpha_1\beta_1} & A_{\alpha_1\beta_2} & \cdots & A_{\alpha_1\beta_k} \\ \cdot & \cdot & \cdots & \cdot \\ \cdot & \cdot & \cdots & \cdot \\ \cdot & \cdot & \cdots & \cdot \\ A_{\alpha_k\beta_1} & \cdot & \cdots & A_{\alpha_k\beta_k} \end{bmatrix},$$

where $1 \leq \alpha_1 < \alpha_2 < \cdots < \alpha_k \leq m$, $1 \leq \beta_1 < \beta_2 < \cdots < \beta_k \leq n$, which is nonsingular.

5. Prove that if A is a nonsingular square matrix and $C = A^{-1}$, then $\bar{A}$, $\tilde{A}$, A^* are nonsingular matrices with respective reciprocals $\bar{C}$, $\tilde{C}$, C^*.

6. Prove that if A and B are nonsingular $n \times n$ matrices then AB is nonsingular, and $(AB)^{-1} = B^{-1}A^{-1}$.

7. If A is an $m \times n$ matrix of rank n, show that there exists an $m \times n$ matrix B such that B^*A is nonsingular; in particular, B may be chosen so that $B^*A = E_n$.

Hint. If $m = n$, then A is nonsingular and one may choose $B = (A^*)^{-1}$. If $n < m$, then there exists an $m \times (m - n)$ matrix C such that the $m \times m$ matrix $P = [A \quad C]$ is nonsingular. If we write $(P^*)^{-1}$ as $[B \quad D]$, where B is $m \times n$ and D is $m \times (m - n)$, then $B^*A = E_n$, $B^*C = 0^{(n,m-n)}$, $D^*A = 0^{(m-n,n)}$ and $D^*C = E_{m-n}$.

8. If A is a nonsingular matrix, show that $A^{-1} = A^*(AA^*)^{-1} = (A^*A)^{-1}A^*$.

2. SOLVABILITY THEOREMS FOR A SYSTEM OF LINEAR ALGEBRAIC EQUATIONS

If $A_{\alpha\beta}$, $(\alpha = 1, \ldots, m;\ \beta = 1, \ldots, n)$ and b_α, $(\alpha = 1, \ldots, m)$, are given complex numbers the solvability theorems for the system of m linear algebraic equations

$$\sum_{\beta=1}^{n} A_{\alpha\beta}x_\beta = b_\alpha, \qquad (\alpha = 1, \ldots, m), \tag{2.1'}$$

are immediate consequences of the results of Section A.4 and the discussion of the above section. In matrix notation system (2.1′) may be written as the vector equation

$$(2.1)\qquad Ax = b,$$

with the understanding that $b = (b_\alpha)$ is a vector of $\mathbf{C}_m$ and $x = (x_\beta)$ is a vector of $\mathbf{C}_n$. For $b = 0$ we have the homogeneous vector equation

$$(2.2)\qquad Ax = 0, \qquad \left\{\sum_{\beta=1}^{n} A_{\alpha\beta} x_\beta = 0, \qquad (\alpha = 1, \ldots, m)\right\}.$$

Associated with this latter equation is the homogeneous conjugate transposed system

$$(2.3)\qquad A^* y = 0, \qquad \left\{\sum_{\alpha=1}^{m} \bar{A}_{\alpha\beta} y_\alpha = 0, \qquad (\beta = 1, \ldots, n)\right\},$$

where it is understood that $y = (y_\alpha)$ is a vector of $\mathbf{C}_m$.

For the transformation $T: \mathbf{C}_n \to \mathbf{C}_m$ with functional value (1.2) in terms of the matrix A, the discussion of the preceding section and Theorem A. 4.1 yield the following results.

Theorem 2.1. *If the $m \times n$ coefficient matrix A is of rank r, then the set of vectors x of $\mathbf{C}_n$ satisfying* (2.2) *is a subspace of $\mathbf{C}_n$ of dimension $n - r$, and the set of vectors y in $\mathbf{C}_m$ satisfying* (2.3) *is a subspace of $\mathbf{C}_m$ of dimension $m - r$.*

Corollary. *If $m < n$, then the set of vectors x of $\mathbf{C}_n$ satisfying* (2.2) *is a subspace of $\mathbf{C}_n$ of dimension at least $n - m$.*

Theorem 2.2. *If the $m \times n$ coefficient matrix A has rank r the vector equation* (2.1) *has a solution x if and only if*

$$(2.4)\qquad y^* b = 0, \qquad \left\{\sum_{\alpha=1}^{m} \bar{y}_\alpha b_\alpha = 0\right\},$$

for all solutions $y = (y_\alpha)$ of the homogeneous conjugate transposed system (2.3). *Moreover, when* (2.4) *holds the general solution $x = (x_\beta)$ of* (2.1) *is*

$$(2.5)\quad x = x^{(0)} + \sum_{\gamma=1}^{n-r} c_\gamma x^{(\gamma)}, \qquad \left\{x_\beta = x_\beta^{(0)} + \sum_{\gamma=1}^{n-r} c_\gamma x_\beta^{(\gamma)}, \qquad (\beta = 1, \ldots, n)\right\},$$

where $x^{(0)} = (x_\beta^{(0)})$ is a particular solution of (2.1), *$x^{(\gamma)} = (x_\beta^{(\gamma)})$, $(\gamma = 1, \ldots, n - r)$, are linearly independent solutions of* (2.2), *and $c_1, \ldots, c_{n-r}$ are arbitrary scalars.*

If $r = m$ then (2.4) holds in a trivial fashion, since in this case the only solution of (2.3) is the m-dimensional zero vector. Correspondingly, when

$r = n$ the only solution of (2.2) is the n-dimensional zero vector, and in this case whenever condition (2.4) holds the solution of (2.2) is unique and the terms involving the $x^{(\gamma)}$ do not appear in (2.5).

The non-negative integer $n - r$ is called the *index of compatibility* of (2.2); if this index is zero then (2.2) is also said to be *incompatible*. Correspondingly, the index of compatibility of (2.3) is $m - r$.

For the special case of $m = n$, we have the following result.

Theorem 2.3. *If A is a square $n \times n$ matrix then the index of compatibility of* (2.2) *is equal to the index of compatibility of* (2.3)*; if A is nonsingular then for arbitrary vectors b of $\mathbf{C}_n$ the equation* (2.1) *has a unique solution given by $x = A^{-1}b$.*

3. THE E. H. MOORE GENERALIZED INVERSE OF A MATRIX

Let A be an $m \times n$ matrix of complex numbers, and as in Section 1 above let $T:\mathbf{C}_n \to \mathbf{C}_m$ be the linear transformation with functional value (1.2). If $r = rk\, A$ then the null space $\mathfrak{N}(T)$ has dimension $n - r$, and if $r < n$ we shall denote by X an $n \times (n - r)$ matrix whose column vectors form a basis for $\mathfrak{N}(T)$. Similarly, $T^*:\mathbf{C}_m \to \mathbf{C}_n$ with functional value $x = A^*y$ has null-space $\mathfrak{N}(T^*)$ of dimension $m - r$, and if $r < m$ we shall denote by Y an $m \times (m - r)$ matrix whose column vectors form a basis for $\mathfrak{N}(T^*)$.

An $n \times m$ matrix B is called a *generalized* (*right-hand*) *inverse* of A if for $y \in \mathfrak{R}(T)$ the element $x = By$ is such that $Ax = y$; that is, the linear mapping $T^{\#}:\mathbf{C}_m \to \mathbf{C}_n$ with functional value $x = By$ is such that if $y \in \mathfrak{R}(T)$ then By belongs to the inverse image of y under the mapping T. Now if $x^0 \in \mathbf{C}_n$ then $Ax^0 \in \mathfrak{R}(T)$, so that if B is a generalized right-hand inverse of A then $ABAx^0 = Ax^0$, and hence $(ABA - A)x^0 = 0$. That is, the matrix $ABA - A$ is such that $\mathfrak{N}(ABA - A) = \mathbf{C}_n$, and hence $ABA - A$ is the zero matrix. Conversely, if $ABA = A$ then clearly B is a generalized inverse of A, so that we have the following result.

***Lemma* 3.1.** *A matrix B is a generalized inverse of A if and only if $ABA = A$.*

Corollary. *If B is generalized inverse of A, then $\bar{B}$, $\tilde{B}$, A^* are generalized inverses of the respective matrices $\bar{A}$, $\tilde{A}$, A^*.*

***Lemma* 3.2.** *If A is an $m \times n$ matrix of rank r, while the $n \times (n - r)$ matrix X and $m \times (m - r)$ matrix Y are defined as indicated above, then:* (a) *if $r < n$ and Θ is an $n \times (n - r)$ matrix such that Θ^*X is nonsingular, then the $(m + n - r) \times n$ matrix*

$$A_\Theta = \begin{bmatrix} A \\ \Theta^* \end{bmatrix} \tag{3.1-a}$$

has rank n; (b) *if* $r < m$, *and* Ψ *is an* $m \times (m - r)$ *matrix such that* $Y^*\Psi$ *is nonsingular, then the* $m \times (n + m - r)$ *matrix*

$$A^{\Psi} = [A \quad \Psi] \tag{3.1-b}$$

has rank m; (c) *if* $r < n$ *and* $r < m$, *then the* $(m + n - r) \times (n + m - r)$ *matrix*

$$A_{\Theta}^{\Psi} = \begin{bmatrix} A & \Psi \\ \Theta^* & 0 \end{bmatrix} \tag{3.1-c}$$

is nonsingular.

For brevity, detailed proof is limited to the case (c), as the procedure in case (a) or (b) is the same with certain matrices deleted and corresponding abbreviation in argument. It is to be remarked that the existence of matrices Θ and Ψ such that the matrices Θ^*X and $Y^*\Psi$ are nonsingular is a consequence of the result of Problem 7 of B.1. If σ and τ are vectors of $\mathbf{C}_n$ and $\mathbf{C}_{m-r}$, respectively, and such that $A\sigma + \Psi\tau = 0$, $\Theta^*\sigma = 0$, then $0 = Y^*[A\sigma + \Psi\tau] = Y^*\Psi\tau$, and $\tau = 0$ since $Y^*\Psi$ is nonsingular. Then $A\sigma = 0$ implies that there exists a vector λ of $\mathbf{C}_{n-r}$ such that $\sigma = X\lambda$, and therefore $0 = \Theta^*\sigma = \Theta^*X\lambda$. The nonsingularity of the matrix Θ^*X then implies $\lambda = 0$ and consequently $\sigma = 0$, thus leading to the conclusion that the matrix (3.1-c) is nonsingular.

Theorem 3.1. *If A is an* $m \times n$ *matrix of rank r, and matrices* Θ, Ψ *are specified as in Lemma* 3.2, *then there exists a unique generalized inverse* $B = A^{\#}_{\Theta,\Psi}$ *of A such that*

$$\Theta^*B = 0, \qquad B\Psi = 0. \tag{3.2}$$

Let the matrix C be defined as follows:

$$\begin{aligned} &\text{(a) } C = A, \quad \textit{if} \quad r = m = n; \\ &\text{(b) } C = A_{\Theta}, \quad \textit{if} \quad r = m < n; \\ &\text{(c) } C = A^{\Psi}, \quad \textit{if} \quad r = n < m; \\ &\text{(d) } C = A_{\Theta}^{\Psi}, \quad \textit{if} \quad r < m, r < n. \end{aligned} \tag{3.3}$$

In all cases C is a nonsingular $(n + m - r) \times (n + m - r)$ matrix, and in the respective cases C^{-1} is of the following form, where B is an $n \times m$ matrix.

$$\begin{aligned} &\text{(a) } C^{-1} = B, \\ &\text{(b) } C^{-1} = [B \quad X\Gamma], \quad \textit{where} \quad \Gamma = (\Theta^*X)^{-1}, \\ &\text{(c) } C^{-1} = \begin{bmatrix} B \\ \Delta^*Y^* \end{bmatrix}, \quad \textit{where} \quad \Delta = (\Psi^*Y)^{-1}, \\ &\text{(d) } C^{-1} = \begin{bmatrix} B & X\Gamma \\ \Delta^*Y^* & 0 \end{bmatrix}, \quad \textit{where } \Gamma \textit{ and } \Delta \textit{ are defined as above.} \end{aligned} \tag{3.4}$$

Again, for brevity the detailed discussion is limited to the case (d), as in the other cases the procedure is the same with certain matrices deleted and corresponding abbreviation in detail. In terms of the component matrices, the fact that A_Θ^Ψ has its inverse of the form (3.4-d) yields the following matrix equations,

$$(3.5)\quad \begin{aligned} &(a)\quad \begin{aligned} AB + \Psi\Delta^* Y^* &= E, & AX\Gamma &= 0,\\ \Theta^* B &= 0, & \Theta^* X\Gamma &= E; \end{aligned}\\ &(b)\quad \begin{aligned} BA + X\Gamma\Theta^* &= E, & B\Psi &= 0,\\ \Delta^* Y^* A &= 0, & \Delta^* Y^*\Psi &= E. \end{aligned} \end{aligned}$$

In general, from these equations it follows that

$$(3.6)\quad \begin{aligned} ABA - A &= 0, & BAB - B &= 0,\\ B\Psi &= 0, & AX &= 0,\\ \Theta^* B &= 0, & Y^* A &= 0. \end{aligned}$$

In particular, B is a generalized inverse of A satisfying (3.2).

Now if B and B_1 are generalized inverses of A satisfying the conditions (3.2), then $D = B - B_1$ is such that

$$ADA = 0, \qquad D\Psi = 0, \qquad \Theta^* D = 0.$$

From the equation $ADA = 0$ it follows that there exists an $(n - r) \times n$ matrix Q such that $DA = XQ$, and $\Theta^* D = 0$ implies that $0 = \Theta^* DA = \Theta^* XQ$, and $Q = 0$ since the matrix $\Theta^* X$ is nonsingular. Hence $DA = 0$, and therefore there exists an $n \times (m - r)$ matrix P such that $D = PY^*$. Consequently, $0 = D\Psi = PY^*\Psi$ and in view of the nonsingularity of $Y^*\Psi$ it follows that $P = 0$, and hence $B - B_1 = D = PY^* = 0$.

The following results are immediate consequences of the above definitions of the matrices (3.3), and the form of their inverses in the respective cases, with the usual understanding as to the deletion of certain matrices in cases (a), (b), (c).

***Corollary* 1.** *A and $A^{\#}_{\Theta,\Psi}$ are of the same rank, and $(A^{\#}_{\Psi,\Theta})^{\#}_{\Psi,\Theta} = A$.*

***Corollary* 2.** *If Λ_1 and Λ_2 are nonsingular matrices of dimensions $n - r$ and $m - r$, respectively, then $A^{\#}_{\Theta\Lambda_1,\Psi\Lambda_2} = A^{\#}_{\Theta,\Psi}$.*

Theorem 3.2. *The most general generalized inverse B of A is of the form*

$$(3.7)\quad B = B_0 + XH + KY^*,$$

where B_0 is any particular generalized inverse, H is an $(n - r) \times m$ matrix, and K is an $n \times (m - r)$ matrix.

If B_0 is a particular generalized inverse of A, and B is an arbitrary $n \times m$ matrix, then $D = B - B_0$ satisfies the equation $ADA = ABA - A$, so that B is a generalized inverse of A if and only if $ADA = 0$. If B is of the form (3.7), then $ADA = A[XH + KY^*]A = (AX)HA + AK(Y^*A) = 0 + 0 = 0$. Conversely, if $ADA = 0$ there exists an $(n - r) \times n$ matrix Q such that $DA = XQ$, and from the first equation of (3.5-a) we have $D = DAB + D\Psi\Delta^* Y^* = X(QB) + (D\Psi\Delta^*)Y^*$, which is of the form (3.7) with $H = QB$ and $K = D\Psi\Delta^*$.

Theorem 3.3. *$B = A^{\#}_{X,Y}$ is the unique generalized inverse of A satisfying the conditions*

$$(3.8)\qquad \text{(a) } (AB)^* = AB; \qquad \text{(b) } (BA)^* = BA; \qquad \text{(c) } BAB = B.$$

For $\Theta = X$ and $\Psi = Y$, in the determination of $A^{\#}_{X,Y}$ as in the proof of Theorem 3.1 the matrices Γ and Δ are the respective hermitian matrices $[X^*X]^{-1}$ and $[Y^*Y]^{-1}$. From the first equations in (3.5-a) and (3.5-b) it then follows that

$$(3.9)\qquad AA^{\#}_{X,Y} + Y(Y^*Y)^{-1}Y^* = E, \qquad A^{\#}_{X,Y}A + X(X^*X)^{-1}X^* = E,$$

and consequently $AA^{\#}_{X,Y}$ and $A^{\#}_{X,Y}A$ are hermitian. Moreover, since $A^{\#}_{X,Y}Y = 0$, it follows from the first equation of (3.9) that $B = A^{\#}_{X,Y}$ satisfies condition (3.8-c).

On the other hand, if B_0 and B are generalized inverses of A such that AB_0, B_0A and AB, BA are hermitian, then $D = B - B_0$ satisfies the conditions $ADA = 0$, $AD = (AD)^*$, $DA = (DA)^*$. Therefore, $0 = (ADA)^* = A^*(AD)^* = A^*AD$, so that $D^*A^*AD = (AD)^*(AD) = 0$, and consequently $AD = 0$. Correspondingly, $0 = (ADA)^* = (DA)^*A^* = DAA^*$, so that $(DA)(DA)^* = (DAA^*)D^* = 0$, and $DA = 0$. Moreover, if B_0 and B are individually solutions of (3.8-c) then $D = B - B_0$ satisfies $B_0AD + DAB_0 + DAD = D$, and since $AD = 0$ and $DA = 0$ it follows that $D = 0$ and $B = B_0$.

Theorem 3.4. *If B is a generalized inverse of A, then A is a generalized inverse of B if and only if the matrices H, K in the representation*

$$(3.7')\qquad B = A^{\#}_{X,Y} + XH + KY^*$$

are such that

$$(3.10)\qquad HAK = (X^*X)^{-1}X^*K + HY(Y^*Y)^{-1}.$$

Since A is a generalized inverse of B if and only if $BAB = B$, the stated result follows from the relation

$$(3.11)\qquad BAB - B = (A^{\#}_{X,Y} + XH + KY^*)\,A(A^{\#}_{X,Y} + XH + KY^*) - (A^{\#}_{X,Y} + XH + KY^*),$$

and the fact that the equations (3.9) imply that the right-hand member of (3.11) is equal to

$$X[HAK - (X^*X)^{-1}X^*K - HY(Y^*Y)^{-1}]Y^*. \tag{3.12}$$

Problems B.3

1. Determine the generalized inverse $A^{\#}_{\Theta,\Psi}$, as defined in Theorem B.3.1, for the 2×3 matrix

$$A = \begin{bmatrix} 0 & -1 & 0 \\ 0 & -1 & 0 \end{bmatrix}, \quad \text{with} \quad \Theta = \begin{bmatrix} 1 & 0 \\ 0 & 1 \\ 0 & 1 \end{bmatrix}, \qquad \Psi = \begin{bmatrix} 1 \\ 0 \end{bmatrix}.$$

2. Determine the most general generalized inverse of each of the following matrices

$$\text{(a) } A = \begin{bmatrix} 1 & 2 \\ 0 & 0 \end{bmatrix}, \qquad \text{(b) } A = [0 \quad 0].$$

3. Prove that if $y \in \mathbf{C}_m$, then $x^0 = A^{\#}_{X,Y}y$ is such that

$$|Ax - y| \geq |Ax^0 - y|, \quad \textit{for all} \quad x \in \mathbf{C}_n;$$

moreover, if $x \in \mathbf{C}_n$, $x \neq x^0$, and $|Ax - y| = |Ax^0 - y|$, then $|x| > |x^0|$.

4. NOTES AND REMARKS

The generalized inverse $A^{\#}_{X,Y}$ of Theorem 3.3 was introduced by E. H. Moore [1]; see also [2; p. 8 and pp. 202–209]. Since then it has been rediscovered by many authors, notably by R. Penrose, (*A generalized inverse for matrices*, Proc. Cambridge Philosophical Soc., LI, (1955), 406–413). A rather comprehensive bibliography on generalized inverses is to be found in *Proceedings of a Symposium on the Theory and Application of Generalized Inverses of Matrices*, held at the Texas Technological College, Lubbock, Texas, March 1968, Texas Technological College Mathematics Series, No. 4.

Appendix C

Determinants

1. DEFINITION

The determinant of the $n \times n$ matrix $A = [A_{\alpha\beta}]$, $(\alpha, \beta = 1, \ldots, n)$, written

$$D = \det \begin{bmatrix} A_{11} & \cdots & A_{1n} \\ \cdot & & \\ \cdot & & \\ \cdot & & \\ A_{n1} & \cdots & A_{nn} \end{bmatrix} = \det [A_{\alpha\beta}] = \det A, \tag{1.1}$$

is defined as the sum of the $n!$ terms

$$(-1)^J A_{1\beta_1} A_{2\beta_2} \cdots A_{n\beta_n}, \tag{1.2}$$

where $\beta_1, \beta_2, \ldots, \beta_n$ is an arrangement of $1, 2, \ldots, n$, derived from the latter by J successive interchanges of pairs of elements. In order to prove that the factor $(-1)^J$ in (1.2) is well-defined, note first that any arrangement $\beta_1, \ldots, \beta_n$ of $1, 2, \ldots, n$ is obtainable from the latter by successive interchanges. To show that for a given arrangement the number of such interchanges is either always even or always odd, it may be verified readily that the interchange of two subscripts α and β changes the sign of the function

$$\prod_{\alpha,\beta=1;\alpha<\beta}^{n} (x_\alpha - x_\beta),$$

and the stated result follows from the consideration of the value of this function under the substitution of $\beta_1, \beta_2, \ldots, \beta_n$ for $1, 2, \ldots, n$, respectively.

2. PROPERTIES

Fundamental properties of determinants of $n \times n$ matrices are as follows:

(a) $\det 0^{(n,n)} = 0$, and $\det E_n = 1$.

(b) $\overline{\det A} = \det \bar{A}$, $\det A = \det \tilde{A}$, and $\det A^* = \overline{\det A}$.

(c) If B is obtained from A by the interchange of two rows, then $\det B = -\det A$; in particular, if two rows of A are equal then $\det A = 0$.

(d) If B is obtained from A by multiplying a row of A by a scalar c, then $\det B = c \det A$.

(e) If the elements in the k-th row of A are of the form $A_{k\beta} = A^0_{k\beta} + A^1_{k\beta}$, ($\beta = 1, \ldots, n$), then $\det A = \det A^0 + \det A^1$, where A^0, A^1 are obtained from A by replacing the k-th row by $A^0_{k1}, \ldots, A^0_{kn}$ and $A^1_{k1}, \ldots, A^1_{kn}$, respectively.

(f) If B is obtained from A by adding to one row a linear combination of the other rows, then $\det B = \det A$.

(g) If $C_{\alpha\beta}$ is the *cofactor* of $A_{\alpha\beta}$ in A, that is, $C_{\alpha\beta}$ equals $(-1)^{\alpha+\beta}$ times the determinant of the minor obtained by deleting the α-th row and β-th column of A, then

$$\text{(2.1)}\quad \begin{aligned} &\text{(a)}\ \sum_{\beta=1}^{n} A_{\alpha\beta} C_{\gamma\beta} = \delta_{\alpha\gamma} \det A, \qquad (\alpha, \gamma = 1, \ldots, n); \\ &\text{(b)}\ \sum_{\alpha=1}^{n} A_{\alpha\beta} C_{\alpha\gamma} = \delta_{\beta\gamma} \det A, \qquad (\beta, \gamma = 1, \ldots, n). \end{aligned}$$

(h) $\det AB = (\det A)(\det B)$.

(i) $\det A \neq 0$ if and only if A is nonsingular; for a nonsingular matrix A the elements of the inverse matrix $A^{-1} = [A^{-1}_{\alpha\beta}]$ are given by

$$\text{(2.2)}\qquad A^{-1}_{\alpha\beta} = C_{\beta\alpha}/\det A, \qquad (\alpha, \beta = 1, \ldots, n).$$

Properties (a)–(e) are ready consequences of the definition of a determinant, while property (f) follows from the properties (e) and (c). In order to establish (2.1), one may derive the first relation of (2.1-a) for $\alpha = 1$, $\gamma = 1$ from the definition of a determinant, and use certain of the preceding results to obtain the remaining relations of (2.1).

By repeated application of (e) and (d) it follows that

$$\det AB = \begin{vmatrix} \sum_{\beta_1=1}^{n} A_{1\beta_1} B_{\beta_1 1} & \cdots & \sum_{\beta_1=1}^{n} A_{1\beta_1} B_{\beta_1 n} \\ \vdots & \vdots & \vdots \\ \sum_{\beta_n=1}^{n} A_{n\beta_n} B_{\beta_n 1} & \cdots & \sum_{\beta_n=1}^{n} A_{n\beta_n} B_{\beta_n n} \end{vmatrix},$$

$$= \sum_{\beta_1=1}^{n} \cdots \sum_{\beta_n=1}^{n} A_{1\beta_1} A_{2\beta_2} \cdots A_{n\beta_n} \begin{vmatrix} B_{\beta_1 1} & \cdots & B_{\beta_1 n} \\ \cdot & \cdots & \cdot \\ B_{\beta_n 1} & \cdots & B_{\beta_n n} \end{vmatrix}.$$

In view of (c), a determinant involving elements of B in this last expression is zero unless $\beta_1, \beta_2, \ldots, \beta_n$ are mutually distinct, in which case $\beta_1, \ldots, \beta_n$ is an arrangement of $1, 2, \ldots, n$, and the result (h) is then a consequence of (c).

If A is nonsingular, then by a result of Section 2 of Appendix B there is an inverse matrix B such that $E = AB$; by (a) and (h) it follows that $1 = (\det A)(\det B)$, and hence $\det A \neq 0$. If $\det A \neq 0$ then it follows from (2.1) that A has an inverse A^{-1} with elements given by (2.2), and hence A is nonsingular.

Appendix D

Geometric and Analytic Aspects of Euclidean Space

1. A METRIC FOR $\mathbf{C}_n$

For x, y vectors of $\mathbf{C}_n$ the numerically valued *inner product*

$$(x, y) = \sum_{\alpha=1}^{n} x_\alpha \bar{y}_\alpha = \tilde{x}\bar{y} = y^* x \tag{1.1}$$

is a positive definite hermitian form $\phi(x, y)$ in the sense of Section A.5, and therefore has the following properties, where x, y, z denote arbitrary vectors of $\mathbf{C}_n$ and c is any scalar:

(1.2)
(a) $(cx, y) = c(x, y)$;
(b) $(x + y, z) = (x, z) + (y, z)$;
(c) $(x, y) = \overline{(y, x)}$;
(d) $(x, x) \geq 0$;
(e) $(x, x) = 0$ *if and only if* $x = 0$.

The *absolute value*, or *length*, of a vector x is defined as the non-negative square root of $(x, x) = \sum_{\alpha=1}^{n} |x_\alpha|^2$, and denoted by $|x|$. The following properties are then direct consequences of Theorem A.5.1:

(1.3)
(a) $|cx| = |c|\,|x|$;
(b) $|x| \geq 0$, *and* $|x| = 0$ *if and only if* $x = 0$;
(c) $|(x, y)| \leq |x|\,|y|$, (*Lagrange-Cauchy inequality*);
(d) $|x + y| \leq |x| + |y|$, (*triangle inequality*).

As an immediate consequence of the definition of the length of a vector we have the inequality

$$\max_{\alpha=1}^{n} |x_\alpha| \leq |x| \leq \sqrt{n} \max_{\alpha=1}^{n} |x_\alpha|, \quad \textit{for} \quad x \in \mathbf{C}_n. \tag{1.4}$$

The function $d(x, y) = |x - y|$ is a *distance function* or *metric* for $\mathbf{C}_n$, satisfying the following conditions:

$$\begin{aligned} &\text{(a) } d(x, y) \geq 0; \\ &\text{(b) } d(x, y) = 0 \quad \textit{if and only if} \quad x = y; \\ &\text{(c) } d(x, y) = d(y, x); \\ &\text{(d) } d(x, z) \leq d(x, y) + d(z, y). \end{aligned} \tag{1.5}$$

The linear space $\mathbf{C}_n$ with this metric is commonly called complex n-dimensional Euclidean space; similarly, $\mathbf{R}_n$ with the above defined metric is referred to as real n-dimensional Euclidean space.

As a special instance of ϕ-orthogonality introduced in Section A.5, two vectors x, y of $\mathbf{C}_n$ are said to be orthogonal if $(x, y) = 0$, or equivalently $(y, x) = 0$. Correspondingly, a set of vectors $y^{(1)}, \ldots, y^{(k)}$ is called an *orthonormal* (orthogonal and normed) system in $\mathbf{C}_n$ if $(y^{(\beta)}, y^{(\gamma)}) = \delta_{\beta\gamma}$, $(\beta, \gamma = 1, \ldots, k)$. For y an arbitrary element of $\mathbf{C}_n$ the following relations are special cases of the results of Theorem 5.2 and its Corollary:

$$\left(y - \sum_{\beta=1}^{k} (y, y^{(\beta)}) y^{(\beta)}, y^{(\gamma)}\right) = 0, \qquad (\gamma = 1, \ldots, k); \tag{1.6}$$

$$\left| y - \sum_{\beta=1}^{k} c_\beta y^{(\beta)} \right|^2 = \left| y - \sum_{\beta=1}^{k} (y, y^{(\beta)}) y^{(\beta)} \right|^2 + \sum_{\beta=1}^{k} |c_\beta - (y, y^{(\beta)})|^2, \tag{1.7}$$

for arbitrary complex $c_1, \ldots, c_k$;

$$|y|^2 = \left| y - \sum_{\beta=1}^{k} (y, y^{(\beta)}) y^{(\beta)} \right|^2 + \sum_{\beta=1}^{k} |(y, y^{(\beta)})|^2, \quad \textit{(Bessel equality)}; \tag{1.8}$$

$$|y|^2 \geq \sum_{\beta=1}^{k} |(y, y^{(\beta)})|^2, \quad \textit{(Bessel inequality)}. \tag{1.9}$$

In view of (1.8) the equality sign in (1.9) holds for all vectors y of $\mathbf{C}_n$ if and only if $[y^{(1)}, \ldots, y^{(k)}]$ is a basis for $\mathbf{C}_n$, that is, if and only if $k = n$.

If $\mathbf{M}$ is a nonzero dimensional subspace of $\mathbf{C}_n$, and $x^{(1)}, \ldots, x^{(d)}$ is a basis for $\mathbf{M}$, then the Gram-Schmidt orthonormalization process yields the set of vectors

$$\begin{aligned} y^{(1)} &= \left(\frac{1}{|x^{(1)}|}\right) x^{(1)}, \\ y^{(\gamma)} &= \frac{x^{(\gamma)} - \sum_{\beta=1}^{\gamma-1} (x^{(\gamma)}, y^{(\beta)}) y^{(\beta)}}{[|x^{(\gamma)}|^2 - \sum_{\beta=1}^{\gamma-1} |(x^{(\gamma)}, y^{(\beta)})|^2]^{1/2}}, \qquad (\gamma = 2, \ldots, d), \end{aligned} \tag{1.10}$$

which forms an orthonormal basis for $\mathbf{M}$.

If A is an $m \times n$ matrix, then the supremum of

$$\{|Ax| \mid x \in \mathbf{C}_n, |x| \leq 1\}$$

is called the *norm* of A, and denoted by $\nu[A]$. From Problem 2 of the following set of problems it follows that $\nu[A]$ does not exceed $(\sum_{\alpha=1}^{m} \sum_{\beta=1}^{n} |A_{\alpha\beta}|^2)^{1/2}$, which in turn does not exceed $\sum_{\alpha=1}^{m} \sum_{\beta=1}^{n} |A_{\alpha\beta}|$.

Problems D.1

1. Prove that if the vector y of $\mathbf{C}_n$ is such that $|(y, z)| \leq K|z|$ for all z of $\mathbf{C}_n$, then $|y| \leq K$.

2. If $A = [A_{\alpha\beta}]$, $(\alpha = 1, \ldots, m;\ \beta = 1, \ldots, n)$, and y is a vector of $\mathbf{C}_n$, show that

$$|Ay| \leq |y| \left[\sum_{\alpha=1}^{m} \sum_{\beta=1}^{n} |A_{\alpha\beta}|^2 \right]^{1/2}.$$

Hint. Use (1.3-c) to obtain bounds for the individual components of Ay.

3. If $z^{(\beta)}$, $(\beta = 1, \ldots, k)$, are vectors of $\mathbf{C}_n$, prove the following results for the (*Gram*) *matrix* $G = [G_{\gamma\beta}] = [(z^{(\beta)}, z^{(\gamma)})]$, $(\beta, \gamma = 1, \ldots, k)$:

(a) G is singular if and only if the set $z^{(1)}, \ldots, z^{(k)}$ is linearly dependent;

(b) $0 \leq \det G \leq \prod_{\beta=1}^{k} |z^{(\beta)}|^2$.

Hints. (a) Show that $\sum_{\beta=1}^{k} G_{\gamma\beta} c_\beta = 0$, $(\gamma = 1, \ldots, k)$, if and only if $|c_1 z^{(1)} + \cdots + c_k z^{(k)}| = 0$.

(b) Note that in view of (a) one need consider only the case of the set $z^{(1)}, \ldots, z^{(k)}$ linearly independent. Proceed by mathematical induction, noting that if $1 \leq m < k$ and $y^{(1)}, \ldots, y^{(m)}$ is an orthonormal basis for $[z^{(1)}, \ldots, z^{(m)}]$, then (1.6), (1.8) for $y = z^{(m+1)}$ imply that upon subtracting from the $(m + 1)$-st column of $[(z^{(\beta)}, z^{(\gamma)})]$, $(\beta, \gamma = 1, \ldots, m + 1)$, suitable multiples of the preceding columns, the first m elements of this column reduce to zero, while the new element in the $(m + 1)$-st row and column is non-negative and not greater than $|z^{(m+1)}|^2$.

4. For an $n \times n$ matrix $A = [A_{\alpha\beta}]$ prove the following (*Hadamard*) *determinant inequality*

$$|\det A|^2 \leq \prod_{\beta=1}^{n} \left(\sum_{\alpha=1}^{n} |A_{\alpha\beta}|^2 \right);$$

in particular, if $|A_{\alpha\beta}| \leq c$, $(\alpha, \beta = 1, \ldots, n)$, then $|\det A| \leq n^{n/2} c^n$.

Hint. If $z^{(1)}, \ldots, z^{(n)}$ denote the column vectors of A, show that $|\det A|^2 = \det [A^* A] = \det [(z^{(\beta)}, z^{(\gamma)})]$, and apply result (b) of Problem 3.

5. Prove that if $\mathbf{M}$ is a subspace of $\mathbf{C}_n$ then for each y of $\mathbf{C}_n$ the function of z equal to $|y - z|$ has a unique minimum on $\mathbf{M}$; that is, there is a unique vector $z = z(y)$ of $\mathbf{M}$ such that $|y - z(y)| \leq |y - z|$ for all z of $\mathbf{M}$. Moreover: (i) $y - z(y)$ is orthogonal to every vector of $\mathbf{M}$; (ii) $z(y) = Ay$, where A is an hermitian $n \times n$ matrix such that $A^2 = A$. {The point $z(y)$ is called the (*orthogonal*) *projection* of y on $\mathbf{M}$.}

Hint. For $y^{(1)}, \ldots, y^{(k)}$ an orthonormal basis for $\mathbf{M}$ use (1.7) and (1.6); show that (ii) holds for $A_{\alpha\beta} = \sum_{\gamma=1}^{n} y_\alpha^{(\gamma)} \bar{y}_\beta^{(\gamma)}$.

6. Prove that:

(a) $|x + y|^2 + |x - y|^2 = 2[|x|^2 + |y|^2]$, *for* $x \in \mathbf{C}_n$, $y \in \mathbf{C}_n$;

(b) $4(x, y) = |x + y|^2 - |x - y|^2$, *for* $x \in \mathbf{R}_n$, $y \in \mathbf{R}_n$,

(c) $4(x, y) = |x + y|^2 - |x - y|^2 + i\,|x + iy|^2 - i\,|x - iy|^2$,

for $x \in \mathbf{C}_n, y \in \mathbf{C}_n$.

7. An $n \times n$ matrix is said to be *unitary* if the column vectors of A form an orthonormal system in $\mathbf{C}_n$. Prove that each of the following conditions is equivalent to A being unitary:

(a) $A^* = A^{-1}$.

(b) $(Ax, Ay) = (x, y)$, *for* $x \in \mathbf{C}_n$, $y \in \mathbf{C}_n$;

(c) $|Ax| = |x|$, *for* $x \in \mathbf{C}_n$.

8. If A is an $n \times n$ matrix which is hermitian, show that $A[x] = (Ax, x) = \sum_{\alpha,\beta=1}^{n} \bar{x}_\alpha A_{\alpha\beta} x_\beta$ is an hermitian form in the sense of Section A.5, and for $x \in \mathbf{C}_n$, $y \in \mathbf{C}_n$ we have

$$4(Ax, y) = A[x + y] - A[x - y] + iA[x + iy] - iA[x - iy].$$

Moreover, if A is *non-negative definite*, (i.e., A is hermitian and $A[x] \geq 0$ for arbitrary $x \in \mathbf{C}_n$), in which case we write $A \geq 0$, then:

$$\text{(1.11)} \quad \begin{aligned} &\text{(a)}\ A^k \geq 0, \quad k = 2, 3, \ldots; \\ &\text{(b)}\ A[x] = 0 \quad \text{if and only if} \quad Ax = 0; \\ &\text{(c)}\ |(Ax, y)| \leq \sqrt{(Ax, x)}\,\sqrt{(Ay, y)}, \quad \textit{for} \quad x \in \mathbf{C}_n, \quad y \in \mathbf{C}_n. \end{aligned}$$

Hints. (a) $(A^{2r}x, x) = (A^r x, A^r x)$; $(A^{2r+1}x, x) = (A[A^r x], A^r x)$;

(b), (c) See Theorem A.5.1.

9. If A is a nonsingular hermitian matrix, show that A^{-1} is also hermitian; moreover, if $A > 0$, (i.e., $A[x]$ is positive definite), then $A^{-1} > 0$.

10. If A is an $m \times n$ matrix, show that:

$\nu[cA] = |c|\,\nu[A]$, *for arbitrary scalars* c;

$\nu[A] = \text{supremum}\,\{|y^*Ax| \mid x \in \mathbf{C}_n, |x| \leq 1; y \in \mathbf{C}_m, |y| \leq 1\}$;

$\nu[A] = \nu[\bar{A}] = \nu[\tilde{A}] = \nu[A^*]$.

11. If A and B are $m \times n$ matrices, show that $\nu[A + B] \leq \nu[A] + \nu[B]$. Also, if A is an $m \times n$ matrix and B is an $n \times p$ matrix, then $\nu[AB] \leq \nu[A]\nu[B]$.

12. If A is an hermitian $n \times n$ matrix, show that

$$\nu[A] = \text{supremum } \{|(Ax, x)| \mid x \in \mathbf{C}_n, |x| \leq 1\}.$$

Hint. If $\nu_1[A] = \text{supremum } \{|(Ax, x)| \mid x \in \mathbf{C}_n, |x| \leq 1\}$, then the Lagrange-Cauchy inequality (1.3-c) implies that $\nu_1[A] \leq \nu[A]$. With the aid of certain results of the above Problems 6 and 8 show that for $x \in \mathbf{C}_n$, $y \in \mathbf{C}_n$, $|x| \leq 1$, $|y| \leq 1$, one has

$$4\,|\Re e(Ax, y)| \leq \nu_1[A][|x + y|^2 + |x - y|^2] = 2\nu_1[A][|x|^2 + |y|^2] \leq 4\nu_1[A].$$

Proceed to show that in view of the arbitrariness of x and y one has the inequality $|(Ax, y)| \leq \nu_1[A]$ for x and y elements of $\mathbf{C}_n$ with $|x| \leq 1$, $|y| \leq 1$, and with a result from Problem 10 above conclude that $\nu[A] \leq \nu_1[A]$.

13. If A is an $n \times n$ hermitian matrix and $A > 0$, show that

$$|(x, z)|^2 \leq (Ax, x)(A^{-1}z, z), \quad \text{for} \quad x \in \mathbf{C}_n, \quad z \in \mathbf{C}_n.$$

Hint. Use a result of Problem 9 above, and inequality (1.11-c) with $y = A^{-1}z$.

2. SEQUENCES OF VECTORS AND MATRICES

Suppose that $y = (y_\alpha)$ and $y^{(j)} = (y_\alpha^{(j)})$, $(j = 1, 2, \ldots)$, are vectors of $\mathbf{C}_n$. The sequence $\{y^{(j)}\}$, $(j = 1, 2, \ldots)$, is said to converge to y or to have y as limit, and we write $\lim_{j\to\infty} y^{(j)} = y$, or $\{y^{(j)}\} \to y$ as $j \to \infty$, whenever the sequence of real numbers $\{|y^{(j)} - y|\}$, $(j = 1, 2, \ldots)$, converges to zero. It follows readily that a sequence cannot have two distinct limit vectors; indeed, if $\{y^{(j)}\} \to y$ and $\{y^{(j)}\} \to z$ as $j \to \infty$, then the triangle inequality (1.3-d) implies that $|y - z| \leq |y - y^{(j)}| + |y^{(j)} - z|$, and as $|y - y^{(j)}| \to 0$ and $|y^{(j)} - z| \to 0$ as $j \to \infty$ it follows that $|y - z| = 0$ and $y = z$. From (1.4) for $x = y^{(j)} - y$ it is clear that $\{y^{(j)}\} \to y$ as $j \to \infty$ if and only if for $\alpha = 1, \ldots, n$ the sequences of individual components $\{y_\alpha^{(j)}\}$, $(j = 1, 2, \ldots)$, converge to the corresponding components y_α of the limit vector y. Moreover, since by (1.4),

$$\max_{\alpha=1}^{n} |y_\alpha^{(j)} - y_\alpha^{(k)}| \leq |y^{(j)} - y^{(k)}| \leq \sqrt{n} \max_{\alpha=1}^{n} |y_\alpha^{(j)} - y_\alpha^{(k)}|, \quad (j, k = 1, 2, \ldots),$$

it follows from the corresponding property of the real number system that the complex n-dimensional number space $\mathbf{C}_n$ is *complete*; that is, a given sequence $\{y^{(j)}\}$ of elements of $\mathbf{C}_n$ is convergent if and only if $\{|y^{(j)} - y^{(k)}|\} \to 0$ as

$j \to \infty$, $k \to \infty$. From the properties (1.3) it follows readily that if sequences $\{y^{(j)}\}$ and $\{z^{(j)}\}$ of vectors in $\mathbf{C}_n$ converge to y and z, respectively, then $\{c_1 y^{(j)} + c_2 z^{(j)}\} \to c_1 y + c_2 z$ as $j \to \infty$ for arbitrary scalars c_1, c_2.

If $v^{(j)} \equiv (v^{(j)}_\alpha)$, $(j = 1, 2, \ldots)$, are vectors of $\mathbf{C}_n$, then the vector series

$$v^{(1)} + \cdots + v^{(j)} + \cdots \tag{2.1}$$

is said to be convergent and to have sum w if the sequence of partial sums $\{w^{(j)}\} \equiv \{v^{(1)} + \cdots + v^{(j)}\}$, $(j = 1, 2, \ldots)$, of (2.1) converges to the limit w. Now from the triangle inequality (1.3-d) it follows that for $j > k$,

$$|w^{(j)} - w^{(k)}| = |v^{(k+1)} + \cdots + v^{(j)}| \leq |v^{(k+1)}| + \cdots + |v^{(j)}|.$$

Consequently for vector series (2.1) one has the following comparison test.

If $d_1 + d_2 + \cdots$ is a convergent infinite series of non-negative real terms and $v^{(j)}$, $(j = 1, 2, \ldots)$, are vectors of $\mathbf{C}_n$ such that $|v^{(j)}| \leq d_j$, $(j = 1, 2, \ldots)$, then the vector series (2.1) *is convergent; moreover, if w is the sum of this series then*

$$|w| \leq d_1 + d_2 + \cdots; \; |w - w^{(j)}| \leq d_{j+1} + d_{j+2} + \cdots. \tag{2.2}$$

Since an arbitrary sequence $\{y^{(j)}\}$, $(j = 1, 2, \ldots)$, of vectors of $\mathbf{C}_n$ is the sequence of partial sums of a vector series (2.1), with $v^{(1)} = y^{(1)}$, $v^{(j+1)} = y^{(j+1)} - y^{(j)}$, $(j = 1, 2, \ldots)$, from the above comparison test for series it follows that a sequence $\{y^{(j)}\}$ is convergent whenever there exist non-negative real constants d_j, $(j = 1, 2, \ldots)$, such that $d_1 + d_2 + \cdots$ is convergent and $|y^{(1)}| \leq d_1$, $|y^{(j+1)} - y^{(j)}| \leq d_{j+1}$, $(j = 1, 2, \ldots)$.

If $M = [M_{\alpha\beta}]$, $M^{(j)} = [M^{(j)}_{\alpha\beta}]$, $(\alpha = 1, \ldots, m;\ \beta = 1, \ldots, n;\ j = 1, 2, \ldots)$, are $m \times n$ matrices then the sequence $\{M^{(j)}\}$ is said to converge to M or to have M as limit, and we write $\lim_{j\to\infty} M^{(j)} = M$ or $\{M^{(j)}\} \to M$ as $j \to \infty$, whenever the sequence of vectors $\{y^{(j)}\} = \{M^{(j)}\xi\}$, $(j = 1, 2, \ldots)$ of $\mathbf{C}_m$ is convergent for arbitrary vectors ξ of $\mathbf{C}_n$. From the above discussion of vector sequences it follows readily that $\{M^{(j)}\} \to M$ as $j \to \infty$ if and only if $\{M^{(j)}_{\alpha\beta}\} \to M_{\alpha\beta}$ as $j \to \infty$ for each $\alpha = 1, \ldots, m$; $\beta = 1, \ldots, n$. Correspondingly, if $V^{(j)}$, $(j = 1, 2, \ldots)$, are $m \times n$ matrices the matrix series

$$V^{(1)} + \cdots + V^{(j)} + \cdots \tag{2.3}$$

is said to be convergent and to have sum W if W is an $m \times n$ matrix and the sequence of partial sums $\{W^{(j)}\} = \{V^{(1)} + \cdots + V^{(j)}\}$, $(j = 1, 2, \ldots)$, converges to the limit W. From these definitions it follows readily that the matrix series (2.3) is convergent in case there exist non-negative real constants k_j, $(j = 1, 2, \ldots)$, such that $k_1 + k_2 + \cdots$ is convergent and $|V^{(j)}\xi| \leq k_j\,|\xi|$ for arbitrary ξ of $\mathbf{C}_n$; moreover, if W is the sum of this series then $|W\xi| \leq (k_1 + k_2 + \cdots)\,|\xi|$ and $|[W - W^{(j)}]\xi| \leq (k_{j+1} + k_{j+2} + \cdots)\,|\xi|$ for arbitrary ξ of $\mathbf{C}_n$.

Now suppose that A is an $n \times n$ matrix and k is a non-negative real constant such that

$$\nu(A) \leq k; \tag{2.4}$$

from Problem 2 of D.1 it follows that k may be chosen to not exceed $[\sum_{\alpha,\beta=1}^{n} |A_{\alpha\beta}|^2]^{1/2}$. Then $|A^2\eta| = |A(A\eta)| \leq k\,|A\eta| \leq k^2\,|\eta|$, and by induction it follows that $|A^j\eta| \leq k^j\,|\eta|$, $(j = 1, 2, \ldots)$. From the above results we have that if the scalar power series

$$c_0 + c_1 z + \cdots + c_j z^j + \cdots \tag{2.5}$$

is absolutely convergent for $z = k$, then the matrix power series

$$c_0 E + c_1 A + \cdots + c_j A^j + \cdots \tag{2.6}$$

converges. In particular, if (2.5) is the Taylor expansion about $z = 0$ of a single-valued analytic function $f(z)$ of one complex variable z, then the matrix power series (2.6) converges for all matrices A which satisfy (2.4) with a k less than the radius of convergence of (2.5); for such matrices A the sum of the series (2.6) will be denoted by $f(A)$. For example, the matrix series

$$e^A = E + A + \frac{1}{2!}A^2 + \cdots,$$

$$\sin A = A - \frac{1}{3!}A^3 + \cdots,$$

$$\cos A = E - \frac{1}{2!}A^2 + \cdots,$$

converge for all square matrices A.

Problems D.2

1. If $\{M^{(j)}\}$, $(j = 1, 2, \ldots)$, is a sequence of $m \times n$ matrices such that $\{M^{(j)}\} \to M$ as $j \to \infty$, show that $\{\bar{M}^{(j)}\} \to \bar{M}$, $\{\tilde{M}^{(j)}\} \to \tilde{M}$ and $\{M^{(j)*}\} \to M^*$ as $j \to \infty$.

2. Compute e^A, $\sin A$, $\cos A$ for $A = \begin{bmatrix} 0 & 1 \\ -1 & 0 \end{bmatrix}$, and $A = \begin{bmatrix} 0 & 1 \\ 0 & 0 \end{bmatrix}$.

3. Show that if A is an $n \times n$ matrix such that (2.6) is convergent, and C is a nonsingular matrix, then for $B = CAC^{-1}$ the series $c_0 E + c_1 B + \cdots$ converges and has sum $C[c_0 E + c_1 A + \cdots]C^{-1}$.

4. Show that if A is an $n \times n$ matrix satisfying (2.4) with $k < 1$ then the matrix series $E + A + A^2 + \cdots$ converges and has sum equal to the inverse of $E - A$.

5. Suppose that $\{y^{(j)}\}$, $(j = 1, 2, \ldots)$, is a bounded sequence of vectors of $\mathbf{C}_n$; that is, there exists a constant $\mu > 0$ such that $|y^{(j)}| \leq \mu$, $(j = 1, 2, \ldots)$. Show that there exists a subsequence $\{y^{(j_\gamma)}\}$, $(j_1 < j_2 < \cdots)$, of $\{y^{(j)}\}$ which is convergent.

Hint. Use (1.4), together with the fact that a bounded sequence of real numbers contains a convergent subsequence.

6. If $\{y^{(j)}\}$, $(j = 1, 2, \ldots)$, is a sequence of vectors of $\mathbf{C}_n$ which converges to y, show that $\lim_{j\to\infty} (y^{(j)} - y, x) = 0$ for arbitrary x of $\mathbf{C}_n$.

7. If A is an $n \times n$ matrix with $\nu[A] < 1$, then the matrix series $\sum_{k=1}^{\infty} c_k A^k$, with

$$c_1 = \tfrac{1}{2},\ c_k = \frac{1 \cdot 3 \cdots (2k - 3)}{k!\ 2^k}, \qquad (k = 2, 3, \ldots),$$

converges. Moreover,

$$C = E - \sum_{k=1}^{\infty} c_k A^k$$

is a square root of $E - A$; that is, $C^2 = E - A$.

Hint. As $c_k > 0$, and the sequence $\{k[(c_k/c_{k+1}) - 1]\}$ has limit 3/2, conclude with the aid of the Raabe comparison test that the series $\sum_{k=1}^{\infty} c_k$ converges. Moreover, if $g(z) = \sum_{k=1}^{\infty} c_k z^k$, then $1 - g(z)$ is the Maclaurin expansion for the branch of $(1 - z)^{1/2}$ that is equal to 1 at $z = 0$. If $A_j = \sum_{k=1}^{j} c_k A^k$, from the fact that $c_k > 0$, $(k = 1, 2, \ldots)$, and the relation $g(z) = [z + g^2(z)]/2$ conclude that

$$\tfrac{1}{2}(A + A_j^2) = \sum_{k=1}^{\infty} d_{kj} A^k,$$

where $d_{kj} = c_k$, $(k = 1, \ldots, j + 1)$, $0 < d_{kj} < c_k$, $(j + 1 < k \leq 2j)$, and $d_{kj} = 0$ for $k > 2j$, and hence

$$\tfrac{1}{2}\nu[(E - A_j)^2 - (E - A)] = \nu[\tfrac{1}{2}(A + A_j^2) - A_j] \leq \sum_{k=j+1}^{\infty} c_k.$$

Appendix E

Spectral Properties of Matrices

1. ELEMENTARY PROPERTIES OF LINEAR TRANSFORMATIONS OF $\mathbf{C}_n$ INTO $\mathbf{C}_n$

If T is a linear transformation of $\mathbf{C}_n$ into $\mathbf{C}_n$, then for the particular basis $\mathcal{E} = \{e^{(1)}, \ldots, e^{(n)}\}$, where $e^{(\beta)} = (e_\alpha^{(\beta)})$ with $e_\alpha^{(\beta)} = \delta_{\alpha\beta}$, $(\alpha, \beta = 1, \ldots, n)$, the matrix $A = [A_{\alpha\beta}] = A(T; \mathcal{E}, \mathcal{E})$ of T is such that $y = Tx$ if and only if $y = (y_\alpha)$, $x = (x_\alpha)$ satisfy the algebraic equation

$$(1.1) \qquad y = Ax, \qquad \left\{ y_\alpha = \sum_{\beta=1}^{n} A_{\alpha\beta} x_\beta, \qquad (\alpha = 1, \ldots, n) \right\}.$$

If P is a nonsingular $n \times n$ matrix with column vectors $p^{(1)}, \ldots, p^{(n)}$, the one-to-one transformation of $\mathbf{C}_n$ onto itself defined by

$$(1.2) \qquad x = P\xi$$

may be considered as a transformation from coordinates $(x_1, \ldots, x_n)$ to coordinates $(\xi_1, \ldots, \xi_n)$, with the one-dimensional subspace of elements of the form $x = cp^{(j)}$ called the ξ_j-th coordinate axis of the new coordinate system. If $x = P\xi$ and $y = P\eta$, then clearly (1.1) holds for x, y if and only if

$$(1.3) \qquad \eta = B\xi, \quad \textit{where} \quad B = P^{-1}AP;$$

that is, (1.3) specifies the same transformation $T: \mathbf{C}_n \to \mathbf{C}_n$ in the new coordinate system defined by (1.2). If two matrices B and A are such that there is a nonsingular matrix P satisfying $B = P^{-1}AP$, then B and A are said to be *similar*; this relation is frequently signified by $B \stackrel{s}{=} A$. Clearly the similarity relation is an *equivalence* in the sense that it possesses the three following properties: (i) reflexivity, ($A \overset{s}{=} A$, for arbitrary A); (ii) symmetry, ($A \overset{s}{=} B$ if $B \overset{s}{=} A$); (iii) transitivity, ($C \overset{s}{=} A$ if $B \overset{s}{=} A$ and $C \overset{s}{=} B$).

A complex number λ is said to be a *proper value*, (*characteristic value*, or *eigenvalue*), of a matrix A if there is a nonzero vector x such that

$$Ax = \lambda x, \tag{1.4}$$

and any such nonzero vector x is termed a *proper vector*, (*characteristic vector*, or *eigenvector*), corresponding to the proper value λ. Clearly λ is a proper value of A if and only if it is a root of the *characteristic equation*

$$\Delta_A(\lambda) \equiv \det(\lambda E - A) = 0. \tag{1.5}$$

The set of proper values of A is called the *spectrum* of A, and of the transformation T with functional value (1.1). The purpose of this section is to present some results concerning the spectrum of a matrix, which are basic for certain aspects of the theory of differential equations. Of particular significance are the spectral decomposition for hermitian matrices, (Theorem 2.2), and the Jordan canonical form for a general square matrix, (Theorem 3.3).

For a proper value λ the dimension of the subspace of vectors x satisfying (1.4) is called the (*geometric*) *index* of λ, and the multiplicity of λ as a root of (1.5) is called the (*algebraic*) *multiplicity* of λ. As

$$\lambda E - P^{-1}AP = P^{-1}(\lambda E - A)P,$$

with the aid of the Property (e) of C.2 it follows that $\Delta_{P^{-1}AP}(\lambda) = \Delta_A(\lambda)$, so that λ is a proper value of A of multiplicity m if and only if λ is a proper value of $P^{-1}AP$ of the same multiplicity. Moreover, x satisfies $(\lambda E - A)x = 0$ if and only if $\xi = P^{-1}x$ satisfies $(\lambda E - P^{-1}AP)\xi = 0$, so that the index of λ as a proper value of A is equal to the index of λ as a proper value of $P^{-1}AP$. That is, the characteristic equation and proper values, together with their indices and multiplicities, are similarity invariants, and for a given transformation T of $\mathbf{C}_n$ into $\mathbf{C}_n$ are independent of the particular basis in terms of which the transformation is represented. It is to be noted that since $\Delta_A(\lambda)$ is a polynomial of degree n the sum of the multiplicities of the proper values of A is equal to n. Moreover, in view of Theorem B.2.3, and the relation $\Delta_A(\lambda) = \Delta_{\tilde{A}}(\lambda) = \overline{\Delta_{A^*}(\bar{\lambda})}$, it follows that $\lambda = \lambda_0$ is a proper value for A of index k and multiplicity m if and only if $\lambda = \lambda_0$, $\{\lambda = \bar{\lambda}_0\}$, is a proper value for $\tilde{A}$, $\{A^*\}$, of index k and multiplicity m.

Theorem 1.1. *For a given $n \times n$ matrix A there exists a similar matrix $B = P^{-1}AP$ which is superdiagonal, that is, $B_{\alpha\beta} = 0$ if $\alpha > \beta$, while the* *sequence of diagonal elements* $B_{\alpha\alpha}$, $(\alpha = 1, \ldots, n)$, *is any prescribed arrangement of the proper values* $\lambda_1, \ldots, \lambda_n$ *of* A, *with each proper value occurring a number of times equal to its multiplicity.*

The theorem will be proved by mathematical induction. If $n = 1$ then A is the 1×1 matrix $[A_{11}]$ with the single proper value $\lambda = A_{11}$, and the result of the theorem holds for P the 1×1 matrix with $P_{11} = 1$. Now suppose that the theorem holds for $n = 1, \ldots, k - 1$, and consider a $k \times k$ matrix $A = [A_{\alpha\beta}]$, $(\alpha, \beta = 1, \ldots, k)$, having $\lambda_1, \ldots, \lambda_k$ as its sequence of proper values, with each repeated a number of times equal to its multiplicity. Then there is a nonzero k-dimensional vector u satisfying $(\lambda_1 E - A)u = 0$, and if U is a nonsingular matrix with u its first column vector then $C = U^{-1}AU$ has $C_{11} = \lambda_1$, $C_{\alpha 1} = 0$ for $\alpha = 2, \ldots, k$. Then $\Delta_A(\lambda) = \Delta_C(\lambda) = (\lambda - \lambda_1) \times \Delta_S(\lambda)$, where $S = [S_{\sigma\tau}]$ is the $(k - 1) \times (k - 1)$ matrix with $S_{\sigma\tau} = C_{1+\sigma,1+\tau}$, $(\sigma, \tau = 1, \ldots, k - 1)$, and consequently $\lambda_2, \ldots, \lambda_k$ is the sequence of proper values of S, with each repeated a number of times equal to its multiplicity. By the induction hypothesis there is a nonsingular $(k - 1) \times (k - 1)$ matrix Q such that $C^0 = Q^{-1}SQ$ is superdiagonal with $C^0_{\sigma\sigma} = \lambda_{1+\sigma}$, $(\sigma = 1, \ldots, k - 1)$. Consequently, if R is the $k \times k$ matrix with $R_{1\alpha} = R_{\alpha 1} = \delta_{\alpha 1}$, $(\alpha = 1, \ldots, k)$, and $R_{1+\sigma,1+\tau} = Q_{\sigma\tau}$, $(\sigma, \tau = 1, \ldots, k - 1)$, the matrix $P = UR$ is such that $B = P^{-1}AP$ is superdiagonal with $B_{\alpha\alpha} = \lambda_\alpha$, $(\alpha = 1, \ldots, k)$, thus completing the induction proof of the theorem.

Corollary. *If* λ_0 *is a proper value for* A *of index* k *and multiplicity* m, *then* $k \leq m$; *moreover, each of the following conditions is necessary and sufficient for the equality* $k = m$:

(i) *if* y *is a proper vector for* A *corresponding to the proper value* λ_0, *then there is no vector* x *such that*

$$(\lambda_0 E - A)x = y; \tag{1.6}$$

(ii) *if* $x^{(1)}, \ldots, x^{(k)}$ *form a basis for the subspace of vectors* x *satisfying* $(\lambda_0 E - A)x = 0$, *and* $y^{(1)}, \ldots, y^{(n-k)}$ *form a basis for the subspace spanned by the column vectors of* $\lambda_0 E - A$, *then* $x^{(1)}, \ldots, x^{(k)}, y^{(1)}, \ldots, y^{(n-k)}$ *is a basis for* $\mathbf{C}_n$;

(iii) *if* $x^{(1)}, \ldots, x^{(k)}$ *and* $z^{(1)}, \ldots, z^{(k)}$ *are linearly independent proper vectors corresponding to the proper value* $\lambda = \lambda_0$ *for* A *and* $\lambda = \bar{\lambda}_0$ *for* A^*, *respectively, then the* $k \times k$ *matrix* $[(x^{(\sigma)}, z^{(\tau)})]$, $(\sigma, \tau = 1, \ldots, k)$, *is nonsingular.*

If $\lambda = \lambda_0$ is a proper value for A of multiplicity m, then in view of Theorem 1.1 one may assume that A is superdiagonal and $A_{\alpha\alpha} = \lambda_0$, $(\alpha = 1, \ldots, m)$, and $A_{\alpha\alpha} \neq \lambda_0$, $(\alpha = m + 1, \ldots, n)$. In particular, the $(n - m) \times (n - m)$ matrix $[\lambda_0 \delta_{\sigma\tau} - A_{m+\sigma,m+\tau}]$, $(\sigma, \tau = 1, \ldots, n - m)$, is nonsingular and if

$x = (x_\alpha)$, $(\alpha = 1, \ldots, n)$, is such that $(\lambda_0 E - A)x = 0$ then $x_{m+\sigma} = 0$, $(\sigma = 1, \ldots, n - m)$, so that the dimension k of the subspace of vectors x satisfying $(\lambda_0 E - A)x = 0$ does not exceed m; moreover, $k = m$ if and only if $A_{\alpha\beta} = 0$, $(\alpha < \beta; \alpha, \beta = 1, \ldots, m)$. Hence if $k = m$ then any vector y of the form $y = (\lambda_0 E - A)x$ has $y_{m+\sigma} = 0$, $(\sigma = 1, \ldots, n - m)$, if and only if $y = 0$, so that if y is a proper vector for A corresponding to the proper value λ_0 then there is no vector x satisfying (1.6). On the other hand, in case $k < m$ not all the elements $A_{\alpha\beta}$, $(\alpha < \beta; \alpha, \beta = 1, \ldots, m)$, are zero, and there exists a smallest integer $\gamma \geq 2$ such that $A_{1\gamma}, \ldots, A_{\gamma-1,\gamma}$ are not all zero. Then $y = (y_\alpha)$ with $y_\alpha = \lambda_0 \delta_{\alpha\gamma} - A_{\alpha\gamma}$, $(\alpha = 1, \ldots, n)$, is a proper vector for A corresponding to the proper value λ_0, and for this proper vector y equation (1.6) has a solution $x = e^{(\gamma)}$. In order to show that condition (ii) is equivalent to (i), one need note only that (i) holds if and only if the set of n vectors $x^{(1)}, \ldots, x^{(k)}, y^{(1)}, \ldots, y^{(n-k)}$ is linearly independent. Finally, in view of Theorem B.2.2, condition (iii) is equivalent to condition (i).

Problems E.1

1. If $A^2 = E$, prove that each proper value of A is either 1 or -1.

2. Show that the result of Theorem 1.1 holds with P a unitary matrix. (The existence of a unitary matrix P such that P^*AP is superdiagonal is known as the *Schur transformation*.)

3. Use the result of the preceding problem to show that if A is hermitian then there is a unitary matrix P such that P^*AP is diagonal.

4. Show that if A is a given $n \times n$ matrix, and $\varepsilon > 0$, then there exists a nonsingular matrix Q such that $B = Q^{-1}AQ$ is superdiagonal with $|B_{\alpha\beta}| < \varepsilon$, $(\alpha < \beta; \alpha, \beta = 1, \ldots, n)$.

Hint. If P is such that $C = P^{-1}AP$ is superdiagonal, $d_\alpha > 0$, and $D = [d_\alpha \delta_{\alpha\beta}]$, then $Q = PD$ is such that $B = Q^{-1}AQ$ is superdiagonal with $B_{\alpha\beta} = C_{\alpha\beta}(d_\beta/d_\alpha)$, $(\alpha \leq \beta)$. Conclude that if $|C_{\alpha\beta}| \leq k$, $(\alpha < \beta)$, then $|B_{\alpha\beta}| < \varepsilon$ for $\alpha < \beta$ whenever $\{d_\beta\}$ is a monotone decreasing sequence with $d_{\beta+1}/d_\beta < \varepsilon/k$.

5. Use the result of Theorem 1.1 to conclude that a matrix A is such that there is a unitary matrix P for which P^*AP is diagonal if and only if A is *normal*, that is, $AA^* = A^*A$.

6. If A is an $n \times n$ matrix, and $\lambda_1, \ldots, \lambda_n$ denote its proper values, with each repeated a number of times equal to its multiplicity as a zero of the polynomial $\Delta_A(\lambda)$, show that $\prod_{\alpha=1}^n \lambda_\alpha = \det A$ and $\sum_{\alpha=1}^n \lambda_\alpha = \operatorname{Tr} A$, where $\operatorname{Tr} A$, the *trace* of A, is defined as $\sum_{\alpha=1}^n A_{\alpha\alpha}$. (This result implies that $\det A$ and $\operatorname{Tr} A$ are similarity invariants).

Hint. Expand $\Delta_A(\lambda) = \det(\lambda E - A)$, which is equal to $\prod_{\alpha=1}^n (\lambda - \lambda_\alpha)$.

7. If A is the $n \times n$ matrix

$$A = \begin{bmatrix} -p_{n-1} & -p_{n-2} & \cdots & -p_0 \\ 1 & 0 & \cdots & 0 \\ 0 & 1 & \cdots & 0 \\ \cdot & \cdot & \cdots & \cdot \\ \cdot & \cdot & 1 & 0 \end{bmatrix},$$

show that $\Delta_A(\lambda) = \lambda^n + p_{n-1}\lambda^{n-1} + \cdots + p_0$.

8. If A and B are $n \times n$ matrices, prove that $\Delta_{AB}(\lambda) = \Delta_{BA}(\lambda)$.

Hint. If A is nonsingular, note that $\Delta_{AB}(\lambda) = \det[A\{\lambda A^{-1} - B\}] = \det[\{\lambda A^{-1} - B\}A] = \Delta_{BA}(\lambda)$. If A is singular, consider $A_1 = A - sI$, where s is "small" and not a proper value of A, and use a continuity argument to deduce the desired result from the relation $\Delta_{A_1B}(\lambda) = \Delta_{BA_1}(\lambda)$.

9. If A and B are $n \times n$ matrices, prove that $\operatorname{Tr}(AB)^j = \operatorname{Tr}(BA)^j$, $(j = 1, 2, \ldots)$.

Hint. Show that the result for $j = 1$ is a consequence of Problems 6 and 8, and proceed by mathematical induction.

10. If λ is a proper value of an $n \times n$ matrix A, prove that $|\mathfrak{Re}\,\lambda| \leq \nu[\mathfrak{Re}\,A]$, $|\mathfrak{Im}\,\lambda| \leq \nu[\mathfrak{Im}\,A]$.

Hint. If x is a proper vector for A corresponding to a proper value λ, note that

$$|x|^2\,\mathfrak{Re}\,\lambda = ([\mathfrak{Re}\,A]x, x),\ |x|^2\,\mathfrak{Im}\,\lambda = ([\mathfrak{Im}\,A]x, x).$$

2. HERMITIAN MATRICES AND FORMS

Attention will be directed now to the case of A an $n \times n$ hermitian matrix, so that $A = A^*$; in particular, for A an hermitian matrix the corresponding form $A[x] = (Ax, x)$ is real-valued, since $A[x] = (Ax, x) = (x, Ax) = \overline{A[x]}$.

Theorem 2.1. *If A is an $n \times n$ hermitian matrix, then:*

(i) *all proper values of A are real;*

(ii) *if x and y are proper vectors of A corresponding to distinct proper values λ and μ, respectively, then $(x, y) = 0$;*

(iii) *each proper value of A has index equal to multiplicity.*

If x is a proper vector of A corresponding to a proper value λ, then the equation $(Ax, x) = (\lambda x, x) = \lambda\,|x|^2$, and the reality of (Ax, x), imply that λ is real. Also, if x and y are proper vectors of A corresponding to respective proper values λ and μ with $\lambda \neq \mu$, then conclusion (ii) follows from the relation

$$\lambda(x, y) = (\lambda x, y) = (Ax, y) = (x, Ay) = \mu(x, y).$$

Now if $(\lambda E - A)y = 0$, and $(\lambda E - A)x = y$, then

$$|y|^2 = ([\lambda E - A]x, y) = (x, [\lambda E - A]y) = (x, 0) = 0,$$

and conclusion (iii) is a consequence of conclusion (i) of the Corollary to Theorem 1.1.

Theorem 2.2. *If A is an $n \times n$ hermitian matrix, then there exists a sequence of proper values $\lambda_1 \leqq \lambda_2 \leq \cdots \leq \lambda_n$ with corresponding proper vectors $x = u^{(\beta)}$ such that:*

(i) $(u^{(\alpha)}, u^{(\beta)}) = \delta_{\alpha\beta}$, $(\alpha, \beta = 1, \ldots, n)$;

(ii) $\lambda_1 = A[u^{(1)}]$ *is the minimum of* $A[x]$ *in the class*

$$\Gamma_N = \{x \mid x \in \mathbf{C}_n, |x| = 1\}; \tag{2.1}$$

(iii) *for* $\beta = 2, \ldots, n$ *the class*

$$\Gamma_{N\beta} = \{x \mid x \in \Gamma_N, (x, u^{(j)}) = 0, j = 1, \ldots, \beta - 1\}, \tag{2.2}$$

is nonempty, and $\lambda_\beta = A[u^{(\beta)}]$ *is the minimum of* $A[x]$ *on* $\Gamma_{N\beta}$.

(iv) *If* $y \in \mathbf{C}_n$, *then*

$$\begin{aligned} A[y] &= \sum_{\beta=1}^{n} \lambda_\beta \, |(y, u^{(\beta)})|^2, \\ |y|^2 &= \sum_{\beta=1}^{n} |(y, u^{(\beta)})|^2. \end{aligned} \tag{2.3}$$

If x is a proper vector of A corresponding to a proper value λ, then the relation $A[x] = (Ax, x) = \lambda(x, x)$ implies that $\lambda \geq \lambda_1$, where λ_1 is defined as the infimum (greatest lower bound) of $A[x]$ on Γ_N. Now Γ_N is a compact set in $\mathbf{C}_n$, and as $A[x]$ is a continuous function on $\mathbf{C}_n$ it follows that there exists an element $u^{(1)} \in \Gamma_N$ such that $A[u^{(1)}] = \lambda_1$. Then

$$([A - \lambda_1 E]u^{(1)}, u^{(1)}) = A[u^{(1)}] - \lambda_1 = 0,$$

and the definitive property of λ_1 implies that $([A - \lambda_1 E]x, x) \geq 0$ for all $x \in \mathbf{C}_n$. From the Lagrange-Cauchy type inequality (D.1.11-c) it then follows that $([A - \lambda_1 E]u^{(1)}, x) = 0$ for arbitrary $x \in \mathbf{C}_n$. Consequently $[A - \lambda_1 E]u^{(1)} = 0$, and $x = u^{(1)}$ is a proper vector of A which satisfies with the corresponding proper value λ_1 the conclusion (ii) of the theorem.

Now suppose that $2 \leq m \leq n$, and that proper values λ_β and corresponding proper vectors $u^{(\beta)}$, $\beta = 1, \ldots, m - 1$, have been determined to satisfy the corresponding conditions (i), (ii), (iii) of the theorem for $\alpha, \beta = 1, \ldots, m - 1$. Then the set

$$\Gamma_{Nm} = \{x \mid x \in \Gamma_N, \quad (x, u^{(\beta)}) = 0, \quad (\beta = 1, \ldots, m - 1)\},$$

is a nonempty compact set in $\mathbf{C}_n$, and if λ_m denotes the infimum of $A[x]$ on Γ_{Nm} then there exists an element $u^{(m)} \in \Gamma_{Nm}$ such that $A[u^{(m)}] = \lambda_m$. Moreover, since $\Gamma_{Nm} \subset \Gamma_{N,m-1}$, it follows that $\lambda_m \geq \lambda_{m-1}$. If Γ_m denotes the subspace

$$\Gamma_m = \{x \mid x \in \mathbf{C}_n, \quad (x, u^{(\beta)}) = 0, \quad (\beta = 1, \ldots, m-1)\},$$

then

$$([A - \lambda_m E]x, x) \geq 0, \quad \textit{for} \quad x \in \Gamma_m,$$

while $u^{(m)} \in \Gamma_m$, with $|u^{(m)}| = 1$, and

$$([A - \lambda_m E]u^{(m)}, u^{(m)}) = 0.$$

Consequently, again by the inequality (D.1.11-c), it follows that

$$([A - \lambda_m E]u^{(m)}, y) = 0, \quad \textit{for} \quad y \in \Gamma_m.$$

Now if $x \in \mathbf{C}_n$ then $y = x - \sum_{\beta=1}^{m-1} (x, u^{(\beta)})u^{(\beta)}$ is such that $(y, u^{(\alpha)}) = 0$, $(\alpha = 1, \ldots, m-1)$, and hence $y \in \Gamma_m$. Moreover, since

$$\begin{aligned}([A - \lambda_m E]u^{(m)}, u^{(\beta)}) &= (Au^{(m)}, u^{(\beta)}) - \lambda_m(u^{(m)}, u^{(\beta)}) \\ &= (u^{(m)}, Au^{(\beta)}) - \lambda_m(u^{(m)}, u^{(\beta)}) \\ &= (\lambda_\beta - \lambda_m)(u^{(m)}, u^{(\beta)}) = 0\end{aligned}$$

for $\beta = 1, \ldots, m-1$, it follows that

$$0 = ([A - \lambda_m E]u^{(m)}, y) = ([A - \lambda_m E]u^{(m)}, x) \quad \textit{for arbitrary} \quad x \in \mathbf{C}_n.$$

Consequently, $u^{(m)}$ is an element of Γ_m with $|u^{(m)}| = 1$ and $[A - \lambda_m E]u^{(m)} = 0$, so that $\lambda_1, \ldots, \lambda_m$ are proper values of A with corresponding proper vectors $u^{(1)}, \ldots, u^{(m)}$ such that the conditions (i), (ii), (iii) of the theorem hold for $\alpha, \beta = 1, \ldots, m$. Hence the conclusions (i), (ii), (iii) of the theorem are established by induction.

Finally, since $\{u^{(1)}, \ldots, u^{(n)}\}$ is an orthonormal set of vectors in $\mathbf{C}_n$ this set is linearly independent, and consequently forms a basis for $\mathbf{C}_n$. If $y \in \mathbf{C}_n$ then y is of the form $\sum_{\alpha=1}^{n} c_\alpha u^{(\alpha)}$, from which it follows that $(y, u^{(\beta)}) = c_\beta$, $(\beta = 1, \ldots, n)$, and hence

$$\begin{aligned}A[y] = (Ay, y) &= \left(\sum_{\alpha=1}^{n}(y, u^{(\alpha)})Au^{(\alpha)}, \sum_{\beta=1}^{n}(y, u^{(\beta)})u^{(\beta)}\right), \\ &= \sum_{\alpha,\beta=1}^{n} (y, u^{(\alpha)})(u^{(\beta)}, y)(Au^{(\alpha)}, u^{(\beta)}), \\ &= \sum_{\alpha,\beta=1}^{n} (y, u^{(\alpha)})(u^{(\beta)}, y)\lambda_\alpha \delta_{\alpha\beta}, \\ &= \sum_{\alpha=1}^{n} \lambda_\alpha \, |(y, u^{(\alpha)})|^2.\end{aligned}$$

Correspondingly, $|y|^2 = (y, y)$ is equal to

$$|y|^2 = (y, y) = \sum_{\alpha,\beta=1}^{n} (y, u^{(\alpha)})(u^{(\beta)}, y)\,\delta_{\alpha\beta},$$
$$= \sum_{\alpha=1}^{n} |(y, u^{(\alpha)})|^2.$$

Theorem 2.3. *Suppose that A is an $n \times n$ hermitian matrix, and $\{\lambda_\alpha, u^{(\alpha)}\}$, $(\alpha = 1, \ldots, n)$, is a sequence of proper values and associated proper vectors satisfying the conclusions of Theorem 2.2. Then:*

(a) *if $1 \leq r \leq n$, and $d_1, \ldots, d_r$ are constants such that $|d_1|^2 + \cdots + |d_r|^2 = 1$, then $x = d_1 u^{(1)} + \cdots + d_r u^{(r)}$ is such that $|x| = 1$ and $A[x] \leq \lambda_r$;*

(b) (*Max-min property*). *If $V = \{v^{(1)}, \ldots, v^{(k)}\}$, with $1 \leq k < n$, are given vectors of $\mathbf{C}_n$, and $\lambda[V]$ denotes the minimum of $A[x]$ on the set*

$$\Gamma_N[V] = \{x \mid x \in \mathbf{C}_n, |x| = 1, (x, v^{(j)}) = 0, (j = 1, \ldots, k)\},$$

then λ_{k+1} is the maximum of $\lambda[V]$.

Conclusion (a) is an immediate consequence of the fact that $x = d_1 u^{(1)} + \cdots + d_r u^{(r)}$ is such that $(x, u^{(\alpha)}) = d_\alpha$ for $\alpha = 1, \ldots, r$ and $(x, u^{(\alpha)}) = 0$ for $\alpha = r + 1, \ldots, n$, so that by conclusion (iv) of Theorem 2.2 we have $|x|^2 = (x, x) = \sum_{\alpha=1}^{r} |d_\alpha|^2 = 1$, and

$$A[x] = (Ax, x) = \sum_{\alpha=1}^{r} \lambda_\alpha |d_\alpha|^2 \leq \lambda_r \sum_{\alpha=1}^{r} |d_\alpha|^2 = \lambda_r.$$

If $V = \{v^{(1)}, \ldots, v^{(k)}\}$ are vectors of $\mathbf{C}_n$, with $1 \leq k < n$, then in view of the Corollary to Theorem B.2.1 there exist constants $d_1, \ldots, d_{k+1}$ not all zero and such that $x = d_1 u^{(1)} + \cdots + d_{k+1} u^{(k+1)}$ satisfies the k linear homogeneous conditions

$$(x, v^{(\alpha)}) = \sum_{\beta=1}^{k+1} d_\beta (u^{(\beta)}, v^{(\alpha)}) = 0, \qquad (\alpha = 1, \ldots, k),$$

and hence there exist such constants with $d_1^2 + \cdots + d_{k+1}^2 = 1$. Then $x \in \Gamma_N[V]$, so that $\lambda[V] \leq A[x]$, while the conclusion (a) also implies $A[x] \leq \lambda_{k+1}$. Consequently, $\lambda[V] \leq \lambda_{k+1}$ for each set V of k vectors of $\mathbf{C}_n$. On the other hand, for $v^{(\alpha)} = u^{(\alpha)}$, $(\alpha = 1, \ldots, k)$, the corresponding value of $\lambda[V]$ is equal to λ_{k+1} by Theorem 2.1, so that λ_{k+1} is indeed the maximum value of $\lambda[V]$.

The hermitian matrix A is said to be *non-negative definite*, {*positive definite*}, if the form $A[x]$ possesses the corresponding definiteness property, and we write correspondingly $A \geq 0$, $\{A > 0\}$. Finally, if A and B are individually hermitian matrices and $A - B \geq 0$, $\{A - B > 0\}$, we write $A \geq B$, $\{A > B\}$. In particular, as an immediate consequence of Theorem 2.2, one has that an hermitian matrix A is non-negative definite, {positive definite}, if and only if all proper values of A are non-negative, {positive}. Also the following comparison theorem is an immediate consequence of the minimizing properties of proper values as established in Theorem 2.2, and conclusion (a) of Theorem 2.3.

Theorem 2.4. *Suppose that A and $\hat{A}$ are $n \times n$ hermitian matrices, with corresponding sequences of proper values and associated proper vectors $\{\lambda_\alpha, u^{(\alpha)}\}$ and $\{\hat{\lambda}_\alpha, \hat{u}^{(\alpha)}\}$, satisfying the conditions of Theorem 2.2. If $A \geq \hat{A}$ then $\lambda_\alpha \geq \hat{\lambda}_\alpha$, $(\alpha = 1, \ldots, n)$; moreover, if $A > \hat{A}$ then $\lambda_\alpha > \hat{\lambda}_\alpha$, $(\alpha = 1, \ldots, n)$.*

A more sophisticated comparison theorem is the following result.

Theorem 2.5. *Suppose that A and $\hat{A}$ are $n \times n$ hermitian matrices, with corresponding sequences of proper values and proper vectors $\{\lambda_\alpha, u^{(\alpha)}\}$ and $\{\hat{\lambda}_\alpha, \hat{u}^{(\alpha)}\}$, satisfying the conditions of Theorem 2.2. Moreover, for the hermitian matrix $B = A + \hat{A}$ let $\{\mu_\alpha, w^{(\alpha)}\}$, denote a sequence of proper values and proper vectors satisfying similar conditions. If α and β are positive integers such that $\alpha + \beta \leq n + 1$ and $1 \leq \alpha \leq n$, $1 \leq \beta \leq n$, then*

$$\mu_{\alpha+\beta-1} \geq \lambda_\alpha + \hat{\lambda}_\beta.$$

The proof of this result is also a ready consequence of the results of Theorem 2.2, and (a) of Theorem 2.3, since under the hypotheses of the theorem there exist constants $d_1, \ldots, d_{\alpha+\beta-1}$ such that $|d_1|^2 + \cdots + |d_{\alpha+\beta-1}|^2 = 1$, and the vector $x = d_1 w^{(1)} + \cdots + d_{\alpha+\beta-1} w^{(\alpha+\beta-1)}$ satisfies the $\alpha + \beta - 2$ conditions

$$(x, u^{(i)}) = 0, \quad \textit{for} \quad i < \alpha, \ (x, \hat{u}^{(j)}) = 0 \quad \textit{for} \quad j < \beta;$$

consequently $|x|^2 = 1$, and

$$\mu_{\alpha+\beta-1} \geq B[x] = A[x] + \hat{A}[x] \geq \lambda_\alpha + \hat{\lambda}_\beta.$$

An hermitian form $A[x] = (Ax, x)$ is said to have (*negative*) *index* equal to r if r is the largest non-negative integer such that there is a subspace of $\mathbf{C}_n$ of dimension r on which $A[x]$ is negative definite. Also, an hermitian form $A[x]$ is said to have *nullity* equal to d if the null-space $\{x \mid x \in \mathbf{C}_n, Ax = 0\}$ of A has dimension equal to d.

Theorem 2.6. *Suppose that A is an $n \times n$ hermitian matrix, with a corresponding sequence of proper values and corresponding proper vectors $\{\lambda_\alpha, u^{(\alpha)}\}$ satisfying the conditions of Theorem 2.2. Then $A[x]$ has index equal to r if and only if $\lambda_r < 0$ and $\lambda_\alpha \geq 0$ if $r < \alpha \leq n$. Correspondingly, the sum of the index and the nullity of $A[x]$ is equal to s if and only if $\lambda_s \leq 0$, and $\lambda_\beta > 0$ if $s < \beta \leq n$.*

Problems E.2

1. With the aid of the result of Problem 2 of E.1, give a different proof of Theorem 2.2.

Hint. If U is an $n \times n$ unitary matrix such that the $n \times n$ hermitian matrix A satisfies $U^*AU = \Lambda$, where $\Lambda = [\lambda_\alpha \delta_{\alpha\beta}]$ is a diagonal matrix, show that each λ_β is real and the β-th column vector $u^{(\beta)}$ of U is a proper vector of A for the proper value $\lambda = \lambda_\beta$. Moreover $A = U\Lambda U^*$, and if $y \in \mathbf{C}_n$ then

$$A[y] = (Ay, y) = y^*U\Lambda U^*y = \sum_{\beta=1}^{n} \lambda_\beta \, |(y, u^{(\beta)})|^2,$$

$$|y|^2 = (y, y) = y^*UU^*y = \sum_{\beta=1}^{n} |(y, u^{(\beta)})|^2;$$

that is, the λ_β, $u^{(\beta)}$ are such that conclusions (i) and (iv) of Theorem 2.2 hold. In turn, use these representations for $A[y]$ and $|y|^2$ to establish conclusions (ii) and (iii) of Theorem 2.2.

2. If A and B are hermitian $n \times n$ matrices such that $A \geq B > 0$, $\{A > B > 0\}$, prove that $B^{-1} \geq A^{-1} > 0$, $\{B^{-1} > A^{-1} > 0\}$.

3. Show by an example involving suitable 2×2 matrices that the condition that two hermitian $n \times n$ matrices, $(n \geq 2)$, satisfy $A > B > 0$ does not imply that $A^2 > B^2$.

4. If $A = [A_{\alpha\beta}]$, $(\alpha, \beta = 1, \ldots, n)$, is an hermitian matrix, possessing proper values $\lambda_1 \leq \cdots \leq \lambda_n$, with each repeated a number of times equal to its multiplicity, prove that:

(i) $\nu[A] = \max\{|\lambda_1|, |\lambda_n|\}$;
(ii) $|A_{\alpha\beta}| \leq \nu[A]$, $(\alpha, \beta = 1, \ldots, n)$;
(iii) if $A \geq 0$, then $\nu[A] \leq \operatorname{Tr} A \leq n\nu[A]$.

Hints. (i) Use the inequality $\lambda_1 |y|^2 \leq (Ay, y) \leq \lambda_n |y|^2$ for $y \in \mathbf{C}_n$, which follows from conclusion (iv) of Theorem 2.2, and the result of Problem 12 of D.1.

(ii) Note that the hint for Problem 12 of D.1 yields the inequality $|(Ax, y)| \leq \nu[A]$ for all $x \in \mathbf{C}_n$, $y \in \mathbf{C}_n$, with $|x| \leq 1$, $|y| \leq 1$, and consider the special vectors $x = e^{(\alpha)}$, $y = e^{(\beta)}$, $(\alpha, \beta = 1, \ldots, n)$.

(iii) Use the result of (i) above, with a result of Problem 6 of E.1, to conclude that

$$\nu[A] = \lambda_n \leq \lambda_1 + \cdots + \lambda_n = \operatorname{Tr} A \leq n\lambda_n = n\nu[A].$$

5. *Suppose that A, $\hat{A}$, and the corresponding sequences $\{\lambda_\alpha, u^{(\alpha)}\}$, $\{\hat{\lambda}_\alpha, \hat{u}^{(\alpha)}\}$ satisfy the conditions of Theorem* 2.4. *If*

$$\sum_{\alpha,\beta=1}^{n} |A_{\alpha\beta} - \hat{A}_{\alpha\beta}|^2 \le \varepsilon^2, \tag{2.4}$$

prove that $|\lambda_\alpha - \hat{\lambda}_\alpha| \le \varepsilon$, $(\alpha = 1, 2, \ldots, n)$.

Hint. Show that (2.4) implies that $A + \varepsilon E \ge \hat{A}$, and $\hat{A} \ge A - \varepsilon E$, while $\{\lambda_\alpha \pm \varepsilon, u^{(\alpha)}\}$ is a sequence of proper values and corresponding proper vectors for $A \pm \varepsilon E$, satisfying the conditions of Theorem 2.2. Apply the result of Theorem 2.4 to $A + \varepsilon E$, $\hat{A}$ and $\hat{A}$, $A - \varepsilon E$.

6. Suppose that for $\tau = (\tau_1, \ldots, \tau_q)$ on a domain $\mathfrak{D}$ of q-dimensional space $\mathbf{C}_q$ the elements of the $n \times n$ matrix $A(\tau) = [A_{\alpha\beta}(\tau)]$ are continuous in τ, and $A(\tau)$ is hermitian for $\tau \in \mathfrak{D}$. If $\{\lambda_\alpha(\tau), u^{(\alpha)}(\tau)\}$ is for $\tau \in \mathfrak{D}$ a sequence of proper values and associated proper vectors satisfying the conditions of Theorem 2.2, use the result of the preceding problem to show that each $\lambda_\alpha(\tau)$ is a continuous function of τ on $\mathfrak{D}$. Show by an example, however, that the associated proper vectors $u^{(\alpha)}(\tau)$ need not be continuous functions of τ.

7. Suppose that $\{A_j\}$, $(j = 1, 2, \ldots)$, is a sequence of hermitian matrices which is non-decreasing and bounded above in the sense that $A_{j+1} \ge A_j$, and there exists a constant κ such that $\kappa E \ge A_j$, $(j = 1, 2, \ldots)$. Prove that there exists an hermitian matrix A such that the sequence $\{A_j\}$ converges to A.

Hint. Show that the general case may be reduced to that in which $0 \le A_1 \le A_2 \le \cdots \le E$, and for this case proceed as follows. For $j < k$ let $A_{jk} = A_k - A_j$, so that $0 \le A_{jk} \le E$. With the aid of inequality (D.1.11-c), show that if $x \in \mathbf{C}_n$ then

$$\begin{aligned} |A_{jk}x|^4 = (A_{jk}x, A_{jk}x)^2 &\le (A_{jk}x, x)(A_{jk}^2 x, A_{jk}x), \\ &\le (A_{jk}x, x)\,|x|^2 = [(A_k x, x) - (A_j x, x)]\,|x|^2. \end{aligned}$$

From the monotoneity and boundedness of the sequence $\{(A_j x, x)\}$ conclude that $|A_{jk}x| = |(A_k x) - (A_j x)|$ tends to zero as $j, k \to \infty$, and hence $\{A_j x\}$ is convergent for each $x \in \mathbf{C}_n$. Consequently, $\{A_j\}$ converges to the matrix A whose β-th column vector is the limit of the sequence $\{A_j e^{(\beta)}\}$.

8. Prove that if A is a non-negative hermitian matrix then there is a unique non-negative hermitian square root of A.

Hint. Show that without loss of generality we may suppose $0 \le A \le E$, and in this case proceed as follows. For $B = E - A$, and $R = E - S$, note that $0 \le B \le E$, and $S^2 = A$ is equivalent to

$$R = \tfrac{1}{2}(B + R^2).$$

Let $R_0 = 0$, and define recursively the matrices R_j as

$$R_{j+1} = \tfrac{1}{2}(B + R_j^2), \qquad (j = 0, 1, \ldots).$$

Proceed by induction to establish that:

(i) R_j is a polynomial in B with non-negative real coefficients, $\nu[R_j] \leq 1$, and $0 \leq R_j \leq E$;

(ii) $R_{j+1} - R_j = \tfrac{1}{2}(R_j + R_{j-1})(R_j - R_{j-1})$, and $R_{j+1} - R_j$ is a polynomial in B with non-negative real coefficients, so that $R_{j+1} - R_j \geq 0$, $(j = 1, 2, \ldots)$.

Let R be the limit of the sequence $\{R_j\}$ as assured by the result of the preceding problem. Note that since R_j is a polynomial in A then $R_jM = MR_j$ for every hermitian matrix M such that $AM = MA$, and consequently also $RM = MR$ for every hermitian matrix M such that $AM = MA$. Now for C any non-negative hermitian square root of A one has $AC = (C^2)C = C(C^2) = CA$, and therefore $RC = CR$. Let D_1 and D_2 be the non-negative hermitian square roots of R and C, respectively, determined in each case by the iterative process described above. If $x \in \mathbf{C}_n$ and $y = [R - C]x$, proceed to show that $|D_1y|^2 + |D_2y|^2 = (D_1^2y, y) + (D_2^2y, y) = (Ry, y) + (Cy, y) = ([R + C]y, y) = ([R + C][R - C]x, y) = ([R^2 - C^2]x, y) = ([A - A]x, y) = 0$, so that $D_1y = 0$ and $D_2y = 0$. Moreover, $Ry = D_1(D_1y) = 0$, $Cy = D_2(D_2y) = 0$, so that $0 = [R - C]y = [R - C]^2x$ and $|[R - C]x|^2 = ([R - C]^2x, x) = 0$; that is $[R - C]x = 0$ for arbitrary $x \in \mathbf{C}_n$ and $C = R$.

9. If A is an hermitian matrix satisfying $0 \leq A \leq E$, show that the non-negative hermitian square root of A is given by the series $E - \sum_{k=1}^{\infty} c_k(E - A)^k$, where the coefficients c_k are as in Problem 7 of D.2.

10. If A is an $n \times n$ hermitian matrix, and $\{\lambda_\alpha, u^{(\alpha)}\}$ is a sequence of proper values and proper vectors satisfying the conditions of Theorem 2.2, show that A^2 is a non-negative definite hermitian matrix such that μ is a proper value of A^2 if and only if either $\sqrt{u}$ or $-\sqrt{\mu}$ is a proper value for A, and the index of μ as a proper value for A^2 is equal to the sum of the indices of $\sqrt{\mu}$ and $-\sqrt{\mu}$ as proper values for A; moreover, a basis for the null space of $\mu E - A^2$ is given by the set of proper vectors $u^{(\alpha)}$ belonging to those λ_α satisfying either $\lambda_\alpha = \sqrt{\mu}$ or $\lambda_\alpha = -\sqrt{\mu}$.

11. If A and B are $n \times n$ hermitian matrices such that $A \geq 0$, $B \geq 0$, prove that AB is an hermitian matrix satisfying $AB \geq 0$ if and only if $AB = BA$.

12. If A is an $n \times n$ hermitian matrix, and Ψ is an $n \times q$ matrix of rank q, where $1 \leq q \leq n - 1$, let $M(\lambda)$ denote the $(n + q) \times (n + q)$ matrix

$$M(\lambda) = \begin{bmatrix} \lambda E - A & \Psi \\ \Psi^* & 0 \end{bmatrix}.$$

Show that $P(\lambda) = \det M(\lambda)$ is a polynomial in λ of degree $n - q$, that all zeros of $P(\lambda)$ are real, and if $\lambda = \tilde{\lambda}_0$ is a zero of $P(\lambda)$ of order k then $M(\tilde{\lambda}_0)$ has rank equal to $n + q - k$. Moreover, the zeros of $P(\lambda)$ may be ordered as a sequence $\tilde{\lambda}_1 \leq \tilde{\lambda}_2 \leq \cdots \leq \tilde{\lambda}_{n-q}$ with corresponding $x = u^{(\beta)} = (u_i^{(\beta)})$, $(i = 1, \ldots, n)$, and $\mu = \mu^{(\beta)} = (\mu_\sigma^{(\beta)})$, $(\sigma = 1, \ldots, q)$, such that $(u; \mu) = (u^{(\beta)}; \mu^{(\beta)})$ is a solution of the system

$$(2.4) \qquad \begin{aligned}(\lambda E - A)u + \Psi\mu &= 0,\\ \Psi^* u &= 0,\end{aligned}$$

for $\lambda = \tilde{\lambda}_\beta$, $(\beta = 1, \ldots, n - q)$, with the properties:

(i) $(u^{(\alpha)}, u^{(\beta)}) = \delta_{\alpha\beta}$, $(\alpha, \beta = 1, \ldots, n - q)$;
(ii) $\tilde{\lambda}_1 = A[u^{(1)}]$ is the minimum of $A[x]$ on the class

$$\Gamma_N = \{x \mid x \in \mathbf{C}_n, \Psi^* x = 0, |x| = 1\};$$

(iii) for $\beta = 2, \ldots, n - q$ the class

$$\Gamma_{N\beta} = \{x \mid x \in \Gamma_N, (x, u^{(j)}) = 0, j = 1, \ldots, \beta - 1\}$$

is nonempty, and $\tilde{\lambda}_\beta = A[u^{(\beta)}]$ is the minimum of $A[x]$ on $\Gamma_{N\beta}$;

(iv) $\prod_{\alpha=1}^{n-q} \tilde{\lambda}_\alpha = \det \begin{bmatrix} A & \Psi \\ \Psi^* & 0 \end{bmatrix}$;

(v) if $\lambda_1 \leq \cdots \leq \lambda_n$ denote the proper values of A as in Theorem 2.2, then $\lambda_j \leq \tilde{\lambda}_j \leq \lambda_{j+q}$, $(j = 1, \ldots, n-q)$;

(vi) if $\lambda = 0$ is a proper value of A of multiplicity q, and $A\Psi = 0$, then the zeros of $P(\lambda)$ are the nonzero proper values of A, each repeated a number of times equal to its multiplicity.

Hints. (i), (ii), (iii). Let Φ be an $n \times (n - q)$ matrix such that $\Psi^*\Phi = 0$, $\Phi^*\Phi = E_{n-q}$. For $U = [\Phi \quad \Psi]$, and $V = \operatorname{diag}\{U, E_q\}$, compute $V^* M(\lambda) V$ to show that $P(\lambda)$ is a nonzero multiple of $\det[\lambda E_q - \Phi^* A \Phi]$. Show that $(u; \mu)$ is a nonzero solution of (2.4) for a value λ if and only if there exists a nonzero $(n - q)$-dimensional vector v such that $u = \Phi v$, and $[\lambda E_q - \Phi^* A \Phi] v = 0$, while $\mu = [\Psi^*\Psi]^{-1} \Psi^* A \Phi v$. Obtain the stated results from Theorem 2.2 for the $(n - q) \times (n - q)$ hermitian matrix $\Phi^* A \Phi$.

(iv) Note that $(-1)^{n-q} \prod_{\alpha=1}^{n-q} \tilde{\lambda}_\alpha = \det M(0)$, and use elementary properties of determinants.

(v) Use the extremizing properties of proper values of A, as set forth in Theorems 2.2 and 2.3.

(vi) Show that if $(u; \mu)$ is a solution of (2.4) then $\mu = 0$.

13. If A is an $n \times n$ hermitian matrix $A = [A_{\alpha\beta}]$, $(\alpha, \beta = 1, \ldots, n)$, for $q = 1, \ldots, n$ denote by A_q the $q \times q$ hermitian matrix $A_q = [A_{\alpha\beta}]$, $(\alpha, \beta = 1, \ldots, q)$, with proper values $\lambda_{1,q} \leq \lambda_{2,q} \leq \cdots \leq \lambda_{q,q}$, ordered as in Theorem

2.2. Prove that

$$\lambda_{i,p} \leq \lambda_{i,p-1} \leq \lambda_{i+1,p}, \qquad (i = 1, \ldots, p-1; p = 2, \ldots, n).$$

Hint. Note that it suffices to prove the stated result for $p = n$, and use results from Problem 12 for $q = 1$, and $\Psi = [\delta_{\alpha n}]$, $(\alpha = 1, \ldots, n)$.

3. THE JORDAN CANONICAL FORM

If $A^{(j)}$ is an $n_j \times n_j$ matrix for $j = 1, \ldots, r$, the symbol

$$(3.1) \qquad A = \operatorname{diag}\{A^{(1)}, \ldots, A^{(r)}\} = \begin{bmatrix} A^{(1)} & 0 & \cdots & 0 \\ & \cdot & \cdots & \cdot \\ 0 & 0 & \cdots & A^{(r)} \end{bmatrix},$$

will denote the square matrix A of $n_1 + \cdots + n_r$ rows and columns which when written as a partitioned $r \times r$ matrix $[M_{ij}]$, $(i,j = 1, \ldots, r)$ with M_{ij} an $n_i \times n_j$ matrix, has $M_{ij} = 0$ if $i \neq j$ and $M_{jj} = A^{(j)}$, $(j = 1, \ldots, r)$. For A of the form (3.1) it follows that $\Delta_A(\lambda) = \Pi_{j=1}^{r} \Delta_{A^{(j)}}(\lambda)$. If $A^{(j)}$ and $B^{(j)}$ are $n_j \times n_j$ matrices for $j = 1, \ldots, r$, and $A = \operatorname{diag}\{A^{(1)}, \ldots, A^{(r)}\}$, $B = \operatorname{diag}\{B^{(1)}, \ldots, B^{(r)}\}$, it may be verified readily that $A + B = \operatorname{diag}\{A^{(1)} + B^{(1)}, \ldots, A^{(r)} + B^{(r)}\}$ and $AB = \operatorname{diag}\{A^{(1)}B^{(1)}, \ldots, A^{(r)}B^{(r)}\}$. Moreover, if $A^{(j)} \overset{s}{=} B^{(j)}$, $(j = 1, \ldots, r)$, then $A \overset{s}{=} B$; indeed, if $B^{(j)} = [P^{(j)}]^{-1}A^{(j)}P^{(j)}$, and $P = \operatorname{diag}\{P^{(1)}, \ldots, P^{(r)}\}$, then $B = P^{-1}AP$.

In the subsequent discussion of a canonical form for arbitrary matrices a role of particular significance will be assumed by the special superdiagonal matrices

$$(3.2) \qquad J = J(\lambda; q) = \begin{bmatrix} \lambda & 1 & 0 & \cdots & 0 & 0 \\ 0 & \lambda & 1 & \cdots & 0 & 0 \\ \cdot & \cdot & \cdot & \cdots & \cdot & \cdot \\ 0 & 0 & 0 & \cdots & \lambda & 1 \\ 0 & 0 & 0 & \cdots & 0 & \lambda \end{bmatrix},$$

of q rows and columns with $J_{\alpha\beta} = 0$ if $\alpha > \beta$ or $\beta > 1 + \alpha$, $J_{\alpha\alpha} = \lambda$, $(\alpha = 1, \ldots, q)$, and $J_{\alpha,\alpha+1} = 1$, $(\alpha = 1, \ldots, q-1)$; in particular, for $q = 1$ the matrix $J(\lambda; 1)$ is the 1×1 matrix with element λ. A matrix (3.2) is called an *elementary Jordan matrix of order q*. The result to be established in this section is that in the field of complex numbers a given square matrix is similar to a matrix of the form (3.1), in which each $A^{(j)}$ is an elementary Jordan matrix. For the theory of differential equations, this result is of particular significance in the study of stability properties of solutions.

For a given $n \times n$ matrix A, as in §1, let $\mathfrak{N}(A^q)$, $(q = 1, 2, \ldots)$, be the subspace of vectors x satisfying $A^q x = 0$. The symbol $\mathfrak{N}(A^0)$ is also introduced for the zero-dimensional subspace; that is, $\mathfrak{N}(A^0) = \mathfrak{N}(E_n)$. Clearly $\mathfrak{N}(A^0) \subset \mathfrak{N}(A^1) \subset \mathfrak{N}(A^2) \subset \cdots$, and if $d_q = \dim \mathfrak{N}(A^q)$ then $0 = d_0 \leq d_1 \leq d_2 \leq \cdots$. In particular, if A is nonsingular then $\mathfrak{N}(A^q) = \mathfrak{N}(A^0)$, $(q = 0, 1, 2, \ldots)$. If A is singular, however, then $d_1 > 0$ and as $d_q \leq n$, $(q = 0, 1, \ldots)$, there is a smallest positive integer $p = p(A)$ such that $d_{p+1} = d_p$; then $\mathfrak{N}(A^{p+1}) = \mathfrak{N}(A^p)$, and p is the smallest integer such that $A^p x = 0$ holds for all x satisfying $A^{p+1}x = 0$. Now if $j \geq 1$ and $A^{p+1+j}x = 0$, then $A^{p+1}(A^j x) = 0$, and by the definitive property of p it follows that $0 = A^p(A^j x) = A^{p+j}x$. Hence $\mathfrak{N}(A^q) = \mathfrak{N}(A^{q-1})$ for $q \geq p + 2$, and $p = p(A)$ is characterized as the smallest integer such that $\mathfrak{N}(A^j) = \mathfrak{N}(A^p)$ for $j \geq p$; in particular, $\mathfrak{N}(A^j)$ is a proper subspace of $\mathfrak{N}(A^{j+1})$ for $j < p$. The subspace $\mathfrak{N}(A^p)$ is termed the *generalized null space* of A, and denoted by $\mathfrak{N}_g(A)$.

***Lemma* 3.1.** *If A is singular then the dimension of $\mathfrak{N}_g(A)$ is equal to the multiplicity of $\lambda = 0$ as a zero of $\Delta_A(\lambda)$.*

If $\lambda = 0$ is a zero of $\Delta_A(\lambda)$ of multiplicity m then $\lambda = 0$ is also a zero of $\Delta_{A^q}(\lambda)$, $(q = 1, 2, \ldots)$, of multiplicity m. Indeed, if m_q denotes the multiplicity of $\lambda = 0$ as a zero of $\Delta_{A^q}(\lambda)$, then $\Delta_{A^q}(\lambda^q)$ has $\lambda = 0$ as a zero of multiplicity qm_q. Moreover, for ω the primitive q-th root $\exp\{2\pi i/q\}$ of 1 the algebraic identity $z^q - 1 = \prod_{\beta=1}^q (z - \omega^\beta)$ implies the slightly more general algebraic identity $z^q - \lambda^q = \prod_{\beta=1}^q (z - \omega^\beta \lambda)$, which in turn implies the matrix identity $A^q - \lambda^q E = \prod_{\beta=1}^q (A - \omega^\beta \lambda E)$. Consequently, we have the relations

$$\Delta_{A^q}(\lambda^q) = \det(\lambda^q E - A^q) = (-1)^n \det(A^q - \lambda^q E) = (-1)^n \prod_{\beta=1}^{q} \det(A - \omega^\beta \lambda E),$$

$$\prod_{\beta=1}^{q} \det(A - \omega^\beta \lambda E) = (-1)^{nq} \prod_{\beta=1}^{q} \det(\omega^\beta \lambda E - A) = (-1)^{nq} \prod_{\beta=1}^{q} \Delta_A(\omega^\beta \lambda),$$

and hence $\lambda = 0$ is a zero of $\Delta_{A^q}(\lambda^q)$ of multiplicity qm, so that $qm_q = qm$ and $m_q = m$.

From the definition of $\mathfrak{N}_g(A)$ the dimension of $\mathfrak{N}_g(A)$ is equal to the index of $\lambda = 0$ as a proper value of A^p for $p = p(A)$. If $A^p y = 0$, and there is a vector x such that $A^p x = y$, then $A^{2p}x = 0$ and from the definitive property of p it follows that $0 = A^p x = y$. From conclusion (i) of the Corollary to Theorem 1.1 it then follows that $\lambda = 0$ is a proper value of A^p with index equal to its multiplicity, and consequently the dimension of $\mathfrak{N}_g(A)$ is equal to m, thus completing the proof of the lemma.

Theorem 3.1. *Suppose that the distinct proper values of A are $\lambda_1, \ldots, \lambda_r$, with λ_j of multiplicity m_j, $(j = 1, \ldots, r)$. If $A_j = A - \lambda_j E$ and $\mathfrak{N}(A_j^q)$, $q = 0, 1, \ldots, p_j = p(A_j)$ and $\mathfrak{N}_g(A_j) = \mathfrak{N}(A_j^{p_j})$ are defined as above, then for $j = 1, \ldots, r$:*

(i) $\mathfrak{N}(A_j{}^q)$ *is a proper subspace of* $\mathfrak{N}_g(A_j)$ *for* $q < p_j$, *and* $\mathfrak{N}(A_j{}^q) = \mathfrak{N}_g(A_j)$ *for* $q \geq p_j$;

(ii) $\dim \mathfrak{N}_g(A_j) = m_j$;

(iii) *if* $x \in \mathfrak{N}_g(A_j)$, *then* $A_j{}^{p_j}x = 0$ *and* $A^q x \in \mathfrak{N}_g(A_j)$, $(q = 1, 2, \ldots)$; *moreover, if* $0 \leq q < p_j$ *there exists an* $x \in \mathfrak{N}_g(A_j)$ *with* $A_j{}^q x \neq 0$;

(iv) *if* $x \in \mathfrak{N}_g(A_j)$, $x \neq 0$, *and* $P(\lambda)$ *is a polynomial with* $P(\lambda_j) \neq 0$, *then* $P(A)x \in \mathfrak{N}_g(A_j)$ *and* $P(A)x \neq 0$;

(v) *if* $v^{(\beta,j)}$, $(\beta = 1, \ldots, m_j)$, *is a basis for* $\mathfrak{N}_g(A_j)$, $(j = 1, \ldots, r)$, *then*

$$v^{(1,1)}, \ldots, v^{(m_1,1)}; \ldots; v^{(1,r)}, \ldots, v^{(m_r,r)}, \tag{3.3}$$

is a basis for $\mathbf{C}_n$; *moreover, if* $P = [v^{(1,1)}, \ldots, v^{(m_r,r)}]$ *is the nonsingular matrix whose sequence of column vectors is* (3.3), *then*

(3.3′) $\quad P^{-1}AP = \operatorname{diag}\{B^{(1)}, \ldots, B^{(r)}\}$, *with* $B^{(j)}$ *an* $m_j \times m_j$ *matrix.*

Conclusion (i) follows from the definitive property of $p_j = p(A_j)$, and the above Lemma 3.1 implies conclusion (ii). The first statement of (iii) follows from the fact that $\mathfrak{N}_g(A_j)$ is the totality of vectors x satisfying $A_j{}^{p_j}x = 0$, and the identity $A^q A_j{}^{p_j} x \equiv A_j{}^{p_j} A^q x$; in view of (i), if $0 \leq q < p_j$ there is an element x of $\mathfrak{N}_g(A_j)$ not belonging to $\mathfrak{N}(A_j{}^q)$, so that $A_j{}^q x \neq 0$.

In order to establish (iv), it clearly suffices to show that if $x \in \mathfrak{N}_g(A_j)$ and $\lambda \neq \lambda_j$, then $(A - \lambda E)x \in \mathfrak{N}_g(A_j)$ and $(A - \lambda E)x = 0$ only if $x = 0$. Now from the first part of (iii) it follows immediately that $(A - \lambda E)x \in \mathfrak{N}_g(A_j)$ whenever $x \in \mathfrak{N}_g(A_j)$; moreover, if $(A - \lambda E)x = 0$ then $A_j x = (\lambda - \lambda_j)x$ and $0 = A_j{}^{p_j}x = (\lambda - \lambda_j)^{p_j}x$ implies that $x = 0$ in case $\lambda \neq \lambda_j$.

For the proof of conclusion (v) it is to be noted first that the relation $m_1 + \cdots + m_r = n$ implies that the vectors of (3.3) are n in number. Now if $c_{\beta j}$, $(\beta = 1, \ldots, m_j; j = 1, \ldots, r)$, are constants such that

$$\sum_{j=1}^{r} \sum_{\beta=1}^{m_j} v^{(\beta,j)} c_{\beta j} = 0,$$

then

$$x^{(j)} = \textstyle\sum_{\beta=1}^{m_j} v^{(\beta,j)} c_{\beta j} \in \mathfrak{N}_g(A_j), \qquad (j = 1, \ldots, r),$$

and $x^{(1)} + \cdots + x^{(r)} = 0$. If $P_k(\lambda)$ is the polynomial $\prod_{j=1; j\neq k}^{r} (\lambda - \lambda_j)^{p_j}$, then in view of the relations $A_j{}^{p_j}x^{(j)} = 0$, $(j = 1, \ldots, r)$, it follows that $0 = p_k(A)[x^{(1)} + \cdots + x^{(r)}] = P_k(A)x^{(k)}$, $(k = 1, \ldots, r)$. On the other hand, since $x^{(k)} \in \mathfrak{N}_g(A_k)$ and $P_k(\lambda_k) \neq 0$, conclusion (iv) implies that $P_k(A)x^{(k)} = 0$ only if $x^{(k)} = 0$, in which case $c_{\beta k} = 0$, $(\beta = 1, \ldots, m_k)$. Hence the n vectors (3.3) are linearly independent, and consequently form a basis for $\mathbf{C}_n$. Finally, if $P^{(j)}$ is the $n \times m_j$ matrix $[v^{(1,j)}, \ldots, v^{(m_j,j)}]$ it follows from condition (iii) that there exists an $m_j \times m_j$ matrix $B^{(j)}$ such that $AP^{(j)} = P^{(j)}B^{(j)}$, $(j = 1, \ldots, r)$, and hence (3.3′) holds.

***Corollary* 1.** *If $P(\lambda)$ is a polynomial in λ which does not contain $(\lambda - \lambda_j)^{p_j}$ as a factor, then there exists an $x \in \mathfrak{N}_g(A_j)$ for which $P(A)x \neq 0$.*

If $P(\lambda) = P_1(\lambda)(\lambda - \lambda_j)^q$, $0 \leq q < p_j$ and $P_1(\lambda_j) \neq 0$, then by (i) there is a vector $x \in \mathfrak{N}_g(A_j)$ for which $y = A_j{}^q x \neq 0$. Now $y \in \mathfrak{N}_g(A_j)$ in view of (iii), and from (iv) it follows that $P(A)x = P_1(A)y \neq 0$.

In view of (v), each $x \in \mathbf{C}_n$ is of the form $x^{(1)} + \cdots + x^{(r)}$, where $x^{(j)} \in \mathfrak{N}_g(A_j)$, $(j = 1, \ldots, r)$. As $A_j{}^{p_j}x = 0$ for $x \in \mathfrak{N}_g(A_j)$, it then follows from the above corollary that $P(A) = 0$, {i.e., $P(A)x = 0$ for arbitrary $x \in \mathbf{C}_n$}, if and only if $P(\lambda)$ is a multiple of the particular polynomial

$$\prod_{j=1}^{r} (\lambda - \lambda_j)^{p_j}. \tag{3.4}$$

The polynomial (3.4) is called the *minimal polynomial* for A. In particular, $p_j \leq m_j$, $(j = 1, \ldots, r)$, and the characteristic polynomial $\Delta_A(\lambda) = \prod_{j=1}^{r} (\lambda - \lambda_j)^{m_j}$ is a multiple of (3.4). Hence we have the following result, which is known as the *Cayley-Hamilton Theorem.*

***Corollary* 2.** $\Delta_A(A) = 0$.

Theorem 3.2. *If the distinct proper values of A are $\lambda_1, \ldots, \lambda_r$, with λ_j of multiplicity m_j, $(j = 1, \ldots, r)$, and P is the $n \times n$ matrix whose sequence of column vectors is* (3.3) *with $v^{(\beta, j)}$, $(\beta = 1, \ldots, m_j)$, a basis for $\mathfrak{N}_g(A_j)$, then $P^{-1}AP = \text{diag}\{B^{(1)}, \ldots, B^{(r)}\}$ where $B^{(j)}$, $(j = 1, \ldots, r)$, is an $m_j \times m_j$ matrix, and:*

(i) $[B^{(j)} - \lambda_j E_{m_j}]^{p_j} = 0$, *and* $[B^{(j)} - \lambda_j E_{m_j}]^q \neq 0$ *for* $q < p_j$;
(ii) $B^{(j)} - \lambda E_{m_j}$ *is nonsingular for* $\lambda \neq \lambda_j$;
(iii) $\Delta_{B^{(j)}}(\lambda) = (\lambda - \lambda_j)^{m_j}$.

In view of conclusion (v) of Theorem 3.1, it follows that for arbitrary λ we have $(A - \lambda E)^q P = PR_\lambda{}^q$, $(q = 1, \ldots)$, with

$$R_\lambda = \text{diag}\{B^{(1)} - \lambda E_{m_1}, \ldots, B^{(r)} - \lambda E_{m_r}\},$$

and conclusion (i) is an immediate consequence of conclusion (i) of Theorem 3.1. In turn, if $\lambda \neq \lambda_j$ and $[B^{(j)} - \lambda E_{m_j}]\xi = 0$, then $0 = [B^{(j)} - \lambda_j E_{m_j}]^{p_j}\xi = (\lambda - \lambda_j)^{p_j}\xi$ so that $\xi = 0$ and consequently $B^{(j)} - \lambda E_{m_j}$ is nonsingular. Finally, conclusion (iii) follows from the relation $\prod_{j=1}^{r} (\lambda - \lambda_j)^{m_j} = \Delta_A(\lambda) = \prod_{j=1}^{r} \Delta_{B^{(j)}}(\lambda)$, together with the results (i), (ii).

Although the results of the above theorems suffice for many of the applications to differential equations, for certain applications one desires the more precise choice for the matrices $B^{(j)}$ that will be established in Theorem 3.3.

A subspace $\mathbf{M}$ of $\mathbf{C}_n$ is said to be *invariant* under a transformation T defined by $y = Ax$ for $x \in \mathbf{C}_n$ if $Ax \in \mathbf{M}$ whenever $x \in \mathbf{M}$. If $\mathbf{M}$ is invariant and there is a positive integer p such that $A^p x = 0$ for every $x \in \mathbf{M}$, then T is said to be *nilpotent* on $\mathbf{M}$, and the smallest such integer p is termed the *degree* (*of nilpotency*) of the transformation T, or of the matrix A, on $\mathbf{M}$. For simplicity, the phrase "A is nilpotent" will be used as synonymous with "A is nilpotent on $\mathbf{C}_n$." The result to be established is a direct consequence of the following lemma, and the fact that the matrix A_j of Theorem 3.1 is nilpotent of degree p_j on $\mathfrak{N}_g(A_j)$.

Lemma 3.2. *If* $\mathbf{M}$ *is a subspace of* $\mathbf{C}_n$ *with* $d = \dim \mathbf{M} \geq 1$, *and* A *is nilpotent on* $\mathbf{M}$, *then* $\mathbf{M}$ *has a basis of the form*

$$A^{q_1-1}u^{(1)}, \ldots, Au^{(1)}, u^{(1)}; \ldots ; A^{q_h-1}u^{(h)}, \ldots, Au^{(h)}, u^{(h)}, \tag{3.5}$$

with

$$A^{q_\gamma}u^{(\gamma)} = 0, \qquad (\gamma = 1, \ldots, h). \tag{3.6}$$

If $d_0 = 0$, *and* d_γ, $(\gamma = 1, 2, \ldots)$, *is the dimension of the subspace* $\mathbf{M}_\gamma$ *consisting of all* $x \in \mathbf{M}$ *satisfying* $A^\gamma x = 0$, *then* $h = d_1$, *and if* n_k *denotes the number of* q_γ *equal to* k *then*

$$n_k = -d_{k-1} + 2d_k - d_{k+1}, \qquad (k = 1, 2, \ldots). \tag{3.7}$$

Consequently, h *is uniquely determined, and the sequence* $q_1, \ldots, q_h$ *is specified except for permutation; in particular, if* p *is the degree of nilpotency of* A *on* $\mathbf{M}$ *then* $d_\gamma = d$ *for* $\gamma \geq p$, $q_\gamma \leq p$ *for* $\gamma = 1, \ldots, h$, *and* $n_p = d - d_{p-1} > 0$.

The result of this lemma will be proved by induction. If $d = \dim \mathbf{M} = 1$ and A is nilpotent on $\mathbf{M}$, then for u a nonzero element of $\mathbf{M}$ the condition that $\mathbf{M}$ is invariant under the transformation (1.1) implies that there is a constant λ such that $Au = \lambda u$, and consequently $A^\gamma u = \lambda^\gamma u$, $(\gamma = 1, 2, \ldots)$. The condition $A^p u = 0$ then implies that $\lambda = 0$, and hence $\mathbf{M}$ has a basis (3.5) satisfying (3.6), with $u^{(1)} = u$, $1 = h = q_1 = p = d_\gamma$, $(\gamma = 1, 2, \ldots)$; in this case the stated properties of the integers $h, q_1, \ldots, q_h$ are clearly valid.

Now suppose that the result of the lemma holds for $\dim \mathbf{M} = 1, 2, \ldots, d - 1 < n$, and consider a transformation (1.1) that is nilpotent on a d-dimensional subspace $\mathbf{M}$ of $\mathbf{C}_n$, If p is the degree of nilpotency of A on $\mathbf{M}$, let u be an element of $\mathbf{M}$ for which $A^{p-1}u \neq 0$, and consider the elements $v^{(\gamma)} = A^{p-\gamma}u$, $(\gamma = 1, \ldots, p)$. As $\mathbf{M}$ is invariant the elements $v^{(\gamma)}$ are in $\mathbf{M}$; moreover, if $c_1 v^{(1)} + \cdots + c_p v^{(p)} = 0$ then the relations $A^{p-1-\gamma}(c_1 v^{(1)} + \cdots + c_p v^{(p)}) = 0$, $(\gamma = 0, 1, \ldots, p - 1)$, and $A^q u = 0$ for $q \geq p$, are seen to imply $c_1 = 0 = c_2 = \cdots = c_p$, so that $v^{(1)}, \ldots, v^{(p)}$ are linearly independent. The subspace $\mathbf{M}^0 = [v^{(1)}, \ldots, v^{(p)}]$ spanned by $v^{(1)}, \ldots, v^{(p)}$ is then of dimension $d^0 = p$, and $\mathbf{M}^0$ is invariant under (1.1) since $Av^{(1)} = 0$ and $Av^{(\gamma)} = v^{(\gamma-1)}$, $(\gamma = 2, \ldots, p)$. As $A^p v^{(\gamma)} = 0$, $(\gamma = 1, \ldots, p)$, and

$$A^{p-1}v^{(p)} = v^{(1)} \neq 0,$$

it follows that A is nilpotent of degree p on $\mathbf{M}^0$. Moreover, if $d_0^0 = 0$ and d_γ^0, $(\gamma = 1, 2, \ldots)$, is the dimension of the subspace $\mathbf{M}_\gamma^0$ consisting of all $x \in \mathbf{M}^0$ satisfying $A^\gamma x = 0$, then it is readily seen that $d_\gamma^0 = \gamma$, $(\gamma = 0, 1, \ldots, p)$, and $d_\gamma^0 = p$ for $\gamma > p$. Now $v^{(1)}, \ldots, v^{(p)}$ is a set of elements of $\mathbf{M}$ of the form (3.5) which satisfies (3.6), with $h = 1$, $q_1 = p$. Also, if n_k^0, $(k = 1, 2, \ldots)$, signifies the number of q_γ equal to k, then each of the quantities n_k^0 and $-d_{k-1}^0 + 2d_k^0 - d_{k+1}^0$ is zero for $k \neq p$, and equal to one in case $k = p$, and hence

$$n_k^0 = -d_{k-1}^0 + 2d_k^0 - d_{k+1}^0, \qquad (k = 1, 2, \ldots). \tag{3.8}$$

In particular, from these properties of $\mathbf{M}^0$ it follows that if $p = d$ then the result of the lemma holds with (3.5) given by $v^{(1)}, \ldots, v^{(p)}$.

On the other hand, in case $p < d$ let v be a vector such that $(A^{p-\gamma}u, v) = \delta_{1\gamma}$, $(\gamma = 1, \ldots, p)$. Then

$$(A^{p+\beta-\gamma-1}u, v) = \delta_{\beta\gamma}, \qquad (\beta, \gamma = 1, \ldots, p), \tag{3.9}$$

and if $w^{(\sigma)}$, $(\sigma = 1, \ldots, d - p)$, are such that $v^{(1)}, \ldots, v^{(p)}, w^{(1)}, \ldots, w^{(d-p)}$ is a basis for $\mathbf{M}$, then the vectors defined by

$$v^{(p+\sigma)} = w^{(\sigma)} - \sum_{\beta=1}^{p} (A^{\beta-1}w^{(\sigma)}, v)v^{(\beta)}, \qquad (\sigma = 1, \ldots, d - p),$$

are such that $v^{(1)}, \ldots, v^{(d)}$ is a basis for $\mathbf{M}$ and

$$(A^{\beta-1}v^{(p+\sigma)}, v) = 0, \qquad (\beta = 1, \ldots, p; \sigma = 1, \ldots, d - p). \tag{3.10}$$

The subspace $\mathbf{M}' = [v^{(p+1)}, \ldots, v^{(d)}]$ spanned by $v^{(p+1)}, \ldots, v^{(d)}$ is then of dimension $d' = d - p$, and in view of (3.9), (3.10) it follows that $\mathbf{M}'$ may be characterized as the set of $x \in \mathbf{M}$ satisfying

$$(A^{\beta-1}x, v) = 0, \qquad (\beta = 1, \ldots, p). \tag{3.11}$$

Now for $x \in \mathbf{M}'$ and $\beta = 1, \ldots, p - 1$ we have $(A^{\beta-1}[Ax], v) = (A^\beta x, v) = 0$ by (3.11), while $(A^{p-1}[Ax], v) = (A^p x, v) = 0$ since $A^p x = 0$. Consequently $Ax \in \mathbf{M}'$ whenever $x \in \mathbf{M}'$, so that $\mathbf{M}'$ is invariant; moreover, as $\mathbf{M}' \subset \mathbf{M}$ it follows that A is nilpotent of degree $p' \leq p$ on $\mathbf{M}'$. By the induction hypothesis it then follows that $\mathbf{M}'$ admits a basis of the form

$$A^{g_1-1}y^{(1)}, \ldots, Ay^{(1)}, y^{(1)}; \ldots; A^{g_s-1}y^{(s)}, \ldots, Ay^{(s)}, y^{(s)},$$

with $A^{g_\gamma}y^{(\gamma)} = 0$, $(\gamma = 1, \ldots, s)$; moreover, if $d_0' = 0$ and d_γ', $(\gamma = 1, 2, \ldots)$, is the dimension of the subspace $\mathbf{M}_\gamma'$ consisting of all $x \in \mathbf{M}'$ satisfying $A^\gamma x = 0$, then $s = d_1'$, and if n_k' denotes the number of g_γ equal to k then

$$n_k' = -d_{k-1}' + 2d_k' - d_{k+1}', \qquad (k = 1, 2, \ldots). \tag{3.12}$$

It then follows that $\mathbf{M}$ has a basis (3.5) satisfying (3.6), with $h = s + 1$, $q_1 = p$, $u^{(1)} = u$, and $q_\gamma = g_\gamma - 1$, $u^{(\gamma)} = y^{(\gamma-1)}$ for $\gamma = 2, \ldots, h$; moreover, $n_k = n_k^0 + n_k'$, $(k = 1, 2, \ldots)$. As each $x \in \mathbf{M}$ is represented uniquely in the form $x = x^{(0)} + x^{(1)}$ with $x^{(0)} \in \mathbf{M}^0$, $x^{(1)} \in \mathbf{M}'$, and $A^\gamma x^{(0)} \in \mathbf{M}^0$, $A^\gamma x^{(1)} \in \mathbf{M}'$, $(\gamma = 1, 2, \ldots)$, it follows that $d_\gamma = d_\gamma^0 + d_\gamma'$, $(\gamma = 0, 1, \ldots)$, and (3.7) is an immediate consequence of (3.8), (3.12).

Theorem 3.3. *A given matrix A is similar to a matrix of the form*

$$B = \operatorname{diag}\{J(\Lambda_1; \tau_1), \ldots, J(\Lambda_\rho; \tau_\rho)\}; \tag{3.13}$$

moreover, for any B of the form (3.13) *that is similar to A the array* $\Lambda_1, \ldots, \Lambda_\rho$ *is the sequence of proper values of* Λ *with each repeated a number of times equal to its index, and for* λ *a proper value of A of index k and multiplicity m the set* $\beta_1, \ldots, \beta_k$ *of values* β *for which* $\Lambda_\beta = \lambda$ *is such that* $\tau_{\beta_1} + \cdots + \tau_{\beta_k} = m$ *and the sequence* $\tau_{\beta_1}, \ldots, \tau_{\beta_k}$ *is unique except for permutation.*

Let $\lambda_1, \ldots, \lambda_r$ denote the distinct proper values of A, with λ_j of index k_j and multiplicity m_j. For $A_j = A - \lambda_j E$ it follows from conclusion (ii) of Theorem 3.1 that $\mathfrak{N}_g(A_j)$ is of dimension m_j, and as A_j is nilpotent on $\mathfrak{N}_g(A_j)$ it follows from Lemma 3.2 that $\mathfrak{N}_g(A_j)$ has a basis $v^{(\beta,j)}$, $(\beta = 1, \ldots, m_j;$ $j = 1, \ldots, r)$, of the form (3.5), and satisfying (3.6), with $A = A_j$, $h = k_j$, and $q_\gamma = q_{j\gamma}$, $(\gamma = 1, \ldots, k_j)$, positive integers such that $q_{j1} + \cdots + q_{jk_j} = m_j$. If P is the matrix with sequence of column vectors the thus determined set (3.3), it then follows from Theorem 3.2 that $P^{-1}AP = B$, where $B = \operatorname{diag}\{B^{(1)}, \ldots, B^{(r)}\}$, and $B^{(j)}$ is the $m_j \times m_j$ matrix $\operatorname{diag}\{J(\lambda_j; q_{j1}), \ldots, J(\lambda_j; q_{jk_j})\}$, so that B is of the form (3.13).

On the other hand, if $P^{-1}AP = B$, with B of the form (3.13), then $P^{-1}(A - \lambda E)P = B_\lambda$ with $B_\lambda = \operatorname{diag}\{J(\Lambda_1 - \lambda; \tau_1), \ldots, J(\Lambda_\rho - \lambda; \tau_\rho)\}$. As $J(\Lambda; \tau)$ is nonsingular for $\Lambda \neq 0$, of rank $\tau - 1$ for $\Lambda = 0$, and $\Delta_{J(\Lambda;\tau)}(\lambda) = (\lambda - \Lambda)^\tau$, it follows that if λ is a proper value of A of index k and multiplicity m then exactly k values $\beta_1, \ldots, \beta_k$ of the index β satisfy $\Lambda_\beta = \lambda$, and $\tau_{\beta_1} + \cdots + \tau_{\beta_k} = m$. Finally, if $P^{(\beta)} = [p^{(1,\beta)}, \ldots, p^{(\tau_\beta,\beta)}]$, $(\beta = 1, \ldots, \rho)$, is the $n \times \tau_\beta$ matrix such that $P = [P^{(1)}, \ldots, P^{(\rho)}]$, then the condition that $P^{-1}AP = B$, with B of the form (3.13), is equivalent to the matrix equations

$$AP^{(\beta)} = P^{(\beta)} J(\Lambda_\beta; \tau_\beta), \qquad (\beta = 1, \ldots, \rho).$$

Consequently $[A - \Lambda_\beta E]P^{(\beta)} = P^{(\beta)} J(0; \tau_\beta)$, and for $u^{(\beta)} = p^{(\tau_\beta,\beta)}$ we have $p^{(j,\beta)} = (A - \Lambda_\beta E)p^{(j+1,\beta)} = (A - \Lambda_\beta E)^{\tau_\beta - j}u^{(\beta)}$, $(j = 1, \ldots, \tau_\beta - 1)$, and $(A - \Lambda_\beta E)p^{(1;\beta)} = 0$, so that

$$P^{(\beta)} = [(A - \Lambda_\beta E)^{\tau_\beta - 1}u^{(\beta)}, \ldots, (A - \Lambda_\beta E)u^{(\beta)}, u^{(\beta)}]$$

and $(A - \Lambda_\beta E)^{\tau_\beta}u^{(\beta)} = 0$. Therefore, if λ is a proper value of A of index k and multiplicity m, then the sequence of column vectors of the matrices $P^{(\beta)}$

with β equal to those values $\beta_1, \ldots, \beta_k$ for which $\Lambda_\beta = \lambda$ is a basis for the generalized null space $\mathfrak{N}_g(A - \lambda E)$ that is of the form (3.5) and satisfies (3.6), with A replaced by $A - \lambda E$, $h = k$, and $q_\gamma = \tau_{\beta_\gamma}$, $(\gamma = 1, \ldots, k)$, and from Lemma 3.2 it follows that the sequence $\tau_{\beta_1}, \ldots, \tau_{\beta_k}$ is unique except for permutation.

A matrix (3.13) which satisfies the properties listed in Theorem 3.3 is said to be a *Jordan canonical form* for A.

Problems E.3

1. For $J = J(\lambda; q)$ show that the matrix series

$$c_0 E + c_1 J + \cdots + c_m J^m + \cdots,$$

converges if and only if the infinite series $f(z) = c_0 + c_1 z + \cdots + c_m z^m + \cdots$, and each of the derivative series $f^{(j)}(z)$, $(j = 1, \ldots, q-1)$, converge for $z = \lambda$.

2. Show that $\exp\{tJ(\rho; q)\} = e^{t\rho} E(t; q)$, where

$$E(t; q) = \exp\{tJ(0; q)\} = \begin{bmatrix} 1 & t & \cdots & \dfrac{t^{q-1}}{(q-1)!} \\ 0 & 1 & \cdots & \dfrac{t^{q-2}}{(q-2)!} \\ \cdot & \cdot & \cdots & \cdot \\ 0 & 0 & \cdots & 1 \end{bmatrix}.$$

3. If $\lambda \neq 0$ and $L(\lambda; q) = [\lambda^{1-\alpha}\delta_{\alpha\beta}]$, $(\alpha, \beta = 1, \ldots, q)$, show that $L(\lambda; q)\, J(\lambda; q)\, L^{-1}(\lambda; q) = \lambda\, J(1; q)$.

4. From Theorem 3.3 conclude that if $t \neq 0$ there is a nonsingular matrix $G(t; q)$ such that $G^{-1}(t; q)\, E(t; q)\, G(t; q) = J(1; q)$.

5. Use the results of Problems 2, 3, and 4 to show that if $\lambda \neq 0$, $t \neq 0$, and ρ is such that $e^{t\rho} = \lambda$ then

$$\exp\{tL^{-1}(\lambda; q)\, G^{-1}(t; q)\, J(\rho; q)\, G(t; q)\, L(\lambda; q)\} = J(\lambda; q).$$

6. From Theorem 3.3 and Problem 5 conclude that if A is a nonsingular matrix then there exists a matrix R such that $A = \exp R$.

7. Suppose that R is an $n \times n$ matrix with distinct proper values $\rho_1, \ldots, \rho_r$, where ρ_j is of index k_j and multiplicity m_j. Show that if $\alpha_1, \ldots, \alpha_g$ are those values of α for which e^{ρ_α} has a common value λ, then λ is a proper value of $\exp R$ of index $k_{\alpha_1} + \cdots + k_{\alpha_g}$ and multiplicity $m_{\alpha_1} + \cdots + m_{\alpha_g}$; in particular, λ is a proper value with index equal to multiplicity if and only if ρ_{α_γ}, $(\gamma = 1, \ldots, g)$, is a proper value of R with index equal to multiplicity.

8. Show that the matrix

$$A = \begin{bmatrix} -2 & 0 & 2 \\ 2 & 1 & -1 \\ -6 & 0 & 5 \end{bmatrix},$$

is similar to diag$\{J(1; 2), J(2; 1)\}$. Find a matrix R such that $\exp R = A$.

9. If X is a nilpotent matrix of degree p, show that $E + X = \exp Z$, where $Z = X - \frac{1}{2}X^2 + \cdots + [(-1)^p/(p-1)]X^{p-1}$.

Hint. Note that the Maclaurin expansion for $\ln(1 + z)$ is

$$z - \tfrac{1}{2}z^2 + \cdots + (-1)^{m+1}\frac{z^m}{m} + \cdots.$$

4. NOTES AND REMARKS

Results in this appendix have been presented in a manner that is of direct use in the study of differential equations, and in many instances the particular suggested proof has been selected because of its analogy to, or identity with, the proof of corresponding results for differential operators. This comment is particularly true for the results of Section 2 on hermitian matrices and forms.

RELATED REFERENCES AND COMMENTS FOR SPECIFIC PROBLEMS

E.2:7,8 The proof suggested for these problems is a well-known one for establishing the generalizations of these matrix results for symmetric operators in Hilbert space, (see, for example, F. Riesz and Béla Sz.-Nagy, *Functional Analysis*, Frederick Unger Publishing Co., New York, 1955, pp. 263–265).

E.2:12 For the significance of the results of this problem for the treatment of ordinary maxima and minima, and also for problems of the calculus of variations, the reader is referred to Carathéodory [2, Ch. XI].

Appendix F

Matrix and Vector Functions

1. PRELIMINARY CONCEPTS

If $A_{\alpha\beta}(t)$, $(\alpha = 1, \ldots, m;\ \beta = 1, \ldots, n)$, are functions of t the $m \times n$ matrix function $[A_{\alpha\beta}(t)]$ is denoted by $A(t)$. If all the elements of a matrix possess a property such as continuity, differentiability, or integrability, for brevity it is said that the matrix has this property. In particular, if $A(t)$ is differentiable the matrix of derivatives $[A'_{\alpha\beta}(t)]$ is signified by $A'(t)$, and the symbol $\bar{A}'(t)$ is employed for the matrix that is the common value of $\overline{A'(t)}$ and $[\overline{A(t)}]'$, with similar conventions for the symbols $\tilde{A}'(t)$ and $A^{*\prime}(t)$.

If $A(t)$ is integrable on a compact interval $[a, b]$ on the real line the matrix of integrals $[\int_a^b A_{\alpha\beta}(t)\, dt$ is denoted by $\int_a^b A(t)\, dt$. It may be verified readily that if $C = \int_a^b A(t)\, dt$ then $\bar{C} = \int_a^b \bar{A}(t)\, dt$, $\tilde{C} = \int_a^b \tilde{A}(t)\, dt$, and $C^* = \int_a^b A^*(t)\, dt$. If $A(t)$, $B(t)$ are $m \times n$ matrices which are differentiable {integrable} on $[a, b]$, then for arbitrary scalars c_1, c_2 the matrix $c_1 A(t) + c_2 B(t)$ is differentiable {integrable} on $[a, b]$ and $[c_1 A(t) + c_2 B(t)]' = c_1 A'(t) + c_2 B'(t)$, $\{\int_a^b [c_1 A(t) + c_2 B(t)]\, dt = c_1 \int_a^b A(t)\, dt + c_2 \int_a^b B(t)\, dt\}$.

In particular, for a given interval I on the real line, and fixed positive integers m and n, let $\mathfrak{D}_{mn}(I)$ denote the set of $m \times n$ matrix functions which are continuous and differentiable on I, and in case I_0 is a compact interval let $\mathfrak{J}_{mn}(I_0)$ denote the set of $m \times n$ matrix functions which are integrable on I_0. It follows readily that the thus specified sets $\mathfrak{D}_{mn}(I)$ and $\mathfrak{J}_{mn}(I_0)$ are vector spaces. In the above statements it has been implicitly assumed that the field of scalars is the field of complex numbers, although analogous statements are valid in case the scalar field is the field of real numbers and the elements of the matrices are restricted to be real-valued.

If $A(t)$ and $B(t)$ are differentiable matrix functions of t on a given interval I, and $A(t)B(t)$ is defined for $t \in I$, then on this interval this product matrix is differentiable and

$$(1.1) \qquad [A(t)B(t)]' = A(t)B'(t) + A'(t)B(t), \qquad t \in I.$$

Whenever the vector function $y(t) = (y_\alpha(t))$, $(\alpha = 1, \ldots, n)$, is integrable on the compact interval $[a, b]$, then the integrability of the individual component functions $y_\alpha(t)$ implies that $|y_\alpha(t)|$, $(\alpha = 1, \ldots, n)$, is also integrable, and in view of the inequality (D.1.4) for arbitrary $t \in [a, b]$ it follows that $|y(t)|$ is integrable on $[a, b]$. For integrals of vector functions the following result is basic.

Theorem 1.1. *If the vector function $y(t)$ is integrable on the compact interval $[a, b]$, then*

$$(1.2) \qquad \left| \int_a^b y(t)\, dt \right| \leq \int_a^b |y(t)|\, dt.$$

For a given $y(t)$ integrable on $[a, b]$, let $z = \int_a^b y(t)\, dt$. As (1.2) clearly holds if z is the zero vector, attention may be restricted to the case in which $|z| > 0$. Now $|z|^2 = (\int_a^b y(t)\, dt, z)$ is real, so that

$$|z|^2 = \Re\left(\int_a^b y(t)\, dt, z \right) = \int_a^b \Re(y(t), z)\, dt.$$

Moreover, $|\Re(y(t), z)| \leq |(y(t), z)| \leq |y(t)|\, |z|$, the last relation following from the Lagrange-Cauchy inequality (D.1.3-c), and hence

$$|z|^2 \leq \left[\int_a^b |y(t)|\, dt \right] |z|,$$

which is equivalent to (1.2) for $|z| > 0$.

The class $\mathfrak{J}_n^2[a, b]$ of n-dimensional vector functions $x(t) = (x_\alpha(t))$, $(\alpha = 1, \ldots, n)$, which are integrable on $[a, b]$ and such that $|x(t)|^2$ is integrable on this interval, is a vector space over the scalar field of complex numbers, and on this vector space the integral

$$(1.3) \qquad ((x, y)) = ((x, y))_n = \int_a^b (x(t), y(t))\, dt,$$

defines an hermitian form in the sense of Section A.5. Therefore, if $x(t)$, $y(t)$, $z(t)$ are vector functions of $\mathfrak{J}_n^2[a, b]$, and c is any complex scalar, then

(a) $((cx, y)) = c((x, y))$;

(b) $((x + y, z)) = ((x, z)) + ((y, z))$;

(1.4) (c) $((x, y)) = \overline{((y, x))}$;

(d) $((x, x)) \geq 0$;

(e) $((x, x)) = 0$ *if and only if* $((x, z)) = 0$ *for arbitrary* $z \in \mathfrak{J}_n^2[a, b]$.

Moreover, if $\|x\|$ is the non-negative square root of $((x, x))$, then

(1.5)
(a) $\|cx\| = |c| \, \|x\|$;
(b) $\|x\| \geq 0$, *and* $\|x\| = 0$ *if and only if* $((x, z)) = 0$ *for arbitrary* $z \in \mathfrak{J}_n^2[a, b]$;
(c) $|((x, y))| \leq \|x\| \cdot \|y\|$, (Bunyakovsky-Schwarz inequality);
(d) $\|x + y\| \leq \|x\| + \|y\|$, (triangle, or Minkowski inequality).

For the class $\mathfrak{J}_n^2[a, b]$ of vector functions one may proceed to obtain other results that are analogous to those of Appendix D for complex n-dimensional number space. For example, if for elements x, y of $\mathfrak{J}_n^2[a, b]$ we define the "distance function" or "metric" $d(x, y)$ as equal to $\|x - y\|$, then on $\mathfrak{J}_n^2[a, b]$ this distance function has properties (a), (c), (d) of (D.1.5), while (D.1.5-b) is replaced by the condition that $d(x, y) = 0$ if and only if $((y - x, z)) = 0$ for arbitrary z of $\mathfrak{J}_n^2[a, b]$. If y, $y^{(j)}$, $(j = 1, 2, \ldots)$, are vector functions of $\mathfrak{J}_n^2[a, b]$ and the sequence $\{y^{(j)}\}$ has with respect to the metric $\|x - y\|$ the limit y, that is, if $\|y^{(j)} - y\| \to 0$ as $j \to \infty$, then from the triangle inequality it follows that

$$\|y^{(j)} - y^{(k)}\| \to 0 \quad as \quad j \to \infty, k \to \infty. \tag{1.6}$$

There arises immediately the question as to whether or not the class $\mathfrak{J}_n^2[a, b]$ with metric $\|x - y\|$ is *complete*, that is, whether or not a sequence $\{y^{(j)}\}$ of elements of $\mathfrak{J}_n^2[a, b]$ satisfying (1.6) is such that there is a y of $\mathfrak{J}_n^2[a, b]$ for which $\|y^{(j)} - y\| \to 0$ as $j \to \infty$. At this point, it is to be emphasized that *in the above statements there has not been a specification of the sense in which the integration procedure is to be understood.* For most of the treatment presented in this work the hypotheses are such that the involved integrals exist in the sense of Riemann, while for other parts of the work, (notably Chapters II and VII), the integrals are taken in the sense of Lebesgue. All properties of integrals discussed above hold with the interpretation of the integration procedure as that of the Riemann integral, or as that of the Lebesgue integral. For the question just posed, however, the answer is dependent upon the integration procedure occurring in the definition of $\mathfrak{J}_n^2[a, b]$. If this procedure is that of the Riemann integral the answer is "no," whereas if the procedure is that of the Lebesgue integral the answer is "yes." That is, the class $\mathfrak{J}_n^2[a, b]$ with $d(x, y) = \|x - y\|$, and integration in the sense of Lebesgue, is complete in this sense, whereas the class $\mathfrak{J}_n^2[a, b]$ with $d(x, y) = \|x - y\|$, and integration in the sense of Riemann, is not complete.

Again, it is important to determine whether or not the class $\mathfrak{J}_n^2[a, b]$ with distance function $\|x - y\|$ has a property corresponding to the result for $\mathbf{C}_n$ given in Problem 5 of D.2. That is, if $\{y^{(j)}\}$, $(j = 1, 2, \ldots)$, is a sequence of elements of $\mathfrak{J}_n^2[a, b]$ for which there is a $\mu > 0$ satisfying $\|y^{(j)}\| \leq \mu$, $(j = 1, 2, \ldots)$, does there exist a subsequence $z^{(\gamma)} = y^{(j_\gamma)}$, $(j_1 < j_2 < \cdots)$, which is such that $\|z^{(\beta)} - z^{(\gamma)}\| \to 0$ as $\beta \to \infty$, $\gamma \to \infty$? The answer to this question is "no," and this result may be established by showing that in $\mathfrak{J}_n^2[a, b]$ there are systems of infinitely many elements $y^{(j)}$, $(j = 1, 2, \ldots)$, which are orthonormal with respect to the inner product functional $((x, y))$ in the sense that $((y^{(j)}, y^{(k)})) = \delta_{jk}$, $(j, k = 1, 2, \ldots)$; for such sequences we have $\|y^{(j)} - y^{(k)}\|^2 = 2$, $(j, k = 1, 2, \ldots)$. For example, in $\mathfrak{J}_n^2[0, \pi]$ the system $y^{(j)}(t)$ with $y_\beta^{(j)}(t) \equiv 0$, $(\beta = 2, \ldots, n; j = 1, 2, \ldots)$, and $y_1^{(j)}(t) = \sqrt{2/\pi}\sin jt$, $(j = 1, 2, \ldots)$, has the stated property.

Problems F.1

1. If $y(t)$ is a vector function which is differentiable on an open interval (a, b), and $s \in (a, b)$, show that $r(t) = |y(t)|$ has the following properties:

(i) if $r(s) \neq 0$, then $r(t)$ has a derivative at $t = s$, and $|r'(s)| \leq |y'(s)|$;

(ii) if $r(s) = 0$, then $r(t)$ has at $t = s$ a right-hand derivative $r'_+(s)$, and a left-hand derivative $r'_-(s)$; moreover, $|r'_\pm(s)| = |y'(s)|$.

2. Suppose that for t on the compact interval $[a, b]$ on the real line the $n \times n$ matrix $A(t) = [A_{\alpha\beta}(t)]$ is hermitian, and let $\lambda_1(t) \leq \cdots \leq \lambda_n(t)$ denote the sequence of proper values of $A(t)$, each repeated a number of times equal to its multiplicity. Prove that:

(i) if $A(t)$ is continuous on $[a, b]$, then each $\lambda_j(t)$, $(j = 1, \ldots, n)$, is continuous on this interval;

(ii) if $A(t)$ is Riemann integrable on $[a, b]$, then each $\lambda_j(t)$, $(j = 1, \ldots, n)$, is Riemann integrable on this interval;

(iii) if $A(t)$ is Lebesgue integrable on $[a, b]$, then each $\lambda_j(t)$, $(j = 1, \ldots, n)$, is Lebesgue integrable on this interval.

Hints. Note that if $s \in [a, b]$, and $A(t)$ is continuous at $t = s$, then the result of Problem 4 of E.2 implies that each $\lambda_j(t)$ is continuous at $t = s$. Moreover, from the result of Problem 2 of D.1 it follows that

$$(1.7) \qquad |\lambda_j(t)| \leq \sum_{\alpha,\beta=1}^{n} |A_{\alpha\beta}(t)|, \quad \textit{for} \quad t \in [a, b], \qquad j = 1, \ldots, n.$$

(i) From the above remarks, conclude that if $A(t)$ is continuous on $[a, b]$ then each $\lambda_j(t)$ is continuous on this interval.

(ii) If $A(t)$ is Riemann integrable on $[a, b]$, then the set of points on this interval where some element of $A(t)$ is discontinuous has Lebesgue measure zero, and hence by the initial remark above the set of points at which some $\lambda_j(t)$ is discontinuous has Lebesgue measure zero. Moreover, since Riemann integrability of $A(t)$ implies that each $|A_{\alpha\beta}(t)|$ is bounded on $[a, b]$, with the aid of (1.7) it follows that each $\lambda_j(t)$ is on $[a, b]$ a bounded function whose points of discontinuity form a set of Lebesgue measure zero, and hence each $\lambda_j(t)$ is Riemann integrable on $[a, b]$.

(iii) If $A(t)$ is Lebesgue integrable on $[a, b]$, extend the definition of $A(t)$ to $(-\infty, \infty)$ by setting $A(t) = A(a)$ for $t < a$, $A(t) = A(b)$ for $t > b$, and consider the matrix function

$$A(t, h) = [2h]^{-1}\int_{t-h}^{t+h} A(s)\,ds = [2h]^{-1}\int_{-h}^{h} A(t+\sigma)\,d\sigma, \quad \textit{for} \quad h > 0, \quad t \in (-\infty, \infty);$$

$$A(t, 0) = A(t), \qquad t \in (-\infty, \infty).$$

Show that $A(t, h)$ is an hermitian matrix for $(t, h) \in (-\infty, \infty) \times [0, \infty)$, which is continuous in (t, h) on $(-\infty, \infty) \times (0, \infty)$. If $\lambda_1(t, h) \leq \cdots \leq \lambda_n(t, h)$ denotes the sequence of proper values of $A(t, h)$, with each repeated a number of times equal to its multiplicity, conclude from Problem 5 of E.2 that each $\lambda_j(t, h)$ is continuous in (t, h) on $(-\infty, \infty) \times (0, \infty)$; in particular, for $h \in (0, \infty)$ each $\lambda_j(t, h)$ is continuous in t on $(-\infty, \infty)$. Moreover, $\lambda_j(t, 0) = \lambda_j(t)$ for $t \in [a, b]$, and on the set

$$T = \{t \mid t \in (-\infty, \infty), \lim_{h\to 0} A(t, h) = A(t)\}$$

we have $\lambda_j(t, h) \to \lambda_j(t)$ as $h \to 0$, in view of the result of Problem 4 of E.2. From the fact that $(-\infty, \infty) - T$ is a set of measure zero, conclude that each $\lambda_j(t)$ is measurable on T, and consequently also measurable on $[a, b]$. Finally, since the Lebesgue integrability of $A(t)$ implies that the right-hand member of (1.7) is integrable on $[a, b]$, conclude that each $\lambda_j(t)$ is a Lebesgue integrable function on this interval.

3. If $A(t) = [A_{\alpha\beta}(t)]$, $(\alpha = 1, \ldots, m; \beta = 1, \ldots, n)$ is an $m \times n$ matrix function on a compact interval $[a, b]$ of the real line, show that $\nu[A(t)]$ is continuous, Riemann integrable, or Lebesgue integrable according as $A(t)$ is continuous, Riemann integrable, or Lebesgue integrable. Moreover,

$$\nu\left[\int_a^b A(s)\,ds\right] \leq \int_a^b \nu[A(s)]\,ds, \tag{1.8}$$

where it is understood that the integrals in (1.8) are in the sense of Riemann or Lebesgue in accordance with the integrability condition possessed by $A(t)$.

Hint. Use the results of Problem 2, noting that $\mu(t) = \nu[A(t)]$ is non-negative, and $\lambda(t) = \mu^2(t)$ is the greatest proper value of the hermitian matrix $A^*(t)A(t)$. Establish the inequality (1.8) by using (1.2), and the result of Problem 1 of D.1, noting that if $z \in \mathbf{C}_n$ then

$$\left|\left(\int_a^b A(s)\,ds\right)z\right| = \left|\int_a^b [A(s)z]\,ds\right| \leq \int_a^b |A(s)z|\,ds \leq \left(\int_a^b \nu[A(s)]\,ds\right)|z|.$$

4. Suppose that $A(t) = [A_{\alpha\beta}(t)]$, $(\alpha, \beta = 1, \ldots, n)$, is an hermitian matrix function on the compact interval $[a, b]$, and that $A(t)$ is integrable on this interval, where integration may be either in the sense of Riemann or in the sense of Lebesgue. If $\lambda_1(t) \leq \cdots \leq \lambda_n(t)$ denote the proper values of $A(t)$, each repeated a number of times equal to its multiplicity, show that:

(i) $\int_a^b A(s)\,ds$ is an hermitian matrix, and if $\lambda_1{}^{ab} \leq \cdots \leq \lambda_n{}^{ab}$ denote its proper values, with each repeated a number of times equal to its multiplicity, then

$$\int_a^b \lambda_1(s)\,ds \leq \lambda_1{}^{ab} \leq \lambda_n{}^{ab} \leq \int_a^b \lambda_n(s)\,ds;$$

(ii) $$\nu\left[\int_a^b A(s)\,ds\right] \leq \max\left\{\left|\int_a^b \lambda_1(s)ds\right|, \left|\int_a^b \lambda_n(s)\,ds\right|\right\},$$
$$\leq \max\left\{\int_a^b |\lambda_1(s)|\,ds, \int_a^b |\lambda_n(s)|\,ds\right\};$$

(iii) *if* $A(t) \geq 0$ *for* $t \in [a, b]$, *and* $a \leq t_1 < t_2 \leq b$, *then*

$$\int_a^{t_2} A(s)\,ds \geq \int_a^{t_1} A(s)\,ds.$$

Hints. (i) For $z \in \mathbf{C}_n$, estimate $z^*(\int_a^b A(s)\,ds)z = \int_a^b [z^*A(s)\,z]\,ds$, using the extremizing properties of the proper values of the involved hermitian matrices.

(ii) Use (i) of Problem 4 of E.2.

5. If $A(t)$ is a non-negative definite hermitian matrix function on the compact interval $[a, b]$ on the real line, and $A^{1/2}(t)$ is the unique non-negative definite hermitian square root matrix of $A(t)$, show that $A^{1/2}(t)$ is continuous, Riemann integrable, or Lebesgue integrable on $[a, b]$ according as $A(t)$ is continuous, Riemann integrable or Lebesgue integrable on this interval.

Hints. Consider first the case in which there is a constant $\kappa > 0$ such that $0 \leq A(t) \leq \kappa E$ for $t \in [a, b]$; in particular, this condition holds if $A(t)$ is either continuous or Riemann integrable on $[a, b]$. In this case, by replacing $A(t)$ by $(1/\kappa)\,A(t)$, one may without loss of generality assume that $0 \leq A(t) \leq E$, so that also $0 \leq E - A(t) \leq E$ for $t \in [a, b]$. Note that for the positive constants c_k defined in Problem 7 of D.2 the matrix series $\sum_{k=1}^{\infty} c_k M^k$ converges uniformly on the class of $n \times n$ matrices which are hermitian and satisfy $0 \leq M \leq E$. Conclude that $A^{1/2}(t)$, which by Problem

9 of E.2 is the sum of the matrix series $E - \sum_{k=1}^{\infty} c_k[E - A(t)]^k$, is continuous at each value on $[a, b]$ at which $A(t)$ is continuous. In particular, $A^{1/2}(t)$ is continuous on $[a, b]$ if $A(t)$ is continuous on this interval. Moreover, if $A(t)$ is Riemann integrable on $[a, b]$ then the set of points of discontinuity of $A(t)$ on $[a, b]$ has Lebesgue measure zero, and consequently the set of points of discontinuity of $A^{1/2}(t)$ on $[a, b]$ has Lebesgue measure zero, so that $A^{1/2}(t)$ is also Riemann integrable.

In the general case of $A(t)$ Lebesgue integrable on $[a, b]$, for $m = 1, 2, \ldots$ let $T_m = \{t \mid t \in [a, b], mE \geq A(t)\}$, and define

$$A_m(t) = A(t), \quad \textit{for} \quad t \in T_m; \qquad A_m(t) = 0, \quad \textit{for} \quad t \in [a, b] - T_m.$$

Show that each nonvacuous T_m is measurable, and $A_m(t)$ is a non-negative definite hermitian matrix which is Lebesgue integrable and satisfies $mE \geq A_m(t)$ for $t \in [a, b]$. Moreover, for each $t \in [a, b]$ we have $\{A_m(t)\} \to A(t)$ and $\{A_m^{1/2}(t)\} \to A^{1/2}(t)$. If $\lambda_n(t)$ denotes the greatest proper value of $A(t)$, and $\mu_{nm}(t)$ denotes the maximum proper value of $A_m^{1/2}(t)$, then use results of Problems 4 and 10 of E.2 to show that

$$\nu[A(t)] = \lambda_n(t) = \mu_{nm}^2(t) = (\nu[A_m^{1/2}(t)])^2, \quad \textit{for} \quad t \in T_m;$$

also $\mu_{nm}(t) = 0$ for $t \in [a, b] - T_m$, and $\mu_{nm}(t) \leq m^{1/2}$ for $t \in [a, b]$. From the representation of $A_m^{1/2}(t)$ as the sum of a matrix series in $E - A_m(t)$, conclude that each $A_m^{1/2}(t)$ is measurable on $[a, b]$. Moreover, $m^{1/2}E \geq A_m^{1/2}(t)$ for $t \in [a, b]$, and with the aid of the bounded convergence theorem of Lebesgue it follows that $A_m^{1/2}(t)$ is integrable on $[a, b]$. Finally, since $\{A_m^{1/2}(t)\} \to A^{1/2}(t)$, $[1 + \lambda_n(t)]E \geq \lambda_n^{1/2}(t)E \geq A^{1/2}(t)$ for $t \in [a, b]$, and $\lambda_n(t) = \nu[A(t)]$ is integrable on $[a, b]$, use the dominated convergence theorem of Lebesgue to show that $A^{1/2}(t)$ is Lebesgue integrable on $[a, b]$.

6. Suppose that $A(t)$ is an $n \times n$ continuous matrix function which is hermitian and positive definite on the compact interval $[a, b]$, and let $A^{1/2}(t)$ be the unique positive definite hermitian square root matrix of $A(t)$. Show that:

(i) if $A(t)$ is continuously differentiable on $[a, b]$, then $A^{1/2}(t)$ is also continuously differentiable on $[a, b]$;

(ii) if $A(t)$ is absolutely continuous on $[a, b]$, then $A^{1/2}(t)$ is also absolutely continuous on $[a, b]$.

Hints. Note that without loss of generality one may assume that there is a positive constant κ such that

$$\kappa E \leq A(t) \leq E, \quad \text{for} \quad t \in [a, b],$$

and hence $B(t) = E - A(t)$ is such that $0 \leq B(t) \leq (1 - \kappa)E$ and $\nu[B(t)] \leq 1 - \kappa$ for $t \in [a, b]$.

(i) If $A(t)$ is continuously differentiable on $[a, b]$, note that $\{B^k(t)\}'$ is the sum of k terms $B^r(t)B'(t)B^s(t)$ with $s = k - 1 - r$, $r = 0, 1, \ldots, k - 1$, and thus $\nu[\{B^k(t)\}'] \le k(\nu[B(t)])^{k-1}\nu[B'(t)]$, $(k = 1, 2, \ldots)$. For c_k, $(k = 1, 2, \ldots)$, defined as in the preceding problem and in Problem 9 of E.2, note that the series $\sum_{k=1}^{\infty} kc_k z^{k-1}$ converges uniformly for $|z| \le 1 - \kappa$. Conclude that the derivative matrix series $\sum_{k=1}^{\infty} c_k\{B^k(t)\}'$ converges uniformly on $[a, b]$, and that $A^{1/2}(t) = E - \sum_{k=1}^{\infty} c_k B^k(t)$ is continuously differentiable on $[a, b]$, with derivative given by $-\sum_{k=1}^{\infty} c_k\{B^k(t)\}'$.

(ii) If $A(t)$ is absolutely continuous on $[a, b]$, use an argument similar to that suggested for (i), now noting that each $B_j(t) = \sum_{k=1}^{j} c_k B^k(t)$ is absolutely continuous, with $\nu[B_j'(t)] \le \nu[B'(t)]\sum_{k=1}^{\infty} kc_k(1 - \kappa)^k$ almost everywhere on $[a, b]$. Use the dominated convergence theorem of Lebesgue to conclude that $A^{1/2}(t)$ is an absolutely continuous matrix function whose derivative is given by $-\sum_{k=1}^{\infty} c_k\{B^k(t)\}'$ almost everywhere on $[a, b]$.

7. Suppose that $y^{(0)}(t), y^{(1)}(t), \ldots, y^{(k)}(t)$ are elements of $\mathfrak{J}_n^2[a, b]$ such that $((y^{(0)}, x)) = 0$ if $x \in \mathfrak{J}_n^2[a, b]$ and $((y^{(\alpha)}, x)) = 0$, $(\alpha = 1, \ldots, k)$. Prove that there exist constants $\lambda_1, \ldots, \lambda_k$ such that $\|y^{(0)} - \sum_{\beta=1}^{k} \lambda_\beta y^{(\beta)}\| = 0$.

Hint. Prove that without loss of generality one may restrict attention to sets $y^{(1)}(t), \ldots, y^{(k)}(t)$ which are linearly independent elements of $\mathfrak{J}_n^2[a, b]$ in the sense that if $c_1, \ldots, c_k$ are constants then $\|c_1 y^{(1)} + \cdots + c_k y^{(k)}\| = 0$ only if $c_\alpha = 0$, $(\alpha = 1, \ldots, k)$, and that in this case the matrix $[((y^{(\alpha)}, y^{(\beta)}))]$ is nonsingular. Proceed to show that the result holds if $\lambda_1, \ldots, \lambda_k$ are determined to satisfy the equations

$$\left(\left(y^{(\alpha)}, y^{(0)} - \sum_{\beta=1}^{k} \lambda_\beta y^{(\beta)}\right)\right) = 0, \qquad (\alpha = 1, \ldots, k).$$

8. If $A(t)$, $t \in [a, \infty)$, is an $n \times n$ hermitian matrix such that $A(s) - A(r) \ge 0$ for $a \le r < s < \infty$, show that the smallest proper value $\lambda[A(t)]$ of $A(t)$ is such that $\lim_{t\to\infty} \lambda[A(t)] = \infty$ if and only if $\lim_{t\to\infty} v^*A(t)v = \infty$ for arbitrary non-null vectors $v \in \mathbf{C}_n$.

Hint. Note that $\lambda[A(t)]$ is monotone nondecreasing, and $v^*A(t)v \ge \lambda[A(t)]v^*v$ for arbitrary $v \in \mathbf{C}_n$. Moreover, if $|v(t)| = 1$, $A(t)v(t) = \lambda[A(t)]\,v(t)$ for $t \in [a, \infty)$, then $v^*(t)A(s)v(t) \le v^*(t)A(t)v(t) = \lambda[A(t)]$ for $s \in [a, t]$. Conclude that if $v \in \mathbf{C}_n$, $|v| = 1$ and there exists a sequence $\{t_j\}$ such that $\{t_j\} \to \infty$, $\{v(t_j)\} \to v$, then $v^*A(t_i)v \le \lim_{t\to\infty} \lambda[A(t)]$, $(i = 1, 2, \ldots)$.

2. CONVERGENCE THEOREMS

For the study of differential equations, certain theorems on the convergence of sequences of vector-valued functions are of basic significance, especially such theorems that present sufficient conditions for uniform convergence. In particular, the following two theorems are of fundamental importance, and will be presented here for specific reference.

Theorem 2.1. (Dini) *Suppose that the functions $f(t)$, $f_j(t)$, $(j = 1, 2, \ldots)$, are real-valued and continuous on a given compact interval $[a, b]$ on the real line, while for each $t \in [a, b]$ the sequence $\{f_j(t)\}$ is monotonic and converges to $f(t)$. Then the sequence $\{f_j(t)\}$ converges to $f(t)$ uniformly on $[a, b]$.*

Under the hypotheses of the theorem the functions $g_j(t) = f_j(t) - f(t)$, $(j = 1, 2, \ldots)$, are continuous on $[a, b]$, for each t on this interval the sequence $\{g_j(t)\}$ converges monotonically to zero, and the result of the theorem is equivalent to the convergence of $\{g_j(t)\}$ being uniform on $[a, b]$. Now for $\varepsilon > 0$ and $t_0 \in [a, b]$ there exists an integer $j_\varepsilon = j_\varepsilon(t_0)$ such that $|g_{j_\varepsilon}(t_0)| < \varepsilon/2$, and a $\delta_\varepsilon > 0$ such that $|g_{j_\varepsilon}(t) - g_{j_\varepsilon}(t_0)| < \varepsilon/2$ for $t \in [a, b] \cap (t_0 - \delta_\varepsilon, t_0 + \delta_\varepsilon)$. As the monotoneity of the sequence $\{g_j(t)\}$ for each fixed t implies that $|g_j(t)| \leq |g_{j_\varepsilon}(t)|$ for $j \geq j_\varepsilon$ and $t \in [a, b]$, it follows, therefore, that for $\varepsilon > 0$ and $t_0 \in [a, b]$ there exists an integer $j_\varepsilon(t_0)$ and an open interval $I(t_0, \varepsilon)$ which contains t_0 and is such that if $t \in I(t_0, \varepsilon) \cap [a, b]$ then $|g_j(t)| < \varepsilon$ for $j > j_\varepsilon(t_0)$. The aggregate of such intervals $I(t_0, \varepsilon)$ forms an open covering of the compact interval $[a, b]$, and by the Heine-Borel Theorem there exists a finite number of such intervals $I(t_0^1, \varepsilon), \ldots, I(t_0^q, \varepsilon)$ whose union covers $[a, b]$. If $j(\varepsilon)$ denotes the greater of the integers $j_\varepsilon(t_0^1), \ldots, j_\varepsilon(t_0^q)$, then for $j > j(\varepsilon)$ we have $|g_j(t)| < \varepsilon$ for all $t \in [a, b]$, thus completing the proof of the uniform convergence of $\{g_j(t)\}$ to zero on $[a, b]$.

A sequence of n-dimensional vector functions $y^{(j)}(t) = (y_\alpha^{(j)}(t))$, $(\alpha = 1, \ldots, n; j = 1, 2, \ldots)$ is said to be *uniformly equi-continuous on a set T* on the real line if to each $\varepsilon > 0$ there corresponds a $\delta(\varepsilon) > 0$ such that

$$(2.1)\quad |y^{(j)}(t_2) - y^{(j)}(t_1)| < \varepsilon, \qquad (j = 1, 2, \ldots), \text{ if } t_1 \in T, \quad t_2 \in T, \quad |t_2 - t_1| < \delta(\varepsilon).$$

For example, $\{y^{(j)}(t)\}$, $(j = 1, 2, \ldots)$, is uniformly equi-continuous on a given set T whenever these vector functions satisfy on this set the uniform *Hölder condition*

$$(2.2)\qquad |y^{(j)}(t_2) - y^{(j)}(t_1)| \leq M\, |t_2 - t_1|^\rho,$$

where the real constants M, ρ satisfy $M \geq 0$, $\rho > 0$. In particular, if each $y^{(j)}(t)$ is continuous and piecewise differentiable on $[a, b]$, and there is a constant $M_1 \geq 0$ satisfying

$$(2.3)\qquad \int_a^b |y^{(j)\prime}(t)|^2\, dt \leq M_1, \qquad (j = 1, 2, \ldots),$$

then (2.2) holds for $M = M_1^{1/2}$ and $\rho = \frac{1}{2}$. Indeed, with the aid of Theorem 1.1 of the preceding section and the Bunyakovsky-Schwarz inequality (1.5-c) for scalar-valued functions it follows that for $y(t) = y^{(j)}(t)$, $(j = 1, 2, \ldots)$, we have

$$|y(t_2) - y(t_1)|^2 = \left|\int_{t_1}^{t_2} y'(t)\, dt\right|^2 \leq \left|\int_{t_1}^{t_2} |y'(t)|\, dt\right|^2$$
$$\leq |t_2 - t_1| \left|\int_{t_1}^{t_2} |y'(t)|^2\, dt\right| \leq M_1\, |t_2 - t_1|\,.$$

It is to be remarked that the same argument implies the uniform equi-continuity of the functions $\{y^{(j)}(t)\}$ on $[a, b]$ if each vector function $y^{(j)}(t)$ is absolutely continuous on this interval, while $|y^{(j)\prime}(t)|^2$ is integrable in the sense of Lebesgue and (2.3) holds.

The following *selection theorem of Ascoli* will be of specific use in the treatment of differential equations.

Theorem 2.2. *Suppose that* $y^{(j)}(t) = (y_\alpha^{(j)}(t))$, $(\alpha = 1, \ldots, n; j = 1, 2, \ldots)$ *is a sequence of n-dimensional vector functions that is uniformly equi-continuous on a compact interval* $[a, b]$, *and such that* $\{|y^{(j)}(t)|\}$, $(j = 1, 2, \ldots)$, *is bounded for each* $t \in [a, b]$. *Then there exists a subsequence* $z^{(\gamma)}(t) = y^{(j_\gamma)}(t)$, $(j_1 < j_2 < \cdots)$, *which converges uniformly on* $[a, b]$ *to a continuous vector function* $z(t)$.

Let $\{t_\beta\}$, $(\beta = 1, 2, \ldots)$, be an infinite sequence of points on $[a, b]$ which is dense on this interval; for example, one might choose the distinct points of the form $a + r(b - a)/2^s$, $(r = 0, 1, \ldots, 2^s;\ s = 1, 2, \ldots)$, ordered as a simple sequence. As the sequence $\{|y^{(j)}(t_1)|\}$ is bounded there is a subsequence $\{y^{(j,1)}(t)\}$ of $\{y^{(j)}(t)\}$ such that $\{y^{(j,1)}(t_1)\}$, $(j = 1, 2, \ldots)$, converges. By mathematical induction there exist sequences $\{y^{(j,\gamma)}(t)\}$, $(\gamma = 1, 2, \ldots)$, such that $\{y^{(j,\gamma)}(t)\}$ is a subsequence of $\{y^{(j,\gamma-1)}(t)\}$, $(\gamma = 2, 3, \ldots)$, and $\{y^{(j,\gamma)}(t)\}$, $(j = 1, 2, \ldots)$, converges for $t = t_\beta$, $(\beta = 1, 2, \ldots, \gamma)$. Then the *diagonal sequence* $z^{(\gamma)}(t) = y^{(j_\gamma)}(t) = y^{(\gamma,\gamma)}(t)$, $(\gamma = 1, 2, \ldots)$, is such that for $k = 1, 2, \ldots$ the sequence $\{z^{(k)}, z^{(k+1)}, \ldots\}$ is a subsequence of $\{y^{(j,k)}\}$, $(j = 1, 2, \ldots)$, and therefore $\{z^{(\gamma)}(t)\}$, $(\gamma = 1, 2, \ldots)$, converges for each $t = t_\beta$, $(\beta = 1, 2, \ldots)$.

For $\varepsilon > 0$ and $\delta(\varepsilon)$ as in (2.1), let $r = r(\varepsilon)$ be a positive integer such that there are r points $s_1, \ldots, s_r$ of the sequence $\{t_\beta\}$ such that for each $t \in [a, b]$ there is a point $s_{[t]}$ of the set $s_1, \ldots, s_r$ satisfying $|t - s_{[t]}| < \delta(\varepsilon/3)$. Let γ_ε be such that $|z^{(\gamma)}(s_k) - z^{(\beta)}(s_k)| < \varepsilon/3$, $(k = 1, \ldots, r(\varepsilon))$, whenever $\gamma > \gamma_\varepsilon$, $\beta > \gamma_\varepsilon$. Since for arbitrary $t \in [a, b]$ we have

$$\begin{aligned}(2.4)\qquad |z^{(\gamma)}(t) - z^{(\beta)}(t)| \leq |z^{(\gamma)}(t) - z^{(\gamma)}(s_{[t]})| &+ |z^{(\gamma)}(s_{[t]}) - z^{(\beta)}(s_{[t]})| \\ &+ |z^{(\beta)}(s_{[t]}) - z^{(\beta)}(t)|\,,\end{aligned}$$

it follows that if $\gamma > \gamma_\varepsilon$, $\beta > \gamma_\varepsilon$ the right hand member of (2.4) is less than ε. Hence $\{z^{(\gamma)}(t)\}$ converges uniformly on $[a, b]$, and if $z(t)$ denotes the limit vector function of this sequence then the continuity of $z(t)$ is a consequence of the uniform convergence of the sequence. Indeed, under the hypotheses of the present theorem, for t_1 and t_2 points on $[a, b]$ satisfying $|t_2 - t_1| < \delta(\varepsilon)$ we have $|z^{(\gamma)}(t_2) - z^{(\gamma)}(t_1)| < \varepsilon$, $(\gamma = 1, 2, \ldots)$, and hence $|z(t_2) - z(t_1)| \leq \varepsilon$ for such values t_1, t_2.

3. NOTES AND REMARKS

The elements of the theory of functions of a real variable needed for the material of this section, aside from an acquaintance with the Lebesgue integral, are to be found in most books of "Advanced Calculus" level. For example, such a background is provided by R. G. Bartle, *The Elements of Real Analysis*, John Wiley and Sons, 1964. A discussion of the basic properties of the Lebesgue integral on the real line is all that is required for the material of this Appendix and the treatment of differential equations in Chapters II and VII of the text, and such material is present in almost any book dealing with the Lebesgue integral. In particular the prerequisite material is to be found in Graves [1, Chs. X, XI].

REFERENCES FOR SPECIFIC PROBLEMS

F.1:5,6 Reid [18, Sec. 4].

Bibliography

ABBREVIATIONS FOR MATHEMATICAL PUBLICATIONS MOST FREQUENTLY LISTED

ACMT	*Acta Mathematica*
AMJM	*American Journal of Mathematics*
AMMM	*American Mathematical Monthly*
AMPA	*Annali di Matematica Pura ed Applicata*
ANLR	*Atti della Accademia Nazionale dei Lincei. Rendiconti. Classe di Scienze, Fisiche, Mathematiche e Naturali*
ANNM	*Annals of Mathematics*
ASEN	*Annales Scientifiques de l'École Normale Supérieure*
ASNP	*Annali della Scuola Normale Superiore di Pisa*
BAMS	*Bulletin of the American Mathematical Society*
BCMS	*Bulletin of the Calcutta Mathematical Society*
BSMF	*Bulletin de la Société Mathématique de France*
BSMT	*Bulletin des Sciences Mathématiques*
BUMI	*Bolletino della Unione Matematica Italiana*
CDJM	*Canadian Journal of Mathematics*
CTCV	*Contributions to the Calculus of Variations*
CTNO	*Contributions to the Theory of Non-linear Oscillations*
DKMJ	*Duke Mathematical Journal*
DOKL	*Doklady Akademii Nauk SSSR, Mathematics Section*
EMTG	*Ergebnisse der Mathematik und ihrer Grenzgebiete*
EMTW	*Enzyklopädie der Mathematischen Wissenschaften*
ESMT	*Encyclopédie des Sciences Mathématiques*
GMTW	*Die Grundlehren der Mathematischen Wissenschaften*
ILJM	*Illinois Journal of Mathematics*
INDM	*Indigationes Mathematicae*
JDEQ	*Journal of Differential Equations*
JFSH	*Journal of the Faculty of Science, Hokkaido University, Series I. Mathematics*
JIMS	*The Journal of the Indian Mathematical Society*
JJMT	*Japanese Journal of Mathematics*
JLMS	*The Journal of the London Mathematical Society*

JMAA	*Journal of Mathematical Analysis and Applications*
JMMC	*Journal of Mathematics and Mechanics*
JMPA	*Journal de Mathématiques Pures et Appliquées*
JMPH	*Journal of Mathematics and Physics*
JNBS	*Journal of Research of the National Bureau of Standards, Section B*
MCMJ	*Michigan Mathematical Journal*
MOMT	*Monatshefte für Mathematik*
MTAN	*Mathematische Annalen*
MTSK	*Matematičeskiǐ Sbornik*
MTZT	*Mathematische Zeitschrift*
NAWK	*Nieuw Archief voor Wiskunde*
PAMS	*Proceedings of the American Mathematical Society*
PFJM	*Pacific Journal of Mathematics*
PIAJ	*Proceedings of the Imperial Academy of Japan*
PKMM	*Prikladnaja Matematika i Mehanika*
PTGM	*Portugaliae Mathematica*
QAMT	*Quarterly of Applied Mathematics*
RCMP	*Rendiconti del Circolo Matematico di Palermo*
RMUP	*Rivista di Matematica della Università di Parma*
SJAM	*SIAM Journal on Applied Mathematics*
STMT	*Studia Mathematica*
TAMS	*Transactions of the American Mathematical Society*

Ahlbrandt, C. D.

1. *Disconjugacy criteria for self-adjoint differential systems*, (Dissertation, Univ. of Oklahoma, 1968), JDEQ, 6(1969), 271–295.

Antosiewicz, H. A.

1. *A survey of Lyapunov's second method*, CTNO, 4(1958), 141–166.

Ascoli, G.

1. *Osservazioni sopra alcune questioni di stabilità*, I. ANLR (8)9(1950), 129–134.

Atkinson, F. V.

1. *Discrete and Continuous Boundary Problems*, Academic Press, New York, 1964.

Barrett, J. H.

1. *Matrix systems of second order differential equations*, PTGM, 14(1955), 79–89.
2. *Behavior of solutions of second order self-adjoint differential equations*, PAMS, 6(1955), 247–251.
3. *A Prüfer transformation for matrix differential equations*, PAMS, 8(1957), 510–518.
4. *Systems disconjugacy of a fourth-order differential equation*, PAMS, 12(1961), 205–213.
5. *Disconjugacy of a self-adjoint differential equation of the fourth order*, PFJM, 11(1961), 25–37.
6. *Two-point boundary problems for linear self-adjoint differential equations of the fourth order with middle term*, DKMJ, 29(1962), 543–554.
7. *Oscillation Theory of Ordinary Linear Differential Equations*. Lectures delivered at the Associated Western Universities Differential Equations Symposium, Boulder, Colorado, Summer 1967, published posthumously in *Advances in Mathematics* 3(1969), 415–509.

Bell, E. T.

1. *Men of Mathematics*, Dover, New York, 1937.
2. *Development of Mathematics*, McGraw-Hill, New York, 1940.

Bellman, R.

1. *The stability of solutions of linear differential equations*, DKMJ, 10(1943), 643–647.
2. *On an application of a Banach-Steinhaus theorem to the study of the boundedness of solutions of non-linear differential and difference equations*, ANNM, 49(1948), 515–522.
3. *Stability Theory of Differential Equations*, McGraw-Hill, New York, 1953.

Birkhoff, G. D.

1. *On the asymptotic character of the solutions of certain linear differential equations containing a parameter*, TMAS, 9(1908), 219–231.
2. *Boundary value and expansion problems of ordinary linear differential equations*, TAMS, 9(1908), 373–395.

Birkhoff, G. D., and M. R. Hestenes

1. *Natural isoperimetric conditions in the calculus of variations*, DKMJ, 1(1935), 198–286.

Birkhoff, G. D., and R. E. Langer

1. *The boundary problems and developments associated with a system of ordinary linear differential equations of the first order*, Proc. Amer. Acad. Arts and Sciences, 58(1923), 51–128.

Birkhoff, G., and G.-C. Rota

1. *Ordinary Differential Equations*, Ginn and Co., Boston, 1962.

Bliss, G. A.

1. *The solutions of differential equations of the first order as functions of their initial values*, ANNM, 6(1904–1905), 49–68.
2. *Differential equations containing arbitrary functions*, TAMS, 21(1920), 79–92.
3. *A boundary value problem for a system of ordinary linear differential equations of the first order*, TAMS, 28(1926), 561–584.
4. *Definitely self-adjoint boundary value problems*, TAMS, 44 (1938), 413–428.
5. *Lectures on the Calculus of Variations*, Univ. of Chicago Press, Chicago, 1946.

Bobonis, A.

1. *Differential systems with boundary conditions involving the characteristic parameter* (Dissertation, Univ. of Chicago, 1939), CTCV, (1938–41), 101–138.

Bôcher, M.

1. *Randwertaufgaben bei gewöhnlichen Differentialgleichungen*, EMTW, II.A 7a (1900).
2. *Green's functions in space of one dimension*, BAMS, (2)7(1901), 297–299.
3. *Boundary problems and Green's functions for linear differential and difference equations*, ANNM, 13(1912), 71–88.
4. *Boundary problems in one dimension*, Proc. of the Fifth International Congress of Mathematicians, I, Cambridge, 1912, 163–195.
5. *Applications and generalizations of the concept of adjoint systems*, TAMS, 14(1913), 403–420.
6. *Leçons sur les méthodes de Sturm dans la théorie des équations différentielles linéaires, et leurs développements modernes*, Gauthier-Villars, Paris, 1917.

Bounitzky, E.

1. *Sur la fonction de Green des équations différentielles linéaires ordinaires*, JMPA, 5(1909), 65–125.

Bourbaki, N.

1. *Éléments de Mathématique*, Livre IV, *Fonctions d'une variable réelle*, Chapter IV, *Équations Différentielles*, Actualites Scientifiques et Industrielles 1132, Hermann, Paris, 1961.

Bradley, J. S.

1. *Adjoint quasi-differential operators of Euler type*, (Dissertation, Univ. of Iowa, 1964), PFJM, 16(1966), 213–237.
2. *Generalized Green's matrices for compatible differential systems*, MCMJ, 13(1966), 97–108.

Brauer, F.

1. *Perturbations of nonlinear systems of differential equations*, JMAA, 14(1966), 198–206.

Brauer, F. and S. Sternberg

1. *Local uniqueness, existence in the large, and the convergence of succsseive approximations*, AMJM, 80(1958), 421–430; Errata, ibid. 81(1959), 797.

Burkhardt, H.

1. *Sur les fonctions de Green relatives à un domaine d'une dimension*, BSMF, 22(1894), 71–75.

Cajori, F.,

1. *A History of Mathematics*, Macmillan Co., New York, 1893; 2nd edition, 1919.

Caligo, D.

1. *Un criterio sufficiente di stabilità per le soluzioni dei sistemi di equazioni integrali lineari e sue applicazioni ai sistemi di equazioni lineari*, Atti 2° Congresso Un. Mat. Ital., Bologna, 1940, 177–185.

Carathéodory, C.

1. *Vorlesungen über reelle Funktionen*, 2nd edition, Teubner, Berlin, 1927.
2. *Variationsrechnung und partielle Differentialgleichungen erster Ordnung*, Teubner, Berlin, 1935.

Cesari, L.

1. *Un nuovo criterio di stabilità per le soluzioni delle equazioni differenziali lineari*, ASNP, 9(1940), 163–186.
2. *Asymptotic Behavior and Stability Problems in Ordinary Differential Equations*, EMTG, 16; 2nd edition, Academic Press, New York, 1963.

Coddington, E. A. and N. Levinson

1. *Uniqueness and convergence of successive approximations*, JIMS, 16(1952), 75–81.
2. *Theory of Ordinary Differential Equations*, McGraw-Hill, New York, 1955.

Cole, R. H.

1. *The expansion problem with boundary conditions at a finite set of points*, CDJM, 13(1961), 462–479.
2. *General boundary conditions for an ordinary linear differential system*, TAMS, 111 (1964), 521–550.

Coles, W. J.

1. *A general Wirtinger-type inequality*, DKMJ, 27(1960), 133–138.
2. *Wirtinger-type integral inequalities*, PJMT, 11(1961), 871–877.

Collatz, L.

1. *Eigenwertprobleme und ihre numerische Behandlung*, Akademische Verlagsgesellschaft, Berlin, 1943; reprinted Chelsea, New York, 1948.

Conti, R.

1. *Sulla stabilità dei sistemi di equazioni differenziali lineari*, RMUP, 6(1955), 3–35.
2. *Sulla "equivalenza asintotica" dei sistemi di equazioni differenziali*, AMPA, (4) 41(1955), 95–104.
3. *Sulla t_∞-similitudine tra matrici e l'equivalenza asintotica dei sistemi differenziali lineari*, RMUP, 8(1957), 43–47.
4. *Recent trends in the theory of boundary value problems for ordinary differential equations*, BUMI, (3) 22(1967), 135–178.

Coppel, W. A.

1. *On the stability of ordinary differential equations*, JLMS, 38(1963), 255–260.
2. *Stability and Asymptotic Behavior of Differential Equations*, Heath, Boston, 1965.

Diaz, J. B., and F. T. Metcalf

1. *Variations of Wirtinger's inequality*, Inequalities, Academic Press, New York, 1967, 79–103.

Diliberto, S. P.

1. *On systems of ordinary differential equations*, CTNO, 1(1950), 1–48.
2. *A note on linear ordinary differential equations*, PAMS, 8(1957), 462–464.

Elliott, W. W.

1. *Generalized Green's functions for compatible differential systems*, AMJM, 50(1928), 243–258.
2. *Green's functions for differential systems containing a parameter*, AMJM, 51(1929), 397–416.

von Escherich, G.

1. *Die zweite Variation der einfachen Integrale*, Wiener Sitzungsberichte, (8)107(1898), 1191–1250.

Etgen, G. J.

1. *Oscillation properties of certain nonlinear matrix differential equations of second order*, TAMS, 122(1966), 289–310.
2. *A note on trigonometric matrices*, PAMS, 17(1966), 1226–1232.
3. *On the determinants of second order matrix differential systems*, JMAA, 18(1967), 585–598.

Ettlinger, H. J.

1. *Existence theorems for the general real self-adjoint linear system of the second order*, TAMS, 19(1918), 79–96.
2. *Oscillation theorems for the real self-adjoint linear system of the second order*, TAMS, 22(1921), 136–143.

Floquet, G.

1. *Sur les équations différentielles linéaires à coefficients périodiques*, ASEN, (2)12(1883), 47–89.

Fukuhara, M.

(see Hukuhara, M.)

Giuliano, L.

1. *Sull'unicitá delle soluzioni dei sistemi di equazioni differenziali ordinarie*, BUMI, (2)2(1940), 221–227.
2. *Su un notevale teorema di confronto e su un teorema di unicitá per i sistemi di equazioni differenziali ordinarie*, Rend. R. Accad. Italia, (7)1(1940) 330–336.
3. *Generalizzazione di un lemma di Gronwall e di una diseguaglianza di Peano*, ANLR, (8)1(1946), 1264–1271.

Gould, S. H.

1. *Variational Methods for Eigenvalue Problems*, 2nd edition, Mathematical Expositions No. 10, Univ. of Toronto Press, Toronto, 1966.

Graves, L. M.

1. *The Theory of Functions of Real Variables*, 2nd edition, McGraw-Hill, New York, 1956.

Green, G.

1. *Essay on the Application of Mathematical Analysis to the Theory of Electricity and Magnetism*, Nottingham, 1828.

Greub, W. and W. C. Rheinboldt

1. *Non-self-adjoint boundary value problems in ordinary differential equations*, JNBS, 64B(1960), 83–90.

Gronwall, T. H.

1. *Note on the derivatives with respect to a parameter of the solutions of a system of differential equations*, ANNM, (2)20(1919), 292–296.

Hahn, W.

1. *Theory and Application of Liapunov's Direct Method*, Prentice-Hall, Englewood Cliffs, N. J., 1963.
2. *Stability of Motion*, GMTW, 138, Springer-Verlag, New York, 1967.

Harris, V. C.

1. *A system of linear difference equations and an associated boundary value problem*, (Dissertation, Northwestern Univ., 1950).

Hartman, P.

1. *On a theorem of Milloux*, AMJM, 70(1948), 395–399.
2. *On the linear logarithmico-exponential differential equation of the second order*, AMJM, 70(1948), 764–779.
3. *Differential equations with non-oscillatory solutions*, DKMJ, 15(1948), 697–709.
4. *On the zeros of solutions of second order linear differential equations*, JLMS, 27(1952), 492–496.
5. *Self-adjoint, nonoscillatory systems of ordinary, second order, linear differential equations*, DKMJ, 24(1957), 25–36.
6. *A differential equation with nonunique solutions*, AMMM, 70(1963), 255–259.
7. *Ordinary Differential Equations*, John Wiley and Sons, New York, 1964.

Hartman, P., and A. Wintner

1. *On non-conservative linear oscillators of low frequency*, AMJM, 70(1948), 529–539.
2. *Oscillatory and nonoscillatory linear differential equations*, AMJM, 71(1949), 627–649.
3. *On an oscillation criterion of Liapunoff*, AMJM, 73(1951), 885–890.
4. *On non-oscillatory linear differential equations*, AMJM, 75(1953), 717–730.
5. *On the assignment of asymptotic values for the solutions of linear differential equations of the second order*, AMJM, 77(1955), 475–483.
6. *On disconjugate differential systems*, CDJM, 8(1956), 72–81.

Hestenes, M. R.

1. *Applications of the theory of quadratic forms in Hilbert space to the calculus of variations*, PFJM, 1(1951), 525–581.

Hilbert, D.

1. *Grundzüge einer allgemeiner Theorie der linearen Integralgleichungen*, Teubner, Leipzig, 1912; reprinted by Chelsea, New York, 1952.

Hille, E.

1. *Nonoscillation theorems*, TAMS, 64(1948), 234–252.
2. *Remarks on a paper by Zeev Nehari*, BAMS, 55(1949), 552–553.
3. *Analytic Function Theory*, Vol. I. Ginn and Co., Boston, 1962.
4. *Lectures on Ordinary Differential Equations*, Addison-Wesley, Reading, 1969.

Hinton, D. B.

1. *Clamped end boundary conditions for fourth order self-adjoint differential equations*, DKMJ, 34(1967), 131–138.

Hölder, E.

1. *Entwicklungssätze aus der Theorie der zweiten Variation, allgemeine Randbedingungen*, ACMT, 70(1939), 193–242.

2. *Einordnung besonderer Eigenwertprobleme in die Eigenwerttheorie kanonischer Differentialgleichungssysteme*, MTAN, 119(1943), 21–66.
3. *Über den Aufbau eines erweiterten Greenschen Tensors kanonischer Differentialgleichungen aus assozierten Lösungssystemen*, Ann. Soc. Polon. Math., 25(1952), 115–121.

Hooker, J. W.
1. *Existence and oscillation theorems for a class of non-linear second order differential equations*, (Dissertation, Univ. of Oklahoma, 1967); JDEQ, 5(1969), 283–306.

Howard, H. C.
1. *Oscillation criteria for fourth-order linear differential equations*, TAMS, 96(1960), 296–311.

Hu, K.-S.
1. *The problem of Bolza and its accessory boundary value problem* (Dissertation, Univ. of Chicago, 1932), CTCV (1931–32), 361–445.

Hukuhara, M.
1. *Sur le théorème d'existence des intégrales des équations différentielles du premier ordre*, JJMT, 5(1928), 239–251.
2. *Sur les systèmes des équations différentielles ordinaires*, PIAJ, 4(1928), 448–449.
3. *Sur l'ensemble des courbes intégrales d'un système d'équations différentielles ordinaires*, I, II, III, PIAJ, 6(1930), 360–362; 7(1931), 37–39; 298–299.
4. *Sur les points singuliers des équations différentielles linéaires, Domaine réel*, JFSH, 2(1934), 13–88.

Hunt, R. W.
1. *The behavior of solutions of ordinary, self-adjoint differential equations of arbitrary even order*, PFJM, 12(1962), 945–961.

Ince, E. L.
1. *Ordinary Differential Equations*, Longmans, Green and Co., London, 1927.

Jackson, D.
1. *Algebraic properties of self-adjoint systems*, TAMS, 17(1916), 418–424.

Jones, W. R.
1. *Differential systems with integral boundary conditions*, JDEQ, 3(1967), 191–202.

Kamke, E.
1. *Differentialgleichungen reeller Funktionen*, Akademische Verlagsgesellschaft, Leipzig, 1930; reprinted by Chelsea, New York, 1947.
2. *Zur Theorie der Systeme gewöhnlicher Differentialgleichungen*, II, ACMT, 58(1932), 57–85.
3. *A new proof of Sturm's comparison theorems*, AMMM, 46(1939), 417–421.
4. *Über die definiten selbstadjungierten Eigenwertaufgaben bei gewöhnlichen linearen Differentialgleichungen*, I, II, III, MTZT, 45(1939), 759–787; 46(1940), 231–250; 251–286.
5. *Differentialgleichungen Lösungsmethoden und Lösungen*, I (*Gewöhnliche Differentialgleichungen*), (6th edition), Akademische Verlagsgesellschaft, Leipzig, 1959.

Kaufman, H., and R. L. Sternberg
1. *Application of the theory of systems of differential equations to multiple non-uniform transmission lines*, JMPH, 31(1952–3), 244–252.
2. *A two-point boundary problem for ordinary self-adjoint differential equations of even order*, DKMJ, 20(1953), 527–531.

Kneser, A.
1. *Untersuchung über die reellen Nullstellen der Integrale linearer Differentialgleichungen*, MTAN, 42(1893), 409–435.

Kneser, H.

1. *Ueber die Lösungen eines Systems gewöhnlicher Differentialgleichungen das der Lipschitzschen Bedingung nicht genügt*, Sitzungsberichte Preuss. Akad. Wiss., Phys.-Math., (1923), 171–174.

Krein, M.

1. *Sur les opérateurs différentielles autoadjoints et leurs fonctions de Green symétriques*, MTSK, 2(44), 1937, 1023–1070.

LaSalle, J., and S. Lefschetz

1. *Stability by Liapunov's Direct Method*, Academic Press, New York, 1961.

Latshaw, V. V.

1. *The algebra of self-adjoint boundary problems*, BAMS, 39(1933), 969–978.

Lefschetz, S.

1. *Differential Equations: Geometric Theory*, 2nd Edition, Interscience, New York, 1963.

Leighton, W.

1. *Principal quadratic functionals*, TAMS, 67(1949), 253–274.
2. *A substitute for the Picone formula*, BAMS, 55(1949), 325–328.
3. *The detection of the oscillation of solutions of a second order linear differential equation*, DKMJ, 17(1950), 57–62.
4. *On self-adjoint differential equations of second order*, JLMS, 27(1952), 37–47.

Leighton, W., and M. Morse

1. *Singular quadratic functionals*, TAMS, 40(1936), 252–286.

Leighton, W., and Z. Nehari

1. *On the oscillation of solutions of self-adjoint linear differential equations of the fourth order*, TAMS, 89(1958), 325–377.

Levin, J. J.

1. *On the matrix Riccati equation*, PAMS, 10(1959), 519–524.

Levinson, N.

1. *The asymptotic behavoir of a system of linear differential equations*, AMJM, 68(1946), 1–6.
2. *The asymptotic nature of solutions of linear systems of differential equations*, DKMJ, 15(1948), 111–126.

Liapunov, A. M.

1. *Problème Général de la Stabilité du Mouvement*, (French translation of a Russian paper dated 1893), Ann. Fac. Sci. Toulouse, (2)9(1907), 27–247; reprinted as Annals of Mathematics Studies, No. 17, Princeton, 1949.

Lindelöf, E.

1. *Sur l'application des méthodes d'approximations successives a l'etude des intégrales réelles des équations différentielles ordinaire*, JMPA, (4)10(1894), 117–128.

Liouville, J.

1. *Sur le développement des fonctions ou parties de fonctions en séries dont les divers termes sont assujettis à satisfaire à une meme équation différentielles du second ordre contenant un paramètre variable*, JMPA, 1(1836), 253–265; 2(1837), 16–35; 418–436.

Lipschitz, R.

1. *Disamina della possibilità di integrare completamente un dato sistema di equazioni differenziali ordinarie*, AMPA, (2)2(1868–69), 288–302. Reproduced in BSMT, 10(1876), 149–159, in French.

Loud, W. S.

1. *Generalized inverses and generalized Green's functions*, SJAM, 14(1966), 342–369.

Malkin, I. G.

1. *On the question of the converse of Lyapunov's theorem on asymptotic stability*, PKMM, 18(1954), 129–138. (Russian).

Mansfield, R.

1. *Differential systems involving k-point boundary conditions*, CTCV, (1938-41), 413–452.

Markus, L.

1. *Continuous matrices and the stability of differential systems*, MTZT, 62(1955), 310–319.

Mason, M.

1. *On the boundary value problems of linear ordinary differential equations of the second order*, TAMS, 7(1906), 337–360.

Massera, J. L., and J. J. Schäffer

1. *Linear Differential Equations and Function Spaces*, Academic Press, New York, 1966.

McShane, E. J.

1. *On the uniqueness of the solutions of differential equations*, BAMS, 45(1939), 755–757.
2. *Integration*, Princeton Univ. Press, Princeton, N.J., 1944.

Moore, E. H.

1. *On the reciprocal of the general algebraic matrix*, Abstract, BAMS, 26(1919–20), 394.
2. *General Analysis*, Part I, Memoirs Amer. Philosophical Soc., 1 (1935).

Moroney, R. M.

1. *A class of characteristic-value problems*, TAMS, 102(1962), 446–470.

Morse, M.

1. *A generalization of the Sturm separation and comparison theorems in n-space*, MTAN, 103(1930), 72–91.
2. *The Calculus of Variations in the Large*, AMS Colloquium Publication, XVIII, (1934).

Müller, M.

1. *Ueber das fundamental Theorem in der Theorie der gewöhnlichen Differentialgleichungen*, MTZT, 26(1927), 619–645.
2. *Beweis eines Satzes des Herrn H. Kneser über die Gesamtheit der Lösungen, die ein System gewöhnlicher Differentialgleichungen durch einen Punkt schickt*, MTZT, 28(1928), 349–355.

Nagumo, M.

1. *Eine hinreichende Bedingung für die Unität der Lösung von Differentialgleichungen erster Ordnung*, JJMT, 3(1926), 107–112.

Nehari, Z.

1. *The Schwarzian derivative and schlicht functions*, BAMS, 55(1949), 545–551.
2. *Oscillation criteria for second-order linear differential equations*, TAMS, 85(1957), 428–445.
3. *Characteristic values associated with a class of non-linear second-order differential equations*, ACMT, 105(1961), 141–175.

Nemytskiǐ, V. V. and V. V. Stepanov

1. *Qualitative Theory of Differential Equations*, (translation of the Russian 2nd edition, 1949), Princeton Univ. Press, Princeton, 1960.

Osgood, W. F.

1. *Beweis der Existenz einer Lösungen der Differentialgleichungen $dy/dx = f(x, y)$ ohne Hinzunahme der Cauchy-Lipschitzschen Bedingung*, MOMT, 9(1898), 331–345.

Painlevé, P.

1. *Gewöhnliche Differentialgleichungen; Existenz der Lösungen*, EMTW, II.A 4a(1900).
2. *Existence de l'intégrale générale. Determination d'une intégrale particulière par ses valeurs initiales*, ESMT, II.16 (1910).

Peano, G.

1. *Sull' integrabilità delle equazioni differenziali del primo ordine*, Atti R. Accad. Sci., Torino, 21(1885–1886), 677–685.
2. *Intégration par séries des équations différentielles linéaires*, MTAN, 32(1888), 450–456.
3. *Démonstration de l'integrabilité des équations différentielles ordinaires*, MTAN, 37(1890), 182–228.

Perron, O.

1. *Ein neuer Existenzbeweis für die Integrale der Differentialgleichungen* $y' = f(x, y)$, MTAN, 76(1915), 471–484.
2. *Ueber Ein- und Mehrdeutigkeit des Integrales eines Systems von Differentailgleichungen*, MTAN, 95(1926), 98–101.
3. *Die Ordnungszahlen Differentialgleichungensysteme*, MTZT, 31(1929), 748–766.
4. *Über eine Matrixtransformation*, MTZT, 32(1930), 465–473.
5. *Die Stabilitätsfrage bei Differentialgleichungen*, MTZT, 32(1930), 703–728.

Persidskiĭ, K.

1. *On the stability of motion specified by the first approximation*, MTSK, (1)40(1933), 284–293, (Russian-German summary).

Picard, É.

1. *Mémoire sur la théorie des équations aux dérivées partielles et la méthode des approximations successives*, JMPA, (5)6(1890), 423–441.

Potter, Ruth L.

1. *On self-adjoint differential equations of second order*, PFJM, 3(1953), 467–491.

Prüfer, H.

1. *Neue Herleitung der Sturm-Liouvilleschen Reihenentwicklung stetiger Funktionen*, MTAN, 95(1926), 499–578.

Putnam, C. R.

1. *An oscillation criterion involving a minimum principle*, DKMJ, 16(1949), 633–636.

Radon, J.

1. *Über die Oszillationstheoreme der konjugierten Punkte beim Probleme von Lagrange*, Münchener Sitzungsberichte, 57(1927), 243–257.
2. *Zum Problem von Lagrange*, Hamburger Mathematische Einzelschriften. 6 Heft., (1928).

Redheffer, R. M.

1. *On solutions of Riccati's equation as functions of initial values*, J. Rat. Mech. Analysis, 5(1956), 835–848.
2. *The Riccati equation: initial values and inequalities*, MTAN, 133 (1957), 235–250.
3. *Inequalities for a matrix Riccati equation*, JMMC, 8(1959), 349–367.
4. *The Mycielski-Paszkowski diffusion problem*, JMMC, 9(1960), 607–622.

Reid, W. T.

1. *Propretiės of solutions of an infinite system of ordinary linear differential equations of the first order with auxiliary boundary conditions* (Dissertation, Univ. of Texas, 1929), TAMS, 32(1930), 284–318.
2. *Generalized Green's matrices for compatible systems of differential equations*, AMJM, 53(1931), 443–459.
3. *A boundary value problem associated with the calculus of variations*, AMJM, 54(1932), 769–790.
4. *The theory of the second variation for the non-parametric problem of Bolza*, AMJM, 57(1935), 573–586.
5. *Boundary value problems of the calculus of variations*, BAMS, 42(1937), 633–666.

6. *An integro-differential boundary problem*, AMJM, 60(1938), 257–292.
7. *A system of ordinary linear differential equations with two-point boundary conditions*, TAMS, 44(1938), 508–521.
8. *Some remarks on linear differential systems*, BAMS, 45(1939), 414–419.
9. *A new class of self-adjoint boundary value problems*, TAMS, 52(1942), 381–425.
10. *A matrix differential equation of Riccati type*, AMJM, 68(1946), 237–246; "Addendum," 70(1948), 250.
11. *Symmetrizable completely continuous linear transformations in Hilbert space*, DKMJ, 18(1951), 41–56.
12. *Oscillation criteria for linear differential systems with complex coefficients*, PFJM. 6(1956), 733–751.
13. *A comparison theorem for self-adjoint differential equations of second order*, ANNM, 65(1957), 197–202.
14. *Adjoint linear differential operators*, TAMS, 85(1957), 446–461.
15. *Remarks on a matrix transformation for linear differential equations*, PAMS, 8(1957), 708–712.
16. *Principal solutions of non-oscillatory self-adjoint linear differential systems*, PFJM, 8(1958), 147–169.
17. *A class of two-point boundary problems*, ILJM, 2(1958), 434–453.
18. *A Prüfer transformation for differential systems*, PFJM, 8(1958), 575–584.
19. *Solutions of a Riccati matrix differential equation as functions of initial values*, JMMC, 8(1959), 221–230.
20. *Generalized linear differential systems*, JMMC, 8(1959), 705–726.
21. *Properties of solutions of a Riccati matrix differential equation*, JMMC, 9(1960), 749–770.
22. *Oscillation criteria for self-adjoint differential systems*, TAMS, 101 (1961), 91–106.
23. *Riccati matrix differential equations and non-oscillation criteria for associated linear differential systems*, PFJM, 13(1963), 665–685.
24. *Principal solutions of non-oscillatory linear differential systems*, JMAA, 9(1964), 397–423.
25. *A matrix equation related to a non-oscillation criterion and Liapunov stability*, QAMT, 23(1965), 83–87.
26. *A class of monotone Riccati matrix differential operators*, DKMJ, 32(1965), 689–696.
27. *Generalized linear differential systems and related Riccati matrix integral equations*, ILJM, 10(1966), 701–722.
28. *Some limit theorems for ordinary differential systems*, JDEQ, 3(1967), 423–439.
29. *Generalized Green's matrices for two-point boundary problems*, SJAM, 15(1967), 853–870.
30. *Variational methods and boundary problems for ordinary linear differential systems*, Proc. US-Japan Sem. on Differential and Functional Equations, Univ. of Minnesota, Minneapolis, Minn., June 26–30, 1967, W. A. Benjamin, Inc., 267–299.
31. *Generalized inverses of differential and integral operators*, Proc. of Symposium on Theory and Application of Generalized Inverses of Matrices, Texas Technological College, Lubbock, Texas, March, 1968, 1–25.
32. *Generalized polar coordinate transformations for differential systems*, to appear in Rocky Mountain Journal of Mathematics, 1(1970).

Richardson, R. G. D.

1. *Contributions to the study of oscillation properties of the solutions of linear differential equations of the second order*, AMJM, 40(1918), 283–316.

Sandor, S.

1. *Sur l'équation différentielle matricielle de type Riccati*, Bull. Math. Soc. Sci. Math. Phys. R. P. Roumaine (N.S.), 3(51), (1959), 229–249.

Sansone, G.

1. *Equazioni differenziali nel campo reale*, 2nd edizione, Zanichelli, Bologna, Parte I (1956); Parte II (1949).

Sansone, G., and R. Conti

1. *Equazioni differenziali non lineari*, Cremonese, Rome, 1956.

Schmidt, E.

1. *Entwicklung willkürlicher Funktionen nach Systemen vorgeschreibener* (Dissertation, Göttingen, 1905), MTAN, 63(1907), 433–476.

Schubert, H.

1. *Über die Entwicklung zulässiger Funktionen nach den Eigenfunktionen bei definiten selbstadjungierten Eigenwertaufgaben*, Sitzungsberichte Heidelberger Akad. Wiss., (1948), 178–192.

Schwarz, H. A.

1. *Ueber ein die Flächen kleinsten Flächeninhalts betreffendes Problem der Variationsrechnung*, Acta societatis Fennicae, 15(1885) 315–362 = Gesammelte Mathematische Abhandlungen, I, Springer, Berlin (1890), 223–269.

Sherman, T. L.

1. *Properties of solutions of n-th order linear differential equations*, PFJM, 15(1965), 1045–1060.
2. *Properties of solutions of quasi-differential equations*, DKMJ, 32(1965), 297–304.

Sloss, F. B.

1. *A self-adjoint boundary value problem with end conditions involving the characteristic parameter* (Dissertation, Northwestern Univ., 1955).

Späth, H.

1. *Über das asymptotische Verhalten der Lösungen nichthomogener linearer Differentialgleichungen*, MTZT, 30(1929), 487–513.

Sternberg, Helen M. and R. L. Sternberg

1. *A two-point boundary problem for ordinary self-adjoint differential equations of fourth order*, CDJM, 6(1954), 416–419.

Sternberg, R. L.

1. *Variational methods and non-oscillation theorems for systems of differential equations* (Dissertation, Northwestern Univ., 1951), DKMJ, 19(1952), 311–322.
2. *A theorem on hermitian solutions for related matrix differential and integral equations*, PTGM, 12(1953), 135–139.

Sturm, J. C. F.

1. *Mémoire sur les équations différentielles linéaires du second ordre*, JMPA, 1(1836), 106–186.

Taam, C.-T.

1. *Non-oscillatory differential equations*, DKMJ, 19(1952), 493–497.
2. *Non-oscillation and comparison theorems of linear differential equations with complex-valued coefficients*, PTGM, 12(1953), 57–72.
3. *On the solution of second order linear differential equations*, PAMS, 4(1953), 876–879.

Tamarkin, J. D.

1. *Sur quelques points de la théorie des équations différentielles linéaires ordinaires et sur la generalization de la série de Fourier*, RCMP, 34(1912), 345–382.
2. *Some general problems of the theory of ordinary linear differential equations and*

expansion of an arbitrary function in series of fundamental solutions, Petrograd, 1917 (Russian); MTZT, 27(1927), 1–54.

Tomastik, E. C.

1. *Singular quadratic functionals of n dependent variables*, TAMS, 124(1966), 60–76.

Tonelli, L.

1. *Sulle equazioni funzionali del tipo di Volterra*, BCMS, 20(1928), 31–48.

Trjitzinsky, W. J.

1. *Properties of growth for solutions of differential equations of dynamic type*, TAMS, 50(1941), 252–294.

Vessiot, E.

1. *Gewöhnliche Differentialgleichungen; Elementare Integrationsmethoden*, EMTW, II.A 4b (1900).
2. *Méthodes d'intégration élémentaires. Étude des équations différentielles ordinaires au point de vue formel*, ESMT, II.16(1910).

Vinograd, R. E.

1. *On a criterion of instability in the sense of Lyapunov for the solutions of a linear system of ordinary differential equations*, DOKL, 84(1952), 201–204 (Russian).

Ważewski, T.

1. *Sur la limitation des intégrales des systèmes d'équations différentielles linéaires ordinaires*, STMT, 10(1948), 48–59.

Weinberger, H.

1. *An extension of the classical Sturm-Liouville theory*, DKMJ, 22(1955), 1–14.

Weyl, H.

1. *Comment on the preceding paper*, AMJM, 68(1946), 7–12.

Whyburn, W. M.

1. *Second-order differential systems with integral and k-point boundary conditions*, TAMS, 30(1928), 630–640.
2. *Existence and oscillation theorems for non-linear differential systems of the second order*, TAMS, 30(1928), 848–854.
3. *On the fundamental existence theorems for differential systems*, ANNM, 30(1929), 31–38.
4. *Differential equations with general boundary conditions*, BAMS, 48(1942), 692–704.
5. *Differential systems with general boundary conditions*, Seminar Reports in Math. (Los Angeles) Univ. of California Publications in Math., N. S., 2(1944), 45–61.

Wintner, A.

1. *The infinities in the non-local existence problem of ordinary differential equations*, AMJM, 68(1946), 173–178.
2. *Asymptotic integration constants*, AMJM, 68(1946), 553–559.
3. *On the Laplace-Fourier transcendents occurring in mathematical physics*, AMJM, 69(1947), 87–98.
4. *A norm criterion for non-oscillatory differential equations*, QAMT, 6(1948), 183–185.
5. *On linear repulsive forces*, AMJM, 71(1949), 362–366.
6. *On the nonexistence of conjugate points*, AMJM, 73(1951), 368–380.
7. *Ordinary differential equations and Laplace transforms*, AMJM, 79(1957), 265–294.

Wyler, O.

1. *Green's operators*, AMPA, 66(1964), 251–264.
2. *On two-point boundary-value problems*, AMPA, 67(1965), 127–142.

Zaanen, A. C.

1. *Ueber vollstetige symmetrische und symmetrisierbare Operatoren*, NAWK, (2)22(1947), 57–80.

2. *On the theory of linear integral equations*, I, II, III, IV, IVa, V, VI, VII, VIII, VIIIa, Nederlandsche Akad. Wet., Proc., 49(1946), 194–212, 292–301, 409–423, 571–585, 608–621; 50(1947), 357–368, 465–473, 612–617, and INDM, 8(1946), 91–109, 161–170, 264–278, 352–380; 9(1947), 215–226, 271–279, 320–325.

Zimmerberg, H. J.

1. *A class of definite boundary value problems* (Dissertation, Univ. of Chicago, 1945).
2. *Two-point boundary problems involving a parameter linearly*, ILJM, 4(1960), 593–608.
3. *Two-point boundary conditions linear in a parameter*, PFJM, 12(1962), 385–393.

Author Index

Abel, N. H., 3, 88
Ahlbrandt, C. D., 89, 399, 400, 530
Alekseev, V. M., 89
d'Alembert, J., 2
Antosiewicz, H. A., 459, 530
Aronszajn, N., 400
Ascoli, G., 459, 530
Atkinson, F. V., 300, 399, 530

Banach, S., 88
Barrett, J. H., 108, 261, 399, 400, 530
Barrow, I., 1
Bartle, R. G., 528
Bell, E. T., 3, 530
Bellman, R., 87, 459, 460, 531
Bernoulli, Daniel, 2
Bernoulli, James, 1, 2
Bernoulli, John, 1, 2
Bessel, F. W., 2
Birkhoff, G. D., 4, 172, 189, 217, 398, 531
Birkhoff, G., 86, 260, 261, 300, 531
Bliss, G. A., 4, 86, 88, 89, 172, 173, 217, 300, 302, 304, 391, 398, 399, 400, 531
Bobonis, A., 399, 531
Bôcher, M., 3, 4, 172, 260, 261, 400, 531
Bounitzky, E., 172, 173, 531
Bouquet, J. C., 3
Bourbaki, N., 108, 531
Bradley, J. S., 173, 532
Brauer, F., 89, 532
Briot, A. A., 3
Burkhardt, H., 172

Cajori, F., 3, 532
Caligo, D., 460, 532
Caratheodory, C., 5, 87, 91, 108, 398, 517, 532
Cauchy, A. L., 2, 3, 16, 87
Cesari, L., 459, 460, 532
Clairaut, A. C., 2
Clebsch, R. F. A., 398
Coddington, E. A., 86, 89, 108, 172, 217, 459, 460, 532
Cole, R. H., 217, 532
Coles, W. J., 217, 532
Collatz, L., 172, 217, 532
Conti, R., 173, 459, 460, 532, 540
Coppel, W. A., 459, 460, 533

Descartes, R., 1
Diaz, J. B., 217, 533
Diliberto, S. P., 460, 533

Elliott, W. W., 173, 533
von Escherich, G., 398, 533
Etgen, G. J., 108, 399, 400, 533
Ettlinger, H. J., 261, 533
Euler, L., 2, 18

Floquet, G., 442, 533
Fourier, J. B. J., 2
Fréchet, M., 88
Fredholm, I., 4
Friedrichs, K., 4
Frobenius, G., 3
Fuchs, L., 3, 88, 172
Fukuhara, M., (*see* Hukuhara, M.) 533

Giuliano, L., 87, 89, 108, 533
Gould, S. H., 400, 533
Graves, L. M., 86, 87, 88, 528, 533
Green, G., 172, 533
Gregory, J., 1
Greub, W., 173, 534
Gronwall, T. H., 86, 534

Hahn, W., 459, 460, 534
Halmos, P. R., 461
Harris, V. C., 399, 534
Hartman, P., 86, 87, 88, 89, 261, 300, 316, 398, 399, 400, 534
Hestenes, M. R., 398, 531, 534
Hilbert, D., 4, 173, 300, 534
Hille, E., 86, 89, 108, 261, 300, 400, 426, 460, 534
Hinton, D. B., 399, 534
Hoffman, R., 461
Hölder, E., 173, 534, 535
Hooker, J. W., 300, 535
l'Hospital, G. F. A., 1
Howard, H. C., 399, 400, 535
Hu, K.-S., 398, 535
Hukuhara, M., 87, 89, 460, 535
Hunt, R. W., 399, 535

Ince, E. L., 1, 3, 88, 89, 217, 260, 261, 300, 460, 535

Jackson, D., 300, 399, 535
Jacobi, C. G. J., 3, 88, 172, 398
Jones, W. R., 173, 535

Kamke, E., 86, 87, 89, 172, 217, 261, 300, 399, 535
Kaufman, H., 399, 535
Kneser, A., 261, 535
Kneser, H., 89, 536
Kodaira, K., 4
Krein, M., 217, 536
Kunze, R., 461

Lagrange, J. L., 2, 88, 172
Langer, R. E., 172, 217, 531
Laplace, P. S., 2
LaSalle, J., 459, 460, 536
Latshaw, V. V., 300, 399, 536
Lavrientieff, M., 88
Lefschetz, S., 459, 460, 536
Legendre, A.-M., 2, 398
Leibniz, G. W., 1
Leighton, W., 261, 316, 342, 398, 399, 400, 536
Levin, J. J., 108, 398, 536
Levinson, N., 86, 89, 108, 172, 217, 459, 460, 532, 536
Liapunov, A. M., 4, 261, 411, 455, 459, 460, 536
Lie, S., 3
Lindelöf, E., 48, 536
Liouville, J., 3, 48, 88, 189, 261, 300, 536
Lipschitz, R., 2, 32, 87, 536
Loud, W. S., 173, 536

Malkin, I. G., 459, 537
Mansfield, R., 173, 537
Markus, L., 460, 537
Mason, M., 300, 537
Massera, J. L., 108, 459, 460, 537
McShane, E. J., 87, 108, 537
Metcalf, F. T., 217, 533
Moore, E. H., 173, 484, 537
Moroney, R. M., 300, 537
Morse, M., 4, 5, 6, 217, 260, 261, 300, 301, 302, 316, 342, 398, 399, 400, 537
Müller, M., 89, 537

Nagumo, M., 89, 537
Nehari, Z., 89, 300, 301, 399, 536, 537
Nehring, E. D., 461
Nemytskiĭ, V. V., 459, 560, 537
von Neumann, J., 4
Newton, I., 1

Olech, C., 89
Osgood, W. F., 89, 537

Painlevé, P., 3, 537
Peano, G., 4, 16, 86, 87, 88, 538
Penrose, R., 484

Perron, O., 87, 89, 409, 460, 538
Persidskiĭ, K., 459, 538
Picard, É., 3, 48, 172, 538
Poincaré, H., 4
Potter, Ruth L., 261, 538
Prüfer, H., 220, 261, 538
Putnam, C. R., 261, 538

Radon, J., 108, 398, 538
Redheffer, R. M., 108, 538
Reid, W. T., 87, 108, 172, 173, 217, 220, 261, 300, 301, 316, 341, 398, 399, 400, 460, 528, 538, 539
Rheinboldt, W. C., 173, 534
Riccati, Count J., 2
Richardson, R. G. D., 261, 539
Riemann, G. F. B., 3
Riesz, F., 517
Rota, G.-C., 86, 261, 300, 531

Sandor, S., 108, 398, 540
Sansone, G., 86, 89, 108, 172, 260, 261, 300, 459, 540
Schäffer, J. J., 108, 459, 460, 537
Schmidt, E., 217, 540
Schubert, H., 217, 399, 540
Schwarz, H. A., 217, 540
Sherman, T. L., 399, 540
Sloss, F. B., 399, 540
Späth, H., 460, 540
Stepanov, V. V., 459, 460, 537
Sternberg, Helen M., 399, 400, 540
Sternberg, R. L., 399, 400, 535, 540
Sternberg, S., 89, 532
Stone, M. H., 4
Sturm, J. C. F., 3, 242, 260, 300, 540
Sz-Nagy, B., 517

Taam, C.-T., 261, 540
Tamarkin, J. D., 4, 172, 173, 217, 540
Taylor, B., 2
Tomastik, E. C., 399, 541
Tonelli, L., 20, 87, 541
Trjitzinsky, W. J., 460, 541

Vessiot, E., 3, 541
Vinograd, R. E., 460, 541
Volterra, V., 4, 57, 87

Ważewski, T., 460, 541
Weierstrass, K., 3
Weinberger, H., 399, 541
Weinstein, A., 400
Weyl, H., 4, 460, 541
Whyburn, W. M., 108, 173, 220, 300, 541
Wintner, A., 89, 261, 398, 399, 400, 460, 534, 541
Wyler, O., 173, 541

Zaanen, A. C., 217, 541
Zimmerberg, H. J., 399, 542

Subject Index

Abbreviations: a = adjoint; b = boundary; d = differential; e = equation; l = linear m = matrix; p = problem; s = self; v = vector.

Abnormality, of s.-a. l. v. d. e., 313; (Probs. VII.3:3; VII.4:7)
Adjoint b. p., 175
Adjoint l. v. d. e., 111; (Probs. III.2:6, 8)
Adjoint scalar d. e., 117-122; (Probs. III.3:4, 5, 6)
Admissible comparison function, 454, (Probs. VIII.8:2-5)
Algebraic l. e., 478-480, (Probs. I.7:1; VII.5:12; VIII.8:1)
Annihilator, 467, 468
Ascoli theorem, 18, 26, 29, 527
Asymptotic behavior of solutions, of l. v. d. e., 3, 4, 403-438; (Probs. VIII.2:1, 2, 7, 8, 9, 11, 12, 14-18; VIII.3:3, 4; VIII.4:3, 4; VIII.6:1, 3-6; VIII.8:2)
  of non-l. v. d. e., 447-457; (Probs. VIII.2:3, 4; VIII.8:3-7)
  of n-th order d. e., (Probs. VIII.4:1; VIII.5:3, 7)
  of second order d. e., (Probs. VIII.2:5, 6; VIII.3:1, 2; VIII.4:2; VIII.5:1, 2, 4, 5, 6)
Asymptotic one-to-one correspondence, 432, 433, 450; (Probs. VIII.5:1, 2, 5; VIII.6:5)
Asymptotic stability, 403, 404, 408; (Probs. VIII.2:1-4,10; VIII.4:1; VIII.8:4)
  uniformly, 403, 404, 408
Autonomous d. e.; (Prob. VIII.1:1)
Basis, conjoined, 307; (Probs. VII.3:1; VII.4:1, 3, 10; VII.5:13,15; VII.7:1-4)
Bessel equality, 473, 489
  equation; (Prob. VIII.5:6)
  inequality, 473, 489
Bilinear concomitant, 120; (Prob. III.3:3)
Boundary conditions, adjoint, 112; (Prob. III.6:3)
  canonical form of, for s.-a. b. p., 262-267; 368-370; (Prob. VI.1:3)
  homogeneous, 126-128
  in parametric form, 127, 130
  non-homogeneous, 129-131
Boundary problems, Chs. IV, VI, VII
  adjoint, 175
  definite, 200-211; (Probs. IV.6:1-8)
  fully s.-a., 197
  involving first order l. v. d. e., 174-182, 196-198, 200-211, 374-379, 381-388, 390-392; (Probs. IV.2:5, 7; IV.5:1, 2, 3; IV.6:1-9; VII.10:1, 2, 3; VII.11:1, 2, 3; VII.12:1-8)
  involving fourth order l. d. e.; (Probs. IV.2:4; IV.6:11)
  involving n-th order l. v. d. e., 190-196; (Prob. IV.6:10)
  involving second order scalar l. d. e., 244-252, Ch. VI; (Probs. I.10:3; IV.2:1, 2, 3, 6; V.7:1-4; VI.3:1-5; VI.4:1-8; VI.5:1-8)

Boundary problems *(Continued)*
s.-a., 196-198, 200-211, 374-379, 381-388, 390-392; (Probs. IV.5:1, 2, 3; IV.6:1-8; VI.1:1, 2; VII.10:1, 2, 3; VII.11:1, 2, 3; VII.12:1-8)
Sturm-Liouville, 3, 250-252; (Probs. V.7:1, 2)
Bunyakovsky-Schwarz inequality, 520

Calculus of variations, fundamental lemma of; (Probs. III.2:1-7)
significance for b. p., 4, 5, 217, 301, 302, 398
Carathéodory d. systems, Chs. II, VII
Cauchy-Bunyakovsky-Schwarz inequality, 472
Cauchy-Lipschitz method, 2, 87
Cauchy-Peano existence theorem, 16-24; (Probs. I.3:2, 3, 4, 6)
Cauchy polygonal functions, 2, 87
Cauchy sequence, 54
Cayley-Hamilton theorem, 512
Center; (Prob. VIII.2:18)
Characteristic e. of m., 497; (Probs. E.1:6, 7, 8)
exponent, 443
multiplier, 441; (Probs. VIII.6:1-5)
Class ΓL, locally, 95; (Prob. II.3:3)
$\Gamma^{(q)}$L, locally, 105
Lip, locally, 96
Lip [D; $\kappa(t)$], 96; (Prob. II.3:2)
Clebsch transformation, 398
Comparison theorems, for s.-a. l. v. d. e., 354-356, 362-366, 390-392; (Probs. VII.7:1-4; VII.12:1, 2)
for second-order scalar l. d. e., 238-243, 279-285; (Probs. V.6:1-4; VI.4:2-8)
Sturmian, 3, 4, 241-243; (Probs. V.6:1-4)
Compatibility, index of, 126-128, 134
Complex d. e., 80-85; (Probs. I.11:1-4)
Conjoined basis, 307; (Probs. VII.3:1; VII.4:1, 3, 10; VII.5:13, 15; VII.7:1-4)
family of solutions, 306
Conjugate point, 233, 313-315, 342; (Probs. VI.3:5; VII.3:1, 2; VII.4:3; VII.5:1; VII.11:3)
order of, 313
Constant of irregularity, 412
Contraction mapping theorem, 55
Courant-Hilbert Max-Min Principle, 277, 383
Cross ratio, 102, 232; (Probs. I.9:16; II.3:10)

Definiteness, of b. p., 200
of hermitian matrices, 491, 503; (Probs. D.1:8, 9, 13; E.2:2, 3, 8-11)
Dependence of solutions of d. e., on initial values, 38, 72-76, 97-99, 104-106; (Probs. I.5:4, 5; I.6:2; I.9:13, 18; I.10:1, 2, 4; I.11:1; II.3:6, 9, 11; II.4:2)
on parameters, 39, 70-75, 98, 99, 104-106; (Probs. I.6:3; I.9:13; I.10:4, 5)
Determinants, 488-490
Hadamard inequality for; (Prob. D.1:4)
Differential inequalities, 15
Differential operator, 109, 110
a., 111, 112; (Probs. III.3:4, 5, 6)
domain of, 109, 126
inverse of, 138, 140
nullspace of, 126, 136
quasi-, 118; (Probs. VII.2:1,2)
range of, 109, 136
s.-a; (Prob. III.8:4)
symmetric, 122; (Probs. III.3:1-3)
Differential systems, two-point, Chs. III, V, VII
a., 132-137; (Prob. III.6:1)
compatible, 126
incompatible, 126
index of compatibility, 126, 128, 134; (Prob. III.4:1)
non-homogeneous, 129-131; (Probs. III.5:1-3; III.6:2)
s.-a., Ch. VII

Differential systems *(Continued)*
involving n-th order l. v. d. e., 152-154; (Probs. III.9:1-6)
involving scalar n-th order l. d. e., 143-150
a., 148
non-homogeneous; (Prob. III.8:1)
s.-a; (Probs. III.8:4; III.11:3)
Differential systems, with non-two-point b. conditions; (Probs. III.10:6-9)
Direction field, 9
Disconjugacy, 233, 313
criteria, for second order l. d. e., 233-236; (Probs. III.8:6; V.3:1, 2, 3; V.5:1, 5, 6, 9)
criteria, for s.-a. l. d. e. of order 2n; (Prob. VII.12:5)
criteria, for s.-a. l. v. d. e., 313, 337-344; (Probs. VII.4:1, 3; VII.5:2-4, 6-9, 12-15; VII.12:2-4)

ϵ-approximate solution, 16, 35, 36; (Prob. I.3:2)
Eigenvalue (*see* Proper value)
Equation of variation, 71; (Probs. I.10:1; II.4:2)
Equicontinuity, uniform, 526-528
Equivalence, of d. systems, 165-170, 427; (Probs. III.11:1-6; VII.2:3)
relation, 407, 496, 497
Euclidean space $\mathbf{C}_n$, 488-490; (Probs. D.1:1, 2, 5, 6)
Lagrange-Cauchy inequality in, 488
Existence theorem for d. e., 2, 4, Chs. I. II
as fixed point theorem, 54-56; (Prob. I.7:4)
by successive approximations, 48-50, 59, 81, 96, 97; (Probs. I.6:1, 4, 5; I.8:1; I.10:4)
of Carathéodory type, 92-99; (Probs. II.3:5-8)

Existence theorem *(Continued)*
of Cauchy-Peano type, 16-24; (Probs. I.3:2, 3, 4, 6)
of Tonelli type, 20-23; 94, 95
Expansion theorem, for definite b. p., 206, 207, 210, 211; (Probs. IV.6:6, 7, 8, 10)
for s.-a. l. v. b. p., 385-389
for s.-a. second-order b. p., 288-298; (Probs. VI.5:1-7)
formulation of, 180
Exp$\{tA\}$; (Probs. I.9:3-6, 14; E.3:2, 4, 5)
Extension of solutions, 23, 24, 27-30, 36-40, 95, 98; (Probs. I.3:1; I.4:4, 5, 6; I.5:9, 17, 18; I.8:3; II.3:5, 6, 7)

Floquet theory, 440-445; (Probs. VIII.6:1-6)
Focal point, 240-243, 258-260, 355-358, 364-366; (Probs. V.6:1-4; V.8:3, 5; VII.7:1-4)
Focus; (Prob. VIII.2:18)
Form, bilinear, 471
hermitian, 471-474, 500-504, 519; (Probs. A.5:2, 3; D.1:8, 9, 12, 13; E.2:1)
Morse fundamental, 253-260, 356-366; (Probs. V.8:1-5)
non-negative definite hermitian, 471; (Prob. D.1:8)
positive definite hermitian, 471; (Probs. D.1:9, 13)
sesquilinear, 471; (Probs. A.5:1, 2)
Fourier coefficients, 180, 206, 210
Functional, hermitian, 322-331, 337-339, 354-366, 374-379, 381-388, 390-392; (Probs. VII.4:1, 2, 4, 6, 8, 9)
quadratic, 224-229, 233-236, 238-243, 258-260, 288-295; (Probs. IV.6:9, 12, 13; V.3:2-6) V.4:1; V.5:1, 2, 3; V.8:1-5; VI.3:1, 3; VI.5:8)
Fundamental partition, 254, 358

Fundamental lemma of the calculus
of variations; (Probs. III.2:1-7)
Fundamental matrix solution, 62, 111; (Probs. I.9:11, 12, 17; I.10:4; I.11:2)
Fundamental set of solutions, 62, 88
normal, 411; (Prob. VIII.2:17)

Generalized inverse, of a matrix, 330, 480-484; (Probs. B.3:1, 2, 3)
Gram matrix; (Prob. D.1:3)
Gram-Schmidt process of orthonormalization, 473, 474, 489
Green's function, expansion for, 295, 296; (Prob. VI.5:4)
for higher order l. d. e., 146, 147, 150; (Probs. III.8:2, 3, 4)
for second order l. d. e., 147; (Probs. III.8:2, 5, 6; VI.3:2)
Green's matrix, expansion for, 183-189, 388
for higher order l. v. d. e., 153-155; (Probs. III.9:2, 6)
for systems with non-two-point b. conditions; (Probs. III.10:6-9)
generalized, 156-160; (Probs. III.10:1-5)
of adjoint, 142, 169; (Prob. III.7:2)
poles of, 184-188, 198
properties of, 138-142, 371; (Probs. III.7:1, 2; IV.3:1, 2)
Gronwall type inequality, 86

Hermitian form, 471-474, 500-504, 519; (Probs. A.5:2, 3; D.1:8, 9, 12, 13; E.2:1)
(negative) index of, 504
nullity of, 504
Hermitian matrix, 476
non-negative square root of; (Probs. E.2:8-10)
properties of, 500-504; (Probs. 3.2:1-13)
Hermitian matrix functions, properties of; (Probs. F.1:2-8)

Index (negative) of hermitian form, 504
Index, of compatibility, 126-128, 134
of proper value of b. p., 174, 175
of proper value of m., 497
Inequality, Bessel, 473
Cauchy-Bunyakovsky-Schwarz, 472
Infinitesimal upper bound, 456
Inner product, 488
Initial value p., existence theorems for, 16-24, 27-30, 48-50, 54-56, 58-60, 80-85, 92-99; (Probs. I.3:2, 3, 4, 6; I.6:1, 4, 5; I.7:4; I.8:1; I.10:4, 5; II.3:5-8)
uniqueness theorem for, 32-40, 47-50, 58-60, 80-83, 96-98; (Probs. I.5:3, 6, 7, 8, 12-16; I.8:2; II.3:4)
Integral, first, 76, 77, 110; (Probs. I.10:6; VIII.8:5, 6)
Integral equation, Fredholm, 4; (Prob. I.7:3)
Volterra, 4; (Prob. I.7:2)
Inverse, of a matrix, 477; (Probs. B.1:5, 6, 8; D.1:9; D.2:4; E.2:2)
of a d. operator, 138
Isocline, 10

Jacobi condition, 398
Jordan matrix, elementary, 509; (Probs. E.3:1-5)
canonical form, 509-516; (Prob. E.3:8)

Kinematic similarity, 406, 411; (Prob. VIII.2:12)

Legendre transformation, 398
Liapunov, direct (second) method, 454-457; (Probs. VIII.8:1-7)
function, 456
similarity, 413
transformation, 413

Linear algebraic system, solvability theorems, 478-480
Linear dependence and independence of vectors, 464-466
Linear functionals, conjugate space of, 466-469
Linearly independent solutions of l. v. d. e., 62
Linear transformation, 469-471
  inverse of, 469
  matrix of, 470, 476
  null-space of, 469
  of $\mathbf{C}_n$ into $\mathbf{C}_m$, 476, 477
  product, 471
  rank of, 469
Liouville transformation; (Prob. V.1:3)
Lipschitz condition, 32
  unilateral, 35
Lipschitz constant, 32; (Prob. I.5:1, 2)
Lipschitzian, locally, 33
Local uniqueness, point of, 37; (Prob. I.5:3)
  region of, 38

Manifold, linear, 464
Matrices, algebra of, 475-477; (Probs. B.1:1-8)
  sequences and series of, 492-494; (Probs. D.2:1-7; E.3:1)
  similar, 496
Matrix, characteristic e. of, 497; (Probs. E.1:1, 7, 8)
  conjugate, 476
  conjugate transpose, 476
  determinant of, 485-487
  exponential of, 494; (Probs. D.2:2; E.3:2, 5-9)
  generalized inverse of, 173, 370, 480-484; (Probs. B.3:1-3)
  generalized null-space of, 510
  Gram; (Prob. D.1:3)
  hermitian, 476, 500-504; (Probs. VIII.8.1; E.1:3; E.2:1-13)
  inverse of, 477; (Probs. B.1:5, 6, 8; D.1:9; D.2:4; E.2:2)
Matrix *(Continued)*
  Jordan canonical form of, 509-516; (Prob. E.3:8)
  logarithm of; (Probs. E.3:5, 6, 9)
  minimal polynomial for, 512
  non-singular, 477; (Probs. B.1:5-8)
  normal; (Prob. E.1:5)
  norm for, 490; (Probs. D.1:10, 11, 12; E.1:10; E.2:4)
  proper values of, 496-498, 500-504; (Probs. E.1:1; E.2:1, 4, 5, 6, 10, 12, 13)
  pure imaginary part of, 476; (Prob. E.1:10)
  rank of, 477; (Probs. B.1:2, 3, 4)
  real part of, 476; (Prob. E.1:10)
  singular, 477
  skew-hermitian, 476
  skew-symmetric, 476
  spectrum of, 497
  square root of; (Probs. D.2:7; E.2:8, 9, 10)
  super-diagonal, 497; (Probs. E.1:2. 4)
  symmetric, 476
  trace (tr) of; (Probs. E.1:5, 6, 9)
  transpose, 476
  unitary; (Probs. D.1:7; E.1:2, 5)
  Wronskian, 64; (Probs. I.9:7, 8)
Matrix functions, Appendix F; (Probs. F.1:2-8)
Matrizant, 63
Maximal interval of existence of solutions of d. e., 23, 24, 27-30, 36-40, 95, 98; (Probs. I.3:5; I.4:5, 6; I.5:9, 17, 18; II.3:5, 6, 7)
Maximal solution, 27-30; (Probs. I.4:1, 2, 4, 5)
Metric space, 54; (Probs. I.7:5, 6)
Minimal polynomial, 512
Minimal solution, 27; (Probs. I.4:1, 3, 4, 5)
Monotoneity Principle A, 13
Morse fundamental form, for conjugate points, 253-258,

Morse *(Continued)*
356-363; (Probs. V.8:1, 2, 4, 5)
for focal points, 258-260, 363-366; (Probs. V.8:3, 5)
Multiplicity of proper value, of a m., 497
of a b. p., 176-182, 184-189, 198; (Probs. IV.2:1, 2, 6, 7)

Nilpotency, degree of, 513
Node; (Prob. VIII.2:18)
Non-oscillation criteria, for s.-a. l. d. e. of order 2n; (Prob. VII.12:5)
for s.-a. l. v. d. e., 313, 337-344; (Probs. VII.4:1, 3; VII.5:2, 3, 4, 6-9, 12-15; VII.12:2, 3, 4)
for second order l. d. e., 233-236; (Probs. V.3:1, 2, 3; V.5:1, 5, 6, 9)
Normal fundamental set of solutions, 411; (Prob. VIII.2:17)
Normality, identical, 313, 340-344, 357-366; (Probs. VII.5:5, 7, 8, 9, 12, 13. 15; VII.7:1, 2, 4)
of b. p., 372-374, 376-379, 381-388, 390-393
Null space of d. operator, 126, 137
Nullity of hermitian form, 504

Orthogonal vectors, 472, 489
Orthonormal systems of proper solutions, 206, 210, 273-275, 381-383
Orthonormal systems of vectors, 472, 489
Oscillation criteria, for s.-a. l. v. d. e., 337-344; (Probs. VII.5:4, 6, 14, 15; VII.11:3; VII.12:7, 8)
for second order l. d. e., 233-236; (Probs. V.1:2; V.5:1, 4, 5, 7, 8)
Oscillation theorem of Sturm, 3, 4, 244-249; (Probs. V.7:2, 3; V.8:5)

Periodic end-point conditions; (Probs. III.4:2, 3, 4; III.5:4; III.6:4)
Perturbation e., 71
Picard-Lindelöf method of iterations, 48-50, 59, 81, 96, 97; (Probs. I.6:1, 4, 5; I.8:1; I.10:4)
Polar coordinate transformation, of second order l. d. e., 220; (Probs. V.1:1, 2)
of s.-a. l. m. d. e., (Probs. VII.2:6, 7, 8; VII.5:14, 15; VII.7:3, 4)
Principal solution, of second order l. d. e., 350, 400; (Prob. VII.5:10, 11)
of s.-a. l. m. d. e., 316-321, 341-344; (Probs. VII.5:7-10)
Proper values for b. p., Chs. IV, VI, VII
asymptotic behavior of; (Probs. VI.4:1, 7; VII.11:2)
comparison theorems for, 279-285, 390-392; (Probs. VI.4:2-8; VII.12:1, 2)
existence of, 203-205, 209, 210, 272-278, 381-389; (Probs. IV.5:1; IV.6:1, 2, 8; VI.3:1, 3; VI.4:1; VI.5:8; VII.11:2, 3; VII.12:2, 8)
extremum properties of, 203, 209, 210, 273-275, 381-383; (Probs. IV.6:7, 10; VI.3:3, 4)
index of, 174, 177-182, 188-189, 197-198; (Probs. IV.2:1, 2, 7)
multiplicity of, 176-182, 184-189, 198; (Probs. IV.2:1, 2, 6, 7)
Proper values for m., Appendix E
comparison theorems for, 503, 504; (Prob. E.2:13)
continuity of, (Prob. E.2:6)
existence of, 500-503
extremum properties of, 503, 504
index of, 497-499
multiplicity of, 497-499
properties of, (Probs. E.1:1, 10; E.2:4, 5, 10, 12, 13)

Prüfer transformation (*see* Polar coordinate transformation)

Reducible v. d. e., 407, 414
Region of local uniqueness, 38
Regular l. v. d. e., 412-414; (Probs. VIII.2:16, 17)
Riccati d. inequality, m., 340; (Prob. VII.5:2)
  scalar, 235; (Prob. V.5:6)
Riccati m. d. e., 101, 317-321;
  distinguished solution of, 319-321; (Prob. VII.5:12)
  properties of solutions of; (Probs. I.9:15; II.3:8-11; II.4:2)
  relation to oscillation criteria, 317-321, 338; (Probs. VII.5:6, 8, 12)
Riccati scalar d. e., 231, 232; (Probs. I.9:16; V.4:1, 2

Saddle point; (Prob. VIII.2:18)
Schur transformation; (Prob. E.1:2)
Schwarz constants, 203, 209
Schwarzian derivative; (Probs. I.11:3,4)
Sequences of vectors and matrices, 492-494; (Probs. I.9:3-6, 14; D.2:1-7; E.2:7-10)
Similarity, of matrices, 496, 497; (Probs. E.1:2-6; E.3:3, 4, 5, 9)
  Liapunov, 413
  of l. v. d. e., kinematic, (t-), 406, 407, 411; (Prob. E.2:12)
Slope field, 9
Solutions, of m. d. e., (Probs. III.2:9; VII.2:3, 4, 5)
  of scalar d. e., (Probs. I.1:1; I.9:9; III.8:5)
Spectral theory, for matrices, Appendix E
  for d. p. p., Chs. IV, VI, VII
Stability, 4, Ch. VIII; (Probs. VIII.2:1, 2, 3, 5, 6, 11, 18; VIII.3:1, 4; VIII.4:1, 2, 3; VIII.5:1-7)
  asymptotic, 403, 404, 408, 422, 425, 428, 433, 444, 445, 448,

Stability, asymptotic *(Continued)* 449, 453, 456; (Probs. VIII.2:1-4, 10; VIII.4:1; VIII.8:4)
  strict, 402, 404, 405, 408, 425, 444; (Probs. VIII.2:4, 7, 8, 9, 12; VIII.4:3)
  uniform, 402, 404, 408, 421, 422, 424, 428, 432, 433, 444, 445, 448, 449, 450, 453, 456; (Probs. VIII.2:10; VIII.3:3; VIII.4:3; VIII.8:4)
  uniformly asymptotic, 403, 404, 408, 425; (Prob. VIII.1:1)
Sturm, comparison theorems of, 3, 4, 241-243; (Probs. V.6:1-4)
  oscillation theorem of, 3, 4, 244-249; (Probs. V.7:2, 3; V.8:5)
Subproblem, 282, 283, 373, 392; (Probs. VI.4:8; VII.12:2)
Subspace, 464
Successive approximations, method of, 48-50, 59, 81, 96, 97; (Probs. I.6:1, 4, 5; I.8:1; I.10:4)
  non-convergence of; (Prob. I.6:4)
Symmetric d. expression, 122; (Probs. III.3:1-3)
Symmetric matrix, 476

Tonelli, method of, 20-23, 94, 95
Transformation, Legendre, or Clebsch, 398
  Liapunov; 413
  Liouville; (Prob. V.1:3)
  Schur; (Prob. E.1:2)
Trigonometric functions, as solutions of d. e., (Prob. I.9:10)
t-similarity, 406; (Prob. E.2:12)
t_∞-similarity; (Prob. VIII.4:4)
Type number, 410

Uniqueness, 32-40, 47-50, 58-60, 80-83, 96-98
  local, 38
  n-th order scalar d. e., 58-60

Uniqueness *(Continued)*
theorem, Kamke's general (Prob. I.5:12)
theorems, 34-36; (Probs. I.5:3, 7, 8)
unilateral, 34, 35; (Probs. I.5:6, 12-16; I.8:2; II.3:4)

Variation, e. of, 71; (Probs. I.10:1; II.4:12)
Variation of parameters, for l. v. d. e., 63, 64; (Probs. I.9:1)
Variation of parameters *(Continued)*
for n-th order scalar l. d. e., 65
Variational principle, 302, 398
Vector and matrix functions, 518-521; (Probs. F.1:1-8)
Vector spaces, 463-474

Wronskian m., 64; (Prob. I.9:7)

Zeros of solutions, of second order d. e., (*see* Disconjugacy, non-oscillation, oscillation, Sturm)